**UPDATED**
**to reflect the revised C**



**for the AP® Exam**

# The Practice of Statistics

SIXTH EDITION

**Daren S. Starnes**
The Lawrenceville School

**Josh Tabor**
Canyon del Oro High School

bedford, freeman & worth
high school publishers
Boston | New York

**Updated The Practice of Statistics** for the AP® Exam
**Sixth Edition**

Vice President, Social Science: Charles Linsmeier
Senior Program Director, High School: Ann Heath
Senior Development Editor: Donald Gecewicz
Senior Media Editor: Kim Morte
Marketing Manager: Thomas Menna
Marketing Assistant: Kelly Noll
Director of Digital Production: Keri deManigold
Senior Media Producer: Alison Lorber
Associate Editor: Corrina Santos
Assistant Editor: Carla Duval
Director of Design, Content Management: Diana Blume
Cover and Interior Designer: Lumina Datamatics, Inc.
Director, Content Management Enhancement: Tracey Kuehn
Senior Managing Editor: Lisa Kinne
Senior Content Project Manager: Vivien Weiss
Senior Workflow Supervisor: Susan Wein
Illustrations: Lumina Datamatics, Inc.
Senior Photo Editor: Robin Fadool
Photo Researcher: Candice Cheesman
Art Manager: Matthew McAdams
Composition: Lumina Datamatics, Inc.
Printing and Binding: LSC Communications
Cover Photo: Prasit Chansareekorn/Moment/Getty Images



Library of Congress Control Number: 2019938329
ISBN-13: 978-1-319-26929-6
ISBN-10: 1-319-26929-X


Printed in the United States of America
1 2 3 4 5 6 23 22 21 20 19

W. H. Freeman and Company
Bedford, Freeman & Worth High School Publishers
One New York Plaza
Suite 4500
New York, NY 10004-1562
highschool.bfwpub.com/catalog

# Contents

This Preliminary Volume includes Chapters 1–6 and selected appendices. The complete Contents are listed here as a preview of the full book to come in January.

## Additional Online Chapters

# About the Authors

Ann Heath

**DAREN S. STARNES** is Mathematics Department Chair and holds the Robert S. and Christina Seix Dow Distinguished Master Teacher Chair in Mathematics at The Lawrenceville School near Princeton, New Jersey. He earned his MA in Mathematics from the University of Michigan and his BS in Mathematics from the University of North Carolina at Charlotte. Daren is also an alumnus of the North Carolina School of Science and Mathematics. Daren has led numerous one-day and weeklong AP® Statistics institutes for new and experienced teachers, and he has been a Reader, Table Leader, and Question Leader for the AP® Statistics exam since 1998. Daren is a frequent speaker at local, state, regional, national, and international conferences. He has written articles for *The Mathematics Teacher* and *CHANCE* magazine. From 2004 to 2009, Daren served on the ASA/NCTM Joint Committee on the Curriculum in Statistics and Probability (which he chaired in 2009). While on the committee, he edited the *Guidelines for Assessment and Instruction in Statistics Education* (GAISE) pre-K–12 report and coauthored (with Roxy Peck) *Making Sense of Statistical Studies*, a capstone module in statistical thinking for high school students. Daren is also coauthor of the popular on-level text *Statistics and Probability with Applications*.

Ann Heath

**JOSH TABOR** has enjoyed teaching on-level and AP® Statistics to high school students for more than 22 years, most recently at his alma mater, Canyon del Oro High School in Oro Valley, Arizona. He received a BS in Mathematics from Biola University, in La Mirada, California. In recognition of his outstanding work as an educator, Josh was named one of the five finalists for Arizona Teacher of the Year in 2011. He is a past member of the AP® Statistics Development Committee (2005–2009) and has been a Reader, Table Leader, and Question Leader at the AP® Statistics Reading since 1999. In 2013, Josh was named to the SAT® Mathematics Development Committee. Each year, Josh leads one-week AP® Summer Institutes and one-day College Board workshops around the country and frequently speaks at local, national, and international conferences. In addition to teaching and speaking, Josh has authored articles in *The American Statistician*, *The Mathematics Teacher*, *STATS Magazine*, and *The Journal of Statistics Education*. He is the author of the *Annotated Teacher's Edition* and *Teacher's Resource Materials* for *The Practice of Statistics*, Fourth Edition and Fifth Edition. Combining his love of statistics and love of sports, Josh teamed with Christine Franklin to write *Statistical Reasoning in Sports*, an innovative textbook for on-level statistics courses. Josh is also coauthor of the popular on-level text *Statistics and Probability with Applications*.

# Content Advisory Board and Supplements Team

**Ann Cannon,** Cornell College, Mount Vernon, IA
*Content Advisor, Accuracy Checker*

Ann has served as Reader, Table Leader, Question Leader, and Assistant Chief Reader for the AP® Statistics exam for the past 17 years. She is also the 2017 recipient of the Mu Sigma Rho William D. Warde Statistics Education Award for a lifetime devotion to the teaching of statistics. Ann has taught introductory statistics at the college level for 25 years and is very active in the Statistics Education Section of the American Statistical Association, currently serving as secretary/treasurer. She is coauthor of *STAT2: Building Models for a World of Data* (W. H. Freeman and Company).

**Luke Wilcox,** East Kentwood High School, Kentwood, MI *Content Advisor, Teacher's Edition, Teacher's Resource Materials*

Luke has been a math teacher for 15 years and is currently teaching Intro Statistics and AP® Statistics. Luke recently received the Presidential Award for Excellence in Mathematics and Science Teaching and was named Michigan Teacher of the Year 2016–2017. He facilitates professional development for teachers in curriculum, instruction, assessment, and strategies for motivating students. Lindsey Gallas and Luke are the co-bloggers at The Stats Medic (www.statsmedic.com), a site dedicated to improving statistics education, which includes activities and lessons for this textbook.

**Erica Chauvet,** Waynesburg University, PA
*Solutions, Tests and Quizzes, Test Bank, Online Homework*

Erica has more than 15 years of experience in teaching high school and college statistics and has served as an AP® Statistics Reader for the past 10 years. Erica famously hosts the two most highly anticipated events at the Reading: the Fun Run and the Closing Ceremonies. She has also worked as a writer, consultant, and reviewer for statistics and calculus textbooks for the past 10 years.

**Doug Tyson,** Central York High School, York, PA
*Content Advisor, Videos and Video Program Manager, Lecture Slide Presentations*

Doug has taught mathematics and statistics to high school and undergraduate students for more than 25 years. He has taught AP® Statistics for 11 years and has been active as an AP® Reader and Table Leader for a decade. Doug is the coauthor of a curriculum module for the College Board and the *Teacher's Edition* for *Statistics and Probability with Applications*, Third Edition. He conducts student review sessions around the country and leads workshops on teaching statistics. Doug also serves on the NCTM/ASA Joint Committee on Curriculum in Statistics and Probability.

**Beth Benzing,** Strath Haven High School, Wallingford/Swarthmore School District, Wallingford, PA
*Activity Videos*

Beth has taught AP® Statistics since 2000 and has served as a Reader for the AP® Statistics exam for the past 7 years. She served as president, and is a current board member, of the regional affiliate for NCTM in the Philadelphia area and has been a moderator for an online course, Teaching Statistics with Fathom. Beth has an MA in Applied Statistics from George Mason University.

**Paul Buckley,** Gonzaga College High School, Washington, DC
*Videos, Online Homework*

Paul has taught high school math for 24 years and AP® Statistics for 16 years. He has been an AP® Statistics Reader for 10 years and a Table Leader for the past 4 years. Paul has presented at Conferences for AP®, NCTM, NCEA (National Catholic Education Association), and JSEA (Jesuit Secondary Education Association) and has served as a representative for the American Statistical Association at the American School Counselors Association annual conference.

**James Bush,** Waynesburg University, Waynesburg, PA
*Test Bank, Videos*

James has taught introductory and advanced courses in statistics for over 35 years. He is currently a Professor of Mathematics at Waynesburg University and is the recipient of the Lucas Hathaway Teaching Excellence Award. James serves as an AP® Table Leader, leads AP® Statistics preparation workshops through the National Math and Science Initiative, and has been a speaker at NCTM, USCOTS, and the Advance Kentucky Fall Forum.

**Monica DeBold,** Harrison High School, Harrison, NY
*Videos*
Monica has taught for 10 years at both the high school and college levels. She is experienced in probability and statistics, as well as AP® Statistics and International Baccalaureate math courses. Monica has served as a mentor teacher in her home district and, more recently, as an AP® Statistics Reader.

**Lindsey Gallas,** East Kentwood High School, Kentwood, MI
*Videos*
Lindsey has recently begun teaching AP® Statistics after spending many years teaching introductory statistics and algebra. Together with Luke Wilcox, Lindsey has created www.statsmedic.com, a site about how to teach high school statistics effectively—which includes daily lesson planning and activities for this textbook.

**Vicki Greenberg,** Atlanta Jewish Academy, Atlanta, GA
*Videos*
Vicki has taught mathematics and statistics to high school and undergraduate students for more than 18 years. She has taught AP® Statistics for 10 years and served as an AP® Reader for 7 years. She is the co-author of an AP® Statistics review book and conducts student review sessions and workshops for teachers. Her educational passion is making mathematics fun and relevant to enhance students' mathematical and statistical understanding.

**DeAnna McDonald,** University of Arizona, Tucson, AZ
*Videos*
DeAnna has taught introductory and AP® Statistics courses for 20 years. She currently teaches statistics as an adjunct instructor at the University of Arizona in the Mathematics Department and taught AP® Statistics at University High School in Tucson for many years. DeAnna has served as an AP® Statistics Reader for 12 years, including 4 years as a Table Leader.

**Leigh Nataro,** Moravian Academy, Bethlehem, PA
*Technology Corner Videos, TI-Nspire Technology Corners*
Leigh has taught AP® Statistics for 13 years and has served as an AP® Statistics Reader for the past 8 years. She enjoys the challenge of writing multiple-choice questions for the College Board for use on the AP® Statistics exam. Leigh is a National Board Certified Teacher in Adolescence and Young Adulthood Mathematics and was previously named a finalist for the Presidential Award for Excellence in Mathematics and Science Teaching in New Jersey.

**Jonathan Osters,** The Blake School, Minneapolis, MN
*Videos*
Jonathan has taught high school mathematics for 12 years. He teaches AP® Statistics, Probability & Statistics, and Geometry and has been a reader for the AP® Statistics exam for 9 years. Jonathan writes a blog about teaching at experiencefirstmath.org and tweets at @callmejosters.

**Tonya Adkins,** Charlotte, NC
*Accuracy Checker, Online Homework*
Tonya has been teaching math and statistics courses for more than 20 years in high schools and colleges in Alabama and North Carolina. She taught AP® Statistics for 10 years and has served as an AP® Reader for the past four years. Tonya also works as a reviewer, consultant, and subject matter expert on mathematics and statistics projects for publishers.

**Robert Lochel,** Hatboro-Horsham High School, Horsham PA
*Online Homework, Desmos Projects*
Bob has served as a high school math teacher and curriculum coach in his district for 21 years, and has taught AP® Statistics for 13 years. He has been an AP® Statistics Reader for the past 6 years. Bob has a passion for developing lessons that leverage technology in math classrooms, and he has shared his ideas at national conferences for NCTM and ISTE. He has served as a section editor for NCTM Mathematics Teacher "Tech Tips" for the last 3 years.

**Sandra Lowell,** Brandeis High School, San Antonio, TX
*Online Homework*
Sandra was a software engineer for 8 years and has taught high school math for 24 years, serving as mathematics coordinator and lead AP® Statistics teacher. She has taught AP® Statistics for 19 years and has been an AP® Statistics Reader for 15 years, serving as a Table Leader for the last 3 years. Sandra is currently teaching at Brandeis

High School and is an adjunct professor at Northwest Vista College.

**Dori Peterson,** Northwest Vista College, San Antonio, TX
*Online Homework*

Dori taught high school math for 28 years and AP® Statistics for 12 years. She served as a mathematics coordinator and statistics lead instructor for 4 years. Dori is currently an adjunct professor of math and statistics at Northwest Vista College. She has been an AP® Statistics Reader for 9 years, serving as a Table Leader for the last 3 years. Dori is a member of the American Statistical Association and served as a project competition judge for 2 years.

**Mary Simons,** Midlothian, VA
*Online Homework*

Mary has taught high school math for 15 years and AP® Statistics for 9 years. She has been an AP® Statistics Reader for the past 5 years. Mary is a member of the American Statistical Association, serving as a project competition judge for the past 4 years. She has also worked as a member of the Delaware Department of Education Mathematics Assessment Committee, the Delaware Mathematics Coalition, and has served as a cooperating teacher for the University of Delaware.

**Jason Molesky** *Strive for a 5 Guide*

Jason served as an AP® Statistics Reader and Table Leader since 2006. After teaching AP® Statistics for 8 years and developing the FRAPPY system for AP® Statistics exam preparation, Jason served as the Director of Program Evaluation and Accountability for the Lakeville Area Public Schools. He has recently settled into his dream job as an educational consultant for Apple. Jason maintains the "Stats Monkey" website, a clearinghouse for AP® Statistics resources.

**Michael Legacy,** Greenhill School, Dallas, TX
*Strive for a 5 Guide*

Michael is a past member of the AP® Statistics Development Committee (2001–2005) and a former Table Leader at the Reading. He currently reads the Alternate Exam and is a presenter at many AP® Summer Institutes. Michael is the author of the 2007 College Board AP® Statistics Teacher's Guide and was named the Texas 2009–2010 AP® Math/Science Teacher of the Year by the Siemens Corporation.

# Acknowledgments

First and foremost, we owe a tremendous debt of gratitude to David Moore and Dan Yates. Professor Moore reshaped the college introductory statistics course through publication of three pioneering texts: *Introduction to the Practice of Statistics (IPS), The Basic Practice of Statistics (BPS)*, and *Statistics: Concepts and Controversies*. He was also one of the original architects of the AP® Statistics course. When the course first launched in the 1996–1997 school year, there were no textbooks written specifically for the high school student audience that were aligned to the AP® Statistics topic outline. Along came Dan Yates. His vision for such a text became reality with the publication of *The Practice of Statistics (TPS)* in 1998. Over a million students have used one of the first five editions of *TPS* for AP® Statistics! Dan also championed the importance of developing high-quality resources for AP® Statistics teachers, which were originally provided in a *Teachers' Resource Binder*. We stand on the shoulders of two giants in statistics education as we carry forward their visions in this and future editions.

*The Practice of Statistics* has continued to evolve, thanks largely to the support of our longtime editor and team captain, Ann Heath. Her keen eye for design is evident throughout the pages of the student and teacher's editions. More importantly, Ann's ability to oversee all of the complex pieces of this project while maintaining a good sense of humor is legendary. Ann has continually challenged everyone involved with *TPS* to innovate in ways that benefit AP® Statistics students and teachers. She is a good friend and an inspirational leader.

*Teamwork* is the secret sauce of *TPS*. We have been blessed to collaborate with many talented AP® Statistics teachers and introductory statistics college professors over the years we have been working on this project. We sincerely appreciate their willingness to give us candid feedback about early drafts of the student edition, and to assist with the development of an expanding cadre of resources for students and teachers.

On the sixth edition, we are especially grateful to the individuals who played lead roles in key components of the project. Ann Cannon did yeoman's work once again in reading, reviewing, and accuracy checking every line in the student edition. Her sage advice and willingness to ask tough questions were much appreciated throughout the writing of *TPS* 6e. Luke Wilcox took on the herculean task of producing the *Teacher's Edition* (TE). We know teachers will appreciate his careful thinking about effective pedagogy and the importance of engaging students with relevant context throughout the TE chapters. Working with his colleague, Lindsay Gallas, Luke also oversaw creation of the fabulous "150 Days of AP® Statistics" resource for teachers at his StatsMedic site.

Erica Chauvet wrote all of the solutions for *TPS* 6e exercises. Her thorough attention to matching the details in worked examples was exceeded only by her remarkable speed in completing this burdensome task. Erica also agreed to manage a substantial revision of the test bank, including crafting prototype quizzes and tests, and has assisted with the online homework content.

Doug Tyson is overseeing production of the vast collection of *TPS* 6e tutorial videos for students and teachers. We are thankful for Doug's expertise in video creation and for his willingness to pitch in wherever we need him. Tonya Adkins kindly agreed to spearhead our new online homework system for this edition. Welcome to the team, Tonya!

Every member of the *TPS* 6e Content Advisory Board and Supplements Team is an experienced teacher with significant involvement in the AP® Statistics program. In addition to the individuals above, we offer our heartfelt thanks to the following list of superstars for their tireless work and commitment to excellence: Beth Benzing, Don Brechlin, Paul Buckley, James Bush, Monica Debold, Kathleen Dickensheets, Lindsay Gallas, Vicki Greenberg, Michael Legacy, Bob Lochel, Sandra Lowell, DeAnna McDonald, Stephen Miller, Jason Molesky, Leigh Nataro, Jonathan Osters, Dori Peterson, Al Reiff, and Mary Simons.

Sincere gratitude also goes to everyone at Bedford, Freeman, and Worth (BFW) involved in *TPS* 6e. Don Gecewicz returned partway through the manuscript writing to offer helpful developmental edits. Vivien Weiss and Susan Wein oversaw the production process with their usual care and professionalism. Louise Ketz kept us clear and consistent with her thoughtful copyediting. Corrina Santos and Kaitlyn Swygard worked behind the scenes to carefully prepare manuscript chapters for production. Diana Blume ensured that the design of the finished book exceeded our expectations. Special thanks go to all the dedicated people on the high school sales and marketing team at BFW who promote *TPS* 6e with enthusiasm. We also offer our thanks to Murugesh Rajkumar Namasivayam and the team at Lumina Datamatics for turning a complex manuscript into good-looking page proofs.

Thank you to all the reviewers who offered encouraging words and thoughtful suggestions for improvement in this and previous editions. And to the many outstanding statistics educators who have taken the time to share their questions and insights with us online, at conferences and workshops, at the AP® Reading, and in assorted other venues, we express our appreciation.

Daren Starnes and Josh Tabor

*A final note from Daren:* I feel extremely fortunate to have partnered with Josh Tabor in writing *TPS* 6e. He is a gifted teacher and talented author in his own right. Josh's willingness to take on half of the chapters in this edition pays tribute to his unwavering commitment to excellence. He enjoys exploring new possibilities, which ensures that *TPS* will keep evolving in future editions. Josh is a good friend and trusted colleague.

My biggest thank you goes to my wife, Judy. She has made incredible sacrifices throughout my years as a textbook author. For Judy's unconditional love and support, I would like to dedicate this edition to her. She is my inspiration.

*A final note from Josh:* I have greatly enjoyed working with Daren Starnes on this edition of *TPS*. No one I know works harder and holds himself to a higher standard than he does. His wealth of experience and vision for this edition made him an excellent writing partner. For your friendship, encouragement, and support—thanks!

I especially want to thank the two most important people in my life. To my wife, Anne, your patience while I spent countless hours working on this project is greatly appreciated. I couldn't have survived without your consistent support and encouragement. To my daughter, Jordan, I can't believe how quickly you are growing up. It won't be long until you are reading *TPS* as a student! I love you both very much.

## Sixth Edition Survey Participants and Reviewers

Paul Bastedo, *Viewpoint School, Calabasas, CA*
Emily Beal, *Chagrin Falls High School, Chagrin Falls, OH*
Raquel Bocast, *Hamilton Union High School, Hamilton City, CA*
Lisa Bonar, *Fossil Ridge High School, Keller, TX*
Robert Boone, *First Coast High School, Jacksonville, FL*
John Bowman, *Hinsdale Central High School, Hinsdale, IL*
Brigette Brankin, *St. Viator High School, Arlington Heights, IL*
Chris Burke, *Hingham High School, Hingham, MA*
Kenny Contreras, *Wilcox High School, Santa Clara, CA*
Nancy Craft, *Chestnut Hill Academy, Philadelphia, PA*
Aimee Davenport, *Heritage High School, Frisco, TX*
Gabrielle Dedrick, *Ernest McBride High School, Long Beach, CA*
Michael Ditzel, *Hampden Academy, Hampden, ME*
Robin Dixon, *Panther Creek High School, Cary, NC*
Parisa Foroutan, *Renaissance School for the Arts, Long Beach, CA*
Rebecca Gaillot, *Metairie Park Country Day School, Metairie, LA*
Roger Gale, *Waukegan High School, Waukegan, IL*
Becky Gerek, *Abraham Lincoln High School, San Francisco, CA*
Lisa Haney, *Monticello High School, Charlottesville, VA*
Kellie Hodge, *Jordan High School, Long Beach, CA*
Susan Knott, *The Oakridge School, Arlington, TX*
Lauren Kriczky, *Clewiston High School, Clewiston, FL*
William Ladley, *Bell High School, Hurst, TX*
Cathy Lichodziejewski, *Fountain Valley High School, Fountain Valley, CA*
Veronica Lunde, *Apollo High School, Saint Cloud, MN*
Shannon McBriar, *Central High School, Macon, GA*
Maureen McMichael, *Seneca High School, Tabernacle, NJ*
Victor Mirrer, *Fairfield Ludlowe High School, Fairfield, CT*
Jose Molina, *St. Mary's Hall, San Antonio, TX*
Kevin Morgan, *Central Columbia High School, Bloomsburg, PA*
Karin Munro, *Marcus High School, Flower Mound, TX*
Leigh Nataro, *Moravian Academy, Bethlehem, PA*
Cindy Parliament, *Klein Oak High School, Spring, TX*
Juliet Pender, *New Egypt High School, New Egypt, NJ*
John Powers, *Cardinal Gibbons High School, Fort Lauderdale, FL*
Jessica Quinn, *Mayfield Senior School, Pasadena, CA*
Gary Remiker, *Cathedral Catholic High School, San Diego, CA*
Michael Rice, *Rainer Beach High School, Seattle, WA*
Laura Ringwood, *Westlake High School, Austin, TX*

Gina Ruth, *Woodrow Wilson High School, Beckley, WV*
Dan Schmidt, *Rift Valley Academy, Kijabe, Kenya*
Ned Smith, *Southwest High School, Fort Worth, TX*
Joseph Tanzosh, *Marian High School, Mishawaka, IN*
Rachael Thiele, *Polytechnic High School, Long Beach, CA*
Jenny Thom-Carroll, *West Essex Senior High School, North Caldwell, NJ*
Tara Truesdale, *Ben Lippen School, Columbia, SC*
Alethea Trundy, *Montachusett Regional Vocational Technical School, Fitchburg, MA*
Crystal Vesperman, *Prairie School, Racine, WI*
Kristine Witzel, *Duchesne High School, Saint Charles, MO*

## Fifth Edition Survey Participants and Reviewers

Blake Abbott, *Bishop Kelley High School, Tulsa, OK*
Maureen Bailey, *Millcreek Township School District, Erie, PA*
Kevin Bandura, *Lincoln County High School, Stanford, KY*
Elissa Belli, *Highland High School, Highland, IN*
Jeffrey Betlan, *Yough School District, Herminie, PA*
Nancy Cantrell, *Macon County Schools, Franklin, NC*
Julie Coyne, *Center Grove High School, Greenwood, IN*
Mary Cuba, *Linden Hall, Lititz, PA*
Tina Fox, *Porter-Gaud School, Charleston, SC*
Ann Hankinson, *Pine View, Osprey, FL*
Bill Harrington, *State College Area School District, State College, PA*
Ronald Hinton, *Pendleton Heights High School, Pendleton, IN*
Kara Immonen, *Norton High School, Norton, MA*
Linda Jayne, *Kent Island High School, Stevensville, MD*
Earl Johnson, *Chicago Public Schools, Chicago, IL*
Christine Kashiwabara, *Mid-Pacific Institute, Honolulu, HI*
Melissa Kennedy, *Holy Names Academy, Seattle, WA*
Casey Koopmans, *Bridgman Public Schools, Bridgman, MI*
David Lee, *Sun Prairie High School, Sun Prairie, WI*
Carolyn Leggert, *Hanford High School, Richland, WA*
Jeri Madrigal, *Ontario High School, Ontario, CA*
Tom Marshall, *Kents Hill School, Kents Hill, ME*
Allen Martin, *Loyola High School, Los Angeles, CA*
Andre Mathurin, *Bellarmine College Preparatory, San José, CA*
Brett Mertens, *Crean Lutheran High School, Irvine, CA*
Sara Moneypenny, *East High School, Denver, CO*
Mary Mortlock, *The Harker School, San José, CA*
Mary Ann Moyer, *Hollidaysburg Area School District, Hollidaysburg, PA*
Howie Nelson, *Vista Murrieta High School, Murrieta, CA*
Shawnee Patry, *Goddard High School, Wichita, KS*
Sue Pedrick, *University High School, Hartford, CT*
Shannon Pridgeon, *The Overlake School, Redmond, WA*
Sean Rivera, *Folsom High, Folsom, CA*
Alyssa Rodriguez, *Munster High School, Munster, IN*
Sheryl Rodwin, *West Broward High School, Pembroke Pines, FL*
Sandra Rojas, *Americas High School, El Paso, TX*
Amanda Schneider, *Battle Creek Public Schools, Charlotte, MI*
Christine Schneider, *Columbia Independent School, Boonville, MO*
Steve Schramm, *West Linn High School, West Linn, OR*
Katie Sinnott, *Revere High School, Revere, MA*
Amanda Spina, *Valor Christian High School, Highlands Ranch, CO*
Julie Venne, *Pine Crest School, Fort Lauderdale, FL*
Dana Wells, *Sarasota High School, Sarasota, FL*
Luke Wilcox, *East Kentwood High School, Grand Rapids, MI*
Thomas Young, *Woodstock Academy, Putnam, CT*

## Fourth Edition Focus Group Participants and Reviewers

Gloria Barrett, *Virginia Advanced Study Strategies, Richmond, VA*
David Bernklau, *Long Island University, Brookville, NY*
Patricia Busso, *Shrewsbury High School, Shrewsbury, MA*
Lynn Church, *Caldwell Academy, Greensboro, NC*
Steven Dafilou, *Springside High School, Philadelphia, PA*
Sandra Daire, *Felix Varela High School, Miami, FL*
Roger Day, *Pontiac High School, Pontiac, IL*
Jared Derksen, *Rancho Cucamonga High School, Rancho Cucamonga, CA*
Michael Drozin, *Munroe Falls High School, Stow, OH*
Therese Ferrell, *I. H. Kempner High School, Sugar Land, TX*
Sharon Friedman, *Newport High School, Bellevue, WA*
Jennifer Gregor, *Central High School, Omaha, NE*
Julia Guggenheimer, *Greenwich Academy, Greenwich, CT*
Dorinda Hewitt, *Diamond Bar High School, Diamond Bar, CA*
Dorothy Klausner, *Bronx High School of Science, Bronx, NY*
Robert Lochel, *Hatboro-Horsham High School, Horsham, PA*
Lynn Luton, *Duchesne Academy of the Sacred Heart, Houston, TX*
Jim Mariani, *Woodland Hills High School, Greensburgh, PA*
Stephen Miller, *Winchester Thurston High School, Pittsburgh, PA*
Jason Molesky, *Lakeville Area Public Schools, Lakeville, MN*
Mary Mortlock, *The Harker School, San José, CA*
Heather Nichols, *Oak Creek High School, Oak Creek, WI*
Jamis Perrett, *Texas A&M University, College Station, TX*
Heather Pessy, *Mount Lebanon High School, Pittsburgh, PA*
Kathleen Petko, *Palatine High School, Palatine, IL*
Todd Phillips, *Mills Godwin High School, Richmond, VA*
Paula Schute, *Mount Notre Dame High School, Cincinnati, OH*
Susan Stauffer, *Boise High School, Boise, ID*
Doug Tyson, *Central York High School, York, PA*
Bill Van Leer, *Flint High School, Oakton, VA*
Julie Verne, *Pine Crest High School, Fort Lauderdale, FL*
Steve Willot, *Francis Howell North High School, St. Charles, MO*
Jay C. Windley, *A. B. Miller High School, Fontana, CA*

# To the Student

## Statistical Thinking and You

The purpose of this book is to give you a working knowledge of the big ideas of statistics and of the methods used in solving statistical problems. Because data always come from a real-world context, doing statistics means more than just manipulating data. *The Practice of Statistics* (*TPS*), Sixth Edition, is full of data. Each set of data has some brief background to help you understand where the data come from. We deliberately chose contexts and data sets in the examples and exercises to pique your interest.

*TPS* 6e is designed to be easy to read and easy to use. This book is written by current high school AP® Statistics teachers, for high school students. We aimed for clear, concise explanations and a conversational approach that would encourage you to read the book. We also tried to enhance both the visual appeal and the book's clear organization in the layout of the pages.

Be sure to take advantage of all that *TPS* 6e has to offer. You can learn a lot by reading the text, but you will develop deeper understanding by doing the Activities and Projects and answering the Check Your Understanding questions along the way. The walkthrough guide on pages xvi–xxii gives you an inside look at the important features of the text.

You learn statistics best by doing statistical problems. This book offers many different types of problems for you to tackle.

- **Section Exercises** include paired odd- and even-numbered problems that test the same skill or concept from that section. There are also some multiple-choice questions to help prepare you for the AP® Statistics exam. Recycle and Review exercises at the end of each exercise set involve material you studied in preceding sections.
- **Chapter Review Exercises** consist of free-response questions aligned to specific learning targets from the chapter. Go through the list of learning targets summarized in the Chapter Review and be sure you can say of each item on the list, "I can do that." Then prove it by solving some problems.
- The **AP® Statistics Practice Test** at the end of each chapter will help you prepare for in-class exams. Each test has about 10 multiple-choice questions and 3 free-response problems, very much in the style of the AP® Statistics exam.
- Finally, the **Cumulative AP® Practice Tests** after Chapters 4, 7, 11, and 12 provide challenging, cumulative multiple-choice and free-response questions like those you might find on a midterm, final, or the AP® Statistics exam.

The main ideas of statistics, like the main ideas of any important subject, took a long time to discover and thus take some time to master. The basic principle of learning them is to be persistent. Once you put it all together, statistics will help you make informed decisions based on data in your daily life.

## *TPS* and AP® Statistics

*The Practice of Statistics* (*TPS*) was the first book written specifically for the Advanced Placement (AP®) Statistics course. This updated version of *TPS* 6e is organized to closely follow the AP® Statistics Course Framework (CF). Every learning objective and essential knowledge statement in the CF is covered thoroughly in the text. Visit the book's website at highschool.bfwpub.com/updatedtps6e for a detailed alignment guide, including "The Nitty Gritty Alignment" guide to EKs and LOs in Updated *TPS* 6e. The few topics in the book that go beyond the AP® Statistics syllabus are marked with an asterisk (*).

Most importantly, *TPS* 6e is designed to prepare you for the AP® Statistics exam. The author team has been involved in the AP® Statistics program since its early days. We have more than 40 years' combined experience teaching AP® Statistics and grading the AP® exam! Both of us have served as Question Leaders for more than 10 years, helping to write scoring rubrics for free-response questions. Including our Content Advisory Board and Supplements Team (page vii), we have extensive knowledge of how the AP® Statistics exam is developed and scored.

*TPS* 6e will help you get ready for the AP® Statistics exam throughout the course by:

- **Using terms, notation, formulas, and tables consistent with those found in the Course Framework and on the AP® Statistics exam.** Key terms are shown in bold in the text, and they are defined in the Glossary. Key terms also are cross-referenced in the Index. See page F-1 to find "Formulas for the AP® Statistics Exam," as well as Tables A, B, and C in the back of the book for reference.
- **Following accepted conventions from AP® Statistics exam rubrics when presenting model solutions.** Over the years, the scoring guidelines for free-response questions have become fairly consistent. We kept these guidelines in mind when writing the solutions that appear throughout *TPS* 6e. For example, the four-step State–Plan–Do–Conclude process that we use to complete inference problems in Chapters 8–12 closely matches the four-point AP® scoring rubrics.
- **Including AP® Exam Tips in the margin where appropriate.** We place exam tips in the margins as "on-the-spot" reminders of common mistakes and how to avoid them. These tips are collected and summarized in the About the AP® Exam and AP® Exam Tips appendix.
- **Providing over 1600 AP®-style exercises throughout the book.** Each chapter contains a mix of free-response and multiple-choice questions that are similar to those found on the AP® Statistics exam. At the start of each Chapter Wrap-Up, you will find a FRAPPY (Free Response AP® Problem, Yay!). Each FRAPPY gives you the chance to solve an AP®-style free-response problem based on the material in the chapter. After you finish, you can view and critique

two example solutions from the book's Student Site (highschool.bfwpub.com/updatedtps6e). Then you can score your own response using a rubric provided by your teacher.

- **Developing the Course Skills for AP® Statistics through frequent repetition.** See the inside back cover for more details.

Turn the page for a tour of the text. See how to use the book to realize success in the course and on the AP® Statistics exam.

# READ THE TEXT and use the book's features to help you grasp the big ideas.

Section 3.1 Scatterplots and Correlation 153

## SECTION 3.1 Scatterplots and Correlation

Read the **LEARNING TARGETS** at the beginning of each section. Focus on mastering these skills and concepts as you work through the chapter.

**LEARNING TARGETS** *By the end of the section, you should be able to:*

- Distinguish between explanatory and response variables for quantitative data.
- Make a scatterplot to display the relationship between two quantitative variables.
- Describe the direction, form, and strength of a relationship displayed in a scatterplot and identify unusual features.
- Interpret the correlation.
- Understand the basic properties of correlation, including how the correlation is influenced by unusual points.
- Distinguish correlation from causation.

A one-variable data set is sometimes called *univariate data*. A data set that describes the relationship between two variables is sometimes called *bivariate data*.

Most statistical studies examine data on more than one variable for a group of individuals. Fortunately, analysis of relationships between two variables builds on the same tools we used to analyze one variable. The principles that guide our work also remain the same:

- Plot the data, then look for overall patterns and departures from those patterns.
- Add numerical summaries.
- When there's a regular overall pattern, use a simplified model to describe it.

### Explanatory and Response Variables

In the "Candy grab" activity, the number of candies is the **response variable**. Hand span is the **explanatory variable** because we anticipate that knowing a student's hand span will help us predict the number of candies that student can grab.

It's important to learn the language of statistics. Take note of the green **DEFINITION** boxes that explain important vocabulary. Flip back to them to review key terms and their definitions, or turn to the Glossary/Glosario at the back of the book.

**DEFINITION Response variable, Explanatory variable**

A **response variable** measures an outcome of a study. An **explanatory variable** may help predict or explain changes in a response variable.

Scan the margins for the green notes, which represent the "voice of the teacher" giving helpful hints for being successful in the course. Many of these notes include important reminders from the AP® Statistics Course Framework.

You will often see explanatory variables called *independent variables* and response variables called *dependent variables*. Because the words *independent* and *dependent* have other meanings in statistics, we won't use them here.

It is easiest to identify explanatory and response variables when we initially specify the values of one variable to see how it affects another variable. For instance, to study the effect of alcohol on body temperature, researchers gave several different amounts of alcohol to mice. Then they measured the change in each mouse's body temperature 15 minutes later. In this case, amount of alcohol is the explanatory variable, and change in body temperature is the response variable. When we don't specify the values of either variable before collecting the data, there may or may not be a clear explanatory variable.

Read the **AP® EXAM TIPS.** They give advice on how to be successful on the AP® Statistics exam.

**AP® EXAM TIP**

When you are asked to *describe* the association shown in a scatterplot, you are expected to discuss the direction, form, and strength of the association, along with any unusual features, *in the context of the problem*. This means that you need to use both variable names in your description.

**HOW TO DESCRIBE A SCATTERPLOT**

To describe a scatterplot, make sure to address the following four characteristics in the context of the data:

- **Direction:** A scatterplot can show a positive association, negative association, or no association.
- **Form:** A scatterplot can show a linear form or a nonlinear form. The form is linear if the overall pattern follows a straight line. Otherwise, t form is nonlinear.
- **Strength:** A scatterplot can show a weak, moderate, or strong associati An association is strong if the points don't deviate much from the forn identified. An association is weak if the points deviate quite a bit from the form identified.

Look for the boxes with the green bands. Some explain how to make graphs or set up calculations, while others recap important concepts.

Watch for **CAUTION ICONS.** They alert you to common mistakes that students often make.

Few relationships are linear for all values of the explanatory variable. **Don't make predictions using values of $x$ that are much larger or much smaller than those that actually appear in your data.**

# LEARN STATISTICS BY *DOING* STATISTICS

Every chapter begins with a hands-on **ACTIVITY** that introduces the content of the chapter. Many of these activities involve collecting data and drawing conclusions from the data. In other activities, you'll use dynamic applets to explore statistical concepts.

**ACTIVITY** Candy grab

In this activity, you will investigate if students with a larger hand span can grab more candy than students with a smaller hand span.[1]

1. Measure the span of your dominant hand to the nearest half-centimeter (cm). Hand span is the distance from the tip of the thumb to the tip of the pinkie finger on your fully stretched-out hand.
2. One student at a time, go to the front of the class and use your dominant hand to grab as many candies as possible from the container. You must grab the candies with your fingers pointing down (no scooping!) and hold the candies for 2 seconds before counting them. After counting, put the candy back into the container.
3. On the board, record your hand span and number of candies in a table with the following headings:

| Hand span (cm) | Number of candies |
|---|---|

4. While other students record their values on the board, copy the table onto a piece of paper and make a graph. Begin by constructing a set of coordinate axes. Label the horizontal axis … the vertical axis "Number of candies." … scale for each axis and plot each point … e as accurately as you can on the graph.

…ell you about the relationship between …r of candies? Summarize your observa-…o.

**ACTIVITY** Investigating properties of the least-squares regression line

APPLET

In this activity, you will use the *Correlation and Regression* applet to explore some properties of the least-squares regression line.

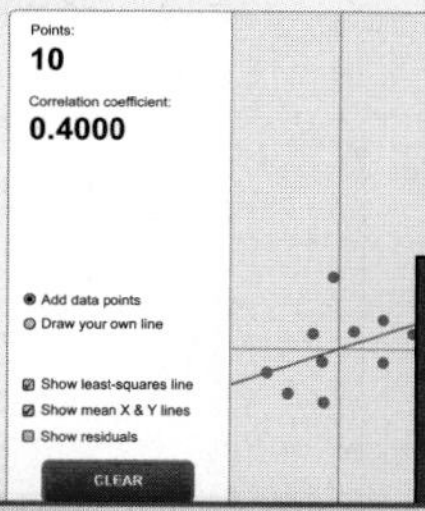


1. Launch the applet at highschool.bfwpub.com/updatedtps6e.
2. Click on the graphing area to add 10 points in the lower-left corner so that the correlation is about $r = 0.40$. Also, check the boxes to show the "Least-Squares Line" and the "Mean X & Y"

Chapters 1, 3, 4, 11, and 12 conclude with a **CHAPTER PROJECT.** Three of the projects (Chapters 1, 3, and 11) provide an opportunity to think like a statistician by analyzing larger data sets with multiple variables of interest. The other two (Chapters 4 and 12) are longer-term projects that require you to engage in the statistical problem-solving process: Ask Questions, Collect Data, Analyze Data, Interpret Results.

## Chapter 3 Project Investigating Relationships in Baseball

What is a better predictor of the number of wins for a baseball team, the number of runs scored by the team or the number of runs they allow the other team to score? What variables can we use to predict the number of runs a team scores? To predict the number of runs it allows the other team to score? In this project, you will use technology to help answer these questions by exploring a large set of data from Major League Baseball.

### Part 1

1. Download the "MLB Team Data 2012–2016" Excel file from the book's website, along with the "Glossary for MLB Team Data file," which explains each of the variables included in the data set.[58] Import the data into th… cal software package you prefer.
2. Create a scatterplot to investigate the relationship … runs scored per game (R/G) and wins (W). The… late the equation of the least-squares regression … standard deviation of the residuals, and $r^2$. *Note:* … the section for hitting statistics and W is in the se… pitching statistics.

5. Because the number of wins a team has is dependent on both how many runs they score and how many runs they allow, we can use a combination of both variables to predict the number of wins. Add a column in your data table for a new variable, run differential. Fill in the values using the formula R/G – RA/G.
6. Create a scatterplot to investigate the relationship between run differential and wins. Then calculate the equation of the least-squares regression line, the standard deviation of the residuals, and $r^2$.
7. Is run differential a better predictor than the variable you chose in Question 4? Explain your reasoning.

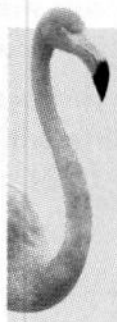

**CHECK YOUR UNDERSTANDING**

In Exercises 3 and 7, we asked you to make and describe a scatterplot for the hiker data shown in the table.

| Body weight (lb) | 120 | 187 | 109 | 103 | 131 | 165 | 158 | 116 |
|---|---|---|---|---|---|---|---|---|
| Backpack weight (lb) | 26 | 30 | 26 | 24 | 29 | 35 | 31 | 28 |

1. Calculate the equation of the least-squares regression line.
2. Make a residual plot for the linear model in Question 1.
3. What does the residual plot indicate about the appropriateness of the linear model? Explain your answer.

**CHECK YOUR UNDERSTANDING** questions appear throughout the section. They help clarify definitions, concepts, and procedures. Be sure to check your answers in the back of the book.

# EXAMPLES: Model statistical problems and how to solve them

Read through each **EXAMPLE**, and then try out the concept yourself by working the **FOR PRACTICE, TRY** exercise in the Section Exercises.

Need extra help? Examples and exercises marked with the **PLAY ICON** are supported by short video clips prepared by experienced AP® Statistics teachers. The video guides you through each step in the example and solution and provides additional explanation when you need it.

## EXAMPLE Old Faithful and fertility
### Describing a scatterplot

BigshotD3/iStock/Getty Images

**PROBLEM:** Describe the relationship in each of the following contexts.

(a) The scatterplot on the left shows the relationship between the duration (in minutes) of an eruption and the interval of time until the next eruption (in minutes) of Old Faithful during a particular month.

(b) The scatterplot on the right shows the relationship between the average income (gross domestic product per person, in dollars) and fertility rate (number of children per woman) in 187 countries.[4]

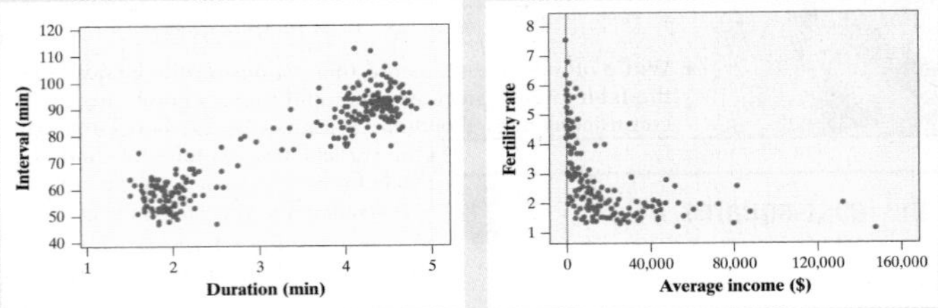


**SOLUTION:**

(a) *There is a strong, positive linear relationship between the duration of an eruption and the interval of time until the next eruption. There are two main clusters of points: one cluster has durations around 2 minutes with in…*
*has durations aro…*

(b) *There is a modera…*
*between average …*
*is a country outsi…*
*$30,000 and a f…*

Even with the clusters, the overall direction is still positive. In some cases, however, the points in a cluster …

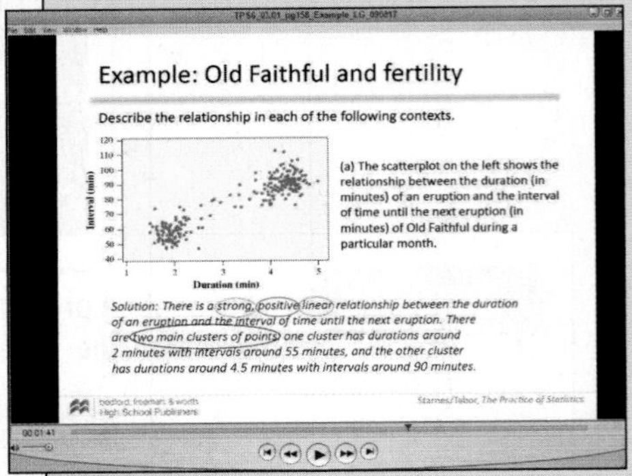


## EXAMPLE Caffeine and pulse rates
### How random assignment works

**PROBLEM:** A total of 20 students have agreed to participate in an experiment comparing the effects of caffeinated cola and caffeine-free cola on pulse rates. Describe how you would randomly assign 10 students to each of the two treatments:

(a) Using 20 identical slips of paper

(b) Using technology

(c) Using Table D

Zaia Snively

**SOLUTION:**

(a) *On 10 slips of paper, write the letter "A"; on the remaining 10 slips, write the letter "B." Shuffle the slips of paper and hand out one slip of paper to each volunteer. Students who get an "A" slip receive the cola with caffeine and students who get a "B" slip receive the cola without caffeine.*

When describing a method of random assignment, don't stop after creating the groups. Make sure to identify which group gets which treatment.

(b) *Label each student with a different integer from 1 to 20. Then randomly generate 10 different integers from 1 to 20. The students with these labels receive the cola with caffeine. The remaining 10 students receive the cola without caffeine.*

When using a random number generator or a table of random digits to assign treatments, make sure to account for the possibility of repeated numbers when describing your method.

(c) *Label each student with a different integer from 01 to 20. Go to a line of Table D and read two-digit groups moving from left to right. The first 10 different labels between 01 and 20 identify the 10 students who receive cola with caffeine. The remaining 10 students receive the caffeine-free cola. Ignore groups of digits from 21 to 00.*

**FOR PRACTICE, TRY EXERCISE 63**

The **SOLUTION** is presented in a special font and models the style, steps, and language that you should use to earn full credit on the AP® Statistics exam.

**THE VOICE OF THE TEACHER.** Study the worked examples and pay special attention to the carefully placed **"Teacher Talk" comment boxes** that guide you step by step through the solution. These comments offer lots of good advice—as if your teacher is working directly with you to solve a problem.

The blue page number icon next to an exercise points you back to the page on which the model example appears.

63. **Layoffs and "survivor guilt"** Workers who survive a layoff of other employees at their location may suffer from "survivor guilt." A study of survivor guilt and its effects used as subjects 120 students who were offered an opportunity to earn extra course credit by doing proofreading. Each subject worked in the same cubicle as another student, who was an accomplice of the experimenters. At a break midway through the work,
pg 251

# EXERCISES: Practice makes perfect!

Start by reading the **SECTION SUMMARY** to be sure that you understand the big ideas and key concepts.

## Section 3.1 Summary

- A **scatterplot** displays the relationship between two quantitative variables measured on the same individuals. Mark values of one variable on the horizontal axis ($x$ axis) and values of the other variable on the vertical axis ($y$ axis). Plot each individual's data as a point on the graph.
- If we think that a variable $x$ may help predict, explain, or even cause changes in another variable $y$, we call $x$ an **explanatory variable** and $y$ a **response variable.** Always plot the explanatory variable on the $x$ axis of a scatterplot. Plot the response variable on the $y$ axis.
- When describing a scatterplot, look for an overall pattern (direction, form, strength) and departures from the pattern (unusual features) and always answer in cor
  - Directio variable

Practice! Work the **EXERCISES** assigned by your teacher. Compare your answers to those in the Solutions appendix at the back of the book. Short solutions to the exercises numbered in red are found in the appendix.

## Section 3.1 Exercises

Most of the **exercises are paired**, meaning that odd- and even-numbered exercises test the same skill or concept. If you answer an assigned exercise incorrectly, try to figure out your mistake. Then see if you can solve the paired exercise.

1. **Coral reefs and cell phones** Identify the explanatory variable and the response variable for the following relationships, if possible. Explain your reasoning. pg 154
   (a) The weight gain of corals in aquariums where the water temperature is controlled at different levels
   (b) The number of text messages sent and the number of phone calls made in a sample of 100 students
2. **Teenagers and corn yield** Identify the explanatory variable and the response variable for the following relationships, if possible. Explain your reasoning.
   (a) The height and arm span of a sample of 50 teenagers
   (b) The yield of corn in bushels per acre and the amount of rain in the growing season
3. **Heavy backpacks** Ninth-grade students at the Webb Schools go on a backpacking trip each fall. Students are divided into hiking groups of size 8 by selecting names from a hat. Before leaving, students and their backpacks are weighed. The data here are from one hiking group. Make a scatterplot by hand that shows how backpack weight relates to body weight. pg 155

| | | | | | | | | |
|---|---|---|---|---|---|---|---|---|
| Body weight (lb) | 120 | 187 | 109 | 103 | 131 | 165 | 158 | 116 |
| Backpack weight (lb) | 26 | 30 | 26 | 24 | 29 | 35 | 31 | 28 |

4. **Putting success** How well do professional golfers putt from various distances to the hole? The data show various distances to the hole (in feet) and the percent of putts mad
   Make a scatt
   of putts mad

| Distance (ft) |
|---|

Track and Field team.[10] Describe the relationship between height and weight for these athletes.

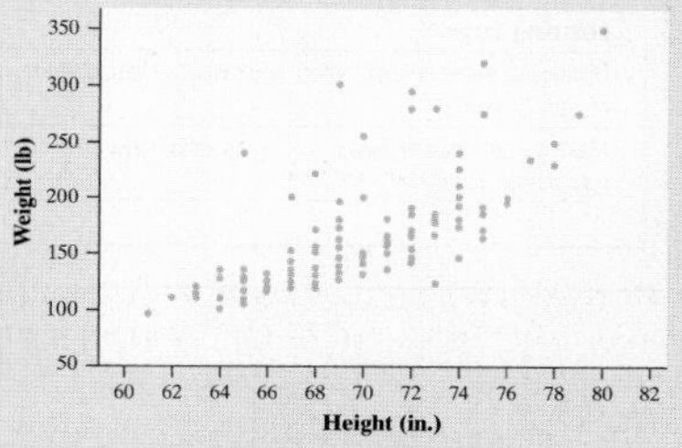

Weight (lb)
Height (in.)

6. **Starbucks** The scatterplot shows the relationship between the amount of fat (in grams) and number of calories in products sold at Starbucks.[11] Describe the relationship between fat and calories for these products.

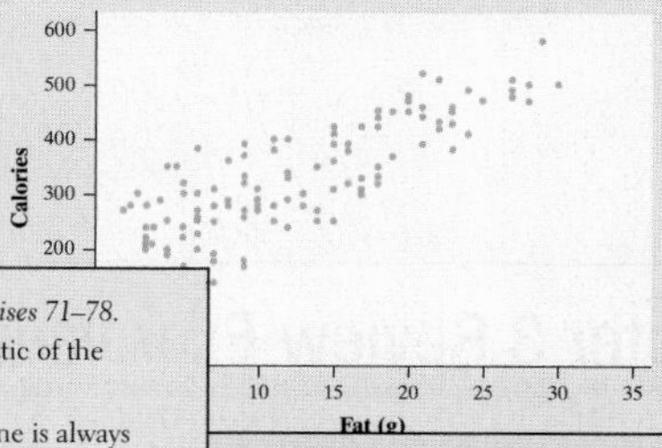

Calories
Fat (g)

Look for **ICONS** that appear next to selected **EXERCISES**. They will guide you to
- the Example that models the exercise.
- videos that provide step-by-step instructions for solving the exercise.

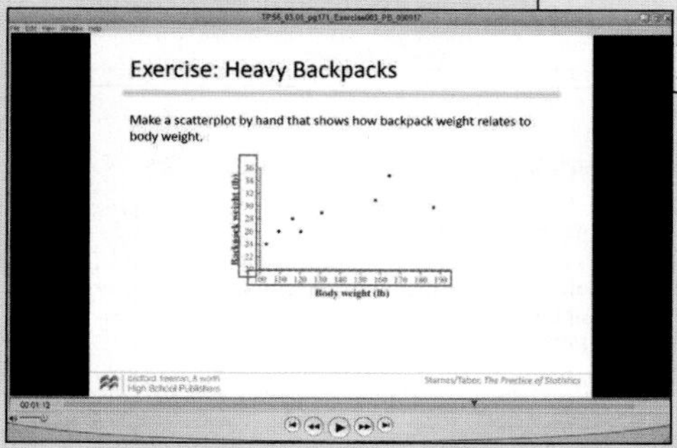

Exercise: Heavy Backpacks
Make a scatterplot by hand that shows how backpack weight relates to body weight.

**Multiple Choice:** *Select the best answer for Exercises 71–78.*

71. Which of the following is *not* a characteristic of the least-squares regression line?
   (a) The slope of the least-squares regression line is always between −1 and 1.
   (b) The least-squares regression line always goes through the point $(\bar{x}, \bar{y})$.
   (c) The least-squares regression line minimizes the sum of squared residuals.
   (d) The slope of the least-squares regression line w always have the same sign as the correlation.
   (e) The least-squares regression line is not resistant outliers.

**Recycle and Review**

79. **Fuel economy** (2.2) In its recent *Fuel Economy Guide*, the Environmental Protection Agency (EPA) gives data on 1152 vehicles. There are a number of outliers, mainly vehicles with very poor gas mileage or hybrids with very good gas mileage. If we ignore the outliers, however, the combined city and highway gas mileage of the other 1120 or so vehicles is approximately Normal with mean 18.7 miles per gallon (mpg) and standard deviation 4.3 mpg.
   (a) The Chevrolet Malibu with a four-cylinder engine has a combined gas mileage of 25 mpg. What percent of the 1120 vehicles have worse gas mileage than the Malibu?

Various types of problems in the Section Exercises let you practice solving many different types of questions, including AP®-style **multiple-choice** and **free-response**. The **Recycle and Review** exercises refer back to concepts and skills learned in an earlier section, noted in purple after the problem title.

# REVIEW and PRACTICE for quizzes and tests

Study the **CHAPTER REVIEW** to be sure that you understand the key concepts in each section.

# Chapter 3 Wrap-Up

## Chapter 3 Review

**Section 3.1: Scatterplots and Correlation**

In this section, you learned how to explore the relationship between two quantitative variables. As with distributions of a single variable, the first step is always to make a graph. A scatterplot is the appropriate type of graph to investigate relationships between two quantitative variables. To describe a scatterplot, be sure to discuss four characteristics: direction, form, strength, and unusual features. The direction of a

Use the **WHAT DID YOU LEARN?** table that directs you to examples and exercises to verify your mastery of each **LEARNING TARGET**.

### What Did You Learn?

| Learning Target | Section | Related Example on Page(s) | Relevant Chapter Review Exercise(s) |
|---|---|---|---|
| Distinguish between explanatory and response variables for quantitative data. | 3.1 | 154 | R3.4 |
| Make a scatterplot to display the relationship between two quantitative variables. | 3.1 | 155 | R3.4 |

**SUMMARY TABLES** in Chapters 8–12 review important details of each inference procedure, including conditions and formulas.

**Comparing significance tests for a proportion and a mean**

| | Significance test for $p$ | Significance test for $\mu$ |
|---|---|---|
| **Name (TI-83/84)** | One-sample $z$ test for $p$ (1-PropZTest) | One-sample $t$ test for $\mu$ (TTest) |
| **Formula** | $z = \dfrac{\hat{p} - p_0}{\sqrt{\dfrac{p_0(1 - p_0)}{n}}}$<br>$P$-value from standard Normal distribution | $t = \dfrac{\bar{x} - \mu_0}{\dfrac{s_x}{\sqrt{n}}}$<br>$P$-value from $t$ distribution with df $= n - 1$ |
| **Conditions** | • **Random:** The data come from a random sample from the population of interest.<br>○ **10%:** When sampling without replacement, $n < 0.10N$.<br>• **Large Counts** Both $np_0$ and $n(1 - p_0)$ are at least 10. That is, the expected number of successes and the expected number of failures in the sample are both at least 10. | • **Random:** The data come from a random sample from the population of interest.<br>○ **10%:** When sampling without replacement, $n < 0.10N$.<br>• **Normal/Large Sample:** The population has a Normal distribution or the sample size is large ($n \geq 30$). If the population distribution has unknown shape and $n < 30$, use a graph of the sample data to assess the Normality of the population. Do not use $t$ procedures if the graph shows strong skewness or outliers. |

## Chapter 3 Review Exercises

*These exercises are designed to help you review the important ideas and methods of the chapter.*

**R3.1 Born to be old?** Is there a relationship between the gestational period (time from conception to birth) of an animal and its average life span? The figure shows a scatterplot of the gestational period and average life span for 43 species of animals.[53]

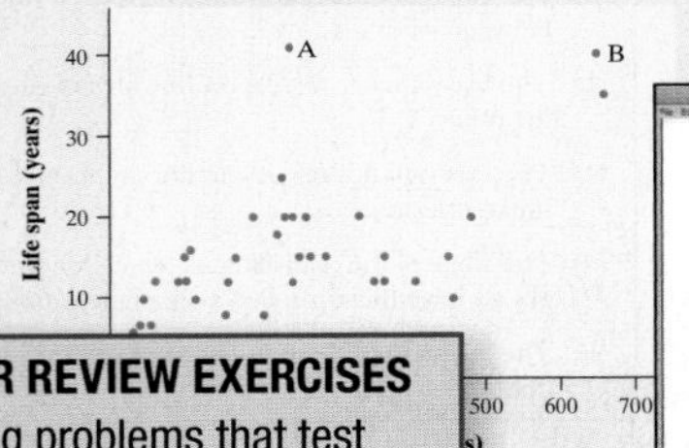
Life span (years)
40
30
20
10
A
B
500
600
700

Tackle the **CHAPTER REVIEW EXERCISES** for practice in solving problems that test concepts from throughout the chapter. Need more help or just want additional insights before you take the practice test? Watch the **Chapter Review Exercise Videos**.

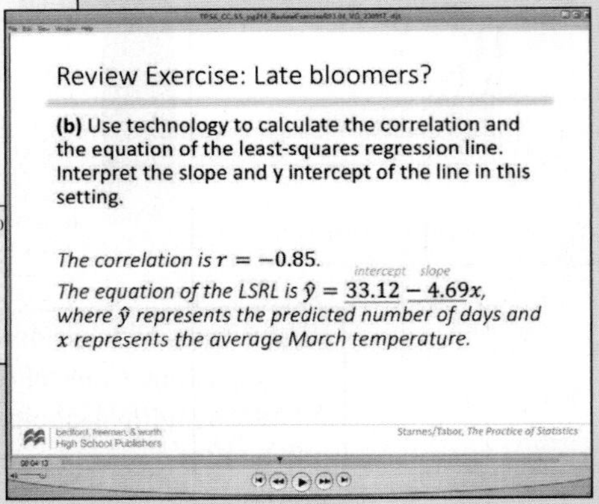
Review Exercise: Late bloomers?
(b) Use technology to calculate the correlation and the equation of the least-squares regression line. Interpret the slope and y intercept of the line in this setting.
The correlation is r = −0.85.
intercept slope
The equation of the LSRL is ŷ = 33.12 − 4.69x, where ŷ represents the predicted number of days and x represents the average March temperature.
Starnes/Tabor, The Practice of Statistics

# and the AP® STATISTICS EXAM

## Chapter 3 AP® Statistics Practice Test

**Section I: Multiple Choice** *Select the best answer for each question.*

**T3.1** A school guidance counselor examines how many extracurricular activities students participate in and their grade point average. The guidance counselor says, "The evidence indicates that the correlation between the number of extracurricular activities a student participates in and his or her grade point average is close to 0." Which of the following is the most appropriate conclusion?

(a) Students involved in many extracurricular activities tend to be students with poor grades.

(b) Students with good grades tend to be students who are not involved in many extracurricular activities.

(c) Students involved in many extracurricular activities are just as likely to get good grades as bad grades.

(d) Students with good grades tend to be students who are involved in many extracurricular activities.

(e) No conclusion should be made based on the correlation without looking at a scatterplot of the data.

*Questions T3.3–T3.5 refer to the following setting.* Scientists examined the activity level of 7 fish at different temperatures. Fish activity was rated on a scale of 0 (no activity) to 100 (maximal activity). The temperature was measured in degrees Celsius. A computer regression printout and a residual plot are provided. Notice that the horizontal axis on the residual plot is labeled "Fitted value," which means the same thing as "predicted value."

Each chapter concludes with an **AP® STATISTICS PRACTICE TEST**. This test includes about 10 AP®-style multiple-choice questions and 3 free-response questions.

## Cumulative AP® Practice Test 1

**Section I: Multiple Choice** *Choose the best answer for Questions AP1.1–AP1.14.*

**AP1.1** You look at real estate ads for houses in Sarasota, Florida. Many houses have prices from \$200,000 to \$400,000. The few houses on the water, however, have prices up to \$15 million. Which of the following statements best describes the distribution of home prices in Sarasota?

(a) The distribution is most likely skewed to the left, and the mean is greater than the median.

(b) The distribution is most likely skewed to the left, and the mean is less than the median.

(c) The distribution is roughly symmetric with a few high outliers, and the mean is approximately equal to the median.

(d) The distribution is most likely skewed to the right, and the mean is greater than the median.

(e) The distribution is most likely skewed to the right, and the mean is less than the median.

**AP1.2** A child is 40 inches tall, which places her at the 90th percentile of all children of similar age. The heights for children of this age form an approximately Normal distribution with a mean of 38 inches. Based on this information, what is the standard deviation of the heights of all children of this age?

(a) 0.20 inch

(b) 0.31 inch

(c) 0.65 inch

(d) 1.21 inches

(e) 1.56 inches

Four **CUMULATIVE AP® PRACTICE TESTS** simulate the real exam. They are placed after Chapters 4, 7, 11, and 12. The tests expand in length and content coverage as you work through the book. The last test models a full AP® Statistics exam.

## FRAPPY! FREE RESPONSE AP® PROBLEM, YAY!

The following problem is modeled after actual AP® Statistics exam free response questions. Your task is to generate a complete, concise response in 15 minutes.

*Directions: Show all your work. Indicate clearly the methods you use, because you will be scored on the correctness of your methods as well as on the accuracy and completeness of your results and explanations.*

Two statistics students went to a flower shop and randomly selected 12 carnations. When they got home, the students prepared 12 identical vases with exactly the same amount of water in each vase. They put one tablespoon of sugar in 3 vases, two tablespoons of sugar in 3 vases, and three tablespoons of sugar in 3 vases. In the remaining 3 vases, they put no sugar. After the vases were prepared, the students randomly assigned 1 carnation to each vase and observed how many hours each flower continued to look fresh. A scatterplot of the data is shown below.

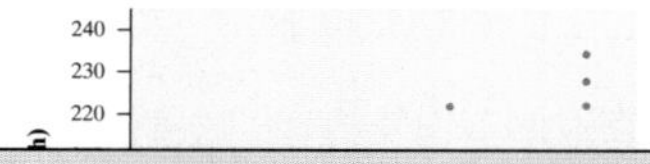


(a) Briefly describe the association shown in the scatterplot.

(b) The equation of the least-squares regression line for these data is $\hat{y} = 180.8 + 15.8x$. Interpret the slope of the line in the context of the study.

(c) Calculate and interpret the residual for the flower that had 2 tablespoons of sugar and looked fresh for 204 hours.

(d) Suppose that another group of students conducted a similar experiment using 12 flowers, but included different varieties in addition to carnations. Would you expect the value of $r^2$ for the second group's data to be greater than, less than, or about the same as the value of $r^2$ for the first group's data? Explain.

After you finish, you can view two example solutions on the book's website (highschool.bfwpub.com/updatedtps6e). Determine whether you think each solution is "complete," "substantial," "developing," or "minimal." If the solution is not complete, what improvements would you suggest to the student who wrote it? Finally, your teacher will provide you with a scoring rubric. Score your response and note what, if anything, you would do differently

Learn how to answer free response questions successfully by working the **FRAPPY!**—the Free Response AP® Problem, Yay!—that begins the Chapter Wrap-Up in every chapter.

# Use TECHNOLOGY to discover and analyze

## 3. Technology Corner COMPUTING NUMERICAL SUMMARIES

*TI-Nspire and other technology instructions are on the book's website at highschool.bfwpub.com/updatedtps6e.*

Let's find numerical summaries for the boys' shoes data from the example on page 64. We'll start by showing you how to compute summary statistics on the TI-83/84 and then look at output from computer software.

I. **One-variable statistics on the TI-83/84**

1. Enter the data in list L1.
2. Find the summary statistics for the shoe data.

- Press STAT (CALC); choose 1-VarStats.
  **OS 2.55 or later:** In the dialog box, press 2nd 1 (L1) and ENTER to specify L1 as the List. Leave FreqList blank. Arrow down to Calculate and press ENTER.
  **Older OS:** Press 2nd 1 (L1) and ENTER.
- Press ▼ to see the rest of the one-variable statistics.

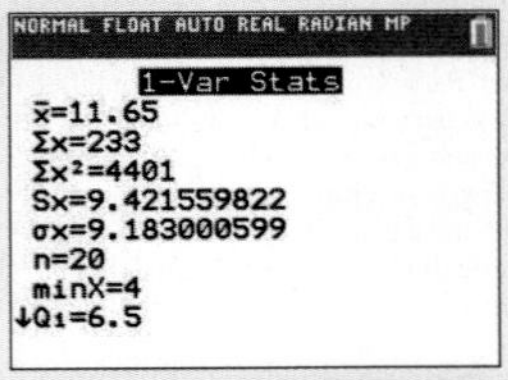


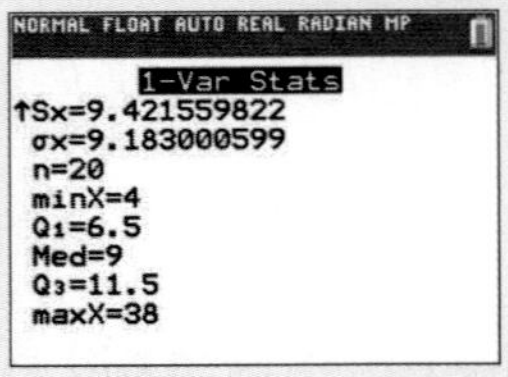


II. **Output from statistical software** We used Minitab statistical software to calculate descriptive statistics for the boys' shoes data. Minitab allows you to choose which numerical summaries are included in the output.

**Descriptive Statistics: Shoes**

| Variable | N | Mean | StDev | Minimum | $Q_1$ | Median | $Q_3$ | Maximum |
|---|---|---|---|---|---|---|---|---|
| Shoes | 20 | 11.65 | 9.42 | 4.00 | 6.25 | 9.00 | 11.75 | 38.00 |

*Note:* The TI-83/84 gives the first and third quartiles of the boys' shoes distribution as $Q_1 = 6.5$ and $Q_3 = 11.5$. Minitab reports that $Q_1 = 6.25$ and $Q_3 = 11.75$. What happened? Minitab and some other software use slightly different rules for locating quartiles. Results from the various rules are usually close to each other. Be aware of possible differences when calculating quartiles as they may affect more than just the *IQR*.

Use technology as a tool for discovery and analysis. **TECHNOLOGY CORNERS** give step-by-step instructions for using the TI-83/84 calculator. Instructions for the TI-Nspire and other calculators are on the book's Student Site (highschool.bfwpub.com/updatedtps6e) and in the e-Book platform.

**Technology Corner videos** are also available to walk you through the key strokes needed to perform each analysis.

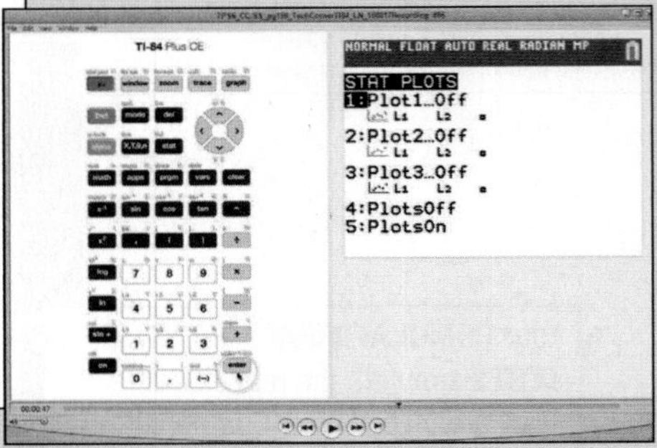


Although the Technology Corners focus on the TI-83/84 graphing calculator, output from multiple programs—including Minitab and JMP—is used in the book's Examples and Exercises to help you become familiar with reading and interpreting many different kinds of statistical summaries.

## 3.2 Technology Corners

*TI-Nspire and other technology instructions are on the book's website at highschool.bfwpub.com/updatedtps6e.*



Find the Technology Corners easily by consulting the summary table at the end of each section or the complete table at the back of the book.

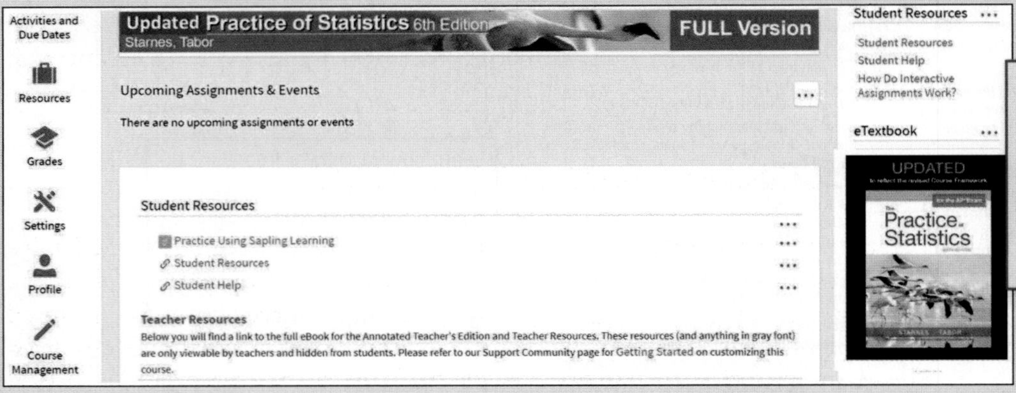


Read, practice, access the resources, and do homework assignments online with the new **Online Homework and e-Book Platform** that may be purchased to enhance your learning experience.

# OVERVIEW: What Is Statistics?

Does listening to music while studying help or hinder learning? If an athlete fails a drug test, how sure can we be that she took a banned substance? Does having a pet help people live longer? How well do SAT scores predict college success? Do most people recycle? Which of two diets will help obese children lose more weight and keep it off? Can a new drug help people quit smoking? How strong is the evidence for global warming?

These are just a few of the questions that statistics can help answer. But what is statistics? And why should you study it?

## Statistics Is the Science of Learning from Data

istockphoto

Data are usually numbers, but they are not "just numbers." *Data are numbers with a context.* The number 10.5, for example, carries no information by itself. But if we hear that a family friend's new baby weighed 10.5 pounds at birth, we congratulate her on the healthy size of the child. The context engages our knowledge about the world and allows us to make judgments. We know that a baby weighing 10.5 pounds is quite large, and that a human baby is unlikely to weigh 10.5 ounces or 10.5 kilograms. The context makes the number meaningful.

In your lifetime, you will be bombarded with data and statistical information. Poll results, television ratings, music sales, gas prices, unemployment rates, medical study outcomes, and standardized test scores are discussed daily in the media. Using data effectively is a large and growing part of most professions. A solid understanding of statistics will enable you to make sound, data-based predictions, decisions, and conclusions in your career and everyday life.

## Data Beat Personal Experiences

It is tempting to base conclusions on your own experiences or the experiences of those you know. But our experiences may not be typical. In fact, the incidents that stick in our memory are often the unusual ones.

Bloomberg via Getty Images

### Do Cell Phones Cause Brain Cancer?

Italian businessman Innocente Marcolini developed a brain tumor at age 60. He also talked on a cellular phone up to 6 hours per day for 12 years as part of his job. Mr. Marcolini's physician suggested that the brain tumor may have been caused by cell-phone use. So Mr. Marcolini decided to file suit in the Italian court system. A court ruled in his favor in October 2012.

Several statistical studies have investigated the link between cell-phone use and brain cancer. One of the largest was conducted by the Danish Cancer Society.

Over 350,000 residents of Denmark were included in the study. Researchers compared the brain-cancer rate for the cell-phone users with the rate in the general population. The result: no statistical difference in brain-cancer rates.[1] In fact, most studies have produced similar conclusions. In spite of the evidence, many people (like Mr. Marcolini) are still convinced that cell phones can cause brain cancer.

In the public's mind, the compelling story wins every time. A statistically literate person knows better. ***Data are more reliable than personal experiences because they systematically describe an overall picture, rather than focus on a few incidents.***

## Where the Data Come from Matters

### Are You Kidding Me?

The famous advice columnist Ann Landers once asked her readers, "If you had it to do over again, would you have children?" A few weeks later, her column was headlined "70% OF PARENTS SAY KIDS NOT WORTH IT." Indeed, 70% of the nearly 10,000 parents who wrote in said they would not have children if they could make the choice again. Do you believe that 70% of all parents regret having children?

istockphoto

You shouldn't. The people who took the trouble to write to Ann Landers are not representative of all parents. Their letters showed that many of them were angry with their children. All we know from these data is that there are some unhappy parents out there. A statistically designed poll, unlike Ann Landers's appeal, targets specific people chosen in a way that gives all parents the same chance to be asked. Such a poll showed that 91% of parents *would* have children again.

Where data come from matters a lot. If you are careless about how you get your data, you may announce 70% "No" when the truth is close to 90% "Yes."

### Who Talks More—Women or Men?

According to Louann Brizendine, author of *The Female Brain,* women say nearly 3 times as many words per day as men. Skeptical researchers devised a study to test this claim. They used electronic devices to record the talking patterns of 396 university students from Texas, Arizona, and Mexico. The device was programmed to record 30 seconds of sound every 12.5 minutes without the carrier's knowledge. What were the results?

According to a published report of the study in *Scientific American,* "Men showed a slightly wider variability in words uttered. . . . But in the end, the sexes came out just about even in the daily averages: women at 16,215 words and men at 15,669."[2] When asked where she got her figures, Brizendine admitted that she used unreliable sources.[3]

*The most important information about any statistical study is how the data were produced.* Only carefully designed studies produce results that can be trusted.

## Always Plot Your Data

Yogi Berra, a famous New York Yankees baseball player known for his unusual quotes, had this to say: "You can observe a lot just by watching." That's a motto for learning from data. *A carefully chosen graph helps us describe patterns in data and identify important departures from those patterns.*

## Do People Live Longer in Wealthier Countries?

The Gapminder website, www.gapminder.org, provides loads of data on the health and well-being of the world's inhabitants. The graph below displays some data from Gapminder.[4] The individual points represent all the world's nations for which data are available. Each point shows the income per person and life expectancy for one country, along with the region (color of point) and population (size of point).

Honggi Zhang/Alamy

We expect people in richer countries to live longer. The overall pattern of the graph does show this, but the relationship has an interesting shape. Life expectancy rises very quickly as personal income increases and then levels off. People in very rich countries like the United States live no longer than people in poorer but not extremely poor nations. In some less wealthy countries, people live longer than in the United States. Several other nations stand out in the graph. What's special about each of these countries?

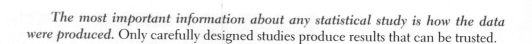

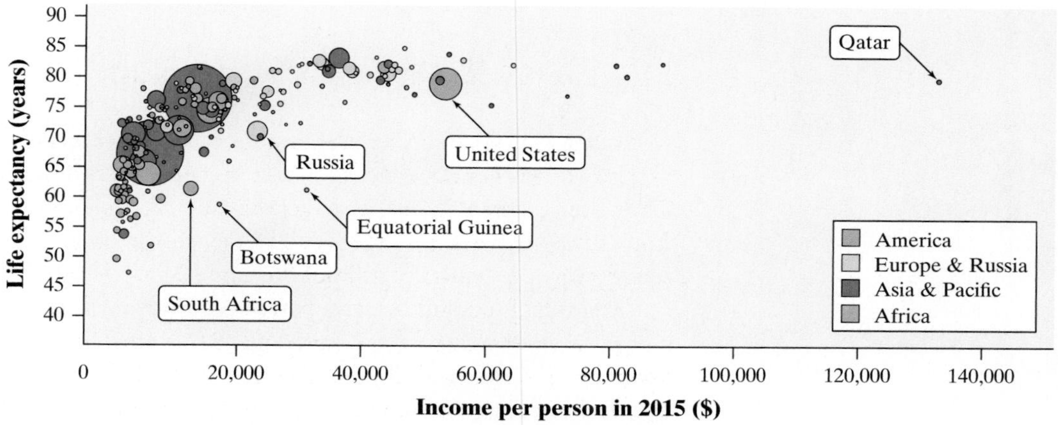


Graph of the life expectancy of people in many nations against each nation's income per person in 2015.

# Variation Is Everywhere

Individuals vary. Repeated measurements on the same individual vary. Chance outcomes—like spins of a roulette wheel or tosses of a coin—vary. Almost everything varies over time. Statistics provides tools for understanding variation.

## Have Most Students Cheated on a Test?

Commercial Eye/Getty Images

Researchers from the Josephson Institute were determined to find out. So they surveyed about 23,000 students from 100 randomly selected schools (both public and private) nationwide. The question was: "How many times have you cheated during a test at school in the past year?" Fifty-one percent said they had cheated at least once.[5]

If the researchers had asked the same question of *all* high school students, would exactly 51% have answered "Yes"? Probably not. If the Josephson Institute had selected a different sample of about 23,000 students to respond to the survey, they would probably have gotten a different estimate. *Variation is everywhere!*

Fortunately, statistics provides a description of how the sample results will vary in relation to the actual population percent. Based on the sampling method that this study used, we can say that the estimate of 51% is very likely to be within 1% of the true population value. That is, we can be quite confident that between 50% and 52% of *all* high school students would say that they have cheated on a test.

***Because variation is everywhere, conclusions are uncertain. Statistics gives us the tools to quantify our uncertainty, allowing for valid, data-based predictions, decisions, and conclusions.***

# UNIT 1

## Exploring One-Variable Data

# Chapter 1

# Data Analysis

AfriPics.com/Alamy Stock Photo

# INTRODUCTION Statistics: The Science and Art of Data

**LEARNING TARGETS** *By the end of the section, you should be able to:*

- Identify the individuals and variables in a set of data.
- Classify variables as categorical or quantitative.

We live in a world of *data*. Every day, the media report poll results, outcomes of medical studies, and analyses of data on everything from stock prices to standardized test scores to global warming. The data are trying to tell us a story. To understand what the data are saying, you need to learn more about **statistics.**

**DEFINITION Statistics**

**Statistics** is the science and art of collecting, analyzing, and drawing conclusions from data.

A solid understanding of statistics will help you make good decisions based on data in your daily life.

## Organizing Data

Rudy Sulgan/Corbis Documentary/Getty Images

Every year, the U.S. Census Bureau collects data from over 3 million households as part of the American Community Survey (ACS). The table displays some data from the ACS in a recent year.

| Household | Region | Number of people | Time in dwelling (years) | Response mode | Household income | Internet access? |
|---|---|---|---|---|---|---|
| 425 | Midwest | 5 | 2–4 | Internet | 52,000 | Yes |
| 936459 | West | 4 | 2–4 | Mail | 40,500 | Yes |
| 50055 | Northeast | 2 | 10–19 | Internet | 481,000 | Yes |
| 592934 | West | 4 | 2–4 | Phone | 230,800 | No |
| 545854 | South | 9 | 2–4 | Phone | 33,800 | Yes |
| 809928 | South | 2 | 30 + | Internet | 59,500 | Yes |
| 110157 | Midwest | 1 | 5–9 | Internet | 80,000 | Yes |
| 999347 | South | 1 | <1 | Mail | 8,400 | No |

Most data tables follow this format—each row describes an **individual** and each column holds the values of a **variable**.

Sometimes the individuals in a data set are called *cases* or *observational units*.

**DEFINITION Individual, Variable**

An **individual** is an object described in a set of data. Individuals can be people, animals, or things.

A **variable** is an attribute that can take different values for different individuals.

For the American Community Survey data set, the *individuals* are households. The *variables* recorded for each household are region, number of people, time in current dwelling, survey response mode, household income, and whether the dwelling has Internet access. Region, time in dwelling, response mode, and Internet access status are **categorical variables**. Number of people and household income are **quantitative variables**.

Note that household is *not* a variable. The numbers in the household column of the data table are just labels for the individuals in this data set. Be sure to look for a column of labels—names, numbers, or other identifiers—in any data table you encounter.

**AP® EXAM TIP**

If you learn to distinguish categorical from quantitative variables now, it will pay big rewards later. You will be expected to analyze categorical and quantitative variables correctly on the AP® exam.

**DEFINITION Categorical variable, Quantitative variable**

A **categorical variable** assigns labels that place each individual into a particular group, called a category.

A **quantitative variable** takes number values that are quantities—counts or measurements.

arithmetic operations make sense

caution

**Not every variable that takes number values is quantitative.** Zip code is one example. Although zip codes are numbers, they are neither counts of anything, nor measurements of anything. They are simply labels for a regional location, making zip code a categorical variable. Some variables—such as gender, race, and occupation—are categorical by nature. Time in dwelling from the ACS data set is also a categorical variable because the values are recorded as intervals of time, such as 2–4 years. If time in dwelling had been recorded to the nearest year for each household, this variable would be quantitative.

To make life simpler, we sometimes refer to *categorical data* or *quantitative data* instead of identifying the variable as categorical or quantitative.

## EXAMPLE

## Census At School
## Individuals and Variables

**PROBLEM:** Census At School is an international project that collects data about primary and secondary school students using surveys. Hundreds of thousands of students from Australia, Canada, Ireland, Japan, New Zealand, South Africa, South Korea, the United Kingdom, and the United States have taken part in the project. Data from the surveys are available online. We used the site's "Random Data Selector" to choose 10 Canadian students who completed the survey in a recent year. The table displays the data.

Garry Black/Alamy

| Province | Gender | Number of languages spoken | Handedness | Height (cm) | Wrist circumference (mm) | Preferred communication |
|---|---|---|---|---|---|---|
| Saskatchewan | Male | 1 | Right | 175.0 | 180 | In person |
| Ontario | Female | 1 | Right | 162.5 | 160 | In person |
| Alberta | Male | 1 | Right | 178.0 | 174 | Facebook |
| Ontario | Male | 2 | Right | 169.0 | 160 | Cell phone |
| Ontario | Female | 2 | Right | 166.0 | 65 | In person |
| Nunavut | Male | 1 | Right | 168.5 | 160 | Text messaging |
| Ontario | Female | 1 | Right | 166.0 | 165 | Cell phone |
| Ontario | Male | 4 | Left | 157.5 | 147 | Text messaging |
| Ontario | Female | 2 | Right | 150.5 | 187 | Text messaging |
| Ontario | Female | 1 | Right | 171.0 | 180 | Text messaging |

(a) Identify the individuals in this data set.

(b) What are the variables? Classify each as categorical or quantitative.

We'll see in Chapter 4 why choosing at random, as we did in this example, is a good idea.

**SOLUTION:**

(a) 10 randomly selected Canadian students who participated in the Census At School survey.

(b) Categorical: Province, gender, handedness, preferred communication method

Quantitative: Number of languages spoken, height (cm), wrist circumference (mm)

There is at least one suspicious value in the data table. We doubt that the girl who is 166 cm tall really has a wrist circumference of 65 mm (about 2.6 inches). Always look to be sure the values make sense!

**FOR PRACTICE, TRY EXERCISE 1**

There are two types of quantitative variables: *discrete* and *continuous*. Most **discrete variables** result from counting something, like the number of languages spoken in the preceding example. **Continuous variables** typically result from measuring something, like height or wrist circumference. Be sure to report the units of measurement (like centimeters for height and millimeters for wrist circumference) for a continuous variable.

**DEFINITION Discrete variable, Continuous variable**

A quantitative variable that takes a fixed set of possible values with gaps between them is a **discrete variable.**

A quantitative variable that can take any value in an interval on the number line is a **continuous variable.**

The proper method of data analysis depends on whether a variable is categorical or quantitative. For that reason, it is important to distinguish these two types of variables. The type of data determines what kinds of graphs and which numerical summaries are appropriate.

**ANALYZING DATA** A variable generally takes values that vary (hence the name *variable*!). Categorical variables sometimes have similar counts in each category and sometimes don't. For instance, we might have expected similar numbers of males and females in the Census At School data set. But we aren't surprised to see that most students are right-handed. Quantitative variables may take values that are very close together or values that are quite spread out. We call the pattern of variation of a variable its **distribution**.

**DEFINITION Distribution**

The **distribution** of a variable tells us what values the variable takes and how often it takes those values.

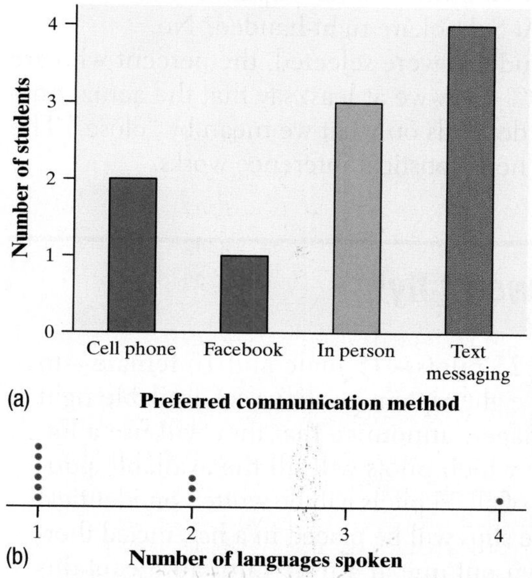


**FIGURE 1.1** (a) Bar graph showing the distribution of preferred communication method for the sample of 10 Canadian students. (b) Dotplot showing the distribution of number of languages spoken by these students.

Let's return to the data for the sample of 10 Canadian students from the preceding example. Figure 1.1(a) shows the distribution of preferred communication method for these students in a *bar graph*. We can see how many students chose each method from the heights of the bars: cell phone (2), Facebook (1), in person (3), text messaging (4). Figure 1.1(b) shows the distribution of number of languages spoken in a *dotplot*. We can see that 6 students speak one language, 3 students speak two languages, and 1 student speaks four languages.

Section 1.1 begins by looking at how to describe the distribution of a single categorical variable and then examines relationships between categorical variables. Sections 1.2 and 1.3 and all of Chapter 2 focus on describing the distribution of a quantitative variable. Chapter 3 investigates relationships between two quantitative variables. In each case, we begin with graphical displays, then add numerical summaries for a more complete description.

**HOW TO ANALYZE DATA**

- Begin by examining each variable by itself. Then move on to study relationships among the variables.
- Start with a graph or graphs. Then add numerical summaries.

This process of exploratory data analysis is known as *descriptive statistics*.

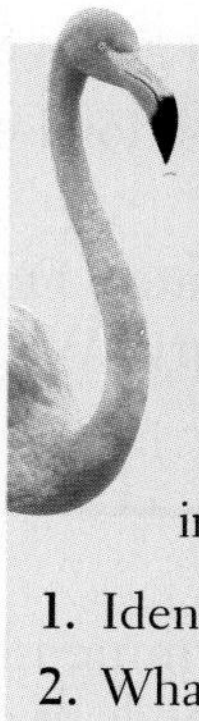

## CHECK YOUR UNDERSTANDING

Jake is a car buff who wants to find out more about the vehicles that his classmates drive. He gets permission to go to the student parking lot and record some data. Later, he does some Internet research on each model of car he found. Finally, Jake makes a spreadsheet that includes each car's license plate, model, number of cylinders, color, highway gas mileage, weight, and whether it has a navigation system. variable

1. Identify the individuals in Jake's study. cars
2. What are the variables? Classify each as categorical or quantitative.
   c- model, color, navigation, cylinders; milleage, weight
3. Identify each quantitative variable as discrete or continuous.
   both continuous

## From Data Analysis to Inference

Sometimes we're interested in drawing conclusions that go beyond the data at hand. That's the idea of *inferential statistics*. In the "Census At School" example, 9 of the 10 randomly selected Canadian students are right-handed. That's 90% of the *sample*. Can we conclude that exactly 90% of the *population* of Canadian students who participated in Census At School are right-handed? No.

If another random sample of 10 students were selected, the percent who are right-handed might not be exactly 90%. Can we at least say that the actual population value is "close" to 90%? That depends on what we mean by "close." The following activity gives you an idea of how statistical inference works.

### ACTIVITY Hiring discrimination—it just won't fly!

An airline has just finished training 25 pilots—15 male and 10 female—to become captains. Unfortunately, only eight captain positions are available right now. Airline managers announce that they will use a lottery to determine which pilots will fill the available positions. The names of all 25 pilots will be written on identical slips of paper. The slips will be placed in a hat, mixed thoroughly, and drawn out one at a time until all 8 captains have been identified.

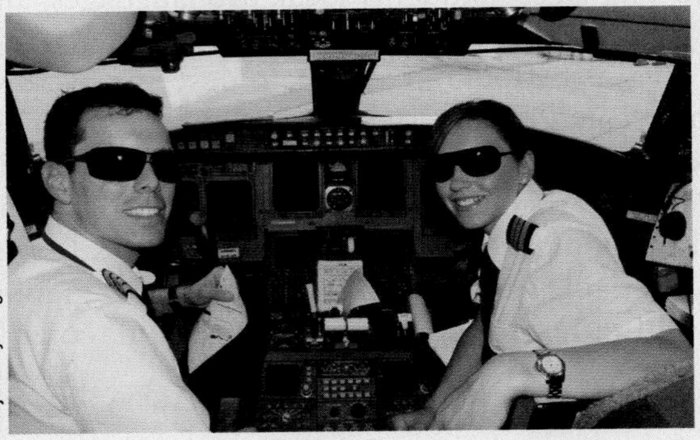

Choja/Getty Images

A day later, managers announce the results of the lottery. Of the 8 captains chosen, 5 are female and 3 are male. Some of the male pilots who weren't selected suspect that the lottery was not carried out fairly. One of these pilots asks your statistics class for advice about whether to file a grievance with the pilots' union.

The key question in this possible discrimination case seems to be: *Is it plausible (believable) that these results happened just by chance?* To find out, you and your classmates will *simulate* the lottery process that airline managers said they used.

1. Your teacher will give you a bag with 25 beads (15 of one color and 10 of another) or 25 slips of paper (15 labeled "M" and 10 labeled "F") to represent the 25 pilots. Mix the beads/slips thoroughly. Without looking, remove 8 beads/slips from the bag. Count the number of female pilots selected. Then return the beads/slips to the bag.
2. Your teacher will draw and label a number line for a class *dotplot*. On the graph, plot the number of females you got in Step 1.
3. Repeat Steps 1 and 2 if needed to get a total of at least 40 simulated lottery results for your class.
4. Discuss the results with your classmates. Does it seem plausible that airline managers conducted a fair lottery? What advice would you give the male pilot who contacted you?

Our ability to do inference is determined by how the data are produced. Chapter 4 discusses the two main methods of data production—sampling and experiments—and the types of conclusions that can be drawn from each. As the activity illustrates, the logic of inference rests on asking, "What are the chances?" *Probability*, the study of chance behavior, is the topic of Chapters 5–7. We'll introduce the most common inference techniques in Chapters 8–12.

## Introduction Summary

- **Statistics** is the science and art of collecting, analyzing, and drawing conclusions from data.
- A data set contains information about a number of **individuals.** Individuals may be people, animals, or things. For each individual, the data give values for one or more **variables.** A variable describes some characteristic of an individual, such as a person's height, gender, or salary.
- A **categorical variable** assigns a label that places each individual in one of several groups, such as male or female. A **quantitative variable** has numerical values that count or measure some characteristic of each individual, such as number of siblings or height in meters.
- There are two types of quantitative variables: discrete and continuous. A **discrete variable** has a fixed set of possible numeric values with gaps between them. A **continuous variable** can take any value in an interval on the number line. Discrete variables usually result from counting something; continuous variables usually result from measuring something.
- The **distribution** of a variable describes what values the variable takes and how often it takes them.

## Introduction Exercises

The solutions to all exercises numbered in red may be found in the Solutions Appendix, starting on page S-1.

1. **A class survey** Here is a small part of the data set that describes the students in an AP® Statistics class. The data come from anonymous responses to a questionnaire filled out on the first day of class.

pg 3

| Gender | Grade level | GPA | Children in family | Homework last night (min) | Android or iPhone? |
|---|---|---|---|---|---|
| F | 9 | 2.3 | 3 | 0–14 | iPhone |
| M | 11 | 3.8 | 6 | 15–29 | Android |
| M | 10 | 3.1 | 2 | 15–29 | Android |
| F | 10 | 4.0 | 1 | 45–59 | iPhone |
| F | 10 | 3.4 | 4 | 0–14 | iPhone |
| F | 10 | 3.0 | 3 | 30–44 | Android |
| M | 9 | 3.9 | 2 | 15–29 | iPhone |
| M | 12 | 3.5 | 2 | 0–14 | iPhone |

(a) Identify the individuals in this data set.

(b) What are the variables? Classify each as categorical or quantitative.

2. **Coaster craze** Many people like to ride roller coasters. Amusement parks try to increase attendance by building exciting new coasters. The following table displays data on several roller coasters that were opened in a recent year.[1]

| Roller coaster | Type | Height (ft) | Design | Speed (mph) | Duration (sec) |
|---|---|---|---|---|---|
| Wildfire | Wood | 187.0 | Sit down | 70.2 | 120 |
| Skyline | Steel | 131.3 | Inverted | 50.0 | 90 |
| Goliath | Wood | 165.0 | Sit down | 72.0 | 105 |
| Helix | Steel | 134.5 | Sit down | 62.1 | 130 |
| Banshee | Steel | 167.0 | Inverted | 68.0 | 160 |
| Black Hole | Steel | 22.7 | Sit down | 25.5 | 75 |

(a) Identify the individuals in this data set.

(b) What are the variables? Classify each as categorical or quantitative.

3. **Hit movies** According to the Internet Movie Database, *Avatar* is tops based on box-office receipts worldwide as of January 2017. The following table displays data on several popular movies. Identify the individuals

and variables in this data set. Classify each variable as categorical or quantitative.

| Movie | Year | Rating | Time (min) | Genre | Box office ($) |
|---|---|---|---|---|---|
| Avatar | 2009 | PG-13 | 162 | Action | 2,783,918,982 |
| Titanic | 1997 | PG-13 | 194 | Drama | 2,207,615,668 |
| Star Wars: The Force Awakens | 2015 | PG-13 | 136 | Adventure | 2,040,375,795 |
| Jurassic World | 2015 | PG-13 | 124 | Action | 1,669,164,161 |
| Marvel's The Avengers | 2012 | PG-13 | 142 | Action | 1,519,479,547 |
| Furious 7 | 2015 | PG-13 | 137 | Action | 1,516,246,709 |
| The Avengers: Age of Ultron | 2015 | PG-13 | 141 | Action | 1,404,705,868 |
| Harry Potter and the Deathly Hallows: Part 2 | 2011 | PG-13 | 130 | Fantasy | 1,328,111,219 |
| Frozen | 2013 | PG | 108 | Animation | 1,254,512,386 |
| Iron Man 3 | 2013 | PG-13 | 129 | Action | 1,172,805,920 |

4. **Skyscrapers** Here is some information about the tallest buildings in the world as of February 2017. Identify the individuals and variables in this data set. Classify each variable as categorical or quantitative.

| Building | Country | Height (m) | Floors | Use | Year completed |
|---|---|---|---|---|---|
| Burj Khalifa | United Arab Emirates | 828.0 | 163 | Mixed | 2010 |
| Shanghai Tower | China | 632.0 | 121 | Mixed | 2014 |
| Makkah Royal Clock Tower Hotel | Saudi Arabia | 601.0 | 120 | Hotel | 2012 |
| Ping An Finance Center | China | 599.0 | 115 | Mixed | 2016 |
| Lotte World Tower | South Korea | 554.5 | 123 | Mixed | 2016 |
| One World Trade Center | United States | 541.0 | 104 | Office | 2013 |
| Taipei 101 | Taiwan | 509.0 | 101 | Office | 2004 |
| Shanghai World Financial Center | China | 492.0 | 101 | Mixed | 2008 |
| International Commerce Center | China | 484.0 | 118 | Mixed | 2010 |
| Petronas Tower 1 | Malaysia | 452.0 | 88 | Office | 1998 |

5. **Protecting wood** What measures can be taken, especially when restoring historic wooden buildings, to help wood surfaces resist weathering? In a study of this question, researchers prepared wooden panels and then exposed them to the weather. Some of the variables recorded were type of wood (yellow poplar, pine, cedar); type of water repellent (solvent-based, water-based); paint thickness (millimeters); paint color (white, gray, light blue); weathering time (months). Classify each variable as categorical or quantitative.

6. **Medical study variables** Data from a medical study contain values of many variables for each subject in the study. Some of the variables recorded were gender (female or male); age (years); race (Asian, Black, White, or other); smoker (yes or no); systolic blood pressure (millimeters of mercury); level of calcium in the blood (micrograms per milliliter). Classify each variable as categorical or quantitative.

7. **College life** A college admissions office collects data from each incoming freshman on several quantitative variables: distance from their home to campus, number of siblings, how many books they have read in the past month, and how long it took them to complete an online survey. Classify each variable as discrete or continuous.

8. **Social media** A social media company records data from each of its users on several quantitative variables: time spent on the site, how many times they visited the site, number of likes received, and how long since they created a member profile. Classify each variable as discrete or continuous.

**Multiple Choice:** *Select the best answer.*

*Exercises 9 and 10 refer to the following setting.* At the Census Bureau website www.census.gov, you can view detailed data collected by the American Community Survey. The following table includes data for 10 people chosen at random from the more than 1 million people in households contacted by the survey. "School" gives the highest level of education completed.

| Weight (lb) | Age (years) | Travel to work (min) | School | Gender | Income last year ($) |
|---|---|---|---|---|---|
| 187 | 66 | 0 | Ninth grade | 1 | 24,000 |
| 158 | 66 | n/a | High school grad | 2 | 0 |
| 176 | 54 | 10 | Assoc. degree | 2 | 11,900 |
| 339 | 37 | 10 | Assoc. degree | 1 | 6000 |
| 91 | 27 | 10 | Some college | 2 | 30,000 |
| 155 | 18 | n/a | High school grad | 2 | 0 |
| 213 | 38 | 15 | Master's degree | 2 | 125,000 |
| 194 | 40 | 0 | High school grad | 1 | 800 |
| 221 | 18 | 20 | High school grad | 1 | 2500 |
| 193 | 11 | n/a | Fifth grade | 1 | 0 |

9. The individuals in this data set are

(a) households. (b) people. (c) adults.
(d) 120 variables. (e) columns.

10. This data set contains

(a) 7 variables, 2 of which are categorical.
(b) 7 variables, 1 of which is categorical.
(c) 6 variables, 2 of which are categorical.
(d) 6 variables, 1 of which is categorical.
(e) None of these.

# SECTION 1.1 Analyzing Categorical Data

**LEARNING TARGETS** *By the end of the section, you should be able to:*

- Make and interpret bar graphs for categorical data.
- Identify what makes some graphs of categorical data misleading.
- Calculate marginal and joint relative frequencies from a two-way table.
- Calculate conditional relative frequencies from a two-way table.
- Use bar graphs to compare distributions of categorical data.
- Describe the nature of the association between two categorical variables.

Here are the data on preferred communication method for the 10 randomly selected Canadian students from the example on page 3:

| | | | | |
|---|---|---|---|---|
| In person | In person | Facebook | Cell phone | In person |
| Text messaging | Cell phone | Text messaging | Text messaging | Text messaging |

We can summarize the distribution of this categorical variable with a **frequency table** or a **relative frequency table**.

Some people use the terms *frequency distribution* and *relative frequency distribution* instead.

**DEFINITION Frequency table, Relative frequency table**

A **frequency table** shows the number of individuals having each value.

A **relative frequency table** shows the proportion or percent of individuals having each value.

To make either kind of table, start by tallying the number of times that the variable takes each value. **Note that the frequencies and relative frequencies listed in these tables are not data.** The tables summarize the data by telling us how many (or what proportion or percent of) students in the sample said "Cell phone," "Facebook," "In person," and "Text messaging."

Frequency vs Relative
count vs proportion

| Preferred method | Tally |
|---|---|
| Cell phone | II |
| Facebook | I |
| In person | III |
| Text messaging | IIII |

**Frequency table**

| Preferred method | Frequency |
|---|---|
| Cell phone | 2 |
| Facebook | 1 |
| In person | 3 |
| Text messaging | 4 |

**Relative frequency table**

| Preferred method | Relative frequency |
|---|---|
| Cell phone | 2/10 = 0.20 or 20% |
| Facebook | 1/10 = 0.10 or 10% |
| In person | 3/10 = 0.30 or 30% |
| Text messaging | 4/10 = 0.40 or 40% |

The same process can be used to summarize the distribution of a quantitative variable. Of course, it would be hard to make a frequency table or a relative frequency table for quantitative data that take many different values, like the ages of people attending a Major League Baseball game. We'll look at a better option for quantitative variables with many possible values in Section 1.2.

# Displaying Categorical Data: Bar Graphs and Pie Charts

A frequency table or relative frequency table summarizes a variable's distribution with numbers. To display the distribution more clearly, use a graph. You can make a **bar graph** or a **pie chart** for categorical data.

Bar graphs are sometimes called *bar charts.* Pie charts are sometimes called *circle graphs.*

**DEFINITION Bar graph, Pie chart**

A **bar graph** shows each category as a bar. The heights of the bars show the category frequencies or relative frequencies.

A **pie chart** shows each category as a slice of the "pie." The areas of the slices are proportional to the category frequencies or relative frequencies.

Figure 1.2 shows a bar graph and a pie chart of the data on preferred communication method for the random sample of Canadian students. Note that the percents for each category come from the relative frequency table.

**Relative frequency table**

| Preferred method | Relative frequency |
|---|---|
| Cell phone | 2/10 = 0.20 or 20% |
| Facebook | 1/10 = 0.10 or 10% |
| In person | 3/10 = 0.30 or 30% |
| Text messaging | 4/10 = 0.40 or 40% |

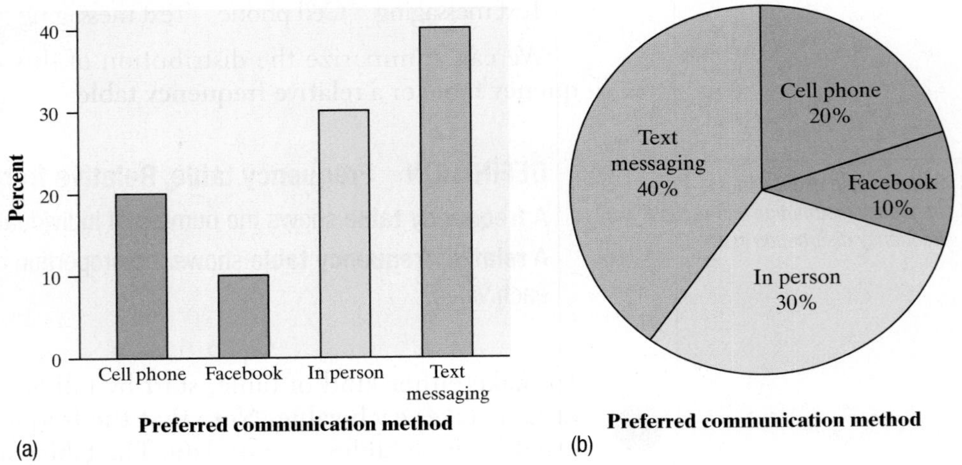


**FIGURE 1.2** (a) Bar graph and (b) pie chart of the distribution of preferred communication method for a random sample of 10 Canadian students.

It is fairly easy to make a bar graph by hand. Here's how you do it.

**HOW TO MAKE A BAR GRAPH**

- **Draw and label the axes.** Put the name of the categorical variable under the horizontal axis. To the left of the vertical axis, indicate whether the graph shows the frequency (count) or relative frequency (percent or proportion) of individuals in each category.
- **"Scale" the axes.** Write the names of the categories at equally spaced intervals under the horizontal axis. On the vertical axis, start at 0 and place tick marks at equal intervals until you exceed the largest frequency or relative frequency in any category.
- **Draw bars above the category names.** Make the bars equal in width and leave gaps between them. Be sure that the height of each bar corresponds to the frequency or relative frequency of individuals in that category.

Making a graph is not an end in itself. The purpose of a graph is to help us understand the data. When looking at a graph, always ask, "What do I see?" We can see from both graphs in Figure 1.2 that the most preferred communication method for these students is text messaging.

## EXAMPLE

### What's on the radio?

### Making and interpreting bar graphs

**PROBLEM:** Arbitron, the rating service for radio audiences, categorizes U.S. radio stations in terms of the kinds of programs they broadcast. The frequency table summarizes the distribution of station formats in a recent year.[2]

(a) Identify the individuals in this data set.

(b) Make a frequency bar graph of the data. Describe what you see.

| Format | Number of stations | Format | Number of stations |
|---|---|---|---|
| Adult contemporary | 2536 | Religious | 3884 |
| All sports | 1274 | Rock | 1636 |
| Contemporary hits | 1012 | Spanish language | 878 |
| Country | 2893 | Variety | 1579 |
| News/Talk/Information | 4077 | Other formats | 4852 |
| Oldies | 831 | **Total** | **25,452** |

**SOLUTION:**

(a) U.S. radio stations

(b)

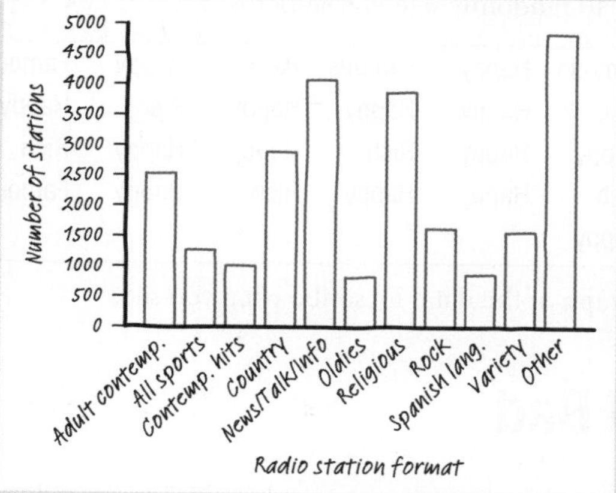


To make the bar graph:

- **Draw and label the axes.**
- **"Scale" the axes.** The largest frequency is 4852. So we choose a vertical scale from 0 to 5000, with tick marks 500 units apart.
- **Draw bars above the category names.**

On U.S. radio stations, the most frequent formats are Other (4852), News/talk/information (4077), and Religious (3884), while the least frequent are Oldies (831), Spanish language (878), and Contemporary hits (1012).

**FOR PRACTICE, TRY EXERCISE 11**

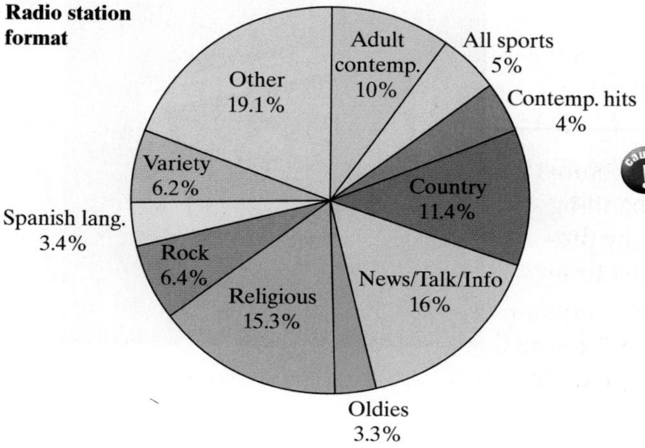


Here is a pie chart of the radio station format data from the preceding example. You can use a pie chart when you want to emphasize each category's relation to the whole. Pie charts are challenging to make by hand, but technology will do the job for you. Note that **a pie chart must include all categories that make up a whole**, which might mean adding an "other" category, as in the radio station example.

## CHECK YOUR UNDERSTANDING

The American Statistical Association sponsors a web-based project that collects data about primary and secondary school students using surveys. We used the site's "Random Sampler" to choose 40 U.S. high school students who completed the survey in a recent year.[3] One of the questions asked:

Which would you prefer to be? Select one.

_______ Rich _______ Happy _______ Famous _______ Healthy

Here are the responses from the 40 randomly selected students:

| | | | | | | | | |
|---|---|---|---|---|---|---|---|---|
| Famous | Healthy | Healthy | Famous | Happy | Famous | Happy | Happy | Famous |
| Rich | Happy | Happy | Rich | Happy | Happy | Happy | Rich | Happy |
| Famous | Healthy | Rich | Happy | Happy | Rich | Happy | Happy | Rich |
| Healthy | Happy | Happy | Rich | Happy | Happy | Rich | Happy | Famous |
| Famous | Happy | Happy | Happy | | | | | |

Make a relative frequency bar graph of the data. Describe what you see.

## Graphs: Good and Bad

Bar graphs are a bit dull to look at. It is tempting to replace the bars with pictures or to use special 3-D effects to make the graphs seem more interesting. Don't do it! Our eyes react to the area of the bars as well as to their height. When all bars have the same width, the area (width × height) varies in proportion to the height, and our eyes receive the right impression about the quantities being compared.

**EXAMPLE**

### Who buys iMacs?
### Beware the pictograph!

Justin Sullivan/Getty Images

**PROBLEM:** When Apple, Inc., introduced the iMac, the company wanted to know whether this new computer was expanding Apple's market share. Was the iMac mainly being bought by previous Macintosh owners, or was it being purchased by first-time computer buyers and by previous PC users who were switching over? To find out, Apple hired a firm to conduct a survey of 500 iMac customers. Each customer was categorized as a new computer purchaser, a previous PC owner, or a previous Macintosh owner. The table summarizes the survey results.[4]

| Previous ownership | Count | Percent (%) |
|---|---|---|
| None | 85 | 17.0 |
| PC | 60 | 12.0 |
| Macintosh | 355 | 71.0 |
| Total | 500 | 100.0 |

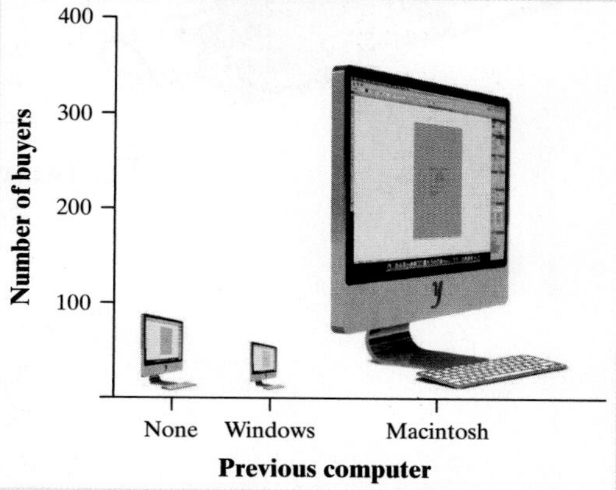


(a) To the right is a clever graph of the data that uses pictures instead of the more traditional bars. How is this pictograph misleading?

(b) Two possible bar graphs of the data are shown below. Which one could be considered deceptive? Why?

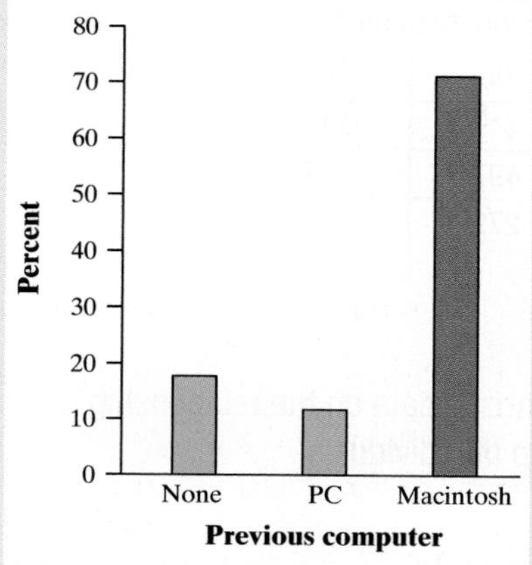


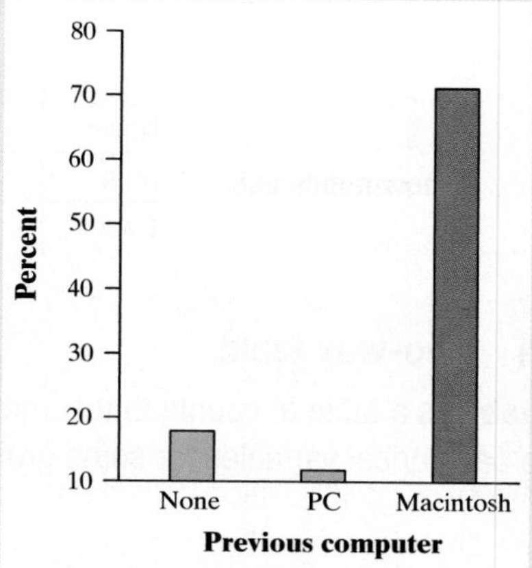


**SOLUTION:**

(a) *The pictograph makes it look like the percentage of iMac buyers who are former Mac owners is at least 10 times larger than either of the other two categories, which isn't true.*

(b) *The bar graph on the right is misleading. By starting the vertical scale at 10 instead of 0, it looks like the percentage of iMac buyers who previously owned a PC is less than half the percentage who are first-time computer buyers, which isn't true.*

In part (a), the *heights* of the images are correct. But the *areas* of the images are misleading. The Macintosh image is about 6 times as tall as the PC image, but its area is about 36 times as large!

FOR PRACTICE, TRY EXERCISE 19

There are two important lessons to be learned from this example: (1) beware the pictograph, and (2) watch those scales.

## Analyzing Data on Two Categorical Variables

You have learned some techniques for analyzing the distribution of a single categorical variable. What should you do when a data set involves two categorical variables? For example, Yellowstone National Park staff surveyed a random sample of 1526 winter visitors to the park. They asked each person whether he or she belonged to an environmental club (like the Sierra Club). Respondents were also

franz12/Shutterstock

asked whether they owned, rented, or had never used a snowmobile. The data set looks something like the following:

| Respondent | Environmental club? | Snowmobile use |
|---|---|---|
| 1 | No | Own |
| 2 | No | Rent |
| 3 | Yes | Never |
| 4 | Yes | Rent |
| 5 | No | Never |
| ⋮ | ⋮ | ⋮ |

The **two-way table** summarizes the survey responses.

| | | Environmental club member? | |
|---|---|---|---|
| | | No | Yes |
| Snowmobile use | Never | 445 | 212 |
| | Rent | 497 | 77 |
| | Own | 279 | 16 |

A two-way table is sometimes called a *contingency table.*

**DEFINITION Two-way table**

A **two-way table** is a table of counts that summarizes data on the relationship between two categorical variables for some group of individuals.

It's easier to grasp the information in a two-way table if row and column totals are included, like the one shown here.

| | | Environmental club | | |
|---|---|---|---|---|
| | | No | Yes | Total |
| Snowmobile use | Never used | 445 | 212 | 657 |
| | Snowmobile renter | 497 | 77 | 574 |
| | Snowmobile owner | 279 | 16 | 295 |
| | Total | 1221 | 305 | 1526 |

Now we can quickly answer questions like:

- What percent of people in the sample are environmental club members?

$$\frac{305}{1526} = 0.200 = 20.0\%$$

- What proportion of people in the sample never used a snowmobile?

$$\frac{657}{1526} = 0.431$$

These percents or proportions are known as **marginal relative frequencies** because they are calculated using values in the margins of the two-way table.

**DEFINITION Marginal relative frequency**

A **marginal relative frequency** gives the percent or proportion of individuals that have a specific value for one categorical variable.

We can compute marginal relative frequencies for the *column* totals to give the distribution of environmental club membership in the entire sample of 1526 park visitors:

> We could call this distribution the *marginal distribution* of environmental club membership.

$$\text{No: } \frac{1221}{1526} = 0.800 \text{ or } 80.0\% \qquad \text{Yes: } \frac{305}{1526} = 0.200 \text{ or } 20.0\%$$

We can compute marginal relative frequencies for the *row* totals to give the distribution of snowmobile use for all the individuals in the sample:

$$\text{Never: } \frac{657}{1526} = 0.431 \text{ or } 43.1\%$$

$$\text{Rent: } \frac{574}{1526} = 0.376 \text{ or } 37.6\%$$

$$\text{Own: } \frac{295}{1526} = 0.193 \text{ or } 19.3\%$$

> We could call this distribution the *marginal distribution* of snowmobile use.

Note that we could use a bar graph or a pie chart to display either of these distributions.

A marginal relative frequency tells you about only *one* of the variables in a two-way table. It won't help you answer questions like these, which involve values of *both* variables:

- What percent of people in the sample are environmental club members and own snowmobiles?

$$\frac{16}{1526} = 0.010 = 1.0\%$$

- What proportion of people in the sample are not environmental club members and never use snowmobiles?

$$\frac{445}{1526} = 0.292$$

These percents or proportions are known as **joint relative frequencies**.

**DEFINITION Joint relative frequency**

A **joint relative frequency** gives the percent or proportion of individuals that have a specific value for one categorical variable and a specific value for another categorical variable.

> In the body, a joint is where two bones come together. In a two-way table, a *joint frequency* is shown where a row and column come together.

## EXAMPLE

### A *Titanic* disaster
### Calculating marginal and joint relative frequencies

Blank Archives/Getty Images

**PROBLEM:** In 1912 the luxury liner *Titanic*, on its first voyage across the Atlantic, struck an iceberg and sank. Some passengers got off the ship in lifeboats, but many died. The two-way table gives information about adult passengers who survived and who died, by class of travel.

(a) What proportion of adult passengers on the *Titanic* survived?

(b) Find the distribution of class of travel for adult passengers on the *Titanic* using relative frequencies.

(c) What percent of adult *Titanic* passengers traveled in third class and survived?

| | | Class of travel | | |
|---|---|---|---|---|
| | | First | Second | Third |
| **Survival status** | Survived | 197 | 94 | 151 |
| | Died | 122 | 167 | 476 |

**SOLUTION:**

Start by finding the marginal totals.

| | | Class of travel | | | |
|---|---|---|---|---|---|
| | | First | Second | Third | Total |
| **Survival status** | Survived | 197 | 94 | 151 | 442 |
| | Died | 122 | 167 | 476 | 765 |
| | Total | 319 | 261 | 627 | 1207 |

(a) $\frac{442}{1207} = 0.366$

(b) First: $\frac{319}{1207} = 0.264 = 26.4\%$

Second: $\frac{261}{1207} = 0.216 = 21.6\%$

Third: $\frac{627}{1207} = 0.519 = 51.9\%$

Remember that a distribution lists the possible values of a variable and how often those values occur.

(c) $\frac{151}{1207} = 0.125 = 12.5\%$

Note that the three percentages for class of travel in part (b) do not add to exactly 100% due to roundoff error.

**FOR PRACTICE, TRY EXERCISE 23**

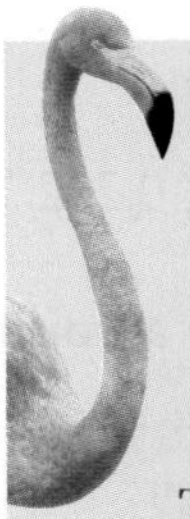

## CHECK YOUR UNDERSTANDING

An article in the *Journal of the American Medical Association* reports the results of a study designed to see if the herb St. John's wort is effective in treating moderately severe cases of depression. The study involved 338 patients who were being treated for major depression. The subjects were randomly assigned to receive one of three treatments: St. John's wort, Zoloft (a prescription drug), or placebo (an inactive treatment) for an 8-week period. The two-way table summarizes the data from the experiment.[5]

| | | Treatment | | |
|---|---|---|---|---|
| | | St. John's wort | Zoloft | Placebo |
| **Change in depression** | Full response | 27 | 27 | 37 |
| | Partial response | 16 | 26 | 13 |
| | No response | 70 | 56 | 66 |

1. What proportion of subjects in the study were randomly assigned to take St. John's wort? Explain why this value makes sense.
2. Find the distribution of change in depression for the subjects in this study using relative frequencies.
3. What percent of subjects took Zoloft and showed a full response?

## Relationships Between Two Categorical Variables

Let's return to the data from the Yellowstone National Park survey of 1526 randomly selected winter visitors. Earlier, we calculated marginal and joint relative frequencies from the two-way table. These values do not tell us much about the *relationship* between environmental club membership and snowmobile use for the people in the sample.

| | | Environmental club | | |
|---|---|---|---|---|
| | | No | Yes | Total |
| Snowmobile use | Never used | 445 | 212 | 657 |
| | Snowmobile renter | 497 | 77 | 574 |
| | Snowmobile owner | 279 | 16 | 295 |
| | Total | 1221 | 305 | 1526 |

We can also use the two-way table to answer questions like:

- What percent of environmental club members in the sample are snowmobile owners?

$$\frac{16}{305} = 0.052 = 5.2\%$$

- What proportion of snowmobile renters in the sample are not environmental club members?

$$\frac{497}{574} = 0.866$$

These percents or proportions are known as **conditional relative frequencies.**

**DEFINITION Conditional relative frequency**

A **conditional relative frequency** gives the percent or proportion of individuals that have a specific value for one categorical variable among individuals who share the same value of another categorical variable (the condition).

**EXAMPLE**

### A *Titanic* disaster

#### Conditional relative frequencies

**PROBLEM:** In 1912 the luxury liner *Titanic,* on its first voyage across the Atlantic, struck an iceberg and sank. Some passengers made it off the ship in lifeboats, but many died. The two-way table gives information about adult passengers who survived and who died, by class of travel.

| | | Class of travel | | | |
|---|---|---|---|---|---|
| | | First | Second | Third | Total |
| Survival status | Survived | 197 | 94 | 151 | 442 |
| | Died | 122 | 167 | 476 | 765 |
| | Total | 319 | 261 | 627 | 1207 |

(a) What proportion of survivors were third-class passengers?

(b) What percent of first-class passengers survived?

Note that a proportion is always a number between 0 and 1, whereas a percent is a number between 0 and 100. To get a percent, multiply the proportion by 100.

**SOLUTION:**

(a) $\frac{151}{442} = 0.342$ (b) $\frac{197}{319} = 0.618 = 61.8\%$

**FOR PRACTICE, TRY EXERCISE 27**

We can study the snowmobile use habits of environmental club members by looking only at the "Yes" column in the two-way table.

| | | Environmental club | | |
|---|---|---|---|---|
| | | No | Yes | Total |
| Snowmobile use | Never used | 445 | 212 | 657 |
| | Snowmobile renter | 497 | 77 | 574 |
| | Snowmobile owner | 279 | 16 | 295 |
| | Total | 1221 | 305 | 1526 |

It is easy to calculate the proportions or percents of environmental club members who never use, rent, and own snowmobiles:

$$\text{Never: } \frac{212}{305} = 0.695 \text{ or } 69.5\% \qquad \text{Rent: } \frac{77}{305} = 0.252 \text{ or } 25.2\%$$

$$\text{Own: } \frac{16}{305} = 0.052 \text{ or } 5.2\%$$

We could also refer to this distribution as the *conditional distribution* of snowmobile use among environmental club members.

This is the distribution of snowmobile use among environmental club members.

We can find the distribution of snowmobile use among the survey respondents who are not environmental club members in a similar way. The table summarizes the conditional relative frequencies for both groups.

| Snowmobile use | Not environmental club members | Environmental club members |
|---|---|---|
| Never | $\frac{445}{1221} = 0.364$ or 36.4% | $\frac{212}{305} = 0.695$ or 69.5% |
| Rent | $\frac{497}{1221} = 0.407$ or 40.7% | $\frac{77}{305} = 0.252$ or 25.2% |
| Own | $\frac{279}{1221} = 0.229$ or 22.9% | $\frac{16}{305} = 0.052$ or 5.2% |

**AP® EXAM TIP**

When comparing groups of different sizes, be sure to use relative frequencies (percents or proportions) instead of frequencies (counts) when analyzing categorical data. Comparing only the frequencies can be misleading, as in this setting. There are many more people who never use snowmobiles among the non-environmental club members in the sample (445) than among the environmental club members (212). However, the *percentage* of environmental club members who never use snowmobiles is much higher (69.5% to 36.4%). Finally, make sure to avoid statements like "More club members never use snowmobiles" when you mean "A greater percentage of club members never use snowmobiles."

Figure 1.3 compares the distributions of snowmobile use for Yellowstone National Park visitors who are environmental club members and those who are not environmental club members with (a) a **side-by-side bar graph**, (b) a **segmented bar graph**, and (c) a **mosaic plot**. Notice that the segmented bar graph can be obtained by stacking the bars in the side-by-side bar graph for each of the two environmental club membership categories (no and yes). The bar widths in the mosaic plot are proportional to the number of survey respondents who are (305) and are not (1221) environmental club members.

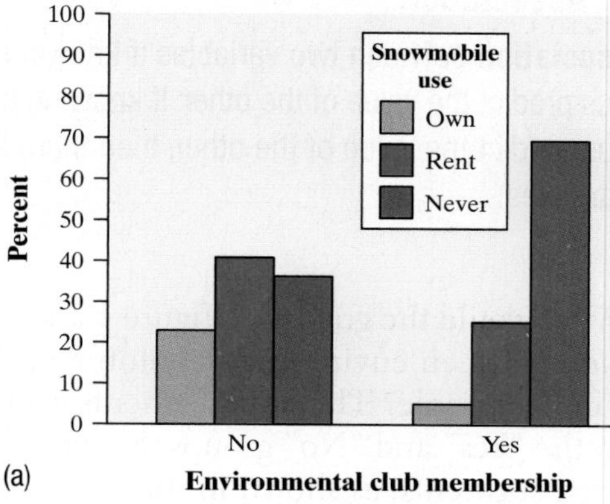


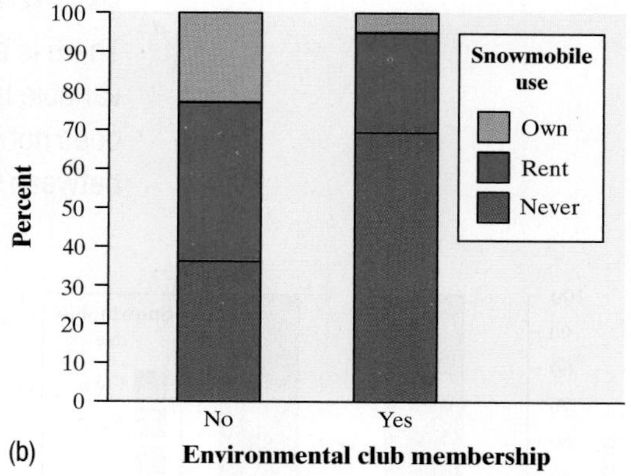


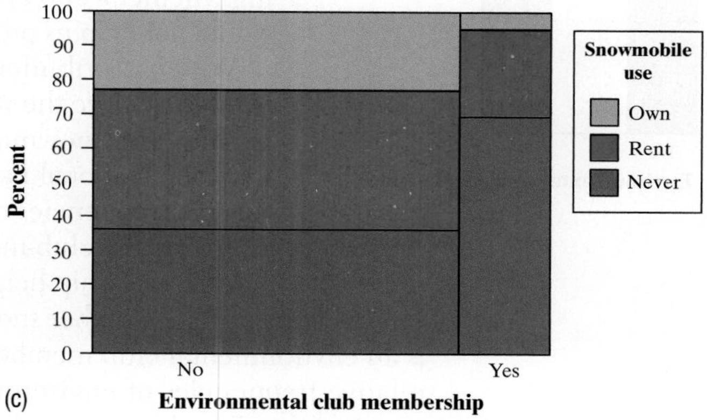


**FIGURE 1.3** (a) Side-by-side bar graph, (b) segmented bar graph, and (c) mosaic plot displaying the distribution of snowmobile use among environmental club members and among non-environmental club members from the 1526 randomly selected winter visitors to Yellowstone National Park.

**DEFINITION Side-by side bar graph, Segmented bar graph, Mosaic plot**

A **side-by-side bar graph** displays the distribution of a categorical variable for each value of another categorical variable. The bars are grouped together based on the values of one of the categorical variables and placed side by side.

A **segmented bar graph** displays the distribution of a categorical variable as segments of a rectangle, with the area of each segment proportional to the percent of individuals in the corresponding category.

A **mosaic plot** is a modified segmented bar graph in which the width of each rectangle is proportional to the number of individuals in the corresponding category.

All three graphs in Figure 1.3 show a clear **association** between environmental club membership and snowmobile use in this random sample of 1526 winter visitors to Yellowstone National Park. The environmental club members were much less likely to rent (25.2% versus 40.7%) or own (5.2% versus 29.0%) snowmobiles than non-club-members and more likely to never use a snowmobile (69.5% versus 36.4%). Knowing whether or not a person in the sample is an environmental club member helps us predict that individual's snowmobile use.

**DEFINITION Association**

There is an **association** between two variables if knowing the value of one variable helps us predict the value of the other. If knowing the value of one variable does not help us predict the value of the other, then there is no association between the variables.

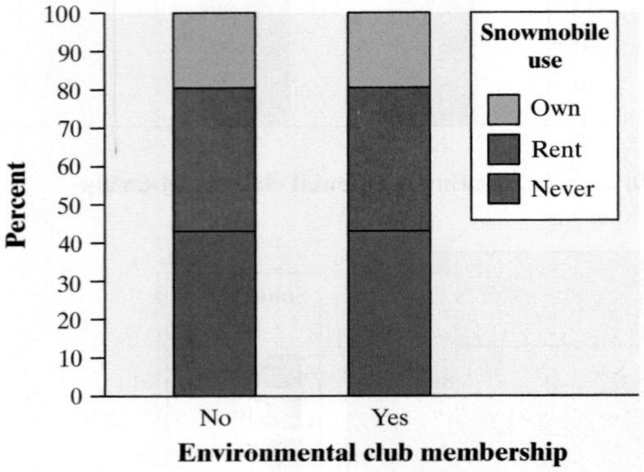


What would the graphs in Figure 1.3 look like if there was *no association* between environmental club membership and snowmobile use in the sample? The blue segments would be the same height for both the "Yes" and "No" groups. So would the green segments and the red segments, as shown in the graph at left. In that case, knowing whether a survey respondent is an environmental club member would *not* help us predict his or her snowmobile use.

Which distributions should we compare? Our goal all along has been to analyze the relationship between environmental club membership and snowmobile use for this random sample of 1526 Yellowstone National Park visitors. We decided to calculate conditional relative frequencies of snowmobile use among environmental club members and among non-club-members. Why? Because we wanted to see if environmental club membership helped us predict snowmobile use. What if we had wanted to determine whether snowmobile use helps us predict whether a person is an environmental club member? Then we would have calculated conditional relative frequencies of environmental club membership among snowmobile owners, renters, and non-users. *In general, you should calculate the distribution of the variable that you want to predict for each value of the other variable.*

Can we say that there is an association between environmental club membership and snowmobile use in the *population* of all winter visitors to Yellowstone National Park? Making this determination requires formal inference, which will have to wait until Chapter 12.

## EXAMPLE A *Titanic* disaster
### Conditional relative frequencies and association

**PROBLEM:** In 1912 the luxury liner *Titanic*, on its first voyage across the Atlantic, struck an iceberg and sank. Some passengers made it off the ship in lifeboats, but many died. The two-way table gives information about adult passengers who survived and who died, by class of travel.

(a) Find the distribution of survival status for each class of travel. Make a segmented bar graph to compare these distributions.

(b) Describe what the graph in part (a) reveals about the association between class of travel and survival status for adult passengers on the *Titanic*.

| | | Class of travel | | | |
|---|---|---|---|---|---|
| | | First | Second | Third | Total |
| Survival status | Survived | 197 | 94 | 151 | 442 |
| | Died | 122 | 167 | 476 | 765 |
| | Total | 319 | 261 | 627 | 1207 |

## SOLUTION:

(a) First class Survived: $\frac{197}{319} = 0.618 = 61.8\%$ Died: $\frac{122}{319} = 0.382 = 38.2\%$

Second class Survived: $\frac{94}{261} = 0.360 = 36.0\%$ Died: $\frac{167}{261} = 0.640 = 64.0\%$

Third class Survived: $\frac{151}{627} = 0.241 = 24.1\%$ Died: $\frac{476}{627} = 0.759 = 75.9\%$

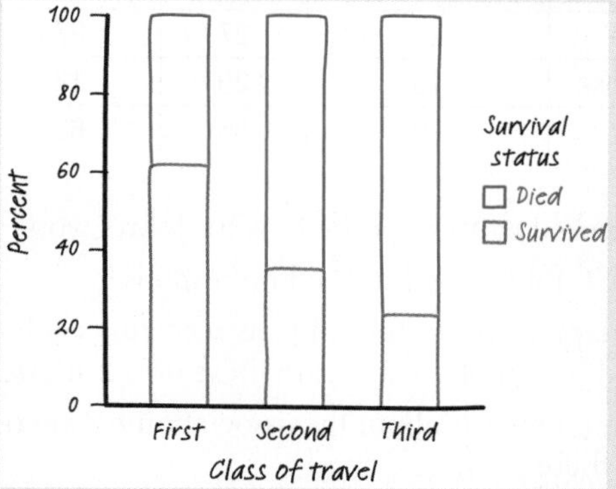


To make the segmented bar graph:

- **Draw and label the axes.** Put class of travel on the horizontal axis and percent on the vertical axis.
- **"Scale" the axes.** Use a vertical scale from 0 to 100%, with tick marks every 20%.
- **Draw bars.** Make each bar have a height of 100%. Be sure the bars are equal in width and leave spaces between them. Segment each bar based on the conditional relative frequencies you calculated. Use different colors or shading patterns to represent the two possible statuses—survived and died. Add a key to the graph that tells us which color (or shading) represents which status.

(b) Knowing a passenger's class of travel helps us predict his or her survival status. First class had the highest percentage of survivors (61.8%), followed by second class (36.0%), and then third class (24.1%).

**FOR PRACTICE, TRY EXERCISE 29**

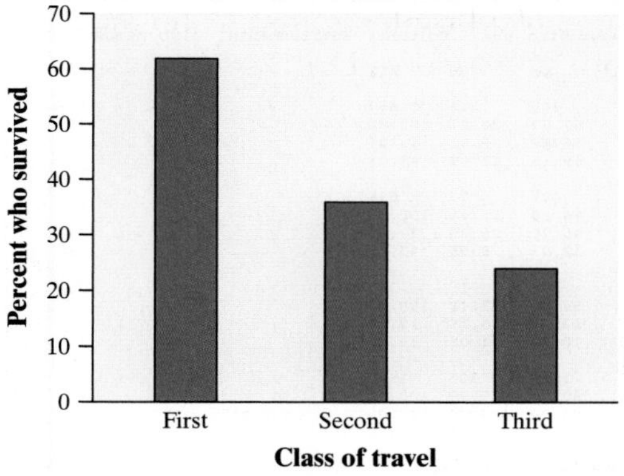


**FIGURE 1.4** Bar graph comparing the percents of passengers who survived among each of the three classes of travel on the *Titanic*.

Because the variable "Survival status" has only two possible values, comparing the three distributions displayed in the segmented bar graph amounts to comparing the percent of passengers in each class of travel who survived. The bar graph in Figure 1.4 shows this comparison. Note that the bar heights do *not* add to 100%, because each bar represents a different group of passengers on the *Titanic*.

caution

We offer a final caution about studying the relationship between two variables: **association does not imply causation.** It may be true that being in a higher class of travel on the *Titanic* increased a passenger's chance of survival. However, there isn't always a cause-and-effect relationship between two variables even if they are clearly associated. For example, a recent study proclaimed that people who are overweight are less likely to die within a few years than are people of normal weight. Does this mean that gaining weight will cause you to live longer? Not at all. The study included smokers, who tend to be thinner and also much more likely to die in a given period than non-smokers. Smokers increased the death rate for the normal-weight category, making it appear as if being overweight is better.[6] The moral of the story: *beware other variables!*

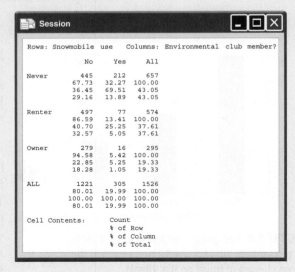

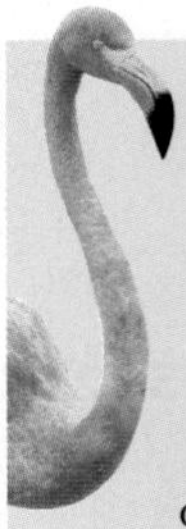

## CHECK YOUR UNDERSTANDING

An article in the *Journal of the American Medical Association* reports the results of a study designed to see if the herb St. John's wort is effective in treating moderately severe cases of depression. The study involved 338 subjects who were being treated for major depression. The subjects were randomly assigned to receive one of three treatments: St. John's wort, Zoloft (a prescription drug), or placebo (an inactive treatment) for an 8-week period. The two-way table summarizes the data from the experiment.

| | | Treatment | | |
|---|---|---|---|---|
| | | **St. John's wort** | **Zoloft** | **Placebo** |
| **Change in depression** | Full response | 27 | 27 | 37 |
| | Partial response | 16 | 26 | 13 |
| | No response | 70 | 56 | 66 |

1. What proportion of subjects who showed a full response took St. John's wort?
2. What percent of subjects who took St. John's wort showed no response?
3. Find the distribution of change in depression for the subjects receiving each of the three treatments. Make a segmented bar graph to compare these distributions.
4. Describe what the graph in Question 3 reveals about the association between treatment and change in depression for these subjects.

## 1. Technology Corner ANALYZING TWO-WAY TABLES

Statistical software will provide marginal relative frequencies, joint relative frequencies, and conditional relative frequencies for data summarized in a two-way table. Here is output from Minitab for the data on snowmobile use and environmental club membership. Use the information on cell contents at the bottom of the output to help you interpret what each value in the table represents.

Session

```
Rows: Snowmobile use    Columns: Environmental club member?

            No      Yes      All

Never      445      212      657
         67.73    32.27   100.00
         36.45    69.51    43.05
         29.16    13.89    43.05

Renter     497       77      574
         86.59    13.41   100.00
         40.70    25.25    37.61
         32.57     5.05    37.61

Owner      279       16      295
         94.58     5.42   100.00
         22.85     5.25    19.33
         18.28     1.05    19.33

ALL       1221      305     1526
         80.01    19.99   100.00
        100.00   100.00   100.00
         80.01    19.99   100.00

Cell Contents:      Count
                    % of Row
                    % of Column
                    % of Total
```

# Section 1.1 Summary

- The distribution of a categorical variable lists the categories and gives the **frequency** (count) or **relative frequency** (percent or proportion) of individuals that fall in each category.
- You can use a **pie chart** or **bar graph** to display the distribution of a categorical variable. When examining any graph, ask yourself, "What do I see?"
- Beware of graphs that mislead the eye. Look at the scales to see if they have been distorted to create a particular impression. Avoid making graphs that replace the bars of a bar graph with pictures whose height and width both change.
- A **two-way table** of counts summarizes data on the relationship between two categorical variables for some group of individuals.
- You can use a two-way table to calculate three types of relative frequencies:
  - A **marginal relative frequency** gives the percent or proportion of individuals that have a specific value for one categorical variable. Use the appropriate row total or column total in a two-way table when calculating a marginal relative frequency.
  - A **joint relative frequency** gives the percent or proportion of individuals that have a specific value for one categorical variable and a specific value for another categorical variable. Use the value from the appropriate cell in the two-way table when calculating a joint relative frequency.
  - A **conditional relative frequency** gives the percent or proportion of individuals that have a specific value for one categorical variable among individuals who share the same value of another categorical variable (the condition). Use conditional relative frequencies to compare distributions of a categorical variable for two or more groups.
- Use a **side-by-side bar graph**, a **segmented bar graph**, or a **mosaic plot** to compare the distribution of a categorical variable for two or more groups.
- There is an **association** between two variables if knowing the value of one variable helps predict the value of the other. To see whether there is an association between two categorical variables, find the distribution of one variable for each value of the other variable by calculating an appropriate set of conditional relative frequencies.

## 1.1 Technology Corner

*TI-Nspire and other technology instructions are on the book's website at highschool.bfwpub.com/updatedtps6e.*

**1. Analyzing two-way tables** **Page 22**

# Section 1.1 Exercises

11. **Birth days** The frequency table summarizes data on the numbers of babies born on each day of the week in the United States in a recent week.[7]

pg 11

| Day | Births |
|---|---|
| Sunday | 7374 |
| Monday | 11,704 |
| Tuesday | 13,169 |
| Wednesday | 13,038 |
| Thursday | 13,013 |
| Friday | 12,664 |
| Saturday | 8459 |

(a) Identify the individuals in this data set.

(b) Make a frequency bar graph to display the data. Describe what you see.

12. **Going up?** As of 2015, there were over 75,000 elevators in New York City. The frequency table summarizes data on the number of elevators of each type.[8]

| Type | Count |
|---|---|
| Passenger elevator | 66,602 |
| Freight elevator | 4140 |
| Escalator | 2663 |
| Dumbwaiter | 1143 |
| Sidewalk elevator | 943 |
| Private elevator | 252 |
| Handicap lift | 227 |
| Manlift | 73 |
| Public elevator | 45 |

(a) Identify the individuals in this data set.

(b) Make a frequency bar graph to display the data. Describe what you see.

13. **Buying cameras** The brands of the last 45 digital single-lens reflex (SLR) cameras sold on a popular Internet auction site are listed here. Make a relative frequency bar graph for these data. Describe what you see.

| | | | | |
|---|---|---|---|---|
| Canon | Sony | Canon | Nikon | Fujifilm |
| Nikon | Canon | Sony | Canon | Canon |
| Nikon | Canon | Nikon | Canon | Canon |
| Canon | Nikon | Fujifilm | Canon | Nikon |
| Nikon | Canon | Canon | Canon | Canon |
| Olympus | Canon | Canon | Canon | Nikon |
| Olympus | Sony | Canon | Canon | Sony |
| Canon | Nikon | Sony | Canon | Fujifilm |
| Nikon | Canon | Nikon | Canon | Sony |

14. **Disc dogs** Here is a list of the breeds of dogs that won the World Canine Disc Championships from 1975 through 2016. Make a relative frequency bar graph for these data. Describe what you see.

| | | |
|---|---|---|
| Whippet | Mixed breed | Australian shepherd |
| Whippet | Australian shepherd | Australian shepherd |
| Whippet | Border collie | Australian shepherd |
| Mixed breed | Australian shepherd | Border collie |
| Mixed breed | Mixed breed | Border collie |
| Other purebred | Mixed breed | Australian shepherd |
| Labrador retriever | Mixed breed | Border collie |
| Mixed breed | Border collie | Border collie |
| Mixed breed | Border collie | Other purebred |
| Border collie | Australian shepherd | Border collie |
| Mixed breed | Border collie | Border collie |
| Mixed breed | Australian shepherd | Border collie |
| Labrador retriever | Border collie | Mixed breed |
| Labrador retriever | Mixed breed | Australian shepherd |

15. **Cool car colors** The most popular colors for cars and light trucks change over time. Silver advanced past green in 2000 to become the most popular color worldwide, then gave way to shades of white in 2007. Here is a relative frequency table that summarizes data on the colors of vehicles sold worldwide in a recent year.[9]

| Color | Percent of vehicles | Color | Percent of vehicles |
|---|---|---|---|
| Black | 19 | Red | 9 |
| Blue | 6 | Silver | 14 |
| Brown/beige | 5 | White | 29 |
| Gray | 12 | Yellow/gold | 3 |
| Green | 1 | Other | ?? |

(a) What percent of vehicles would fall in the "Other" category?

(b) Make a bar graph to display the data. Describe what you see.

(c) Would it be appropriate to make a pie chart of these data? Explain.

16. **Spam** Email spam is the curse of the Internet. Here is a relative frequency table that summarizes data on the most common types of spam:[10]

| Type of spam | Percent |
|---|---|
| Adult | 19 |
| Financial | 20 |
| Health | 7 |
| Internet | 7 |
| Leisure | 6 |
| Products | 25 |
| Scams | 9 |
| Other | ?? |

(a) What percent of spam would fall in the "Other" category?

(b) Make a bar graph to display the data. Describe what you see.

(c) Would it be appropriate to make a pie chart of these data? Explain.

17. **Hispanic origins** Here is a pie chart prepared by the Census Bureau to show the origin of the more than 50 million Hispanics in the United States in 2010.[11] About what percent of Hispanics are Mexican? Puerto Rican?

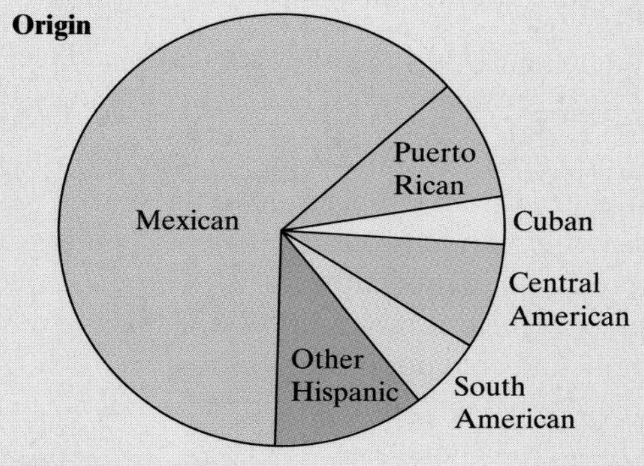


18. **Which major?** About 3 million first-year students enroll in colleges and universities each year. What do they plan to study? The pie chart displays data on the percent of first-year students who plan to major in several disciplines.[12] About what percent of first-year students plan to major in business? In social science?

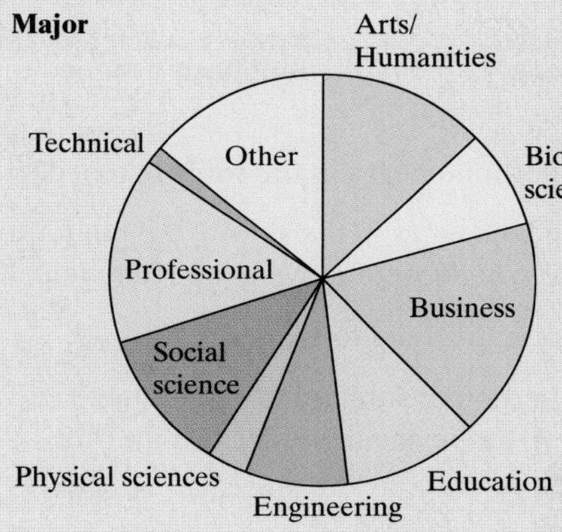


19. **Going to school** Students in a high school statistics class were given data about the main method of transportation to school for a group of 30 students. They produced the pictograph shown. Explain how this graph is misleading.

pg 12

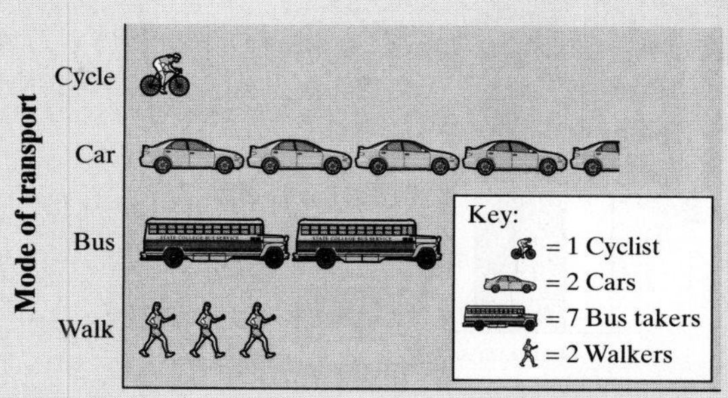


20. **Social media** The Pew Research Center surveyed a random sample of U.S. teens and adults about their use of social media. The following pictograph displays some results. Explain how this graph is misleading.

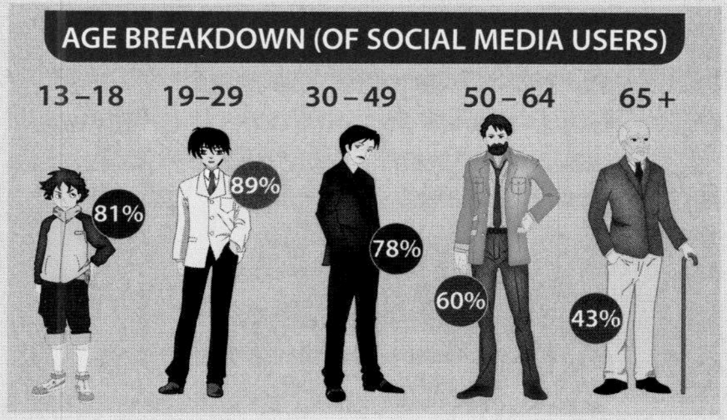


21. **Binge-watching** Do you "binge-watch" television series by viewing multiple episodes of a series at one sitting? A survey of 800 people who binge-watch were asked how many episodes is too many to watch in one viewing session. The results are displayed in the bar graph.[13] Explain how this graph is misleading.

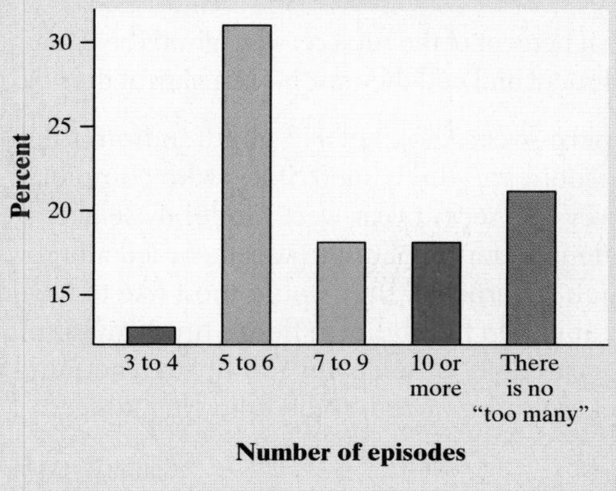


22. **Support the court?** A news network reported the results of a survey about a controversial court decision. The network initially posted on its website a bar graph of the data similar to the one that follows. Explain how this graph is misleading. (*Note:* When notified about

the misleading nature of its graph, the network posted a corrected version.)

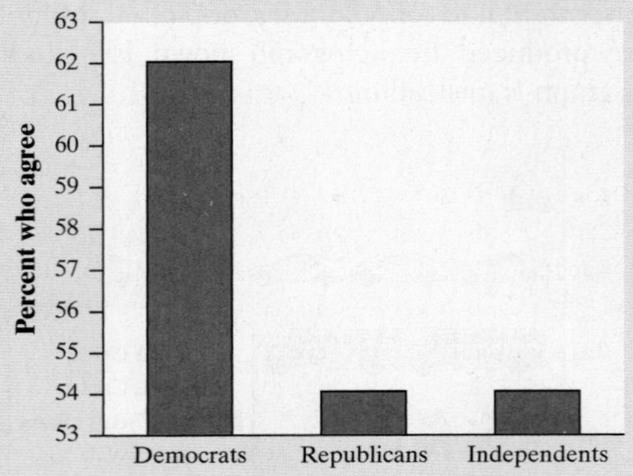


**23. A smash or a hit?** Researchers asked 150 subjects to recall the details of a car accident they watched on video. Fifty subjects were randomly assigned to be asked, "About how fast were the cars going when they smashed into each other?" For another 50 randomly assigned subjects, the words "smashed into" were replaced with "hit." The remaining 50 subjects—the control group—were not asked to estimate speed. A week later, all subjects were asked if they saw any broken glass at the accident (there wasn't any). The table shows each group's response to the broken glass question.[14]

pg 15

| | | Treatment | | |
|---|---|---|---|---|
| | | "Smashed into" | "Hit" | Control |
| Response | Yes | 16 | 7 | 6 |
| | No | 34 | 43 | 44 |

(a) What proportion of subjects were given the control treatment?

(b) Find the distribution of responses about whether there was broken glass at the accident for the subjects in this study using relative frequencies.

(c) What percent of the subjects were given the "smashed into" treatment and said they saw broken glass at the accident?

**24. Superpowers** A total of 415 children from the United Kingdom and the United States who completed a survey in a recent year were randomly selected. Each student's country of origin was recorded along with which superpower they would most like to have: the ability to fly, ability to freeze time, invisibility, superstrength, or telepathy (ability to read minds). The data are summarized in the following table.[15]

| | | Country | |
|---|---|---|---|
| | | U.K. | U.S. |
| Superpower | Fly | 54 | 45 |
| | Freeze time | 52 | 44 |
| | Invisibility | 30 | 37 |
| | Superstrength | 20 | 23 |
| | Telepathy | 44 | 66 |

(a) What proportion of students in the sample are from the United States?

(b) Find the distribution of superpower preference for the students in the sample using relative frequencies.

(c) What percent of students in the sample are from the United Kingdom and prefer telepathy as their superpower preference?

**25. Body image** A random sample of 1200 U.S. college students was asked, "What is your perception of your own body? Do you feel that you are overweight, underweight, or about right?" The two-way table summarizes the data on perceived body image by gender.[16]

| | | Gender | | |
|---|---|---|---|---|
| | | Female | Male | Total |
| Body image | About right | 560 | 295 | 855 |
| | Overweight | 163 | 72 | 235 |
| | Underweight | 37 | 73 | 110 |
| | Total | 760 | 440 | 1200 |

(a) What percent of respondents feel that their body weight is about right?

(b) What proportion of the sample is female?

(c) What percent of respondents are males and feel that they are overweight or underweight?

**26. Python eggs** How is the hatching of water python eggs influenced by the temperature of the snake's nest? Researchers randomly assigned newly laid eggs to one of three water temperatures: hot, neutral, or cold. Hot duplicates the extra warmth provided by the mother python, and cold duplicates the absence of the mother. The two-way table summarizes the data on whether or not the eggs hatched.[17]

| | | Water temperature | | | |
|---|---|---|---|---|---|
| | | Cold | Neutral | Hot | Total |
| Hatched? | Yes | 16 | 38 | 75 | 129 |
| | No | 11 | 18 | 29 | 58 |
| | Total | 27 | 56 | 104 | 187 |

(a) What percent of eggs were randomly assigned to hot water?

(b) What proportion of eggs in the study hatched?

(c) What percent of eggs in the study were randomly assigned to cold or neutral water and hatched?

**27. A smash or a hit** Refer to Exercise 23.

pg 17

(a) What proportion of subjects who said they saw broken glass at the accident received the "hit" treatment?

(b) What percent of subjects who received the "smashed into" treatment said they did not see broken glass at the accident?

28. **Superpower** Refer to Exercise 24.

(a) What proportion of students in the sample who prefer invisibility as their superpower are from the United States?

(b) What percent of students in the sample who are from the United Kingdom prefer superstrength as their superpower?

29. **A smash or a hit** Refer to Exercise 23.

pg 20

(a) Find the distribution of responses about whether there was broken glass at the accident for each of the three treatment groups. Make a segmented bar graph to compare these distributions.

(b) Describe what the graph in part (a) reveals about the association between response about broken glass at the accident and treatment received for the subjects in the study.

30. **Superpower** Refer to Exercise 24.

(a) Find the distribution of superpower preference for the students in the sample from each country (i.e., the United States and the United Kingdom). Make a segmented bar graph to compare these distributions.

(b) Describe what the graph in part (a) reveals about the association between country of origin and superpower preference for the students in the sample.

31. **Body image** Refer to Exercise 25.

(a) Of the respondents who felt that their body weight was about right, what proportion were female?

(b) Of the female respondents, what percent felt that their body weight was about right?

(c) The mosaic plot displays the distribution of perceived body image by gender. Describe what this graph reveals about the association between these two variables for the 1200 college students in the sample.

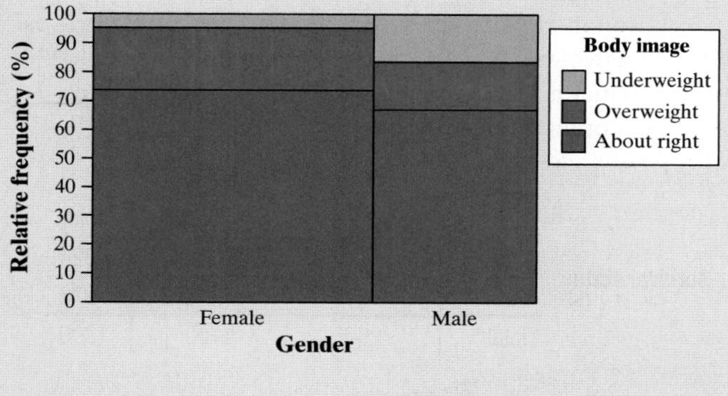


32. **Python eggs** Refer to Exercise 26.

(a) Of the eggs that hatched, what proportion were randomly assigned to hot water?

(b) Of the eggs that were randomly assigned to hot water, what percent hatched?

(c) The mosaic plot displays the distribution of hatching status by water temperature. Describe what this graph reveals about the association between these two variables for the python eggs in this experiment.

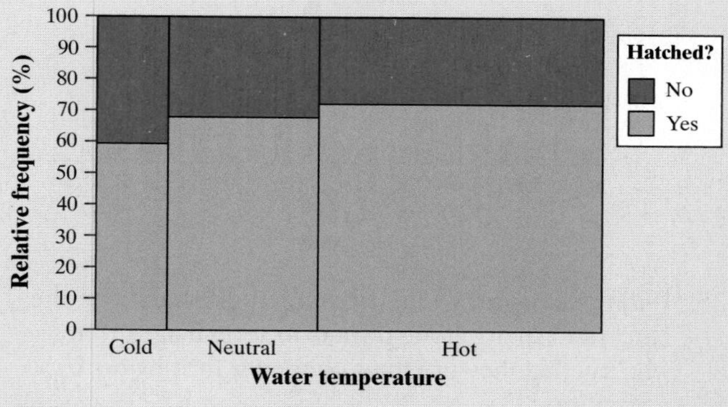


33. **Far from home** A survey asked first-year college students, "How many miles is this college from your permanent home?" Students had to choose from the following options: 5 or fewer, 6 to 10, 11 to 50, 51 to 100, 101 to 500, or more than 500. The side-by-side bar graph shows the percentage of students at public and private 4-year colleges who chose each option.[18] Write a few sentences comparing the distributions of distance from home for students from private and public 4-year colleges who completed the survey.

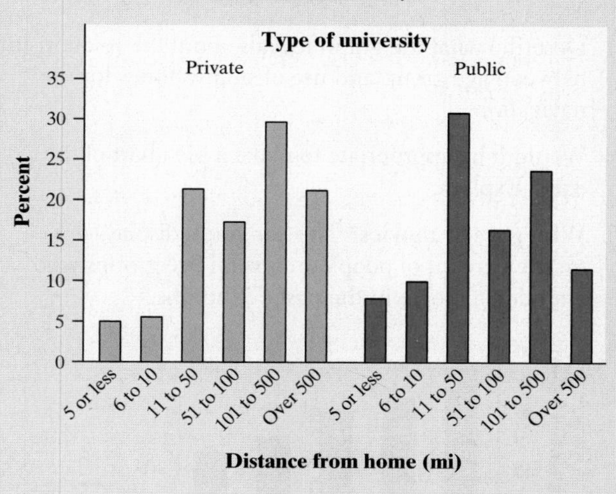

34. **Popular car colors** Favorite car colors may differ among countries. The side-by-side bar graph displays data on the most popular car colors in a recent year for North America and Asia. Write a few sentences comparing the distributions.[19]

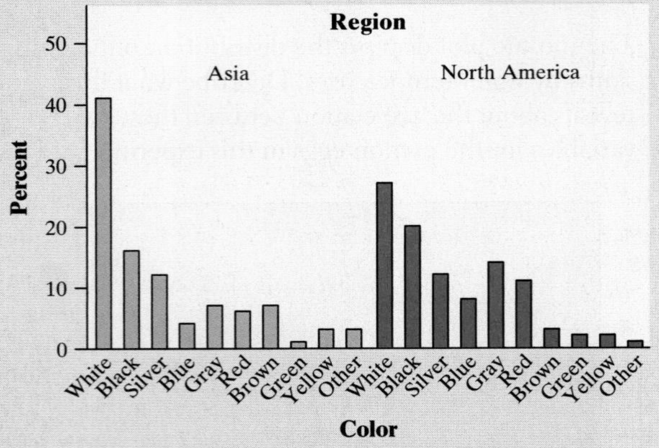


35. **Phone navigation** The bar graph displays data on the percent of smartphone owners in several age groups who say that they use their phone for turn-by-turn navigation.[20]

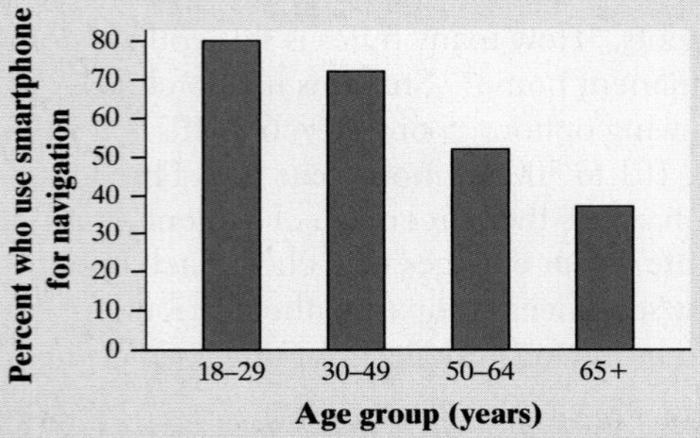


(a) Describe what the graph reveals about the relationship between age group and use of smartphones for navigation.

(b) Would it be appropriate to make a pie chart of the data? Explain.

36. **Who goes to movies?** The bar graph displays data on the percent of people in several age groups who attended a movie in the past 12 months.[21]

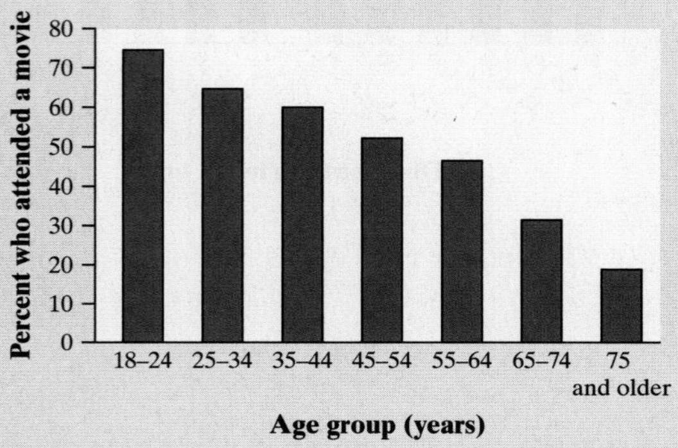


(a) Describe what the graph reveals about the relationship between age group and movie attendance.

(b) Would it be appropriate to make a pie chart of the data? Explain.

37. **Marginal totals aren't the whole story** Here are the row and column totals for a two-way table with two rows and two columns:

| | | |
|---|---|---|
| *a* | *b* | 50 |
| *c* | *d* | 50 |
| 60 | 40 | 100 |

Find *two different* sets of counts *a, b, c,* and *d* for the body of the table that give these same totals. This shows that the relationship between two variables cannot be obtained from the two individual distributions of the variables.

38. **Women and children first?** Here's another table that summarizes data on survival status by gender and class of travel on the *Titanic*:

| | Class of travel | | | | | |
|---|---|---|---|---|---|---|
| | First class | | Second class | | Third class | |
| Survival status | Female | Male | Female | Male | Female | Male |
| Survived | 140 | 57 | 80 | 14 | 76 | 75 |
| Died | 4 | 118 | 13 | 154 | 89 | 387 |

(a) Find the distributions of survival status for males and for females within each class of travel. Did women survive the disaster at higher rates than men? Explain.

(b) In an earlier example, we noted that survival status is associated with class of travel. First-class passengers had the highest survival rate, while third-class passengers had the lowest survival rate. Does this same relationship hold for both males and females in all three classes of travel? Explain.

39. **Simpson's paradox** Accident victims are sometimes taken by helicopter from the accident scene to a hospital. Helicopters save time. Do they also save lives? The two-way table summarizes data from a sample of patients who were transported to the hospital by helicopter or by ambulance.[22]

| | | Method of transport | | |
|---|---|---|---|---|
| | | Helicopter | Ambulance | Total |
| Survival status | Died | 64 | 260 | 324 |
| | Survived | 136 | 840 | 976 |
| | Total | 200 | 1100 | 1300 |

(a) What percent of patients died with each method of transport?

Here are the same data broken down by severity of accident:

**Serious accidents**

| | | Method of transport | | |
|---|---|---|---|---|
| | | Helicopter | Ambulance | Total |
| Survival status | Died | 48 | 60 | 108 |
| | Survived | 52 | 40 | 92 |
| | Total | 100 | 100 | 200 |

**Less serious accidents**

| | | Method of transport | | |
|---|---|---|---|---|
| | | Helicopter | Ambulance | Total |
| Survival status | Died | 16 | 200 | 216 |
| | Survived | 84 | 800 | 884 |
| | Total | 100 | 1000 | 1100 |

(b) Calculate the percent of patients who died with each method of transport for the serious accidents. Then calculate the percent of patients who died with each method of transport for the less serious accidents. What do you notice?

(c) See if you can explain how the result in part (a) is possible given the result in part (b).

*Note:* This is an example of *Simpson's paradox*, which states that an association between two variables that holds for each value of a third variable can be changed or even reversed when the data for all values of the third variable are combined.

**Multiple Choice:** *Select the best answer for Exercises 40–43.*

**40.** For which of the following would it be *inappropriate* to display the data with a single pie chart?

(a) The distribution of car colors for vehicles purchased in the last month

(b) The distribution of unemployment percentages for each of the 50 states

(c) The distribution of favorite sport for a sample of 30 middle school students

(d) The distribution of shoe type worn by shoppers at a local mall

(e) The distribution of presidential candidate preference for voters in a state

**41.** The following bar graph shows the distribution of favorite subject for a sample of 1000 students. What is the most serious problem with the graph?

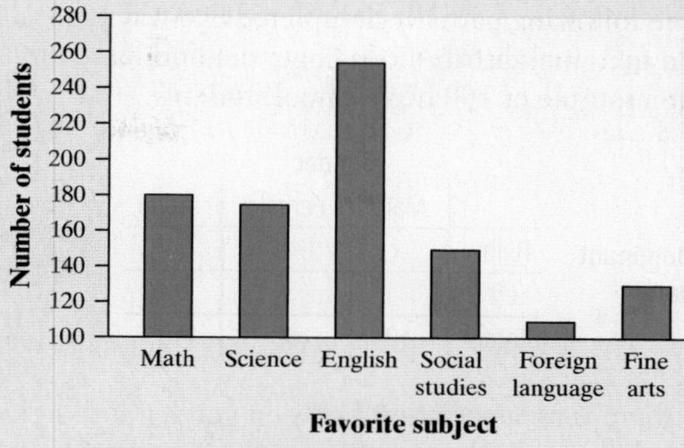


(a) The subjects are not listed in the correct order.

(b) This distribution should be displayed with a pie chart.

(c) The vertical axis should show the percent of students.

(d) The vertical axis should start at 0 rather than 100.

(e) The foreign language bar should be broken up by language.

**42.** The Dallas Mavericks won the NBA championship in the 2010–2011 season. The two-way table displays the relationship between the outcome of each game in the regular season and whether the Mavericks scored at least 100 points.

| | | Points scored | | |
|---|---|---|---|---|
| | | 100 or more | Fewer than 100 | Total |
| Outcome of game | Win | 43 | 14 | 57 |
| | Loss | 4 | 21 | 25 |
| | Total | 47 | 35 | 82 |

Which of the following is the best evidence that there is an association between the outcome of a game and whether or not the Mavericks scored at least 100 points?

(a) The Mavericks won 57 games and lost only 25 games.

(b) The Mavericks scored at least 100 points in 47 games and fewer than 100 points in only 35 games.

(c) The Mavericks won 43 games when scoring at least 100 points and only 14 games when scoring fewer than 100 points.

(d) The Mavericks won a higher proportion of games when scoring at least 100 points (43/47) than when they scored fewer than 100 points (14/35).

(e) The combination of scoring 100 or more points and winning the game occurred more often (43 times) than any other combination of outcomes.

43. The following partially completed two-way table shows the marginal distributions of gender and handedness for a sample of 100 high school students.

| | | Gender | | |
|---|---|---|---|---|
| | | Male | Female | Total |
| Dominant hand | Right | $x$ | | 90 |
| | Left | | | 10 |
| | Total | 40 | 60 | 100 |

If there is no association between gender and handedness for the members of the sample, which of the following is the correct value of $x$?

(a) 20
(b) 30
(c) 36
(d) 45
(e) Impossible to determine without more information.

**Recycle and Review**

44. **Hotels** (Introduction) A high school lacrosse team is planning to go to Buffalo for a three-day tournament. The tournament's sponsor provides a list of available hotels, along with some information about each hotel. The following table displays data about hotel options. Identify the individuals and variables in this data set. Classify each variable as categorical or quantitative.

| Hotel | Pool | Exercise room? | Internet ($/day) | Restaurants | Distance to site (mi) | Room service? | Room rate ($/day) |
|---|---|---|---|---|---|---|---|
| Comfort Inn | Out | Y | 0.00 | 1 | 8.2 | Y | 149 |
| Fairfield Inn & Suites | In | Y | 0.00 | 1 | 8.3 | N | 119 |
| Baymont Inn & Suites | Out | Y | 0.00 | 1 | 3.7 | Y | 60 |
| Chase Suite Hotel | Out | N | 15.00 | 0 | 1.5 | N | 139 |
| Courtyard | In | Y | 0.00 | 1 | 0.2 | Dinner | 114 |
| Hilton | In | Y | 10.00 | 2 | 0.1 | Y | 156 |
| Marriott | In | Y | 9.95 | 2 | 0.0 | Y | 145 |

# SECTION 1.2 Displaying Quantitative Data with Graphs

**LEARNING TARGETS** *By the end of the section, you should be able to:*

- Make and interpret dotplots, stemplots, and histograms of quantitative data.
- Identify the shape of a distribution from a graph.
- Describe the overall pattern (shape, center, and variability) of a distribution and identify any major departures from the pattern (outliers).
- Compare distributions of quantitative data using dotplots, stemplots, and histograms.

To display the distribution of a categorical variable, use a bar graph or a pie chart. How can we picture the distribution of a quantitative variable? In this section, we present several types of graphs that can be used to display quantitative data.

## Dotplots

One of the simplest graphs to construct and interpret is a **dotplot**.

**DEFINITION Dotplot**

A **dotplot** shows each data value as a dot above its location on a number line.

Note that goals scored is a discrete variable.

Here are data on the number of goals scored in 20 games played by the 2016 U.S. women's soccer team:

5 5 1 10 5 2 1 1 2 3 3 2 1 4 2 1 2 1 9 3

Figure 1.5 shows a dotplot of these data.

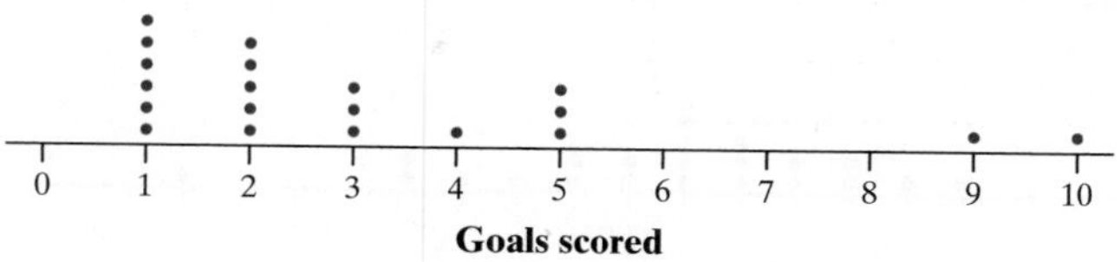


**FIGURE 1.5** Dotplot of goals scored in 20 games by the 2016 U.S. women's soccer team.

It is fairly easy to make a dotplot by hand for small sets of quantitative data.

### HOW TO MAKE A DOTPLOT

- **Draw and label the axis.** Draw a horizontal axis and put the name of the quantitative variable underneath. Be sure to include units of measurement for continuous variables.
- **Scale the axis.** Look at the smallest and largest values in the data set. Start the horizontal axis at a convenient number equal to or less than the smallest value and place tick marks at equal intervals until you equal or exceed the largest value.
- **Plot the values.** Mark a dot above the location on the horizontal axis corresponding to each data value. Try to make all the dots the same size and space them out equally as you stack them.

Remember what we said in Section 1.1: Making a graph is not an end in itself. When you look at a graph, always ask, "What do I see?" From Figure 1.5, we see that the 2016 U.S. women's soccer team scored 4 or more goals in $6/20 = 0.30$ or 30% of its games. That's quite an offense! Unfortunately, the team lost to Sweden on penalty kicks in the 2016 Summer Olympics.

## EXAMPLE

### Give it some gas!

### Making and interpreting dotplots

**PROBLEM:** The Environmental Protection Agency (EPA) is in charge of determining and reporting fuel economy ratings for cars. To estimate fuel economy, the EPA performs tests on several vehicles of the same make, model, and year. Here are data on the highway fuel economy ratings for a sample of 25 model year 2018 Toyota 4Runners tested by the EPA:

| | | | | | | | | | | | | |
|---|---|---|---|---|---|---|---|---|---|---|---|---|
| 22.4 | 22.4 | 22.3 | 23.3 | 22.3 | 22.3 | 22.5 | 22.4 | 22.1 | 21.5 | 22.0 | 22.2 | 22.7 |
| 22.8 | 22.4 | 22.6 | 22.9 | 22.5 | 22.1 | 22.4 | 22.2 | 22.9 | 22.6 | 21.9 | 22.4 | |

(a) Make a dotplot of these data.

(b) Toyota reports the highway gas mileage of its 2018 model year 4Runners as 22 mpg. Do these data give the EPA sufficient reason to investigate that claim?

**SOLUTION:**

(a)

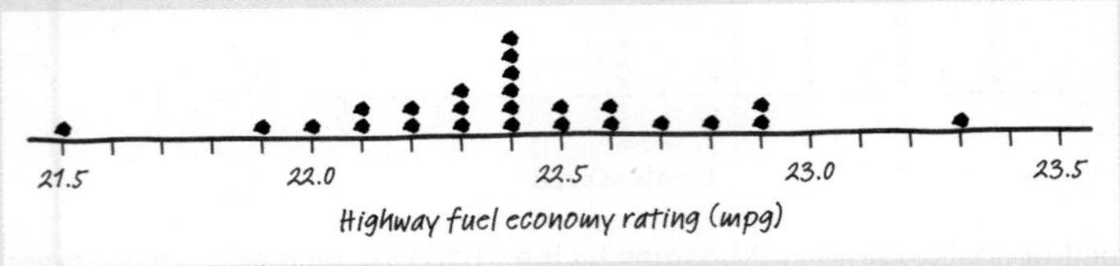


To make the dotplot:

- **Draw and label the axis.** Note the name and units for this continuous variable.
- **Scale the axis.** The smallest value is 21.5 and the largest value is 23.3. So we choose a scale from 21.5 to 23.5 with tick marks 0.1 units apart.
- **Plot the values.**

(b) No. 23 of the 25 cars tested had an estimated highway fuel economy of 22 mpg or greater.

**FOR PRACTICE, TRY EXERCISE 45**

## Describing Shape

When you describe the shape of a dotplot or another graph of quantitative data, focus on the main features. Look for major *peaks*, not for minor ups and downs in the graph. Look for *clusters* of values and obvious *gaps*. Decide if the distribution is roughly **symmetric** or clearly **skewed**.

We could also describe a distribution with a long tail to the left as "skewed toward negative values" or "negatively skewed" and a distribution with a long right tail as "positively skewed."

**DEFINITION Symmetric and skewed distributions**

A distribution is roughly **symmetric** if the right side of the graph (containing the half of observations with the largest values) is approximately a mirror image of the left side.

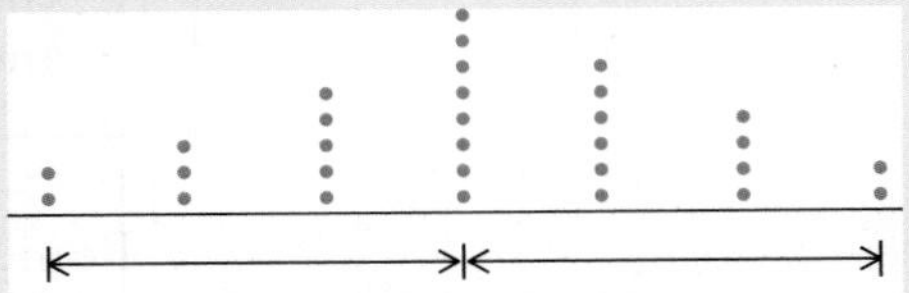

**Roughly symmetric**

A distribution is **skewed to the right** if the right side of the graph is much longer than the left side.

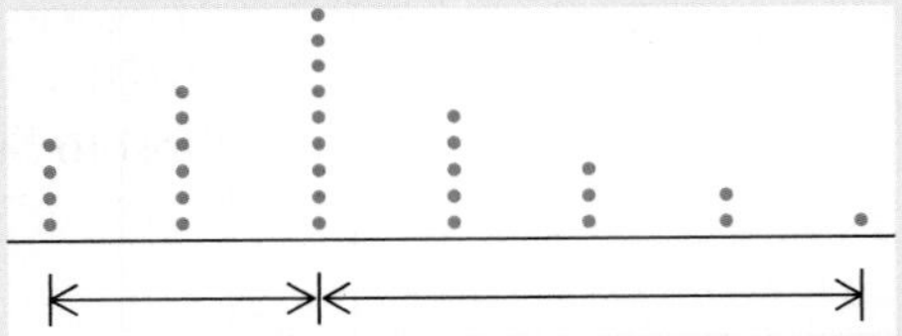

**Skewed to the right**

A distribution is **skewed to the left** if the left side of the graph is much longer than the right side.

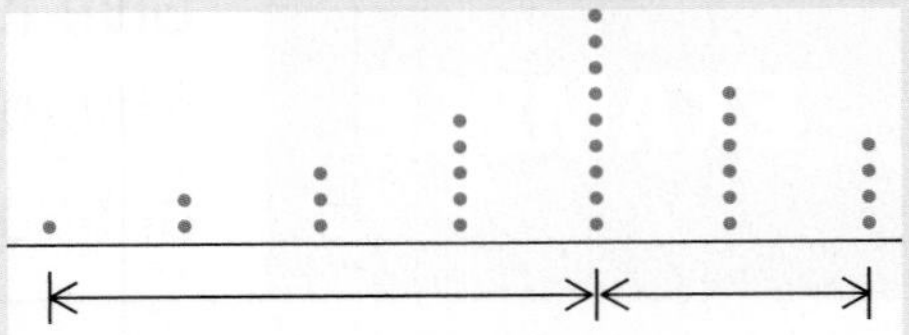

**Skewed to the left**

For ease, we sometimes say "left-skewed" instead of "skewed to the left" and "right-skewed" instead of "skewed to the right." **The direction of skewness is toward the long tail, not the direction where most observations are clustered.** The drawing is a cute but corny way to help you keep this straight. To avoid danger, Mr. Starnes skis on the gentler slope—in the direction of the skewness.

# EXAMPLE

## Quiz scores and die rolls
### Describing shape

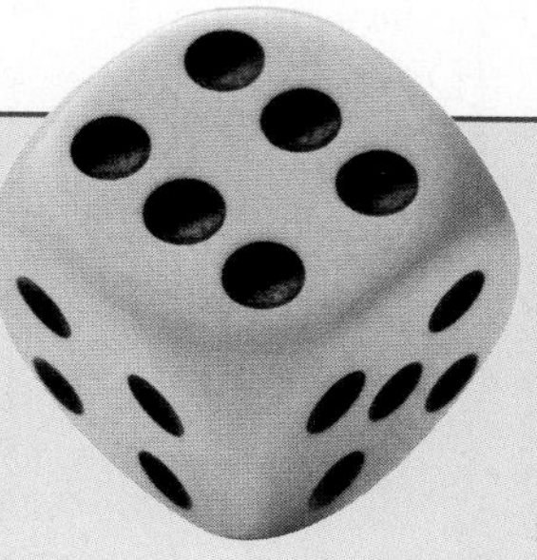

Malerapaso/Getty Images

**PROBLEM:** The dotplots display two different sets of quantitative data. Graph (a) shows the scores of 21 statistics students on a 20-point quiz. Graph (b) shows the results of 100 rolls of a 6-sided die. Describe the shape of each distribution.

(a)

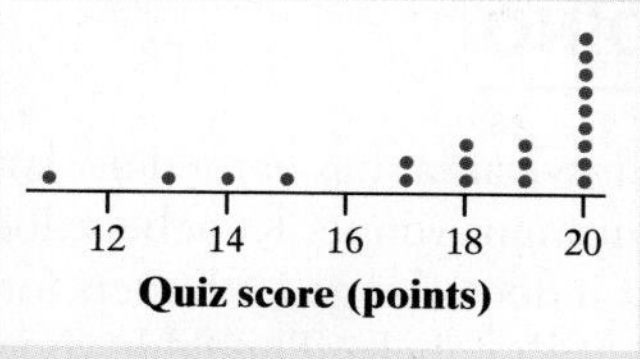


(b)

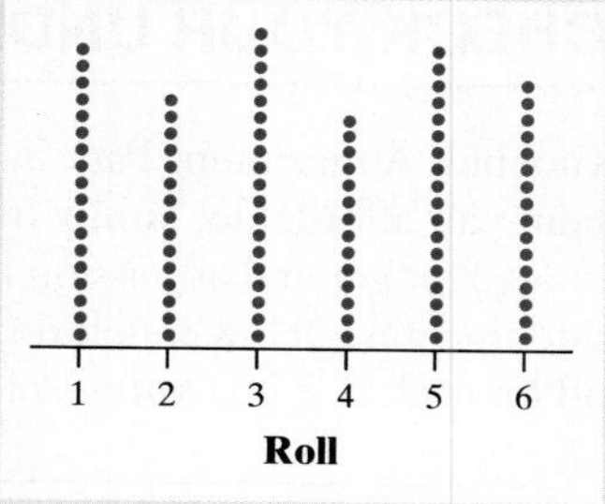


**SOLUTION:**

(a) The distribution of statistics quiz scores is skewed to the left, with a single peak at 20 (a perfect score). There are two small gaps at 12 and 16.

(b) The distribution of die rolls is roughly symmetric. It has no clear peak.

> We can describe the shape of the distribution in part (b) as *approximately uniform* because the frequencies are about the same for all possible rolls.

**FOR PRACTICE, TRY EXERCISE 49**

Some people refer to graphs with a single peak, like the dotplot of quiz scores in part (a) of the example, as *unimodal*. Figure 1.6 is a dotplot of the duration (in minutes) of 220 eruptions of the Old Faithful geyser. The graph has two distinct clusters and two peaks: one at about 2 minutes and one at about 4.5 minutes. We would describe this distribution's shape as roughly symmetric and *bimodal*. (Although we could continue the pattern with "trimodal" for three peaks and so on, it's more common to refer to distributions with more than two clear peaks as *multimodal*.) When you examine a graph of quantitative data, describe any pattern you see as clearly as you can.

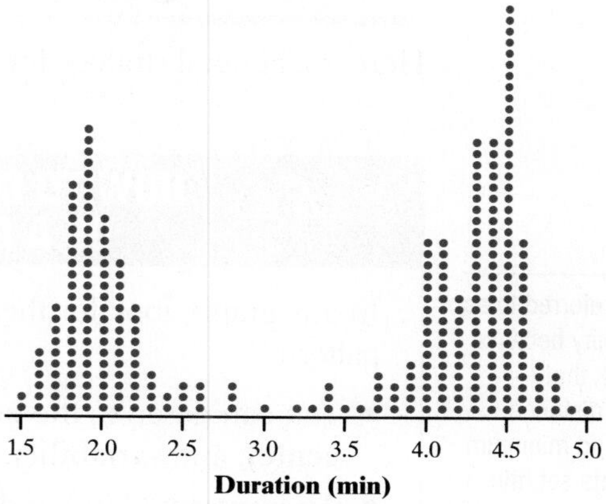


**FIGURE 1.6** Dotplot displaying duration (in minutes) of 220 Old Faithful eruptions. This graph has two distinct clusters and two clear peaks (bimodal).

Some quantitative variables have distributions with predictable shapes. Many biological measurements on individuals from the same species and gender—lengths of bird bills, heights of young women—have roughly symmetric distributions. Salaries and home prices, on the other hand, usually have right-skewed distributions. There are many moderately priced houses, for example, but the few very expensive mansions give the distribution of house prices a strong right skew.

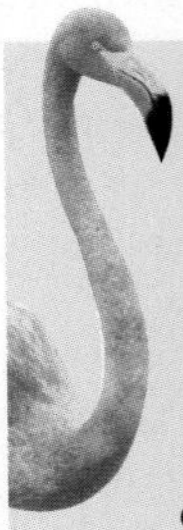

## CHECK YOUR UNDERSTANDING

Knoebels Amusement Park in Elysburg, Pennsylvania, has earned acclaim for being an affordable, family-friendly entertainment venue. Knoebels does not charge for general admission or parking, but it does charge customers for each ride they take. How much do the rides cost at Knoebels? The table shows the cost for each ride in a sample of 22 rides in a recent year.

| Name | Cost | Name | Cost |
|---|---|---|---|
| Merry Mixer | \$1.50 | Looper | \$1.75 |
| Italian Trapeze | \$1.50 | Flying Turns | \$3.00 |
| Satellite | \$1.50 | Flyer | \$1.50 |
| Galleon | \$1.50 | The Haunted Mansion | \$1.75 |
| Whipper | \$1.25 | StratosFear | \$2.00 |
| Skooters | \$1.75 | Twister | \$2.50 |
| Ribbit | \$1.25 | Cosmotron | \$1.75 |
| Roundup | \$1.50 | Paratrooper | \$1.50 |
| Paradrop | \$1.25 | Downdraft | \$1.50 |
| The Phoenix | \$2.50 | Rockin' Tug | \$1.25 |
| Gasoline Alley | \$1.75 | Sklooosh! | \$1.75 |

1. Make a dotplot of the data.
2. Describe the shape of the distribution.

# Describing Distributions

Here is a general strategy for describing a distribution of quantitative data.

Variability is sometimes referred to as *spread.* We prefer variability because students sometimes think that spread refers only to the distance between the maximum and minimum value of a quantitative data set (the *range*). There are several ways to measure the variability (spread) of a distribution, including the range.

### HOW TO DESCRIBE THE DISTRIBUTION OF A QUANTITATIVE VARIABLE

In any graph, look for the *overall pattern* and for clear *departures* from that pattern.

- You can describe the overall pattern of a distribution by its **shape, center,** and **variability.**
- An important kind of departure is an **outlier,** an observation that falls outside the overall pattern.

**AP® EXAM TIP**

Always be sure to include context when you are asked to describe a distribution. This means using the variable name, not just the units the variable is measured in.

We will discuss more formal ways to measure center and variability and to identify outliers in Section 1.3. For now, just use the *median* (middle value in the ordered data set) when describing center and the *minimum* and *maximum* when describing variability.

Let's practice with the dotplot of goals scored in 20 games played by the 2016 U.S. women's soccer team.

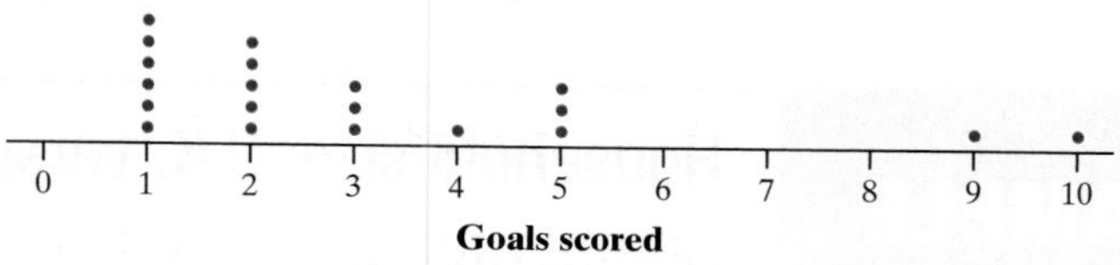


When describing a distribution of quantitative data, don't forget: **S**tatistical **O**pinions **C**an **V**ary (**S**hape, **O**utliers, **C**enter, **V**ariability).

*Shape:* The distribution of goals scored is skewed to the right, with a single peak at 1 goal. There is a gap between 5 and 9 goals.

*Outliers:* The games when the team scored 9 and 10 goals appear to be outliers.

*Center:* The median is 2 goals scored.

*Variability:* The data vary from 1 to 10 goals scored.

## EXAMPLE

### Give it some gas!
### Describing a distribution

**PROBLEM:** Here is a dotplot of the highway fuel economy ratings for a sample of 25 model year 2018 Toyota 4Runners tested by the EPA. Describe the distribution.

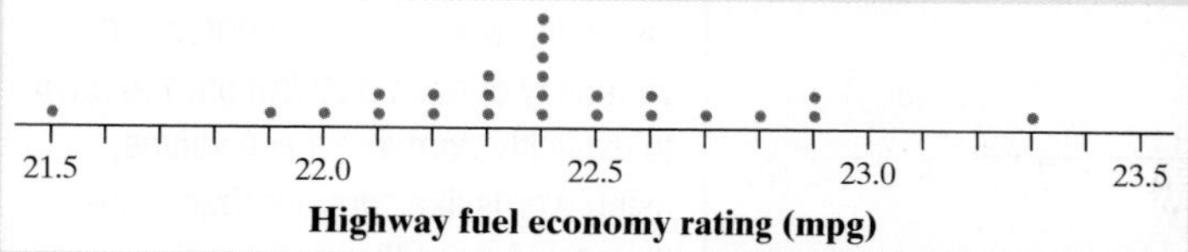

Daren Starnes

**SOLUTION:**

*Shape:* The distribution of highway fuel economy ratings is roughly symmetric, with a single peak at 22.4 mpg. There are two clear gaps: between 21.5 and 21.9 mpg and between 22.9 and 23.3 mpg.

*Outliers:* The cars with 21.5 mpg and 23.3 mpg ratings are possible outliers.

*Center:* The median rating is 22.4 mpg.

*Variability:* The ratings vary from 21.5 to 23.3 mpg.

Be sure to include context by discussing the variable of interest, highway fuel economy ratings. And give the units of measurement: miles per gallon (mpg).

**FOR PRACTICE, TRY EXERCISE 53**

## Comparing Distributions

Some of the most interesting statistics questions involve comparing two or more groups. Which of two popular diets leads to greater long-term weight loss? Who texts more—males or females? As the following example suggests, you should always discuss shape, outliers, center, and variability whenever you compare distributions of a quantitative variable.

### EXAMPLE Household size: U.K. versus South Africa
#### Comparing distributions

**PROBLEM:** How do the numbers of people living in households in the United Kingdom (U.K.) and South Africa compare? To help answer this question, we used Census At School's "Random Data Selector" to choose 50 students from each country. Here are dotplots of the household sizes reported by the survey respondents. Compare the distributions of household size for these two countries.

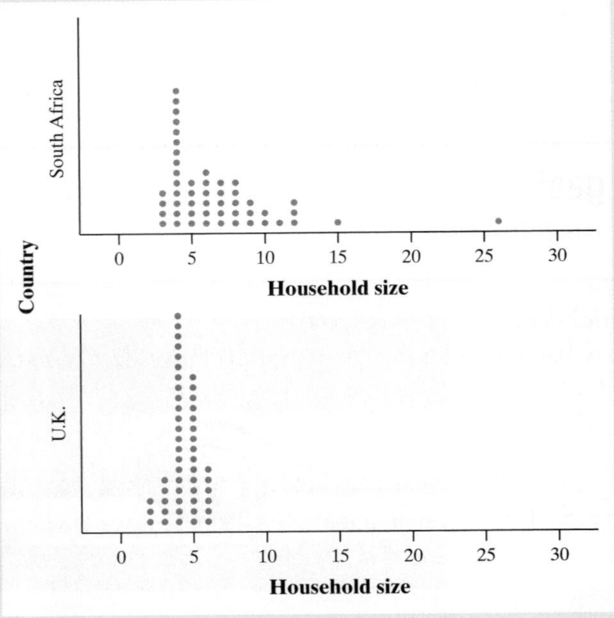

**AP® EXAM TIP**

When comparing distributions of quantitative data, it's not enough just to list values for the center and variability of each distribution. You have to explicitly *compare* these values, using words like "greater than," "less than," or "about the same as."

**SOLUTION:**

*Shape:* The distribution of household size for the U.K. sample is roughly symmetric, with a single peak at 4 people. The distribution of household size for the South Africa sample is skewed to the right, with a single peak at 4 people and a clear gap between 15 and 26.

*Outliers:* There don't appear to be any outliers in the U.K. distribution. The South African distribution seems to have two outliers: the households with 15 and 26 people.

*Center:* Household sizes for the South African students tend to be larger (median = 6 people) than for the U.K. students (median = 4 people).

*Variability:* The household sizes for the South African students vary more (from 3 to 26 people) than for the U.K. students (from 2 to 6 people).

Don't forget to include context! It isn't enough to refer to the U.K. distribution or the South Africa distribution. You need to mention the variable of interest, household size.

FOR PRACTICE, TRY EXERCISE 55

Notice that in the preceding example, we discussed the distributions of household size only for the two *samples* of 50 students. We might be interested in whether the sample data give us convincing evidence of a difference in the *population* distributions of household size for South Africa and the United Kingdom. We'll have to wait a few chapters to decide whether we can reach such a conclusion, but our ability to make such an inference later will be helped by the fact that the students in our samples were chosen at random.

## CHECK YOUR UNDERSTANDING

For a statistics class project, Jonathan and Crystal hosted an ice-cream-eating contest. Each student in the contest was given a small cup of ice cream and instructed to eat it as fast as possible. Jonathan and Crystal then recorded each contestant's gender and time (in seconds), as shown in the dotplots. Compare the distributions of eating times for males and females.

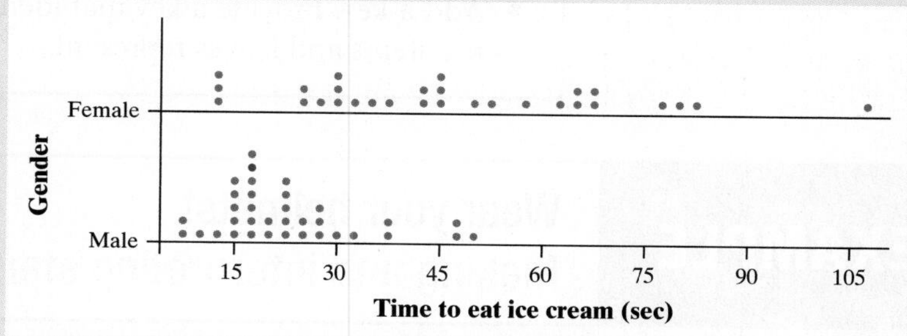


## Stemplots

Another simple type of graph for displaying quantitative data is a **stemplot**.

A stemplot is also known as a *stem-and-leaf plot*.

**DEFINITION Stemplot**

A **stemplot** shows each data value separated into two parts: a *stem,* which consists of all but the final digit, and a *leaf,* the final digit. The stems are ordered from lowest to highest and arranged in a vertical column. The leaves are arranged in increasing order out from the appropriate stems.

Here are data on the resting pulse rates (beats per minute) of 19 middle school students:

71 104 76 88 78 71 68 86 70 90 74 76 69 68 88 96 68 82 120

Figure 1.7 shows a stemplot of these data.

```
 6 | 8889
 7 | 0114668
 8 | 2688
 9 | 06
10 | 4
11 |
12 | 0
```

Key: 8|2 is a student whose resting pulse rate is 82 beats per minute.

**FIGURE 1.7** Stemplot of the resting pulse rates of 19 middle school students.

According to the American Heart Association, a resting pulse rate above 100 beats per minute is considered high for this age group. We can see that $2/19 = 0.105 = 10.5\%$

of these students have high resting pulse rates by this standard. Also, the distribution of pulse rates for these 19 students is skewed to the right (toward the larger values).

Stemplots give us a quick picture of a distribution that includes the individual observations in the graph. It is fairly easy to make a stemplot by hand for small sets of quantitative data.

### HOW TO MAKE A STEMPLOT

- **Make stems.** Separate each observation into a stem, consisting of all but the final digit, and a leaf, the final digit. Write the stems in a vertical column with the smallest at the top. Draw a vertical line at the right of this column. Do not skip any stems, even if there is no data value for a particular stem.
- **Add leaves.** Write each leaf in the row to the right of its stem.
- **Order leaves.** Arrange the leaves in increasing order out from the stem.
- **Add a key.** Provide a key that identifies the variable and explains what the stems and leaves represent.

## EXAMPLE

### Wear your helmets!
### Making and interpreting stemplots

Pete Saloutos/AGE Fotostock

**PROBLEM:** Many athletes (and their parents) worry about the risk of concussions when playing sports. A football coach plans to obtain specially made helmets for his players that are designed to reduce the chance of getting a concussion. Here are the measurements of head circumference (in inches) for the 30 players on the team:

| | | | | | | | | | | | | | | |
|---|---|---|---|---|---|---|---|---|---|---|---|---|---|---|
| 23.0 | 22.2 | 21.7 | 22.0 | 22.3 | 22.6 | 22.7 | 21.5 | 22.7 | 25.6 | 20.8 | 23.0 | 24.2 | 23.5 | 20.8 |
| 24.0 | 22.7 | 22.6 | 23.9 | 22.5 | 23.1 | 21.9 | 21.0 | 22.4 | 23.5 | 22.5 | 23.9 | 23.4 | 21.6 | 23.3 |

(a) Make a stemplot of these data.

(b) Describe the shape of the distribution. Are there any obvious outliers?

**SOLUTION:**

(a)

| | |
|---|---|
| 20 | 88 |
| 21 | 05679 |
| 22 | 02345566777 |
| 23 | 001345599 |
| 24 | 02 |
| 25 | 6 |

Key: 23|5 is a player with a head circumference of 23.5 inches.

To make the stemplot:
- **Make stems.** The smallest head circumference is 20.8 inches and the largest is 25.6 inches. We use the first two digits as the stem and the final digit as the leaf. So we need stems from 20 to 25.
- **Add leaves.**
- **Order leaves.**
- **Add a key.**

(b) **The distribution of head circumferences for the 30 players on the team is roughly symmetric, with a single peak on the 22-inch stem. There are no obvious outliers.**

FOR PRACTICE, TRY EXERCISE 59

We can get a better picture of the head circumference data by *splitting stems*. In Figure 1.8(a), leaf values from 0 to 9 are placed on the same stem. Figure 1.8(b) shows another stemplot of the same data. This time, values with leaves from 0 to 4 are placed on one stem, while those with leaves from 5 to 9 are placed on another stem. Now we can see the shape of the distribution even more clearly—including the possible outlier at 25.6 inches.

```
20 | 88
21 | 05679
22 | 02345566777
23 | 001345599
24 | 02
25 | 6
```

Key: 23|5 is a player with a head circumference of 23.5 inches.

(a)

```
20 | 88
21 | 0
21 | 5679
22 | 0234
22 | 5566777
23 | 00134
23 | 5599
24 | 02
24 |
25 |
25 | 6
```

(b)

**FIGURE 1.8** Two stemplots showing the head circumference data. The graph in (b) improves on the graph in (a) by splitting stems.

Here are a few tips to consider before making a stemplot:

- There is no magic number of stems to use. Too few or too many stems will make it difficult to see the distribution's shape. Five stems is a good minimum.
- If you split stems, be sure that each stem is assigned an equal number of possible leaf digits.
- When the data have too many digits, you can get more flexibility by rounding or truncating the data. See Exercises 61 and 62 for an illustration of rounding data before making a stemplot.

You can use a *back-to-back stemplot* with common stems to compare the distribution of a quantitative variable in two groups. The leaves are placed in order on each side of the common stem. For example, Figure 1.9 shows a back-to-back stemplot of the 19 middle school students' resting pulse rates and their pulse rates after 5 minutes of running.

```
 Resting       After exercise
    9888 |  6 |
 8664110 |  7 |
    8862 |  8 | 6788
      60 |  9 | 02245899
       4 | 10 | 044
         | 11 | 8
       0 | 12 | 44
         | 13 |
         | 14 | 6
```

Key: 8|2 is a student whose pulse rate is 82 beats per minute.

**FIGURE 1.9** Back-to-back stemplot of 19 middle school students' resting pulse rates and their pulse rates after 5 minutes of running.

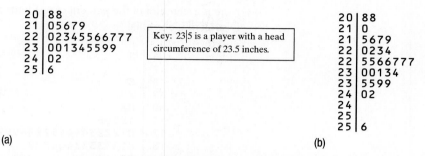

## CHECK YOUR UNDERSTANDING

1. Write a few sentences comparing the distributions of resting and after-exercise pulse rates in Figure 1.9.

Multiple Choice: *Select the best answer for Questions 2–4.*

Here is a stemplot of the percent of residents aged 65 and older in the 50 states and the District of Columbia:

| | |
|---|---|
| 6 | 8 |
| 7 | |
| 8 | 8 |
| 9 | 79 |
| 10 | 08 |
| 11 | 15566 |
| 12 | 01222344445788899 |
| 13 | 01233333444899 |
| 14 | 02666 |
| 15 | 23 |
| 16 | 8 |

Key: 8|8 represents a state in which 8.8% of residents are 65 and older.

2. The low outlier is Alaska. What percent of Alaska residents are 65 or older?
   (a) 0.68 (b) 6.8 (c) 8.8 (d) 16.8 (e) 68
3. Ignoring the outlier, the shape of the distribution is
   (a) skewed to the right.
   (b) skewed to the left.
   (c) skewed to the middle.
   (d) double-peaked (bimodal).
   (e) roughly symmetric.
4. The center of the distribution is close to
   (a) 11.6%. (b) 12.0%. (c) 12.8%. (d) 13.3%. (e) 6.8% to 16.8%.

# Histograms

You can use a dotplot or stemplot to display quantitative data. Both graphs show every individual data value. For large data sets, this can make it difficult to see the overall pattern in the graph. We often get a clearer picture of the distribution by grouping together nearby values. Doing so allows us to make a new type of graph: a **histogram**.

**DEFINITION Histogram**

A **histogram** shows each interval of values as a bar. The heights of the bars show the frequencies or relative frequencies of values in each interval.

Figure 1.10 shows a dotplot and a histogram of the durations (in minutes) of 220 eruptions of the Old Faithful geyser. Notice how the histogram groups together nearby values.

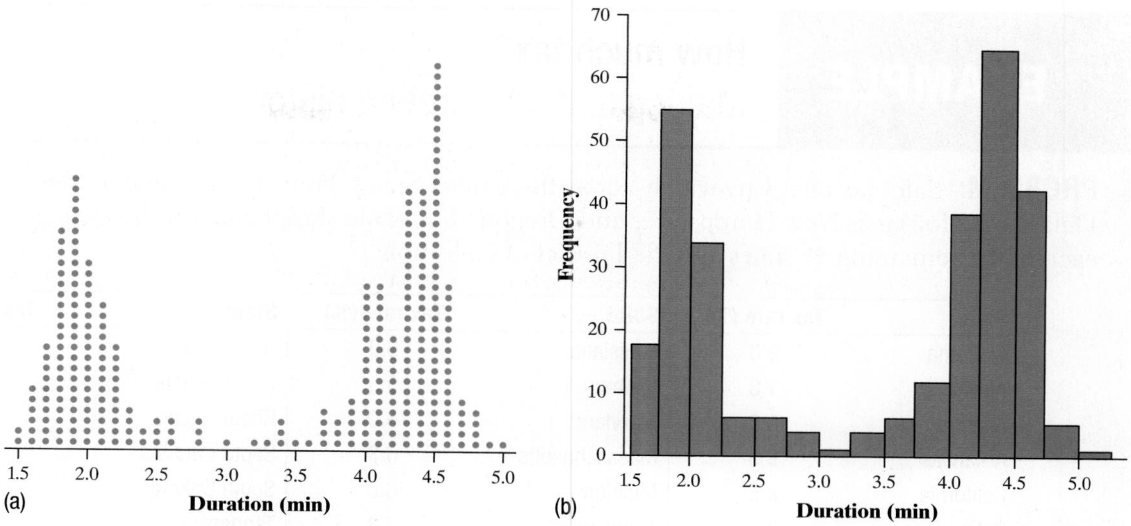


**FIGURE 1.10** (a) Dotplot and (b) histogram of the duration (in minutes) of 220 eruptions of the Old Faithful geyser.

It is fairly easy to make a histogram by hand. Here's how you do it.

## HOW TO MAKE A HISTOGRAM

- **Choose equal-width intervals** that span the data. Five intervals is a good minimum.
- **Make a table** that shows the frequency (count) or relative frequency (percent or proportion) of individuals in each interval. Put values that fall on an interval boundary in the interval containing larger values.
- **Draw and label the axes.** Draw horizontal and vertical axes. Put the name of the quantitative variable under the horizontal axis. To the left of the vertical axis, indicate whether the graph shows the frequency (count) or relative frequency (percent or proportion) of individuals in each interval.
- **Scale the axes.** Place equally spaced tick marks at the smallest value in each interval along the horizontal axis. On the vertical axis, start at 0 and place equally spaced tick marks until you exceed the largest frequency or relative frequency in any interval.
- **Draw bars** above the intervals. Make the bars equal in width and leave no gaps between them. Be sure that the height of each bar corresponds to the frequency or relative frequency of individuals in that interval. An interval with no data values will appear as a bar of height 0 on the graph.

It is possible to choose intervals of unequal widths when making a histogram. Such graphs are beyond the scope of this book.

# EXAMPLE

## How much tax?

### Making and interpreting histograms

**PROBLEM:** Sales tax rates vary widely across the United States. Four states charge no state or local sales tax: Delaware, Montana, New Hampshire, and Oregon. The table shows data on the average total tax rate for each of the remaining 46 states and the District of Columbia.[23]

| State | Tax rate (%) | State | Tax rate (%) | State | Tax rate (%) |
|---|---|---|---|---|---|
| Alabama | 9.0 | Louisiana | 9.0 | Oklahoma | 8.8 |
| Alaska | 1.8 | Maine | 5.5 | Pennsylvania | 6.3 |
| Arizona | 8.3 | Maryland | 6.0 | Rhode Island | 7.0 |
| Arkansas | 9.3 | Massachusetts | 6.3 | South Carolina | 7.2 |
| California | 8.5 | Michigan | 6.0 | South Dakota | 5.8 |
| Colorado | 7.5 | Minnesota | 7.3 | Tennessee | 9.5 |
| Connecticut | 6.4 | Mississippi | 7.1 | Texas | 8.2 |
| Florida | 6.7 | Missouri | 7.9 | Utah | 6.7 |
| Georgia | 7.0 | Nebraska | 6.9 | Vermont | 6.2 |
| Hawaii | 4.4 | Nevada | 8.0 | Virginia | 5.6 |
| Idaho | 6.0 | New Jersey | 7.0 | Washington | 8.9 |
| Illinois | 8.6 | New Mexico | 7.5 | West Virginia | 6.2 |
| Indiana | 7.0 | New York | 8.5 | Wisconsin | 5.4 |
| Iowa | 6.8 | North Carolina | 6.9 | Wyoming | 5.4 |
| Kansas | 8.6 | North Dakota | 6.8 | District of Columbia | 5.8 |
| Kentucky | 6.0 | Ohio | 7.1 | | |

(a) Make a frequency histogram to display the data.
(b) What percent of values in the distribution are less than 6.0? Interpret this result in context.

**SOLUTION:**

(a)

| Interval | Frequency |
|---|---|
| 1.0 to <2.0 | 1 |
| 2.0 to <3.0 | 0 |
| 3.0 to <4.0 | 0 |
| 4.0 to <5.0 | 1 |
| 5.0 to <6.0 | 6 |
| 6.0 to <7.0 | 15 |
| 7.0 to <8.0 | 11 |
| 8.0 to <9.0 | 9 |
| 9.0 to <10.0 | 4 |

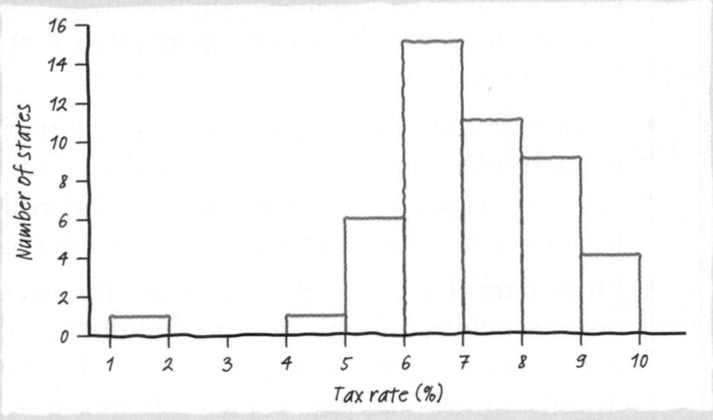


To make the histogram:

- **Choose equal-width intervals** that span the data. The data vary from 1.8 percent to 9.5 percent. So we choose intervals of width 1.0, starting at 1.0%.
- **Make a table.** Record the number of states in each interval to make a frequency histogram.
- **Draw and label the axes.** Don't forget units (percent) for the variable (tax rate).
- **Scale the axes.**
- **Draw bars.**

(b) 8/47 = 0.170 = 17.0%; 17% of the states (including the District of Columbia) have tax rates less than 6%.

**FOR PRACTICE, TRY EXERCISE 67**

Figure 1.11 shows two different histograms of the state sales tax data. Graph (a) uses the intervals of width 1% from the preceding example. The distribution has a single peak in the 6.0 to <7.0 interval. Graph (b) uses intervals half as wide: 1.0 to <1.5, 1.5 to <2.0, and so on. Now we see a distribution with more than one distinct peak. **The choice of intervals in a histogram can affect the appearance of a distribution.** Histograms with more intervals show more detail but may have a less clear overall pattern.

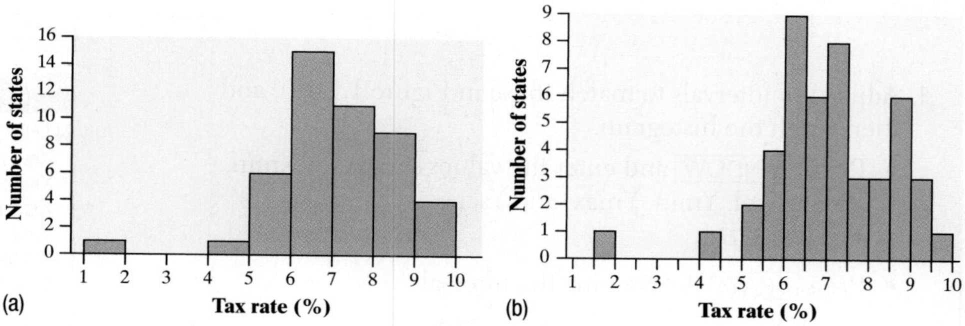


**FIGURE 1.11** (a) Frequency histogram of the sales tax rate in the states that have local or state sales taxes and the District of Columbia with intervals of width 1.0%, from the preceding example. (b) Frequency histogram of the data with intervals of width 0.5%.

You can use a graphing calculator, statistical software, or an applet to make a histogram. The technology's default choice of intervals is a good starting point, but you should adjust the intervals to fit with common sense.

## 2. Technology Corner MAKING HISTOGRAMS

*TI-Nspire and other technology instructions are on the book's website at highschool.bfwpub.com/updatedtps6e.*

1. Enter the data from the sales tax example in your Statistics/List Editor.
   - Press STAT and choose Edit...
   - Type the values into list L1.

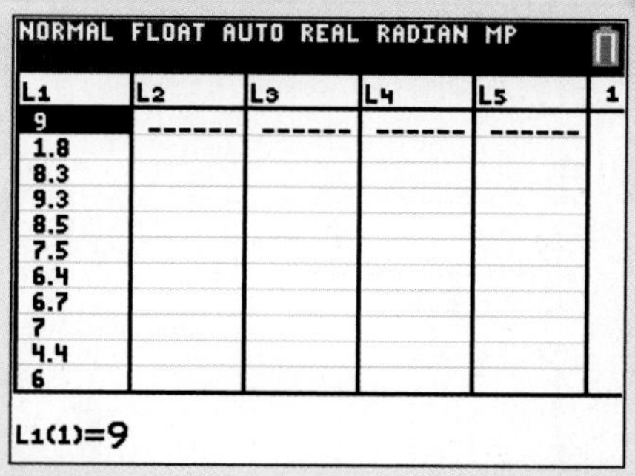


2. Set up a histogram in the Statistics Plots menu.
   - Press 2nd Y= (STAT PLOT).
   - Press ENTER or 1 to go into Plot1.
   - Adjust the settings as shown.

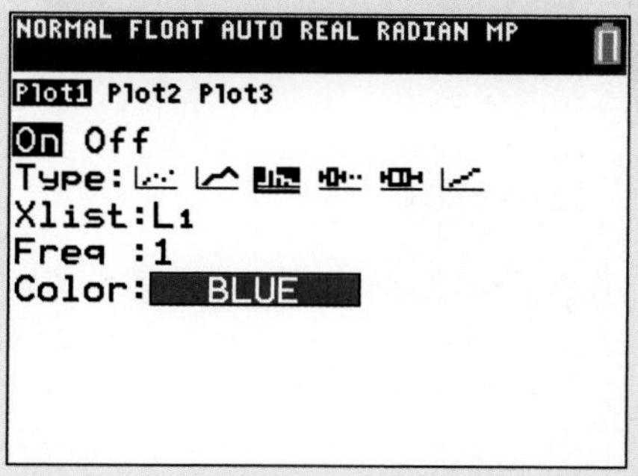

3. Use ZoomStat to let the calculator choose intervals and make a histogram.
   - Press ZOOM and choose ZoomStat.
   - Press TRACE to examine the intervals.

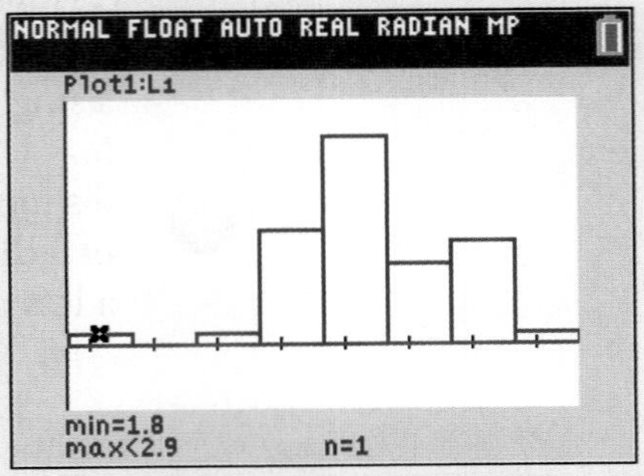


4. Adjust the intervals to match those in Figure 1.11(a), and then graph the histogram.
   - Press WINDOW and enter the values shown for Xmin, Xmax, Xscl, Ymin, Ymax, and Yscl.
   - Press GRAPH.
   - Press TRACE to examine the intervals.

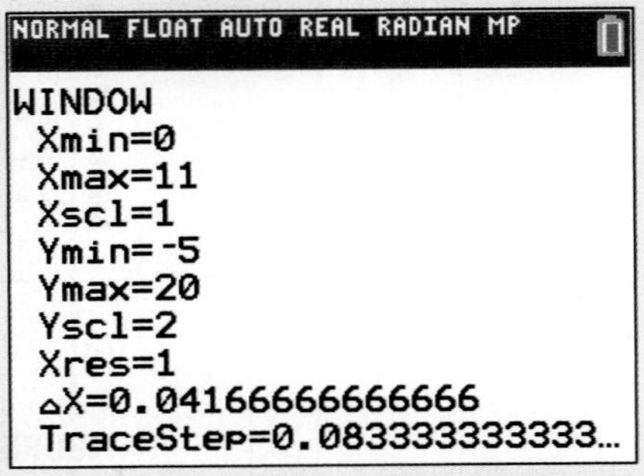


5. See if you can match the histogram in Figure 1.11(b).

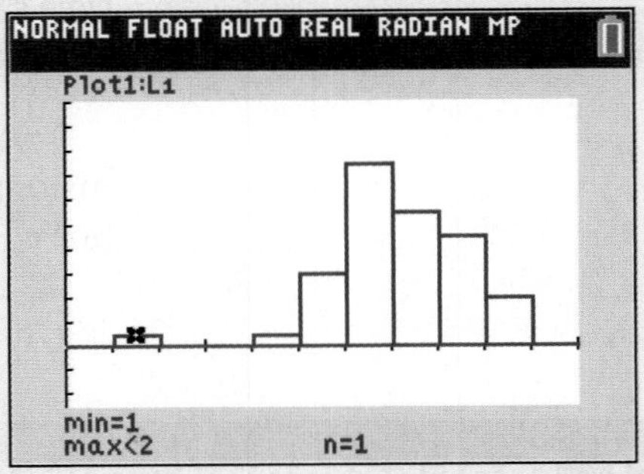


**AP® EXAM TIP**

If you're asked to make a graph on a free-response question, be sure to label and scale your axes. Unless your calculator shows labels and scaling, don't just transfer a calculator screen shot to your paper.

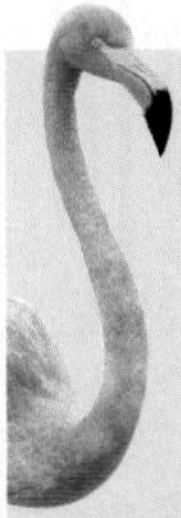

## CHECK YOUR UNDERSTANDING

Many people believe that the distribution of IQ scores follows a "bell curve," like the one shown. But is this really how such scores are distributed? The IQ scores of 60 fifth-grade students chosen at random from one school are shown here.[24]

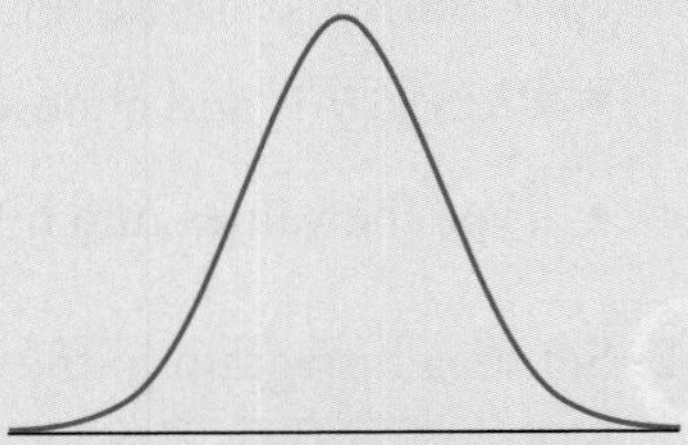

| | | | | | | | | | |
|---|---|---|---|---|---|---|---|---|---|
| 145 | 139 | 126 | 122 | 125 | 130 | 96 | 110 | 118 | 118 |
| 101 | 142 | 134 | 124 | 112 | 109 | 134 | 113 | 81 | 113 |
| 123 | 94 | 100 | 136 | 109 | 131 | 117 | 110 | 127 | 124 |
| 106 | 124 | 115 | 133 | 116 | 102 | 127 | 117 | 109 | 137 |
| 117 | 90 | 103 | 114 | 139 | 101 | 122 | 105 | 97 | 89 |
| 102 | 108 | 110 | 128 | 114 | 112 | 114 | 102 | 82 | 101 |

1. Construct a histogram that displays the distribution of IQ scores effectively.
2. Describe what you see. Is the distribution bell-shaped?

## Using Histograms Wisely

We offer several cautions based on common mistakes students make when using histograms.

1. **Don't confuse histograms and bar graphs.** Although histograms resemble bar graphs, their details and uses are different. A histogram displays the distribution of a quantitative variable. Its horizontal axis identifies intervals of values that the variable takes. A bar graph displays the distribution of a categorical variable. Its horizontal axis identifies the categories. Be sure to draw bar graphs with blank space between the bars to separate the categories. Draw histograms with no space between bars for adjacent intervals. For comparison, here is one of each type of graph from earlier examples:

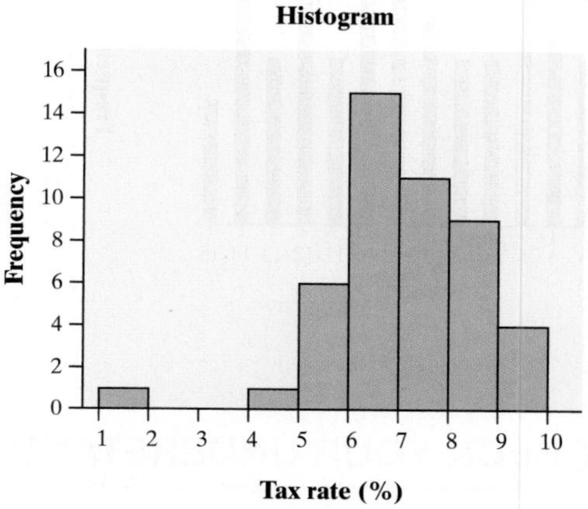


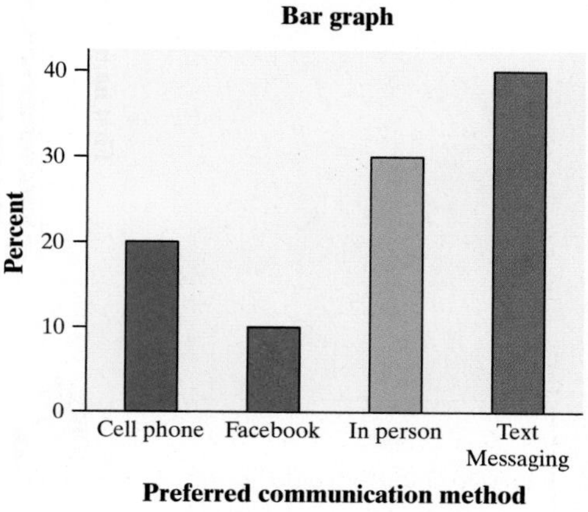


2. **Use percents or proportions instead of counts on the vertical axis when comparing distributions with different numbers of observations.** Mary was interested in comparing the reading levels of a biology journal and an airline magazine. She counted the number of letters in the first 400 words of an article in the journal and of the first 100 words of an article in the airline magazine. Mary then used statistical software to produce the histograms shown in Figure 1.12(a). This figure is misleading—it compares frequencies, but the two samples were of very different sizes (400 and 100). Using the same data, Mary's teacher produced the histograms in Figure 1.12(b). By using relative frequencies, this figure makes the comparison of word lengths in the two samples much easier.

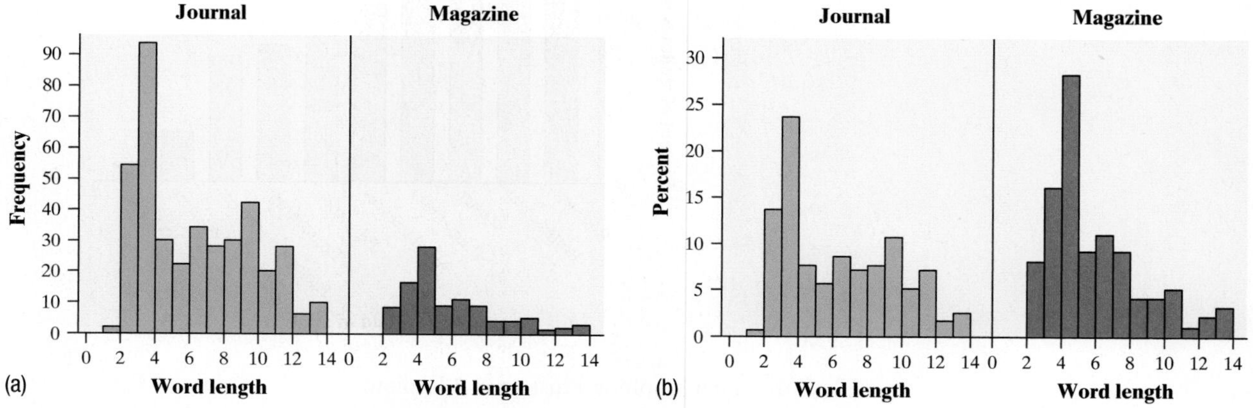


**FIGURE 1.12** Two sets of histograms comparing word lengths in articles from a biology journal and from an airline magazine. In graph (a), the vertical scale uses frequencies. Graph (b) fixes the problem of different sample sizes by using percents (relative frequencies) on the vertical scale.

3. **Just because a graph looks nice doesn't make it a meaningful display of data.** The 15 students in a small statistics class recorded the number of letters in their first names. One student entered the data into an Excel spreadsheet and then used Excel's "chart maker" to produce the graph shown on the left. What kind of graph is this? It's a bar graph that compares the raw data values. But first-name length is a quantitative variable, so a bar graph is not an appropriate way to display its distribution. The histogram on the right is a much better choice because the graph makes it easier to identify the shape, center, and variability of the distribution of name length.

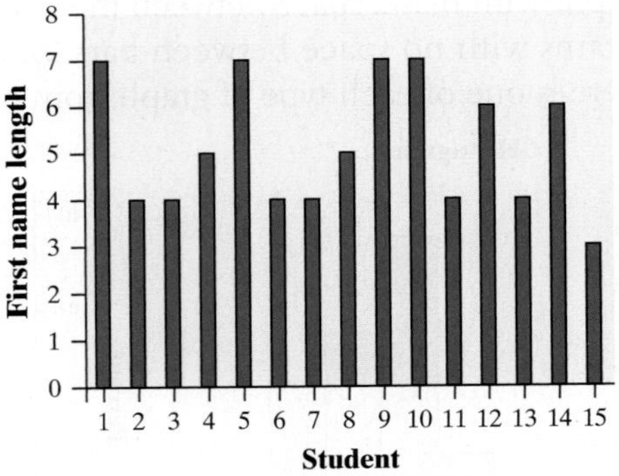


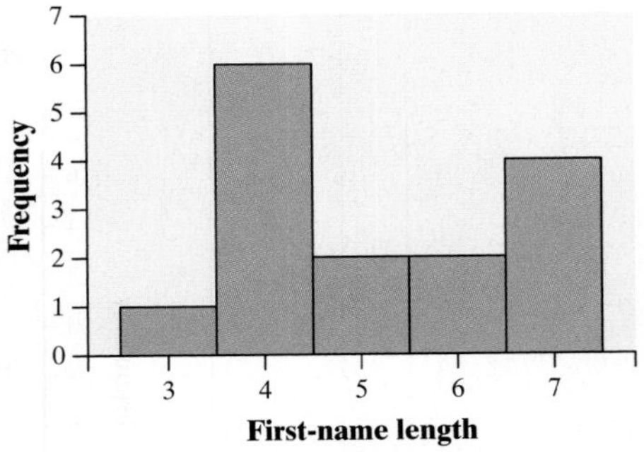

## CHECK YOUR UNDERSTANDING

1. Write a few sentences comparing the distributions of word length shown in Figure 1.12(b).

*Questions 2 and 3 refer to the following setting.* About 3 million first-year students enroll in colleges and universities each year. What do they plan to study? The graph displays data on the percent of first-year students who plan to major in several disciplines.[25]

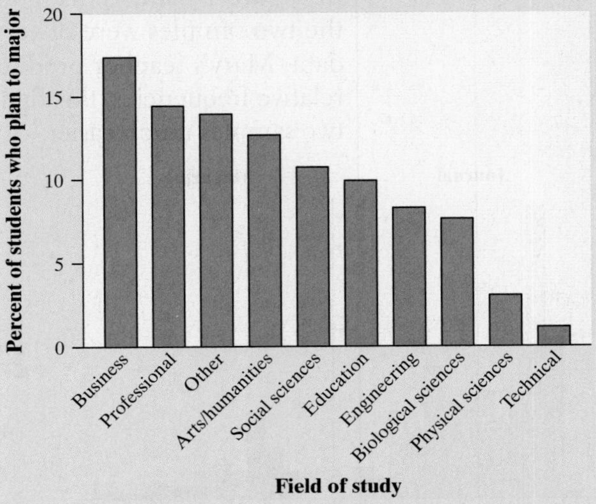


2. Is this a bar graph or a histogram? Explain.
3. Would it be correct to describe this distribution as right-skewed? Why or why not?

# Section 1.2 Summary

- You can use a **dotplot, stemplot,** or **histogram** to show the distribution of a quantitative variable. A dotplot displays individual values on a number line. Stemplots separate each observation into a stem and a one-digit leaf. Histograms plot the frequencies (counts) or relative frequencies (proportions or percents) of values in equal-width intervals.
- Some distributions have simple shapes, such as **symmetric, skewed to the left,** or **skewed to the right.** The number of peaks is another aspect of overall shape. So are distinct clusters and gaps.
- A single-peaked graph is sometimes called *unimodal,* and a double-peaked graph is sometimes called *bimodal.*
- When examining any graph of quantitative data, look for an *overall pattern* and for clear *departures* from that pattern. **Shape, center,** and **variability** describe the overall pattern of the distribution of a quantitative variable. **Outliers** are observations that lie outside the overall pattern of a distribution.
- When comparing distributions of quantitative data, be sure to compare shape, center, variability, and possible outliers.
- Remember: histograms are for quantitative data; bar graphs are for categorical data. Be sure to use relative frequencies when comparing data sets of different sizes.

## 1.2 Technology Corner

*TI-Nspire and other technology instructions are on the book's website at highschool.bfwpub.com/updatedtps6e.*

**2. Making histograms** **Page 43**

# Section 1.2 Exercises

**45. Feeling sleepy?** Students in a high school statistics class responded to a survey designed by their teacher. One of the survey questions was "How much sleep did you get last night?" Here are the data (in hours):

pg 31

| | | | | | | | | | | | | | |
|---|---|---|---|---|---|---|---|---|---|---|---|---|---|
| 9 | 6 | 8 | 7 | 8 | 8 | 6 | 6.5 | 7 | 7 | 9.0 | 4 | 3 | 4 |
| 5 | 6 | 11 | 6 | 3 | 7 | 6 | 10.0 | 7 | 8 | 4.5 | 9 | 7 | 7 |

(a) Make a dotplot to display the data.

(b) Experts recommend that high school students sleep at least 9 hours per night. What proportion of students in this class got the recommended amount of sleep?

**46. Easy reading?** Here are data on the lengths of the first 25 words on a randomly selected page from Toni Morrison's *Song of Solomon*:

| | | | | | | | | | | | | |
|---|---|---|---|---|---|---|---|---|---|---|---|---|
| 2 | 3 | 4 | 10 | 2 | 11 | 2 | 8 | 4 | 3 | 7 | 2 | 7 |
| 5 | 3 | 6 | 4 | 4 | 2 | 5 | 8 | 2 | 3 | 4 | 4 | |

(a) Make a dotplot of these data.

(b) Long words can make a book hard to read. What percentage of words in the sample have 8 or more letters?

**47. U.S. women's soccer—2016** Earlier, we examined data on the number of goals scored by the 2016 U.S. women's soccer team in 20 games played. The following dotplot displays the goal differential for those same games, computed as U.S. goals scored minus opponent goals scored.

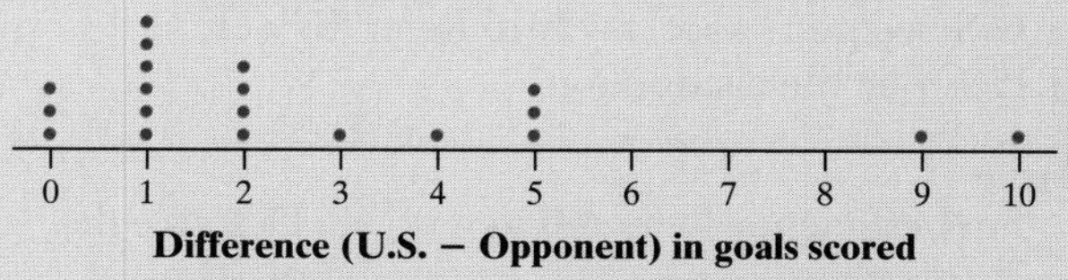


(a) Explain what the dot above 3 represents.

(b) What does the graph tell us about how well the team did in 2016? Be specific.

**48. Fuel efficiency** The dotplot shows the difference (Highway – City) in EPA mileage ratings, in miles per gallon (mpg) for each of 24 model year 2018 cars.

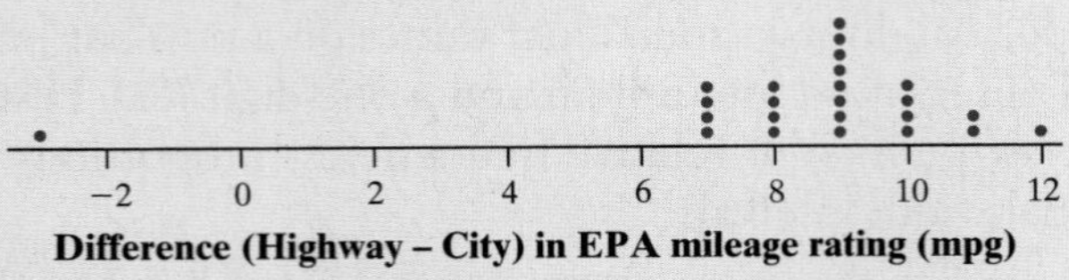


(a) Explain what the dot above –3 represents.

(b) What does the graph tell us about fuel economy in the city versus on the highway for these car models? Be specific.

**49. Getting older** How old is the oldest person you know? pg 33 Prudential Insurance Company asked 400 people to place a blue sticker on a huge wall next to the age of the oldest person they have ever known. An image of the graph is shown here. Describe the shape of the distribution.

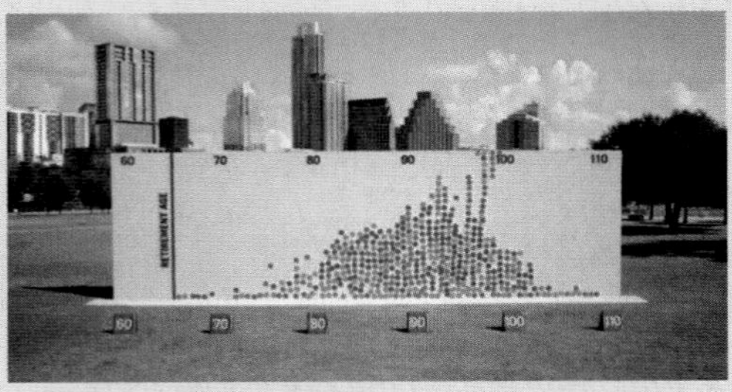

**50. Pair-o-dice** The dotplot shows the results of rolling a pair of fair, six-sided dice and finding the sum of the up-faces 100 times. Describe the shape of the distribution.

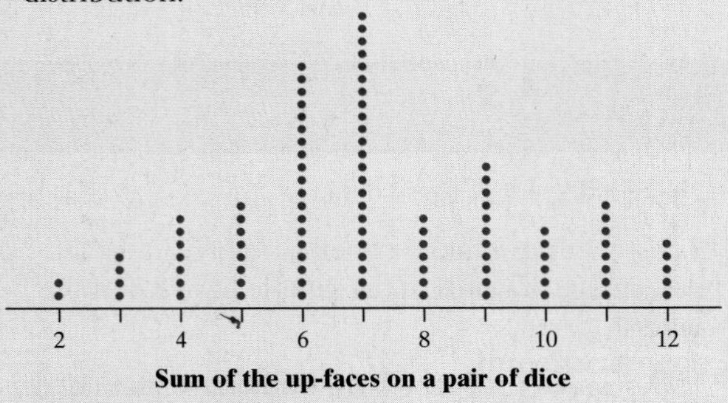


**51. Feeling sleepy?** Refer to Exercise 45. Describe the shape of the distribution.

**52. Easy reading?** Refer to Exercise 46. Describe the shape of the distribution.

**53. U.S. women's soccer—2016** Refer to Exercise 47. pg 35 Describe the distribution.

**54. Fuel efficiency** Refer to Exercise 48. Describe the distribution.

**55. Making money** The parallel dotplots show the pg 36 total family income of randomly chosen individuals from Indiana (38 individuals) and New Jersey (44 individuals). Compare the distributions of total family incomes in these two samples.

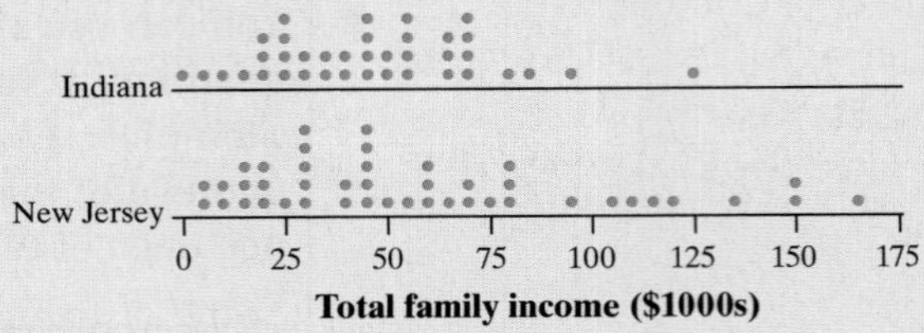


**56. Healthy streams** Nitrates are organic compounds that are a main ingredient in fertilizers. When those fertilizers run off into streams, the nitrates can have a toxic effect on fish. An ecologist studying nitrate pollution in two streams measures nitrate concentrations at 42 places on Stony Brook and 42 places on Mill Brook. The parallel dotplots display the data. Compare the distributions of nitrate concentration in these two streams.

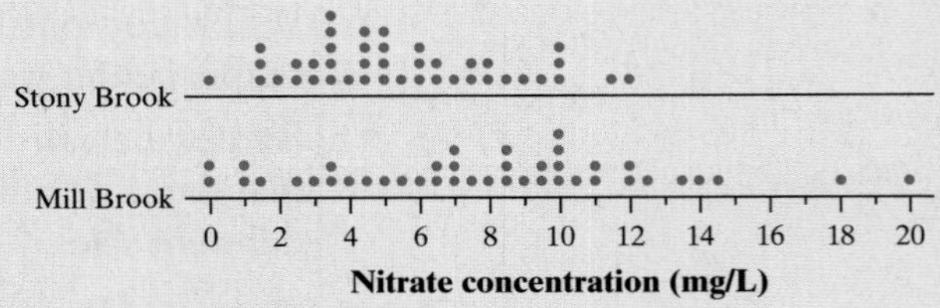


**57. Enhancing creativity** Do external rewards—things like money, praise, fame, and grades—promote creativity? Researcher Teresa Amabile recruited 47 experienced creative writers who were college students and divided them at random into two groups. The students in one group were given a list of statements about external reasons (E) for writing, such as public recognition, making money, or pleasing their parents. Students in the other group were given a list of statements about internal reasons (I) for writing, such as expressing yourself and enjoying wordplay. Both groups were then instructed to write a poem about laughter. Each student's poem was rated separately by 12 different poets using a creativity scale.[26] These ratings were averaged to obtain an overall creativity score for each poem. Parallel dotplots of the two groups' creativity scores are shown here.

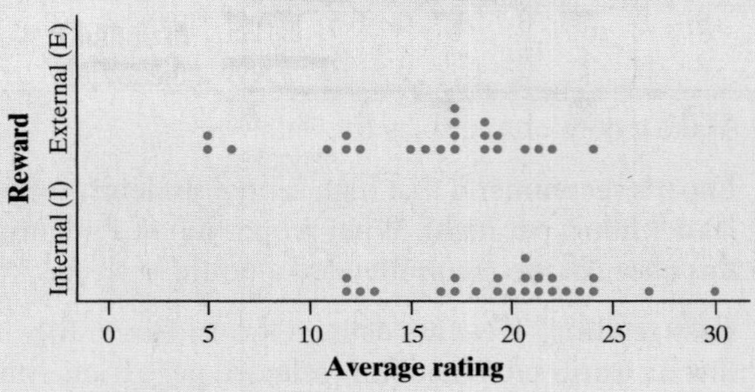


(a) Is the variability in creativity scores similar or different for the two groups? Justify your answer.

(b) Do the data suggest that external rewards promote creativity? Justify your answer.

**58. Healthy cereal?** Researchers collected data on 76 brands of cereal at a local supermarket.[27] For each brand, the sugar content (grams per serving) and the shelf in the store on which the cereal was located (1 = bottom, 2 = middle, 3 = top) were recorded. A dotplot of the data is shown here.

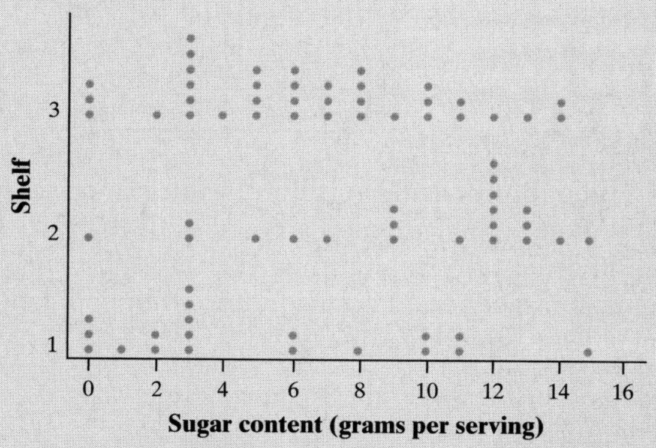


(a) Is the variability in sugar content of the cereals on the three shelves similar or different? Justify your answer.

(b) Critics claim that supermarkets tend to put sugary cereals where kids can see them. Do the data from this study support this claim? Justify your answer. (Note that Shelf 2 is at about eye level for kids in most supermarkets.)

**59. Snickers® are fun!** Here are the weights (in grams) of 17 Snickers Fun Size bars from a single bag:

pg 38

| | | | | | | | | |
|---|---|---|---|---|---|---|---|---|
| 17.1 | 17.4 | 16.6 | 17.4 | 17.7 | 17.1 | 17.3 | 17.7 | 17.8 |
| 19.2 | 16.0 | 15.9 | 16.5 | 16.8 | 16.5 | 17.1 | 16.7 | |

(a) Make a stemplot of these data.

(b) What interesting feature does the graph reveal?

(c) The advertised weight of a Snickers Fun Size bar is 17 grams. What proportion of candy bars in this sample weigh less than advertised?

**60. Eat your beans!** Beans and other legumes are a great source of protein. The following data give the protein content of 30 different varieties of beans, in grams per 100 grams of cooked beans.[28]

| | | | | | | | | | |
|---|---|---|---|---|---|---|---|---|---|
| 7.5 | 8.2 | 8.9 | 9.3 | 7.1 | 8.3 | 8.7 | 9.5 | 8.2 | 9.1 |
| 9.0 | 9.0 | 9.7 | 9.2 | 8.9 | 8.1 | 9.0 | 7.8 | 8.0 | 7.8 |
| 7.0 | 7.5 | 13.5 | 8.3 | 6.8 | 10.6 | 8.3 | 7.6 | 7.7 | 8.1 |

(a) Make a stemplot of these data.

(b) What interesting feature does the graph reveal?

(c) What proportion of these bean varieties contain more than 9 grams of protein per 100 grams of cooked beans?

**61. South Carolina counties** Here is a stemplot of the areas of the 46 counties in South Carolina. Note that the data have been rounded to the nearest 10 square miles (mi²).

| | |
|---|---|
| 3 | 9999 |
| 4 | 0116689 |
| 5 | 01115566778 |
| 6 | 47899 |
| 7 | 01245579 |
| 8 | 0011 |
| 9 | 13 |
| 10 | 8 |
| 11 | 233 |
| 12 | 2 |

Key: 6|4 represents a county with an area of 635 to 644.99 square miles.

(a) What is the area of the largest South Carolina county?

(b) Describe the distribution of area for the 46 South Carolina counties.

**62. Shopping spree** The stemplot displays data on the amount spent by 50 shoppers at a grocery store. Note that the values have been rounded to the nearest dollar.

| | |
|---|---|
| 0 | 399 |
| 1 | 1345677889 |
| 2 | 000123455668888 |
| 3 | 25699 |
| 4 | 1345579 |
| 5 | 0359 |
| 6 | 1 |
| 7 | 0 |
| 8 | 366 |
| 9 | 3 |

Key: 9|3 = $92.50 to $93.49 spent

(a) What was the smallest amount spent by any of the shoppers?

(b) Describe the distribution of amount spent by these 50 shoppers.

**63. Where do the young live?** Here is a stemplot of the percent of residents aged 25 to 34 in each of the 50 states:

| | |
|---|---|
| 11 | 44 |
| 11 | 66778 |
| 12 | 0134 |
| 12 | 666778888 |
| 13 | 0000001111444 |
| 13 | 7788999 |
| 14 | 0044 |
| 14 | 567 |
| 15 | 11 |
| 15 | |
| 16 | 0 |

(a) Why did we split stems?

(b) Give an appropriate key for this stemplot.

(c) Describe the shape of the distribution. Are there any outliers?

64. **Watch that caffeine!** The U.S. Food and Drug Administration (USFDA) limits the amount of caffeine in a 12-ounce can of carbonated beverage to 72 milligrams. That translates to a maximum of 48 milligrams of caffeine per 8-ounce serving. Data on the caffeine content of popular soft drinks (in milligrams per 8-ounce serving) are displayed in the stemplot.

| | |
|---|---|
| 1 | 556 |
| 2 | 033344 |
| 2 | 55667778888899 |
| 3 | 113 |
| 3 | 55567778 |
| 4 | 33 |
| 4 | 77 |

(a) Why did we split stems?

(b) Give an appropriate key for this graph.

(c) Describe the shape of the distribution. Are there any outliers?

65. **Acorns and oak trees** Of the many species of oak trees in the United States, 28 grow on the Atlantic Coast and 11 grow in California. The back-to-back stemplot displays data on the average volume of acorns (in cubic centimeters) for these 39 oak species.[29] Write a few sentences comparing the distributions of acorn size for the oak trees in these two regions.

| Atlantic Coast | | California |
|---:|---|---|
| 998643 | 0 | 4 |
| 88864211111 | 1 | 06 |
| 50 | 2 | 06 |
| 6640 | 3 | |
| 8 | 4 | 1 |
| | 5 | 59 |
| 8 | 6 | 0 |
| | 7 | 1 |
| 1 | 8 | |
| 1 | 9 | |
| 5 | 10 | |
| | 11 | |
| | 12 | |
| | 13 | |
| | 14 | |
| | 15 | |
| | 16 | |
| | 17 | 1 |

Key: 2|6 = An oak species whose acorn volume is 2.6 $cm^3$.

66. **Who studies more?** Researchers asked the students in a large first-year college class how many minutes they studied on a typical weeknight. The back-to-back stemplot displays the responses from random samples of 30 women and 30 men from the class, rounded to the nearest 10 minutes. Write a few sentences comparing the male and female distributions of study time.

| Women | | Men |
|---:|---|---|
| | 0 | 03333 |
| 96 | 0 | 56668999 |
| 22222222 | 1 | 02222222 |
| 888888888875555 | 1 | 558 |
| 4440 | 2 | 00344 |
| | 2 | |
| | 3 | 0 |
| 6 | 3 | |

Key: 2|3 = 225–234 minutes

67. **Carbon dioxide emissions** Burning fuels in power plants and motor vehicles emits carbon dioxide ($CO_2$), which contributes to global warming. The table displays $CO_2$ emissions per person from countries with populations of at least 20 million.[30]
pg 42

(a) Make a histogram of the data using intervals of width 2, starting at 0.

(b) Describe the shape of the distribution. Which countries appear to be outliers?

| Country | $CO_2$ | Country | $CO_2$ |
|---|---|---|---|
| Algeria | 3.3 | Mexico | 3.8 |
| Argentina | 4.5 | Morocco | 1.6 |
| Australia | 16.9 | Myanmar | 0.2 |
| Bangladesh | 0.4 | Nepal | 0.1 |
| Brazil | 2.2 | Nigeria | 0.5 |
| Canada | 14.7 | Pakistan | 0.9 |
| China | 6.2 | Peru | 2.0 |
| Colombia | 1.6 | Philippines | 0.9 |
| Congo | 0.5 | Poland | 8.3 |
| Egypt | 2.6 | Romania | 3.9 |
| Ethiopia | 0.1 | Russia | 12.2 |
| France | 5.6 | Saudi Arabia | 17.0 |
| Germany | 9.1 | South Africa | 9.0 |
| Ghana | 0.4 | Spain | 5.8 |
| India | 1.7 | Sudan | 0.3 |
| Indonesia | 1.8 | Tanzania | 0.2 |
| Iran | 7.7 | Thailand | 4.4 |
| Iraq | 3.7 | Turkey | 4.1 |
| Italy | 6.7 | Ukraine | 6.6 |
| Japan | 9.2 | United Kingdom | 7.9 |
| Kenya | 0.3 | United States | 17.6 |
| Korea, North | 11.5 | Uzbekistan | 3.7 |
| Korea, South | 2.9 | Venezuela | 6.9 |
| Malaysia | 7.7 | Vietnam | 1.7 |

68. **Traveling to work** How long do people travel each day to get to work? The following table gives the average travel times to work (in minutes) for workers in each state and the District of Columbia who are at least 16 years old and don't work at home.[31]

| | | | | | |
|---|---|---|---|---|---|
| AL | 23.6 | LA | 25.1 | OH | 22.1 |
| AK | 17.7 | ME | 22.3 | OK | 20.0 |
| AZ | 25.0 | MD | 30.6 | OR | 21.8 |
| AR | 20.7 | MA | 26.6 | PA | 25.0 |
| CA | 26.8 | MI | 23.4 | RI | 22.3 |
| CO | 23.9 | MN | 22.0 | SC | 22.9 |
| CT | 24.1 | MS | 24.0 | SD | 15.9 |
| DE | 23.6 | MO | 22.9 | TN | 23.5 |
| FL | 25.9 | MT | 17.6 | TX | 24.6 |
| GA | 27.3 | NE | 17.7 | UT | 20.8 |
| HI | 25.5 | NV | 24.2 | VT | 21.2 |
| ID | 20.1 | NH | 24.6 | VA | 26.9 |
| IL | 27.9 | NJ | 29.1 | WA | 25.2 |
| IN | 22.3 | NM | 20.9 | WV | 25.6 |
| IA | 18.2 | NY | 30.9 | WI | 20.8 |
| KS | 18.5 | NC | 23.4 | WY | 17.9 |
| KY | 22.4 | ND | 15.5 | DC | 29.2 |

(a) Make a histogram to display the travel time data using intervals of width 2 minutes, starting at 14 minutes.

(b) Describe the shape of the distribution. What is the most common interval of travel times?

69. **DRP test scores** There are many ways to measure the reading ability of children. One frequently used test is the Degree of Reading Power (DRP). In a research study on third-grade students, the DRP was administered to 44 students.[32] Their scores were as follows.

| | | | | | | | | | | |
|---|---|---|---|---|---|---|---|---|---|---|
| 40 | 26 | 39 | 14 | 42 | 18 | 25 | 43 | 46 | 27 | 19 |
| 47 | 19 | 26 | 35 | 34 | 15 | 44 | 40 | 38 | 31 | 46 |
| 52 | 25 | 35 | 35 | 33 | 29 | 34 | 41 | 49 | 28 | 52 |
| 47 | 35 | 48 | 22 | 33 | 41 | 51 | 27 | 14 | 54 | 45 |

Make a histogram to display the data. Write a few sentences describing the distribution of DRP scores.

70. **Country music** The lengths, in minutes, of the 50 most popular mp3 downloads of songs by country artist Dierks Bentley are given here.

| | | | | | |
|---|---|---|---|---|---|
| 4.2 | 4.0 | 3.9 | 3.8 | 3.7 | 4.7 |
| 3.4 | 4.0 | 4.4 | 5.0 | 4.6 | 3.7 |
| 4.6 | 4.4 | 4.1 | 3.0 | 3.2 | 4.7 |
| 3.5 | 3.7 | 4.3 | 3.7 | 4.8 | 4.4 |
| 4.2 | 4.7 | 6.2 | 4.0 | 7.0 | 3.9 |
| 3.4 | 3.4 | 2.9 | 3.3 | 4.0 | 4.2 |
| 3.2 | 3.4 | 3.7 | 3.5 | 3.4 | 3.7 |
| 3.9 | 3.7 | 3.8 | 3.1 | 3.7 | 3.6 |
| 4.5 | 3.7 | | | | |

Make a histogram to display the data. Write a few sentences describing the distribution of song lengths.

71. **Returns on common stocks** The return on a stock is the change in its market price plus any dividend payments made. Return is usually expressed as a percent of the beginning price. The figure shows a histogram of the distribution of monthly returns for the U.S. stock market over a 273-month period.[33]

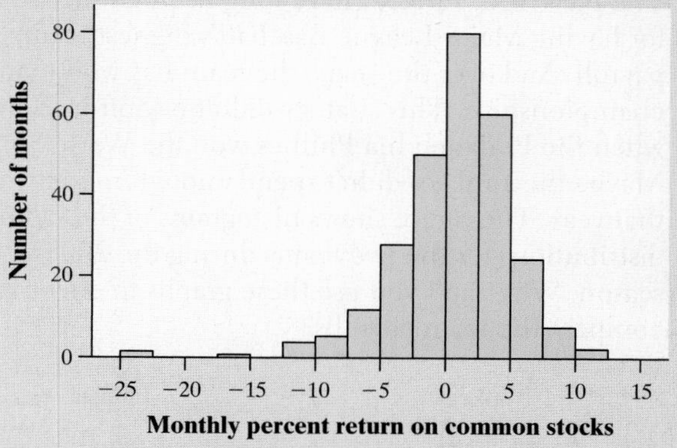


(a) Describe the overall shape of the distribution of monthly returns.

(b) What is the approximate center of this distribution?

(c) Explain why you cannot find the exact value for the minimum return. Between what two values does it lie?

(d) A return less than 0 means that stocks lost value in that month. About what percent of all months had returns less than 0?

72. **Healthy cereal?** Researchers collected data on calories per serving for 77 brands of breakfast cereal. The histogram displays the data.[34]

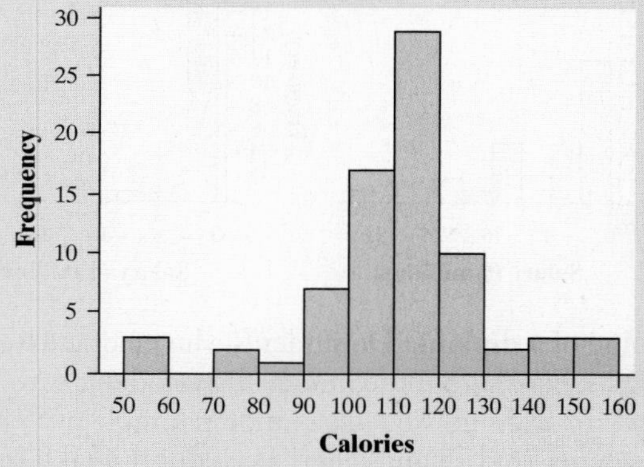


(a) Describe the overall shape of the distribution of calories.

(b) What is the approximate center of this distribution?

(c) Explain why you cannot find the exact value for the maximum number of calories per serving for these cereal brands. Between what two values does it lie?

(d) About what percent of the cereal brands have 130 or more calories per serving?

73. **Paying for championships** Does paying high salaries lead to more victories in professional sports? The New York Yankees have long been known for having Major League Baseball's highest team payroll. And over the years, the team has won many championships. This strategy didn't pay off in 2008, when the Philadelphia Phillies won the World Series. Maybe the Yankees didn't spend enough money that year. The figure shows histograms of the salary distributions for the two teams during the 2008 season. Why can't you use these graphs to effectively compare the team payrolls?

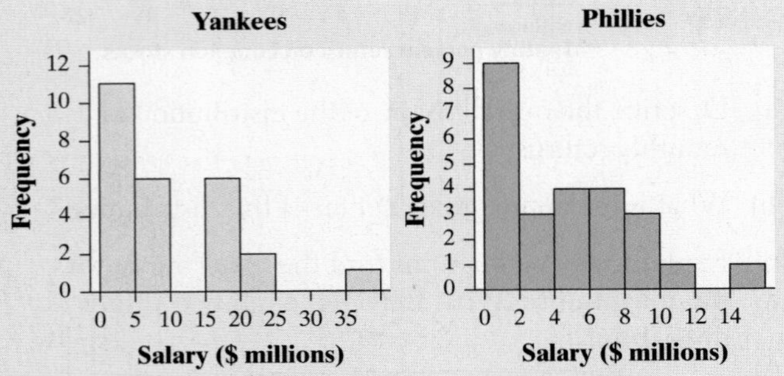


74. **Paying for championships** Refer to Exercise 73. Here is a better graph of the 2008 salary distributions for the Yankees and the Phillies. Write a few sentences comparing these two distributions.

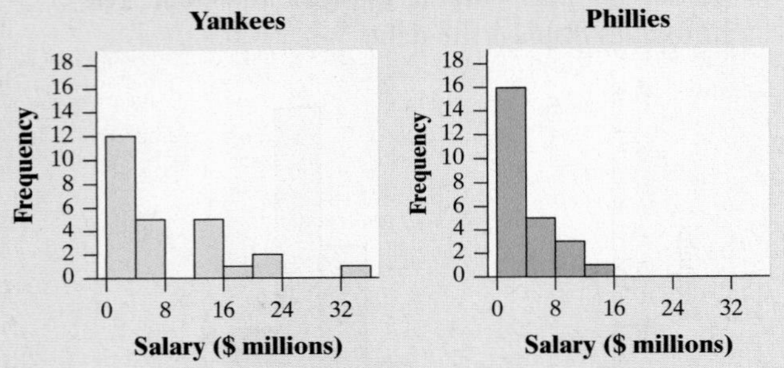


75. **Value of a diploma** Do students who graduate from high school earn more money than students who do not? To find out, we took a random sample of 371 U.S. residents aged 18 and older. The educational level and total personal income of each person were recorded. The data for the 57 non-graduates (No) and the 314 graduates (Yes) are displayed in the relative frequency histograms.

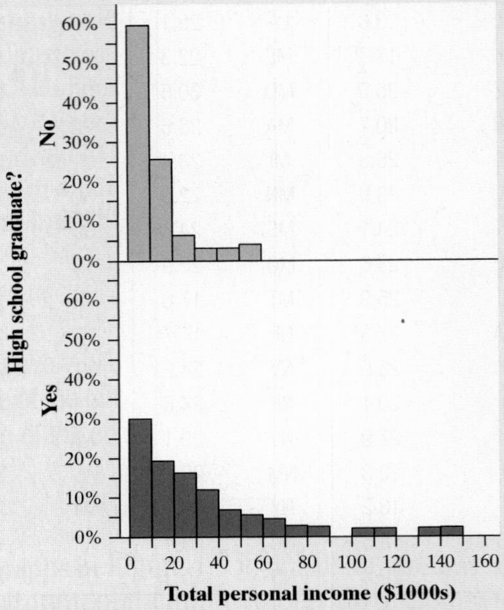


(a) Would it be appropriate to use frequency histograms instead of relative frequency histograms in this setting? Explain why or why not.

(b) Compare the distributions of total personal income for the two groups.

76. **Strong paper towels** In commercials for Bounty paper towels, the manufacturer claims that they are the "quicker picker-upper," but are they also the stronger picker-upper? Two of Mr. Tabor's statistics students, Wesley and Maverick, decided to find out. They selected a random sample of 30 Bounty paper towels and a random sample of 30 generic paper towels and measured their strength when wet. To do this, they uniformly soaked each paper towel with 4 ounces of water, held two opposite edges of the paper towel, and counted how many quarters each paper towel could hold until ripping, alternating brands. The data are displayed in the relative frequency histograms. Compare the distributions.

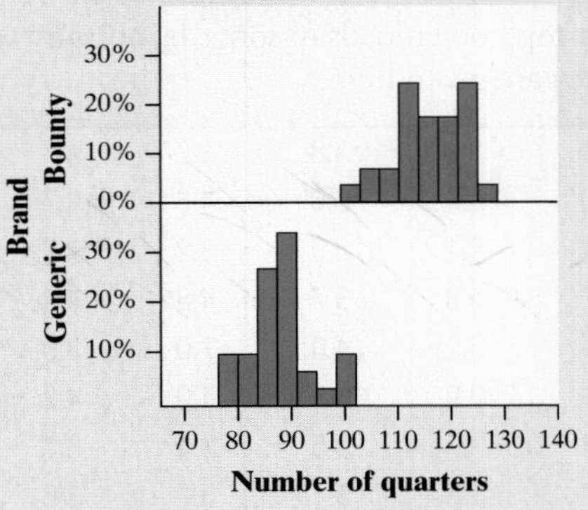


(a) Would it be appropriate to use frequency histograms instead of relative frequency histograms in this setting? Explain why or why not.

(b) Compare the distributions of number of quarters until breaking for the two paper towel brands.

77. **Birth months** Imagine asking a random sample of 60 students from your school about their birth months. Draw a plausible (believable) graph of the distribution of birth months. Should you use a bar graph or a histogram to display the data?

78. **Die rolls** Imagine rolling a fair, six-sided die 60 times. Draw a plausible graph of the distribution of die rolls. Should you use a bar graph or a histogram to display the data?

79. **AP® exam scores** The table gives the distribution of grades earned by students taking the AP® Calculus AB and AP® Statistics exams in 2016.[35]

| | Grade | | | | | |
|---|---|---|---|---|---|---|
| | 5 | 4 | 3 | 2 | 1 | Total |
| Calculus AB | 76,486 | 53,467 | 53,533 | 30,017 | 94,712 | 308,215 |
| Statistics | 29,627 | 44,884 | 51,367 | 32,120 | 48,565 | 206,563 |

(a) Make an appropriate graphical display to compare the grade distributions for AP® Calculus AB and AP® Statistics.

(b) Write a few sentences comparing the two distributions of exam grades.

**Multiple Choice:** *Select the best answer for Exercises 80–85.*

80. Here are the amounts of money (cents) in coins carried by 10 students in a statistics class: 50, 35, 0, 46, 86, 0, 5, 47, 23, 65. To make a stemplot of these data, you would use stems

(a) 0, 2, 3, 4, 6, 8.
(b) 0, 1, 2, 3, 4, 5, 6, 7, 8.
(c) 0, 3, 5, 6, 7.
(d) 00, 10, 20, 30, 40, 50, 60, 70, 80, 90.
(e) None of these.

81. The histogram shows the heights of 300 randomly selected high school students. Which of the following is the best description of the shape of the distribution of heights?

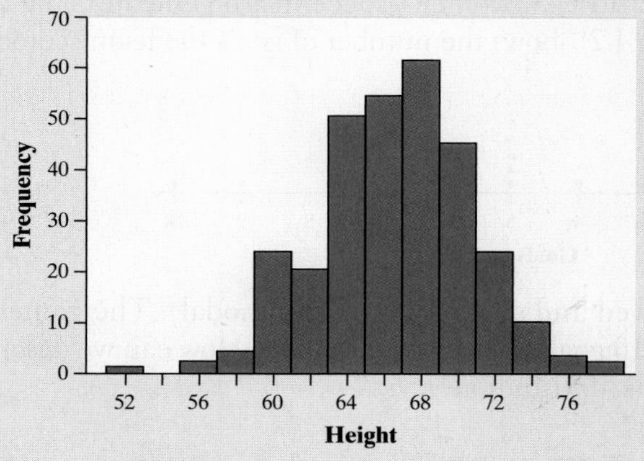


(a) Roughly symmetric and single-peaked (unimodal)
(b) Roughly symmetric and double-peaked (bimodal)
(c) Roughly symmetric and multi-peaked (multimodal)
(d) Skewed to the left
(e) Skewed to the right

82. You look at real estate ads for houses in Naples, Florida. There are many houses ranging from \$200,000 to \$500,000 in price. The few houses on the water, however, are priced up to \$15 million. The distribution of house prices will be

(a) skewed to the left.
(b) roughly symmetric.
(c) skewed to the right.
(d) single-peaked.
(e) approximately uniform.

83. The histogram shows the distribution of the percents of women aged 15 and over who have never married in each of the 50 states and the District of Columbia. Which of the following statements about the histogram is correct?

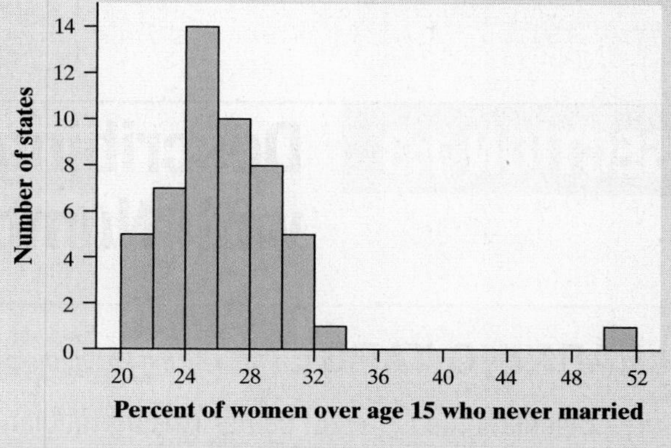


(a) The center (median) of the distribution is about 36%.
(b) There are more states with percentages above 32 than there are states with percentages less than 24.
(c) It would be better if the values from 34 to 50 were deleted on the horizontal axis so there wouldn't be a large gap.
(d) There was one state with a value of exactly 33%.
(e) About half of the states had percentages between 24% and 28%.

84. When comparing two distributions, it would be best to use relative frequency histograms rather than frequency histograms when

(a) the distributions have different shapes.
(b) the distributions have different amounts of variability.
(c) the distributions have different centers.
(d) the distributions have different numbers of observations.
(e) at least one of the distributions has outliers.

85. Which of the following is the best reason for choosing a stemplot rather than a histogram to display the distribution of a quantitative variable?

(a) Stemplots allow you to split stems; histograms don't.
(b) Stemplots allow you to see the values of individual observations.
(c) Stemplots are better for displaying very large sets of data.
(d) Stemplots never require rounding of values.
(e) Stemplots make it easier to determine the shape of a distribution.

**Recycle and Review**

86. **Risks of playing soccer** (1.1) A study in Sweden looked at former elite soccer players, people who had played soccer but not at the elite level, and people of the same age who did not play soccer. Here is a two-way table that classifies these individuals by whether or not they had arthritis of the hip or knee by their mid-fifties:[36]

| | | Soccer level | | |
|---|---|---|---|---|
| | | Elite | Non-elite | Did not play |
| Whether person developed arthritis | Yes | 10 | 9 | 24 |
| | No | 61 | 206 | 548 |

(a) What percent of the people in this study were elite soccer players? What percent of the people in this study developed arthritis?

(b) What percent of the elite soccer players developed arthritis? What percent of those who got arthritis were elite soccer players?

(c) Researchers suspected that the more serious soccer players were more likely to develop arthritis later in life. Do the data confirm this suspicion? Calculate appropriate percentages to support your answer.

# SECTION 1.3 Describing Quantitative Data with Numbers

**LEARNING TARGETS** *By the end of the section, you should be able to:*

- Calculate measures of center (mean, median) for a distribution of quantitative data.
- Calculate and interpret measures of variability (range, standard deviation, *IQR*) for a distribution of quantitative data.
- Explain how outliers and skewness affect measures of center and variability.
- Identify outliers using the $1.5 \times IQR$ rule.
- Make and interpret boxplots of quantitative data.
- Use boxplots and numerical summaries to compare distributions of quantitative data.

How much offense did the 2016 U.S. women's soccer team generate? The dotplot (reproduced from Section 1.2) shows the number of goals the team scored in 20 games played.

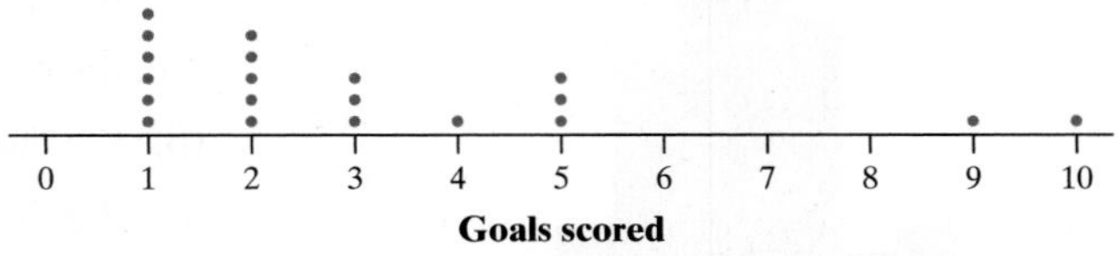


The *mode* of a data set is the most frequently occurring value. For the distribution of goals scored by the 2016 U.S. women's soccer team, the mode is 1. As you can see, the mode is rarely a good way to describe the center of the distribution.

The distribution is right-skewed and single-peaked (unimodal). The games in which the team scored 9 and 10 goals appear to be outliers. How can we describe the center and variability of this distribution?

## Measuring Center: The Mean

The most common measure of center is the **mean**.

**DEFINITION The mean**

The **mean** of a distribution of quantitative data is the average of all the individual data values. To find the mean, add all the values and divide by the total number of data values.

If the $n$ observations are $x_1, x_2, \ldots, x_n$, the sample mean $\bar{x}$ (pronounced "$x$-bar") is given by the formula

$$\bar{x} = \frac{\text{sum of data values}}{\text{number of data values}} = \frac{x_1 + x_2 + \cdots + x_n}{n} = \frac{\sum x_i}{n}$$

The $\Sigma$ (capital Greek letter sigma) in the formula is short for "add them all up." The subscripts on the observations $x_i$ are just a way of keeping the $n$ data values distinct. They do not necessarily indicate order or any other special facts about the data.

## EXAMPLE

### How many goals? Calculating the mean

Kyodo News/ Getty Images

**PROBLEM:** Here are the data on the number of goals scored in 20 games played by the 2016 U.S. women's soccer team:

5 5 1 10 5 2 1 1 2 3 3 2 1 4 2 1 2 1 9 3

(a) Calculate the mean number of goals scored per game by the team. Show your work.

(b) The earlier description of these data (page 35) suggests that the games in which the team scored 9 and 10 goals are possible outliers. Calculate the mean number of goals scored per game by the team in the other 18 games that season. What do you notice?

**SOLUTION:**

(a) $$\bar{x} = \frac{5+5+1+10+5+2+1+1+2+3+3+2+1+4+2+1+2+1+9+3}{20}$$

$$= \frac{63}{20} = 3.15 \text{ goals}$$

$$\bar{x} = \frac{\sum x_i}{n}$$

(b) The mean for the other 18 games is

$$\bar{x} = \frac{5+5+1+5+2+1+1+2+3+3+2+1+4+2+1+2+1+3}{18}$$

$$= \frac{44}{18} = 2.44 \text{ goals}$$

These two games increased the team's mean number of goals scored per game by 0.71 goals.

FOR PRACTICE, TRY EXERCISE 87

The notation $\bar{x}$ refers to the mean of a *sample*. Most of the time, the data we encounter can be thought of as a sample from some larger population. When we need to refer to a *population mean*, we'll use the symbol $\mu$ (Greek letter mu, pronounced "mew"). If you have the entire population of data available, then you calculate $\mu$ in just the way you'd expect: add the values of all the observations, and divide by the number of observations.

We call $\bar{x}$ a **statistic** and $\mu$ a **parameter**. Remember **s** and **p**: **s**tatistics come from **s**amples and **p**arameters come from **p**opulations.

**DEFINITION Statistic, Parameter**

A **statistic** is a number that describes some characteristic of a sample.

A **parameter** is a number that describes some characteristic of a population.

The preceding example illustrates an important weakness of the mean as a measure of center: **the mean is sensitive to extreme values in a distribution.** These may be outliers, but a skewed distribution that has no outliers will also pull the mean toward its long tail. We say that the mean is not a **resistant** measure of center.

**DEFINITION Resistant**

A statistical measure is **resistant** if it isn't sensitive to extreme values.

The mean of a distribution also has a physical interpretation, as the following activity shows.

## ACTIVITY Mean as a "balance point"

In this activity, you'll investigate an important property of the mean.

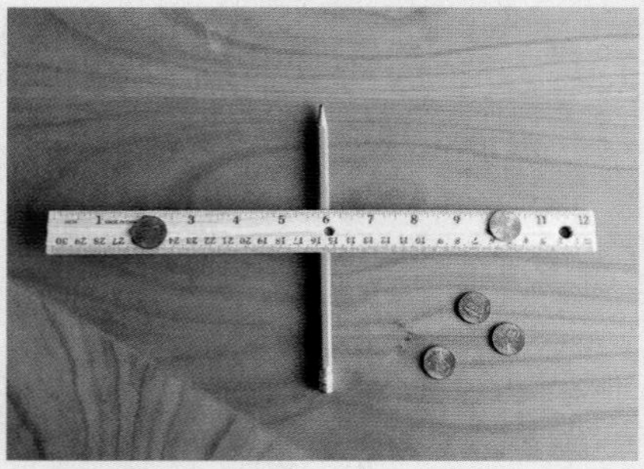

Ann Heath

1. Stack 5 pennies on top of the 6-inch mark on a 12-inch ruler. Place a pencil under the ruler to make a "seesaw" on a desk or table. Move the pencil until the ruler balances. What is the relationship between the location of the pencil and the mean of the five data values 6, 6, 6, 6, and 6?
2. Move one penny off the stack to the 8-inch mark on your ruler. Now move one other penny so that the ruler balances again without moving the pencil. Where did you put the other penny? What is the mean of the five data values represented by the pennies now?
3. Move one more penny off the stack to the 2-inch mark on your ruler. Now move both remaining pennies from the 6-inch mark so that the ruler still balances with the pencil in the same location. Is the mean of the data values still 6?
4. Discuss with your classmates: Why is the mean called the "balance point" of a distribution?

The activity gives a physical interpretation of the mean as the balance point of a distribution. For the data on goals scored in each of 20 games played by the 2016 U.S. women's soccer team, the dotplot balances at $\bar{x} = 3.15$ goals.

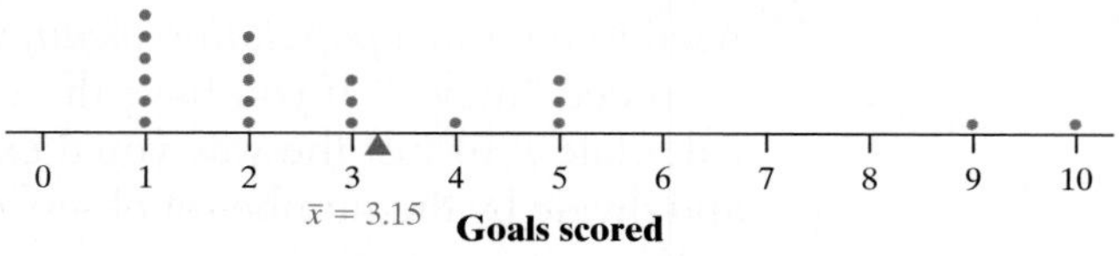

## Measuring Center: The Median

We could also report the value in the "middle" of a distribution as its center. That's the idea of the **median**.

**DEFINITION Median**

The **median** is the midpoint of a distribution, the number such that about half the observations are smaller and about half are larger.

To find the median, arrange the data values from smallest to largest.

- If the number $n$ of data values is odd, the median is the middle value in the ordered list.
- If the number $n$ of data values is even, use the average of the two middle values in the ordered list as the median.

The median is easy to find by hand for small sets of data. For instance, here are the data from Section 1.2 on the highway fuel economy ratings for a sample of 25 model year 2018 Toyota 4Runners tested by the EPA:

22.4 22.4 22.3 23.3 22.3 22.3 22.5 22.4 22.1 21.5 22.0 22.2 22.7
22.8 22.4 22.6 22.9 22.5 22.1 22.4 22.2 22.9 22.6 21.9 22.4

Start by sorting the data values from smallest to largest:

21.5 21.9 22.0 22.1 22.1 22.2 22.2 22.3 22.3 22.3 22.4 22.4 **22.4**
22.4 22.4 22.4 22.5 22.5 22.6 22.6 22.7 22.8 22.9 22.9 23.3

There are $n = 25$ data values (an odd number), so the median is the middle (13th) value in the ordered list, the bold **22.4**.

**EXAMPLE**

### How many goals?
### Finding the median

Icon Sports Wire/Getty Images

**PROBLEM:** Here are the data on the number of goals scored in 20 games played by the 2016 U.S. women's soccer team:

5 5 1 10 5 2 1 1 2 3 3 2 1 4 2 1 2 1 9 3

Find the median.

**SOLUTION:**

1 1 1 1 1 1 2 2 2 (2 | 2) 3 3 3 4 5 5 5 9 10

The median is $\frac{2+2}{2} = 2$.

To find the median, sort the data values from smallest to largest. Because there are $n = 20$ data values (an even number), the median is the average of the middle two values in the ordered list.

**FOR PRACTICE, TRY EXERCISE 89**

# Comparing the Mean and the Median

Which measure—the mean or the median—should we report as the center of a distribution? That depends on both the shape of the distribution and whether there are any outliers.

- **Shape:** Figure 1.13 shows the mean and median for dotplots with three different shapes. Notice how these two measures of center compare in each case. The mean is pulled in the direction of the long tail in a skewed distribution.

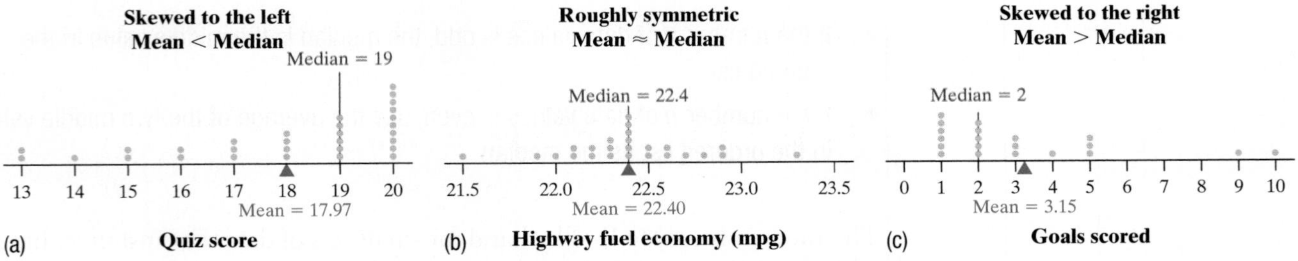


**FIGURE 1.13** Dotplots that show the relationship between the mean and median in distributions with different shapes: (a) Scores of 30 statistics students on a 20-point quiz, (b) highway fuel economy ratings for a sample of 25 model year 2018 Toyota 4Runners, and (c) number of goals scored in 20 games played by the 2016 U.S. women's soccer team.

You can compare how the mean and median behave by using the *Mean and Median* applet at the book's website, highschool.bfwpub.com/updatedtps6e.

- **Outliers:** We noted earlier that the mean is sensitive to extreme values. If we remove the two possible outliers (9 and 10) in Figure 1.13(c), the mean number of goals scored per game decreases from 3.15 to 2.44. The median number of goals scored is 2 whether we include these two games or not. The median is a resistant measure of center, but the mean is not.

**EFFECT OF SKEWNESS AND OUTLIERS ON MEASURES OF CENTER**

- If a distribution of quantitative data is roughly symmetric and has no outliers, the mean and median will be similar.
- If the distribution is strongly skewed, the mean will be pulled in the direction of the skewness but the median won't. For a right-skewed distribution, we expect the mean to be greater than the median. For a left-skewed distribution, we expect the mean to be less than the median.
- The median is resistant to outliers but the mean isn't.

The mean and median measure center in different ways, and both are useful. In Major League Baseball (MLB), the distribution of player salaries is strongly skewed to the right. Most players earn close to the minimum salary (which was $507,500 in 2016), while a few earn more than $20 million. The median salary for MLB players in 2016 was about $1.5 million—but the mean salary was about $4.4 million. Clayton Kershaw, Miguel Cabrera, John Lester, and several other highly paid superstars pulled the mean up but that did not affect the median.

The median gives us a good idea of what a "typical" MLB salary is. If we want to know the total salary paid to MLB players in 2016, however, we would multiply the mean salary by the total number of players: (\$4.4 million)(862) $\approx$ \$3.8 billion!

## CHECK YOUR UNDERSTANDING

Some students purchased pumpkins for a carving contest. Before the contest began, they weighed the pumpkins. The weights in pounds are shown here, along with a histogram of the data.

| | | | | | | | | | | | |
|---|---|---|---|---|---|---|---|---|---|---|---|
| 3.6 | 4.0 | 9.6 | 14.0 | 11.0 | 12.4 | 13.0 | 2.0 | 6.0 | 6.6 | 15.0 | 3.4 |
| 12.7 | 6.0 | 2.8 | 9.6 | 4.0 | 6.1 | 5.4 | 11.9 | 5.4 | 31.0 | 33.0 | |

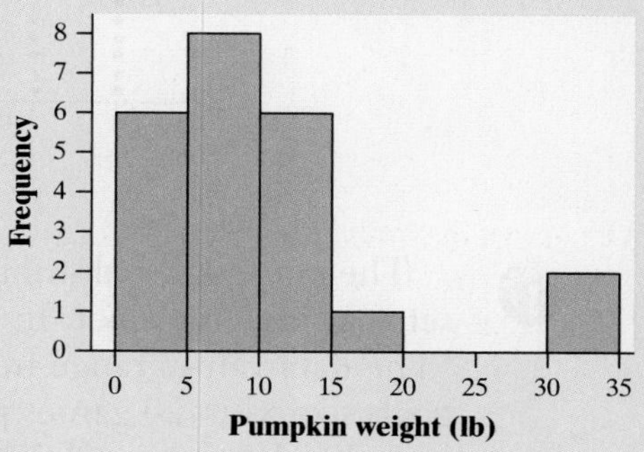


1. Calculate the mean weight of the pumpkins.
2. Find the median weight of the pumpkins.
3. Would you use the mean or the median to summarize the typical weight of a pumpkin in this contest? Explain.

# Measuring Variability: The Range

Being able to describe the shape and center of a distribution is a great start. However, two distributions can have the same shape and center, but still look quite different.

Figure 1.14 shows comparative dotplots of the length (in millimeters) of separate random samples of PVC pipe from two suppliers, A and B.[37] Both distributions are roughly symmetric and single-peaked (unimodal), with centers at about 600 mm, but the variability of these two distributions is quite different. The sample of pipes from Supplier A has much more consistent lengths (less variability) than the sample from Supplier B.

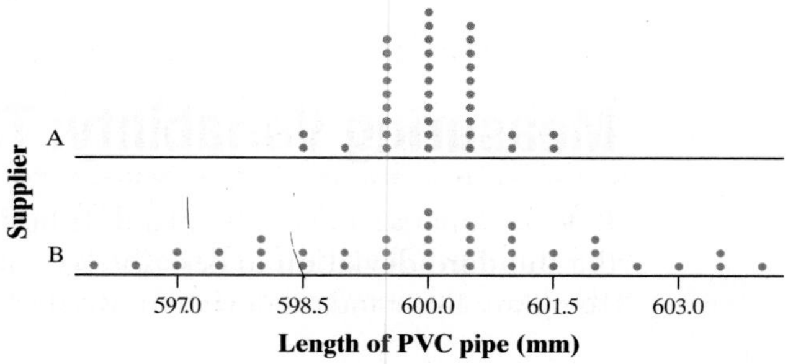


**FIGURE 1.14** Comparative dotplots of the length of PVC pipes in separate random samples from Supplier A and Supplier B.

There are several ways to measure the variability of a distribution. The simplest is the **range**.

> **DEFINITION Range**
>
> The **range** of a distribution is the distance between the minimum value and the maximum value. That is,
>
> $$\text{Range} = \text{Maximum} - \text{Minimum}$$

Here are the data on the number of goals scored in 20 games played by the 2016 U.S. women's soccer team, along with a dotplot:

5 5 1 10 5 2 1 1 2 3 3 2 1 4 2 1 2 1 9 3

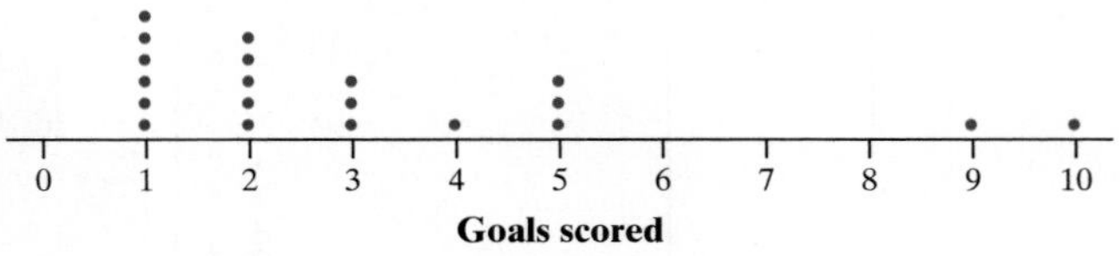


The range of this distribution is $10 - 1 = 9$ goals. Note that **the range of a data set is a single number.** In everyday language, people sometimes say things like, "The data values range from 1 to 10." A correct statement is "The number of goals scored in 20 games played by the 2016 U.S. women's soccer team varies from 1 to 10, a range of 9 goals."

The range is *not* a resistant measure of variability. It depends on only the maximum and minimum values, which may be outliers. Look again at the data on goals scored by the 2016 U.S. women's soccer team. Without the possible outliers at 9 and 10 goals, the range of the distribution would decrease to $5 - 1 = 4$ goals.

The following graph illustrates another problem with the range as a measure of variability. The parallel dotplots show the lengths (in millimeters) of a sample of 11 nails produced by each of two machines.[38] Both distributions are centered at 70 mm and have a range of $72 - 68 = 4$ mm. But the lengths of the nails made by Machine B clearly vary more from the center of 70 mm than the nails made by Machine A.

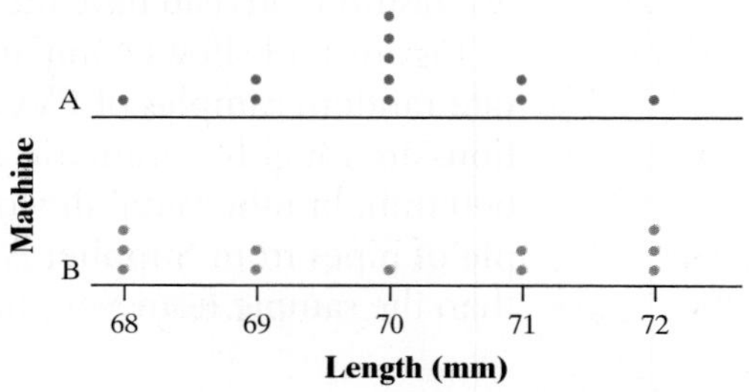


# Measuring Variability: The Standard Deviation

If we summarize the center of a distribution with the mean, then we should use the **standard deviation** to describe the variation of data values around the mean. To obtain the standard deviation, we start by calculating the **variance**.

> **DEFINITION Standard deviation, Variance**
>
> The **standard deviation** measures the typical distance of the values in a distribution from the mean. It is calculated by finding an average of the squared deviations and then taking the square root. This average squared deviation is called the **variance**. If the values in a data set are $x_1, x_2, \ldots, x_n$, the sample variance $s_x^2$ is given by the formula
>
> $$s_x^2 = \frac{(x_1-\bar{x})^2+(x_2-\bar{x})^2+\cdots+(x_n-\bar{x})^2}{n-1} = \frac{\sum(x_i-\bar{x})^2}{n-1}$$
>
> The sample standard deviation $s_x$ is the square root of the variance:
>
> $$s_x = \sqrt{\frac{\sum(x_i-\bar{x})^2}{n-1}}$$

> **AP® EXAM TIP**
>
> The formula sheet provided with the AP® Statistics exam gives the sample standard deviation in the equivalent form
>
> $$s_x = \sqrt{\frac{1}{n-1}\sum(x_i-\bar{x})^2}$$

How do we calculate the standard deviation $s_x$ of a quantitative data set with $n$ values? Here are the steps.

**HOW TO CALCULATE THE SAMPLE STANDARD DEVIATION $s_x$**

- Find the mean of the distribution.
- Calculate the *deviation* of each value from the mean:
  deviation = value − mean
- Square each of the deviations.
- Add all the squared deviations, then divide by $n - 1$ to get the sample variance.
- Take the square root to return to the original units.

The population standard deviation $\sigma$ is calculated by dividing the sum of squared deviations by the population size $N$ (not $N - 1$) before taking the square root.

The notation $s_x$ refers to the standard deviation of a *sample*. When we need to refer to the standard deviation of a population, we'll use the symbol $\sigma$ (Greek lowercase sigma). We often use the sample statistic $s_x$ to estimate the population parameter $\sigma$.

## EXAMPLE How many friends?
### Calculating and interpreting standard deviation

**PROBLEM:** Eleven high school students were asked how many "close" friends they have. Here are their responses, along with a dotplot:

1 2 2 2 3 3 3 3 4 4 6

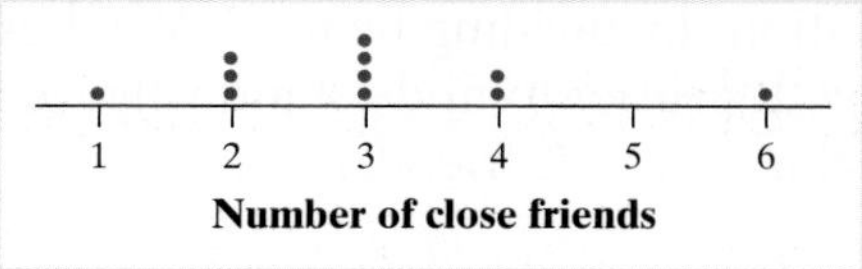

LaraBelova/Getty Images

Calculate the sample standard deviation. Interpret this value.

**SOLUTION:**

$$\bar{x} = \frac{1+2+2+2+3+3+3+3+4+4+6}{11} = 3$$

| $x_i$ | $x_i - \bar{x}$ | $(x_i - \bar{x})^2$ |
|---|---|---|
| 1 | 1−3=−2 | $(-2)^2 = 4$ |
| 2 | 2−3=−1 | $(-1)^2 = 1$ |
| 2 | 2−3=−1 | $(-1)^2 = 1$ |
| 2 | 2−3=−1 | $(-1)^2 = 1$ |
| 3 | 3−3=0 | $0^2 = 0$ |
| 3 | 3−3=0 | $0^2 = 0$ |
| 3 | 3−3=0 | $0^2 = 0$ |
| 3 | 3−3=0 | $0^2 = 0$ |
| 4 | 4−3=1 | $1^2 = 1$ |
| 4 | 4−3=1 | $1^2 = 1$ |
| 6 | 6−3=3 | $3^2 = 9$ |
| | | Sum=18 |

To calculate the sample standard deviation:
- Find the mean of the distribution.
- Calculate the *deviation* of each value from the mean:
  deviation = value − mean
- Square each of the deviations.
- Add all the squared deviations, then divide by $n-1$ to get the sample variance.
- Take the square root to return to the original units.

$$s_x^2 = \frac{18}{11-1} = 1.80$$

$$s_x = \sqrt{1.80} = 1.34 \text{ close friends}$$

$$s_x = \sqrt{\frac{\sum(x_i - \bar{x})^2}{n-1}}$$

**Interpretation: The number of close friends these students have typically varies by about 1.34 close friends from the mean of 3 close friends.**

**FOR PRACTICE, TRY EXERCISE 99**

In the preceding example, the sample variance is $s_x^2 = 1.80$. Unfortunately, the units are "squared close friends." Because variance is measured in squared units, it is not a very helpful way to describe the variability of a distribution.

### Think About It

**WHY IS THE STANDARD DEVIATION CALCULATED IN SUCH A COMPLEX WAY?** Add the deviations from the mean in the preceding example. You should get a sum of 0. Why? Because the mean is the balance point of the distribution. We square the deviations to avoid the positive and negative deviations balancing each other out and adding to 0. It might seem strange to "average" the squared deviations by dividing by $n-1$. We'll explain the reason for doing this in Chapter 7. It's easier to understand why we take the square root: to return to the original units (close friends).

More important than the details of calculating $s_x$ are the properties of the standard deviation as a measure of variability:

- **$s_x$ is always greater than or equal to 0.** $s_x = 0$ only when there is no variability, that is, when all values in a distribution are the same.
- **Larger values of $s_x$ indicate greater variation** from the mean of a distribution. The comparative dotplot shows the lengths of PVC pipe in random samples from two different suppliers. Supplier A's pipe lengths have a standard deviation of 0.681 mm, while Supplier B's pipe lengths have a standard deviation of 2.02 mm. The lengths of pipes from Supplier B are typically farther from the mean than the lengths of pipes from Supplier A.

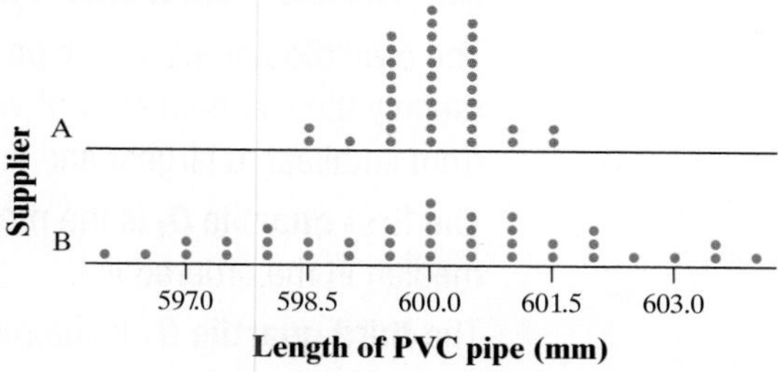


- **$s_x$ is not a resistant measure of variability.** The use of squared deviations makes $s_x$ even more sensitive than $\bar{x}$ to extreme values in a distribution. For example, the standard deviation of the number of goals scored in 20 games played by the 2016 U.S women's soccer team is 2.58 goals. If we omit the possible outliers of 9 and 10 goals, the standard deviation drops to 1.46 goals.

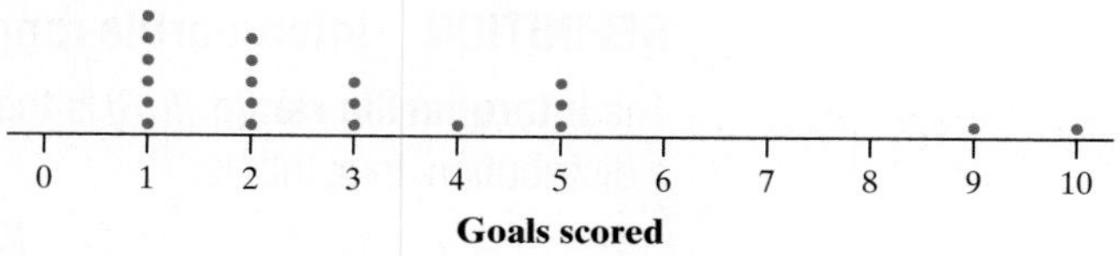


- **$s_x$ measures variation about the mean.** It should be used only when the mean is chosen as the measure of center.

In the close friends example, 11 high school students had an average of $\bar{x}=3$ close friends with a standard deviation of $s_x=1.34$. What if a 12th high school student was added to the sample who had 3 close friends? The mean number of close friends in the sample would still be $\bar{x}=3$. How would $s_x$ be affected? Because the standard deviation measures the typical distance of the values in a distribution from the mean, $s_x$ would *decrease* because this 12th value is at a distance of 0 from the mean. In fact, the new standard deviation would be

$$s_x = \sqrt{\frac{\sum(x_i - \bar{x})^2}{n-1}} = \sqrt{\frac{18}{12-1}} = 1.28 \text{ close friends}$$

## Measuring Variability: The Interquartile Range (*IQR*)

We can avoid the impact of extreme values on our measure of variability by focusing on the middle of the distribution. Start by ordering the data values from smallest to largest. Then find the **quartiles**, the values that divide the distribution into four groups of roughly equal size. The **first quartile** $Q_1$ lies one-quarter of the way up the list. The second quartile is the median, which is halfway up the list. The **third quartile** $Q_3$ lies three-quarters of the way up the list. The first and third quartiles mark out the middle half of the distribution.

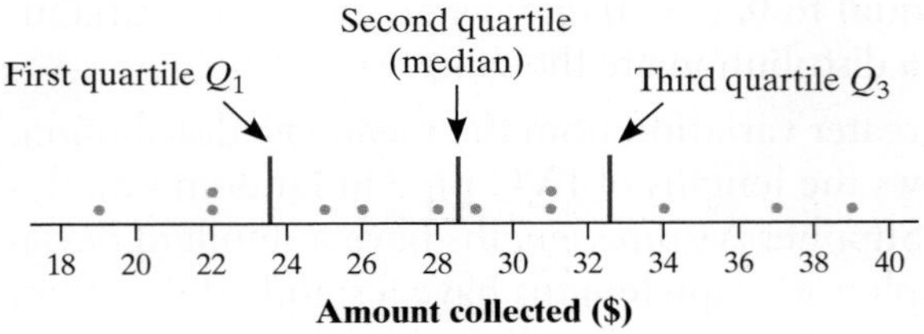


For example, here are the amounts collected each hour by a charity at a local store: \$19, \$26, \$25, \$37, \$31, \$28, \$22, \$22, \$29, \$34, \$39, and \$31. The dotplot displays the data. Because there are 12 data values, the quartiles divide the distribution into 4 groups of 3 values.

**DEFINITION Quartiles, First quartile $Q_1$, Third quartile $Q_3$**

The **quartiles** of a distribution divide the ordered data set into four groups having roughly the same number of values. To find the quartiles, arrange the data values from smallest to largest and find the median.

The **first quartile $Q_1$** is the median of the data values that are to the left of the median in the ordered list.

The **third quartile $Q_3$** is the median of the data values that are to the right of the median in the ordered list.

**The interquartile range** (*IQR*) measures the variability in the middle half of the distribution.

**DEFINITION Interquartile range (*IQR*)**

The **interquartile range** *(IQR)* is the distance between the first and third quartiles of a distribution. In symbols:

$$IQR = Q_3 - Q_1$$

Notice that the *IQR* is simply the range of the "middle half" of the distribution.

## EXAMPLE Boys and their shoes?
### Finding the *IQR*

**PROBLEM:** How many pairs of shoes does a typical teenage boy own? To find out, two AP® Statistics students surveyed a random sample of 20 male students from their large high school and recorded the number of pairs of shoes that each boy owned. Here are the data, along with a dotplot:

14 7 6 5 12 38 8 7 10 10 10 11 4 5 22 7 5 10 35 7

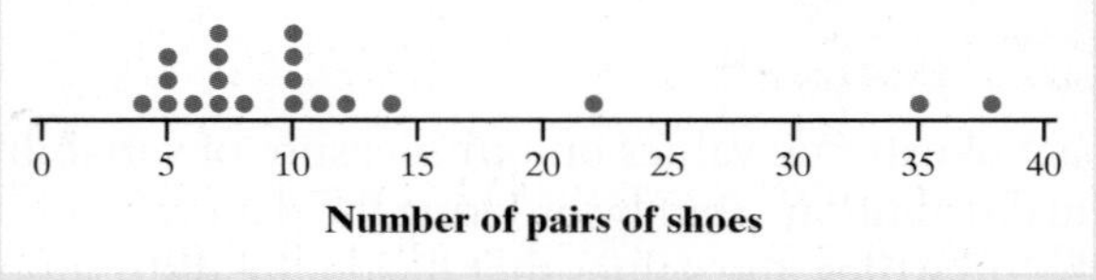

Peter Cade/Getty Images

Find the interquartile range.

**SOLUTION:**

4 5 5 5 6 7 7 7 7 (8 | 10) 10 10 10 11 12 14 22 35 38
Median = 9

Sort the data values from smallest to largest and find the median.

4 5 5 5 (6 | 7) 7 7 7 8 | 10 10 10 10 (11 | 12) 14 22 35 38

$Q_1 = 6.5$ Median $Q_3 = 11.5$

Find the first quartile $Q_1$ and the third quartile $Q_3$.

$IQR = 11.5 - 6.5 = 5$ pairs of shoes

$IQR = Q_3 - Q_1$

**FOR PRACTICE, TRY EXERCISE 105**

The quartiles and the interquartile range are *resistant* because they are not affected by a few extreme values. For the shoe data, $Q_3$ would still be 11.5 and the *IQR* would still be 5 if the maximum were 58 rather than 38.

Be sure to leave out the median when you locate the quartiles. In the preceding example, the median was not one of the data values. For the earlier close friends data set, we ignore the circled median of 3 when finding $Q_1$ and $Q_3$.

1 2 (2) 2 3 (3) 3 3 (4) 4 6

$Q_1$ Median $Q_3$

## CHECK YOUR UNDERSTANDING

Here are data on the highway fuel economy ratings for a sample of 25 model year 2018 Toyota 4Runners tested by the EPA, along with a dotplot:

22.4 22.4 22.3 23.3 22.3 22.3 22.5 22.4 22.1 21.5 22.0 22.2 22.7
22.8 22.4 22.6 22.9 22.5 22.1 22.4 22.2 22.9 22.6 21.9 22.4

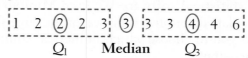

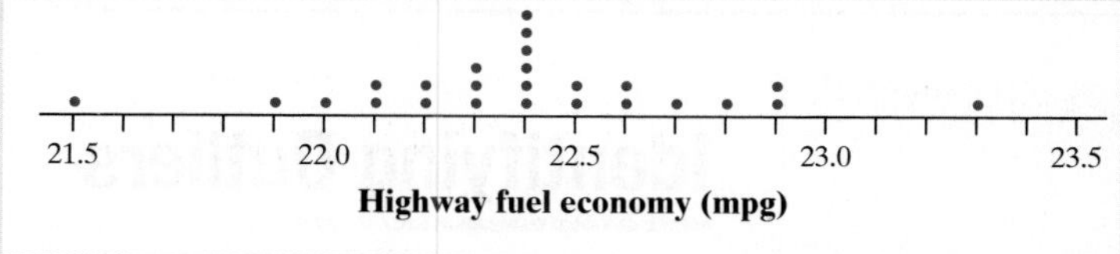


1. Find the range of the distribution.
2. The mean and standard deviation of the distribution are 22.404 mpg and 0.363 mpg, respectively. Interpret the standard deviation.
3. Find the interquartile range of the distribution.
4. Which measure of variability would you choose to describe the distribution? Explain.

# Numerical Summaries with Technology

Graphing calculators and computer software will calculate numerical summaries for you. Using technology to perform calculations will allow you to focus on choosing the right methods and interpreting your results.

## 3. Technology Corner COMPUTING NUMERICAL SUMMARIES

*TI-Nspire and other technology instructions are on the book's website at highschool.bfwpub.com/updatedtps6e.*

Let's find numerical summaries for the boys' shoes data from the example on page 64. We'll start by showing you how to compute summary statistics on the TI-83/84 and then look at output from computer software.

```
NORMAL FLOAT AUTO REAL RADIAN MP
1-Var Stats
x̄=11.65
Σx=233
Σx²=4401
Sx=9.421559822
σx=9.183000599
n=20
minX=4
↓Q1=6.5
```

```
NORMAL FLOAT AUTO REAL RADIAN MP
1-Var Stats
↑Sx=9.421559822
σx=9.183000599
n=20
minX=4
Q1=6.5
Med=9
Q3=11.5
maxX=38
```

I. **One-variable statistics on the TI-83/84**

1. Enter the data in list L1.
2. Find the summary statistics for the shoe data.
   - Press [STAT] (CALC); choose 1-Var Stats.
     **OS 2.55 or later:** In the dialog box, press [2nd] [1] (L1) and [ENTER] to specify L1 as the List. Leave FreqList blank. Arrow down to Calculate and press [ENTER].
     **Older OS:** Press [2nd] [1] (L1) and [ENTER].
   - Press [▼] to see the rest of the one-variable statistics.

II. **Output from statistical software** We used Minitab statistical software to calculate descriptive statistics for the boys' shoes data. Minitab allows you to choose which numerical summaries are included in the output.

**Descriptive Statistics: Shoes**

| Variable | N | Mean | StDev | Minimum | $Q_1$ | Median | $Q_3$ | Maximum |
|---|---|---|---|---|---|---|---|---|
| Shoes | 20 | 11.65 | 9.42 | 4.00 | 6.25 | 9.00 | 11.75 | 38.00 |

*Note:* The TI-83/84 gives the first and third quartiles of the boys' shoes distribution as $Q_1 = 6.5$ and $Q_3 = 11.5$. Minitab reports that $Q_1 = 6.25$ and $Q_3 = 11.75$. What happened? Minitab and some other software use slightly different rules for locating quartiles. Results from the various rules are usually close to each other. Be aware of possible differences when calculating quartiles as they may affect more than just the *IQR*.

# Identifying Outliers

Besides serving as a measure of variability, the interquartile range (*IQR*) is used as a "ruler" for identifying outliers.

There are other rules for determining outliers, such as "any value that is more than 2 (or 3) standard deviations from the mean." We always use the 1.5 × *IQR* rule in this book because it is based on statistics that are resistant to outliers, unlike the mean and standard deviation.

### HOW TO IDENTIFY OUTLIERS: THE 1.5 × *IQR* RULE

Call an observation an outlier if it falls more than $1.5 \times IQR$ above the third quartile or below the first quartile. That is,

$$\text{Low outliers} < Q_1 - 1.5 \times IQR \qquad \text{High outliers} > Q_3 + 1.5 \times IQR$$

Here are sorted data on the highway fuel economy ratings for a sample of 25 model year 2018 Toyota 4Runners tested by the EPA, along with a dotplot:

21.5 21.9 22.0 22.1 22.1 22.2 22.2 22.3 22.3 22.3 22.4 22.4 22.4
22.4 22.4 22.4 22.5 22.5 22.6 22.6 22.7 22.8 22.9 22.9 23.3

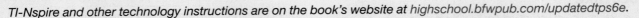

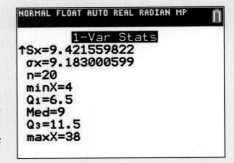

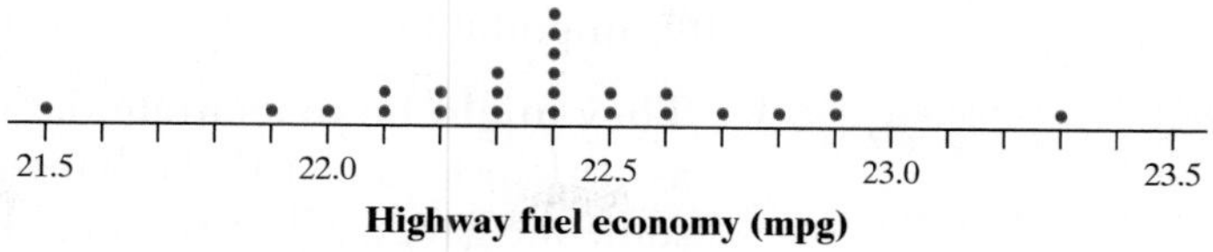


Does the $1.5 \times IQR$ rule identify any outliers in this distribution? If you did the preceding Check Your Understanding, you should have found that $Q_1 = 22.2$ mpg, $Q_3 = 22.6$ mpg, and $IQR = 0.4$ mpg. For these data,

$$\text{High outliers} > Q_3 + 1.5 \times IQR = 22.6 + 1.5 \times 0.4 = 23.2$$

and

$$\text{Low outliers} < Q_1 - 1.5 \times IQR = 22.2 - 1.5 \times 0.4 = 21.6$$

The cars with estimated highway fuel economy ratings of 21.5 and 23.3 are identified as outliers.

**AP® EXAM TIP**

You may be asked to determine whether a quantitative data set has any outliers. Be prepared to state and use the rule for identifying outliers.

## EXAMPLE

### How many goals?
### Identifying outliers

Icon Sports Wire/Getty Images

**PROBLEM:** Here are sorted data on the number of goals scored in 20 games played by the 2016 U.S women's soccer team, along with a dotplot:

1 1 1 1 1 1 2 2 2 2 2 3 3 3 4 5 5 5 9 10

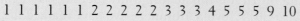

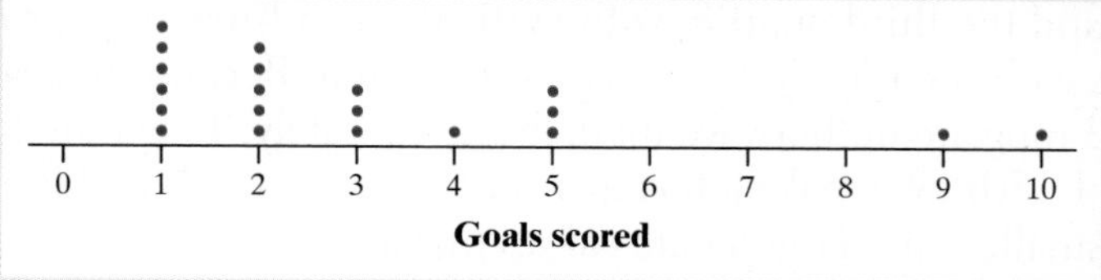


Identify any outliers in the distribution. Show your work.

**SOLUTION:**

1 1 1 1 (1 1) 2 2 2 (2 | 2) 3 3 3 (4 5) 5 5 9 10

$Q_1 = 1$ Median $= 2$ $Q_3 = 4.5$

$IQR = Q_3 - Q_1 = 4.5 - 1 = 3.5$

Low outliers $< 1 - 1.5 \times 3.5 = -4.25$

High outliers $> 4.5 + 1.5 \times 3.5 = 9.75$

There are no data values less than $-4.25$, but the game in which the team scored 10 goals is an outlier.

Low outliers $< Q_1 - 1.5 \times IQR$
High outliers $> Q_3 + 1.5 \times IQR$

The game in which the team scored 9 goals is not identified as an outlier by the $1.5 \times IQR$ rule.

**FOR PRACTICE, TRY EXERCISE 107**

It is important to identify outliers in a distribution for several reasons:

1. **They might be inaccurate data values.** Maybe someone recorded a value as 10.1 instead of 101. Perhaps a measuring device broke down. Or maybe someone gave a silly response, like the student in a class survey who claimed to study 30,000 minutes per night! Try to correct errors like these if possible. If you can't, give summary statistics with and without the outlier.
2. **They can indicate a remarkable occurrence.** For example, in a graph of net worth, Bill Gates is likely to be an outlier.
3. **They can heavily influence the values of some summary statistics,** like the mean, range, and standard deviation.

## Making and Interpreting Boxplots

You can use a dotplot, stemplot, or histogram to display the distribution of a quantitative variable. Another graphical option for quantitative data is a **boxplot**. A boxplot summarizes a distribution by displaying the location of 5 important values within the distribution, known as its **five-number summary**.

A boxplot is sometimes called a *box-and-whisker* plot.

**DEFINITION Five-number summary, Boxplot**

The **five-number summary** of a distribution of quantitative data consists of the minimum, the first quartile $Q_1$, the median, the third quartile $Q_3$, and the maximum.

A **boxplot** is a visual representation of the five-number summary.

Figure 1.15 illustrates the process of making a boxplot. The dotplot in Figure 1.15(a) shows the data on EPA estimated highway fuel economy ratings for a sample of 25 model year 2018 Toyota 4Runners. We have marked the first quartile, the median, and the third quartile with vertical green lines. The process of testing for outliers with the $1.5 \times IQR$ rule is shown in red. Because the values of 21.5 mpg and 23.3 mpg are outliers, we mark these separately. To get the finished boxplot in Figure 1.15(b), we make a box spanning from $Q_1$ to $Q_3$ and then draw "whiskers" to the smallest and largest data values that are not outliers

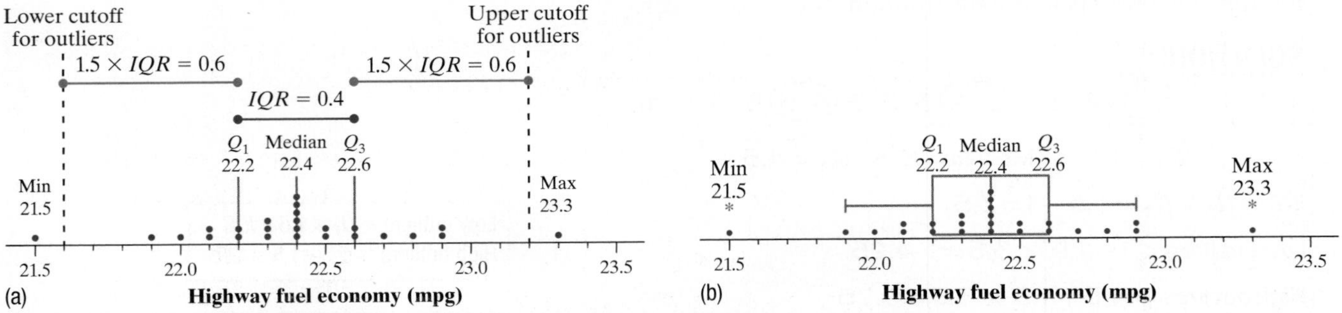


**FIGURE 1.15** A visual illustration of how to make a boxplot for the Toyota 4Runner highway gas mileage data. (a) Dotplot of the data with the five-number summary and $1.5 \times IQR$ marked. (b) Boxplot of the data with outliers identified (*).

As you can see, it is fairly easy to make a boxplot by hand for small sets of data. Here's a summary of the steps.

### HOW TO MAKE A BOXPLOT

- **Find the five-number summary** for the distribution.
- **Identify outliers** using the $1.5 \times IQR$ rule.
- **Draw and label the axis.** Draw a horizontal axis and put the name of the quantitative variable underneath, including units if applicable.
- **Scale the axis.** Look at the minimum and maximum values in the data set. Start the horizontal axis at a convenient number equal to or below the minimum and place tick marks at equal intervals until you equal or exceed the maximum.
- **Draw a box** that spans from the first quartile ($Q_1$) to the third quartile ($Q_3$).
- **Mark the median** with a vertical line segment that's the same height as the box.
- **Draw whiskers**—lines that extend from the ends of the box to the smallest and largest data values that are *not* outliers. Mark any outliers with a special symbol such as an asterisk (*).

We see from the boxplot in Figure 1.15 that the distribution of highway gas mileage ratings for this sample of model year 2018 Toyota 4Runners is roughly symmetric with one high outlier and one low outlier.

## EXAMPLE

### Picking pumpkins
### Making and interpreting boxplots

cscredon/Getty Images

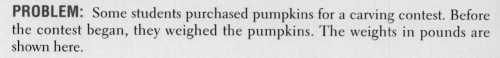

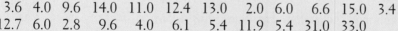

**PROBLEM:** Some students purchased pumpkins for a carving contest. Before the contest began, they weighed the pumpkins. The weights in pounds are shown here.

| | | | | | | | | | | | |
|---|---|---|---|---|---|---|---|---|---|---|---|
| 3.6 | 4.0 | 9.6 | 14.0 | 11.0 | 12.4 | 13.0 | 2.0 | 6.0 | 6.6 | 15.0 | 3.4 |
| 12.7 | 6.0 | 2.8 | 9.6 | 4.0 | 6.1 | 5.4 | 11.9 | 5.4 | 31.0 | 33.0 | |

(a) Make a boxplot of the data.

(b) Explain why the median and *IQR* would be a better choice for summarizing the center and variability of the distribution of pumpkin weights than the mean and standard deviation.

**SOLUTION:**

(a)

Min $\qquad$ $Q_1$ $\qquad$ Median

2.0 2.8 3.4 3.6 4.0 (4.0) 5.4 5.4 6.0 6.0 6.1 (6.6)

9.6 9.6 11.0 11.9 12.4 (12.7) 13.0 14.0 15.0 31.0 33.0

$Q_3$ $\qquad$ Max

$IQR = Q_3 - Q_1 = 12.7 - 4.0 = 8.7$

Low outliers $< Q_1 - 1.5 \times IQR = 4.0 - 1.5 \times 8.7 = -9.05$

High outliers $> Q_3 + 1.5 \times IQR = 12.7 + 1.5 \times 8.7 = 25.75$

The pumpkins that weighed 31.0 and 33.0 pounds are outliers.

To make the boxplot:
- **Find the five-number summary.**
- **Identify outliers.**
- **Draw and label the axis.**
- **Scale the axis.**
- **Draw a box.**
- **Mark the median.**
- **Draw whiskers** to the smallest and largest data values that are *not* outliers. Mark outliers with an asterisk.

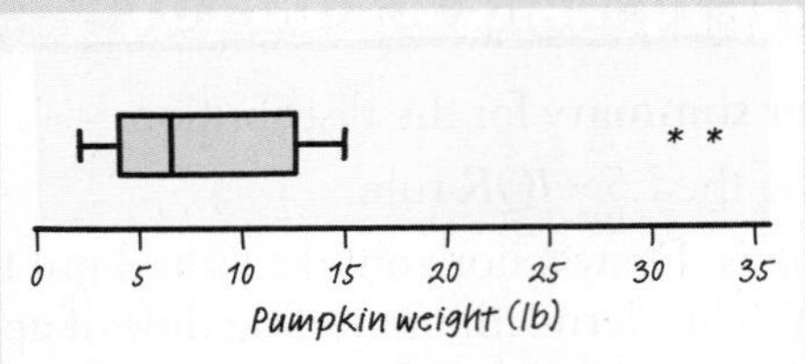


(b) The distribution of pumpkin weights is skewed to the right with two high outliers. Because the mean and standard deviation are sensitive to outliers, it would be better to use the median and *IQR*, which are resistant.

We know the distribution is skewed to the right because the left half of the distribution varies from 2.0 to 6.6 pounds, while the right half of the distribution (excluding outliers) varies from 6.6 to 15.0 pounds.

**FOR PRACTICE, TRY EXERCISE 111**

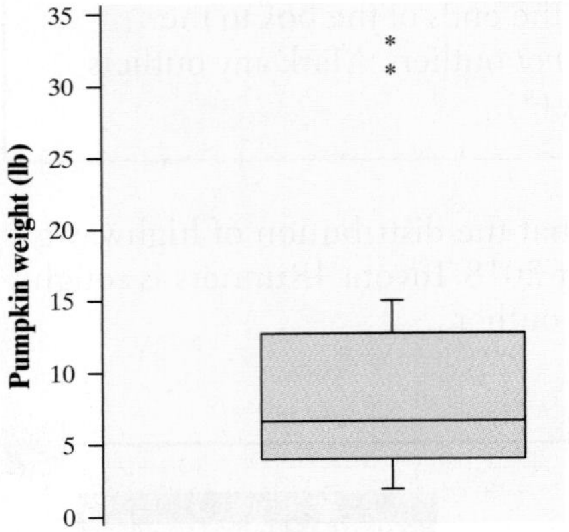


Boxplots provide a quick summary of the center and variability of a distribution. The median is displayed as a line in the central box, the interquartile range is the length of the box, and the range is the length of the entire plot, including outliers. Note that some statistical software orients boxplots vertically. At left is a vertical boxplot of the pumpkin weight data from the preceding example. You can see that the graph is skewed toward the larger values.

Boxplots do not display each individual value in a distribution. And **boxplots don't show gaps, clusters, or peaks**. For instance, the dotplot below left displays the duration, in minutes, of 220 eruptions of the Old Faithful geyser. The distribution of eruption durations is clearly double-peaked (*bimodal*). But a boxplot of the data hides this important information about the shape of the distribution.

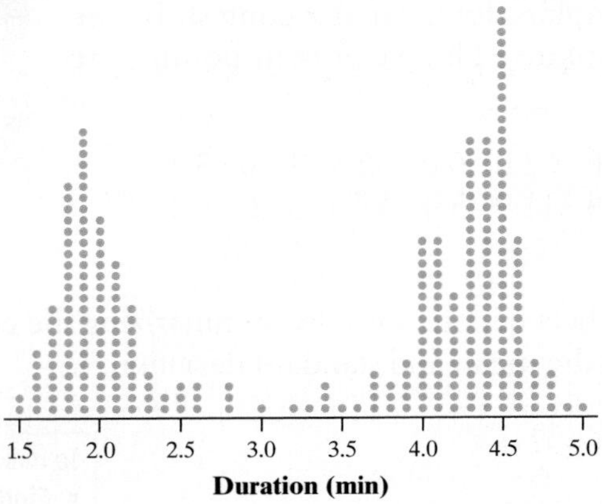


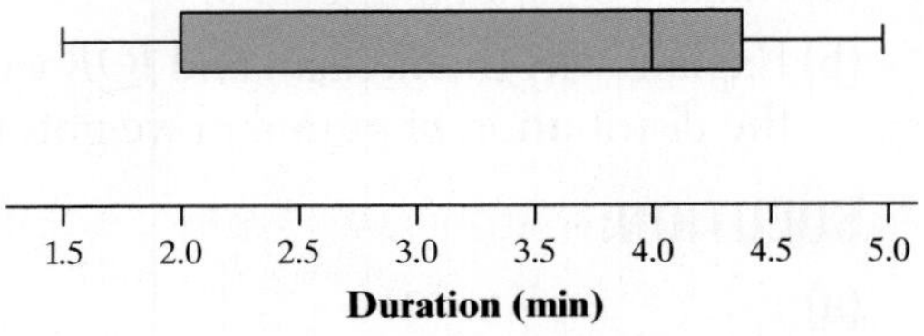

## CHECK YOUR UNDERSTANDING

Ryan and Brent were curious about the amount of french fries they would get in a large order from their favorite fast-food restaurant, Burger King. They went to several different Burger King locations over a series of days and ordered a total of 14 large fries. The weight of each order (in grams) is as follows:

165 163 160 159 166 152 166 168 173 171 168 167 170 170

1. Make a boxplot to display the data.
2. According to a nutrition website, Burger King's large fries weigh 160 grams, on average. Ryan and Brent suspect that their local Burger King restaurants may be skimping on fries. Does the boxplot in Question 1 support their suspicion? Explain why or why not.

## Comparing Distributions with Boxplots

Boxplots are especially effective for comparing the distribution of a quantitative variable in two or more groups, as seen in the following example.

**EXAMPLE**

### Which company makes better tablets?
### Comparing distributions with boxplots

**PROBLEM:** In a recent year, *Consumer Reports* rated many tablet computers for performance and quality. Based on several variables, the magazine gave each tablet an overall rating, where higher scores indicate better ratings. The overall ratings of the tablets produced by Apple and Samsung are given here, along with parallel boxplots and numerical summaries of the data.[39]

Peter Cade/Getty Images

| | | | | | | | | | | | | |
|---|---|---|---|---|---|---|---|---|---|---|---|---|
| **Apple** | 87 | 87 | 87 | 87 | 86 | 86 | 86 | 86 | 84 | 84 | | |
| | 84 | 84 | 83 | 83 | 83 | 83 | 81 | 79 | 76 | 73 | | |
| **Samsung** | 88 | 87 | 87 | 86 | 86 | 86 | 86 | 86 | 84 | 84 | 83 | 83 |
| | 77 | 76 | 76 | 75 | 75 | 75 | 75 | 75 | 74 | 71 | 62 | |

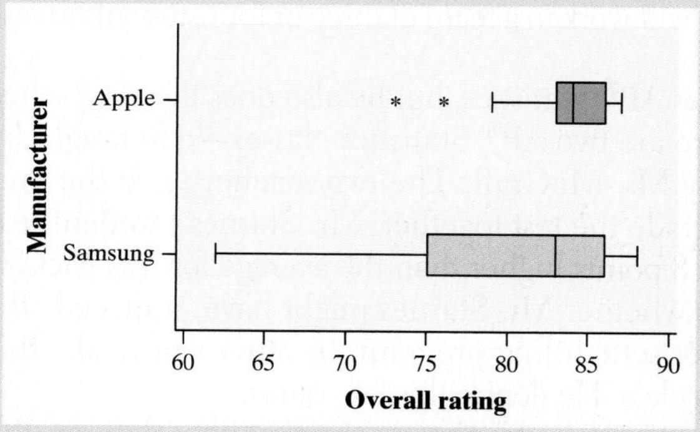


| | $\bar{x}$ | $s_x$ | Min | $Q_1$ | Median | $Q_3$ | Max | *IQR* |
|---|---|---|---|---|---|---|---|---|
| Apple | 83.45 | 3.762 | 73 | 83 | 84 | 86 | 87 | 3 |
| Samsung | 79.87 | 6.74 | 62 | 75 | 83 | 86 | 88 | 11 |

Compare the distributions of overall rating for Apple and Samsung.

**SOLUTION:**

*Shape:* Both distributions of overall ratings are skewed to the left.

*Outliers:* There are two low outliers in the Apple tablet distribution: overall ratings of 73 and 76. The Samsung tablet distribution has no outliers.

*Center:* The Apple tablets had a slightly higher median overall rating (84) than the Samsung tablets (83). More importantly, about 75% of the Apple tablets had overall ratings that were greater than or equal to the median for the Samsung tablets.

*Variability:* There is much more variation in overall rating among the Samsung tablets than the Apple tablets. The *IQR* for Samsung tablets (11) is almost four times larger than the *IQR* for Apple tablets (3).

Remember to compare shape, outliers, center, and variability!

Because of the strong skewness and outliers, use the median and *IQR* instead of the mean and standard deviation when comparing center and variability.

FOR PRACTICE, TRY EXERCISE 115

**AP® EXAM TIP**

Use statistical terms carefully and correctly on the AP® Statistics exam. Don't say "mean" if you really mean "median." Range is a single number; so are $Q_1$, $Q_3$, and *IQR*. Avoid poor use of language, like "the outlier *skews* the mean" or "the median is in the middle of the *IQR*." Skewed is a shape and the *IQR* is a single number, not a region. If you misuse a term, expect to lose some credit.

Here's an activity that gives you a chance to put into practice what you have learned in this section.

## ACTIVITY Team challenge: Did Mr. Starnes stack his class?

In this activity, you will work in a team of three or four students to resolve a dispute.

Mr. Starnes teaches AP® Statistics, but he also does the class scheduling for the high school. There are two AP® Statistics classes—one taught by Mr. Starnes and one taught by Ms. McGrail. The two teachers give the same first test to their classes and grade the test together. Mr. Starnes's students earned an average score that was 8 points higher than the average for Ms. McGrail's class. Ms. McGrail wonders whether Mr. Starnes might have "adjusted" the class rosters from the computer scheduling program. In other words, she thinks he might have "stacked" his class. He denies this, of course.

To help resolve the dispute, the teachers collect data on the cumulative grade point averages and SAT Math scores of their students. Mr. Starnes provides the GPA data from his computer. The students report their SAT Math scores. The following table shows the data for each student in the two classes.

Did Mr. Starnes stack his class? Give appropriate graphical and numerical evidence to support your conclusion. Be prepared to defend your answer.

| Student | Teacher | GPA | SAT-M |
|---|---|---|---|
| 1 | Starnes | 2.900 | 670 |
| 2 | Starnes | 2.860 | 520 |
| 3 | Starnes | 2.600 | 570 |
| 4 | Starnes | 3.600 | 710 |
| 5 | Starnes | 3.200 | 600 |
| 6 | Starnes | 2.700 | 590 |
| 7 | Starnes | 3.100 | 640 |
| 8 | Starnes | 3.085 | 570 |
| 9 | Starnes | 3.750 | 710 |
| 10 | Starnes | 3.400 | 630 |
| 11 | Starnes | 3.338 | 630 |
| 12 | Starnes | 3.560 | 670 |
| 13 | Starnes | 3.800 | 650 |
| 14 | Starnes | 3.200 | 660 |
| 15 | Starnes | 3.100 | 510 |
| 16 | McGrail | 2.900 | 620 |
| 17 | McGrail | 3.300 | 590 |
| 18 | McGrail | 3.980 | 650 |
| 19 | McGrail | 2.900 | 600 |
| 20 | McGrail | 3.200 | 620 |
| 21 | McGrail | 3.500 | 680 |
| 22 | McGrail | 2.800 | 500 |
| 23 | McGrail | 2.900 | 502.5 |
| 24 | McGrail | 3.950 | 640 |
| 25 | McGrail | 3.100 | 630 |
| 26 | McGrail | 2.850 | 580 |
| 27 | McGrail | 2.900 | 590 |
| 28 | McGrail | 3.245 | 600 |
| 29 | McGrail | 3.000 | 600 |
| 30 | McGrail | 3.000 | 620 |
| 31 | McGrail | 2.800 | 580 |
| 32 | McGrail | 2.900 | 600 |
| 33 | McGrail | 3.200 | 600 |

You can use technology to make boxplots, as the following Technology Corner illustrates.

## 4. Technology Corner MAKING BOXPLOTS

*TI-Nspire and other technology instructions are on the book's website at highschool.bfwpub.com/updatedtps6e.*

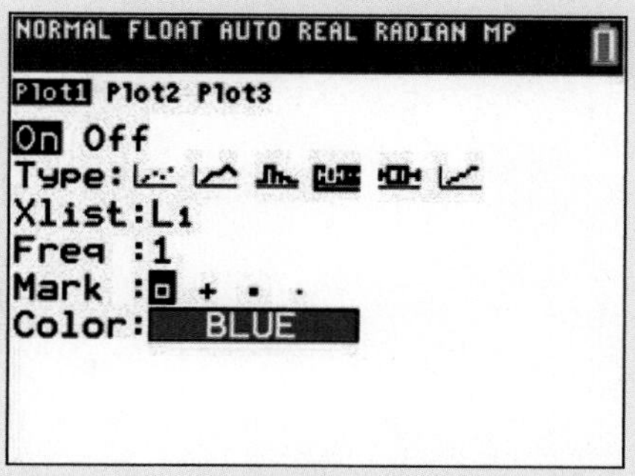


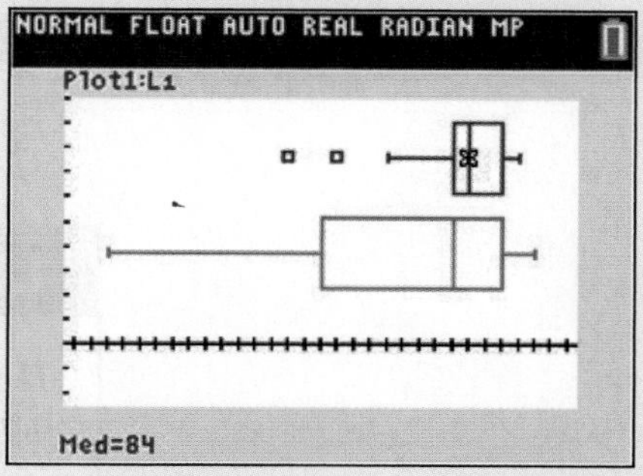


The TI-83/84 can plot up to three boxplots in the same viewing window. Let's use the calculator to make parallel boxplots of the overall rating data for Apple and Samsung tablets.

1. Enter the ratings for Apple tablets in list L1 and for Samsung in list L2.
2. Set up two statistics plots: Plot1 to show a boxplot of the Apple data in list L1 and Plot2 to show a boxplot of the Samsung data in list L2. The setup for Plot1 is shown. When you define Plot2, be sure to change L1 to L2.

*Note:* The calculator offers two types of boxplots: one that shows outliers and one that doesn't. We'll always use the type that identifies outliers.

3. Use the calculator's Zoom feature to display the parallel boxplots. Then Trace to view the five-number summary.
   - Press ZOOM and select ZoomStat.
   - Press TRACE.

## Section 1.3 Summary

- A numerical summary of a distribution should include measures of **center** and **variability.**
- The **mean** and the **median** describe the center of a distribution in different ways. The mean is the average of the observations. In symbols, the sample mean $\bar{x} = \frac{\sum x_i}{n}$. The median is the midpoint of the distribution, the number such that about half the observations are smaller and half are larger.
- A **statistic** is a number that describes a sample. A **parameter** is a number that describes a population. We often use statistics (like the sample mean $\bar{x}$) to estimate parameters (like the population mean $\mu$).
- The simplest measure of variability for a distribution of quantitative data is the **range,** which is the distance from the maximum value to the minimum value.
- When you use the mean to describe the center of a distribution, use the **standard deviation** to describe the distribution's variability. The standard deviation gives the typical distance of the values in a distribution from the mean. In symbols, the sample standard deviation $s_x = \sqrt{\frac{\sum(x_i - \bar{x})^2}{n-1}}$. The standard deviation $s_x$ is 0 when there is no variability and gets larger as variability from the mean increases.
- The **variance** is the average of the squared deviations from the mean. The sample variance $s_x^2$ is the square of the sample standard deviation.
- When you use the median to describe the center of a distribution, use the **interquartile range** (*IQR*) to describe the distribution's variability. The **first quartile** $Q_1$ has about one-fourth of the observations below it, and the **third quartile** $Q_3$ has about three-fourths of the observations below it. The interquartile range measures variability in the middle half of the distribution and is found using $IQR = Q_3 - Q_1$.
- The median is a **resistant** measure of center because it is relatively unaffected by extreme observations. The mean is not resistant. Among measures of variability, the *IQR* is resistant, but the standard deviation and range are not.
- According to the **1.5 × *IQR* rule,** an observation is an outlier if it is less than $Q_1 - 1.5 \times IQR$ or greater than $Q_3 + 1.5 \times IQR$.
- The **five-number summary** of a distribution consists of the minimum, $Q_1$, the median, $Q_3$, and the maximum. A **boxplot** displays the five-number summary, marking outliers with a special symbol. The box shows the variability in the middle half of the distribution. Whiskers extend from the box to the smallest and the largest observations that are not outliers. Boxplots are especially useful for comparing distributions.

### 1.3 Technology Corners

*TI-Nspire and other technology instructions are on the book's website at highschool.bfwpub.com/updatedtps6e.*

**3. Computing numerical summaries** — **Page 66**

**4. Making boxplots** — **Page 73**

# Section 1.3 Exercises

**87. Quiz grades** Joey's first 14 quiz grades in a marking period were as follows:
pg 55

| | | | | | | |
|---|---|---|---|---|---|---|
| 86 | 84 | 91 | 75 | 78 | 80 | 74 |
| 87 | 76 | 96 | 82 | 90 | 98 | 93 |

(a) Calculate the mean. Show your work.

(b) Suppose Joey has an unexcused absence for the 15th quiz, and he receives a score of 0. Recalculate the mean. What property of the mean does this illustrate?

**88. Pulse rates** Here are data on the resting pulse rates (in beats per minute) of 19 middle school students:

| | | | | | | | | | |
|---|---|---|---|---|---|---|---|---|---|
| 71 | 104 | 76 | 88 | 78 | 71 | 68 | 86 | 70 | 90 |
| 74 | 76 | 69 | 68 | 88 | 96 | 68 | 82 | 120 | |

(a) Calculate the mean. Show your work.

(b) The student with a 120 pulse rate has a medical issue. Find the mean pulse rate for the other 18 students. What property of the mean does this illustrate?

**89. Quiz grades** Refer to Exercise 87.
pg 57

(a) Find the median of Joey's first 14 quiz grades.

(b) Find the median of Joey's quiz grades after his unexcused absence. Explain why the 0 quiz grade does not have much effect on the median.

**90. Pulse rates** Refer to Exercise 88.

(a) Find the median pulse rate for all 19 students.

(b) Find the median pulse rate excluding the student with the medical issue. Explain why this student's 120 pulse rate does not have much effect on the median.

**91. Electing the president** To become president of the United States, a candidate does not have to receive a majority of the popular vote. The candidate does have to win a majority of the 538 Electoral College votes. Here is a stemplot of the number of electoral votes in 2016 for each of the 50 states and the District of Columbia:

```
0 | 3333333344444
0 | 55566666677788999
1 | 00001111234
1 | 5668
2 | 00
2 | 99
3 |
3 | 8
4 |
4 |
5 |
5 | 5
```

Key: 1|5 is a state with 15 electoral votes.

(a) Find the median.

(b) Without doing any calculations, explain how the mean and median compare.

(c) Is the value you found in part (a) a statistic or a parameter? Justify your answer.

**92. Birthrates in Africa** One of the important factors in determining population growth rates is the birthrate per 1000 individuals in a population. The dotplot shows the birthrates per 1000 individuals (rounded to the nearest whole number) for all 54 African nations.

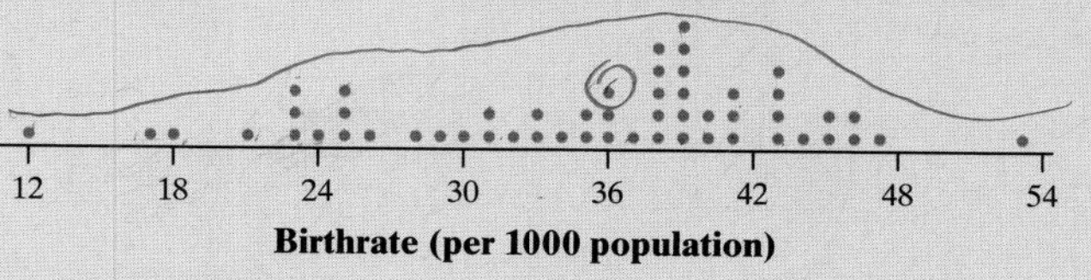


(a) Find the median.

(b) Without doing any calculations, explain how the mean and median compare.

(c) Is the value you found in part (a) a statistic or a parameter? Justify your answer.

**93. House prices** The mean and median selling prices of existing single-family homes sold in September 2016 were \$276,200 and \$234,200.[40] Which of these numbers is the mean and which is the median? Explain your reasoning.

**94. Mean salary?** Last year a small accounting firm paid each of its five clerks \$32,000, two junior accountants \$60,000 each, and the firm's owner \$280,000.

(a) What is the mean salary paid at this firm? How many of the employees earn less than the mean? What is the median salary?

(b) Write a sentence to describe how an unethical recruiter could use statistics to mislead prospective employees.

**95. Do adolescent girls eat fruit?** We all know that fruit is good for us. Here is a histogram of the number of servings of fruit per day claimed by 74 seventeen-year-old girls in a study in Pennsylvania:[41]

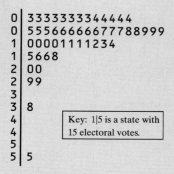

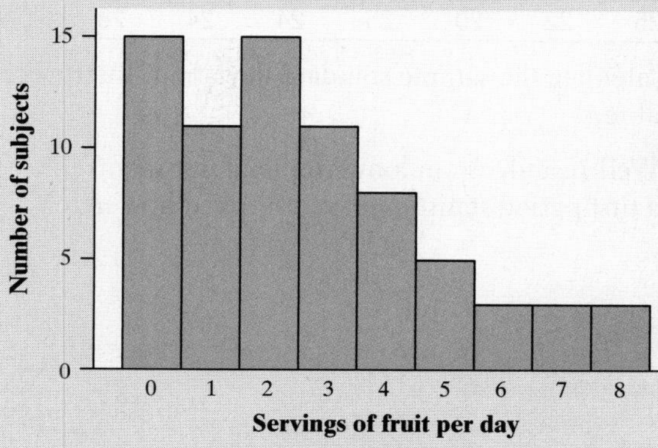

(a) Find the median number of servings of fruit per day from the histogram. Explain your method clearly.

(b) Calculate the mean of the distribution. Show your work.

96. **Shakespeare** The histogram shows the distribution of lengths of words used in Shakespeare's plays.[42]

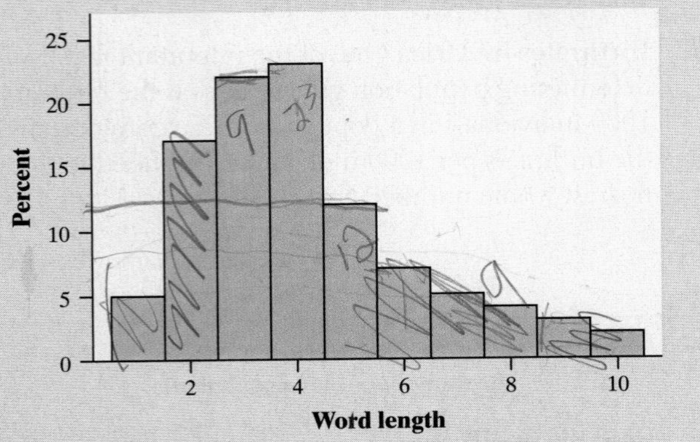


(a) Find the median word length in Shakespeare's plays from the histogram. Explain your method clearly.

(b) Calculate the mean of the distribution. Show your work.

97. **Quiz grades** Refer to Exercise 87.

(a) Find the range of Joey's first 14 quiz grades and the range of Joey's quiz grades after his unexcused absence.

(b) Explain what part (a) suggests about using the range as a measure of variability for a distribution of quantitative data.

98. **Pulse rates** Refer to Exercise 88.

(a) Find the range of the pulse rates for all 19 students and the range of the pulse rates excluding the student with the medical issue.

(b) Explain what part (a) suggests about using the range as a measure of variability for a distribution of quantitative data.

99. **Foot lengths** Here are the foot lengths (in centimeters) for a random sample of seven 14-year-olds from the United Kingdom:

pg 61

| | | | | | | |
|---|---|---|---|---|---|---|
| 25 | 22 | 20 | 25 | 24 | 24 | 28 |

Calculate the sample standard deviation. Interpret this value.

100. **Well rested?** A random sample of 6 students in a first-period statistics class was asked how much sleep (to the nearest hour) they got last night. Their responses were 6, 7, 7, 8, 10, and 10. Calculate the sample standard deviation. Interpret this value.

101. **File sizes** How much storage space does your music use? Here is a dotplot of the file sizes (to the nearest tenth of a megabyte) for 18 randomly selected files on Nathaniel's mp3 player:

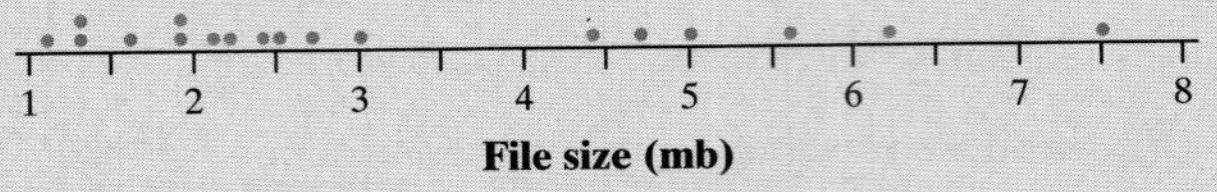


(a) The distribution of file size has a mean of $\bar{x} = 3.2$ megabytes and a standard deviation of $s_x = 1.9$ megabytes. Interpret the standard deviation.

(b) Suppose the music file that takes up 7.5 megabytes of storage space is replaced with another version of the file that only takes up 4 megabytes. How would this affect the mean and the standard deviation? Justify your answer.

102. **Healthy fast food?** Here is a dotplot of the amount of fat (to the nearest gram) in 12 different hamburgers served at a fast-food restaurant:

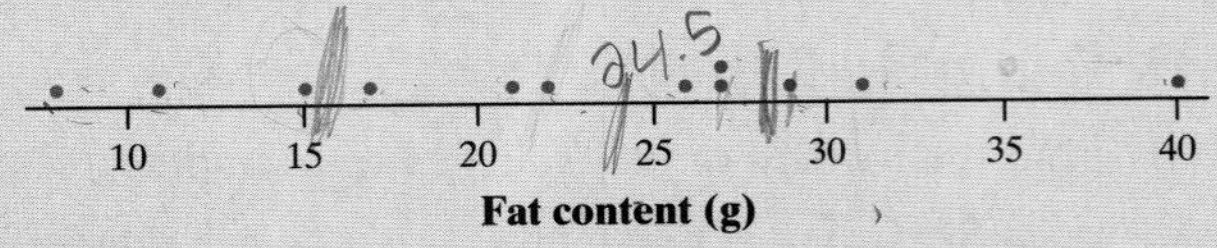


(a) The distribution of fat content has a mean of $\bar{x} = 22.83$ grams and a standard deviation of $s_x = 9.06$ grams. Interpret the standard deviation.

(b) Suppose the restaurant replaces the burger that has 22 grams of fat with a new burger that has 35 grams of fat. How would this affect the mean and the standard deviation? Justify your answer.

103. **Comparing SD** Which of the following distributions has a smaller standard deviation? Justify your answer.

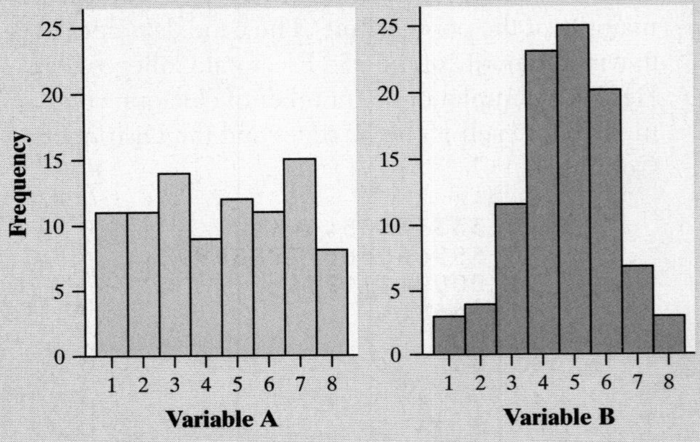

**104. Comparing SD** The parallel dotplots show the lengths (in millimeters) of a sample of 11 nails produced by each of two machines. Which distribution has the larger standard deviation? Justify your answer.

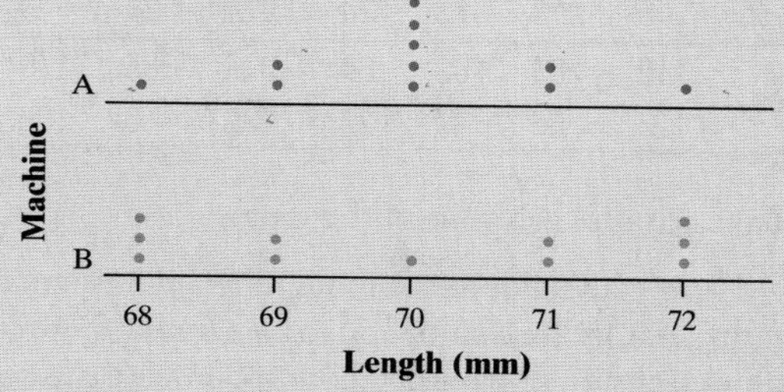


**105. File sizes** Refer to Exercise 101. Find the interquartile range of the file size distribution shown in the dotplot.
pg 64

**106. Healthy fast food?** Refer to Exercise 102. Find the interquartile range of the fat content distribution shown in the dotplot.

**107. File sizes** Refer to Exercises 101 and 105. Identify any outliers in the distribution. Show your work.
pg 67

**108. Healthy fast food?** Refer to Exercises 102 and 106. Identify any outliers in the distribution. Show your work.

**109. Shopping spree** The figure displays computer output for data on the amount spent by 50 grocery shoppers.

| | $\bar{x}$ | $s_x$ | Min | $Q_1$ | Med | $Q_3$ | Max |
|---|---|---|---|---|---|---|---|
| Amount spent | 34.70 | 21.70 | 3.11 | 19.27 | 27.86 | 45.40 | 93.34 |

(a) What would you guess is the shape of the distribution based only on the computer output? Explain.

(b) Interpret the value of the standard deviation.

(c) Are there any outliers? Justify your answer.

**110. C-sections** A study in Switzerland examined the number of cesarean sections (surgical deliveries of babies) performed in a year by samples of male and female doctors. Here are summary statistics for the two distributions:

| | $\bar{x}$ | $s_x$ | Min | $Q_1$ | Med | $Q_3$ | Max |
|---|---|---|---|---|---|---|---|
| Male doctors | 41.333 | 20.607 | 20 | 27 | 34 | 50 | 86 |
| Female doctors | 19.1 | 10.126 | 5 | 10 | 18.5 | 29 | 33 |

(a) Based on the computer output, which distribution would you guess has a more symmetrical shape? Explain your answer.

(b) Explain how the *IQR*s of these two distributions can be so similar even though the standard deviations are quite different.

(c) Does either distribution have any outliers? Justify your answer.

**111. Don't call me** According to a study by Nielsen Mobile, "Teenagers ages 13 to 17 are by far the most prolific texters, sending 1742 messages a month." Mr. Williams, a high school statistics teacher, was skeptical about the claims in the article. So he collected data from his first-period statistics class on the number of text messages they had sent in the past 24 hours. Here are the data:
pg 69

| | | | | | | | | | | | | |
|---|---|---|---|---|---|---|---|---|---|---|---|---|
| 0 | 7 | 1 | 29 | 25 | 8 | 5 | 1 | 25 | 98 | 9 | 0 | 26 |
| 8 | 118 | 72 | 0 | 92 | 52 | 14 | 3 | 3 | 44 | 5 | 42 | |

(a) Make a boxplot of these data.

(b) Use the boxplot you created in part (a) to explain how these data seem to contradict the claim in the article.

**112. Acing the first test** Here are the scores of Mrs. Liao's students on their first statistics test:

| | | | | | | | | | | |
|---|---|---|---|---|---|---|---|---|---|---|
| 93 | 93 | 87.5 | 91 | 94.5 | 72 | 96 | 95 | 93.5 | 93.5 | 73 |
| 82 | 45 | 88 | 80 | 86 | 85.5 | 87.5 | 81 | 78 | 86 | 89 |
| 92 | 91 | 98 | 85 | 82.5 | 88 | 94.5 | 43 | | | |

(a) Make a boxplot of these data.

(b) Use the boxplot you created in part (a) to describe how the students did on Mrs. Liao's first test.

**113. Electing the president** Refer to Exercise 91. Here are a boxplot and some numerical summaries of the electoral vote data:

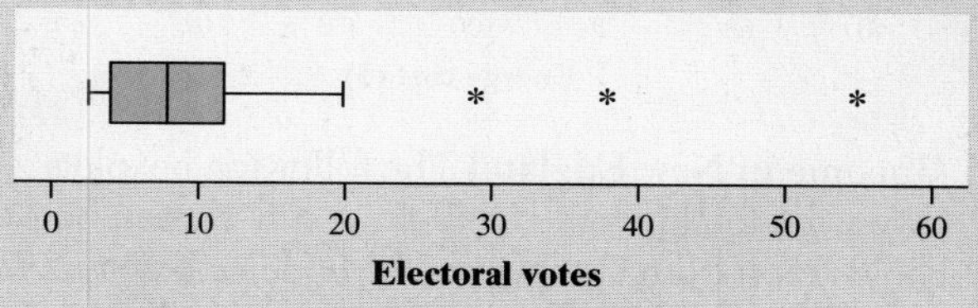


| Variable | $N$ | Mean | SD | Min | $Q_1$ | Median | $Q_3$ | Max |
|---|---|---|---|---|---|---|---|---|
| Electoral votes | 51 | 10.55 | 9.69 | 3 | 4 | 8 | 12 | 55 |

(a) Explain why the median and *IQR* would be a better choice for summarizing the center and variability of the distribution of electoral votes than the mean and standard deviation.

(b) Identify an aspect of the distribution that the stemplot in Exercise 91 reveals that the boxplot does not.

**114. Birthrates in Africa** Refer to Exercise 92. Here are a boxplot and some numerical summaries of the birthrate data:

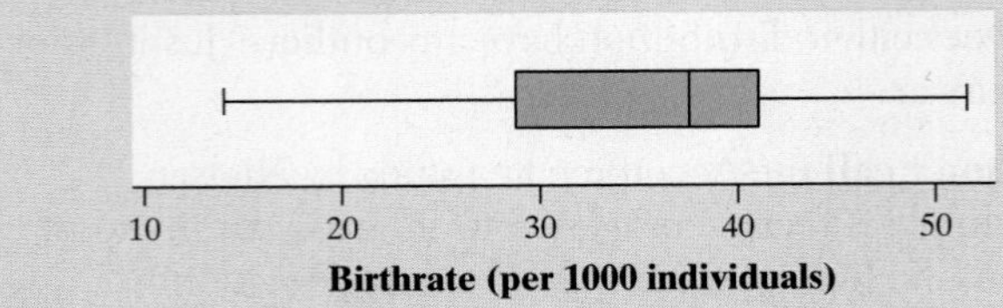


| Variable | $N$ | Mean | SD | Min | $Q_1$ | Median | $Q_3$ | Max |
|---|---|---|---|---|---|---|---|---|
| Birthrate | 54 | 34.91 | 8.57 | 14.00 | 29.00 | 37.50 | 41.00 | 53.00 |

(a) Explain why the median and *IQR* would be a better choice for summarizing the center and variability of the distribution of birthrates in African countries than the mean and standard deviation.

(b) Identify an aspect of the distribution that the dotplot in Exercise 92 reveals that the boxplot does not.

**115. Energetic refrigerators** *Consumer Reports* magazine rated different types of refrigerators, including those with bottom freezers, those with top freezers, and those with side freezers. One of the variables they measured was annual energy cost (in dollars). The following boxplots show the energy cost distributions for each of these types. Compare the energy cost distributions for the three types of refrigerators.

pg 71

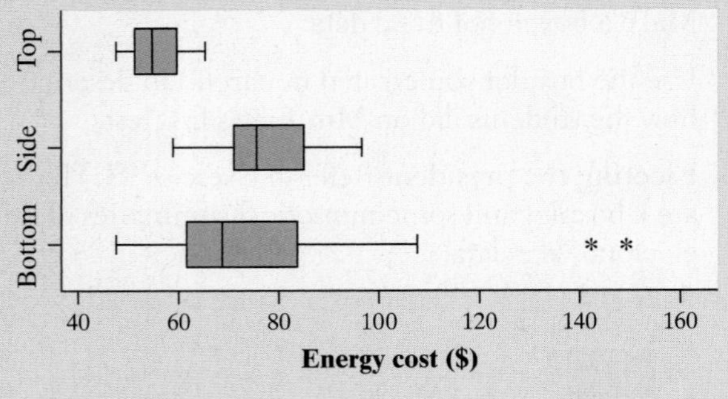


**116. Income in New England** The following boxplots show the total income of 40 randomly chosen households each from Connecticut, Maine, and Massachusetts, based on U.S. Census data from the American Community Survey. Compare the distributions of annual incomes in the three states.

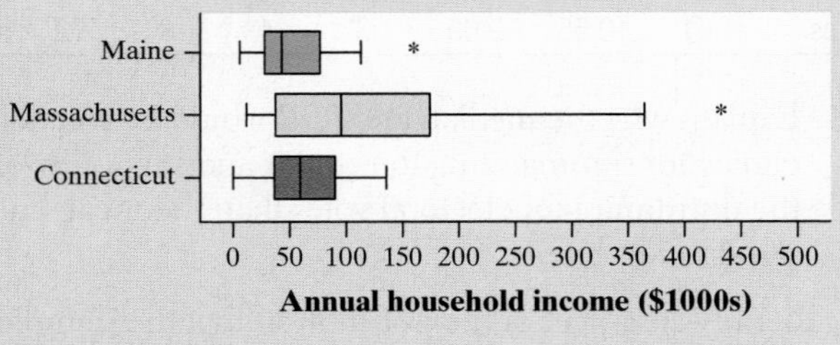


**117. Who texts more?** For their final project, a group of AP® Statistics students wanted to compare the texting habits of males and females. They asked a random sample of students from their school to record the number of text messages sent and received over a two-day period. Here are their data:

| | | | | | | | | |
|---|---|---|---|---|---|---|---|---|
| **Males** | 127 | 44 | 28 | 83 | 0 | 6 | 78 | 6 |
| | 5 | 213 | 73 | 20 | 214 | 28 | 11 | |
| **Females** | 112 | 203 | 102 | 54 | 379 | 305 | 179 | 24 |
| | 127 | 65 | 41 | 27 | 298 | 6 | 130 | 0 |

(a) Make parallel boxplots of the data.

(b) Use your calculator to compute separate numerical summaries for the males and for the females. Are these values statistics or parameters? Explain your answer.

(c) Do these data suggest that males and females at the school differ in their texting habits? Use the results from parts (a) and (b) to support your answer.

**118. SSHA scores** Here are the scores on the Survey of Study Habits and Attitudes (SSHA) for a random sample of 18 first-year college women:

| | | | | | | | | |
|---|---|---|---|---|---|---|---|---|
| 154 | 109 | 137 | 115 | 152 | 140 | 154 | 178 | 101 |
| 103 | 126 | 126 | 137 | 165 | 165 | 129 | 200 | 148 |

Here are the SSHA scores for a random sample of 20 first-year college men:

| | | | | | | |
|---|---|---|---|---|---|---|
| 108 | 140 | 114 | 91 | 180 | 115 | 126 |
| 92 | 169 | 146 | 109 | 132 | 75 | 88 |
| 113 | 151 | 70 | 115 | 187 | 104 | |

Note that high scores indicate good study habits and attitudes toward learning.

(a) Make parallel boxplots of the data.

(b) Use your calculator to compute separate numerical summaries for the women and for the men. Are these values statistics or parameters? Explain your answer.

(c) Do these data support the belief that men and women differ in their study habits and attitudes toward learning? Use your results from parts (a) and (b) to support your answer.

**119. Income and education level** Each March, the Bureau of Labor Statistics compiles an Annual Demographic Supplement to its monthly Current Population Survey.[43] Data on about 71,067 individuals between the ages of 25 and 64 who were employed full-time were collected in one of these surveys. The parallel boxplots compare the distributions of income for people with five levels of education. This figure is a variation of the boxplot idea: because large data sets often contain very extreme observations, we omitted the individuals in each category with the top 5% and bottom 5% of incomes. Also, the whiskers are drawn all the way to the maximum and minimum values of the remaining data for each distribution.

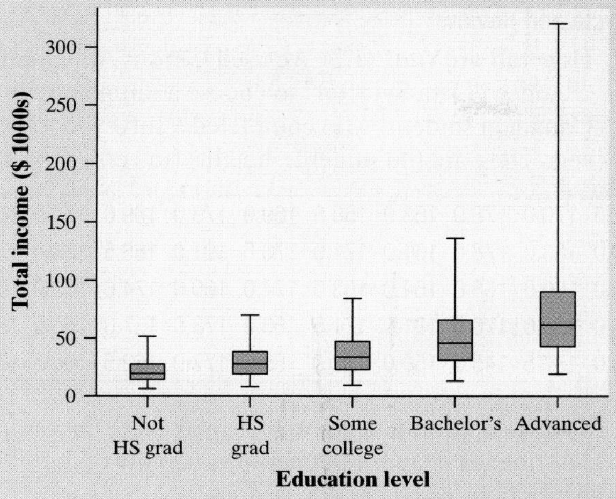


Use the graph to help answer the following questions.

(a) What shape do the distributions of income have?

(b) Explain how you know that there are outliers in the group that earned an advanced degree.

(c) How does the typical income change as the highest education level reached increases? Why does this make sense?

(d) Describe how the variability in income changes as the highest education level reached increases.

120. **Sleepless nights** Researchers recorded data on the amount of sleep reported each night during a week by a random sample of 20 high school students. Here are parallel boxplots comparing the distribution of time slept on all 7 nights of the study:[44]

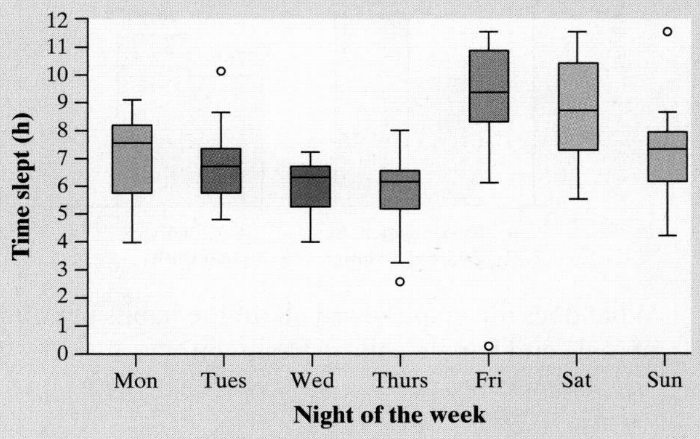


Use the graph to help answer the following questions.

(a) Which distributions have a clear left-skewed shape?

(b) Which outlier stands out the most, and why?

(c) How does the typical amount of sleep that the students got compare on these seven nights?

(d) On which night was there the most variation in how long the students slept? Justify your answer.

121. **SD contest** This is a standard deviation contest. You must choose four numbers from the whole numbers 0 to 10, with repeats allowed.

(a) Choose four numbers that have the smallest possible standard deviation.

(b) Choose four numbers that have the largest possible standard deviation.

(c) Is more than one choice possible in either part (a) or (b)? Explain.

122. **What do they measure?** For each of the following summary statistics, decide (i) whether it could be used to measure center or variability and (ii) whether it is resistant.

(a) $\frac{Q_1 + Q_3}{2}$ (b) $\frac{\text{Max} - \text{Min}}{2}$

**Multiple Choice:** *Select the best answer for Exercises 123–126.*

123. If a distribution is skewed to the right with no outliers, which expression is correct?

(a) mean < median
(b) mean ≈ median
(c) mean = median
(d) mean > median
(e) We can't tell without examining the data.

124. The scores on a statistics test had a mean of 81 and a standard deviation of 9. One student was absent on the test day, and his score wasn't included in the calculation. If his score of 84 was added to the distribution of scores, what would happen to the mean and standard deviation?

(a) Mean will increase, and standard deviation will increase.

(b) Mean will increase, and standard deviation will decrease.

(c) Mean will increase, and standard deviation will stay the same.

(d) Mean will decrease, and standard deviation will increase.

(e) Mean will decrease, and standard deviation will decrease.

125. The stemplot shows the number of home runs hit by each of the 30 Major League Baseball teams in a single season. Home run totals above what value should be considered outliers?

| | |
|---|---|
| 09 | 15 |
| 10 | 3789 |
| 11 | 47 |
| 12 | 19 |
| 13 | |
| 14 | 89 |
| 15 | 34445 |
| 16 | 239 |
| 17 | 223 |
| 18 | 356 |
| 19 | 1 |
| 20 | 3 |
| 21 | 0 |
| 22 | 2 |

Key: 14|8 is a team with 148 home runs.

(a) 173 (b) 210 (c) 222

(d) 229 (e) 257

126. Which of the following boxplots best matches the distribution shown in the histogram?

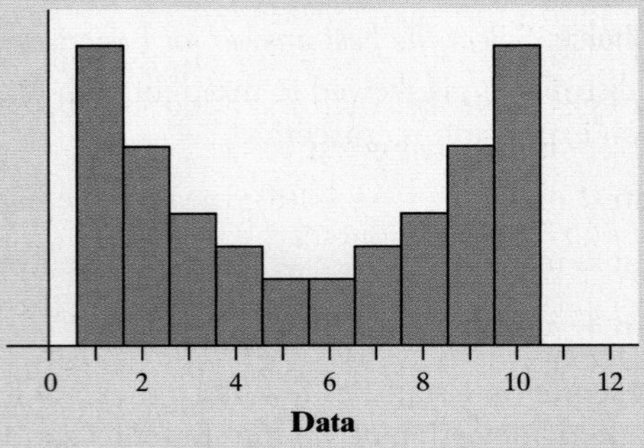


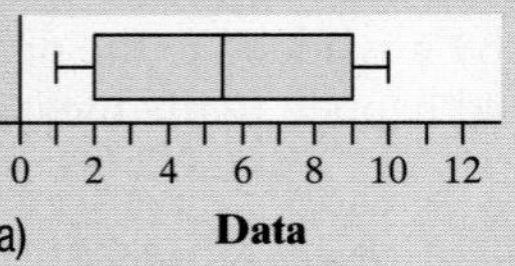


(a)

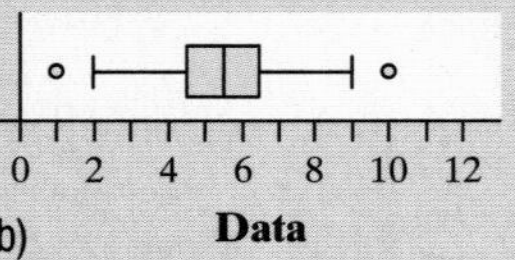


(b)

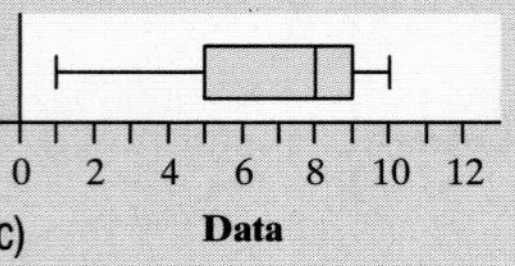


(c)

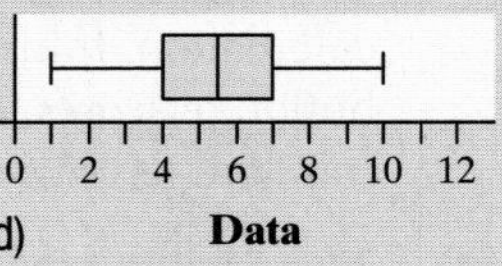


(d)

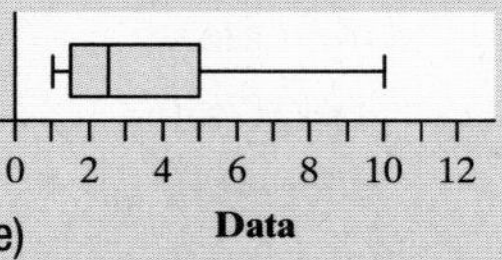


(e)

### Recycle and Review

127. **How tall are you?** (1.2) We used Census At School's "Random Data Selector" to choose a sample of 50 Canadian students who completed a survey in a recent year. Here are the students' heights (in centimeters):

| | | | | | | | | | |
|---|---|---|---|---|---|---|---|---|---|
| 166.5 | 170.0 | 178.0 | 163.0 | 150.5 | 169.0 | 173.0 | 169.0 | 171.0 | 166.0 |
| 190.0 | 183.0 | 178.0 | 161.0 | 171.0 | 170.0 | 191.0 | 168.5 | 178.5 | 173.0 |
| 175.0 | 160.5 | 166.0 | 164.0 | 163.0 | 174.0 | 160.0 | 174.0 | 182.0 | 167.0 |
| 166.0 | 170.0 | 170.0 | 181.0 | 171.5 | 160.0 | 178.0 | 157.0 | 165.0 | 187.0 |
| 168.0 | 157.5 | 145.5 | 156.0 | 182.0 | 168.5 | 177.0 | 162.5 | 160.5 | 185.5 |

Make an appropriate graph to display these data. Describe the shape, center, and variability of the distribution. Are there any outliers?

128. **Success in college** (1.1) The Freshman Survey asked first-year college students about their "habits of mind"—specific behaviors that college faculty have identified as being important for student success. One question asked students, "How often in the past year did you revise your papers to improve your writing?" Another asked, "How often in the past year did you seek feedback on your academic work?" The figure is a bar graph comparing the percent of males and females who answered "frequently" to these two questions.[45]

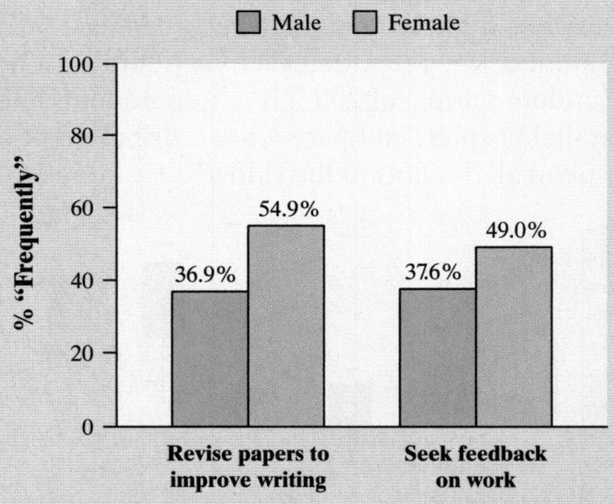


What does the graph reveal about the habits of mind of male and female college freshmen?

# Chapter 1 Wrap-Up

## FRAPPY! FREE RESPONSE AP® PROBLEM, YAY!

**The following problem is modeled after actual AP® Statistics exam free response questions. Your task is to generate a complete, concise response in 15 minutes.**

*Directions: Show all your work. Indicate clearly the methods you use, because you will be scored on the correctness of your methods as well as on the accuracy and completeness of your results and explanations.*

Using data from the 2010 census, a random sample of 348 U.S. residents aged 18 and older was selected. Among the variables recorded were gender (male or female), housing status (rent or own), and marital status (married or not married).

The two-way table below summarizes the relationship between gender and housing status.

| | Male | Female | Total |
|---|---|---|---|
| Own | 132 | 122 | **254** |
| Rent | 50 | 44 | **94** |
| **Total** | **182** | **166** | **348** |

(a) What percent of males in the sample own their home?

(b) Make a graph to compare the distribution of housing status for males and females.

(c) Using your graph from part (b), describe the relationship between gender and housing status.

(d) The two-way table below summarizes the relationship between marital status and housing status.

| | Married | Not Married | Total |
|---|---|---|---|
| Own | 172 | 82 | **254** |
| Rent | 40 | 54 | **94** |
| **Total** | **212** | **136** | **348** |

For the members of the sample, is the relationship between marital status and housing status stronger or weaker than the relationship between gender and housing status that you described in part (c)? Justify your choice using the data provided in the two-way tables.

**After you finish, you can view two example solutions on the book's website (highschool.bfwpub.com/updatedtps6e). Determine whether you think each solution is "complete," "substantial," "developing," or "minimal." If the solution is not complete, what improvements would you suggest to the student who wrote it? Finally, your teacher will provide a scoring rubric. Score your response and note what, if anything, you would do differently to improve your own score.**

# Chapter 1 Review

### Introduction: Data Analysis: Making Sense of Data

In this brief section, you learned several fundamental concepts that will be important throughout the course: the idea of a distribution and the distinction between quantitative and categorical variables. You also learned a strategy for exploring data:

- Begin by examining each variable by itself. Then move on to study relationships between variables.
- Start with a graph or graphs. Then add numerical summaries.

### Section 1.1: Analyzing Categorical Data

In this section, you learned how to display the distribution of a single categorical variable with bar graphs and pie charts and what to look for when describing these displays. Remember to properly label your graphs! Poor labeling is an easy way to lose points on the AP® Statistics exam. You should also be able to recognize misleading graphs and be careful to avoid making misleading graphs yourself.

Next, you learned how to investigate the relationship between two categorical variables. Using a two-way table, you learned how to calculate and display marginal and joint relative frequencies. Conditional relative frequencies and side-by-side bar graphs, segmented bar graphs, or mosaic plots allow you to look for an association between the variables. If there is no association between the two variables, comparative bar graphs of the distribution of one variable for each value of the other variable will be identical. If differences in the corresponding conditional relative frequencies exist, there is an association between the variables. That is, knowing the value of one variable helps you predict the value of the other variable.

### Section 1.2: Displaying Quantitative Data with Graphs

In this section, you learned how to create three different types of graphs for a quantitative variable: dotplots, stemplots, and histograms. Each of the graphs has distinct benefits, but all of them are good tools for examining the distribution of a quantitative variable. Dotplots and stemplots are handy for small sets of data. Histograms are the best choice when there are a large number of observations. On the AP® exam, you will be expected to create each of these types of graphs, label them properly, and comment on their characteristics.

When you are describing the distribution of a quantitative variable, you should look at its graph for the overall pattern (shape, center, variability) and striking departures from that pattern (outliers). Use the acronym SOCV (shape, outliers, center, variability) to help remember these four characteristics. When comparing distributions, you should include explicit comparison words for center and variability such as "is greater than" or "is approximately the same as." When asked to compare distributions, a very common mistake on the AP® exam is describing the characteristics of each distribution separately without making these explicit comparisons.

### Section 1.3: Describing Quantitative Data with Numbers

To measure the center of a distribution of quantitative data, you learned how to calculate the mean and the median of a distribution. You also learned that the median is a resistant measure of center, but the mean isn't resistant because it can be greatly affected by skewness or outliers.

To measure the variability of a distribution of quantitative data, you learned how to calculate the range, standard deviation, and interquartile range. The standard deviation is the most commonly used measure of variability and approximates the typical distance of a value in the data set from the mean. The standard deviation is not resistant—it is heavily affected by extreme values. The interquartile range (*IQR*) is a resistant measure of variability because it ignores the upper 25% and lower 25% of the distribution, but the range isn't resistant because it uses only the minimum and maximum value.

To identify outliers in a distribution of quantitative data, you learned the $1.5 \times IQR$ rule. You also learned that boxplots are a great way to visually summarize a distribution of quantitative data. Boxplots are helpful for comparing the center (median) and variability (range, *IQR*) for multiple distributions. Boxplots aren't as useful for identifying the shape of a distribution because they do not display peaks, clusters, gaps, and other interesting features.

## What Did You Learn?

| Learning Target | Section | Related Example on Page(s) | Relevant Chapter Review Exercise(s) |
|---|---|---|---|
| Identify the individuals and variables in a set of data. | Intro | 3 | R1.1 |
| Classify variables as categorical or quantitative. | Intro | 3 | R1.1 |
| Make and interpret bar graphs for categorical data. | 1.1 | 11 | R1.2 |
| Identify what makes some graphs of categorical data misleading. | 1.1 | 12 | R1.3 |
| Calculate marginal and joint relative frequencies from a two-way table. | 1.1 | 15 | R1.4 |
| Calculate conditional relative frequencies from a two-way table. | 1.1 | 17 | R1.4, R1.5 |
| Use bar graphs to compare distributions of categorical data. | 1.1 | 20 | R1.5 |
| Describe the nature of the association between two categorical variables. | 1.1 | 20 | R1.5 |
| Make and interpret dotplots, stemplots, and histograms of quantitative data. | 1.2 | Dotplots: 31<br>Stemplots: 38<br>Histograms: 42 | R1.6, R1.7 |

| Learning Target | Section | Related Example on Page(s) | Relevant Chapter Review Exercise(s) |
|---|---|---|---|
| Identify the shape of a distribution from a graph. | 1.2 | 33 | R1.6 |
| Describe the overall pattern (shape, center, and variability) of a distribution and identify any major departures from the pattern (outliers). | 1.2 | 35 | R1.6 |
| Compare distributions of quantitative data using dotplots, stemplots, and histograms. | 1.2 | 36 | R1.8 |
| Calculate measures of center (mean, median) for a distribution of quantitative data. | 1.3 | Mean: 55<br>Median: 57 | R1.6 |
| Calculate and interpret measures of variability (range, standard deviation, *IQR*) for a distribution of quantitative data. | 1.3 | SD: 61<br>IQR: 64 | R1.9 |
| Explain how outliers and skewness affect measures of center and variability. | 1.3 | 69 | R1.9 |
| Identify outliers using the $1.5 \times IQR$ rule. | 1.3 | 67 | R1.7, R1.9 |
| Make and interpret boxplots of quantitative data. | 1.3 | 69 | R1.7 |
| Use boxplots and numerical summaries to compare distributions of quantitative data. | 1.3 | 71 | R1.10 |

# Chapter 1 Review Exercises

*These exercises are designed to help you review the important ideas and methods of the chapter.*

**R1.1 Who buys cars?** A car dealer keeps records on car buyers for future marketing purposes. The table gives information on the last 4 buyers.

| Buyer's name | Zip code | Gender | Buyer's distance from dealer (mi) | Car model | Model year | Price |
|---|---|---|---|---|---|---|
| P. Smith | 27514 | M | 13 | Fiesta | 2018 | $26,375 |
| K. Ewing | 27510 | M | 10 | Mustang | 2015 | $39,500 |
| L. Shipman | 27516 | F | 2 | Fusion | 2016 | $38,400 |
| S. Reice | 27243 | F | 4 | F-150 | 2016 | $56,000 |

(a) Identify the individuals in this data set.

(b) What variables were measured? Classify each as categorical or quantitative.

**R1.2 I want candy!** Mr. Starnes bought some candy for his AP® Statistics class to eat on Halloween. He offered the students an assortment of Snickers®, Milky Way®, Butterfinger®, Twix®, and 3 Musketeers® candies. Each student was allowed to choose one option. Here are the data on the type of candy selected. Make a relative frequency bar graph to display the data. Describe what you see.

| | | |
|---|---|---|
| Twix | Snickers | Butterfinger |
| Butterfinger | Snickers | Snickers |
| 3 Musketeers | Snickers | Snickers |
| Butterfinger | Twix | Twix |
| Twix | Twix | Twix |
| Snickers | Snickers | Twix |
| Snickers | Milky Way | Twix |
| Twix | Twix | Butterfinger |
| Milky Way | Butterfinger | 3 Musketeers |
| Milky Way | Butterfinger | Butterfinger |

**R1.3 I'd die without my phone!** In a survey of over 2000 U.S. teenagers by Harris Interactive, 47% said that "their social life would end or be worsened without their cell phone."[46] One survey question asked the teens how important it is for their phone to have certain features. The following figure displays data on the percent who indicated that a particular feature is vital.

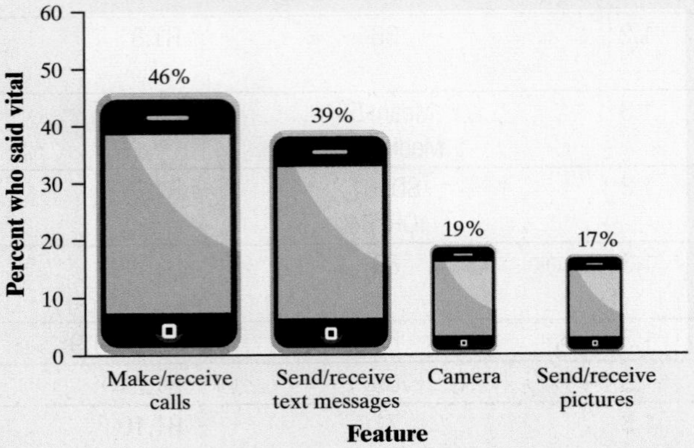


(a) Explain how the graph gives a misleading impression.

(b) Would it be appropriate to make a pie chart to display these data? Why or why not?

**R1.4 Facebook and age** Is there a relationship between Facebook use and age among college students? The following two-way table displays data for the 219 students who responded to the survey.[47]

| | | Age | | |
|---|---|---|---|---|
| | | Younger (18–22) | Middle (23–27) | Older (28 and up) |
| Facebook user? | Yes | 78 | 49 | 21 |
| | No | 4 | 21 | 46 |

(a) What percent of the students who responded were Facebook users?

(b) What percent of the students in the sample were aged 28 or older?

(c) What percent of the students who responded were older Facebook users?

(d) What percent of the Facebook users in the sample were younger students?

**R1.5 Facebook and age** Refer to the preceding exercise.

(a) Find the distribution of Facebook use for each of the three age groups. Make a segmented bar graph to compare these distributions.

(b) Describe what the graph in part (a) reveals about the association between age and Facebook use.

**R1.6 Density of the earth** In 1798, the English scientist Henry Cavendish measured the density of the earth several times by careful work with a torsion balance. The variable recorded was the density of the earth as a multiple of the density of water. Here are Cavendish's 29 measurements:[48]

| | | | | | | | | | |
|---|---|---|---|---|---|---|---|---|---|
| 5.50 | 5.61 | 4.88 | 5.07 | 5.26 | 5.55 | 5.36 | 5.29 | 5.58 | 5.65 |
| 5.57 | 5.53 | 5.62 | 5.29 | 5.44 | 5.34 | 5.79 | 5.10 | 5.27 | 5.39 |
| 5.42 | 5.47 | 5.63 | 5.34 | 5.46 | 5.30 | 5.75 | 5.68 | 5.85 | |

(a) Make a stemplot of the data.

(b) Describe the distribution of density measurements.

(c) The currently accepted value for the density of earth is 5.51 times the density of water. How does this value compare to the mean of the distribution of density measurements?

**R1.7 Guinea pig survival times** Here are the survival times (in days) of 72 guinea pigs after they were injected with infectious bacteria in a medical experiment.[49] Survival times, whether of machines under stress or cancer patients after treatment, usually have distributions that are skewed to the right.

| | | | | | | | | | | | |
|---|---|---|---|---|---|---|---|---|---|---|---|
| 43 | 45 | 53 | 56 | 56 | 57 | 58 | 66 | 67 | 73 | 74 | 79 |
| 80 | 80 | 81 | 81 | 81 | 82 | 83 | 83 | 84 | 88 | 89 | 91 |
| 91 | 92 | 92 | 97 | 99 | 99 | 100 | 100 | 101 | 102 | 102 | 102 |
| 103 | 104 | 107 | 108 | 109 | 113 | 114 | 118 | 121 | 123 | 126 | 128 |
| 137 | 138 | 139 | 144 | 145 | 147 | 156 | 162 | 174 | 178 | 179 | 184 |
| 191 | 198 | 211 | 214 | 243 | 249 | 329 | 380 | 403 | 511 | 522 | 598 |

(a) Make a histogram of the data. Does it show the expected right skew?

(b) Now make a boxplot of the data.

(c) Compare the histogram from part (a) with the boxplot from part (b). Identify an aspect of the distribution that one graph reveals but the other does not.

**R1.8 Household incomes** Rich and poor households differ in ways that go beyond income. Here are histograms that compare the distributions of household size (number of people) for low-income and high-income households.[50] Low-income households had annual incomes less than $15,000, and high-income households had annual incomes of at least $100,000.

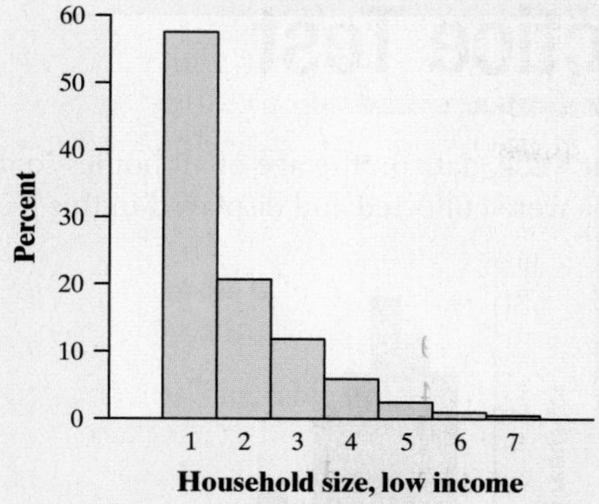


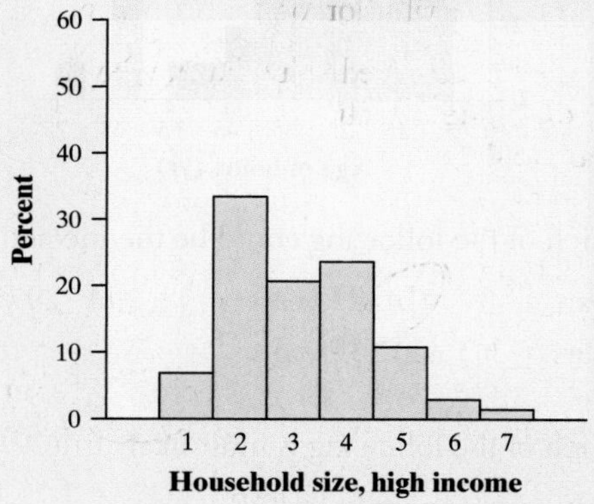


(a) About what percent of each group of households consisted of four or more people?

(b) Describe the similarities and differences in these two distributions of household size.

*Exercises R1.9 and R1.10 refer to the following setting.* Do you like to eat tuna? Many people do. Unfortunately, some of the tuna that people eat may contain high levels of mercury. Exposure to mercury can be especially hazardous for pregnant women and small children. How much mercury is safe to consume? The Food and Drug Administration will take action (like removing the product from store shelves) if the mercury concentration in a 6-ounce can of tuna is 1.00 ppm (parts per million) or higher.

What is the typical mercury concentration in cans of tuna sold in stores? A study conducted by Defenders of Wildlife set out to answer this question. Defenders collected a sample of 164 cans of tuna from stores across the United States. They sent the selected cans to a laboratory that is often used by the Environmental Protection Agency for mercury testing.[51]

**R1.9 Mercury in tuna** Here are a dotplot and numerical summaries of the data on mercury concentration in the sampled cans (in parts per million, ppm):

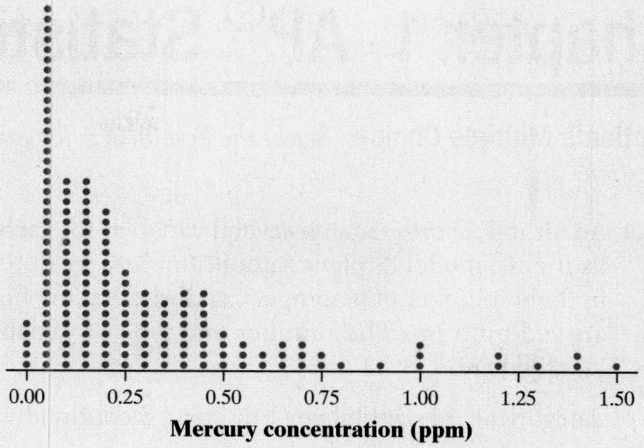


| Variable | $N$ | Mean | SD | Min |
|---|---|---|---|---|
| Mercury | 164 | 0.285 | 0.300 | 0.012 |
| Variable | $Q_1$ | Med | $Q_3$ | Max |
| Mercury | 0.071 | 0.180 | 0.380 | 1.500 |

(a) Interpret the standard deviation.

(b) Determine whether there are any outliers.

(c) Explain why the mean is so much larger than the median of the distribution.

**R1.10 Mercury in tuna** Is there a difference in the mercury concentration of light tuna and albacore tuna? Use the parallel boxplots and the computer output to write a few sentences comparing the two distributions.

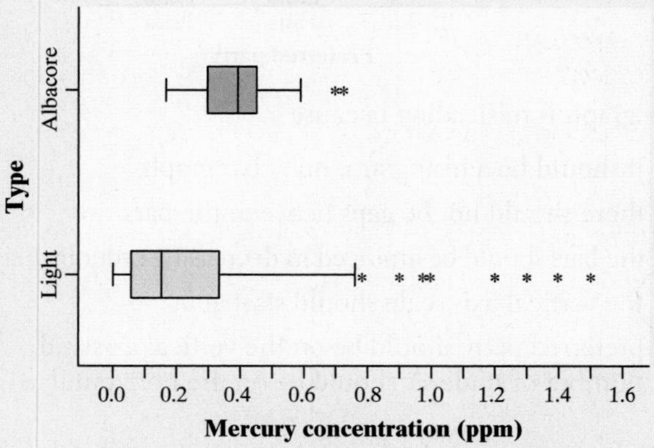


| Type | N | Mean | SD | Min |
|---|---|---|---|---|
| Albacore | 20 | 0.401 | 0.152 | 0.170 |
| Light | 144 | 0.269 | 0.312 | 0.012 |
| Type | $Q_1$ | Med | $Q_3$ | Max |
| Albacore | 0.293 | 0.400 | 0.460 | 0.730 |
| Light | 0.059 | 0.160 | 0.347 | 1.500 |

# Chapter 1 AP® Statistics Practice Test

**Section I: Multiple Choice** *Select the best answer for each question.*

**T1.1** An airline records data on several variables for each of its flights: model of plane, amount of fuel used, time in flight, number of passengers, and whether the flight arrived on time. The number and type of variables recorded are

(a) 1 categorical, 4 quantitative (1 discrete, 3 continuous)
(b) 1 categorical, 4 quantitative (2 discrete, 2 continuous)
(c) 2 categorical, 3 quantitative (1 discrete, 2 continuous)
(d) 2 categorical, 3 quantitative (2 discrete, 1 continuous)
(e) 3 categorical, 2 quantitative (1 discrete, 1 continuous)

**T1.2** The students in Mr. Tyson's high school statistics class were recently asked if they would prefer a pasta party, a pizza party, or a donut party. The following bar graph displays the data.

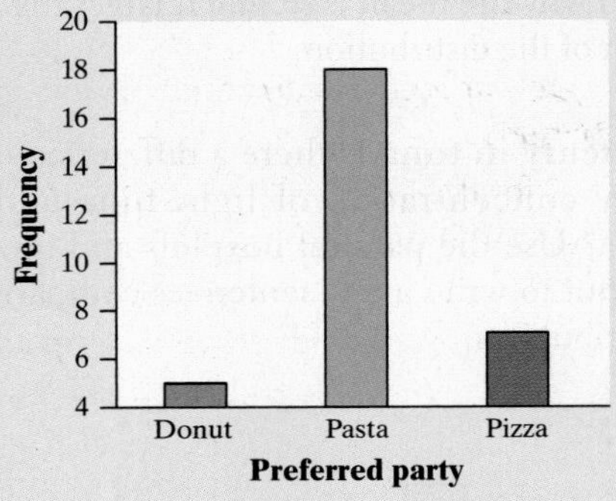


This graph is misleading because

(a) it should be a histogram, not a bar graph.
(b) there should not be gaps between the bars.
(c) the bars should be arranged in decreasing order by height.
(d) the vertical axis scale should start at 0.
(e) preferred party should be on the vertical axis and number of students should be on the horizontal axis.

**T1.3** Forty students took a statistics test worth 50 points. The dotplot displays the data. The third quartile is

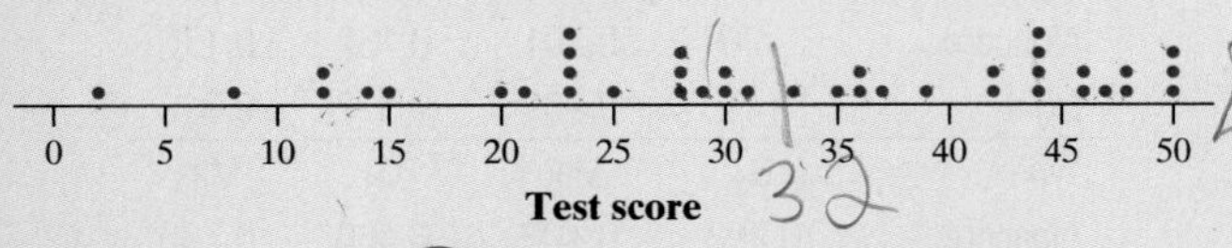


(a) 45.
(b) 44.
(c) 43.
(d) 32.
(e) 23.

*Questions T1.4–T1.6 refer to the following setting.* Realtors collect data in order to serve their clients more effectively. In a recent week, data on the age of all homes sold in a particular area were collected and displayed in this histogram.

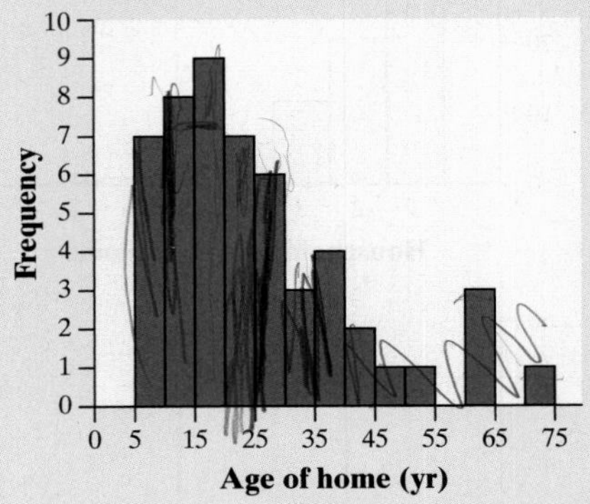


**T1.4** Which of the following could be the median age?

(a) 19 years
(b) 24 years
(c) 29 years
(d) 34 years
(e) 39 years

**T1.5** Which of the following is most likely true?

(a) mean > median, range < *IQR*
(b) mean < median, range < *IQR*
(c) mean > median, range > *IQR*
(d) mean < median, range > *IQR*
(e) mean = median, range > *IQR*

**T1.6** The standard deviation of the distribution of house age is about 16 years. Interpret this value.

(a) The age of all houses in the sample is within 16 years of the mean.
(b) The gap between the youngest and oldest house is 16 years.
(c) The age of all the houses in the sample is 16 years from the mean.
(d) The gap between the first quartile and the third quartile is 16 years.
(e) The age of the houses in the sample typically varies by about 16 years from the mean age.

**T1.7** The mean salary of all female workers at a company is $35,000. The mean salary of all male workers at the company is $41,000. What must be true about the mean salary of all workers at the company?

(a) It must be $38,000.
(b) It must be larger than the median salary.
(c) It could be any number between $35,000 and $41,000.
(d) It must be larger than $38,000.
(e) It cannot be larger than $40,000.

*Questions T1.8 and T1.9 refer to the following setting.* A survey was designed to study how business operations vary by size. Companies were classified as small, medium, or large. Questionnaires were sent to 200 randomly selected businesses of each size. Because not all questionnaires are returned, researchers decided to investigate the relationship between the response rate and the size of the business. The data are given in the following two-way table.

| | | Business size | | |
|---|---|---|---|---|
| | | Small | Medium | Large |
| Response? | Yes | 125 | 81 | 40 |
| | No | 75 | 119 | 160 |

**T1.8** What percent of all small companies receiving questionnaires responded?

(a) 12.5% (b) 20.8% (c) 33.3%
(d) 50.8% (e) 62.5%

**T1.9** Which of the following conclusions seems to be supported by the data?

(a) There are more small companies than large companies in the survey.
(b) Small companies appear to have a higher response rate than medium or big companies.
(c) Exactly the same number of companies responded as didn't respond.
(d) Overall, more than half of companies responded to the survey.
(e) If we combined the medium and large companies, then their response rate would be equal to that of the small companies.

**T1.10** An experiment was conducted to investigate the effect of a new weed killer to prevent weed growth in onion crops. Two chemicals were used: the standard weed killer (S) and the new chemical (N). Both chemicals were tested at high and low concentrations on 50 test plots. The percent of weeds that grew in each plot was recorded. Here are some boxplots of the results.

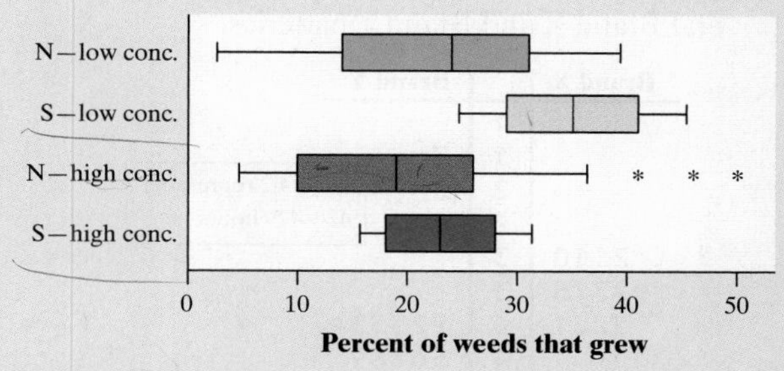


Which of the following is *not* a correct statement about the results of this experiment?

(a) At both high and low concentrations, the new chemical results in better weed control than the standard weed killer.
(b) For both chemicals, a smaller percentage of weeds typically grew at higher concentrations than at lower concentrations.
(c) The results for the standard weed killer are less variable than those for the new chemical.
(d) High and low concentrations of either chemical have approximately the same effects on weed growth.
(e) Some of the results for the low concentration of weed killer show a smaller percentage of weeds growing than some of the results for the high concentration.

## Section II: Free Response

*Show all your work. Indicate clearly the methods you use, because you will be graded on the correctness of your methods as well as on the accuracy and completeness of your results and explanations.*

**T1.11** You are interested in how many contacts older adults have in their smartphones. Here are data on the number of contacts for a random sample of 30 elderly adults with smartphones in a large city:

| | | | | | | | | | |
|---|---|---|---|---|---|---|---|---|---|
| 7 | 20 | 24 | 25 | 25 | 28 | 28 | 30 | 32 | 35 |
| 42 | 43 | 44 | 45 | 46 | 47 | 48 | 48 | 50 | 51 |
| 72 | 75 | 77 | 78 | 79 | 83 | 87 | 88 | 135 | 151 |

(a) Construct a histogram of these data.
(b) Are there any outliers? Justify your answer.
(c) Would it be better to use the mean and standard deviation or the median and *IQR* to describe the center and variability of this distribution? Why?

**T1.12** A study among the Pima Indians of Arizona investigated the relationship between a mother's diabetic status and the number of birth defects in her children. The results appear in the two-way table.

| | | Diabetic status | | |
|---|---|---|---|---|
| | | Nondiabetic | Prediabetic | Diabetic |
| Number of birth defects | None | 754 | 362 | 38 |
| | One or more | 31 | 13 | 9 |

(a) What proportion of the women in this study had a child with one or more birth defects?
(b) What percent of the women in this study were diabetic or prediabetic, and had a child with one or more birth defects?

(c) Make a segmented bar graph to display the distribution of number of birth defects for the women with each of the three diabetic statuses.

(d) Describe the nature of the association between mother's diabetic status and number of birth defects for the women in this study.

**T1.13** The back-to-back stemplot shows the lifetimes of several Brand X and Brand Y batteries.

| Brand X | | Brand Y |
|---|---|---|
| | 1 | |
| | 1 | 7 |
| | 2 | 2 |
| | 2 | 6 |
| 2110 | 3 | |
| 99775 | 3 | |
| 3221 | 4 | 223334 |
| | 4 | 56889 |
| 4 | 5 | 0 |
| 5 | 5 | |

Key: 4|2 represents 420–429 hours.

(a) What is the longest that any battery lasted?

(b) Give a reason someone might prefer a Brand X battery.

(c) Give a reason someone might prefer a Brand Y battery.

**T1.14** Catherine and Ana suspect that athletes (i.e., students who have been on at least one varsity team) typically have a faster reaction time than other students. To test this theory, they gave an online reflex test to separate random samples of 33 varsity athletes and 29 other students at their school. Here are parallel boxplots and numerical summaries of the data on reaction times (in milliseconds) for the two groups of students.

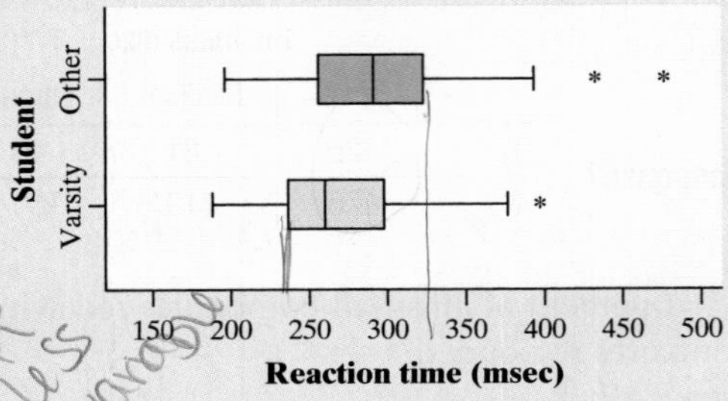


| Student | $n$ | Mean | StDev | Min | $Q_1$ | Med | $Q_3$ | Max |
|---|---|---|---|---|---|---|---|---|
| Other | 29 | 297.3 | 65.9 | 197.0 | 255.0 | 292.0 | 325.0 | 478.0 |
| Athlete | 33 | 270.1 | 57.7 | 189.6 | 236.0 | 261.0 | 300.0 | 398.0 |

(a) Are these numerical summaries statistics or parameters? Explain your answer.

(b) Write a few sentences comparing the distribution of reaction time for the two types of students.

# Chapter 1 Project American Community Survey

Each month, the U.S. Census Bureau selects a random sample of about 300,000 U.S. households to participate in the American Community Survey (ACS). The chosen households are notified by mail and invited to complete the survey online. The Census Bureau follows up on any uncompleted surveys by phone or in person. Data from the ACS are used to determine how the federal government allocates over $400 billion in funding for local communities.

The file **acs survey ch1 project.xls** can be accessed from the book's website at highschool.bfwpub.com/updatedtps6e. It contains data for 3000 randomly selected households in one month's ACS survey. Download the file to a computer for further analysis using the application specified by your teacher.

Each row in the spreadsheet describes a household. A serial number that identifies the household is in the first column. The other columns contain values of several variables. See the code sheet on the book's website for details on how each variable is recorded. Note that all the categorical variables have been coded to have numerical values in the spreadsheet.

Use the files provided to answer the following questions.

1. How many variables are recorded? Classify each one as categorical or quantitative.
2. Examine the distribution of location (division or region) for the households in the sample. Make a bar graph to display the data. Then calculate numerical summaries (counts, percents, or proportions). Describe what you see.
3. Explore the relationship between two categorical variables of interest to you. Summarize the data in a two-way table. Then calculate appropriate conditional relative frequencies and make a side-by-side or segmented bar graph. Write a few sentences comparing the distributions.
4. Analyze the distribution of household income (HINCP) using appropriate graphs and numerical summaries.
5. Compare the distribution of a quantitative variable that interests you in two or more groups. For instance, you might compare the distribution of number of people in a family (NPF) by region. Make appropriate graphs and calculate numerical summaries. Then write a few sentences comparing the distributions.

# UNIT 1

## Exploring One-Variable Data

# Chapter 2

# Modeling Distributions of Quantitative Data

# INTRODUCTION

Suppose Emily earns 43 out of 50 points on a statistics test. Should she be satisfied or disappointed with her performance? That depends on how her score compares with the scores of the other students who took the test. If 43 is the highest score, Emily might be very pleased. Maybe her teacher will "scale" the grades so that Emily's 43 becomes an "A." But if Emily's 43 falls below the class average, she may not be so happy.

Section 2.1 focuses on describing the location of an individual within a distribution of quantitative data. We begin by discussing a familiar measure of location: *percentiles*. Next, we introduce a new type of graph that is useful for displaying percentiles. Then we consider another way to describe an individual's location that is based on the mean and standard deviation. In the process, we examine the effects of transforming data on the shape, center, and variability of a distribution.

Sometimes it is helpful to use graphical models called *density curves* to estimate an individual's location in a distribution, rather than relying on actual data values. Such models are especially helpful when data fall in a bell-shaped pattern called a *Normal distribution*. Section 2.2 examines the properties of Normal distributions and shows you how to perform useful calculations with them.

## ACTIVITY Where do I stand?

In this activity, you and your classmates will explore ways to describe where you stand (literally!) within a distribution.

1. Your teacher will mark out a number line on the floor with a scale running from about 58 to 78 inches.
2. Make a human dotplot. Each member of the class should stand at the appropriate location along the number line scale based on height (to the nearest inch).
3. Your teacher will make a copy of the dotplot on the board for your reference. Describe the class's distribution of heights.
4. What percent of the students in the class have heights less than or equal to yours? This *percentile* is one way to measure your location in the distribution of heights.
5. Calculate the mean and standard deviation of the class's distribution of height from the dotplot. Confirm these values with your classmates.
6. Does your height fall above or below the mean? How far above or below the mean is it? How many standard deviations above or below the mean is it? This *standardized score* (also called a *z-score*) is another way to measure your location in the class's distribution of heights.
7. *Class discussion:* What would happen to the class's distribution of height if you converted each data value from inches to centimeters? (There are 2.54 centimeters in 1 inch.) How would this change of units affect the shape, center, variability, and the measures of location (percentile and *z*-score) that you calculated?

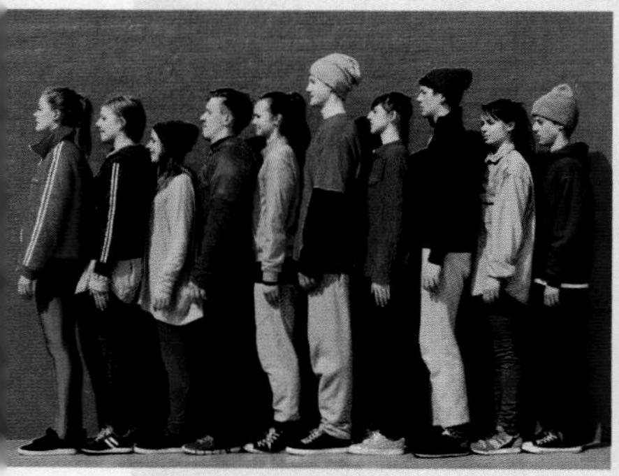

Henrik Sorensen/Getty Images

Want to know more about where you stand—in terms of height, weight, or even body mass index? Do a web search for "Clinical Growth Charts" at the National Center for Health Statistics site, www.cdc.gov/nchs.

# SECTION 2.1 Describing Location in a Distribution

**LEARNING TARGETS** *By the end of the section, you should be able to:*

- Find and interpret the percentile of an individual value in a distribution of data.
- Estimate percentiles and individual values using a cumulative relative frequency graph.
- Find and interpret the standardized score (*z*-score) of an individual value in a distribution of data.
- Describe the effect of adding, subtracting, multiplying by, or dividing by a constant on the shape, center, and variability of a distribution of data.

There are 25 students in Mr. Tabor's statistics class. He gives them a first test worth 50 points. Here are the students' scores:

35 18 37 38 42 41 25 37 36 32 12 43 31
29 32 48 44 45 38 40 45 38 38 40 22

The score marked in red is Emily's 43. How did she perform on this test relative to her classmates?

Figure 2.1 displays a dotplot of the class's test scores, with Emily's score marked in red. The distribution is skewed to the left with some possible low outliers. From the dotplot, we can see that Emily's score is above the mean (balance point) of the distribution. We can also see that Emily did better on the test than most other students in the class.

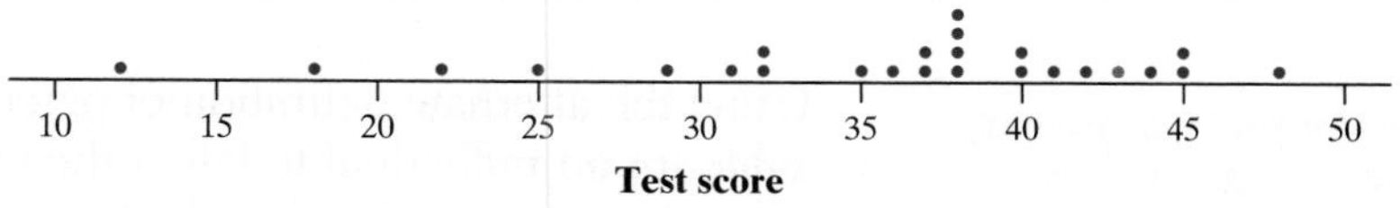


**FIGURE 2.1** Dotplot of scores (out of 50 points) on Mr. Tabor's first statistics test. Emily's score of 43 is marked in red.

## Measuring Location: Percentiles

One way to describe Emily's location in the distribution of test scores is to calculate her **percentile.**

**DEFINITION Percentile**

The *p*th **percentile** of a distribution is the value with *p*% of observations less than or equal to it.

Some people define the *p*th percentile of a distribution as the value with *p*% of observations *less than it*. This distinction matters for discrete variables (like score on Mr. Tabor's first test), but not for continuous variables.

Using the dotplot, we see that four students in the class earned test scores greater than Emily's 43. Because 21 of the 25 observations (84%) are less than or equal to her score, Emily is at the 84th percentile in the class's test score distribution.

caution **Be careful with your language when describing percentiles.** Percentiles are specific locations in a distribution, so an observation isn't "in" the 84th percentile. Rather, it is "at" the 84th percentile.

## EXAMPLE

### Mr. Tabor's first test
### Finding and interpreting percentiles

antoniodiaz/Shutterstock.com

**PROBLEM:** Refer to the dotplot of 25 scores on Mr. Tabor's first statistics test.

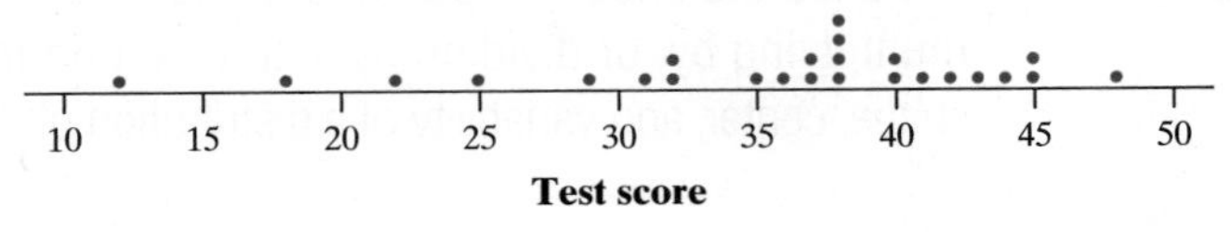


(a) Find the percentile for Jacob, who scored 18 on the test.

(b) Maria's test score is at the 48th percentile of the distribution. Interpret this value in context. What score did Maria earn?

**SOLUTION:**

(a) $2/25 = 0.08$, so Jacob scored at the 8th percentile on this test.

Only 2 of the 25 scores in the class are less than or equal to Jacob's 18.

(b) 48% of students in the class earned a test score less than or equal to Maria's. Because $(0.48)(25) = 12$, Maria scored greater than or equal to 12 of the 25 students. Maria earned a 37 on the test.

Two students in the class scored a 37 on the test. Both these students' scores are at the 48th percentile because 12 of the 25 students in the class earned equal or lower scores.

**FOR PRACTICE, TRY EXERCISE 1**

The three quartiles $Q_1$, $Q_2$ (median), and $Q_3$ divide a distribution into four groups of roughly equal size. Similarly, the 99 percentiles should divide a distribution into 100 groups of roughly equal size. This concept of percentile makes sense for only very large sets of quantitative data.

Using the alternate definition of percentile (see margin note on page 91), it is possible for an individual to fall at the 0th percentile. If we used this definition, Jacob's score of 18 would fall at the 4th percentile (1 of 25 scores were less than 18). Calculating percentiles is not an exact science, especially with small data sets!

The median of a distribution is roughly the 50th percentile because about half the observations are less than or equal to the median. The first quartile $Q_1$ is roughly the 25th percentile of a distribution because it separates the lowest 25% of values from the upper 75%. Likewise, the third quartile $Q_3$ is roughly the 75th percentile.

caution

**A high percentile is not always a good thing.** For example, a man whose cholesterol level is at the 90th percentile for his age group may need treatment for his high cholesterol!

## Cumulative Relative Frequency Graphs

There are some interesting graphs that can be made with percentiles. One of the most common graphs starts with a frequency table for a quantitative variable. For instance, the frequency table on the next page summarizes the ages of the first 45 U.S. presidents when they took office.

| Age | Frequency |
|---|---|
| 40–45 | 2 |
| 45–50 | 7 |
| 50–55 | 13 |
| 55–60 | 12 |
| 60–65 | 7 |
| 65–70 | 3 |
| 70–75 | 1 |

Note that age at inauguration is a *continuous* variable. For instance, Donald Trump was 70 years, 7 months, 7 days (about 70.603 years) old when he took office. The intervals in the frequency table are given as 40–45, 45–50, and so on because each president's inauguration age falls in exactly one of these intervals on the number line.

Let's expand this table to include columns for relative frequency, *cumulative frequency*, and *cumulative relative frequency*.

| Age | Frequency | Relative frequency | Cumulative frequency | Cumulative relative frequency (percentile) |
|---|---|---|---|---|
| 40–45 | 2 | 2/45 = 0.044 = 4.4% | 2 | 2/45 = 0.044 = 4.4% |
| 45–50 | 7 | 7/45 = 0.156 = 15.6% | 9 | 9/45 = 0.200 = 20.0% |
| 50–55 | 13 | 13/45 = 0.289 = 28.9% | 22 | 22/45 = 0.489 = 48.9% |
| 55–60 | 12 | 12/45 = 0.267 = 26.7% | 34 | 34/45 = 0.756 = 75.6% |
| 60–65 | 7 | 7/45 = 0.156 = 15.6% | 41 | 41/45 = 0.911 = 91.1% |
| 65–70 | 3 | 3/45 = 0.067 = 6.7% | 44 | 44/45 = 0.978 = 97.8% |
| 70–75 | 1 | 1/45 = 0.022 = 2.2% | 45 | 45/45 = 1.000 = 100% |

The table reveals that 20.0% of U.S. presidents took office by the time they turned 50. In other words, the 20th percentile of the distribution of inauguration age is 50.000 years. We can display the percentiles from the table in a **cumulative relative frequency graph.**

Some people refer to cumulative relative frequency graphs as *ogives* (pronounced "o-jives") or as *percentile plots.*

**DEFINITION Cumulative relative frequency graph**

A **cumulative relative frequency graph** plots a point corresponding to the percentile of a given value in a distribution of quantitative data. Consecutive points are then connected with a line segment to form the graph.

Figure 2.2 shows the completed cumulative relative frequency graph for the presidential age at inauguration data. Notice the following details:

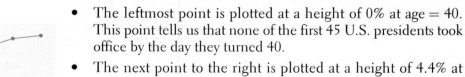

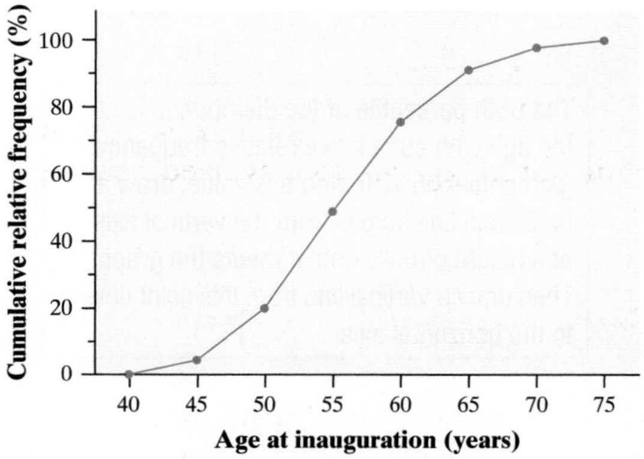


**FIGURE 2.2** Cumulative relative frequency graph of the ages of U.S. presidents when they took office.

- The leftmost point is plotted at a height of 0% at age = 40. This point tells us that none of the first 45 U.S. presidents took office by the day they turned 40.
- The next point to the right is plotted at a height of 4.4% at age = 45. This point tells us that 4.4% of these presidents were inaugurated by their 45th birthday.
- The graph grows very gradually at first because few presidents were inaugurated when they were in their 40s. Then the graph gets very steep beginning at age 50 because most U.S. presidents were in their 50s when they took office. The rapid growth in the graph slows at age 60.
- The rightmost point on the graph is plotted above age 75 and has cumulative relative frequency 100%. That's because 100% of these U.S. presidents took office by age 75.

A cumulative relative frequency graph can be used to describe the position of an individual value in a distribution or to locate a specified percentile of the distribution.

## EXAMPLE

### Ages of U.S. presidents
### Interpreting a cumulative relative frequency graph

Jae C. Hong/AP Photo

**PROBLEM:** Use the graph in Figure 2.2 to help you answer each question.

(a) Was Barack Obama, who was first inaugurated at age 47 years, 169 days (about 47.463 years), unusually young?

(b) Estimate and interpret the 65th percentile of the distribution.

**SOLUTION:**

(a)

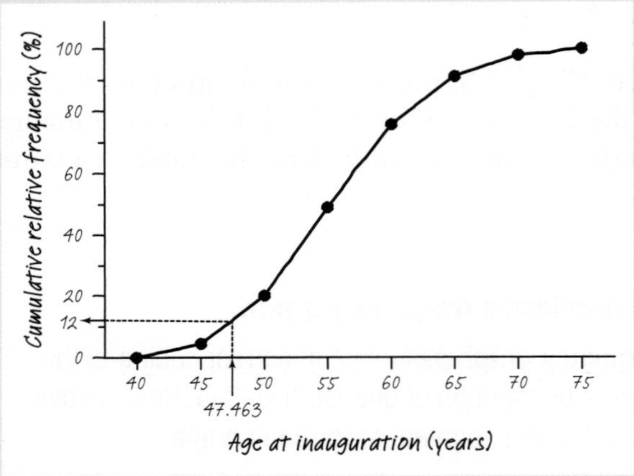


To find President Obama's location in the distribution, draw a vertical line up from his age (47.463) on the horizontal axis until it meets the graph. Then draw a horizontal line from this point to the vertical axis.

Barack Obama's inauguration age places him at about the 12th percentile. About 12% of the first 45 U.S. presidents took office by the time they were 47.463 years old. So Obama was fairly young, but not unusually young, when he took office.

(b)

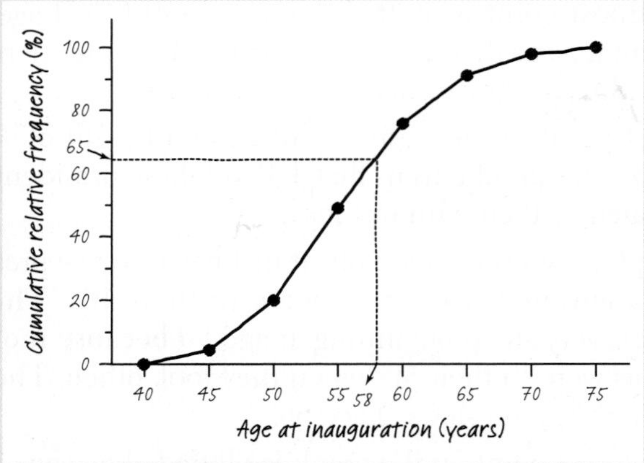


The 65th percentile of the distribution is the age with cumulative relative frequency (percentile) 65%. To find this value, draw a horizontal line across from the vertical axis at a height of 65% until it meets the graph. Then draw a vertical line from this point down to the horizontal axis.

The 65th percentile is about 58 years old. About 65% of the first 45 U.S. presidents took office by the time they turned 58.

**FOR PRACTICE, TRY EXERCISE 9**

## CHECK YOUR UNDERSTANDING

1. *Multiple choice: Select the best answer.* Mark receives a score report detailing his performance on a statewide test. On the math section, Mark earned a raw score of 39, which placed him at the 68th percentile. This means that
   (a) Mark did the same as or better than about 39% of those who took the test.
   (b) Mark did worse than about 39% of those who took the test.
   (c) Mark did the same as or better than about 68% of those who took the test.
   (d) Mark did worse than about 68% of those who took the test.
   (e) Mark got more than half of the questions correct on this test.
2. Mrs. Munson is concerned about how her daughter's height and weight compare with those of other girls of the same age. She uses an online calculator to determine that her daughter is at the 87th percentile for weight and the 67th percentile for height. Explain to Mrs. Munson what these values mean.

*Questions 3 and 4 relate to the following setting.* The graph displays the cumulative relative frequency of the lengths of phone calls made from the mathematics department office at Gabalot High last month.

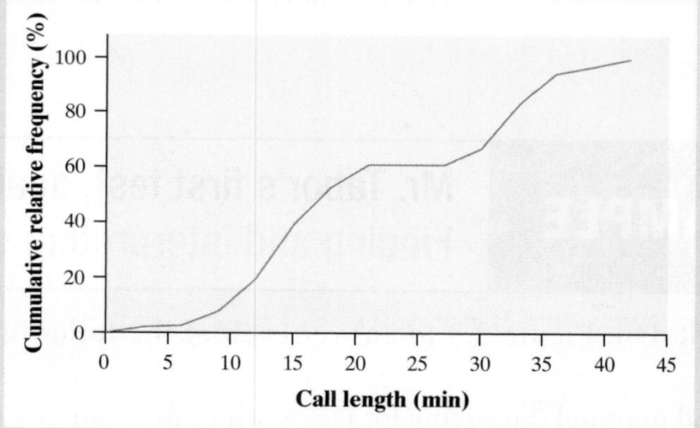


3. About what percent of calls lasted less than or equal to 30 minutes? More than 30 minutes?
4. Estimate $Q_1$, $Q_3$, and the *IQR* of the distribution of phone call length.

# Measuring Location: Standardized Scores

A percentile is one way to describe the location of an individual in a distribution of quantitative data. Another way is to give the **standardized score** (*z*-**score**) for the observed value.

**DEFINITION Standardized score (*z*-score)**

The **standardized score (*z*-score)** for an individual value in a distribution tells us how many standard deviations from the mean the value falls, and in what direction. To find the standardized score (*z*-score), compute

$$z = \frac{\text{value} - \text{mean}}{\text{standard deviation}}$$

Values larger than the mean have positive $z$-scores. Values smaller than the mean have negative $z$-scores.

Let's return to the data from Mr. Tabor's first statistics test. Figure 2.3 shows a dotplot of the data, along with numerical summaries.

The relationship between the mean and the median is about what you'd expect in this left-skewed distribution.

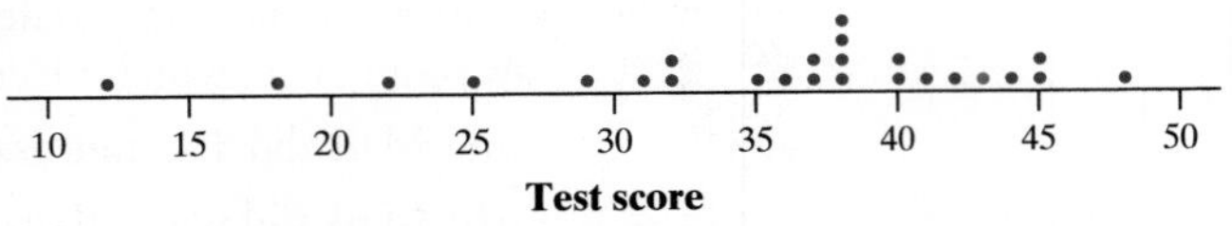


| $n$ | $\bar{x}$ | $s_x$ | Min | $Q_1$ | Med | $Q_3$ | Max |
|---|---|---|---|---|---|---|---|
| 25 | 35.44 | 8.77 | 12 | 31.5 | 38 | 41.5 | 48 |

**FIGURE 2.3** Dotplot and summary statistics of the scores on Mr. Tabor's first statistics test. Emily's score of 43 is marked in red on the dotplot.

Where does Emily's 43 (marked in red on the dotplot) fall in the distribution? Her standardized score ($z$-score) is

$$z = \frac{\text{value} - \text{mean}}{\text{standard deviation}} = \frac{43 - 35.44}{8.77} = 0.86$$

That is, Emily's test score is 0.86 standard deviations above the mean score of the class.

## EXAMPLE

### Mr. Tabor's first test, again
### Finding and interpreting $z$-scores

**PROBLEM:** Use Figure 2.3 to answer each of the following questions.

(a) Find and interpret the $z$-score for Jacob, who earned an 18 on the test.

(b) Tamika had a standardized score of 0.292. Find Tamika's test score.

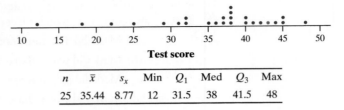

ESB Professional/Shutterstock.com

**SOLUTION:**

(a) $z = \frac{18 - 35.44}{8.77} = -1.99$

Jacob's test score is 1.99 standard deviations below the class mean of 35.44.

Be sure to interpret the magnitude (number of standard deviations) and direction (greater than or less than the mean) of a $z$-score in context.

(b) $0.292 = \frac{\text{value} - 35.44}{8.77}$

$0.292(8.77) + 35.44 = \text{value}$

$38 = \text{value}$

Tamika's test score was 38.

Be sure to show your work when finding the value that corresponds to a given $z$-score.

**FOR PRACTICE, TRY EXERCISE 13**

We often standardize observed values to express them on a common scale. For example, we might compare the heights of two children of different ages or genders by calculating their *z*-scores.

- At age 9, Jordan is 55 inches tall. Her height puts her at a *z*-score of 1. That is, Jordan is 1 standard deviation above the mean height of 9-year-old girls.
- Zayne's height at age 11 is 58 inches. His corresponding *z*-score is 0.5. In other words, Zayne is 1/2 standard deviation above the mean height of 11-year-old boys.

Even though Zayne's height is larger, Jordan is taller relative to girls her age than Zayne is relative to boys his age. The standardized heights tell us where each child stands (pun intended!) in the distribution for his or her age group.

## CHECK YOUR UNDERSTANDING

1. Mrs. Navard's statistics class has just completed the first three steps of the "Where do I stand?" activity (page 90). The figure shows a dotplot of the distribution of height for the class, along with summary statistics from computer output.

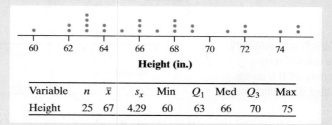

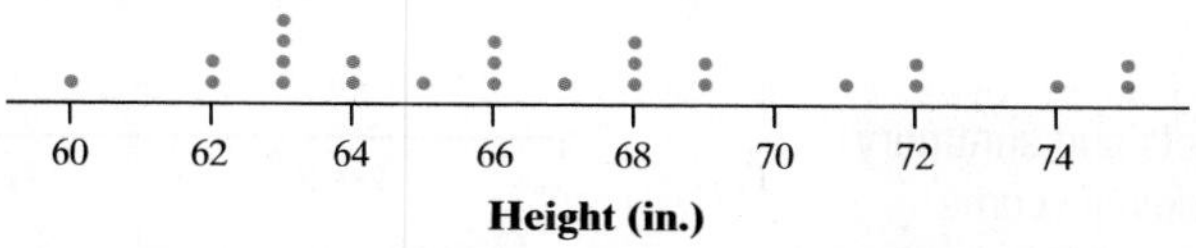


| Variable | $n$ | $\bar{x}$ | $s_x$ | Min | $Q_1$ | Med | $Q_3$ | Max |
|---|---|---|---|---|---|---|---|---|
| Height | 25 | 67 | 4.29 | 60 | 63 | 66 | 70 | 75 |

Lynette, a student in the class, is 62 inches tall. Find and interpret her *z*-score.

2. Brent is a member of the school's basketball team and is 74 inches tall. The mean height of the players on the team is 76 inches. Brent's height translates to a standardized score of −0.85 in the team's distribution of height. What is the standard deviation of the team members' heights?

# Transforming Data

To find the standardized score (*z*-score) for an individual observation, we transform this data value by subtracting the mean and dividing the difference by the standard deviation. Transforming converts the observation from the original units of measurement (e.g., inches) to a standardized scale.

There are other reasons to transform data. We may want to change the units of measurement for a data set from kilograms to pounds (1 kg $\approx$ 2.2 lb), or from Fahrenheit to Celsius $\left[°C = \frac{5}{9}(°F - 32)\right]$. Or perhaps a measuring device is calibrated wrong, so we have to add a constant to each data value to get accurate measurements. What effect do these kinds of transformations—adding or subtracting; multiplying or dividing—have on the shape, center, and variability of a distribution?

**EFFECT OF ADDING OR SUBTRACTING A CONSTANT** Recall that Mr. Tabor gave his class of 25 statistics students a first test worth 50 points. Here is a dotplot of the students' scores along with some numerical summaries.

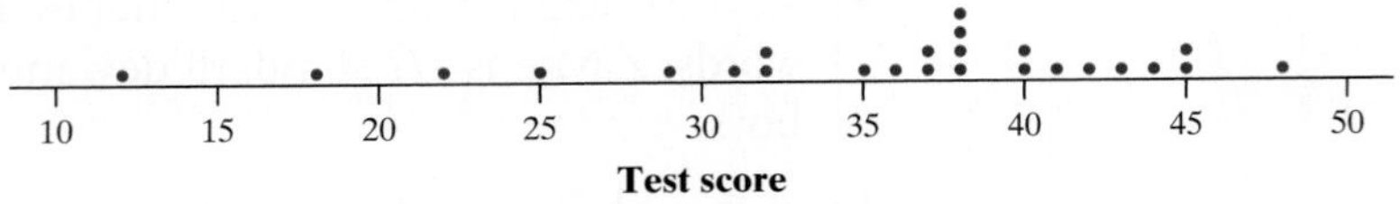

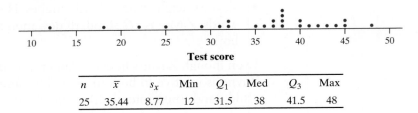

| $n$ | $\bar{x}$ | $s_x$ | Min | $Q_1$ | Med | $Q_3$ | Max |
|---|---|---|---|---|---|---|---|
| 25 | 35.44 | 8.77 | 12 | 31.5 | 38 | 41.5 | 48 |

Suppose Mr. Tabor was nice and added 5 points to each student's test score. How would this affect the distribution of scores? Figure 2.4 shows graphs and numerical summaries for the original test scores and adjusted scores.

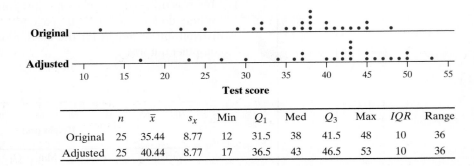

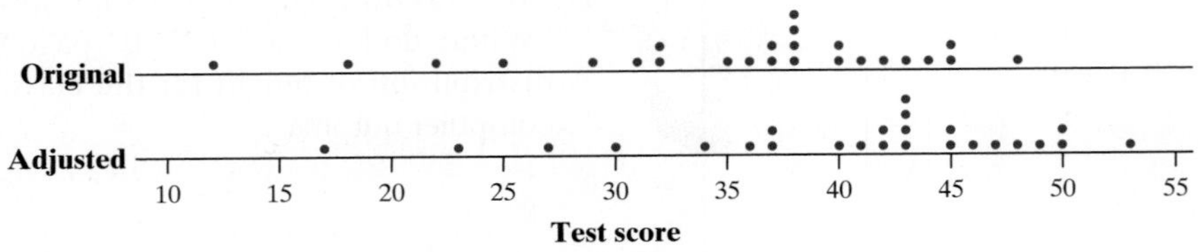


| | $n$ | $\bar{x}$ | $s_x$ | Min | $Q_1$ | Med | $Q_3$ | Max | $IQR$ | Range |
|---|---|---|---|---|---|---|---|---|---|---|
| Original | 25 | 35.44 | 8.77 | 12 | 31.5 | 38 | 41.5 | 48 | 10 | 36 |
| Adjusted | 25 | 40.44 | 8.77 | 17 | 36.5 | 43 | 46.5 | 53 | 10 | 36 |

**FIGURE 2.4** Dotplots and summary statistics for the original scores and adjusted scores (with 5 points added) on Mr. Tabor's statistics test.

From both the graph and summary statistics, we can see that the measures of center (mean and median) and location (min, $Q_1$, $Q_3$, and max) increased by 5 points. The shape of the distribution did not change. Neither did the variability of the distribution—the range, the standard deviation, and the interquartile range ($IQR$) all stayed the same.

As this example shows, adding the same positive number to each value in a data set shifts the distribution to the right by that number. Subtracting a positive constant from each data value would shift the distribution to the left by that constant.

## THE EFFECT OF ADDING OR SUBTRACTING A CONSTANT

Adding the same positive number $a$ to (subtracting $a$ from) each observation:

- Adds $a$ to (subtracts $a$ from) measures of center and location (mean, five-number summary, percentiles)
- Does not change measures of variability (range, $IQR$, standard deviation)
- Does not change the shape of the distribution

# EXAMPLE

## How wide is this room?
### Effect of adding/subtracting a constant

Regissercom/Shutterstock.com

**PROBLEM:** Soon after the metric system was introduced in Australia, a group of students was asked to guess the width of their classroom to the nearest meter. A dotplot of the data and some numerical summaries are shown.[1]

The actual width of the room was 13 meters. We can examine the distribution of students' errors by defining a new variable as follows: error = guess − 13. Note that a negative value for error indicates that a student's guess for the width of the room was too small.

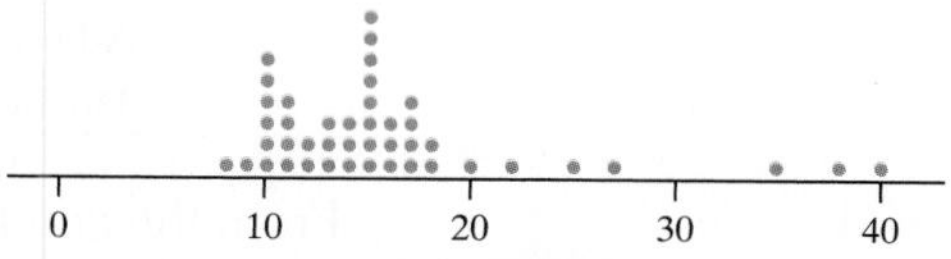


| | $n$ | $\bar{x}$ | $s_x$ | Min | $Q_1$ | Med | $Q_3$ | Max | $IQR$ | Range |
|---|---|---|---|---|---|---|---|---|---|---|
| **Guess (m)** | 44 | 16.02 | 7.14 | 8 | 11 | 15 | 17 | 40 | 6 | 32 |

(a) What shape does the distribution of error have?

(b) Find the mean and the median of the distribution of error.

(c) Find the standard deviation and interquartile range ($IQR$) of the distribution of error.

**SOLUTION:**

(a) The same shape as the original distribution of guesses: skewed to the right with two distinct peaks.

(b) Mean: 16.02 − 13 = 3.02 meters; Median: 15 − 13 = 2 meters

It is not a surprise that the mean is greater than the median in this right-skewed distribution.

(c) The same as for the original distribution of guesses:

Standard deviation = 7.14 meters, $IQR$ = 6 meters

**FOR PRACTICE, TRY EXERCISE 21**

Figure 2.5 confirms the results of the example.

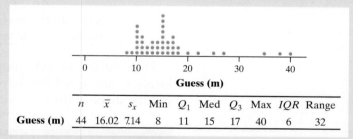

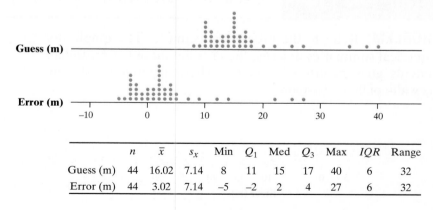

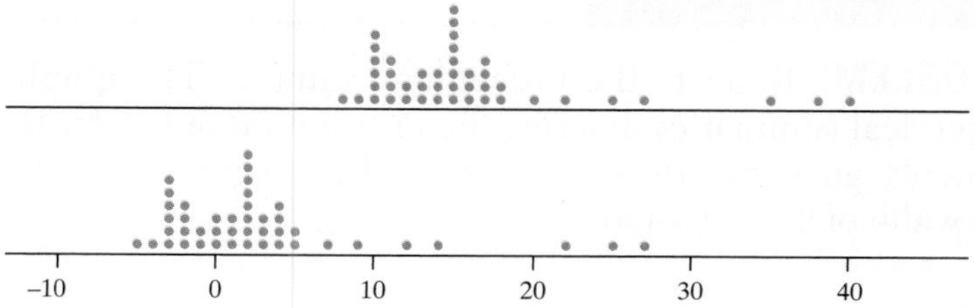


| | $n$ | $\bar{x}$ | $s_x$ | Min | $Q_1$ | Med | $Q_3$ | Max | $IQR$ | Range |
|---|---|---|---|---|---|---|---|---|---|---|
| Guess (m) | 44 | 16.02 | 7.14 | 8 | 11 | 15 | 17 | 40 | 6 | 32 |
| Error (m) | 44 | 3.02 | 7.14 | −5 | −2 | 2 | 4 | 27 | 6 | 32 |

**FIGURE 2.5** Dotplots and summary statistics for the Australian students' guesses of classroom width and the errors in their guesses, in meters.

What about outliers? You can check that the four highest guesses—which are 27, 35, 38, 40 meters—are outliers by the 1.5 × $IQR$ rule. The same individuals will still be outliers in the distribution of error, but each of their values will be decreased by 13 meters: 14, 22, 25, and 27 meters.

**EFFECT OF MULTIPLYING OR DIVIDING BY A CONSTANT** Suppose that Mr. Tabor wants to convert his students' adjusted test scores to percents. Because the test was out of 50 points, he should multiply each score by 2 to be counted out of 100 points instead. Here are graphs and numerical summaries for the adjusted scores and the doubled scores:

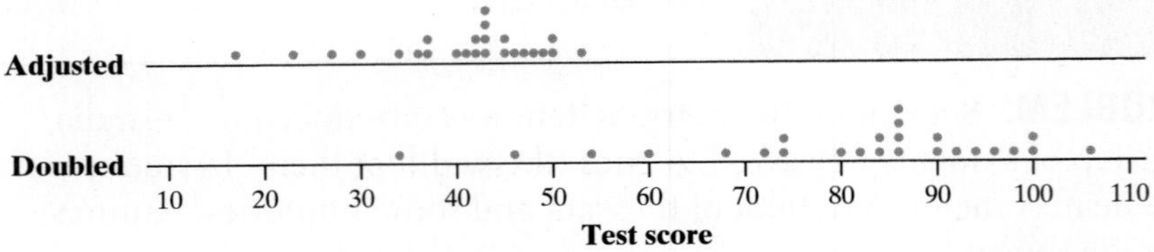


| | $n$ | $\bar{x}$ | $s_x$ | Min | $Q_1$ | Med | $Q_3$ | Max | $IQR$ | Range |
|---|---|---|---|---|---|---|---|---|---|---|
| Adjusted | 25 | 40.44 | 8.77 | 17 | 36.5 | 43 | 46.5 | 53 | 10 | 36 |
| Doubled | 25 | 80.88 | 17.54 | 34 | 73 | 86 | 93 | 106 | 20 | 72 |

From the graphs and summary statistics, we can see that the measures of center, location, and variability have all doubled, just like the individual observations. But the shape of the two distributions is the same.

It is not common to multiply (or divide) each observation in a data set by a *negative* number $b$. Doing so would multiply (or divide) the measures of variability by the *absolute value* of $b$. We can't have a negative amount of variability! Multiplying or dividing by a negative number would also affect the shape of the distribution, as all values would be reflected over the $y$ axis.

### EFFECT OF MULTIPLYING OR DIVIDING BY A CONSTANT

Multiplying (or dividing) each observation by the same positive number $b$:

- Multiplies (divides) measures of center and location (mean, five-number summary, percentiles) by $b$
- Multiplies (divides) measures of variability (range, $IQR$, standard deviation) by $b$
- Does not change the shape of the distribution

## EXAMPLE

### How far off were our guesses?
### Effect of multiplying/dividing by a constant

**PROBLEM:** Refer to the preceding example. The graph and numerical summaries describe the distribution of the Australian students' guessing errors (in meters) when they tried to estimate the width of their classroom.

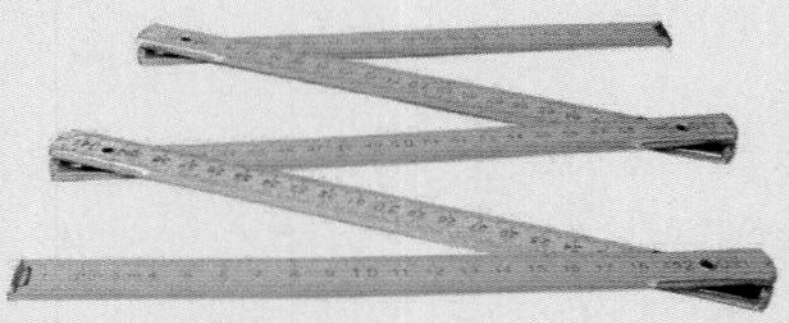

N center/Shutterstock.com

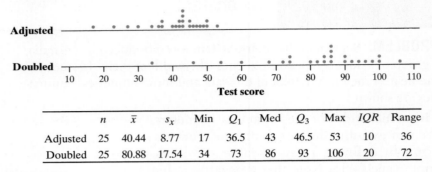

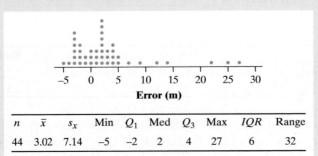

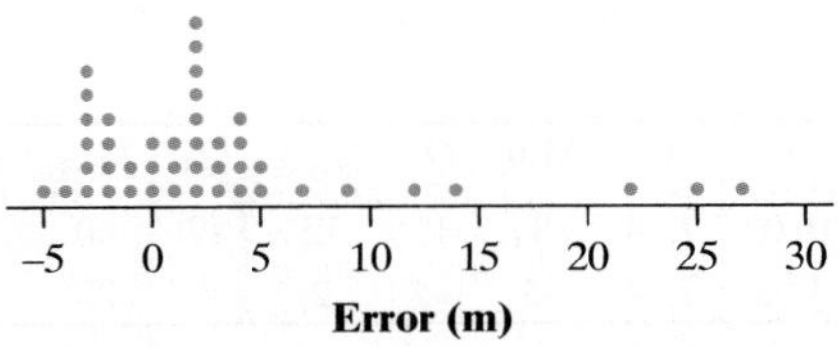


| $n$ | $\bar{x}$ | $s_x$ | Min | $Q_1$ | Med | $Q_3$ | Max | $IQR$ | Range |
|---|---|---|---|---|---|---|---|---|---|
| 44 | 3.02 | 7.14 | −5 | −2 | 2 | 4 | 27 | 6 | 32 |

Because the students are having some difficulty with the metric system, it may not be helpful to tell them that their guesses tended to be about 2 meters too high. Let's convert the error data to feet before we report back to them.

There are roughly 3.28 feet in a meter. For the student whose error was −5 meters, that translates to

$$-5 \text{ meters} \times \frac{3.28 \text{ feet}}{1 \text{ meter}} = -16.4 \text{ feet}$$

To change the units of measurement from meters to feet, we multiply each of the error values by 3.28.

(a) What shape does the resulting distribution of error have?

(b) Find the median of the distribution of error in feet.

(c) Find the interquartile range (*IQR*) of the distribution of error in feet.

**SOLUTION:**

(a) The same shape as the original distribution of guesses: skewed to the right with two distinct peaks.

(b) Median $= 2 \times 3.28 = 6.56$ feet

(c) $IQR = 6 \times 3.28 = 19.68$ feet

**FOR PRACTICE, TRY EXERCISE 25**

Figure 2.6 confirms the results of the example.

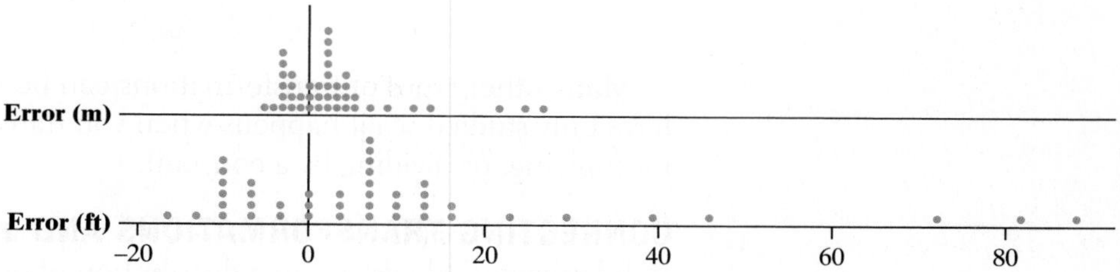


| | $n$ | $\bar{x}$ | $s_x$ | Min | $Q_1$ | Med | $Q_3$ | Max | *IQR* | Range |
|---|---|---|---|---|---|---|---|---|---|---|
| Error (m) | 44 | 3.02 | 7.14 | −5 | −2 | 2 | 4 | 27 | 6 | 32 |
| Error (ft) | 44 | 9.91 | 23.42 | −16.4 | −6.56 | 6.56 | 13.12 | 88.56 | 19.68 | 104.96 |

**FIGURE 2.6** Dotplots and summary statistics for the errors in Australian students' guesses of classroom width, in meters and in feet.

**PUTTING IT ALL TOGETHER: ADDING/SUBTRACTING AND MULTIPLYING/DIVIDING** What happens if we transform a data set by both adding or subtracting a constant and multiplying or dividing by a constant? We just use the facts about transforming data that we've already established and the order of operations.

## EXAMPLE

### Too cool at the cabin?
### Analyzing the effects of transformations

**PROBLEM:** During the winter months, the temperatures at the Starnes's Colorado cabin can stay well below freezing (32°F, or 0°C) for weeks at a time. To prevent the pipes from freezing, Mrs. Starnes sets the thermostat at 50°F. She also buys a digital thermometer that records the indoor temperature each night at midnight. Unfortunately, the thermometer is programmed to measure the temperature in degrees Celsius. Following are a dotplot and numerical summaries of the midnight temperature readings for a 30-day period.

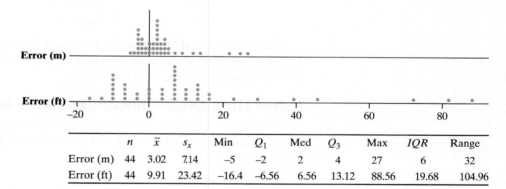

gulfix/Getty Images

Use the fact that °F = (9/5)°C + 32 to help you answer the following questions.

(a) Find the mean temperature in degrees Fahrenheit. Does the thermostat setting seem accurate?

(b) Calculate the standard deviation of the temperature readings in degrees Fahrenheit. Interpret this value in context.

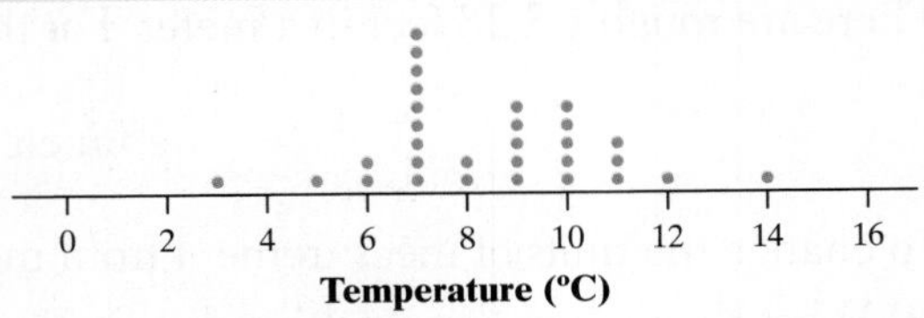


| $n$ | Mean | SD | Min | $Q_1$ | Med | $Q_3$ | Max |
|---|---|---|---|---|---|---|---|
| 30 | 8.43 | 2.27 | 3.00 | 7.00 | 8.50 | 10.00 | 14.00 |

**SOLUTION:**

(a) Mean = (9/5)(8.43) + 32 = 47.17°F. The thermostat doesn't seem to be very accurate. It is set at 50°F, but the mean temperature over the 30-day period is about 47°F.

(b) SD = (9/5)(2.27) = 4.09°F. The temperature readings typically vary from the mean by about 4°F. That's a lot of variation!

Multiplying each observation by 9/5 multiplies the standard deviation by 9/5. However, adding 32 to each observation doesn't affect the variability.

**FOR PRACTICE, TRY EXERCISE 29**

Many other types of transformations can be very useful in analyzing data. We have only studied what happens when you transform data by adding, subtracting, multiplying, or dividing by a constant.

**CONNECTING TRANSFORMATIONS AND *z*-SCORES** What happens if we standardize *all* the values in a distribution of quantitative data? Here is a dotplot of the original test scores for the 25 students in Mr. Tabor's statistics class, along with some numerical summaries:

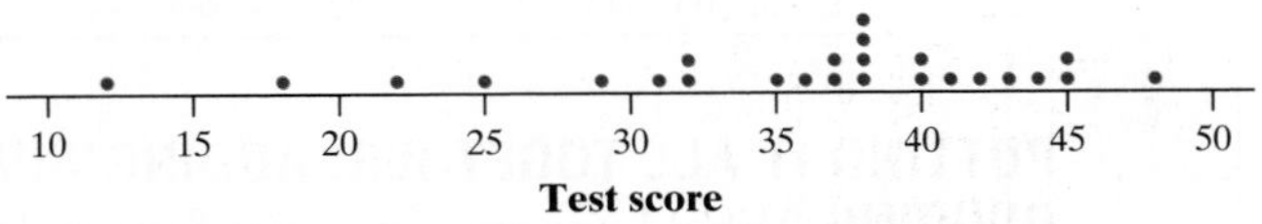


| $n$ | $\bar{x}$ | $s_x$ |
|---|---|---|
| 25 | 35.44 | 8.77 |

We calculate the standardized score for each student using

$$z = \frac{\text{score} - 35.44}{8.77}$$

In other words, we subtract 35.44 from each student's test score and then divide by 8.77. What effect do these transformations have on the shape, center, and variability of the distribution?

Here is a dotplot of the class's *z*-scores. Let's describe the distribution.

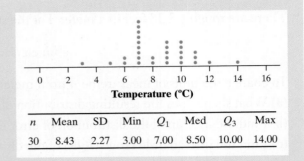

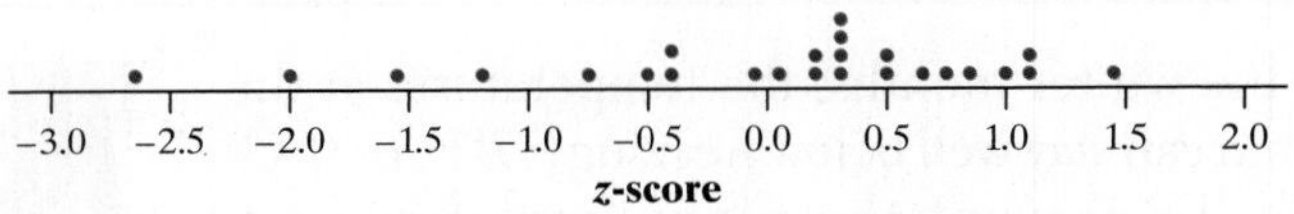


| $n$ | $\bar{x}$ | $s_x$ |
|---|---|---|
| 25 | 0 | 1 |

- **Shape:** *The shape of the distribution of z-scores is the same as the shape of the original distribution of test scores*—skewed to the left. Neither subtracting a constant nor dividing by a constant changes the shape of the graph.
- **Center:** *The mean of the distribution of z-scores is 0.* Subtracting 35.44 from each test score would reduce the mean from 35.44 to 0. Dividing each of

these new data values by 8.77 would divide the new mean of 0 by 8.77, which still yields a mean of 0.

- **Variability:** *The standard deviation of the distribution of z-scores is 1.* Subtracting 35.44 from each test score does not affect the standard deviation. However, dividing all of the resulting values by 8.77 would divide the original standard deviation of 8.77 by 8.77, yielding 1.

## CHECK YOUR UNDERSTANDING

Knoebels Amusement Park in Elysburg, Pennsylvania, has earned acclaim for being an affordable, family-friendly entertainment venue. Knoebels does not charge for general admission or parking, but it does charge customers for each ride they take. How much do the rides cost at Knoebels? The figure shows a dotplot of the cost for each of 22 rides in a recent year, along with summary statistics.

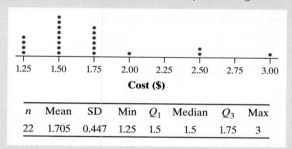

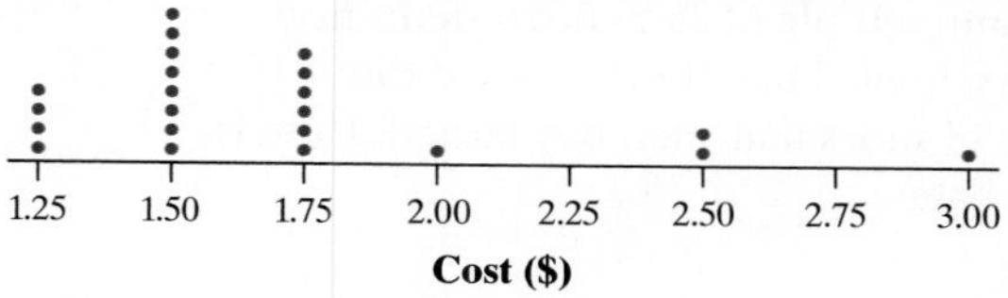


| $n$ | Mean | SD | Min | $Q_1$ | Median | $Q_3$ | Max |
|---|---|---|---|---|---|---|---|
| 22 | 1.705 | 0.447 | 1.25 | 1.5 | 1.5 | 1.75 | 3 |

1. Suppose you convert the cost of the rides from dollars to cents ($1 = 100 cents). Describe the shape, mean, and standard deviation of the distribution of ride cost in cents.
2. Knoebels' managers decide to increase the cost of each ride by 25 cents. How would the shape, center, and variability of this distribution compare with the distribution of cost in Question 1?
3. Now suppose you convert the increased costs from Question 2 to *z*-scores. What would be the shape, mean, and standard deviation of this distribution? Explain your answers.

## Section 2.1 Summary

- Two ways of describing an individual value's location in a distribution are **percentiles** and **standardized scores (*z*-scores).** The *p*th percentile of a distribution is the value with *p* percent of the observations less than or equal to it.
- To standardize any data value, subtract the mean of the distribution and then divide the difference by the standard deviation. The resulting *z*-score

$$z = \frac{\text{value} - \text{mean}}{\text{standard deviation}}$$

measures how many standard deviations the data value lies above or below the mean of the distribution. We can also use percentiles and *z*-scores to compare the relative location of individuals in different distributions.

- A **cumulative relative frequency graph** allows us to examine location in a distribution. The completed graph allows you to estimate the percentile for an individual value, and vice versa.

- It is necessary to **transform data** when changing units of measurement.
  - When you add a positive constant *a* to (subtract *a* from) all the values in a data set, measures of center and location—mean, five-number summary, percentiles—increase (decrease) by *a*. Measures of variability—range, *IQR*, SD—do not change.
  - When you multiply (divide) all the values in a data set by a positive constant *b*, measures of center, location, and variability are multiplied (divided) by *b*.
  - Neither of these transformations changes the shape of the distribution.

## Section 2.1 Exercises

1. **Shoes** How many pairs of shoes does a typical teenage boy own? To find out, two AP® Statistics students surveyed a random sample of 20 male students from their large high school. Then they recorded the number of pairs of shoes that each boy owned. Here is a dotplot of the data:

pg 92

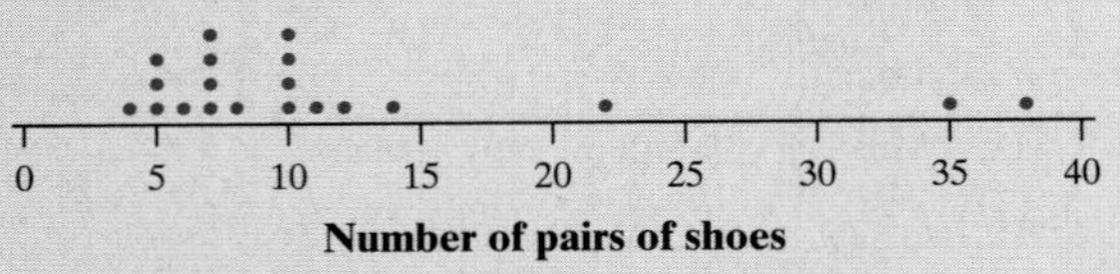


(a) Find the percentile for Jackson, who reported owning 22 pairs of shoes.

(b) Raul's reported number of pairs of shoes is at the 45th percentile of the distribution. Interpret this value. How many pairs of shoes does Raul own?

2. **Old folks** Here is a stemplot of the percents of residents aged 65 and older in the 50 states:

| | |
|---|---|
| 7 | 0 |
| 8 | 8 |
| 9 | 8 |
| 10 | 019 |
| 11 | 16777 |
| 12 | 01122456778999 |
| 13 | 0001223344455689 |
| 14 | 023568 |
| 15 | 24 |
| 16 | 9 |

Key: 15|2 means 15.2% of this state's residents are 65 or older.

(a) Find the percentile for Colorado, where 10.1% of the residents are aged 65 and older.

(b) Rhode Island is at the 80th percentile of the distribution. Interpret this value. What percent of Rhode Island's residents are aged 65 and older?

3. **Wear your helmet!** Many athletes (and their parents) worry about the risk of concussions when playing sports. A football coach plans to obtain specially made helmets for his players that are designed to reduce the chance of getting a concussion. Here are the measurements of head circumference (in inches) for the players on the team:

| | | | | | | | |
|---|---|---|---|---|---|---|---|
| 23.0 | 22.2 | 21.7 | 22.0 | 22.3 | 22.6 | 22.7 | 21.5 |
| 22.7 | 25.6 | 20.8 | 23.0 | 24.2 | 23.5 | 20.8 | 24.0 |
| 22.7 | 22.6 | 23.9 | 22.5 | 23.1 | 21.9 | 21.0 | 22.4 |
| 23.5 | 22.5 | 23.9 | 23.4 | 21.6 | 23.3 | | |

(a) Antawn, the team's starting quarterback, has a head circumference of 22.4 inches. What is Antawn's percentile?

(b) Find the head circumference of the player at the 90th percentile of the distribution.

4. **Don't call me** According to a study by Nielsen Mobile, "Teenagers ages 13 to 17 are by far the most prolific texters, sending 1742 messages a month." Mr. Williams, a high school statistics teacher, was skeptical about the claims in the article. So he collected data from his first-period statistics class on the number of text messages they had sent over the past 24 hours. Here are the data:

| | | | | | | | | | | | | |
|---|---|---|---|---|---|---|---|---|---|---|---|---|
| 0 | 7 | 1 | 29 | 25 | 8 | 5 | 1 | 25 | 98 | 9 | 0 | 26 |
| 8 | 118 | 72 | 0 | 92 | 52 | 14 | 3 | 3 | 44 | 5 | 42 | |

(a) Sunny was the student who sent 42 text messages. What is Sunny's percentile?

(b) Find the number of text messages sent by Joelle, who is at the 20th percentile of the distribution.

5. **Setting speed limits** According to the *Los Angeles Times*, speed limits on California highways are set at the 85th percentile of vehicle speeds on those stretches of road. Explain to someone who knows little statistics what that means.

6. **Blood pressure** Larry came home very excited after a visit to his doctor. He announced proudly to his wife, "My doctor says my blood pressure is at the 90th percentile among men like me. That means I'm better off than about 90% of similar men."

How should his wife, who is a statistician, respond to Larry's statement?

7. **Growth charts** We used an online growth chart to find percentiles for the height and weight of a 16-year-old girl who is 66 inches tall and weighs 118 pounds. According to the chart, this girl is at the 48th percentile for weight and the 78th percentile for height. Explain what these values mean in plain English.

8. **Track star** Peter is a star runner on the track team. In the league championship meet, Peter records a time that would fall at the 80th percentile of all his race times that season. But his performance places him at the 50th percentile in the league championship meet. Explain how this is possible. (Remember that shorter times are better in this case!)

9. **Run fast!** As part of a student project, high school students were asked to sprint 50 yards, and their times (in seconds) were recorded. A cumulative relative frequency graph of the sprint times is shown here.

pg 94

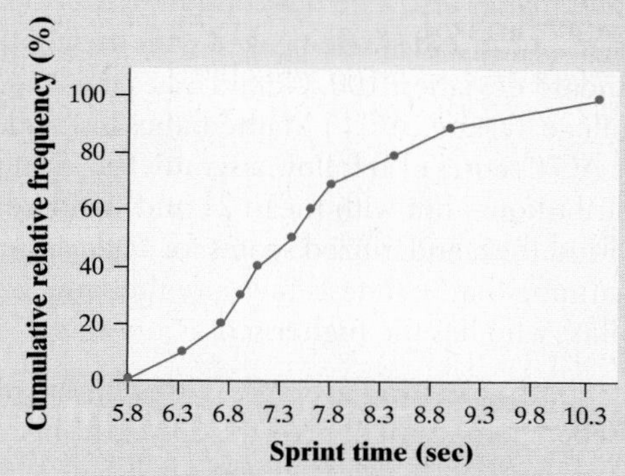


(a) One student ran the 50 yards in 8.05 seconds. Is a sprint time of 8.05 seconds unusually slow?

(b) Estimate and interpret the 20th percentile of the distribution.

10. **Household incomes** The cumulative relative frequency graph describes the distribution of median household incomes in the 50 states in a recent year.[2]

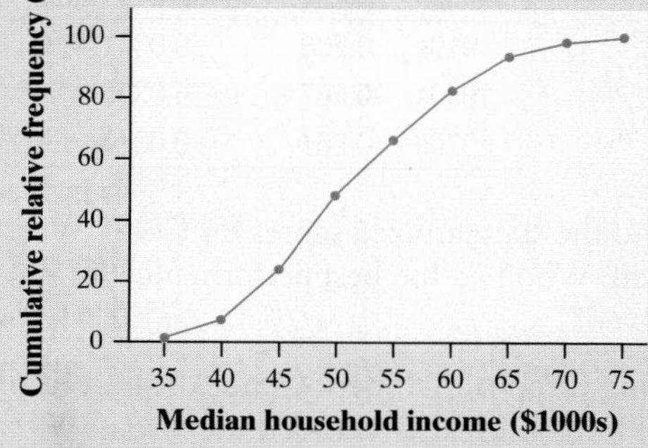


(a) The median household income in North Dakota that year was $55,766. Is North Dakota an unusually wealthy state?

(b) Estimate and interpret the 90th percentile of the distribution.

11. **Foreign-born residents** The cumulative relative frequency graph shows the distribution of the percent of foreign-born residents in the 50 states.

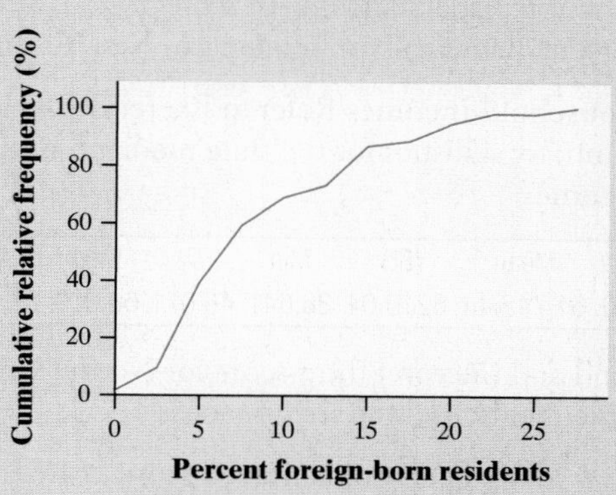


(a) Estimate the interquartile range (*IQR*) of this distribution. Show your method.

(b) Arizona had 15.1% foreign-born residents that year. Estimate its percentile.

(c) Explain why the graph is fairly flat between 20% and 27.5%.

12. **Shopping spree** The figure is a cumulative relative frequency graph of the amount spent by 50 consecutive grocery shoppers at a store.

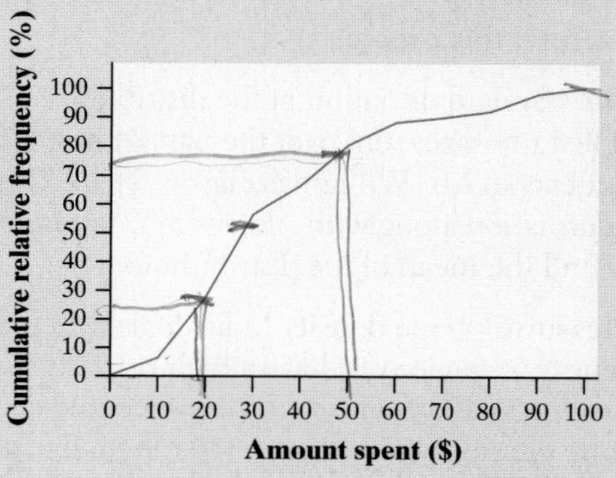


(a) Estimate the interquartile range (*IQR*) of this distribution. Show your method.

(b) One shopper spent $19.50. Estimate this person's percentile.

(c) Explain why the graph is steepest between $10 and $30.

13. **Foreign-born residents** Refer to Exercise 11. Here pg 96 are summary statistics for the percent of foreign-born residents in the 50 states:

| *n* | Mean | SD | Min | $Q_1$ | Med | $Q_3$ | Max |
|---|---|---|---|---|---|---|---|
| 50 | 8.73 | 6.12 | 1.3 | 4.1 | 6.2 | 13.4 | 27.1 |

(a) Find and interpret the *z*-score for Montana, which had 1.9% foreign-born residents.

(b) New York had a standardized score of 2.10. Find the percent of foreign-born residents in New York at that time.

14. **Household incomes** Refer to Exercise 10. Here are summary statistics for the state median household incomes:

| *n* | Mean | SD | Min | $Q_1$ | Med | $Q_3$ | Max |
|---|---|---|---|---|---|---|---|
| 50 | 51,742.44 | 8210.64 | 36,641 | 46,071 | 50,009 | 57,020 | 71,836 |

(a) Find and interpret the *z*-score for North Carolina, with a median household income of $41,553.

(b) New Jersey had a standardized score of 1.82. Find New Jersey's median household income for that year.

15. **Shoes** Refer to Exercise 1. Jackson, who reported owning 22 pairs of shoes, has a standardized score of $z = 1.10$.

(a) Interpret this *z*-score.

(b) The standard deviation of the distribution of number of pairs of shoes owned in this sample of 20 boys is 9.42. Use this information along with Jackson's standardized score to find the mean of the distribution.

16. **Don't call me** Refer to Exercise 4. Alejandro, who sent 92 texts, has a standardized score of $z = 1.89$.

(a) Interpret this *z*-score.

(b) The standard deviation of the distribution of number of text messages sent over the past 24 hours by the students in Mr. Williams's class is 34.15. Use this information along with Alejandro's standardized score to find the mean of the distribution.

17. **Measuring bone density** Individuals with low bone density (osteoporosis) have a high risk of broken bones (fractures). Physicians who are concerned about low bone density in patients can refer them for specialized testing. Currently, the most common method for testing bone density is dual-energy X-ray absorptiometry (DEXA). The bone density results for a patient who undergoes a DEXA test usually are reported in grams per square centimeter (g/cm$^2$) and in standardized units.

Judy, who is 25 years old, has her bone density measured using DEXA. Her results indicate bone density in the hip of 948 g/cm$^2$ and a standardized score of $z = -1.45$. The mean bone density in the hip is 956 g/cm$^2$ in the reference population of 25-year-old women like Judy.[3]

(a) Judy has not taken a statistics class in a few years. Explain to her in simple language what the standardized score reveals about her bone density.

(b) Use the information provided to calculate the standard deviation of bone density in the reference population.

18. **Comparing bone density** Refer to Exercise 17. Judy's friend Mary also had the bone density in her hip measured using DEXA. Mary is 35 years old. Her bone density is also reported as 948 g/cm$^2$, but her standardized score is $z = 0.50$. The mean bone density in the hip for the reference population of 35-year-old women is 944 grams/cm$^2$.

(a) Whose bones are healthier for her age: Judy's or Mary's? Justify your answer.

(b) Calculate the standard deviation of the bone density in Mary's reference population. How does this compare with your answer to Exercise 17(b)? Are you surprised?

19. **SAT versus ACT** Eleanor scores 680 on the SAT Mathematics test. The distribution of SAT Math scores is symmetric and single-peaked with mean 500 and standard deviation 100. Gerald takes the American College Testing (ACT) Mathematics test and scores 29. ACT scores also follow a symmetric, single-peaked distribution—but with mean 21 and standard deviation 5. Find the standardized scores for both students. Assuming that both tests measure the same kind of ability, who has the higher score?

20. **Comparing batting averages** Three landmarks of baseball achievement are Ty Cobb's batting average of 0.420 in 1911, Ted Williams's 0.406 in 1941, and George Brett's 0.390 in 1980. These batting averages cannot be compared directly because the distribution of major league batting averages has changed over the years. The distributions are quite symmetric, except for outliers such as Cobb, Williams, and Brett. While the mean batting average has been held roughly constant by rule changes and the balance between hitting and pitching, the standard deviation has dropped over time. Here are the facts:

| Decade | Mean | Standard deviation |
|---|---|---|
| 1910s | 0.266 | 0.0371 |
| 1940s | 0.267 | 0.0326 |
| 1970s | 0.261 | 0.0317 |

Find the standardized scores for Cobb, Williams, and Brett. Who had the best performance for the decade he played?[4]

21. **Long jump** A member of a track team was practicing pg 99 the long jump and recorded the distances (in centimeters) shown in the dotplot. Some numerical summaries of the data are also provided.

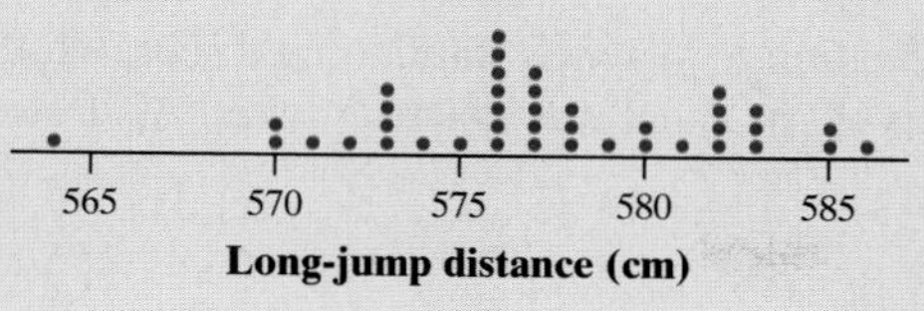


| $n$ | Mean | SD | Min | $Q_1$ | Med | $Q_3$ | Max |
|---|---|---|---|---|---|---|---|
| 40 | 577.3 | 4.713 | 564 | 574.5 | 577 | 581.5 | 586 |

After chatting with a teammate, the jumper realized that he measured his jumps from the back of the board instead of the front. Thus, he had to subtract 20 centimeters from each of his jumps to get the correct measurement for each jump.

(a) What shape would the distribution of corrected long-jump distance have?

(b) Find the mean and median of the distribution of corrected long-jump distance.

(c) Find the standard deviation and interquartile range (*IQR*) of the distribution of corrected long-jump distance.

22. **Step right up!** A dotplot of the distribution of height for Mrs. Navard's class is shown, along with some numerical summaries of the data.

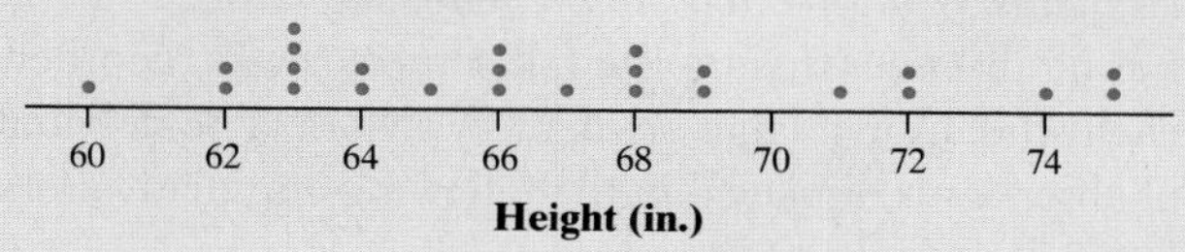


| $n$ | $\bar{x}$ | $s_x$ | Min | $Q_1$ | Med | $Q_3$ | Max |
|---|---|---|---|---|---|---|---|
| 25 | 67 | 4.29 | 60 | 63 | 66 | 70 | 75 |

Suppose that Mrs. Navard has the entire class stand on a 6-inch-high platform and then asks the students to measure the distance from the top of their heads to the ground.

(a) What shape would this distribution of distance have?

(b) Find the mean and median of the distribution of distance.

(c) Find the standard deviation and interquartile range (*IQR*) of the distribution of distance.

23. **Teacher raises** A school system employs teachers at salaries between \$38,000 and \$70,000. The teachers' union and school board are negotiating the form of next year's increase in the salary schedule. Suppose that every teacher is given a \$1000 raise. What effect will this raise have on each of the following characteristics of the resulting distribution of salary?

(a) Shape

(b) Mean and median

(c) Standard deviation and interquartile range (*IQR*)

24. **Used cars, cheap!** A used-car salesman has 28 cars in his inventory, with prices ranging from \$11,500 to \$25,000. For a Labor Day sale, he reduces the price of each car by \$500. What effect will this reduction have on each of the following characteristics of the resulting distribution of price?

(a) Shape

(b) Mean and median

(c) Standard deviation and interquartile range (*IQR*)

25. **Long jump** Refer to Exercise 21. Suppose that the corrected long-jump distances are converted from centimeters to meters (note that 100 cm = 1 m).
pg 100

(a) What shape would the resulting distribution have? Explain your answer.

(b) Find the mean of the distribution of corrected long-jump distance in meters.

(c) Find the standard deviation of the distribution of corrected long-jump distance in meters.

26. **Step right up!** Refer to Exercise 22. Suppose that the distances from the tops of the students' heads to the ground are converted from inches to feet (note that 12 in. = 1 ft).

(a) What shape would the resulting distribution have? Explain your answer.

(b) Find the mean of the distribution of distance in feet.

(c) Find the standard deviation of the distribution of distance in feet.

27. **Teacher raises** Refer to Exercise 23. Suppose each teacher receives a 5% raise instead of a \$1000 raise. What effect will this raise have on each of the following characteristics of the resulting salary distribution?

(a) Shape

(b) Median

(c) Interquartile range (*IQR*)

28. **Used cars, cheap!** Refer to Exercise 24. Suppose each car's price is reduced by 10% instead of \$500. What effect will this discount have on each of the following characteristics of the resulting price distribution?

(a) Shape

(b) Median

(c) Interquartile range (*IQR*)

29. **Cool pool?** Coach Ferguson uses a thermometer to measure the temperature (in degrees Fahrenheit) at 20 different locations in the school swimming pool. An analysis of the data yields a mean of 77°F and a standard deviation of 3°F. (Recall that $°C = \frac{5}{9}°F - \frac{160}{9}$.)
pg 101

(a) Find the mean temperature reading in degrees Celsius.

(b) Calculate the standard deviation of the temperature readings in degrees Celsius.

30. **Measure up** Clarence measures the diameter of each tennis ball in a bag with a standard ruler. Unfortunately, he uses the ruler incorrectly so that each of his measurements is 0.2 inch too large. Clarence's data had a mean of 3.2 inches and a standard deviation of 0.1 inch. (Recall that 1 in. = 2.54 cm.)

(a) Find the mean of the corrected measurements in centimeters.

(b) Calculate the standard deviation of the corrected measurements in centimeters.

31. **Taxi!** In 2016, taxicabs in Los Angeles charged an initial fee of \$2.85 plus \$2.70 per mile. In equation form, Fare = 2.85 + 2.7(miles). At the end of a month, a businessman collects all his taxicab receipts and calculates some numerical summaries. The mean fare he paid was \$15.45 with a standard deviation of \$10.20. What are the mean and standard deviation of the lengths of his cab rides in miles?

32. **Quiz scores** The scores on Ms. Martin's statistics quiz had a mean of 12 and a standard deviation of 3. Ms. Martin wants to transform the scores to have a mean of 75 and a standard deviation of 12. What transformations should she apply to each test score? Explain your answer.

**Multiple Choice:** *Select the best answer for Exercises 33–38.*

33. Jorge's score on Exam 1 in his statistics class was at the 64th percentile of the scores for all students. His score falls

(a) between the minimum and the first quartile.

(b) between the first quartile and the median.

(c) between the median and the third quartile.

(d) between the third quartile and the maximum.

(e) at the mean score for all students.

34. When Sam goes to a restaurant, he always tips the server \$2 plus 10% of the cost of the meal. If Sam's distribution of meal costs has a mean of \$9 and a standard deviation of \$3, what are the mean and standard deviation of his tip distribution?

(a) \$2.90, \$0.30 (d) \$11.00, \$2.00

(b) \$2.90, \$2.30 (e) \$2.00, \$0.90

(c) \$9.00, \$3.00

35. Scores on the ACT college entrance exam follow a bell-shaped distribution with mean 21 and standard deviation 5. Wayne's standardized score on the ACT was −0.6. What was Wayne's actual ACT score?

(a) 3 (b) 13 (c) 16

(d) 18 (e) 24

36. George's average bowling score is 180; he bowls in a league where the average for all bowlers is 150 and the standard deviation is 20. Bill's average bowling score is 190; he bowls in a league where the average is 160 and the standard deviation is 15. Who ranks higher in his own league, George or Bill?

(a) Bill, because his 190 is higher than George's 180.

(b) Bill, because his standardized score is higher than George's.

(c) Bill and George have the same rank in their leagues, because both are 30 pins above the mean.

(d) George, because his standardized score is higher than Bill's.

(e) George, because the standard deviation of bowling scores is higher in his league.

*Exercises 37 and 38 refer to the following setting.* The number of absences during the fall semester was recorded for each student in a large elementary school. The distribution of absences is displayed in the following cumulative relative frequency graph.

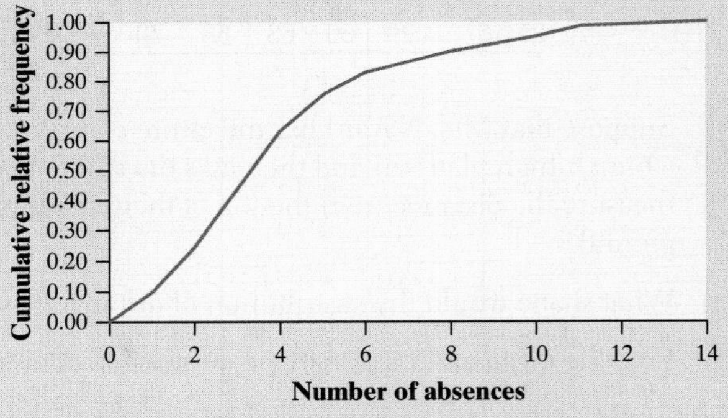


37. What is the interquartile range ($IQR$) for the distribution of absences?

(a) 1 (b) 2 (c) 3

(d) 5 (e) 14

38. If the distribution of absences was displayed in a histogram, what would be the best description of the histogram's shape?

(a) Symmetric (d) Skewed right

(b) Uniform (e) Cannot be determined

(c) Skewed left

**Recycle and Review** *Exercises 39 and 40 refer to the following setting.* We used Census At School's Random Data Selector to choose a sample of 50 Canadian students who completed a survey in a recent year.

39. **Travel time** (1.2) The dotplot displays data on students' responses to the question "How long does it usually take you to travel to school?" Describe the distribution.

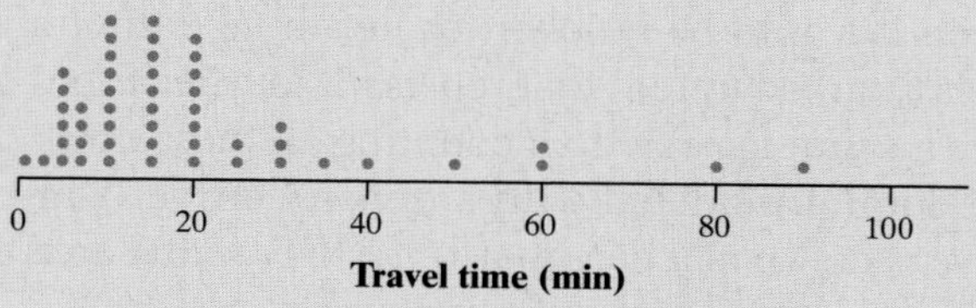


40. **Lefties** (1.1) Students were asked, "Are you right-handed, left-handed, or ambidextrous?" The responses (R = right-handed, L = left-handed, A = ambidextrous) are shown here.

| | | | | | | | | | | | | | |
|---|---|---|---|---|---|---|---|---|---|---|---|---|---|
| R | R | R | R | R | R | R | R | R | R | R | L | R | R |
| R | R | R | R | R | R | R | R | R | R | R | R | R | A |
| R | R | R | R | A | R | R | L | R | R | R | R | L | A |
| R | R | R | R | R | R | R | R | | | | | | |

(a) Make an appropriate graph to display these data.

(b) Over 10,000 Canadian high school students took the Census At School survey that year. What percent of this population would you estimate is left-handed? Justify your answer.

# SECTION 2.2 Density Curves and Normal Distributions

**LEARNING TARGETS** *By the end of the section, you should be able to:*

- Use a density curve to model a distribution of quantitative data.
- Identify the relative locations of the mean and median of a distribution from a density curve.
- Use the empirical rule to estimate (i) the proportion of values in a specified interval, or (ii) the value that corresponds to a given percentile in a Normal distribution.
- Find the proportion of values in a specified interval in a Normal distribution using Table A or technology.
- Find the value that corresponds to a given percentile in a Normal distribution using Table A or technology.
- Determine whether a distribution of data is approximately Normal from graphical and numerical evidence.

In Chapter 1, we developed graphical and numerical tools for describing distributions of quantitative data. Our work gave us a clear strategy for exploring data on a single quantitative variable.

- Always plot your data: make a graph—usually a dotplot, stemplot, or histogram.
- Look for the overall pattern (shape, center, variability) and for striking departures such as outliers.
- Calculate numerical summaries to describe center and variability.

In this section, we add one more step to this strategy.

- When there's a regular overall pattern, use a simplified model called a *density curve* to describe it.

## Density Curves

age footstock/Alamy Stock Photo

Selena works at a bookstore in the Denver International Airport. She takes the airport train from the main terminal to get to work each day. The airport just opened a new walkway that would allow Selena to get from the main terminal to the bookstore in 4 minutes. She wonders if it will be faster to walk or take the train to work.

Figure 2.7(a) shows a dotplot of the amount of time it has taken Selena to get to the bookstore by train each day for the last 1000 days she worked. To estimate the percent of days on which it would be quicker for her to take the train, we could find the percent of dots (marked in red) that represent journey times of less than 4 minutes. Surely, there's a simpler way than counting all those dots!

Figure 2.7(b) shows the dotplot modeled with a **density curve.** You might wonder why the density curve is drawn at a height of 1/3. That's so the area under the density curve between 2 minutes and 5 minutes is equal to

$$3 \times 1/3 = 1.00 = 100\%$$

representing 100% of the observations in the distribution shown in Figure 2.7(a).

**DEFINITION Density curve**

A **density curve** models the distribution of a quantitative variable with a curve that

- Is always on or above the horizontal axis
- Has area exactly 1 underneath it

The area under the curve and above any interval of values on the horizontal axis estimates the proportion of all observations that fall in that interval.

The red shaded area under the density curve in Figure 2.7(b) provides a good approximation for the proportion or percent of red dots. Because the shaded region is rectangular,

$$\text{area} = \text{base} \times \text{height} = 2 \times 1/3 = 2/3 = 0.667 = 66.7\%$$

So we estimate that it would be quicker for Selena to take the train to work on about 66.7% of days. In fact, on 669 of the 1000 days, Selena's journey from the terminal to the bookstore took less than 4 minutes. That's 669/1000 = 0.669 = 66.9%—very close to the estimate we got using the density curve.

Recall from Chapter 1 that we can describe the distribution of journey times in Figure 2.7(a) as *approximately uniform*. The density curve in Figure 2.7(b) is called a *uniform density curve* because it has constant height.

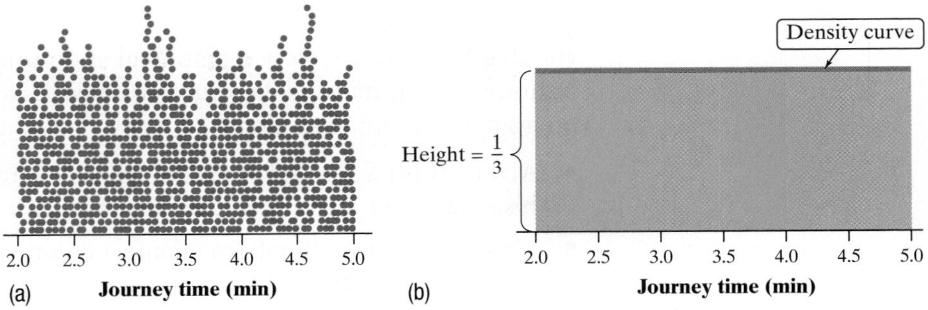


**FIGURE 2.7** (a) Dotplot of Selena's travel time over the past several years via train from the Denver airport main terminal to the bookstore where she works. The red dots indicate times when it took her less than 4 minutes to get to work. (b) Density curve modeling the dotplot in part (a). The red shaded area estimates the proportion of times that it took Selena less than 4 minutes to get to work.

No set of quantitative data is exactly described by a density curve. The curve is an approximation that is easy to use and accurate enough in most cases. The density curve simply smooths out the irregularities in the distribution.

## EXAMPLE

### That's so random!
### Density curves

**PROBLEM:** Suppose you use a calculator or computer random number generator to produce a number between 0 and 2 (like 0.84522 or 1.1111119). The random number generator will spread its output uniformly across the entire interval from 0 to 2 as we allow it to generate a long sequence of random numbers.

(a) Draw a density curve to model this distribution of random numbers. Be sure to include scales on both axes.

(b) About what percent of the randomly generated numbers will fall between 0.87 and 1.55?

(c) Estimate the 65th percentile of this distribution of random numbers.

**SOLUTION:**

(a)

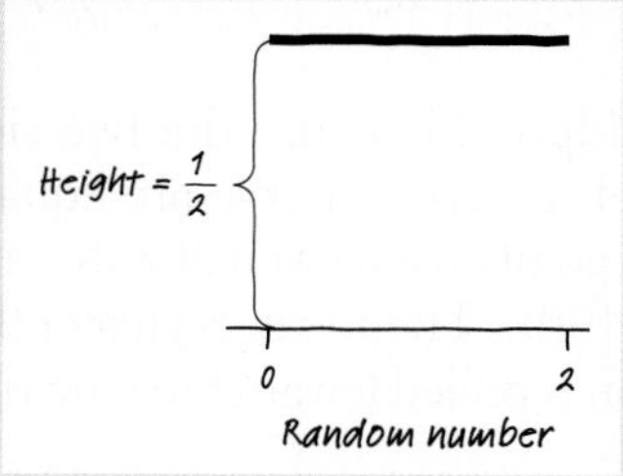


The height of the curve needs to be 1/2 so that

$$\text{area} = \text{base} \times \text{height} = 2 \times 1/2 = 1$$

(b)

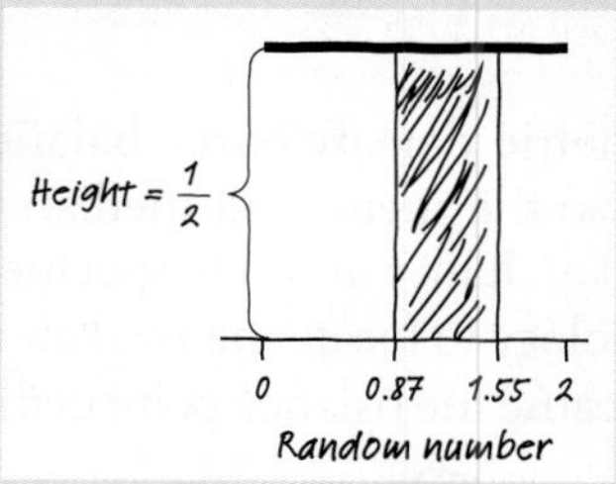


$$\text{Area} = (1.55 - 0.87) \times 1/2 = 0.34 = 34\%$$

(c)

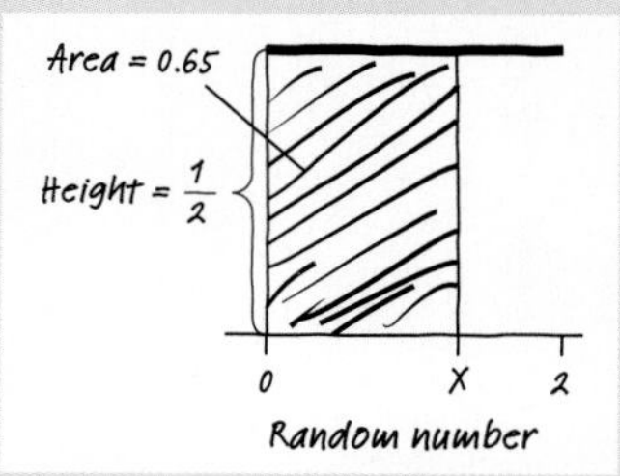


$$0.65 = (x - 0) \times 1/2$$
$$0.65 = (1/2)x$$
$$1.30 = x$$

**FOR PRACTICE, TRY EXERCISE 41**

## Describing Density Curves

Density curves come in many shapes. As with the distribution of a quantitative variable, we start by looking for rough symmetry or clear skewness. Then we identify any clear peaks. Figure 2.8 shows three density curves with distinct shapes.

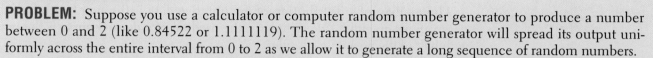

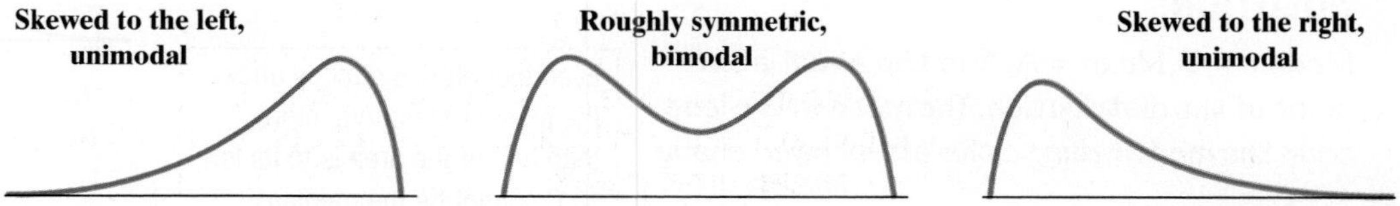


**FIGURE 2.8** Density curves with different shapes.

Our measures of center and variability apply to density curves as well as to distributions of quantitative data. Recall that the mean is the balance point of a distribution. Figure 2.9 illustrates this idea for the **mean of a density curve.**

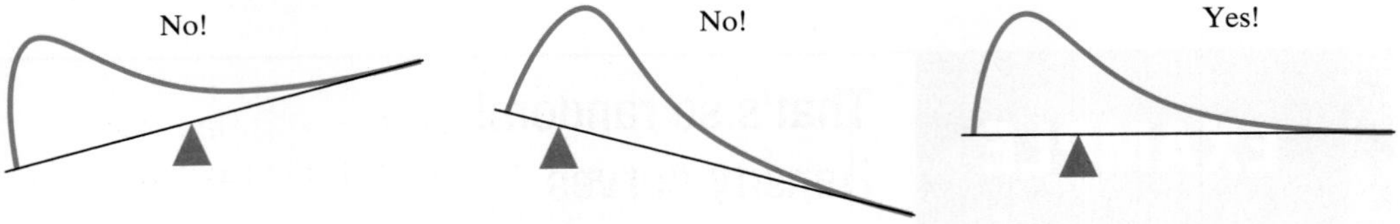


**FIGURE 2.9** The mean of a density curve is its balance point.

The median of a distribution of quantitative data is the point with half the observations on either side. Similarly, the **median of a density curve** is the point with half of the area on each side.

**DEFINITION Mean of a density curve, Median of a density curve**

The **mean of a density curve** is the point at which the curve would balance if made of solid material.

The **median of a density curve** is the equal-areas point, the point that divides the area under the curve in half.

A symmetric density curve balances at its midpoint because the two sides are identical. So the mean and median of a symmetric density curve are equal, as in Figure 2.10(a). It isn't so easy to spot the equal-areas point on a skewed density curve. We used technology to locate the median in Figure 2.10(b). The mean is greater than the median because the balance point of the distribution is pulled toward the long right tail.

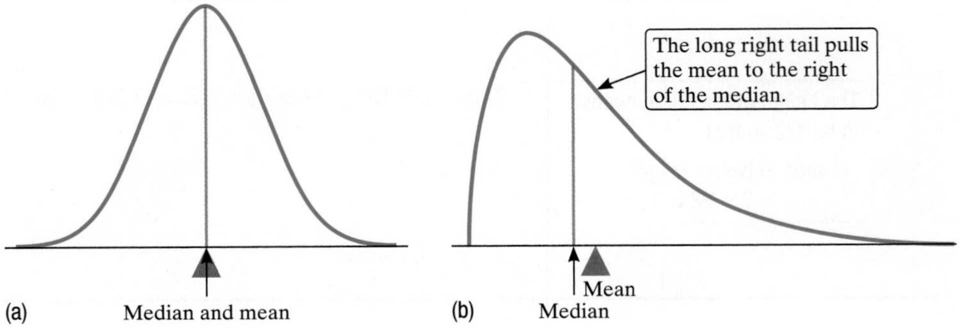


**FIGURE 2.10** (a) Both the median and mean of a symmetric density curve lie at the point of symmetry. (b) In a right-skewed density curve, the mean is pulled away from the median toward the long tail.

## EXAMPLE

### What does the left skew do?
### Mean versus median

**PROBLEM:** The density curve that models a distribution of quantitative data is shown here. Identify the location of the mean and median by letter. Justify your answers.

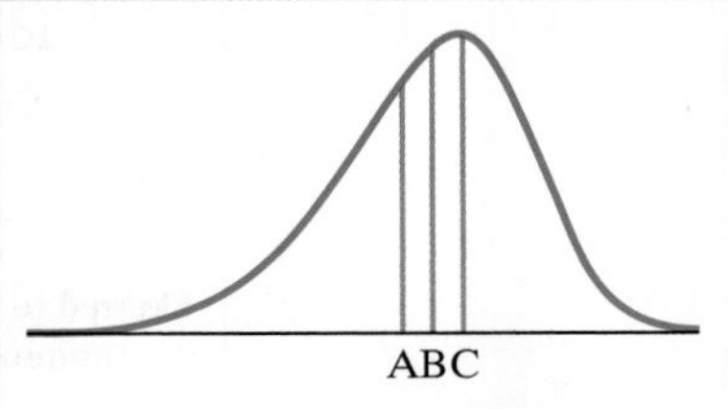


**SOLUTION:**

Median = B, Mean = A. B is the equal-areas point of the distribution. The mean will be less than the median due to the left-skewed shape.

Even though C is directly under the peak of the curve, more than half of the area is to its left, so C cannot be the median.

**FOR PRACTICE, TRY EXERCISE 45**

A density curve is an idealized model for the distribution of a quantitative variable. As a result, we label the mean of a density curve as $\mu$ and the standard deviation of a density curve as $\sigma$. This is the same notation we used for the population mean and standard deviation in Chapter 1. In both cases, we refer to $\mu$ and $\sigma$ as *parameters*.

We can roughly locate the mean $\mu$ of any density curve by eye, as the balance point. No easy way exists to estimate the standard deviation for density curves in general. But there is one family of density curves for which we can estimate the standard deviation by eye.

## Normal Distributions

Blend Images/Shutterstock.com

When we examine a distribution of quantitative data, how does it compare with an idealized density curve? Figure 2.11(a) shows a histogram of the scores of all seventh-grade students in Gary, Indiana, on the vocabulary part of the Iowa Test of Basic Skills (ITBS).[5] The scores are grade-level equivalents, so a score of 6.3 indicates that the student's performance is typical for a student in the third month of grade 6. The histogram is roughly symmetric, and both tails fall off smoothly from a single center peak. There are no large gaps or obvious outliers.

The density curve drawn through the tops of the histogram bars in Figure 2.11(b) is a good description of the overall pattern of the ITBS score distribution. We call it a **Normal curve.** The distributions described by Normal curves are called **Normal distributions.** In this case, the ITBS vocabulary scores of Gary, Indiana, seventh-graders are approximately Normally distributed.

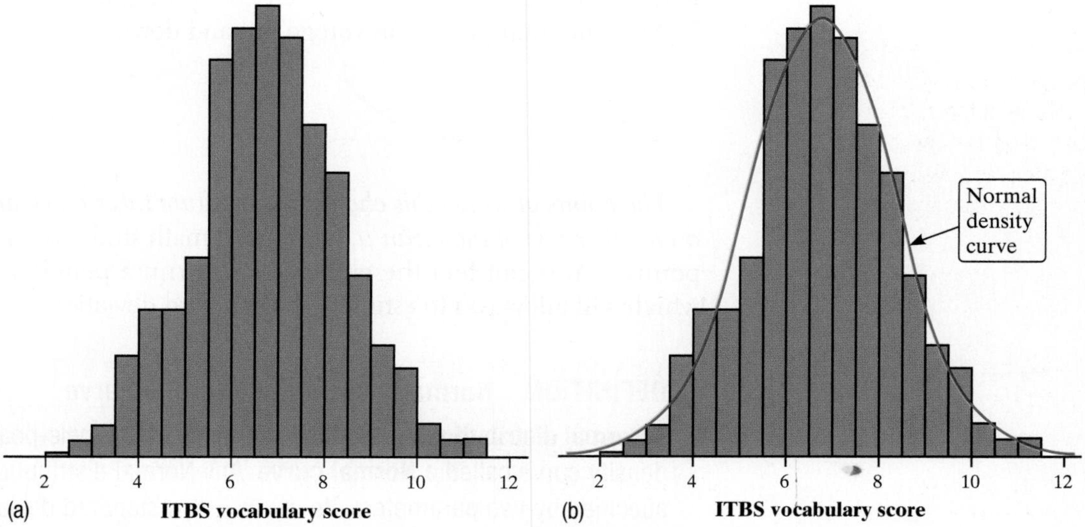


**FIGURE 2.11** (a) Histogram of the Iowa Test of Basic Skills (ITBS) vocabulary scores of all seventh-grade students in Gary, Indiana. (b) The Normal density curve shows the overall shape of the distribution.

Normal distributions play a large role in statistics, but they are rather special and not at all "normal" in the sense of being usual or typical. We capitalize Normal to remind you that these density curves are special.

Look at the two Normal distributions in Figure 2.12. They illustrate several important facts:

- **Shape:** All Normal distributions have the same overall shape: symmetric, single-peaked (unimodal), and bell-shaped.
- **Center:** The mean $\mu$ is located at the midpoint of the symmetric density curve and is the same as the median.
- **Variability:** The standard deviation $\sigma$ measures the variability (width) of a Normal distribution.

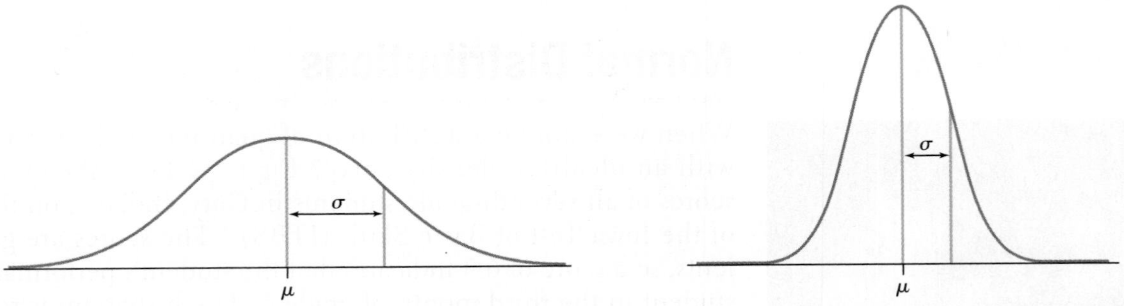


**FIGURE 2.12** Two Normal curves, showing the mean $\mu$ and standard deviation $\sigma$.

You can estimate $\sigma$ by eye on a Normal curve. Here's how. Imagine that you are skiing down a mountain that has the shape of a Normal distribution. At first, you descend at an increasingly steep angle as you go out from the peak.

Fortunately, before you find yourself going straight down, the slope begins to get flatter rather than steeper as you go out and down:

*The points at which this change of curvature takes place are located at a distance $\sigma$ on either side of the mean $\mu$.* (Advanced math students know these as "inflection points.") You can feel the change as you run a pencil along a Normal curve, which will allow you to estimate the standard deviation.

**DEFINITION Normal distribution, Normal curve**

A **Normal distribution** is described by a symmetric, single-peaked, bell-shaped density curve called a **Normal curve**. Any Normal distribution is completely specified by two parameters: its mean $\mu$ and standard deviation $\sigma$.

The distribution of ITBS vocabulary scores for seventh-grade students in Gary, Indiana, is modeled well by a Normal distribution with mean $\mu = 6.84$ and standard deviation $\sigma = 1.55$. The figure shows this distribution with the points 1, 2, and 3 standard deviations from the mean labeled on the horizontal axis.

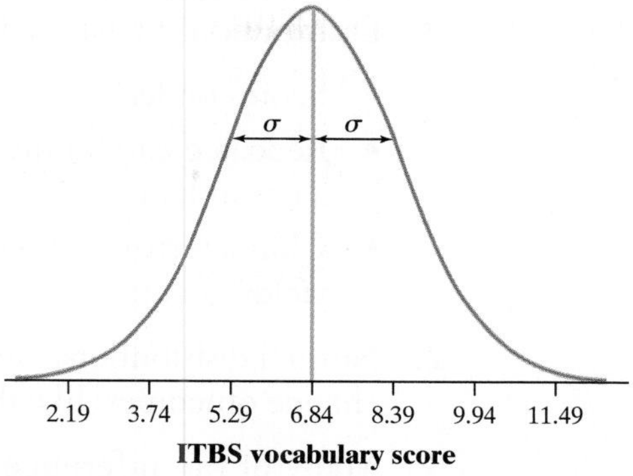


You will be asked to make reasonably accurate sketches of Normal distributions to model quantitative data sets like the ITBS vocabulary scores. The best way to learn is to practice.

## EXAMPLE

### Stop the car!
### Sketching a Normal distribution

**PROBLEM:** Many studies on automobile safety suggest that when automobile drivers make emergency stops, the stopping distances follow an approximately Normal distribution. Suppose that for one model of car traveling at 62 mph under typical conditions on dry pavement, the mean stopping distance is $\mu = 155$ ft with a standard deviation of $\sigma = 3$ ft. Sketch the Normal curve that approximates the distribution of stopping distance. Label the mean and the points that are 1, 2, and 3 standard deviations from the mean.

fstop123/Getty Images

**SOLUTION:**

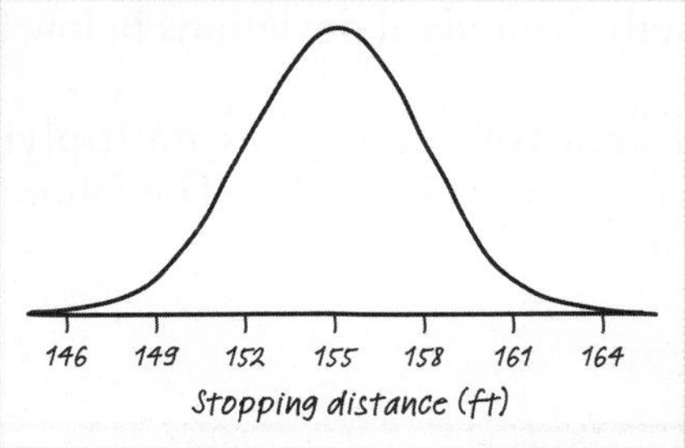


The mean (155) is at the midpoint of the bell-shaped density curve. The standard deviation (3) is the distance from the center to the change-of-curvature points on either side. Label the mean and the points that are 1, 2, and 3 SDs from the mean:

1 SD: $155 - 1(3) = 152$ and $155 + 1(3) = 158$
2 SD: $155 - 2(3) = 149$ and $155 + 2(3) = 161$
3 SD: $155 - 3(3) = 146$ and $155 + 3(3) = 164$

FOR PRACTICE, TRY EXERCISE 47

caution

**Remember that $\mu$ and $\sigma$ alone do not specify the appearance of most distributions.** The shape of density curves, in general, does not reveal $\sigma$. These are special properties of Normal distributions.

Why are Normal distributions important in statistics? Here are three reasons.

1. Normal distributions are good descriptions for some distributions of real data. Distributions that are often close to Normal include:
   - Scores on tests taken by many people (such as SAT exams and IQ tests)
   - Repeated careful measurements of the same quantity (like the diameter of a tennis ball)
   - Characteristics of biological populations (such as lengths of crickets and yields of corn)
2. Normal distributions are good approximations to the results of many kinds of chance outcomes, like the proportion of heads in many tosses of a fair coin.
3. Many of the inference methods in Chapters 8–12 are based on Normal distributions.

DEA/A. DAGLI ORTI/De Agostini/Getty Images

Normal curves were first applied to data by the great mathematician Carl Friedrich Gauss (1777–1855). He used them to describe the small errors made by astronomers and surveyors in repeated careful measurements of the same quantity. You will sometimes see Normal distributions labeled "Gaussian" in honor of Gauss. His image was even featured on a previous German DM 10 bill, along with a sketch of the Normal distribution.

## The Empirical Rule

Earlier, we saw that the distribution of Iowa Test of Basic Skills (ITBS) vocabulary scores for seventh-grade students in Gary, Indiana, is approximately Normal with mean $\mu = 6.84$ and standard deviation $\sigma = 1.55$. How unusual is it for a Gary seventh-grader to get an ITBS score less than 3.74? The figure shows the Normal density curve for this distribution with the area of interest shaded. Note that the boundary value, 3.74, is exactly 2 standard deviations below the mean.

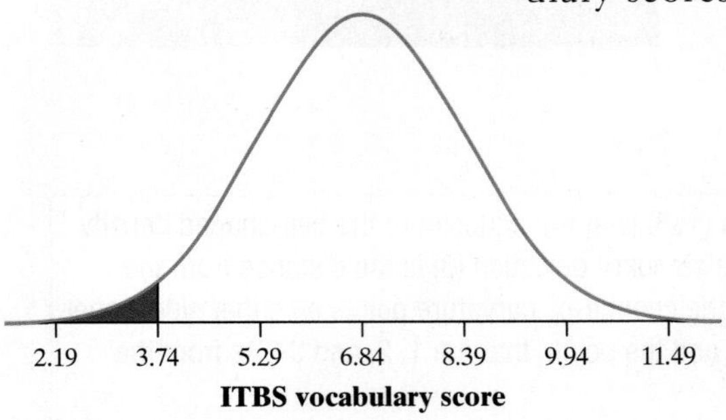


Calculating the shaded area isn't as easy as multiplying base × height, but it's not as hard as you might think. The following activity shows you how to do it.

### ACTIVITY What's so special about Normal distributions?

APPLET

In this activity, you will use the *Normal Density Curve* applet at *highschool.bfwpub.com/updatedtps6e* to explore an interesting property of Normal distributions.

Change the mean to 6.84 and the standard deviation to 1.55, and click on "UPDATE." (These are the values for the distribution of ITBS vocabulary scores of seventh-graders in Gary, Indiana.) A figure like the one that follows should appear:

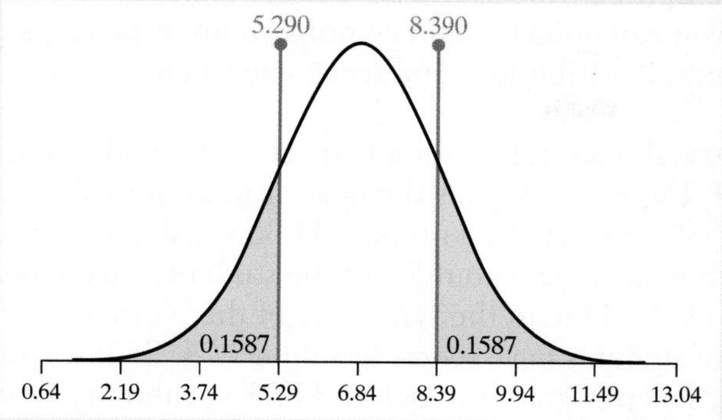


Use the applet to help you answer the following questions.

1. (a) What percent of the area under the Normal curve lies within 1 standard deviation of the mean? That is, about what percent of Gary, Indiana, seventh-graders have ITBS vocabulary scores between 5.29 and 8.39?

   (b) What percent of the area under the Normal curve lies within 2 standard deviations of the mean? Interpret this result in context.

   (c) What percent of the area under the Normal curve lies within 3 standard deviations of the mean? Interpret this result in context.

2. The distribution of IQ scores in the adult population is approximately Normal with mean $\mu = 100$ and standard deviation $\sigma = 15$. Adjust the applet to display this distribution. About what percent of adults have IQ scores within 1, 2, and 3 standard deviations of the mean?

When you hear the phrase "standard Normal distribution," think standardized scores (*z*-scores), which have a mean of 0 and a standard deviation of 1.

3. Adjust the applet to have a mean of 0 and a standard deviation of 1. Then click "UPDATE." The resulting density curve describes the *standard Normal distribution.* What percent of the area under this Normal density curve lies within 1, 2, and 3 standard deviations of the mean?

4. Summarize by completing this sentence: "For any Normal distribution, the area under the Normal curve within 1, 2, and 3 standard deviations of the mean is about ___%, ___%, and ___%."

Although there are many Normal distributions, they all have properties in common. In particular, all Normal distributions obey the **empirical rule.**

*Empirical* means "learned from experience or by observation."

**DEFINITION The empirical rule**

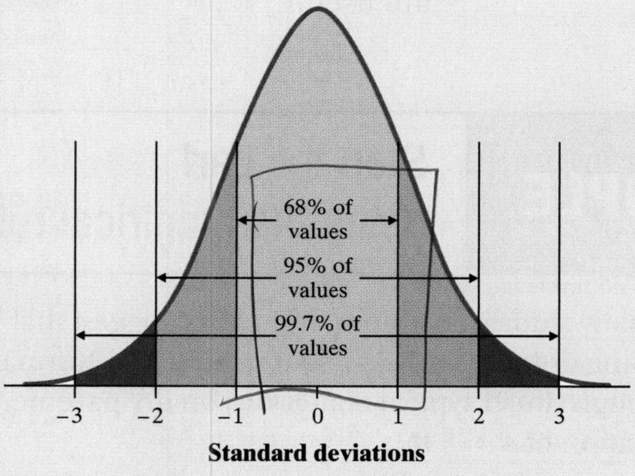


In a Normal distribution with mean $\mu$ and standard deviation $\sigma$:

- Approximately **68%** of the observations fall within $\sigma$ of the mean $\mu$.
- Approximately **95%** of the observations fall within $2\sigma$ of the mean $\mu$.
- Approximately **99.7%** of the observations fall within $3\sigma$ of the mean $\mu$.

This result is known as the **empirical rule.**

Recall from Chapter 1 that there are other rules for determining outliers, such as "any value that is more than 2 (or 3) standard deviations from the mean." This rule makes better sense when we are discussing a Normal distribution!

Some people refer to the empirical rule as the *68–95–99.7 rule*. By remembering these three numbers, you can quickly estimate proportions or percents of observations (areas) using Normal distributions and recognize when an observation is unusual.

Earlier, we asked how unusual it would be for a Gary seventh-grader to get an ITBS score less than 3.74. Figure 2.13 gives the answer in graphical form. By the empirical rule, about 95% of these students have ITBS vocabulary scores between 3.74 and 9.94, which means that about 5% of the students have scores less than 3.74 or greater than 9.94. Due to the symmetry of the Normal distribution, about 5% / 2 = 2.5% of students have scores less than 3.74. So it is quite unusual for a Gary, Indiana, seventh-grader to get an ITBS vocabulary score below 3.74.

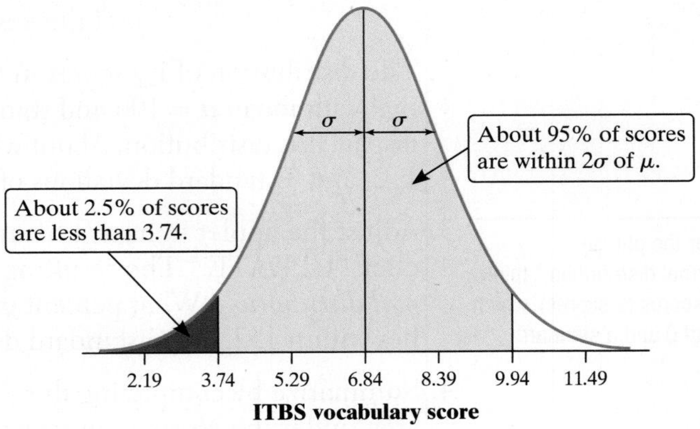


**FIGURE 2.13** Using the empirical rule to estimate the percent of Gary, Indiana, seventh-graders with ITBS vocabulary scores less than 3.74.

How well does the empirical rule describe the distribution of ITBS vocabulary scores for Gary, Indiana, seventh-graders? Well, 900 of the 947 scores are between 3.74 and 9.94. That's 95.04%, which is very accurate indeed. Of the remaining 47 scores, 20 are below 3.74 and 27 are above 9.94. The number of values in each tail is not quite equal, as it would be in an exactly Normal distribution. Normal distributions often describe real data better in the center of the distribution than in the extreme high and low tails. As famous statistician George Box once noted, "All models are wrong, but some are useful!"

## EXAMPLE

### Stop the car!
### Using the empirical rule

**PROBLEM:** Many studies on automobile safety suggest that when automobile drivers must make emergency stops, the stopping distances follow an approximately Normal distribution. Suppose that for one model of car traveling at 62 mph under typical conditions on dry pavement, the mean stopping distance is $\mu = 155$ ft with a standard deviation of $\sigma = 3$ ft.

(a) About what percent of cars of this model would take more than 158 feet to make an emergency stop? Show your method clearly.

(b) A car of this model that takes 158 feet to make an emergency stop would be at about what percentile of the distribution? Justify your answer.

**SOLUTION:**

(a)

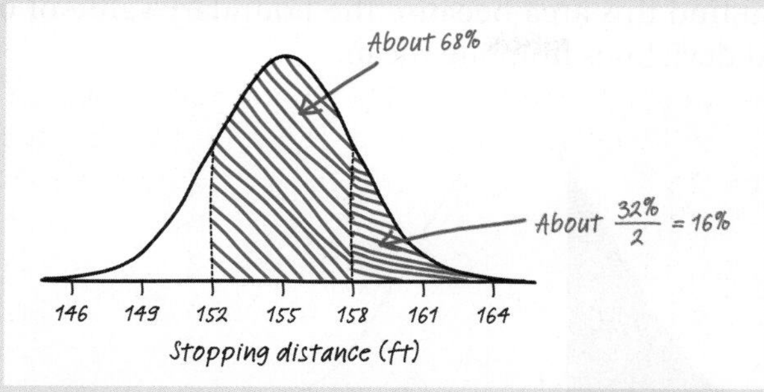


Start by sketching a Normal curve and labeling the values 1, 2, and 3 standard deviations from the mean. Then shade the area of interest.

Use the empirical rule and the symmetry of the Normal distribution to find the desired area.

About 16% of cars of this model would take more than 158 feet to make an emergency stop.

(b) About the 84th percentile because about 100% − 16% = 84% of cars of this model would stop in less than or equal to 158 feet.

FOR PRACTICE, TRY EXERCISE 51

caution

**Note that the empirical rule applies *only* to Normal distributions.** Is there a similar rule that would apply to *any* distribution? Sort of. A result known as *Chebyshev's inequality* says that in any distribution, the proportion of observations falling within $k$ standard deviations of the mean is *at least* $1-\frac{1}{k^2}$. If $k = 2$, for example, Chebyshev's inequality tells us that at least $1-\frac{1}{2^2} = 0.75$ of the observations in *any* distribution are within 2 standard deviations of the mean. For Normal distributions, we know that this proportion is much higher than 0.75. In fact, it's approximately 0.95.

Chebyshev's inequality is an interesting result, but it is not required for the AP® Statistics exam.

## CHECK YOUR UNDERSTANDING

The distribution of heights of young women aged 18 to 24 is approximately Normal with mean $\mu = 64.5$ inches and standard deviation $\sigma = 2.5$ inches.

1. Sketch the Normal curve that approximates the distribution of young women's height. Label the mean and the points that are 1, 2, and 3 standard deviations from the mean.
2. About what percent of young women have heights less than 69.5 inches? Show your work.
3. Is a young woman with a height of 62 inches unusually short? Justify your answer.

# Finding Areas in a Normal Distribution

Let's return to the distribution of ITBS vocabulary scores among all Gary, Indiana, seventh-graders. Recall that this distribution is approximately Normal with mean $\mu = 6.84$ and standard deviation $\sigma = 1.55$. What proportion of these seventh-graders have vocabulary scores that are below sixth-grade level? Figure 2.14 shows the Normal curve with the area of interest shaded. We can't use the empirical rule to find this area because the boundary value of 6 is not exactly 1, 2, or 3 standard deviations from the mean.

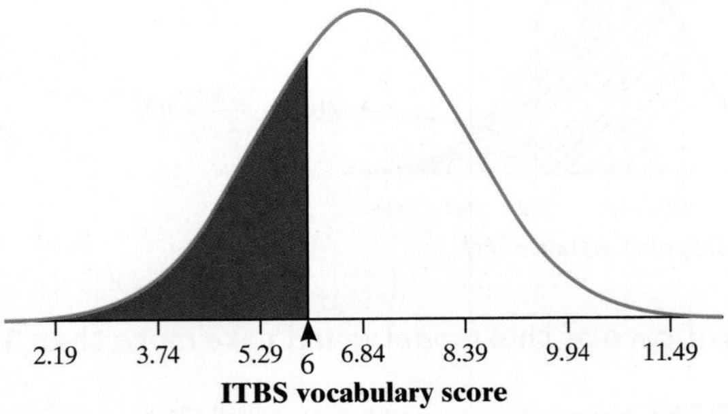


**FIGURE 2.14** Normal curve we would use to estimate the proportion of Gary, Indiana, seventh-graders with ITBS vocabulary scores that are less than 6—that is, below sixth-grade level.

As the empirical rule suggests, all Normal distributions are the same if we measure in units of size $\sigma$ from the mean $\mu$. Changing to these units requires us to standardize, just as we did in Section 2.1:

$$z = \frac{\text{value} - \text{mean}}{\text{standard deviation}} = \frac{x - \mu}{\sigma}$$

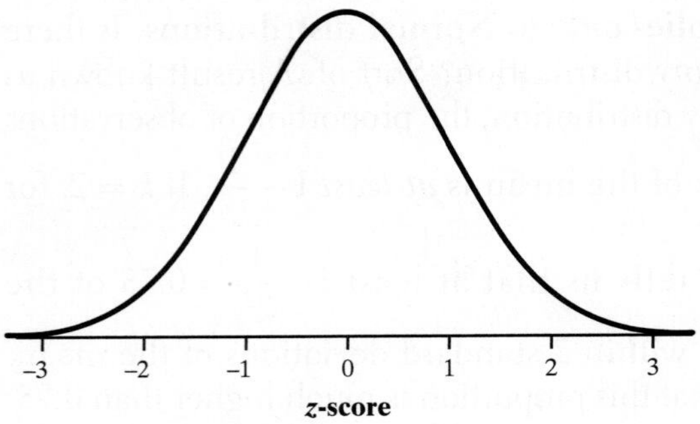


Recall that subtracting a constant and dividing by a constant don't change the shape of a distribution. If the quantitative variable we standardize has an approximately Normal distribution, then so does the new variable $z$. This new distribution of standardized values can be modeled with a Normal curve having mean $\mu = 0$ and standard deviation $\sigma = 1$. It is called the **standard Normal distribution**.

**DEFINITION Standard Normal distribution**

The **standard Normal distribution** is the Normal distribution with mean 0 and standard deviation 1.

Because all Normal distributions are the same when we standardize, we can find areas under any Normal curve using the standard Normal distribution. Table A in the back of the book gives areas under the standard Normal curve. The table entry for each $z$-score is the area under the curve *to the left* of $z$.

For the ITBS test score data, we want to find the area to the left of 6 under the Normal distribution with mean 6.84 and standard deviation 1.55. See Figure 2.15(a). We start by standardizing the boundary value $x = 6$:

$$z = \frac{\text{value} - \text{mean}}{\text{standard deviation}} = \frac{6 - 6.84}{1.55} = -0.54$$

Figure 2.15(b) shows the standard Normal distribution with the area to the left of $z = -0.54$ shaded. Notice that the shaded areas in the two graphs are the same.

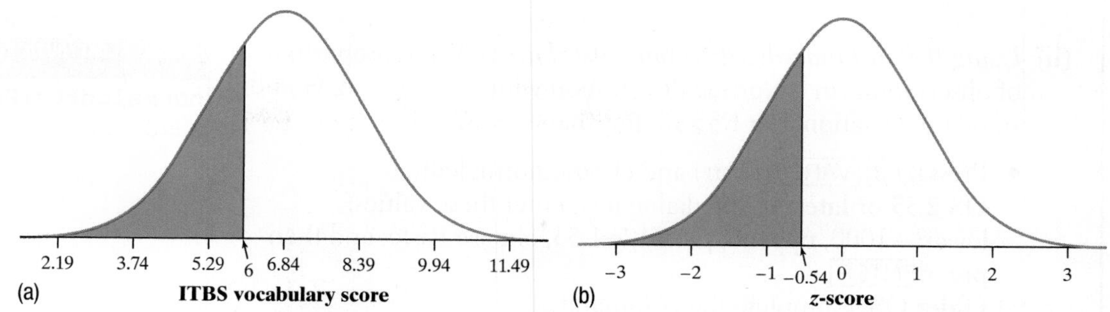


**FIGURE 2.15** (a) Normal distribution estimating the proportion of Gary, Indiana, seventh-graders who earn ITBS vocabulary scores less than sixth-grade level. (b) The corresponding area in the standard Normal distribution.

| *z* | .03 | .04 | .05 |
|---|---|---|---|
| −0.6 | .2643 | .2611 | .2578 |
| −0.5 | .2981 | .2946 | .2912 |
| −0.4 | .3336 | .3300 | .3264 |

To find the area to the left of $z = -0.54$ using Table A, locate −0.5 in the left-hand column, then locate the remaining digit 4 as .04 in the top row. The entry to the right of −0.5 and under .04 is .2946. This is the area we seek. We estimate that about 29.46% of Gary, Indiana, seventh-grader scores fall below the sixth-grade level on the ITBS vocabulary test. *Note that we have made a connection between z-scores and percentiles when the shape of a distribution is approximately Normal.*

It is also possible to find areas under a Normal curve using technology.

## 5. Technology Corner FINDING AREAS FROM VALUES IN A NORMAL DISTRIBUTION

*TI-Nspire and other technology instructions are on the book's website at highschool.bfwpub.com/updatedtps6e.*

The normalcdf command on the TI-83/84 can be used to find areas under a Normal curve. The syntax is normalcdf(lower bound, upper bound, mean, standard deviation). Let's use this command to calculate the proportion of ITBS vocabulary scores in Gary, Indiana, that are less than 6. Note that we can do the area calculation using the standard Normal distribution or the Normal distribution with mean 6.84 and standard deviation 1.55.

(i) *Using the standard Normal distribution:* What proportion of observations in a standard Normal distribution are less than $z = -0.54$? Recall that the standard Normal distribution has mean $\mu = 0$ and standard deviation $\sigma = 1$.

- Press 2nd VARS (Distr) and choose normalcdf(.
  **OS 2.55 or later:** In the dialog box, enter these values: lower: − 1000, upper: − 0.54, $\mu$:0, $\sigma$:1, choose Paste, and then press ENTER.
  **Older OS:** Complete the command normalcdf (−1000,−0.54,0,1) and press ENTER.

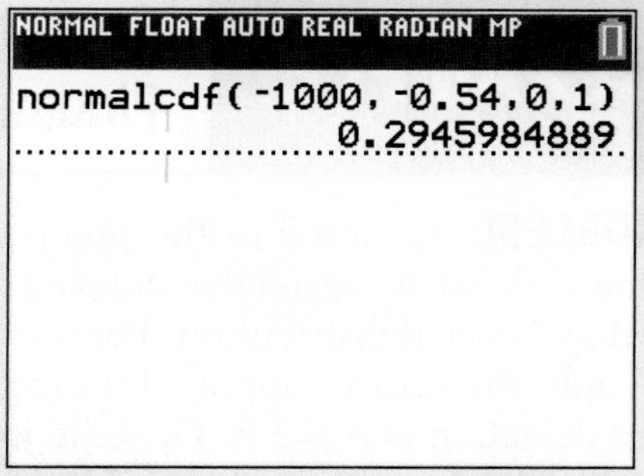


*Note:* We chose −1000 as the lower bound because it's many, many standard deviations less than the mean.

(ii) *Using the unstandardized Normal distribution:* What proportion of observations in a Normal distribution with mean $\mu = 6.84$ and standard deviation $\sigma = 1.55$ are less than $x = 6$?

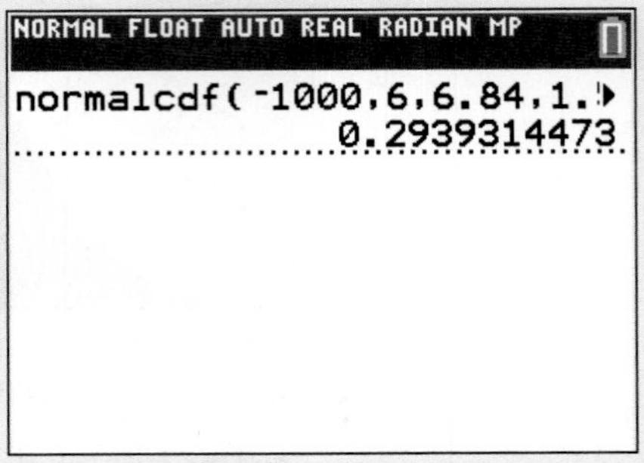


- Press 2nd VARS (Distr) and choose normalcdf(.
  **OS 2.55 or later:** In the dialog box, enter these values: lower: $-1000$, upper:6, $\mu$:6.84, $\sigma$:1.55, choose Paste, and then press ENTER.
  **Older OS:** Complete the command normalcdf($-1000$,6,6.84,1.55) and press ENTER.

This answer differs slightly from the one we got using the standard Normal distribution because we rounded the standardized score to two decimal places: $z = -0.54$.

The following box summarizes the process of finding areas in a Normal distribution. In Step 2, each method of performing calculations has some advantages, so check with your teacher to see which option will be used in your class.

## HOW TO FIND AREAS IN ANY NORMAL DISTRIBUTION

**Step 1: Draw a Normal distribution** with the horizontal axis labeled and scaled using the mean and standard deviation, the boundary value(s) clearly identified, and the area of interest shaded.

**Step 2: Perform calculations—show your work!** Do one of the following:

(i) Standardize each boundary value and use Table A or technology to find the desired area under the standard Normal curve; or

(ii) Use technology to find the desired area without standardizing.

Be sure to answer the question that was asked.

### AP® EXAM TIP

Students often do not get full credit on the AP® Statistics exam because they use option (ii) with "calculator-speak" to show their work on Normal calculation questions—for example, normalcdf(–1000,6,6.84,1.55). This is *not* considered clear communication. To get full credit, follow the two-step process above, making sure to carefully label each of the inputs in the calculator command if you use technology in Step 2: normalcdf(lower: – 1000, upper: 6, mean: 6.84, SD:1.55).

## EXAMPLE

### Stop the car!
### Finding area to the left

**PROBLEM:** As noted in the preceding example, studies on automobile safety suggest that stopping distances follow an approximately Normal distribution. For one model of car traveling at 62 mph, the mean stopping distance is $\mu = 155$ ft with a standard deviation of $\sigma = 3$ ft. Danielle is driving one of these cars at 62 mph when she spots a wreck 160 feet in front of her and needs to make an emergency stop. About what percent of cars of this model when going 62 mph would be able to make an emergency stop in less than 160 feet? Is Danielle likely to stop safely?

fizkes/Shutterstock.com

**SOLUTION:**

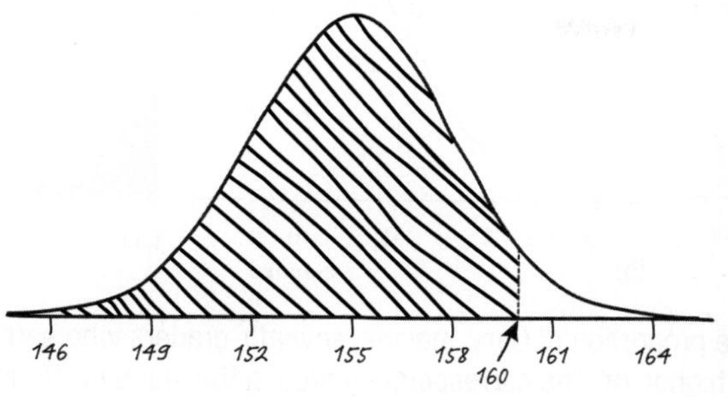


**1. Draw a normal distribution.** Be sure to:
- Scale the horizontal axis.
- Label the horizontal axis with the variable name, including units of measurement.
- Clearly identify the boundary value(s).
- Shade the area of interest.

**2. Perform calculations—show your work!**
(i) Standardize the boundary value and use Table A or technology to find the desired probability; or
(ii) Use technology to find the desired area without standardizing.

(i) $z = \dfrac{160 - 155}{3} = 1.67$

*Using Table A:* Area for $z < 1.67$ is 0.9525.
*Using technology:* normalcdf(lower: −1000, upper: 1.67, mean: 0, SD: 1) = 0.9525

(ii) normalcdf(lower: −1000, upper: 160, mean: 155, SD: 3) = 0.9522

| z | .06 | .07 | .08 |
|---|---|---|---|
| **1.5** | .9406 | .9418 | .9429 |
| **1.6** | .9515 | .9525 | .9535 |
| **1.7** | .9608 | .9616 | .9625 |

About 95% of cars of this model would be able to make an emergency stop within 160 feet. So Danielle is likely to be able to stop safely.

Be sure to answer the question that was asked.

**FOR PRACTICE, TRY EXERCISE 53**

What percent of cars of this model would be able to make an emergency stop in less than 140 feet? The standardized score for $x = 140$ is

$$z = \frac{140 - 155}{3} = -5.00$$

Table A does not go beyond $z = -3.50$ and $z = 3.50$ because it is highly unusual for a value to be more than 3.5 standard deviations from the mean in a Normal distribution. For practical purposes, we can act as if there is approximately zero probability outside the range of Table A. So there is almost no chance that a car of this model going 62 mph would be able to make an emergency stop within 140 feet.

**FINDING AREAS TO THE RIGHT IN A NORMAL DISTRIBUTION** What proportion of Gary, Indiana, seventh-grade scores on the ITBS vocabulary test are *at least* 9? Start with a picture. Figure 2.16(a) on the next page shows the Normal distribution with mean $\mu = 6.84$ and standard deviation $\sigma = 1.55$ with the area of interest shaded. Next, standardize the boundary value:

$$z = \frac{9 - 6.84}{1.55} = 1.39$$

Figure 2.16(b) shows the standard Normal distribution with the area to the right of $z = 1.39$ shaded. Again, notice that the shaded areas in the two graphs are the same.

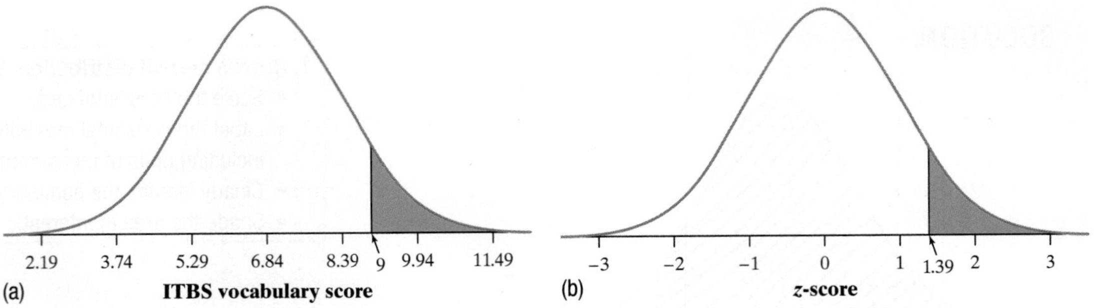


**FIGURE 2.16** (a) Normal distribution estimating the proportion of Gary, Indiana, seventh-graders who earn ITBS vocabulary scores at the ninth-grade level or higher. (b) The corresponding area in the standard Normal distribution.

| *z* | .07 | .08 | .09 |
|---|---|---|---|
| 1.2 | .8980 | .8997 | .9015 |
| 1.3 | .9147 | .9162 | .9177 |
| 1.4 | .9292 | .9306 | .9319 |

To find the area to the right of $z = 1.39$, locate 1.3 in the left-hand column of Table A, then locate the remaining digit 9 as .09 in the top row. The entry to the right of 1.3 and under .09 is .9177. However, this is the area *to the left* of $z = 1.39$. We can use the fact that the total area in the standard Normal distribution is 1 to find that the area *to the right* of $z = 1.39$ is $1 - 0.9177 = 0.0823$. We estimate that about 8.23% of Gary, Indiana, seventh-graders earn scores at the ninth-grade level or above on the ITBS vocabulary test.

caution !

**A common student mistake is to look up a *z*-score in Table A and report the entry corresponding to that *z*-score, regardless of whether the problem asks for the area to the left or to the right of that *z*-score.** This mistake can usually be prevented by drawing a Normal distribution and shading the area of interest. Look to see if the area should be closer to 0 or closer to 1. In the preceding example, for instance, it should be obvious that 0.9177 is *not* the correct area.

## EXAMPLE

## Can Spieth clear the trees?

### Finding area to the right

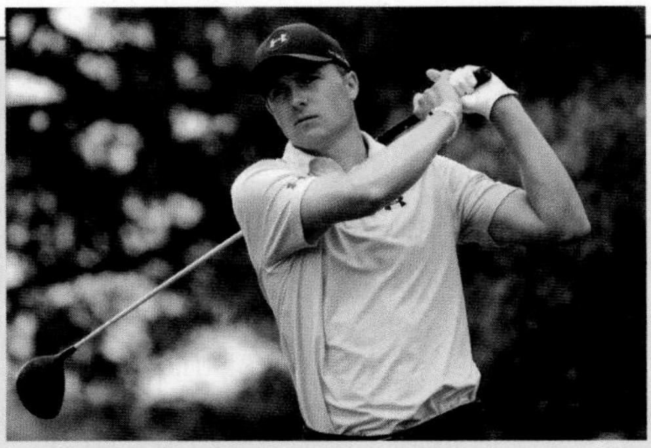
Christian Petersen/Getty Images

**PROBLEM:** When professional golfer Jordan Spieth hits his driver, the distance the ball travels can be modeled by a Normal distribution with mean 304 yards and standard deviation 8 yards. On a specific hole, Jordan would need to hit the ball at least 290 yards to have a clear second shot that avoids a large group of trees. What percent of Spieth's drives travel at least 290 yards? Is he likely to have a clear second shot?

**SOLUTION:**

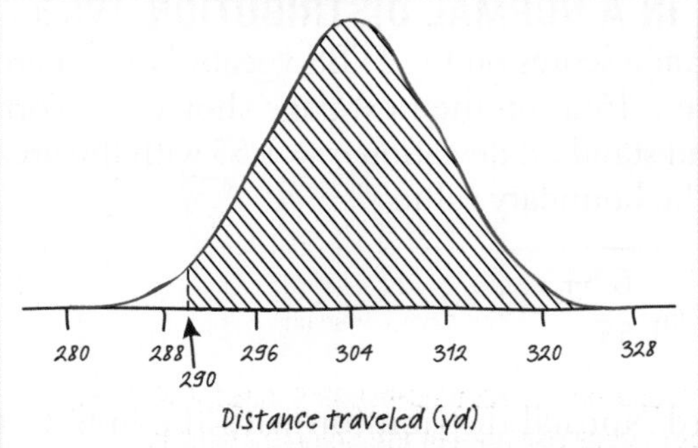


**1. Draw a Normal distribution.**

(i) $z = \dfrac{290 - 304}{8} = -1.75$

*Using Table A:* Area for $z < -1.75$ is 0.0401. Area for $z \geq -1.75$ is $1 - 0.0401 = 0.9599$.

*Using technology:* normalcdf(lower: −1.75, upper: 1000, mean: 0, SD: 1) = 0.9599

(ii) normalcdf(lower: 290, upper: 1000, mean: 304, SD: 8) = 0.9599

**2. Perform calculations—show your work!**
(i) Standardize and use Table A or technology; or
(ii) Use technology without standardizing.

About 96% of Jordan Spieth's drives travel at least 290 yards. So he is likely to have a clear second shot.

Be sure to answer the question that was asked.

**FOR PRACTICE, TRY EXERCISE 55**

### Think About It

**WHAT PROPORTION OF JORDAN SPIETH'S DRIVES GO EXACTLY 290 YARDS?** There is no area under the Normal density curve in the preceding example directly above the point 290.000000000. . . . So the answer to our question based on the Normal distribution is 0. One more thing: the areas under the curve with $x \geq 290$ and $x > 290$ are the same. According to the Normal model, the proportion of Spieth's drives that travel at least 290 yards is the same as the proportion that travel more than 290 yards.

**FINDING AREAS BETWEEN TWO VALUES IN A NORMAL DISTRIBUTION** How do you find the area in a Normal distribution that is between two values? For instance, suppose we want to estimate the proportion of Gary, Indiana, seventh-graders with ITBS vocabulary scores between 6 and 9. Figure 2.17(a) shows the desired area under the Normal curve with mean $\mu = 6.84$ and standard deviation $\sigma = 1.55$. We can use Table A or technology to find the desired area.

**Option (i):** If we standardize each boundary value, we get

$$z = \frac{6 - 6.84}{1.55} = -0.54, \qquad z = \frac{9 - 6.84}{1.55} = 1.39$$

Figure 2.17(b) shows the corresponding area of interest in the standard Normal distribution.

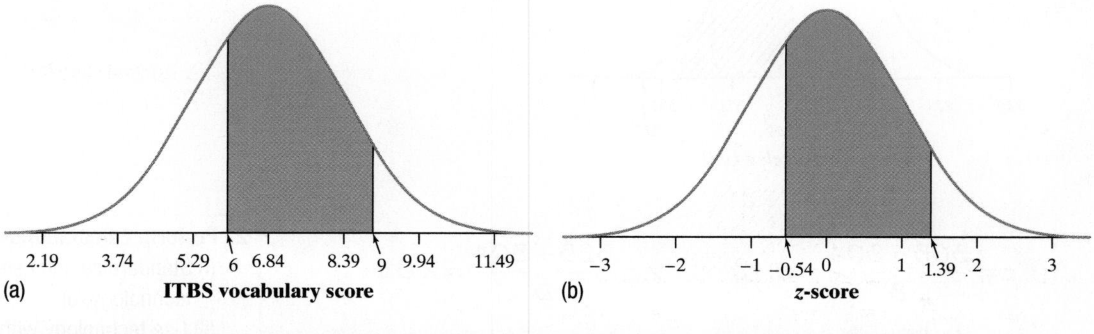


**FIGURE 2.17** (a) Normal distribution approximating the proportion of seventh-graders in Gary, Indiana, with ITBS vocabulary scores between 6 and 9. (b) The corresponding area in the standard Normal distribution.

*Using Table A*: The table makes this process a bit trickier because it only shows areas to the left of a given $z$-score. The visual shows one way to think about the calculation.

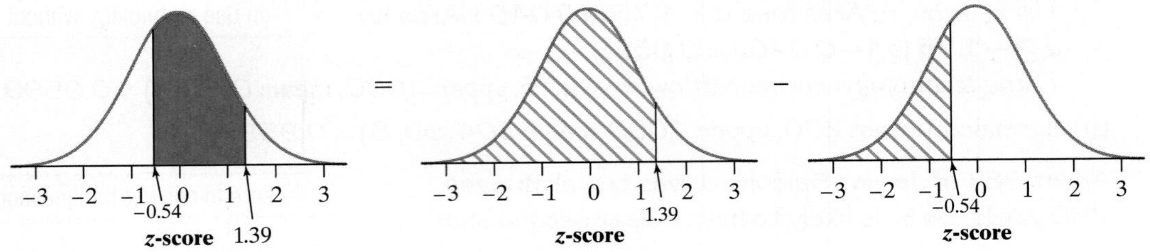


Area between $z = -0.54$ and $z = 1.39$
$= $ Area to the left of $z = 1.39 -$ Area to the left of $z = -0.54$
$= 0.9177 - 0.2946$
$= 0.6231$

*Using technology*: normalcdf(lower: −0.54, upper: 1.39, mean: 0, SD: 1) = 0.6231

**Option (ii)**: normalcdf(lower: 6, upper: 9, mean: 6.84, SD: 1.55) = 0.6243.

About 62% of Gary, Indiana, seventh-graders earned grade-equivalent scores between 6 and 9.

## EXAMPLE

### Can Spieth reach the green?

### Finding areas between two values

Ben Stansall/AFP/Getty Images

**PROBLEM:** When professional golfer Jordan Spieth hits his driver, the distance the ball travels can be modeled by a Normal distribution with mean 304 yards and standard deviation 8 yards. On another golf hole, Spieth has the opportunity to drive the ball onto the green if he hits the ball between 305 and 325 yards. What percent of Spieth's drives travel a distance that falls in the interval? Is he likely to get the ball on the green with his drive?

**SOLUTION:**

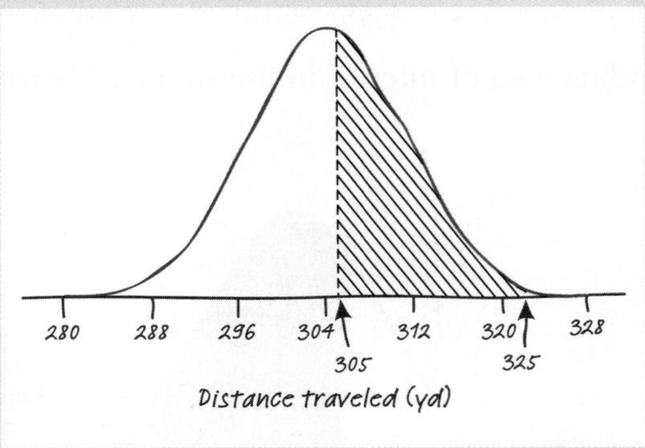


**1. Draw a Normal distribution.**

(i) $z = \dfrac{305 - 304}{8} = 0.13$ $\quad z = \dfrac{325 - 304}{8} = 2.63$

**2. Perform calculations—show your work!**
(i) Standardize and use Table A or technology; or
(ii) Use technology without standardizing.

*Using Table A:* 0.9957 − 0.5517 = 0.4440
*Using technology:* normalcdf(lower:0.13, upper:2.63, mean:0, SD:1) = 0.4440

(ii) normalcdf(lower:305, upper:325, mean:304, SD:8) = 0.4459

About 45% of Spieth's drives travel between 305 and 325 yards. He has a fairly good chance of getting the ball on the green—assuming he hits the shot straight.

Be sure to answer the question that was asked.

**FOR PRACTICE, TRY EXERCISE 57**

## CHECK YOUR UNDERSTANDING

High levels of cholesterol in the blood increase the risk of heart disease. For 14-year-old boys, the distribution of blood cholesterol is approximately Normal with mean $\mu = 170$ milligrams of cholesterol per deciliter of blood (mg/dl) and standard deviation $\sigma = 30$ mg/dl.[6]

1. Cholesterol levels higher than 240 mg/dl may require medical attention. What percent of 14-year-old boys have more than 240 mg/dl of cholesterol?
2. People with cholesterol levels between 200 and 240 mg/dl are at considerable risk for heart disease. What proportion of 14-year-old boys have blood cholesterol between 200 and 240 mg/dl?

# Working Backward: Finding Values from Areas

So far, we have focused on finding areas in Normal distributions that correspond to specific values. What if we want to find the value that corresponds to a particular area? For instance, suppose we want to estimate the 90th percentile of the distribution of ITBS vocabulary scores for Gary, Indiana, seventh-graders. Figure 2.18(a) shows the Normal curve with mean $\mu = 6.84$ and standard deviation $\sigma = 1.55$ that models this distribution. We're looking for the ITBS score $x$ that has 90% of the area under the curve less than or equal to it. Figure 2.18(b) shows the standard Normal distribution with the corresponding area shaded.

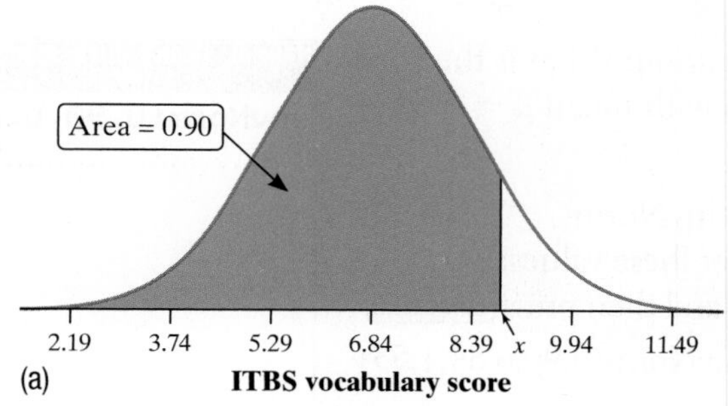

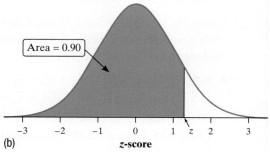

**FIGURE 2.18** (a) Normal distribution showing the 90th percentile of ITBS vocabulary scores for Gary, Indiana, seventh-graders. (b) The 90th percentile in the standard Normal distribution.

We can use Table A or technology to find the $z$-score with an area of 0.90 to its left. Because Table A gives the area to the left of a specified $z$-score, all we have to do is find the value closest to 0.90 in the middle of the table. From the

| $z$ | .07 | .08 | .09 |
|---|---|---|---|
| 1.1 | .8790 | .8810 | .8830 |
| 1.2 | .8980 | .8997 | .9015 |
| 1.3 | .9147 | .9162 | .9177 |

reproduced portion of Table A, you can see that the desired value is $z = 1.28$. Then we "unstandardize" to get the corresponding ITBS vocabulary score $x$.

$$z = \frac{x - \mu}{\sigma}$$

$$1.28 = \frac{x - 6.84}{1.55}$$

$$1.28(1.55) + 6.84 = x$$

$$8.824 = x$$

So the 90th percentile of the distribution of ITBS vocabulary scores for Gary, Indiana, seventh-graders is 8.824.

It is also possible to find the 90th percentile of either distribution in Figure 2.18 using technology.

## 6. Technology Corner FINDING VALUES FROM AREAS IN A NORMAL DISTRIBUTION

*TI-Nspire and other technology instructions are on the book's website at highschool.bfwpub.com/updatedtps6e.*

The TI-83/84 invNorm command calculates the value corresponding to a given percentile in a Normal distribution. The syntax is invNorm(area to the left, mean, standard deviation). Let's use this command to confirm the 90th percentile for the ITBS vocabulary scores in Gary, Indiana. Note that we can do the calculation using the standard Normal distribution or the Normal distribution with mean 6.84 and standard deviation 1.55.

(i) *Using the standard Normal distribution:* What is the 90th percentile of the standard Normal distribution?

- Press 2nd VARS (Distr) and choose invNorm(.
  **OS 2.55 or later:** In the dialog box, enter these values: area:0.90, $\mu$:0, $\sigma$:1, choose Paste, and then press ENTER.
  **Older OS:** Complete the command invNorm(0.90,0,1) and press ENTER.

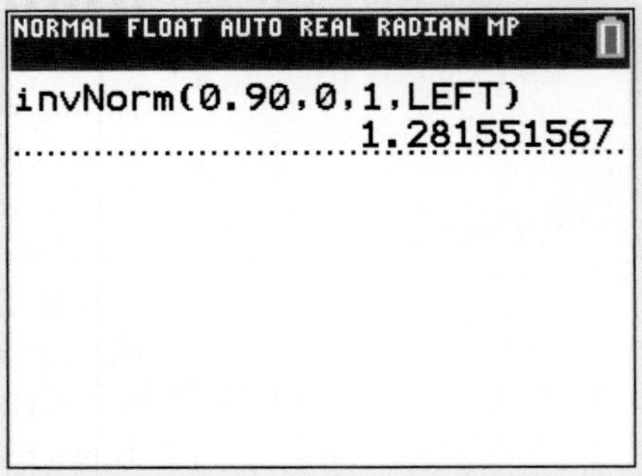


*Note:* The most recent TI-84 Plus CE OS has added an option for specifying area in the LEFT, CENTER, or RIGHT of the distribution. Choose LEFT in this case.

This result matches what we got using Table A. Now "unstandardize" as shown preceding the Technology Corner to get $x = 8.824$.

(ii) *Using the unstandardized Normal distribution:* What is the 90th percentile of a Normal distribution with mean $\mu = 6.84$ and standard deviation $\sigma = 1.55$?

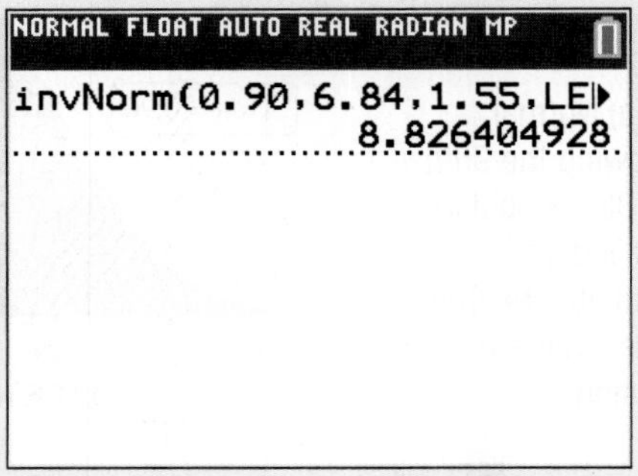


- Press 2nd VARS (Distr) and choose invNorm(.
  **OS 2.55 or later:** In the dialog box, enter these values: area:0.90, $\mu$:6.84, $\sigma$:1.55, choose Paste, and then press ENTER.
  **Older OS:** Complete the command invNorm(0.90,6.84,1.55) and press ENTER.

*Note:* The most recent TI-84 Plus CE OS has added an option for specifying area in the LEFT, CENTER, or RIGHT of the distribution. Choose LEFT in this case.

The following box summarizes the process of finding a value corresponding to a given area in a Normal distribution. In Step 2, each method of performing calculations has some advantages, so check with your teacher to see which option will be used in your class.

## HOW TO FIND VALUES FROM AREAS IN ANY NORMAL DISTRIBUTION

**Step 1: Draw a Normal distribution** with the horizontal axis labeled and scaled using the mean and standard deviation, the area of interest shaded and labeled, and unknown boundary value clearly marked.

**Step 2: Perform calculations—show your work!** Do one of the following:

(i) Use Table A or technology to find the value of $z$ with the indicated area under the standard Normal curve, then "unstandardize" to transform back to the original distribution; or

(ii) Use technology to find the desired value without standardizing.

Be sure to answer the question that was asked.

### AP® EXAM TIP

As noted previously, to make sure that you get full credit on the AP® Statistics exam, do not use "calculator-speak" alone—for example, invNorm(0.90,6.84,1.55). This is *not* considered clear communication. To get full credit, follow the two-step process above, making sure to carefully label each of the inputs in the calculator command if you use technology in Step 2: invNorm(area: 0.90, mean: 6.84, SD:1.55).

## EXAMPLE

### How tall are 3-year-old girls?
### Finding a value from an area

Cultura Creative (RF)/ Alamy Stock Photo

**PROBLEM:** According to www.cdc.gov/growthcharts/, the heights of 3-year-old females are approximately Normally distributed with a mean of 94.5 centimeters and a standard deviation of 4 centimeters. Seventy-five percent of 3-year-old girls are taller than what height?

**SOLUTION:**

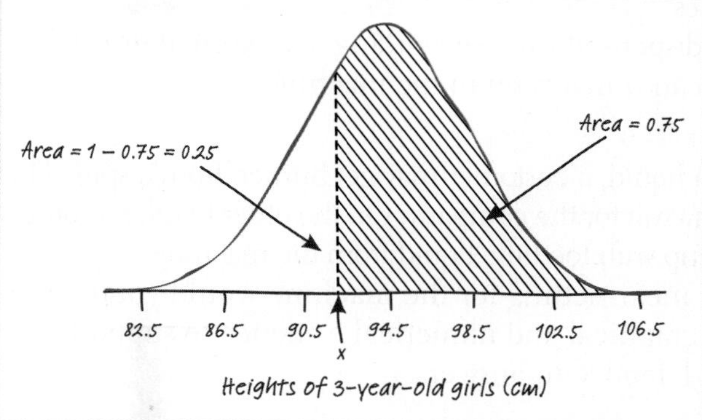


If 75% of 3-year-old girls are taller than a certain height, then 25% of 3-year-old girls are shorter than that height. So we just need to find the 25th percentile of this distribution of height.

**1. Draw a Normal distribution.**
From the empirical rule, we know that about 16% of the observations in a Normal distribution will fall more than 1 standard deviation less than the mean. So the 25th percentile will be located slightly to the right of 90.5, as shown.

(i) *Using Table A:* 0.25 area to the left → z = −0.67

*Using technology:* invNorm(area: 0.25, mean: 0, SD: 1) = −0.67

$$-0.67 = \frac{x - 94.5}{4}$$

$$-0.67(4) + 94.5 = x$$

$$91.82 = x$$

(ii) invNorm (area: 0.25, mean: 94.5, SD: 4) = 91.80

About 75% of 3-year-old girls are taller than 91.80 centimeters.

**2. Perform calculations—show your work!**
(i) Use Table A or technology to find the value of *z* with the indicated area under the standard Normal curve, then "unstandardize" to transform back to the original distribution; or
(ii) Use technology to find the desired value without standardizing.

Be sure to answer the question that was asked.

**FOR PRACTICE, TRY EXERCISE 63**

Here's an activity that gives you a chance to apply what you have learned so far in this section.

## ACTIVITY Team challenge: The vending machine problem

Ed Honowitz/The Image Bank/Getty Images

In this activity, you will work in a team of three or four students to resolve a real-world problem.

Have you ever purchased a hot drink from a vending machine? The intended sequence of events goes something like this: You insert your money into the machine and select your preferred beverage. A cup falls out of the machine, landing upright. Liquid pours out until the cup is nearly full. You reach in, grab the piping-hot cup, and drink happily.

Except sometimes, things go wrong. The machine might swipe your money. Or the cup might fall over. More frequently, everything goes smoothly until the liquid begins to flow. It might stop flowing when the cup is only half full. Or the liquid might keep coming until your cup overflows. Neither of these results leaves you satisfied.

The vending machine company wants to keep its customers happy. So it has decided to hire you as a statistical consultant. The company provides you with the following summary of important facts about the vending machine:

- Cups will hold 8 ounces.
- The amount of liquid dispensed varies according to a Normal distribution centered at the mean $\mu$ that is set in the machine.
- $\sigma = 0.2$ ounces.

If a cup contains too much liquid, a customer may get burned from a spill. This could result in an expensive lawsuit for the company. On the other hand, customers may be irritated if they get a cup with too little liquid from the machine.

Given these issues, what mean setting for the machine would you recommend? Provide appropriate graphical and numerical evidence to support your conclusion. Be prepared to defend your answer.

## CHECK YOUR UNDERSTANDING

High levels of cholesterol in the blood increase the risk of heart disease. For 14-year-old boys, the distribution of blood cholesterol is approximately Normal with mean $\mu = 170$ milligrams of cholesterol per deciliter of blood (mg/dl) and standard deviation $\sigma = 30$ mg/dl. What cholesterol level would place a 14-year-old boy at the 10th percentile of the distribution?

# Assessing Normality

Normal distributions provide good models for some distributions of quantitative data. Examples include SAT and IQ test scores, the highway gas mileage of 2018 Corvette convertibles, weights of 9-ounce bags of potato chips, and heights of 3-year-old girls (see Figure 2.19).

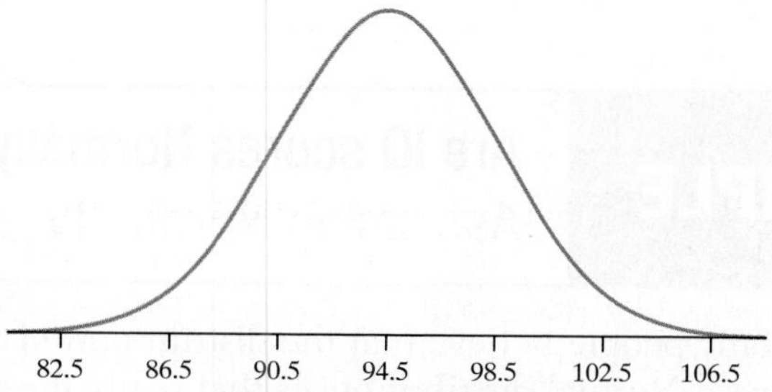


**FIGURE 2.19** The heights of 3-year-old girls are approximately Normally distributed with a mean of 94.5 centimeters and standard deviation of 4 centimeters.

The distributions of other quantitative variables are skewed and therefore distinctly non-Normal. Examples include single-family home prices in a certain city, survival times of cancer patients after treatment, and number of siblings for students in a statistics class.

While experience can suggest whether or not a Normal distribution is a reasonable model in a particular case, it is risky to assume that a distribution is approximately Normal without first analyzing the data. As in Chapter 1, we start with a graph and then add numerical summaries to assess the Normality of a distribution of quantitative data.

If a graph of the data is clearly skewed, has multiple peaks, or isn't bell-shaped, that's evidence the distribution is not Normal. Here is a dotplot of the number of siblings reported by each student in a statistics class. This distribution is skewed to the right and therefore not approximately Normal.

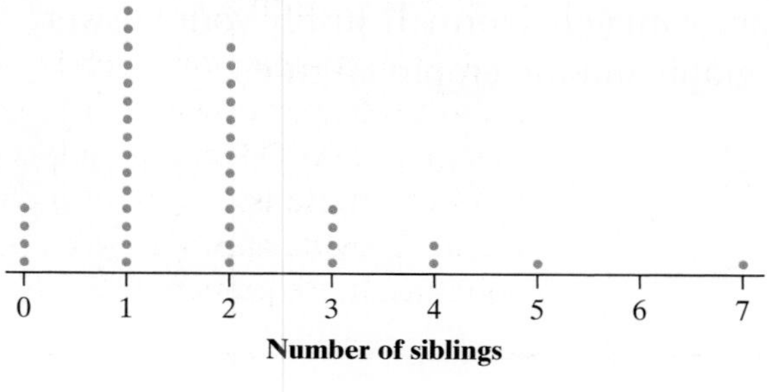

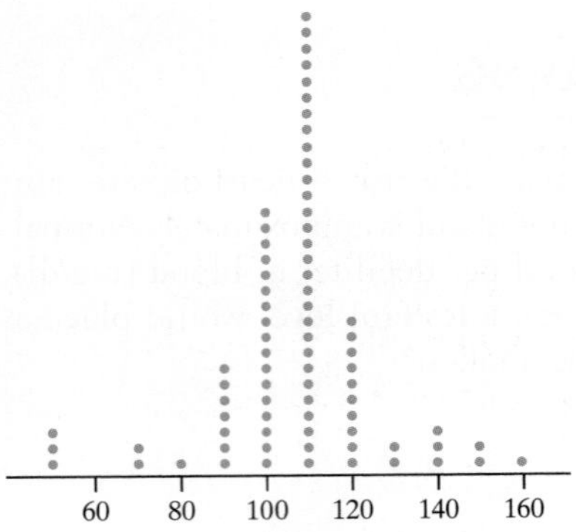


| $n$ | Mean | SD | Min | $Q_1$ | Med | $Q_3$ | Max |
|---|---|---|---|---|---|---|---|
| 77 | 106.883 | 19.484 | 50 | 100 | 110 | 110 | 160 |

**FIGURE 2.20** Dotplot and summary statistics for data on calories per serving in 77 different brands of breakfast cereal.

Even if a graph of the data looks roughly symmetric and bell-shaped, we shouldn't assume that the distribution is approximately Normal. The empirical rule can give additional evidence in favor of or against Normality.

Figure 2.20 shows a dotplot and numerical summaries for data on calories per serving in 77 brands of breakfast cereal.[7] The graph is roughly symmetric, single-peaked (unimodal), and somewhat bell-shaped. Let's count the number of data values within 1, 2, and 3 standard deviations of the mean:

| | | | |
|---|---|---|---|
| Mean $\pm$ 1 SD: | $106.883 \pm 1(19.484)$ | 87.399 to 126.367 | 63 out of 77 = 81.8% |
| Mean $\pm$ 2 SD: | $106.883 \pm 2(19.484)$ | 67.915 to 145.851 | 71 out of 77 = 92.2% |
| Mean $\pm$ 3 SD: | $106.883 \pm 3(19.484)$ | 48.431 to 165.335 | 77 out of 77 = 100.0% |

In a Normal distribution, about 68% of the values fall within 1 standard deviation of the mean. For the cereal data, almost 82% of the brands had between 87.399 and 126.367 calories per serving. These two percentages are far apart. So this distribution of calories in breakfast cereals is not approximately Normal.

## EXAMPLE

### Are IQ scores Normally distributed?
### Assessing Normality

Monkey Business Images/
Shutterstock.com

**PROBLEM:** Many people believe that the distribution of IQ scores follows a Normal distribution. Is that really the case? To find out, researchers obtained the IQ scores of 60 randomly selected fifth-grade students from one school. Here are the data:[8]

| | | | | | | | | | | |
|---|---|---|---|---|---|---|---|---|---|---|
| 81 | 82 | 89 | 90 | 94 | 96 | 97 | 100 | 101 | 101 | 101 |
| 102 | 102 | 102 | 103 | 105 | 106 | 108 | 109 | 109 | 109 | 110 |
| 110 | 110 | 112 | 112 | 113 | 113 | 114 | 114 | 114 | 115 | 116 |
| 117 | 117 | 117 | 118 | 118 | 122 | 122 | 123 | 124 | 124 | 124 |
| 125 | 126 | 127 | 127 | 128 | 130 | 131 | 133 | 134 | 134 | 136 |
| 137 | 139 | 139 | 142 | 145 | | | | | | |

A histogram and summary statistics for the data are shown. Is this distribution of IQ scores of fifth-graders at this school approximately Normal? Justify your answer based on the graph and the empirical rule.

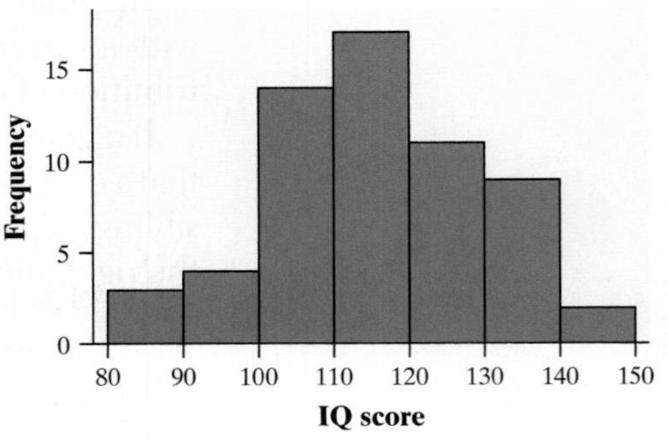


| $n$ | Mean | SD | Min | $Q_1$ | Med | $Q_3$ | Max |
|---|---|---|---|---|---|---|---|
| 60 | 114.983 | 14.801 | 81 | 104 | 114 | 125.5 | 145 |

**SOLUTION:**

The histogram looks roughly symmetric, single-peaked, and somewhat bell-shaped. The percents of values within 1, 2, and 3 standard deviations of the mean are

| | | |
|---|---|---|
| Mean $\pm$ 1 SD: $114.983 \pm 1(14.801)$ | 100.182 to 129.784 | 41 out of 60 = 68.3% |
| Mean $\pm$ 2 SD: $114.983 \pm 2(14.801)$ | 85.381 to 144.585 | 57 out of 60 = 95.0% |
| Mean $\pm$ 3 SD: $114.983 \pm 3(14.801)$ | 70.580 to 159.386 | 60 out of 60 = 100.0% |

These percents are very close to the 68%, 95%, and 99.7% targets for a Normal distribution. The graphical and numerical evidence suggests that this distribution of IQ scores is approximately Normal.

**FOR PRACTICE, TRY EXERCISE 75**

**AP® EXAM TIP**

Never say that a distribution of quantitative data *is* Normal. Real-world data always show at least slight departures from a Normal distribution. The most you can say is that the distribution is "approximately Normal."

Because the IQ data come from a random sample, we can use the sample mean IQ score to make an inference about the population mean IQ score of all fifth-graders at the school. As you will see in later chapters, the methods for inference about a population mean work best when the population distribution is Normal. Because the distribution of IQ scores in the sample is approximately Normal, it is reasonable to believe that the population distribution is approximately Normal.

**NORMAL PROBABILITY PLOTS** A graph called a **Normal probability plot** (or a *Normal quantile plot*) provides a good assessment of whether or not a distribution of quantitative data is approximately Normal.

**DEFINITION Normal probability plot**

A **Normal probability plot** is a scatterplot of the ordered pair (data value, expected $z$-score) for each of the individuals in a quantitative data set. That is, the $x$-coordinate of each point is the actual data value and the $y$-coordinate is the expected $z$-score corresponding to the percentile of that data value in a standard Normal distribution.

**AP® EXAM TIP**

Normal probability plots are not included in the AP® Statistics Course Framework. However, these graphs are very useful for assessing Normality. You may use them on the AP® Statistics exam if you wish—just be sure that you know what you're looking for.

Some software plots the data values on the horizontal axis and the expected $z$-scores on the vertical axis, while other software does just the reverse. The TI-83/84 gives you both options. We prefer the data values on the horizontal axis, which is consistent with other types of graphs we have made.

Technology Corner 7 at the end of this subsection shows you how to make a Normal probability plot. For now, let's focus on how to interpret Normal probability plots.

Figure 2.21 shows dotplots and Normal probability plots for each of the data sets in this subsection.

- Panel (a): We confirmed earlier that the distribution of IQ scores is approximately Normal. Its Normal probability plot has a *linear* form.
- Panel (b): The distribution of number of siblings is clearly right-skewed. Its Normal probability plot has a curved form.
- Panel (c): We determined earlier that the distribution of calories in breakfast cereals is *not* approximately Normal, even though the graph looks roughly symmetric and is somewhat bell-shaped. Its Normal probability plot has a different kind of nonlinear form.

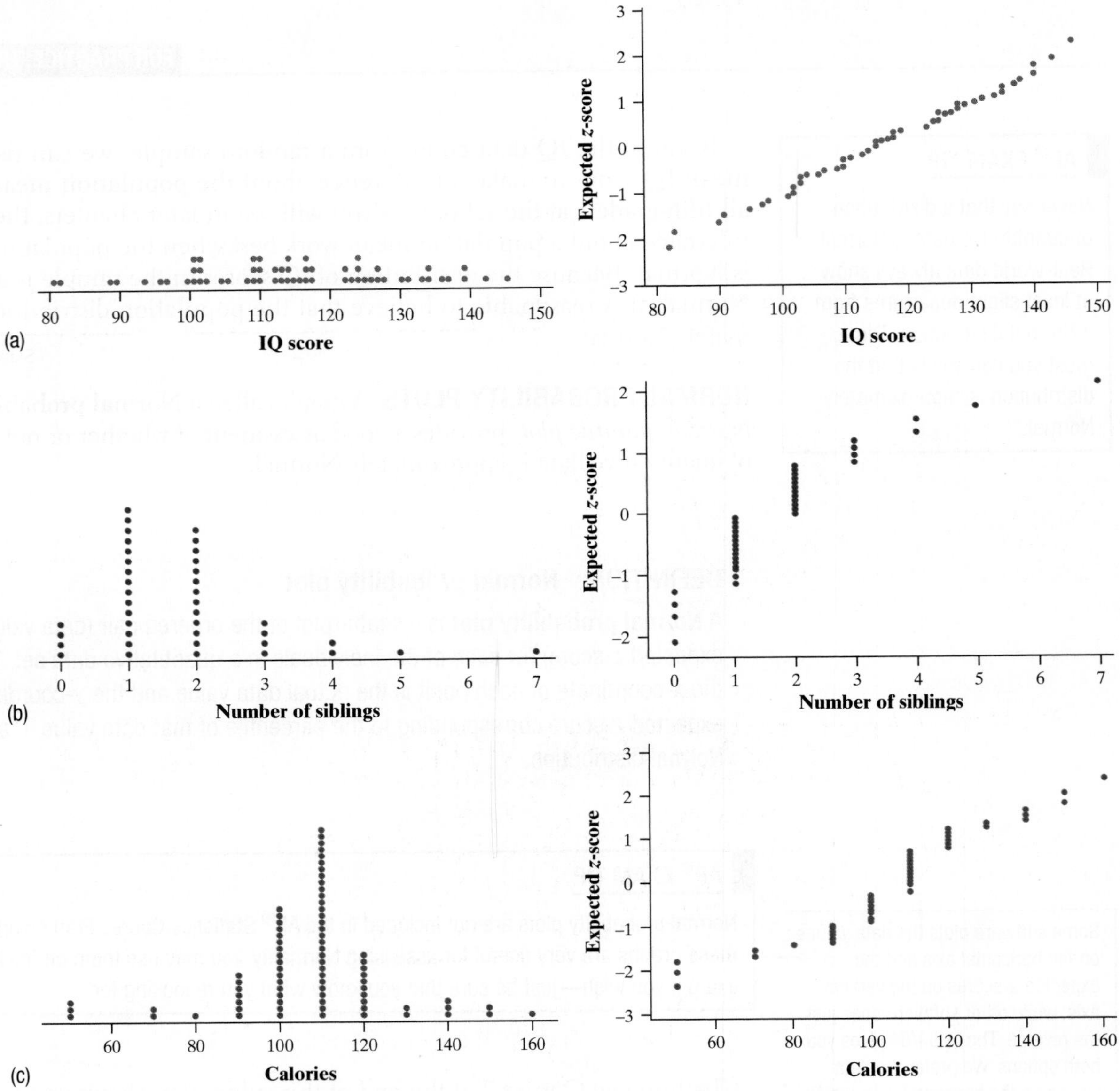


**FIGURE 2.21** Dotplot and Normal probability plot of (a) IQ scores for 60 randomly selected fifth-graders from one school, (b) Number of siblings for each student in a college statistics class, and (c) Calories per serving in 77 brands of breakfast cereal. The distribution of IQ scores in (a) is approximately Normal because the Normal probability plot has a linear form. The nonlinear Normal probability plots in (b) and (c) confirm that neither of these distributions is approximately Normal.

## HOW TO ASSESS NORMALITY WITH A NORMAL PROBABILITY PLOT

If the points on a Normal probability plot lie close to a straight line, the data are approximately Normally distributed. A nonlinear form in a Normal probability plot indicates a non-Normal distribution.

When examining a Normal probability plot, look for shapes that show clear departures from Normality. Don't overreact to minor wiggles in the plot. We used a TI-84 to generate three different random samples of size 20 from a Normal distribution. The screen shots show Normal probability plots for each of the samples. Although none of the plots is perfectly linear, it is reasonable to believe that each sample came from a Normal population.

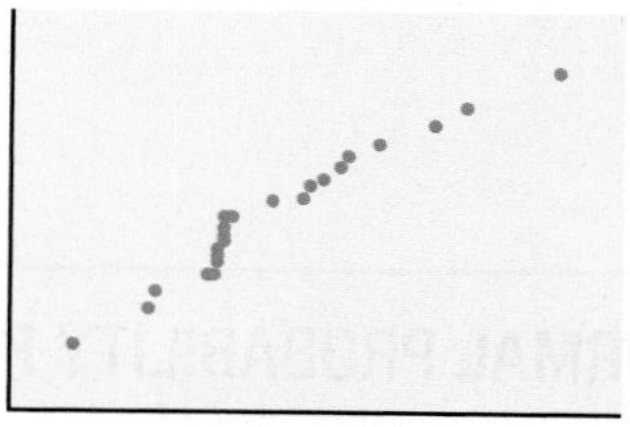
Sample 1

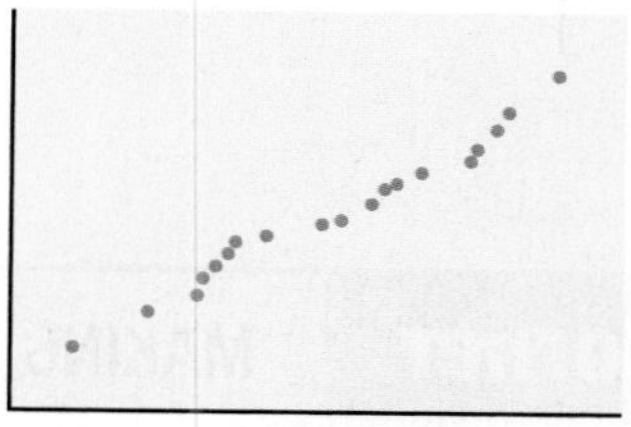
Sample 2

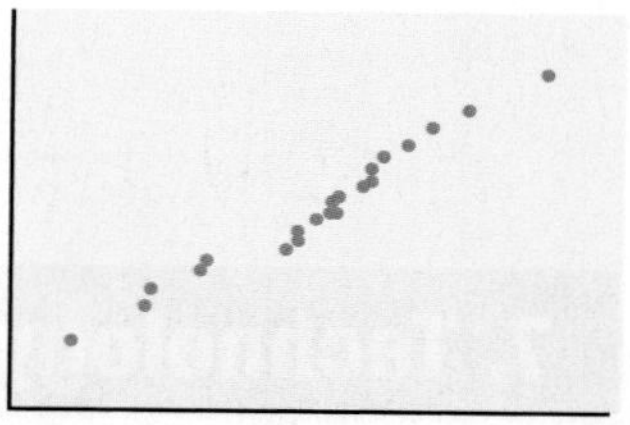
Sample 3

## EXAMPLE

### How Normal are survival times?
### Interpreting Normal probability plots

**PROBLEM:** Researchers recorded the survival times in days of 72 guinea pigs after they were injected with infectious bacteria in a medical experiment.[9] A Normal probability plot of the data is shown. Use the graph to determine if the distribution of survival times is approximately Normal.

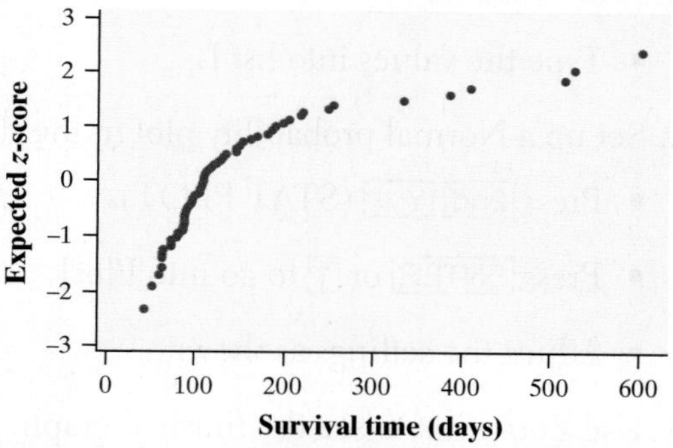


**SOLUTION:**

The Normal probability plot is clearly curved, indicating that the distribution of survival time for the 72 guinea pigs is not approximately Normal.

**FOR PRACTICE, TRY EXERCISE 79**

### Think About It

**HOW CAN WE DETERMINE SHAPE FROM A NORMAL PROBABILITY PLOT?** Look at the Normal probability plot of the guinea pig survival data in the example. Imagine all the points falling down onto the horizontal axis. The resulting dotplot would have many values stacked up between 50 and 150 days, and fewer values that are further spread apart from 150 to about 600 days. The distribution would be skewed to the right due to the greater variability in the upper half of the data set. The dotplot of the data confirms our answer.

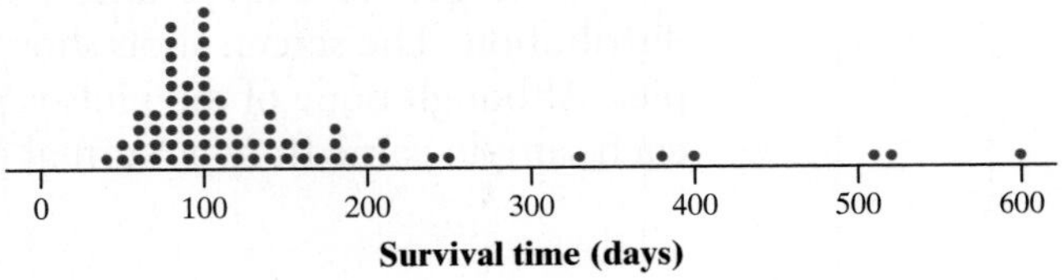


## 7. Technology Corner — MAKING NORMAL PROBABILITY PLOTS

*TI-Nspire and other technology instructions are on the book's website at highschool.bfwpub.com/updatedtps6e.*

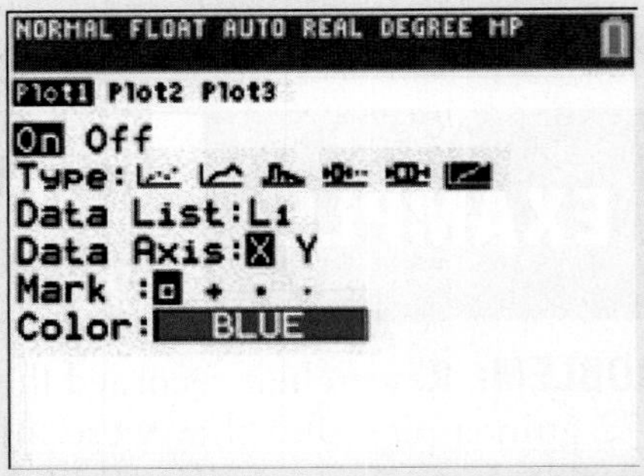


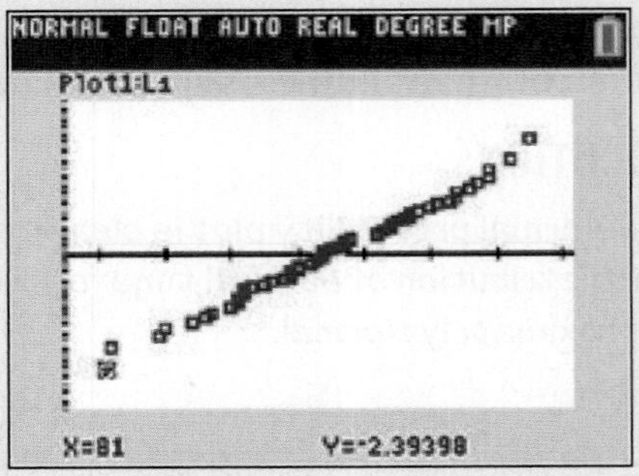


Let's use the TI-83/84 to make a Normal probability plot for the IQ score data (page 132).

1. Enter the data values in list L1.
   - Press STAT and choose Edit....
   - Type the values into list $L_1$.
2. Set up a Normal probability plot in the statistics plots menu.
   - Press 2nd Y= (STAT PLOT).
   - Press ENTER or 1 to go into Plot1.
   - Adjust the settings as shown.
3. Use ZoomStat to see the finished graph.

*Interpretation*: The Normal probability plot is quite linear, which confirms our earlier belief that the distribution of IQ scores is approximately Normal.

# Section 2.2 Summary

- We can describe the overall pattern of the distribution of a quantitative variable with a **density curve.** A density curve always remains on or above the horizontal axis and has total area 1 underneath it. An area under a density curve estimates the proportion of observations that fall in an interval of values.
- A density curve is an idealized description of a distribution that smooths out any irregularities. We write the **mean of a density curve** as $\mu$ and the **standard deviation of a density curve** as $\sigma$. The values of $\mu$ and $\sigma$ are the parameters of the density curve.
- The mean and the median of a density curve can be located by eye. The mean $\mu$ is the balance point of the curve. The median divides the area under the curve in half. The standard deviation $\sigma$ cannot be located by eye on most density curves.
- The mean and median are equal for symmetric density curves. The mean of a skewed density curve is located farther toward the long tail than is the median.
- **Normal distributions** are described by a special family of bell-shaped, symmetric density curves, called **Normal curves.** The mean $\mu$ and standard deviation $\sigma$ completely specify a Normal distribution. The mean is the center of the curve, and $\sigma$ is the distance from $\mu$ to the change-of-curvature points on either side.
- The **empirical rule** describes what percent of observations in any Normal distribution fall within 1, 2, and 3 standard deviations of the mean: about 68%, 95%, and 99%, respectively.
- All Normal distributions are the same when observations are standardized. If $x$ follows a Normal distribution with mean $\mu$ and standard deviation $\sigma$, we can standardize using

$$z = \frac{x - \mu}{\sigma}$$

  Then the variable $z$ has the **standard Normal distribution** with mean 0 and standard deviation 1.
- **Table A** in the back of the book gives percentiles for the standard Normal distribution. You can use Table A or technology to determine area for given values of the variable or the value that corresponds to a given percentile in any Normal distribution.
- To find the area in a Normal distribution corresponding to given values:

  **Step 1: Draw a Normal distribution** with the horizontal axis labeled and scaled using the mean and standard deviation, the boundary value(s) clearly identified, and the area of interest shaded.

  **Step 2: Perform calculations—show your work!** Do one of the following:

  (i) Standardize each boundary value and use Table A or technology to find the desired area under the standard Normal curve; or

  (ii) Use technology to find the desired area without standardizing.

  Be sure to answer the question that was asked.

- To find the value in a Normal distribution corresponding to a given area (percentile):

  **Step 1: Draw a Normal distribution** with the horizontal axis labeled and scaled using the mean and standard deviation, the area of interest shaded and labeled, and unknown boundary value clearly marked.

  **Step 2: Perform calculations—show your work!** Do one of the following:

  (i) Use Table A or technology to find the value of $z$ with the indicated area under the standard Normal curve, then "unstandardize" to transform back to the original distribution; or

  (ii) Use technology to find the desired area without standardizing.

  Be sure to answer the question that was asked.

- To assess Normality for a given set of quantitative data, we first observe the shape of a dotplot, stemplot, or histogram. Then we can check how well the data fit the empirical rule for Normal distributions. Another good method for assessing Normality is to construct a **Normal probability plot.** If the Normal probability plot has a linear form, then we can say that the distribution is approximately Normal.

## 2.2 Technology Corners

*TI-Nspire and other technology instructions are on the book's website at highschool.bfwpub.com/updatedtps6e.*



# Section 2.2 Exercises

41. **Where's the bus?** Sally takes the same bus to work every morning. The amount of time (in minutes) that she has to wait for the bus can be modeled by a uniform distribution on the interval from 0 minutes to 10 minutes.

pg 111

(a) Draw a density curve to model the amount of time that Sally has to wait for the bus. Be sure to include scales on both axes.

(b) On about what percent of days does Sally wait between 2.5 and 5.3 minutes for the bus?

(c) Find the 70th percentile of Sally's wait times.

42. **Still waiting for the server?** How does your web browser get a file from the Internet? Your computer sends a request for the file to a web server, and the web server sends back a response. For one particular web server, the time (in seconds) after the start of an hour at which a request is received can be modeled by a uniform distribution on the interval from 0 to 3600 seconds.

(a) Draw a density curve to model the amount of time after an hour at which a request is received by the web server. Be sure to include scales on both axes.

(b) About what proportion of requests are received within the first 5 minutes (300 seconds) after the hour?

(c) Find the interquartile range of this distribution.

43. **Quick, click!** An Internet reaction time test asks subjects to click their mouse button as soon as a light flashes on the screen. The light is

programmed to go on at a randomly selected time after the subject clicks "Start." The density curve models the amount of time the subject has to wait for the light to flash.

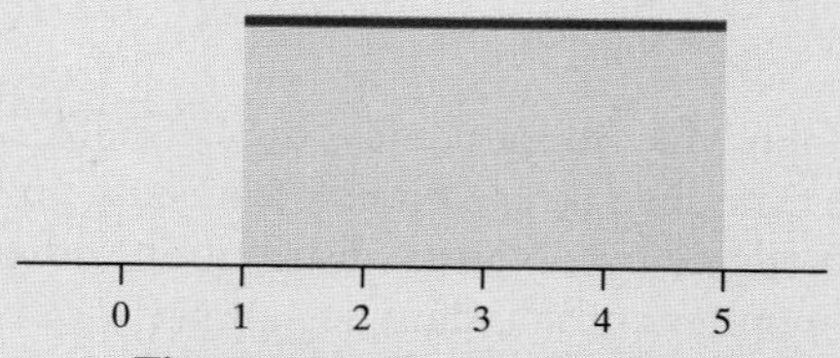


**Time (sec) until the light flashes**

(a) What height must the density curve have? Justify your answer.

(b) About what percent of the time will the light flash more than 3.75 seconds after the subject clicks "Start"?

(c) Calculate and interpret the 38th percentile of this distribution.

**44. Class is over!** Mr. Shrager does not always let his statistics class out on time. In fact, he seems to end class according to his own "internal clock." The density curve models the distribution of the amount of time after class ends (in minutes) when Mr. Shrager dismisses the class. (A negative value indicates he ended class early.)

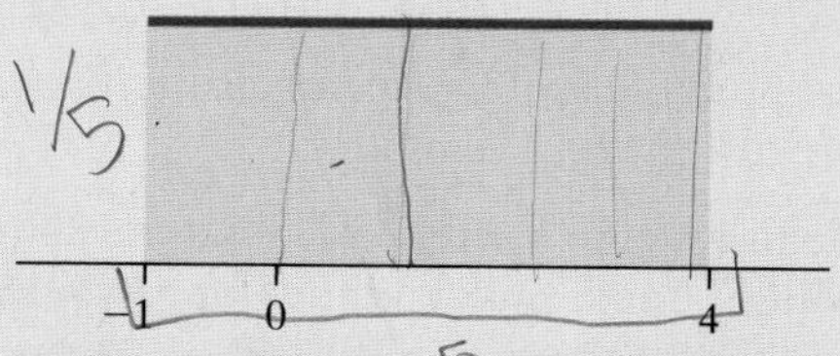


**Dismissal time (minutes after class ends)**

(a) What height must the density curve have? Justify your answer.

(b) About what proportion of the time does Mr. Shrager dismiss class within 1 minute of its scheduled end time?

(c) Calculate and interpret the 20th percentile of the distribution.

**45. Mean and median** The figure displays two density curves that model different distributions of quantitative data. Identify the location of the mean and median by letter for each graph. Justify your answers.

pg 112

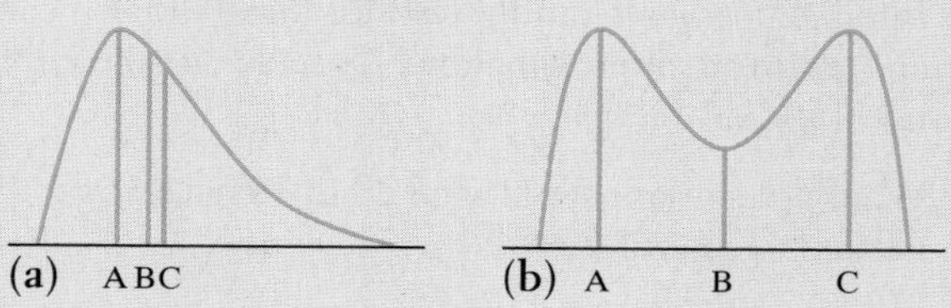


**46. Mean and median** The figure displays two density curves that model different distributions of quantitative data. Identify the location of the mean and median by letter for each graph. Justify your answers.

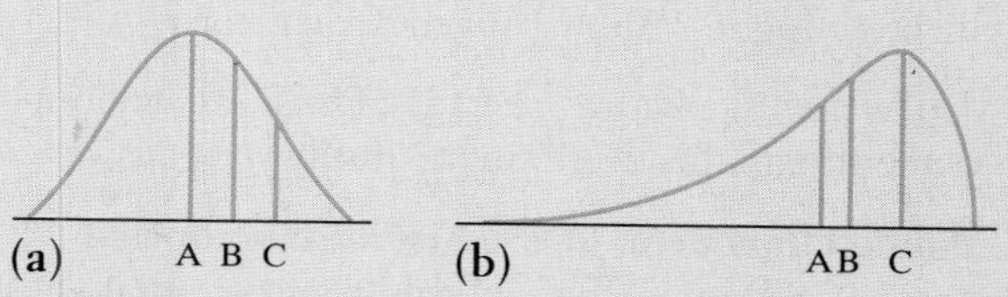


**47. Potato chips** The weights of 9-ounce bags of a particular brand of potato chips can be modeled by a Normal distribution with mean $\mu = 9.12$ ounces and standard deviation $\sigma = 0.05$ ounce. Sketch the Normal density curve. Label the mean and the points that are 1, 2, and 3 standard deviations from the mean.

pg 115

**48. Batter up!** In baseball, a player's batting average is the proportion of times the player gets a hit out of his total number of times at bat. The distribution of batting averages in a recent season for Major League Baseball players with at least 100 plate appearances can be modeled by a Normal distribution with mean $\mu = 0.261$ and standard deviation $\sigma = 0.034$. Sketch the Normal density curve. Label the mean and the points that are 1, 2, and 3 standard deviations from the mean.

**49. Normal curve** Estimate the mean and standard deviation of the Normal density curve below.

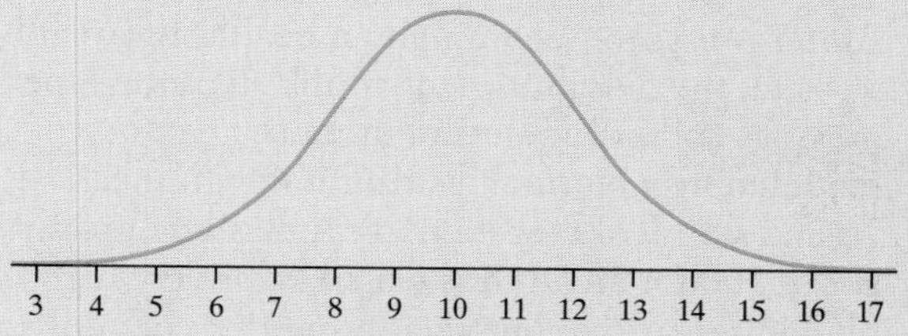


**50. Normal curve** Estimate the mean and standard deviation of the Normal density curve below.

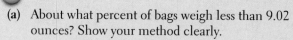

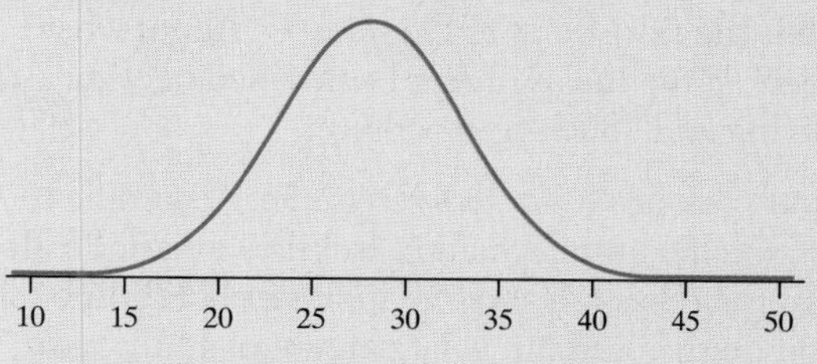


**51. Potato chips** Refer to Exercise 47. Use the empirical rule to answer the following questions.

pg 118

(a) About what percent of bags weigh less than 9.02 ounces? Show your method clearly.

(b) A bag that weighs 9.07 ounces is at about what percentile in this distribution? Justify your answer.

52. **Batter up!** Refer to Exercise 48. Use the empirical rule to answer the following questions.
(a) About what percent of Major League Baseball players with 100 plate appearances had batting averages of 0.363 or higher? Show your method clearly.
(b) A player with a batting average of 0.227 is at about what percentile in this distribution? Justify your answer.

53. **Potato chips** Refer to Exercise 47. About what percent of 9-ounce bags of this brand of potato chips weigh less than the advertised 9 ounces? Is this likely to pose a problem for the company that produces these chips?
pg 122

54. **Batter up!** Refer to Exercise 48. A player with a batting average below 0.200 is at risk of sitting on the bench during important games. About what percent of players are at risk?

55. **Watch the salt!** A study investigated about 3000 meals ordered from Chipotle restaurants using the online site Grubhub. Researchers calculated the sodium content (in milligrams) for each order based on Chipotle's published nutrition information. The distribution of sodium content is approximately Normal with mean 2000 mg and standard deviation 500 mg.[10] About what percent of the meals ordered exceeded the recommended daily allowance of 2400 mg of sodium?
pg 124

56. **Blood pressure** According to a health information website, the distribution of adults' diastolic blood pressure (in millimeters of mercury) can be modeled by a Normal distribution with mean 70 and standard deviation 20. A diastolic pressure above 100 for an adult is classified as very high blood pressure. About what proportion of adults have very high blood pressure according to this criterion?

57. **Watch the salt!** Refer to Exercise 55. About what percent of the meals ordered contained between 1200 mg and 1800 mg of sodium?
pg 126

58. **Blood pressure** Refer to Exercise 56. According to the same health information website, a diastolic blood pressure between 80 and 90 indicates borderline high blood pressure. About what percent of adults have borderline high blood pressure?

59. **Standard Normal areas** Find the proportion of observations in a standard Normal distribution that satisfies each of the following statements.
(a) $z > -1.66$
(b) $-1.66 < z < 2.85$

60. **Standard Normal areas** Find the proportion of observations in a standard Normal distribution that satisfies each of the following statements.
(a) $z < -2.46$
(b) $0.89 < z < 2.46$

61. **Sudoku** Mrs. Starnes enjoys doing Sudoku puzzles. The time she takes to complete an easy puzzle can be modeled by a Normal distribution with mean 5.3 minutes and standard deviation 0.9 minute.
(a) What proportion of the time does Mrs. Starnes finish an easy Sudoku puzzle in less than 3 minutes?
(b) How often does it take Mrs. Starnes more than 6 minutes to complete an easy puzzle?
(c) What percent of easy Sudoku puzzles take Mrs. Starnes between 6 and 8 minutes to complete?

62. **Hit an ace!** Professional tennis player Novak Djokovic hits the ball extremely hard. His first-serve speeds can be modeled by a Normal distribution with mean 112 miles per hour (mph) and standard deviation 5 mph.
(a) How often does Djokovic hit his first serve faster than 120 mph?
(b) What percent of Djokovic's first serves are slower than 100 mph?
(c) What proportion of Djokovic's first serves have speeds between 100 and 110 mph?

63. **Sudoku** Refer to Exercise 61. Find the 20th percentile of Mrs. Starnes's Sudoku times for easy puzzles.
pg 129

64. **Hit an ace!** Refer to Exercise 62. Find the 85th percentile of Djokovic's first-serve speeds.

65. **Deciles** The deciles of any distribution are the values at the 10th, 20th, . . . , 90th percentiles. The first and last deciles are the 10th and the 90th percentiles, respectively. What are the first and last deciles of the standard Normal distribution?

66. **Outliers** The percent of the observations that are classified as outliers by the $1.5 \times IQR$ rule is the same in any Normal distribution. What is this percent? Show your method clearly.

67. **IQ test scores** Scores on the Wechsler Adult Intelligence Scale (an IQ test) for the 20- to 34-year-old age group are approximately Normally distributed with $\mu = 110$ and $\sigma = 25$.
(a) What percent of people aged 20 to 34 have IQs between 125 and 150?

(b) MENSA is an elite organization that admits as members people who score in the top 2% on IQ tests. What score on the Wechsler Adult Intelligence Scale would an individual aged 20 to 34 have to earn to qualify for MENSA membership?

68. **Post office** A local post office weighs outgoing mail and finds that the weights of first-class letters are approximately Normally distributed with a mean of 0.69 ounce and a standard deviation of 0.16 ounce.

(a) Estimate the 60th percentile of first-class letter weights.

(b) First-class letters weighing more than 1 ounce require extra postage. What proportion of first-class letters at this post office require extra postage?

*Exercises 69 and 70 refer to the following setting.* At some fast-food restaurants, customers who want a lid for their drinks get them from a large stack near the straws, napkins, and condiments. The lids are made with a small amount of flexibility so they can be stretched across the mouth of the cup and then snugly secured. When lids are too small or too large, customers can get very frustrated, especially if they end up spilling their drinks. At one particular restaurant, large drink cups require lids with a "diameter" of between 3.95 and 4.05 inches. The restaurant's lid supplier claims that the diameter of its large lids follows a Normal distribution with mean 3.98 inches and standard deviation 0.02 inch. Assume that the supplier's claim is true.

69. **Put a lid on it!**

(a) What percent of large lids are too small to fit?

(b) What percent of large lids are too big to fit?

(c) Compare your answers to parts (a) and (b). Does it make sense for the lid manufacturer to try to make one of these values larger than the other? Why or why not?

70. **Put a lid on it!** The supplier is considering two changes to reduce to 1% the percentage of its large-cup lids that are too small. One strategy is to adjust the mean diameter of its lids. Another option is to alter the production process, thereby decreasing the standard deviation of the lid diameters.

(a) If the standard deviation remains at $\sigma = 0.02$ inch, at what value should the supplier set the mean diameter of its large-cup lids so that only 1% are too small to fit?

(b) If the mean diameter stays at $\mu = 3.98$ inches, what value of the standard deviation will result in only 1% of lids that are too small to fit?

(c) Which of the two options in parts (a) and (b) do you think is preferable? Justify your answer. (Be sure to consider the effect of these changes on the percent of lids that are too large to fit.)

71. **Flight times** An airline flies the same route at the same time each day. The flight time varies according to a Normal distribution with unknown mean and standard deviation. On 15% of days, the flight takes more than an hour. On 3% of days, the flight lasts 75 minutes or more. Use this information to determine the mean and standard deviation of the flight time distribution.

72. **Brush your teeth** The amount of time Ricardo spends brushing his teeth follows a Normal distribution with unknown mean and standard deviation. Ricardo spends less than 1 minute brushing his teeth about 40% of the time. He spends more than 2 minutes brushing his teeth 2% of the time. Use this information to determine the mean and standard deviation of this distribution.

73. **Normal highway driving?** The dotplot shows the EPA highway gas mileage estimates in miles per gallon (mpg) for a random sample of 21 model year 2018 midsize cars.[11] Explain why this distribution of highway gas mileage is not approximately Normal.

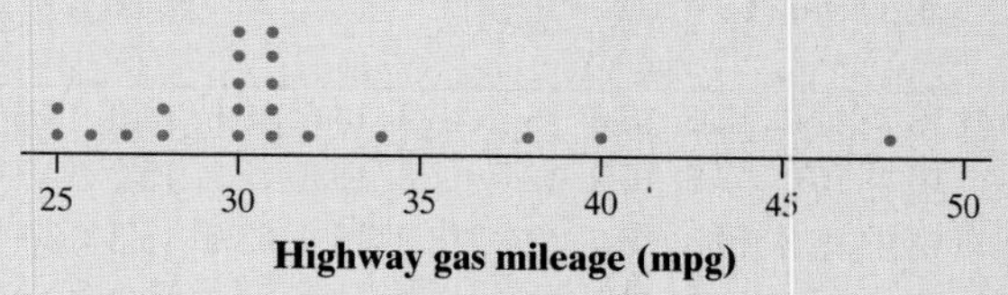


74. **Normal to be foreign born?** The histogram displays the percent of foreign-born residents in each of the 50 states.[12] Explain why this distribution of the percent of foreign-born residents in the states is not approximately Normal.

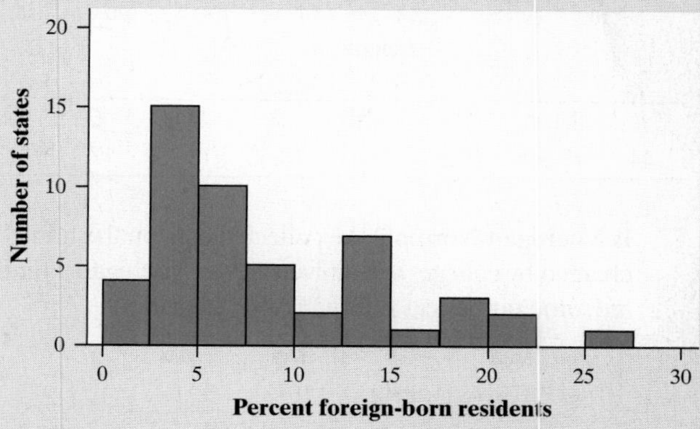

75. **Refrigerators** *Consumer Reports* magazine collected data on the usable capacity (in cubic feet) of a sample of 36 side-by-side refrigerators. Here are the data:

pg 132

| | | | | | | | | | | | |
|---|---|---|---|---|---|---|---|---|---|---|---|
| 12.9 | 13.7 | 14.1 | 14.2 | 14.5 | 14.5 | 14.6 | 14.7 | 15.1 | 15.2 | 15.3 | 15.3 |
| 15.3 | 15.3 | 15.5 | 15.6 | 15.6 | 15.8 | 16.0 | 16.0 | 16.2 | 16.2 | 16.3 | 16.4 |
| 16.5 | 16.6 | 16.6 | 16.6 | 16.8 | 17.0 | 17.0 | 17.2 | 17.4 | 17.4 | 17.9 | 18.4 |

A histogram of the data and summary statistics are shown here. Is this distribution of refrigerator capacities approximately Normal? Justify your answer based on the graph and the empirical rule.

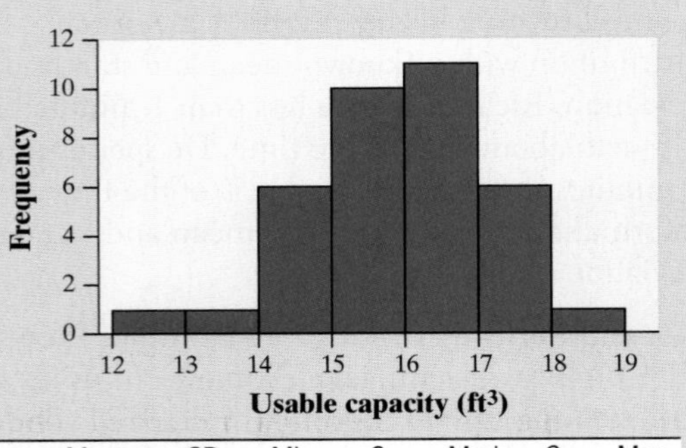


| $n$ | Mean | SD | Min | $Q_1$ | Med | $Q_3$ | Max |
|---|---|---|---|---|---|---|---|
| 36 | 15.825 | 1.217 | 12.9 | 15.15 | 15.9 | 16.6 | 18.4 |

76. **Big sharks** Here are the lengths (in feet) of 44 great white sharks:[13]

| | | | | | | | | | | |
|---|---|---|---|---|---|---|---|---|---|---|
| 18.7 | 12.3 | 18.6 | 16.4 | 15.7 | 18.3 | 14.6 | 15.8 | 14.9 | 17.6 | 12.1 |
| 16.4 | 16.7 | 17.8 | 16.2 | 12.6 | 17.8 | 13.8 | 12.2 | 15.2 | 14.7 | 12.4 |
| 13.2 | 15.8 | 14.3 | 16.6 | 9.4 | 18.2 | 13.2 | 13.6 | 15.3 | 16.1 | 13.5 |
| 19.1 | 16.2 | 22.8 | 16.8 | 13.6 | 13.2 | 15.7 | 19.7 | 18.7 | 13.2 | 16.8 |

A dotplot of the data and summary statistics are shown below. Is this distribution of shark length approximately Normal? Justify your answer based on the graph and the empirical rule.

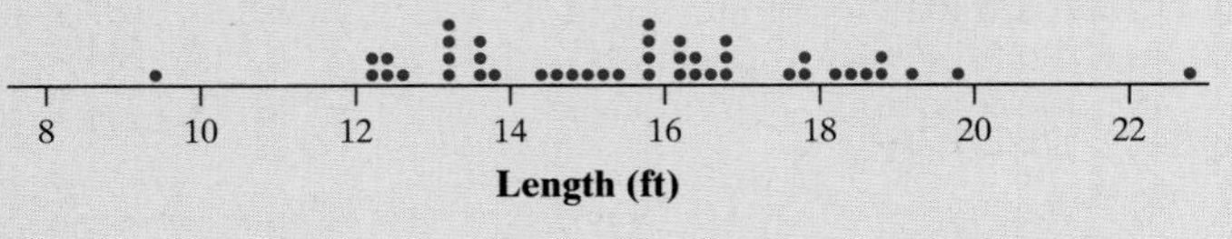


| $n$ | Mean | SD | Min | $Q_1$ | Med | $Q_3$ | Max |
|---|---|---|---|---|---|---|---|
| 44 | 15.586 | 2.55 | 9.4 | 13.55 | 15.75 | 17.2 | 22.8 |

77. **Is Michigan Normal?** We collected data on the tuition charged by colleges and universities in Michigan. Here are some numerical summaries for the data:

| Mean | SD | Min | Max |
|---|---|---|---|
| 10,614 | 8049 | 1873 | 30,823 |

Based on the relationship between the mean, standard deviation, minimum, and maximum, is it reasonable to believe that the distribution of Michigan tuitions is approximately Normal? Explain your answer.

78. **Are body weights Normal?** The heights of people of the same gender and similar ages follow Normal distributions reasonably closely. How about body weights? The weights of women aged 20 to 29 have mean 141.7 pounds and median 133.2 pounds. The first and third quartiles are 118.3 pounds and 157.3 pounds. Is it reasonable to believe that the distribution of body weights for women aged 20 to 29 is approximately Normal? Explain your answer.

79. **Runners' heart rates** The following figure is a Normal probability plot of the heart rates of 200 male runners after 6 minutes of exercise on a treadmill.[14] Use the graph to determine if this distribution of heart rates is approximately Normal.

pg 135

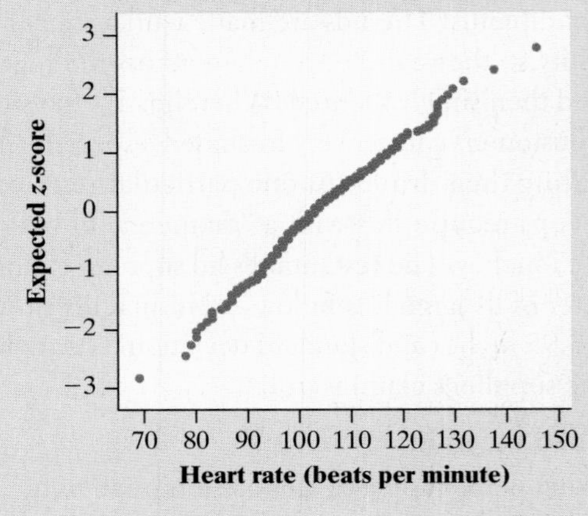


80. **Carbon dioxide emissions** The following figure is a Normal probability plot of the emissions of carbon dioxide ($CO_2$) per person in 48 countries.[15] Use the graph to determine if this distribution of $CO_2$ emissions is approximately Normal.

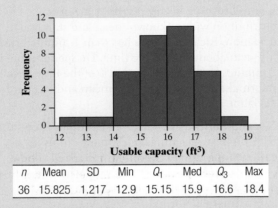

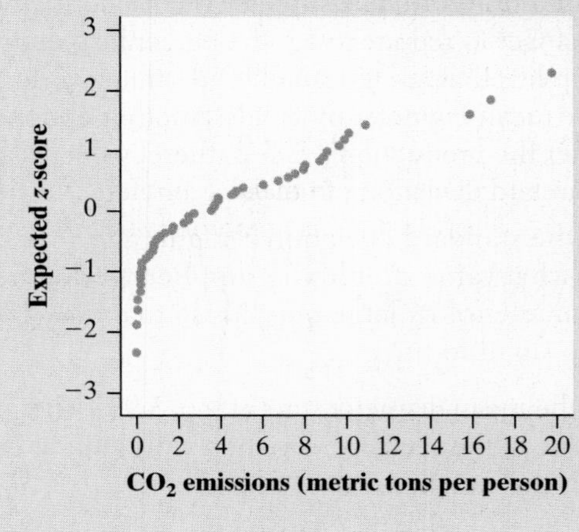

81. **Normal states?** The Normal probability plot displays data on the areas (in thousands of square miles) of each of the 50 states. Use the graph to determine if the distribution of land area is approximately Normal.

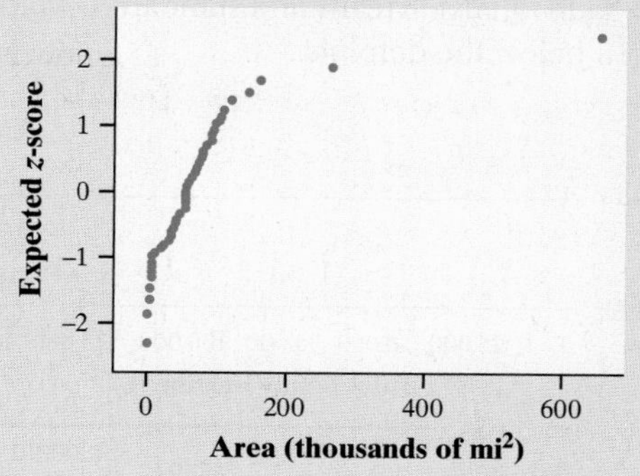


82. **Density of the earth** In 1798, the English scientist Henry Cavendish measured the density of the earth several times by careful work with a torsion balance. The variable recorded was the density of the earth as a multiple of the density of water. A Normal probability plot of the data is shown.[16] Use the graph to determine if this distribution of density measurement is approximately Normal.

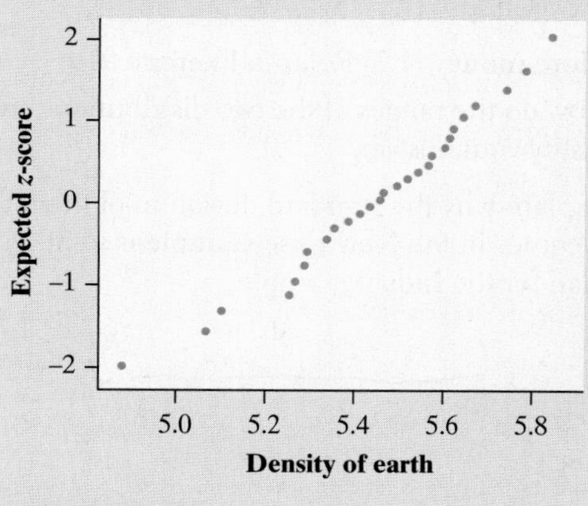


83. **Refrigerators** Refer to Exercise 75.

(a) Use your calculator to make a Normal probability plot of the data. Sketch this graph on your paper.

(b) What does the graph in part (a) imply about whether the distribution of refrigerator capacity is approximately Normal? Explain.

84. **Big sharks** Refer to Exercise 76.

(a) Use your calculator to make a Normal probability plot of the data. Sketch this graph on your paper.

(b) What does the graph in part (a) imply about whether the distribution of shark length is approximately Normal? Explain.

**Multiple Choice:** *Select the best answer for Exercises 85–90.*

85. Two measures of center are marked on the density curve shown. Which of the following is correct?

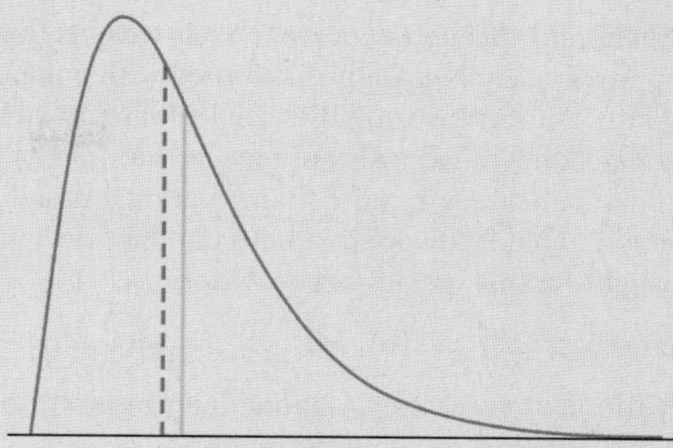

(a) The median is at the yellow line and the mean is at the red line.

(b) The median is at the red line and the mean is at the yellow line.

(c) The mode is at the red line and the median is at the yellow line.

(d) The mode is at the yellow line and the median is at the red line.

(e) The mode is at the red line and the mean is at the yellow line.

*Exercises 86–88 refer to the following setting.* The weights of laboratory cockroaches can be modeled with a Normal distribution having mean 80 grams and standard deviation 2 grams. The following figure is the Normal curve for this distribution of weights.

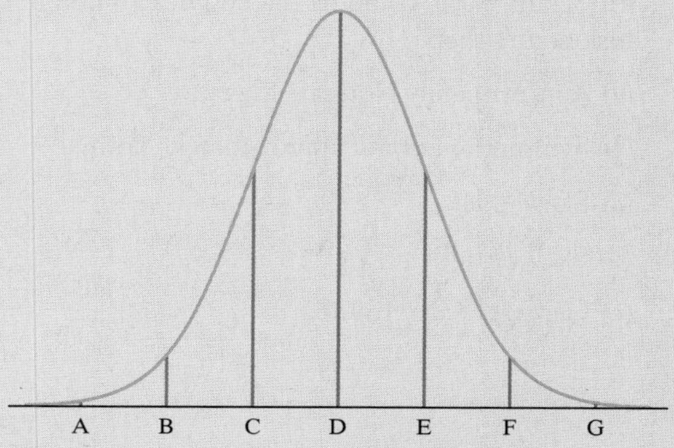


86. Point C on this Normal curve corresponds to

(a) 84 grams. (b) 82 grams. (c) 78 grams.
(d) 76 grams. (e) 74 grams.

87. About what percent of the cockroaches have weights between 76 and 84 grams?

(a) 99.7% (b) 95% (c) 68%
(d) 47.5% (e) 34%

88. About what proportion of the cockroaches will have weights greater than 83 grams?

(a) 0.0228 (b) 0.0668 (c) 0.1587
(d) 0.9332 (e) 0.0772

89. A different species of cockroach has weights that are approximately Normally distributed with a mean of 50 grams. After measuring the weights of many of these cockroaches, a lab assistant reports that 14% of the cockroaches weigh more than 55 grams. Based on this report, what is the approximate standard deviation of weight for this species of cockroach?

(a) 4.6
(b) 5.0
(c) 6.2
(d) 14.0
(e) Cannot determine without more information.

90. The following Normal probability plot shows the distribution of points scored for the 551 players in a single NBA season.

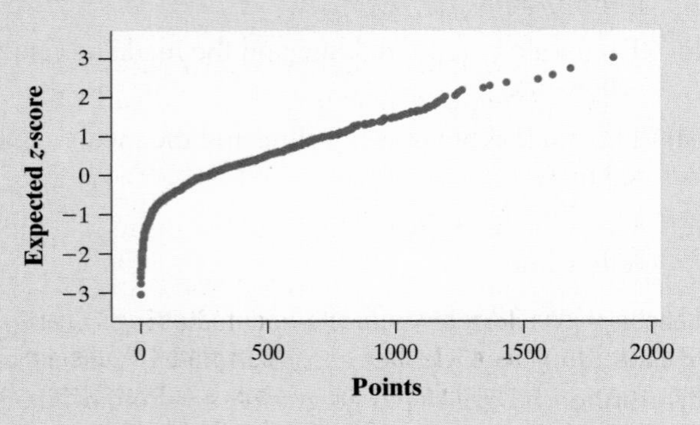


If the distribution of points was displayed in a histogram, what would be the best description of the histogram's shape?

(a) Approximately Normal
(b) Symmetric but not approximately Normal
(c) Skewed left
(d) Skewed right
(e) Cannot be determined

### Recycle and Review

91. **Making money** (2.1) The parallel dotplots show the total family income of randomly chosen individuals from Indiana (38 individuals) and New Jersey (44 individuals). Means and standard deviations are given below the dotplots.

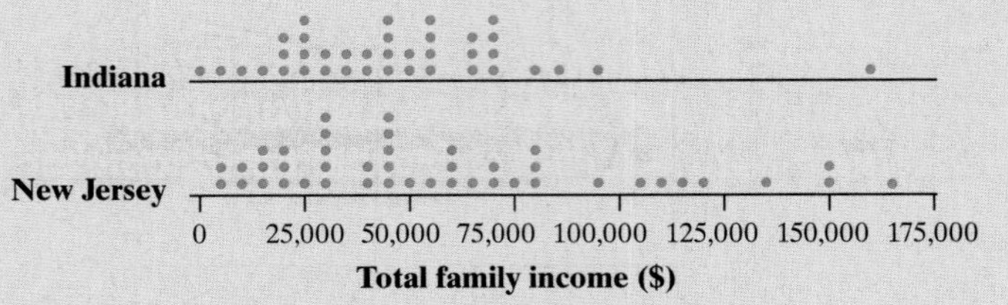


| | Mean | Standard deviation |
|---|---|---|
| Indiana | $47,400 | $29,400 |
| New Jersey | $58,100 | $41,900 |

Consider individuals in each state with total family incomes of $95,000. Which individual has a higher income, relative to others in his or her state? Use percentiles and $z$-scores to support your answer.

92. **More money** (1.3) Refer to Exercise 91.

(a) How do the ranges of the two distributions compare? Justify your answer.

(b) Explain why the standard deviation of the total family incomes in the New Jersey sample is so much larger than for the Indiana sample.

# Chapter 2 Wrap-Up

## FRAPPY! FREE RESPONSE AP® PROBLEM, YAY!

The following problem is modeled after actual AP® Statistics exam free response questions. Your task is to generate a complete, concise response in 15 minutes.

*Directions: Show all your work. Indicate clearly the methods you use, because you will be scored on the correctness of your methods as well as on the accuracy and completeness of your results and explanations.*

The distribution of scores on a recent test closely followed a Normal distribution with a mean of 22 points and a standard deviation of 4 points.

(a) What proportion of the students scored at least 25 points on this test?

(b) What is the 31st percentile of the distribution of test scores?

(c) The teacher wants to transform the test scores so that they have an approximately Normal distribution with a mean of 80 points and a standard deviation of 10 points. To do this, she will use a formula in the form

$$\textit{new score} = a + b(\textit{old score})$$

Find the values of $a$ and $b$ that the teacher should use to transform the distribution of test scores.

(d) Before the test, the teacher gave a review assignment for homework. The maximum score on the assignment was 10 points. The distribution of scores on this assignment had a mean of 9.2 points and a standard deviation of 2.1 points. Would it be appropriate to use a Normal distribution to calculate the proportion of students who scored below 7 points on this assignment? Explain your answer.

After you finish, you can view two example solutions on the book's website (highschool.bfwpub.com/updatedtps6e). Determine whether you think each solution is "complete," "substantial," "developing," or "minimal." If the solution is not complete, what improvements would you suggest to the student who wrote it? Finally, your teacher will provide a scoring rubric. Score your response and note what, if anything, you would do differently to improve your own score.

# Chapter 2 Review

### Section 2.1: Describing Location in a Distribution

In this section, you learned two different ways to describe the location of individuals in a distribution: percentiles and standardized scores (*z*-scores). Percentiles describe the location of an individual value in a distribution by measuring what percent of the observations are less than or equal to that value. A cumulative relative frequency graph is a handy tool for identifying percentiles in a distribution. You can use it to estimate the percentile for a particular value of a variable or to estimate the value of the variable at a particular percentile.

Standardized scores (*z*-scores) describe the location of an individual in a distribution by measuring how many standard deviations the individual is above or below the mean. To find the standardized score for a particular observation, transform the value by subtracting the mean and then dividing the difference by the standard deviation. Besides describing the location of an individual in a distribution, you can also use *z*-scores to compare observations from different distributions—standardizing the values puts them on a standard scale.

You also learned to describe the effects on the shape, center, and variability of a distribution when transforming data from one scale to another. Adding a positive constant to (or subtracting it from) each value in a data set changes the measures of center and location, but not the shape or variability of the distribution. Multiplying or dividing each value in a data set by a positive constant changes the measures of center and location and measures of variability, but not the shape of the distribution.

### Section 2.2: Density Curves and Normal Distributions

In this section, you learned how density curves are used to model distributions of quantitative data. An area under a density curve estimates the proportion of observations that fall in a specified interval of values. The total area under a density curve is 1, or 100%.

The most commonly used density curve is called a Normal curve. The Normal curve is symmetric, single-peaked, and bell-shaped with mean $\mu$ and standard deviation $\sigma$. For any distribution of data that is approximately Normal in shape, about 68% of the observations will be within 1 standard deviation of the mean, about 95% of the observations will be within 2 standard deviations of the mean, and about 99.7% of the observations will be within 3 standard deviations of the mean. This handy result is known as the empirical rule.

When observations do not fall exactly 1, 2, or 3 standard deviations from the mean, you learned how to use Table A or technology to identify the proportion of values in any specified interval under a Normal curve. You also learned how to use Table A or technology to determine the value of an individual that falls at a specified percentile in a Normal distribution. On the AP® Statistics exam, it is extremely important that you clearly communicate your methods when answering questions that involve a Normal distribution. Shading a Normal curve with the mean, standard deviation, and boundaries clearly identified is a great start. If you use technology to perform calculations, be sure to label the inputs of your calculator commands.

Finally, you learned how to determine if a distribution of data is approximately Normal using graphs (dotplots, stemplots, histograms) and the empirical rule. You also learned that a Normal probability plot is a great way to determine whether the shape of a distribution is approximately Normal. The more linear the Normal probability plot, the more Normal the distribution of the data.

## What Did You Learn?

| Learning Target | Section | Related Example on Page(s) | Relevant Chapter Review Exercise(s) |
|---|---|---|---|
| Find and interpret the percentile of an individual value in a distribution of data. | 2.1 | 92 | R2.1, R2.3(c) |
| Estimate percentiles and individual values using a cumulative relative frequency graph. | 2.1 | 94 | R2.2 |
| Find and interpret the standardized score (*z*-score) of an individual value in a distribution of data. | 2.1 | 96 | R2.1 |
| Describe the effect of adding, subtracting, multiplying by, or dividing by a constant on the shape, center, and variability of a distribution of data. | 2.1 | 99, 100, 101 | R2.3 |
| Use a density curve to model a distribution of quantitative data. | 2.2 | 111 | R2.4 |
| Identify the relative locations of the mean and median of a distribution from a density curve. | 2.2 | 112 | R2.4 |
| Use the empirical rule to estimate (i) the proportion of values in a specified interval, or (ii) the value that corresponds to a given percentile in a Normal distribution. | 2.2 | 118 | R2.5 |
| Find the proportion of values in a specified interval in a Normal distribution using Table A or technology. | 2.2 | 122, 124, 126 | R2.5, R2.6, R2.7 |
| Find the value that corresponds to a given percentile in a Normal distribution using Table A or technology. | 2.2 | 129 | R2.5, R2.6, R2.7 |
| Determine whether a distribution of data is approximately Normal from graphical and numerical evidence. | 2.2 | 132, 135 | R2.8, R2.9 |

# Chapter 2 Review Exercises

*These exercises are designed to help you review the important ideas and methods of the chapter.*

**R2.1 Is Paul tall?** According to the National Center for Health Statistics, the distribution of heights for 15-year-old males has a mean of 170 centimeters (cm) and a standard deviation of 7.5 cm. Paul is 15 years old and 179 cm tall.

(a) Find the *z*-score corresponding to Paul's height. Explain what this value means.

(b) Paul's height puts him at the 85th percentile among 15-year-old males. Explain what this means to someone who knows no statistics.

**R2.2 Computer use** Mrs. Causey asked her students how much time they had spent watching television during the previous week. The figure shows a cumulative relative frequency graph of her students' responses.

(a) At what percentile is a student who watched TV for 7 hours last week?

(b) Estimate from the graph the interquartile range (*IQR*) for time spent watching TV.

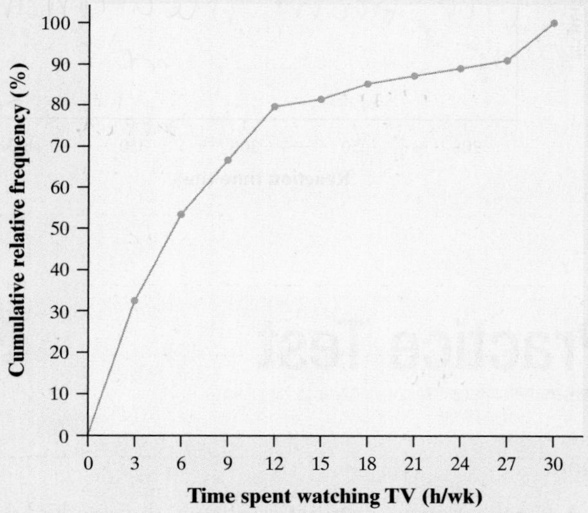


**R2.3 Aussie, Aussie, Aussie** A group of Australian students were asked to estimate the width of their classroom in feet. Use the dotplot and summary statistics to answer the following questions.

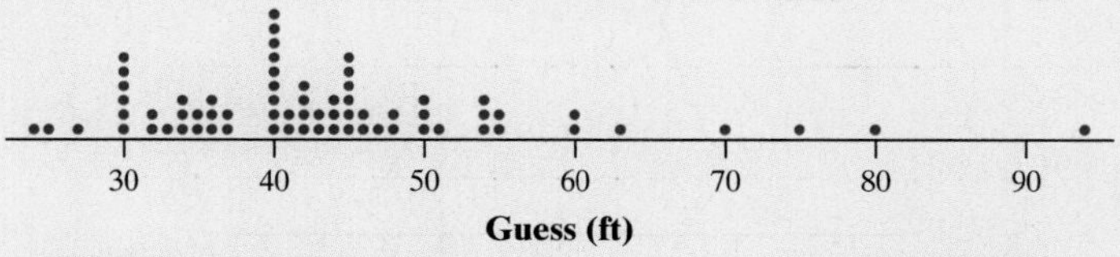


| Variable | $n$ | Mean | SD | Min | $Q_1$ | Med | $Q_3$ | Max |
|---|---|---|---|---|---|---|---|---|
| Guess (ft) | 66 | 43.70 | 12.50 | 24.00 | 35.50 | 42.00 | 48.00 | 94.00 |

(a) Suppose we converted each student's guess from feet to meters (3.28 ft = 1 m). How would the shape of the distribution be affected? Find the mean, median, standard deviation, and *IQR* for the transformed data.

(b) The actual width of the room was 42.6 feet. Suppose we calculated the error in each student's guess as follows: guess − 42.6. Find the mean and standard deviation of the errors in feet.

(c) Find the percentile for the student who estimated the classroom width as 63 feet.

**R2.4 Density curves** The following figure is a density curve that models a distribution of quantitative data. Trace the curve onto your paper.

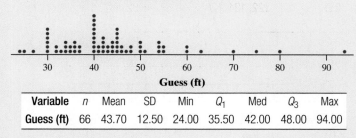

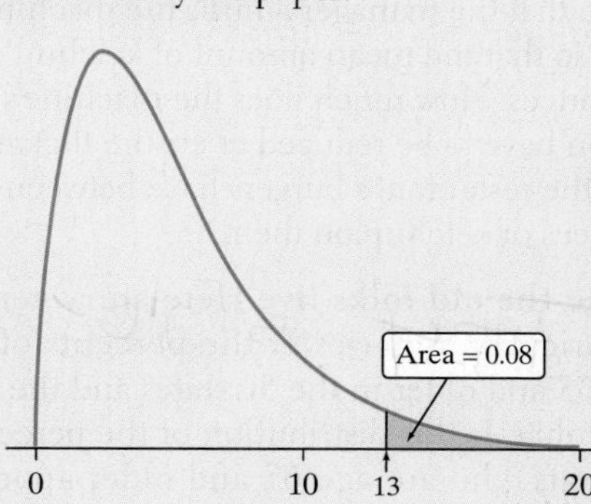


(a) What percent of observations have values less than 13? Justify your answer.

(b) Mark the approximate location of the median. Explain your choice of location.

(c) Mark the approximate location of the mean. Explain your choice of location.

**R2.5 Low-birth-weight babies** Researchers in Norway analyzed data on the birth weights of 400,000 newborns over a 6-year period. The distribution of birth weights is approximately Normal with a mean of 3668 grams and a standard deviation of 511 grams.[17] Babies that weigh less than 2500 grams at birth are classified as "low birth weight."

(a) Fill in the blanks: About 99.7% of the babies had birth weights between ____ and ____ grams.

(b) What percent of babies will be identified as low birth weight?

(c) Find the quartiles of the birth weight distribution.

**R2.6 Acing the GRE** The Graduate Record Examinations (GREs) are widely used to help predict the performance of applicants to graduate schools. The scores on the GRE Chemistry test are approximately Normal with mean = 694 and standard deviation = 112.

(a) Approximately what percent of test takers earn a score less than 500 or greater than 900 on the GRE Chemistry test?

(b) Estimate the 99th percentile score on the GRE Chemistry test.

**R2.7 Ketchup** A fast-food restaurant has just installed a new automatic ketchup dispenser for use in preparing its burgers. The amount of ketchup dispensed by the machine can be modeled by a Normal distribution with mean 1.05 ounces and standard deviation 0.08 ounce.

(a) If the restaurant's goal is to put between 1 and 1.2 ounces of ketchup on each burger, about what percent of the time will this happen?

(b) Suppose that the manager adjusts the machine's settings so that the mean amount of ketchup dispensed is 1.1 ounces. How much does the machine's standard deviation have to be reduced to ensure that at least 99% of the restaurant's burgers have between 1 and 1.2 ounces of ketchup on them?

**R2.8 Where the old folks live** Here are a stemplot and numerical summaries of the percents of residents aged 65 and older in the 50 states and the District of Columbia. Is this distribution of the percent of state residents who are age 65 and older approximately Normal? Justify your answer based on the graph and the empirical rule.

| | |
|---|---|
| 7 | 7 |
| 8 | |
| 9 | 0 |
| 10 | 379 |
| 11 | 44 |
| 12 | 02333445899 |
| 13 | 0223455555577888889 |
| 14 | 012334445689 |
| 15 | 49 |
| 16 | 0 |
| 17 | 3 |

Key: 12|2 represents a state in which 12.2% of residents are 65 and older.

| $n$ | Mean | SD | Min | $Q_1$ | Med | $Q_3$ | Max |
|---|---|---|---|---|---|---|---|
| 51 | 13.255 | 1.668 | 7.7 | 12.4 | 13.5 | 14.3 | 17.3 |

**R2.9 Assessing Normality** Catherine and Ana gave an online reflex test to 33 varsity athletes at their school. The following Normal probability plot displays the data on reaction times (in milliseconds) for these students. Is the distribution of reaction times for these athletes approximately Normal? Why or why not?

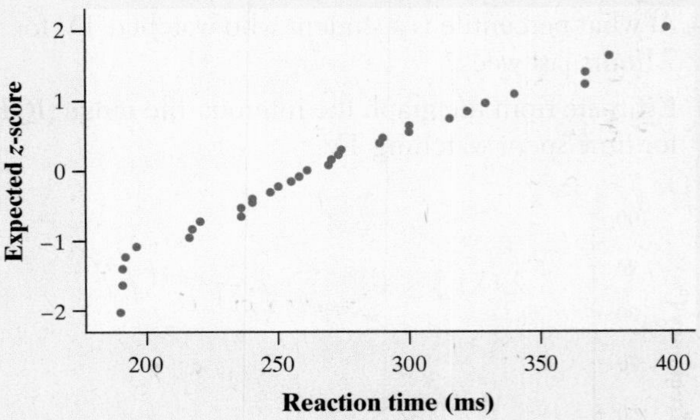


# Chapter 2 AP® Statistics Practice Test

**Section I: Multiple Choice** *Select the best answer for each question.*

**T2.1** Many professional schools require applicants to take a standardized test. Suppose that 1000 students take such a test. Several weeks after the test, Pete receives his score report: he got a 63, which placed him at the 73rd percentile. This means that

(a) Pete's score was below the median.

(b) Pete did worse than about 63% of test takers.

(c) Pete did worse than about 73% of test takers.

(d) Pete did the same as or better than about 63% of test takers.

(e) Pete did the same as or better than about 73% of test takers.

**T2.2** For the Normal distribution shown, the standard deviation is closest to

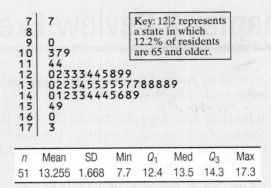

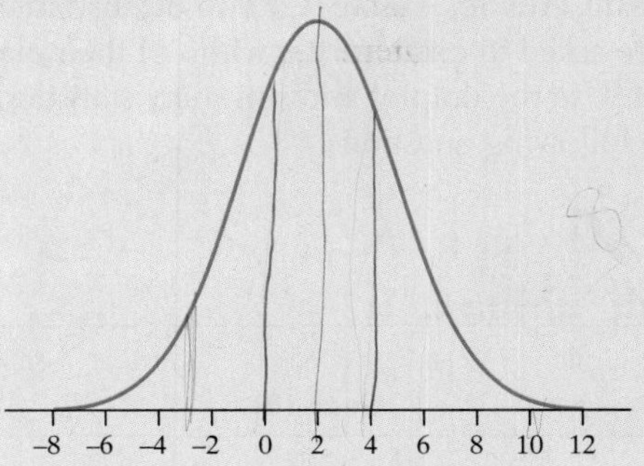


(a) 0.
(b) 1.
(c) 2.
(d) 3.
(e) 5.

**T2.3** Rainwater was collected in water containers at 30 different sites near an industrial complex, and the amount of acidity (pH level) was measured. The mean and standard deviation of the values are 4.60 and 1.10, respectively. When the pH meter was recalibrated back at the laboratory, it was found to be in error. The error can be corrected by adding 0.1 pH unit to all of the values and then multiplying the result by 1.2. What are the mean and standard deviation of the corrected pH measurements?

(a)  5.64, 1.44
(b)  5.64, 1.32
(c)  5.40, 1.44
(d)  5.40, 1.32
(e)  5.64, 1.20

**T2.4** The figure shows a cumulative relative frequency graph of the number of ounces of alcohol consumed per week in a sample of 150 adults who report drinking alcohol occasionally. About what percent of these adults consume between 4 and 8 ounces per week?

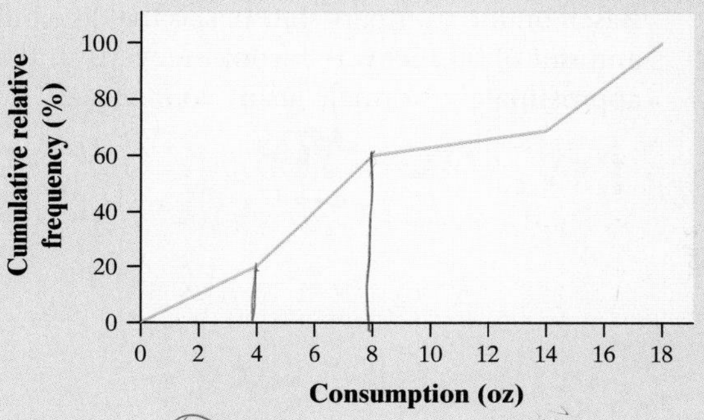

(a)  20%
(b)  40%
(c)  50%
(d)  60%
(e)  80%

**T2.5** The average yearly snowfall in Chillyville is approximately Normally distributed with a mean of 55 inches. If the snowfall in Chillyville exceeds 60 inches in 15% of the years, what is the standard deviation?

(a)  4.83 inches
(b)  5.18 inches
(c)  6.04 inches
(d)  8.93 inches
(e)  The standard deviation cannot be computed from the given information.

**T2.6** The figure shown is the density curve of a distribution. Seven values are marked on the density curve. Which of the following statements is true?

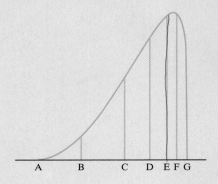

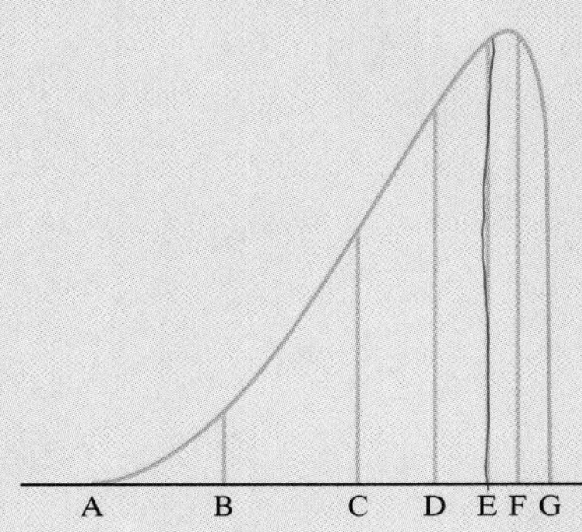

(a)  The mean of the distribution is E.
(b)  The area between B and F is 0.50.
(c)  The median of the distribution is C.
(d)  The 3rd quartile of the distribution is D.
(e)  The area under the curve between A and G is 1.

**T2.7** If the heights of a population of men are approximately Normally distributed, and the middle 99.7% have heights between 5'0" and 7'0", what is the standard deviation of the heights in this population?

(a)  1"
(b)  3"   *3 SD*
(c)  4"
(d)  6"
(e)  12"

**T2.8** The distribution of the time it takes for different people to solve a certain crossword puzzle is strongly skewed to the right with a mean of 30 minutes and a standard deviation of 15 minutes. The distribution of z-scores for those times is

(a)  Normally distributed with mean 30 and SD 15.
(b)  skewed to the right with mean 30 and SD 15.
(c)  Normally distributed with mean 0 and SD 1.
(d)  skewed to the right with mean 0 and SD 1.
(e)  skewed to the right, but the mean and standard deviation cannot be determined without more information.

**T2.9** The Environmental Protection Agency (EPA) requires that the exhaust of each model of motor vehicle be tested for the level of several pollutants. The level of oxides of nitrogen (NOX) in the exhaust of one light truck model was found to vary among individual trucks according to an approximately Normal distribution with mean $\mu = 1.45$ grams per mile driven and standard deviation $\sigma = 0.40$ gram per mile. Which of the following best estimates the proportion of light trucks of this model with NOX levels greater than 2 grams per mile?

(a)  0.0228
(b)  0.0846
(c)  0.4256
(d)  0.9154
(e)  0.9772

**T2.10** Until the scale was changed in 1995, SAT scores were based on a scale set many years ago. For Math scores, the mean under the old scale in the early 1990s was 470 and the standard deviation was 110. In 2016, the mean was 510 and the standard deviation was 103. Gina took the SAT in 1994 and scored 500. Her cousin Colleen took the SAT in 2016 and scored 530. Who did better on the exam, and how can you tell?

(a)  Colleen—she scored 30 points higher than Gina.
(b)  Colleen—her standardized score is higher than Gina's.
(c)  Gina—her standardized score is higher than Colleen's.
(d)  Gina—the standard deviation was larger in 2016.
(e)  The two cousins did equally well—their z-scores are the same.

**Section II: Free Response** *Show all your work. Indicate clearly the methods you use, because you will be graded on the correctness of your methods as well as on the accuracy and completeness of your results and explanations.*

**T2.11** The dotplot gives the sale prices for 40 houses in Ames, Iowa, sold during a recent month. The mean sale price was $203,388 with a standard deviation of $87,609.

Price ($1000s)

*27/40 = 68%*

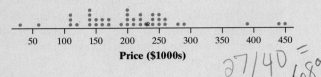

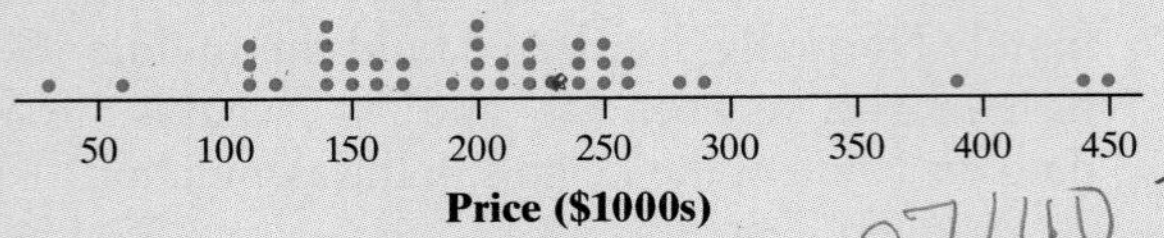

**(a)** Find the percentile of the house indicated in red on the dotplot.

**(b)** Calculate and interpret the standardized score (*z*-score) for the house indicated by the red dot, which sold for $234,000. *falls .303 SD away from mean*

**T2.12** A study of 12,000 able-bodied male students at the University of Illinois found that their times for the mile run were approximately Normally distributed with mean 7.11 minutes and standard deviation 0.74 minute.[18]

*a)* $\frac{6-7.11}{.74} = -1.5$   *0,7.11,0.74   .5120 = .6144*

*0.0068*

*≈ 801*

*lower ≈ 1,000*

**(a)** About how many students ran the mile in less than 6 minutes?

**(b)** Approximately how long did it take the slowest 10% of students to run the mile?   *8.05 m*

**(c)** Suppose that these mile run times are converted from minutes to seconds. Estimate the percent of students who ran the mile in between 400 and 500 seconds.   *M - 426.6   SD = 44.4   68%*

**T2.13** A study recorded the amount of oil recovered from the 64 wells in an oil field, in thousands of barrels. Here are descriptive statistics for that set of data from statistical software.

| Descriptive Statistics: Oilprod | | | | | | | |
|---|---|---|---|---|---|---|---|
| Variable | N | Mean | Median | StDev | Min | Max | $Q_1$ | $Q_3$ |
| Oilprod | 64 | 48.25 | 37.80 | 40.24 | 2.00 | 204.90 | 21.40 | 60.75 |

Based on the summary statistics, is the distribution of amount of oil recovered from the wells in this field approximately Normal? Justify your answer.

*No, bc the SD is big and max is very far from mean making it right skewed. Large dif between mean + median. Dist between min & mean + max & mean would be similar*

*PERCENT → InvNorm*

*$\frac{X - 7.11}{.74} = 6.34$*

*X - 7.11 = .469*

*X - 7.11*

*less = opposite*

# UNIT 2

## Exploring Two-Variable Data

# Chapter 3

# Exploring Two-Variable Quantitative Data

AfriPics.com/Alamy Stock Photo

# INTRODUCTION

Investigating relationships between variables is central to what we do in statistics. When we understand the relationship between two variables, we can use the value of one variable to help us make predictions about the other variable. In Section 1.1, we explored relationships between *categorical* variables, such as membership in an environmental club and snowmobile use for visitors to Yellowstone National Park. The association between these two variables suggests that members of environmental clubs are less likely to own or rent snowmobiles than nonmembers.

In this chapter, we investigate relationships between two *quantitative* variables. What can we learn about the price of a used car from the number of miles it has been driven? What does the length of a fish tell us about its weight? Can students with larger hands grab more candy? The following activity will help you explore the last question.

## ACTIVITY Candy grab

In this activity, you will investigate if students with a larger hand span can grab more candy than students with a smaller hand span.[1]

Josh Tabor

1. Measure the span of your dominant hand to the nearest half-centimeter (cm). Hand span is the distance from the tip of the thumb to the tip of the pinkie finger on your fully stretched-out hand.
2. One student at a time, go to the front of the class and use your dominant hand to grab as many candies as possible from the container. You must grab the candies with your fingers pointing down (no scooping!) and hold the candies for 2 seconds before counting them. After counting, put the candy back into the container.
3. On the board, record your hand span and number of candies in a table with the following headings:

| Hand span (cm) | Number of candies |
|---|---|

4. While other students record their values on the board, copy the table onto a piece of paper and make a graph. Begin by constructing a set of coordinate axes. Label the horizontal axis "Hand span (cm)" and the vertical axis "Number of candies." Choose an appropriate scale for each axis and plot each point from your class data table as accurately as you can on the graph.
5. What does the graph tell you about the relationship between hand span and number of candies? Summarize your observations in a sentence or two.

# SECTION 3.1 Scatterplots and Correlation

**LEARNING TARGETS** *By the end of the section, you should be able to:*

- Distinguish between explanatory and response variables for quantitative data.
- Make a scatterplot to display the relationship between two quantitative variables.
- Describe the direction, form, and strength of a relationship displayed in a scatterplot and identify unusual features.
- Interpret the correlation.
- Understand the basic properties of correlation, including how the correlation is influenced by unusual points.
- Distinguish correlation from causation.

A one-variable data set is sometimes called *univariate data.* A data set that describes the relationship between two variables is sometimes called *bivariate data.*

Most statistical studies examine data on more than one variable for a group of individuals. Fortunately, analysis of relationships between two variables builds on the same tools we used to analyze one variable. The principles that guide our work also remain the same:

- Plot the data, then look for overall patterns and departures from those patterns.
- Add numerical summaries.
- When there's a regular overall pattern, use a simplified model to describe it.

## Explanatory and Response Variables

In the "Candy grab" activity, the number of candies is the **response variable**. Hand span is the **explanatory variable** because we anticipate that knowing a student's hand span will help us predict the number of candies that student can grab.

**DEFINITION Response variable, Explanatory variable**

A **response variable** measures an outcome of a study. An **explanatory variable** may help predict or explain changes in a response variable.

You will often see explanatory variables called *independent variables* and response variables called *dependent variables.* Because the words *independent* and *dependent* have other meanings in statistics, we won't use them here.

It is easiest to identify explanatory and response variables when we initially specify the values of one variable to see how it affects another variable. For instance, to study the effect of alcohol on body temperature, researchers gave several different amounts of alcohol to mice. Then they measured the change in each mouse's body temperature 15 minutes later. In this case, amount of alcohol is the explanatory variable, and change in body temperature is the response variable. When we don't specify the values of either variable before collecting the data, there may or may not be a clear explanatory variable.

## EXAMPLE
### Diamonds and the SAT
### Explanatory or response?

stori/Deposit Photos

**PROBLEM:** Identify the explanatory variable and response variable for the following relationships, if possible. Explain your reasoning.

(a) The weight (in carats) and the price (in dollars) for a sample of diamonds.
(b) The SAT math score and the SAT evidence-based reading and writing score for a sample of students.

**SOLUTION:**

(a) **Explanatory: weight; Response: price. The weight of a diamond helps explain how expensive it is.**
(b) **Either variable could be the explanatory variable because each one could be used to predict or explain the other.**

**FOR PRACTICE, TRY EXERCISE 1**

In many studies, the goal is to show that changes in one or more explanatory variables actually *cause* changes in a response variable. However, other explanatory–response relationships don't involve direct causation. In the alcohol and mice study, alcohol actually *causes* a change in body temperature. But there is no cause-and-effect relationship between SAT math and evidence-based reading and writing scores.

## Displaying Relationships: Scatterplots

Although there are many ways to display the distribution of a single quantitative variable, a **scatterplot** is the best way to display the relationship between two quantitative variables.

**DEFINITION Scatterplot**

A **scatterplot** shows the relationship between two quantitative variables measured on the same individuals. The values of one variable appear on the horizontal axis, and the values of the other variable appear on the vertical axis. Each individual in the data set appears as a point in the graph.

Figure 3.1 shows a scatterplot that displays the relationship between hand span (cm) and number of Starburst™ candies for the 24 students in Mr. Tyson's class

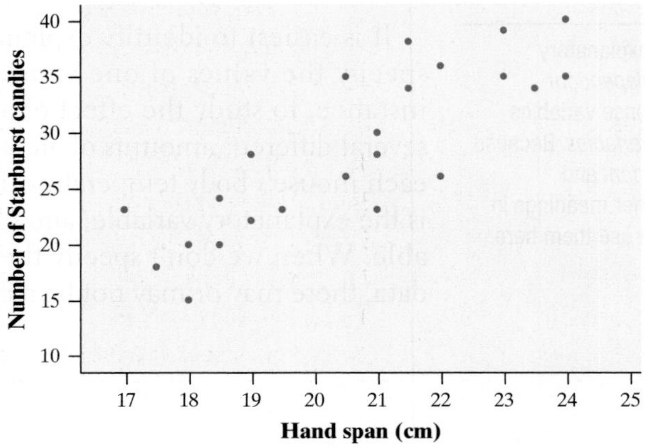


**FIGURE 3.1** Scatterplot of hand span (cm) and number of Starburst candies grabbed by 24 students. Only 23 points appear because two students had hand spans of 19 cm and grabbed 28 Starburst candies.

who did the "Candy grab" activity. As you can see, students with larger hand spans were typically able to grab more candies.

After collecting bivariate quantitative data, it is easy to make a scatterplot.

## HOW TO MAKE A SCATTERPLOT

- **Label the axes.** Put the name of the explanatory variable under the horizontal axis and the name of the response variable to the left of the vertical axis. If there is no explanatory variable, either variable can go on the horizontal axis.
- **Scale the axes.** Place equally spaced tick marks along each axis, beginning at a convenient number just below the smallest value of the variable and continuing until you exceed the largest value.
- **Plot individual data values.** For each individual, plot a point directly above that individual's value for the variable on the horizontal axis and directly to the right of that individual's value for the variable on the vertical axis.

The following example illustrates the process of constructing a scatterplot.

## EXAMPLE Buying wins
### Making a scatterplot

Ezra Shaw/Getty Images

**PROBLEM:** Do baseball teams that spend more money on players also win more games? The table shows the payroll (in millions of dollars) and number of wins for each of the 30 Major League Baseball teams during the 2016 regular season.[2] Make a scatterplot to show the relationship between payroll and wins.

| Team | Payroll | Wins | Team | Payroll | Wins |
|---|---|---|---|---|---|
| Arizona Diamondbacks | 103 | 69 | Milwaukee Brewers | 75 | 73 |
| Atlanta Braves | 122 | 68 | Minnesota Twins | 112 | 59 |
| Baltimore Orioles | 157 | 89 | New York Mets | 150 | 87 |
| Boston Red Sox | 215 | 93 | New York Yankees | 227 | 84 |
| Chicago Cubs | 182 | 103 | Oakland Athletics | 98 | 69 |
| Chicago White Sox | 141 | 78 | Philadelphia Phillies | 117 | 71 |
| Cincinnati Reds | 114 | 68 | Pittsburgh Pirates | 106 | 78 |
| Cleveland Indians | 114 | 94 | San Diego Padres | 127 | 68 |
| Colorado Rockies | 121 | 75 | San Francisco Giants | 181 | 87 |
| Detroit Tigers | 206 | 86 | Seattle Mariners | 155 | 86 |
| Houston Astros | 118 | 84 | St. Louis Cardinals | 167 | 86 |
| Kansas City Royals | 145 | 81 | Tampa Bay Rays | 71 | 68 |
| Los Angeles Angels | 181 | 74 | Texas Rangers | 169 | 95 |
| Los Angeles Dodgers | 274 | 91 | Toronto Blue Jays | 159 | 89 |
| Miami Marlins | 81 | 79 | Washington Nationals | 163 | 95 |

**SOLUTION:**

- **Label the axes.** The explanatory variable is payroll because we think it might help explain the number of wins.
- **Scale the axes.**
- **Plot individual data values.**

**FOR PRACTICE, TRY EXERCISE 3**

## Describing a Scatterplot

To describe a scatterplot, follow the basic strategy of data analysis from Chapter 1: look for patterns and important departures from those patterns.

The scatterplot in Figure 3.2(a) shows a **positive association** between wins and payroll for MLB teams in 2016. That is, teams that spent more money typically won more games. Other scatterplots, such as the one in Figure 3.2(b), show a **negative association**. Teams that allow their opponents to score more runs typically win *fewer* games.

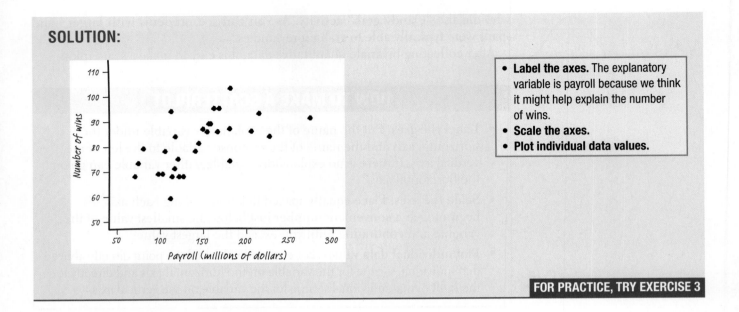

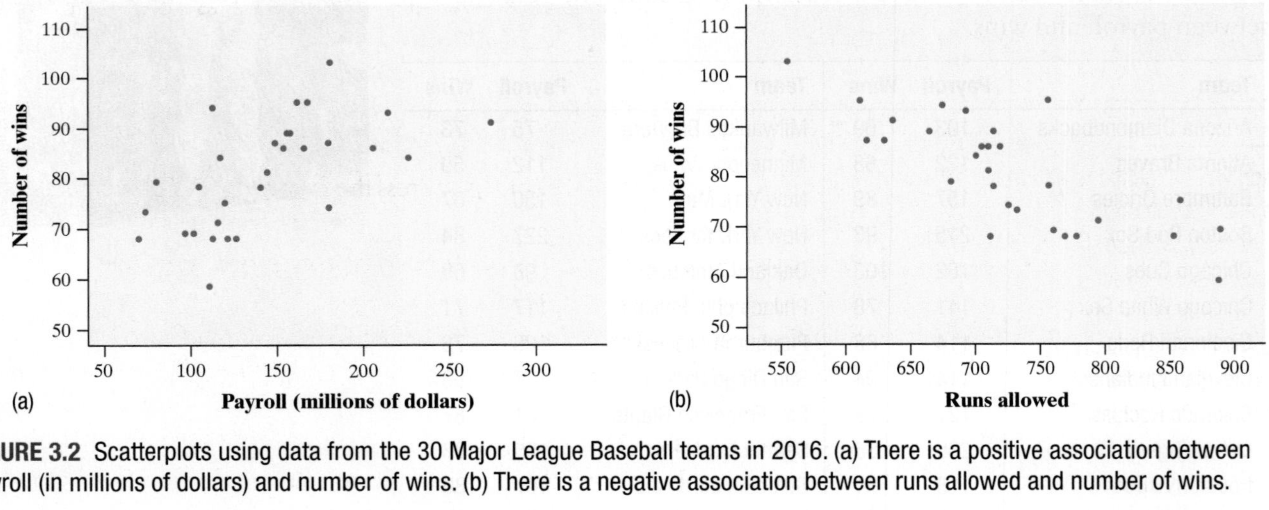


**FIGURE 3.2** Scatterplots using data from the 30 Major League Baseball teams in 2016. (a) There is a positive association between payroll (in millions of dollars) and number of wins. (b) There is a negative association between runs allowed and number of wins.

In some cases, there is **no association** between two variables. For example, the following scatterplot shows the relationship between height (in centimeters) and the typical amount of sleep on a non-school night (in hours) for a sample of students.[3] Knowing the height of a student doesn't help predict how much he or she likes to sleep on Saturday night!

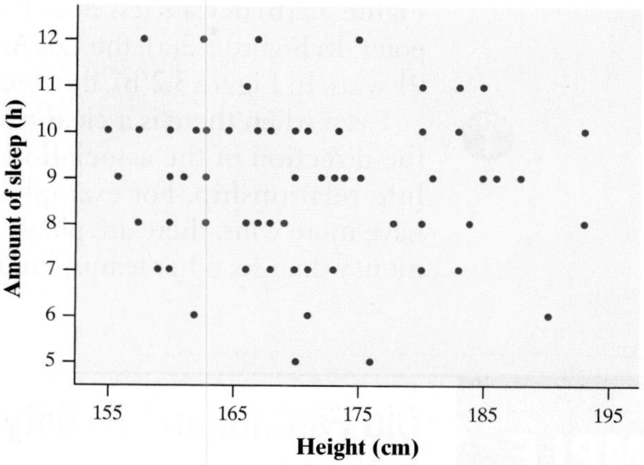


Recall from Section 1.1 that two variables have an *association* if knowing the value of one variable helps us predict the value of the other variable.

**DEFINITION Positive association, Negative association, No association**

Two variables have a **positive association** when values of one variable tend to increase as the values of the other variable increase.

Two variables have a **negative association** when values of one variable tend to decrease as the values of the other variable increase.

There is **no association** between two variables if knowing the value of one variable does not help us predict the value of the other variable.

Identifying the direction of an association in a scatterplot is a good start, but there are several other characteristics that we need to address when describing a scatterplot.

**AP® EXAM TIP**

When you are asked to *describe* the association shown in a scatterplot, you are expected to discuss the direction, form, and strength of the association, along with any unusual features, *in the context of the problem.* This means that you need to use both variable names in your description.

## HOW TO DESCRIBE A SCATTERPLOT

To describe a scatterplot, make sure to address the following four characteristics in the context of the data:

- **Direction:** A scatterplot can show a positive association, negative association, or no association.
- **Form:** A scatterplot can show a linear form or a nonlinear form. The form is linear if the overall pattern follows a straight line. Otherwise, the form is nonlinear.
- **Strength:** A scatterplot can show a weak, moderate, or strong association. An association is strong if the points don't deviate much from the form identified. An association is weak if the points deviate quite a bit from the form identified.
- **Unusual features:** Look for individual points that fall outside the overall pattern and distinct clusters of points.

Even though they have opposite directions, both associations in Figure 3.2 on page 156 have a linear form. However, the association between runs allowed and wins is stronger than the relationship between payroll and wins because the points in Figure 3.2(b) deviate less from the linear pattern. Each scatterplot has one unusual point: In Figure 3.2(a), the Los Angeles Dodgers spent \$274 million and had "only" 91 wins. In Figure 3.2(b), the Texas Rangers gave up 757 runs but had 95 wins.

caution !

**Even when there is a clear relationship between two variables in a scatterplot, the direction of the association describes only the overall trend—not an absolute relationship.** For example, even though teams that spend more generally have more wins, there are plenty of exceptions. The Minnesota Twins spent more money than six other teams, but had fewer wins than any team in the league.

## EXAMPLE

### Old Faithful and fertility
### Describing a scatterplot

BigshotD3/iStock/Getty Images

**PROBLEM:** Describe the relationship in each of the following contexts.

(a) The scatterplot on the left shows the relationship between the duration (in minutes) of an eruption and the interval of time until the next eruption (in minutes) of Old Faithful during a particular month.

(b) The scatterplot on the right shows the relationship between the average income (gross domestic product per person, in dollars) and fertility rate (number of children per woman) in 187 countries.[4]

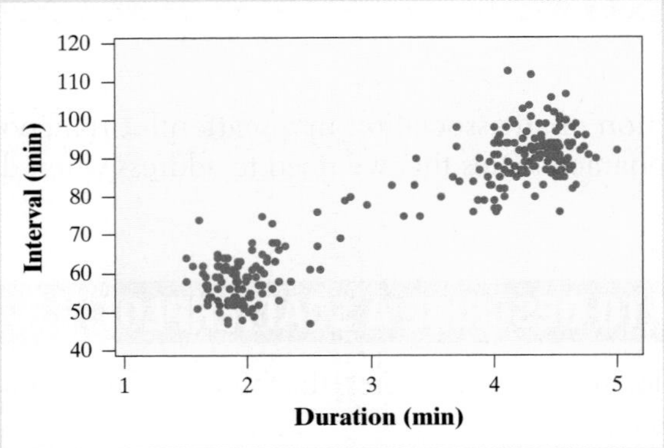


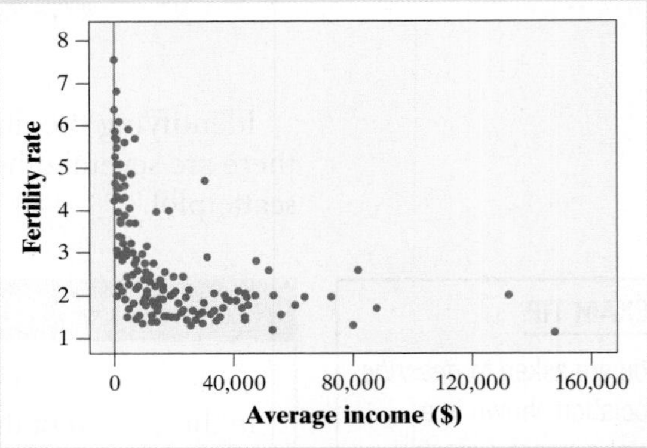


**SOLUTION:**

(a) There is a strong, positive linear relationship between the duration of an eruption and the interval of time until the next eruption. There are two main clusters of points: one cluster has durations around 2 minutes with intervals around 55 minutes, and the other cluster has durations around 4.5 minutes with intervals around 90 minutes.

> Even with the clusters, the overall direction is still positive. In some cases, however, the points in a cluster go in the opposite direction of the overall association.

(b) There is a moderately strong, negative nonlinear relationship between average income and fertility rate in these countries. There is a country outside this pattern with an average income around \$30,000 and a fertility rate around 4.7.

> The association is called "nonlinear" because the pattern of points is clearly curved.

**FOR PRACTICE, TRY EXERCISE 5**

## CHECK YOUR UNDERSTANDING

Is there a relationship between the amount of sugar (in grams) and the number of calories in movie-theater candy? Here are the data from a sample of 12 types of candy.[5]

| Name | Sugar (g) | Calories | Name | Sugar (g) | Calories |
|---|---|---|---|---|---|
| Butterfinger Minis | 45 | 450 | Reese's Pieces | 61 | 580 |
| Junior Mints | 107 | 570 | Skittles | 87 | 450 |
| M&M'S® | 62 | 480 | Sour Patch Kids | 92 | 490 |
| Milk Duds | 44 | 370 | SweeTarts | 136 | 680 |
| Peanut M&M'S® | 79 | 790 | Twizzlers | 59 | 460 |
| Raisinets | 60 | 420 | Whoppers | 48 | 350 |

1. Identify the explanatory and response variables. Explain your reasoning.
2. Make a scatterplot to display the relationship between amount of sugar and the number of calories in movie-theater candy.
3. Describe the relationship shown in the scatterplot.

## 8. Technology Corner MAKING SCATTERPLOTS

*TI-Nspire and other technology instructions are on the book's website at highschool.bfwpub.com/updatedtps6e.*

Making scatterplots with technology is much easier than constructing them by hand. We'll use the MLB data from page 155 to show how to construct a scatterplot on a TI-83/84.

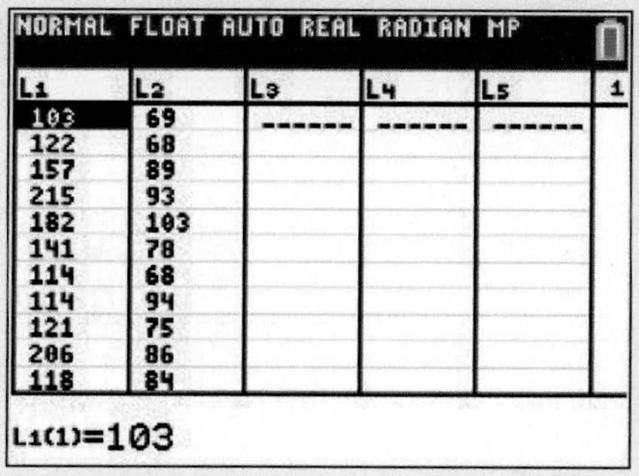


1. Enter the payroll values in L1 and the number of wins in L2.
   - Press STAT and choose Edit....
   - Type the values into L1 and L2.
2. Set up a scatterplot in the statistics plots menu.
   - Press 2nd Y= (STAT PLOT).
   - Press ENTER or 1 to go into Plot1.
   - Adjust the settings as shown.

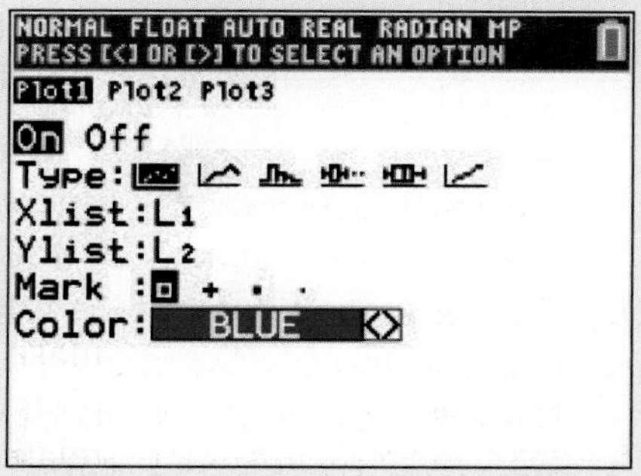

3. Use ZoomStat to let the calculator choose an appropriate window.
   - Press ZOOM and choose ZoomStat.

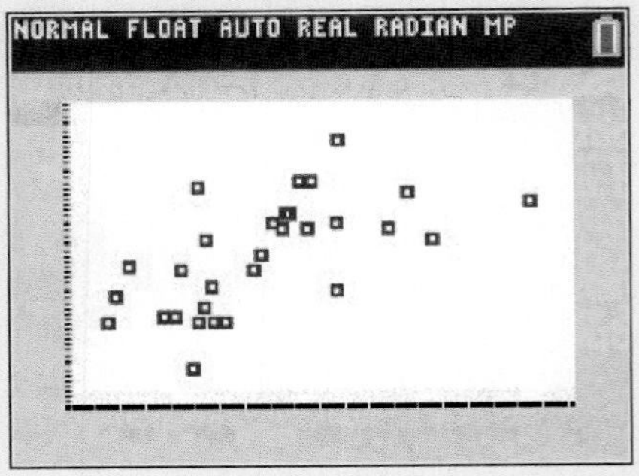


**AP® EXAM TIP**

If you are asked to make a scatterplot, be sure to label and scale both axes. *Don't* just copy an unlabeled calculator graph directly onto your paper.

## Measuring Linear Association: Correlation

A scatterplot displays the direction, form, and strength of a relationship between two quantitative variables. Linear relationships are particularly important because a straight line is a simple pattern that is quite common. A linear relationship is considered strong if the points lie close to a straight line and is considered weak if the points are widely scattered about the line. Unfortunately, our eyes are not the most reliable tools when it comes to judging the strength of a linear relationship. When the association between two quantitative variables is linear, we can use the **correlation** $r$ to help describe the strength and direction of the association.

Some people refer to $r$ as the "correlation coefficient."

**DEFINITION Correlation $r$**

For a linear association between two quantitative variables, the **correlation $r$** measures the direction and strength of the association.

Here are some important properties of the correlation $r$:

- The correlation $r$ is always a number between $-1$ and $1$ $(-1 \leq r \leq 1)$.
- The correlation $r$ indicates the direction of a linear relationship by its sign: $r > 0$ for a positive association and $r < 0$ for a negative association.
- The extreme values $r = -1$ and $r = 1$ occur *only* in the case of a perfect linear relationship, when the points lie exactly along a straight line.
- If the linear relationship is strong, the correlation $r$ will be close to 1 or $-1$. If the linear relationship is weak, the correlation $r$ will be close to 0.

caution **It is only appropriate to use the correlation to describe strength and direction for a linear relationship.** This is why the word *linear* kept appearing in the list above!

Figure 3.3 shows six scatterplots that correspond to various values of $r$. To make the meaning of $r$ clearer, the standard deviations of both variables in these plots are equal, and the horizontal and vertical scales are the same. The correlation $r$ describes the direction and strength of the linear relationship in each scatterplot.

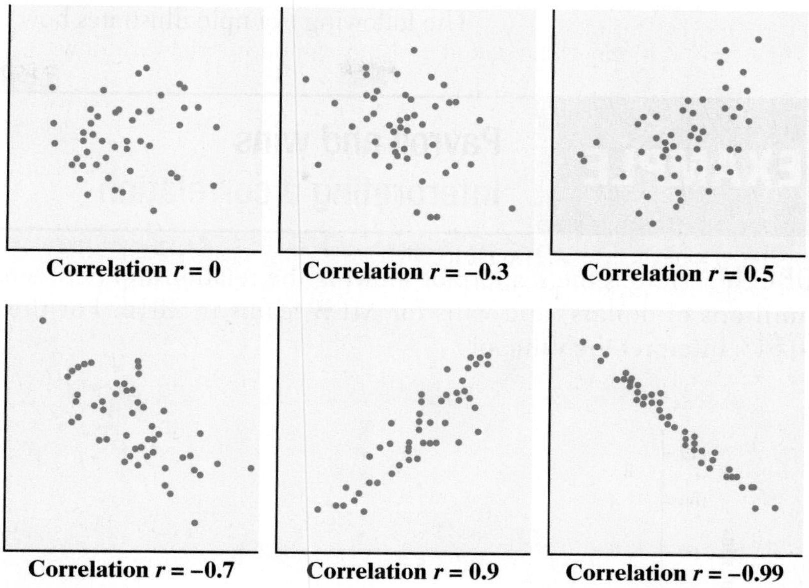


**FIGURE 3.3** How correlation measures the strength and direction of a linear relationship. When the dots are tightly packed around a line, the correlation will be close to 1 or −1.

## ACTIVITY Guess the correlation

APPLET

In this activity, we will have a class competition to see who can best guess the correlation.

1. Load the *Guess the Correlation* applet at www.rossmanchance.com/applets.

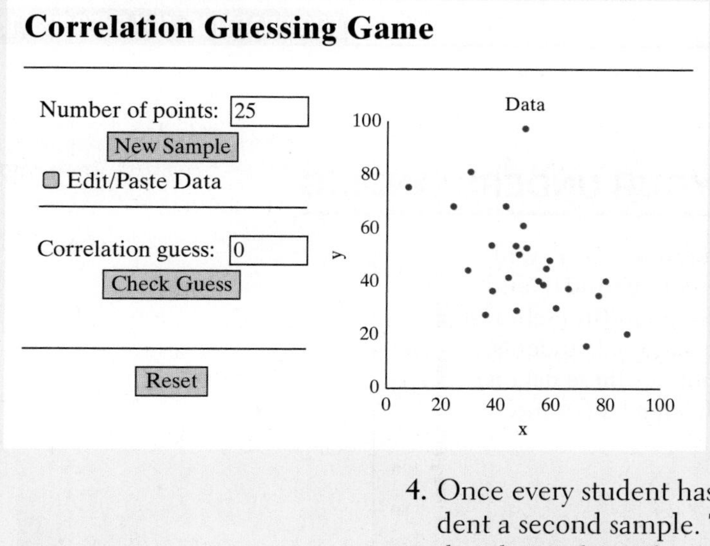


2. The teacher will press the "New Sample" button to see a "random" scatterplot. As a class, try to guess the correlation. Type the guess in the "Correlation guess" box and press "Check Guess" to see how the class did. Repeat several times to see more examples. For the competition, there will be two rounds.
3. Starting on one side of the classroom and moving in order to the other side, the teacher will give each student *one* new sample and have him or her guess the correlation. The teacher will then record how far off the guess was from the true correlation.
4. Once every student has made an attempt, the teacher will give each student a second sample. This time, the students will go in reverse order so that the student who went first in Round 1 will go last in Round 2. The student who has the closest guess in either round wins a prize!

The following example illustrates how to interpret the correlation.

## EXAMPLE

### Payroll and wins
### Interpreting a correlation

jonathansloane/E+/Getty Images

**PROBLEM:** Here is the scatterplot showing the relationship between payroll (in millions of dollars) and wins for MLB teams in 2016. For these data, $r = 0.613$. Interpret the value of $r$.

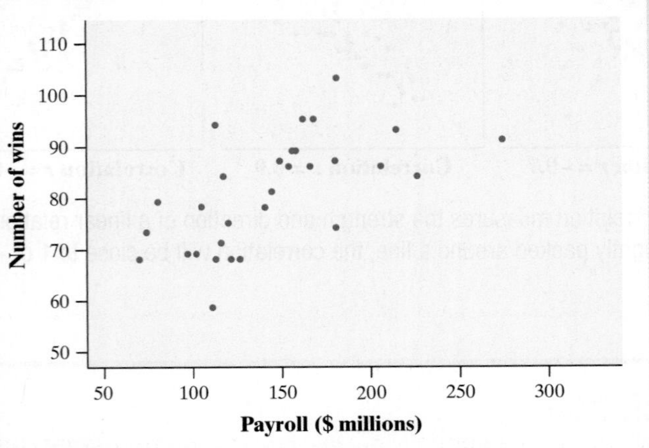


**SOLUTION:**

**The correlation of $r = 0.613$ confirms that the linear association between payroll and number of wins is moderately strong and positive.**

Always include context by using the variable names in your answer.

**FOR PRACTICE, TRY EXERCISE 15**

## CHECK YOUR UNDERSTANDING

The scatterplot shows the 40-yard-dash times (in seconds) and long-jump distances (in inches) for a small class of 12 students. The correlation for these data is $r = -0.838$. Interpret this value.

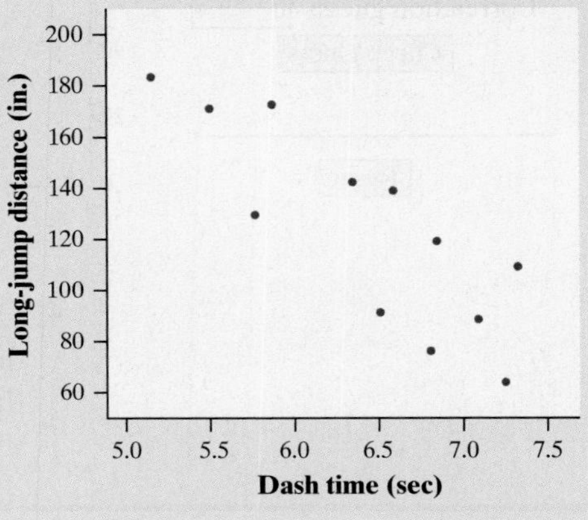

## Cautions about Correlation

While the correlation is a good way to measure the strength and direction of a linear relationship, it has several limitations.

**Correlation doesn't imply causation.** In many cases, two variables might have a strong correlation, but changes in one variable are very unlikely to cause changes in the other variable. Consider the following scatterplot showing total revenue generated by skiing facilities in the United States and the number of people who died by becoming tangled in their bedsheets in 10 recent years.[6] The correlation for these data is $r = 0.97$. Does an increase in skiing revenue *cause* more people to die by becoming tangled in their bedsheets? We doubt it!

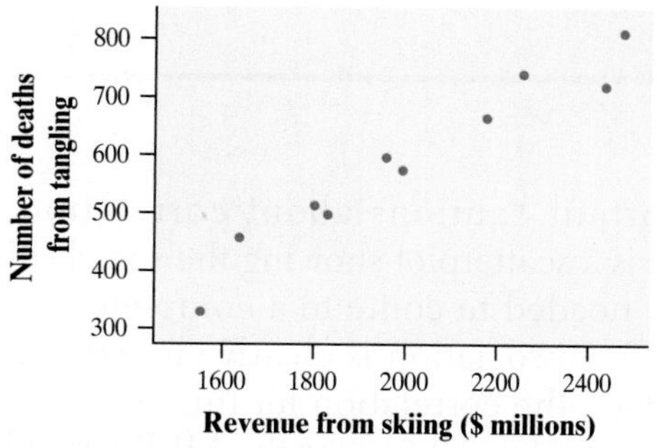


Even though we shouldn't automatically conclude that there is a cause-and-effect relationship between two variables when they have an association, in some cases there might actually be a cause-and-effect relationship. You will learn how to distinguish these cases in Chapter 4.

The following activity helps you explore some additional limitations of the correlation.

## ACTIVITY Correlation and Regression applet

In this activity, you will use an applet to investigate some important properties of the correlation. Go to the book's website at highschool.bfwpub.com/updatedtps6e and launch the *Correlation and Regression* applet.

1. You are going to use the *Correlation and Regression* applet to make several scatterplots that have correlation close to 0.7.
   (a) Start by putting two points on the graph. What's the value of the correlation? Why does this make sense?
   (b) Make a lower-left to upper-right pattern of 10 points with correlation about $r = 0.7$. You can drag points up or down to adjust $r$ after you have 10 points.
   (c) Make a new scatterplot, this time with 9 points in a vertical stack at the left of the plot. Add 1 point far to the right and move it until the correlation is close to 0.7.
   (d) Make a third scatterplot, this time with 10 points in a curved pattern that starts at the lower left and rises to the right. Adjust the points up or down until you have a smooth curve with correlation close to 0.7.

*Summarize:* If you know only that the correlation between two variables is $r = 0.7$, what can you say about the form of the relationship?

2. Click on the scatterplot to create a group of 7 points in a U shape so that there is a strong nonlinear association. What is the correlation?

*Summarize:* If you know only that the correlation between two variables is $r = 0$, what can you say about the strength of the relationship?

3. Click on the scatterplot to create a group of 10 points in the lower-left corner of the scatterplot with a strong linear pattern (correlation about 0.9).
   (a) Add 1 point at the upper right that is in line with the first 10. How does the correlation change?
   (b) Drag this last point straight down. How small can you make the correlation? Can you make the correlation negative?

*Summarize:* What did you learn from Step 3 about the effect of an unusual point on the correlation?

The activity highlighted some important cautions about correlation. **Correlation does not measure form.** Here is a scatterplot showing the speed (in miles per hour) and the distance (in feet) needed to come to a complete stop when a motorcycle's brake was applied.[7] The association is clearly curved, but the correlation is quite large: $r = 0.98$. In fact, the correlation for this *nonlinear* association is much greater than the correlation of $r = 0.613$ for the MLB payroll data, which had a clear linear association.

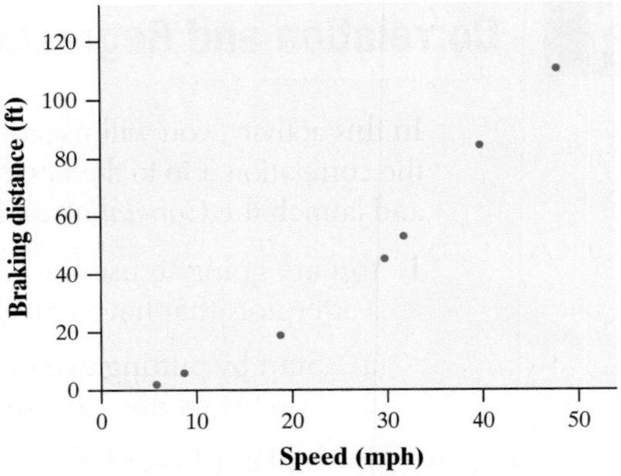


**Correlation should only be used to describe linear relationships.** The association displayed in the following scatterplot is extremely strong, but the correlation is $r = 0$. This isn't a contradiction because correlation doesn't measure the strength of nonlinear relationships.

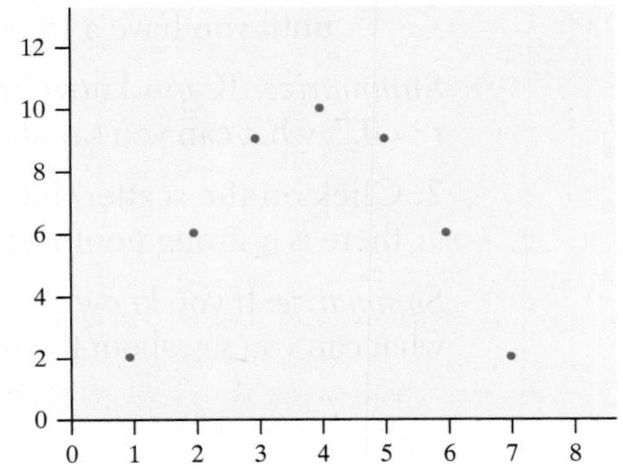

**The correlation is not a resistant measure of strength.** In the following scatterplot, the correlation is $r = -0.13$. But when the unusual point in the lower right corner is excluded, the correlation becomes $r = 0.72$.

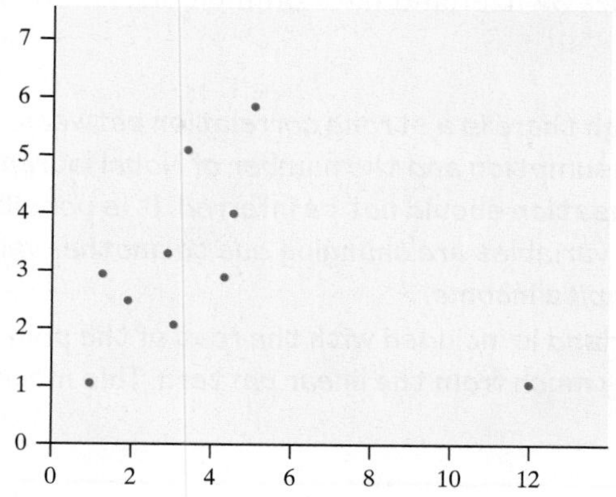

Like the mean and the standard deviation, the correlation can be greatly influenced by unusual points.

## EXAMPLE Nobel chocolate
### Cautions about correlation

amphotora/Getty Images

**PROBLEM:** Most people love chocolate for its great taste. But does it also make you smarter? A scatterplot like this one recently appeared in the *New England Journal of Medicine*.[8] The explanatory variable is the chocolate consumption per person for a sample of countries. The response variable is the number of Nobel Prizes per 10 million residents of that country.

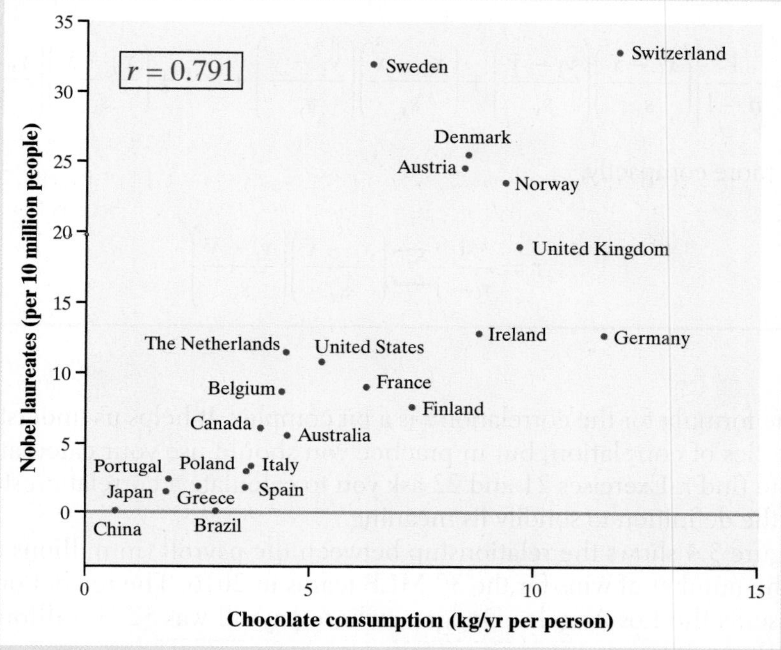

(a) If people in the United States started eating more chocolate, could we expect more Nobel Prizes to be awarded to residents of the United States? Explain.

(b) What effect does Switzerland have on the correlation? Explain.

**SOLUTION**

(a) No; even though there is a strong correlation between chocolate consumption and the number of Nobel laureates in a country, causation should not be inferred. It is possible that both of these variables are changing due to another variable, such as per capita income.

Not all questions about cause and effect include the word *cause*. Make sure to read questions—and reports in the media—very carefully.

(b) When Switzerland is included with the rest of the points, it makes the association stronger because it doesn't vary much from the linear pattern. This makes the correlation closer to 1.

**FOR PRACTICE, TRY EXERCISES 17 AND 19**

## Calculating Correlation

Now that you understand the meaning and limitations of the correlation, let's look at how it's calculated.

### HOW TO CALCULATE THE CORRELATION *r*

Suppose that we have data on variables $x$ and $y$ for $n$ individuals. The values for the first individual are $x_1$ and $y_1$, the values for the second individual are $x_2$ and $y_2$, and so on. The means and standard deviations of the two variables are $\bar{x}$ and $s_x$ for the $x$-values, and $\bar{y}$ and $s_y$ for the $y$-values. The correlation $r$ between $x$ and $y$ is

$$r = \frac{1}{n-1}\left[\left(\frac{x_1-\bar{x}}{s_x}\right)\left(\frac{y_1-\bar{y}}{s_y}\right) + \left(\frac{x_2-\bar{x}}{s_x}\right)\left(\frac{y_2-\bar{y}}{s_y}\right) + \cdots + \left(\frac{x_n-\bar{x}}{s_x}\right)\left(\frac{y_n-\bar{y}}{s_y}\right)\right]$$

or, more compactly,

$$r = \frac{1}{n-1}\sum\left(\frac{x_i-\bar{x}}{s_x}\right)\left(\frac{y_i-\bar{y}}{s_y}\right)$$

The formula for the correlation $r$ is a bit complex. It helps us understand some properties of correlation, but in practice you should use your calculator or software to find $r$. Exercises 21 and 22 ask you to calculate a correlation step by step from the definition to solidify its meaning.

Figure 3.4 shows the relationship between the payroll (in millions of dollars) and the number of wins for the 30 MLB teams in 2016. The red dot on the right represents the Los Angeles Dodgers, whose payroll was $274 million and who won 91 games.

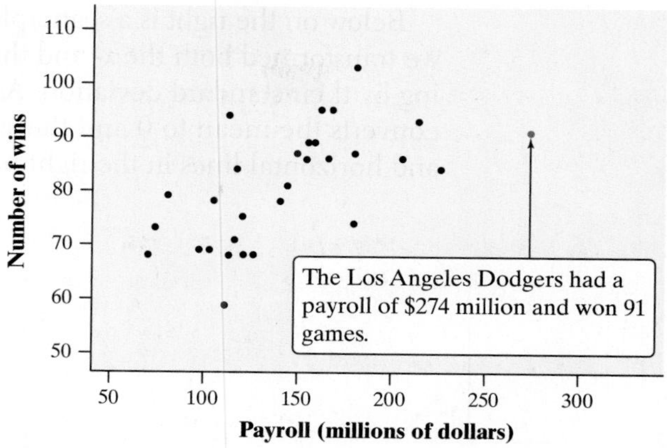


**FIGURE 3.4** Scatterplot showing the relationship between payroll (in millions of dollars) and number of wins for 30 Major League Baseball teams in 2016. The point representing the Los Angeles Dodgers is highlighted in red.

The formula for $r$ begins by standardizing the observations. The value

$$\frac{x_i - \bar{x}}{s_x}$$

in the correlation formula is the standardized payroll ($z$-score) of the $i$th team. In 2016, the mean payroll was $\bar{x} = \$145.033$ million with a standard deviation of $s_x = \$46.879$ million. For the Los Angeles Dodgers, the corresponding $z$-score is

$$z_x = \frac{274 - 145.033}{46.879} = 2.75$$

ZUMA Press Inc/Alamy

The Dodgers' payroll is 2.75 standard deviations above the mean.

Likewise, the value

$$\frac{y_i - \bar{y}}{s_y}$$

in the correlation formula is the standardized number of wins for the $i$th team. In 2016, the mean number of wins was $\bar{y} = 80.9$ with a standard deviation of $s_y = 10.669$. For the Los Angeles Dodgers, the corresponding $z$-score is

$$z_y = \frac{91 - 80.9}{10.669} = 0.95$$

The Dodgers' number of wins is 0.95 standard deviation above the mean.

Multiplying the Dodgers' two $z$-scores, we get a product of $(2.75)(0.95) = 2.6125$. The correlation $r$ is an "average" of the products of the standardized scores for all the teams. Just as in the case of the standard deviation $s_x$, we divide by 1 fewer than the number of individuals to find the average. Finishing the calculation reveals that $r = 0.613$ for the 30 MLB teams.

Some people like to write the correlation formula as

$$r = \frac{1}{n-1}\sum z_x z_y$$

to emphasize the product of standardized scores in the calculation.

To understand what correlation measures, consider the graphs in Figure 3.5 on the next page. At the left is a scatterplot of the MLB data with two lines added—a vertical line at the mean payroll and a horizontal line at the mean number of wins. Most of the points fall in the upper-right or lower-left "quadrants" of the graph. Teams with above-average payrolls tend to have above-average numbers of wins, like the Dodgers. Teams with below-average payrolls tend to have numbers of wins that are below average. This confirms the positive association between the variables.

Below on the right is a scatterplot of the standardized scores. To get this graph, we transformed both the *x*- and the *y*-values by subtracting their mean and dividing by their standard deviation. As we saw in Chapter 2, standardizing a data set converts the mean to 0 and the standard deviation to 1. That's why the vertical and horizontal lines in the right-hand graph are both at 0.

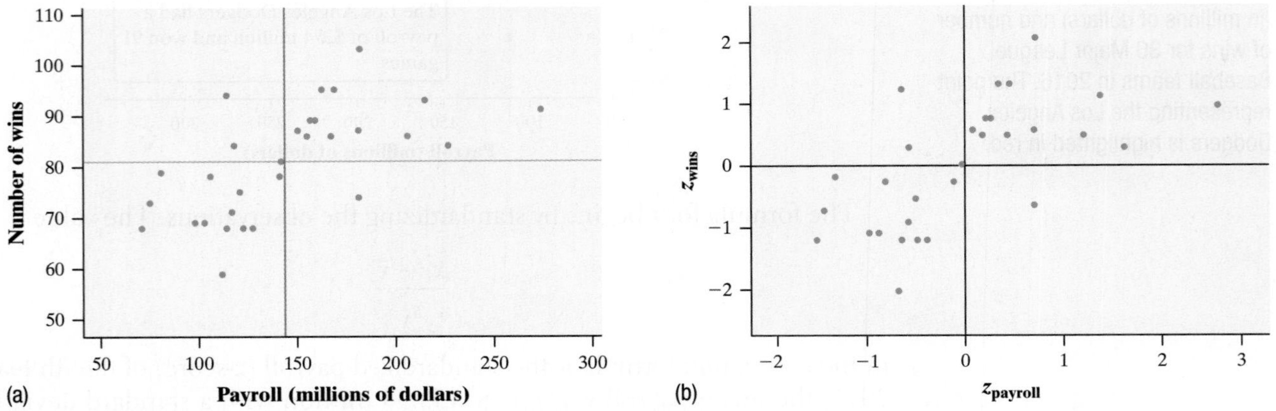


**FIGURE 3.5** (a) Scatterplot showing the relationship between payroll (in millions of dollars) and number of wins for 30 Major League Baseball teams in 2016, with lines showing the mean of each variable. (b) Scatterplot showing the relationship between the standardized values of payroll and the standardized values of number of wins for the same 30 teams.

For the points in the upper-right quadrant and the lower-left quadrant, the products of the standardized values will be positive. Because most of the points are in these two quadrants, the sum of the *z*-score products will also be positive, resulting in a positive correlation *r*.

What if there was a negative association between two variables? Most of the points would be in the upper-left and lower-right quadrants and their *z*-score products would be negative, resulting in a negative correlation.

## Additional Facts about Correlation

Now that you have seen how the correlation is calculated, here are some additional facts about correlation.

1. *Correlation requires that both variables be quantitative,* so that it makes sense to do the arithmetic indicated by the formula for *r*. We cannot calculate a correlation between the incomes of a group of people and what city they live in because city is a categorical variable. When one or both of the variables are categorical, use the term *association* rather than *correlation.*
2. *Correlation makes no distinction between explanatory and response variables.* When calculating the correlation, it makes no difference which variable you call *x* and which you call *y*. Can you see why from the formula?

$$r = \frac{1}{n-1}\sum\left(\frac{x_i - \bar{x}}{s_x}\right)\left(\frac{y_i - \bar{y}}{s_y}\right)$$

3. Because $r$ uses the standardized values of the observations, *$r$ does not change when we change the units of measurement of $x$, $y$, or both.* The correlation between height and weight won't change if we measure height in centimeters rather than inches and measure weight in kilograms rather than pounds.
4. *The correlation $r$ has no unit of measurement.* It is just a number.

## EXAMPLE Long strides
### More about correlation

Monkey Business Images/ Shutterstock.com

**PROBLEM:** The following scatterplot shows the height (in inches) and number of steps needed for a random sample of 36 students to walk the length of a school hallway. The correlation is $r = -0.632$.

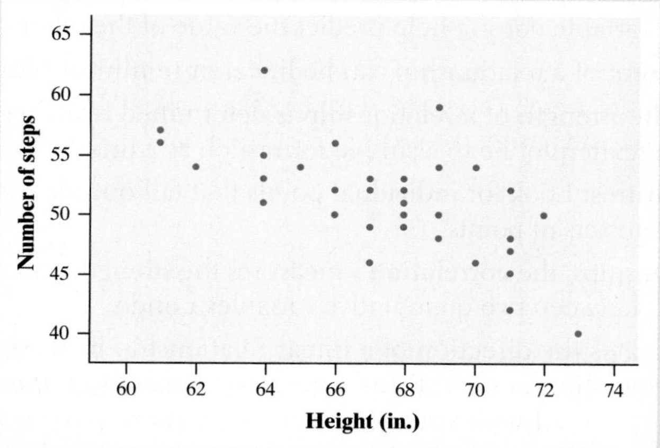


(a) Explain why it isn't correct to say that the correlation is $-0.632$ steps per inch.

(b) What would happen to the correlation if number of steps was used as the explanatory variable and height was used as the response variable?

(c) What would happen to the correlation if height was measured in centimeters instead of inches? Explain.

**SOLUTION:**

(a) Because correlation is calculated using standardized values, it doesn't have units.

(b) The correlation would be the same because correlation doesn't make a distinction between explanatory and response variables.

(c) The correlation would be the same. Because $r$ is calculated using standardized values, changes of units don't affect correlation.

> Although it is unlikely that you will need to calculate the correlation by hand, understanding how the formula works makes it easier to answer questions like these.

> Changing from inches to centimeters won't change the locations of the points, only the numbers on the horizontal scale.

FOR PRACTICE, TRY EXERCISE 23

## Section 3.1 Summary

- A **scatterplot** displays the relationship between two quantitative variables measured on the same individuals. Mark values of one variable on the horizontal axis ($x$ axis) and values of the other variable on the vertical axis ($y$ axis). Plot each individual's data as a point on the graph.
- If we think that a variable $x$ may help predict, explain, or even cause changes in another variable $y$, we call $x$ an **explanatory variable** and $y$ a **response variable.** Always plot the explanatory variable on the $x$ axis of a scatterplot. Plot the response variable on the $y$ axis.
- When describing a scatterplot, look for an overall pattern (direction, form, strength) and departures from the pattern (unusual features) and always answer in context.
  - **Direction:** A relationship has a **positive association** when values of one variable tend to increase as the values of the other variable increase, a **negative association** when values of one variable tend to decrease as the values of the other variable increase, or **no association** when knowing the value of one variable doesn't help predict the value of the other variable.
  - **Form:** The form of a relationship can be linear or nonlinear (curved).
  - **Strength:** The strength of a relationship is determined by how close the points in the scatterplot lie to a simple form such as a line.
  - **Unusual features:** Look for individual points that fall outside the pattern and distinct clusters of points.
- For linear relationships, the **correlation** $r$ measures the strength and direction of the association between two quantitative variables $x$ and $y$.
- Correlation indicates the direction of a linear relationship by its sign: $r > 0$ for a positive association and $r < 0$ for a negative association. Correlation always satisfies $-1 \leq r \leq 1$ with stronger linear associations having values of $r$ closer to 1 and $-1$. Correlations of $r = 1$ and $r = -1$ occur only when the points on a scatterplot lie exactly on a straight line.
- Remember these limitations of $r$: Correlation does not imply causation. The correlation is not resistant, so unusual points can greatly change the value of $r$. The correlation should only be used to describe linear relationships.
- Correlation ignores the distinction between explanatory and response variables. The value of $r$ does not have units and is not affected by changes in the unit of measurement of either variable.

### 3.1 Technology Corner

*TI-Nspire and other technology instructions are on the book's website at highschool.bfwpub.com/updatedtps6e.*

**8. Making scatterplots** **Page 159**

# Section 3.1 Exercises

1. **Coral reefs and cell phones** Identify the explanatory variable and the response variable for the following relationships, if possible. Explain your reasoning. pg 154
   (a) The weight gain of corals in aquariums where the water temperature is controlled at different levels
   (b) The number of text messages sent and the number of phone calls made in a sample of 100 students

2. **Teenagers and corn yield** Identify the explanatory variable and the response variable for the following relationships, if possible. Explain your reasoning.
   (a) The height and arm span of a sample of 50 teenagers
   (b) The yield of corn in bushels per acre and the amount of rain in the growing season

3. **Heavy backpacks** Ninth-grade students at the Webb Schools go on a backpacking trip each fall. Students are divided into hiking groups of size 8 by selecting names from a hat. Before leaving, students and their backpacks are weighed. The data here are from one hiking group. Make a scatterplot by hand that shows how backpack weight relates to body weight. pg 155

| Body weight (lb) | 120 | 187 | 109 | 103 | 131 | 165 | 158 | 116 |
|---|---|---|---|---|---|---|---|---|
| Backpack weight (lb) | 26 | 30 | 26 | 24 | 29 | 35 | 31 | 28 |

4. **Putting success** How well do professional golfers putt from various distances to the hole? The data show various distances to the hole (in feet) and the percent of putts made at each distance for a sample of golfers.[9] Make a scatterplot by hand that shows how the percent of putts made relates to the distance of the putt.

| Distance (ft) | Percent made | Distance (ft) | Percent made |
|---|---|---|---|
| 2 | 93.3 | 12 | 25.7 |
| 3 | 83.1 | 13 | 24.0 |
| 4 | 74.1 | 14 | 31.0 |
| 5 | 58.9 | 15 | 16.8 |
| 6 | 54.8 | 16 | 13.4 |
| 7 | 53.1 | 17 | 15.9 |
| 8 | 46.3 | 18 | 17.3 |
| 9 | 31.8 | 19 | 13.6 |
| 10 | 33.5 | 20 | 15.8 |
| 11 | 31.6 | | |

5. **Olympic athletes** The scatterplot shows the relationship between height (in inches) and weight (in pounds) for the members of the U.S. 2016 Olympic Track and Field team.[10] Describe the relationship between height and weight for these athletes. pg 158

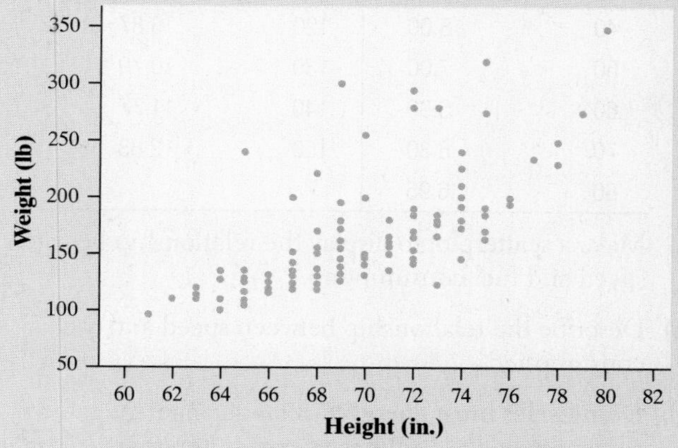


6. **Starbucks** The scatterplot shows the relationship between the amount of fat (in grams) and number of calories in products sold at Starbucks.[11] Describe the relationship between fat and calories for these products.

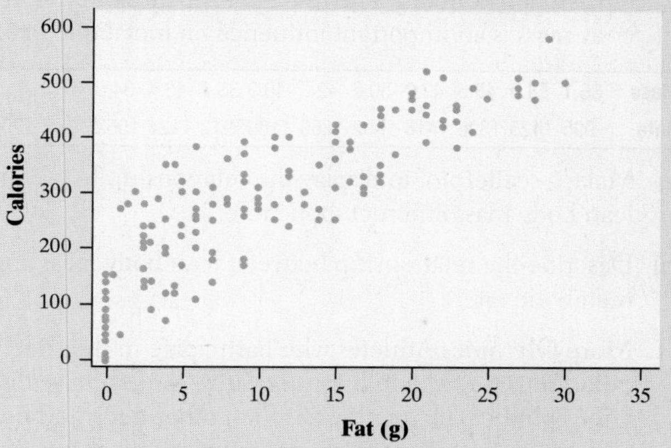


7. **More heavy backpacks** Refer to your graph from Exercise 3. Describe the relationship between body weight and backpack weight for this group of hikers.

8. **More putting success** Refer to your graph from Exercise 4. Describe the relationship between distance from hole and percent of putts made for the sample of professional golfers.

9. **Does fast driving waste fuel?** How does the fuel consumption of a car change as its speed increases? Here are data for a British Ford Escort. Speed is measured in kilometers per hour, and fuel consumption is measured in liters of gasoline used per 100 kilometers traveled.[12]

| Speed (km/h) | Fuel used (L/100 km) | Speed (km/h) | Fuel used (L/100 km) |
|---|---|---|---|
| 10 | 21.00 | 90 | 7.57 |
| 20 | 13.00 | 100 | 8.27 |
| 30 | 10.00 | 110 | 9.03 |
| 40 | 8.00 | 120 | 9.87 |
| 50 | 7.00 | 130 | 10.79 |
| 60 | 5.90 | 140 | 11.77 |
| 70 | 6.30 | 150 | 12.83 |
| 80 | 6.95 | | |

(a) Make a scatterplot to display the relationship between speed and fuel consumption.

(b) Describe the relationship between speed and fuel consumption.

10. **Do muscles burn energy?** Metabolic rate, the rate at which the body consumes energy, is important in studies of weight gain, dieting, and exercise. We have data on the lean body mass and resting metabolic rate for 12 women who are subjects in a study of dieting. Lean body mass, given in kilograms, is a person's weight leaving out all fat. Metabolic rate is measured in calories burned per 24 hours. The researchers believe that lean body mass is an important influence on metabolic rate.

| | | | | | | | | | | | | |
|---|---|---|---|---|---|---|---|---|---|---|---|---|
| **Mass** | 36.1 | 54.6 | 48.5 | 42.0 | 50.6 | 42.0 | 40.3 | 33.1 | 42.4 | 34.5 | 51.1 | 41.2 |
| **Rate** | 995 | 1425 | 1396 | 1418 | 1502 | 1256 | 1189 | 913 | 1124 | 1052 | 1347 | 1204 |

(a) Make a scatterplot to display the relationship between lean body mass and metabolic rate.

(b) Describe the relationship between lean body mass and metabolic rate.

11. **More Olympics** Athletes who participate in the shot put, discus throw, and hammer throw tend to have different physical characteristics than other track and field athletes. The scatterplot shown here enhances the scatterplot from Exercise 5 by plotting these athletes with blue squares. How are the relationships between height and weight the same for the two groups of athletes? How are the relationships different?

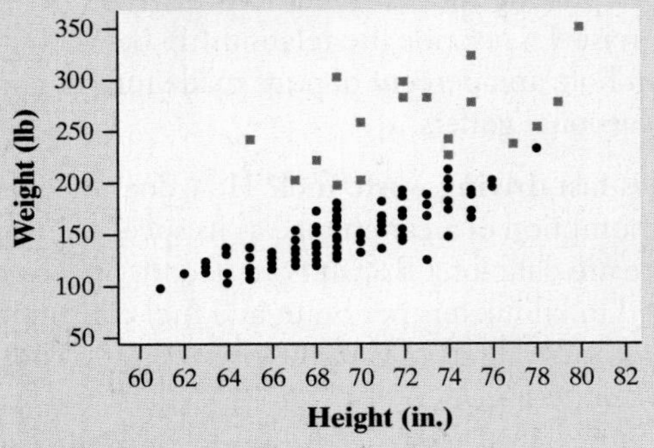


12. **More Starbucks** How do the nutritional characteristics of food products differ from drink products at Starbucks? The scatterplot shown here enhances the scatterplot from Exercise 6 by plotting the food products with blue squares. How are the relationships between fat and calories the same for the two types of products? How are the relationships different?

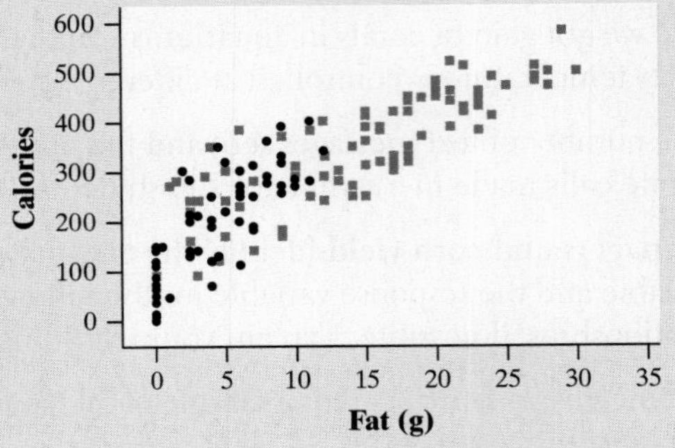


13. **Manatees** Manatees are large, gentle, slow-moving sea creatures found along the coast of Florida. Many manatees are injured or killed by boats. Here is a scatterplot showing the relationship between the number of boats registered in Florida (in thousands) and the number of manatees killed by boats for the years 1977 to 2015.[13] Is $r > 0$ or $r < 0$? Closer to $r = 0$ or $r = \pm 1$? Explain your reasoning.

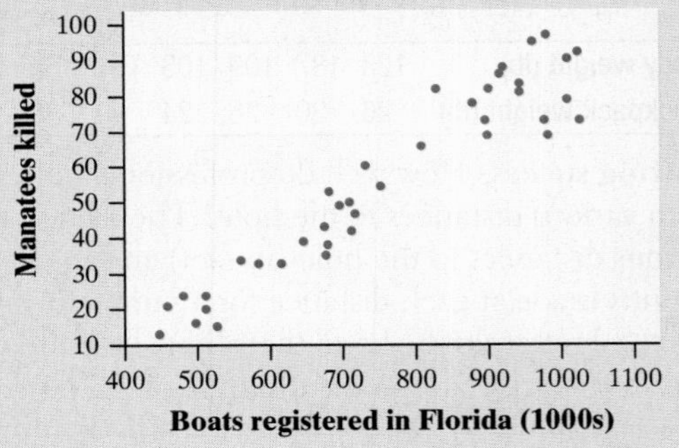


14. **Windy city** Is it possible to use temperature to predict wind speed? Here is a scatterplot showing the average temperature (in degrees Fahrenheit) and average wind speed (in miles per hour) for 365 consecutive days at O'Hare International Airport in Chicago.[14] Is $r > 0$ or $r < 0$? Closer to $r = 0$ or $r = \pm 1$? Explain your reasoning.

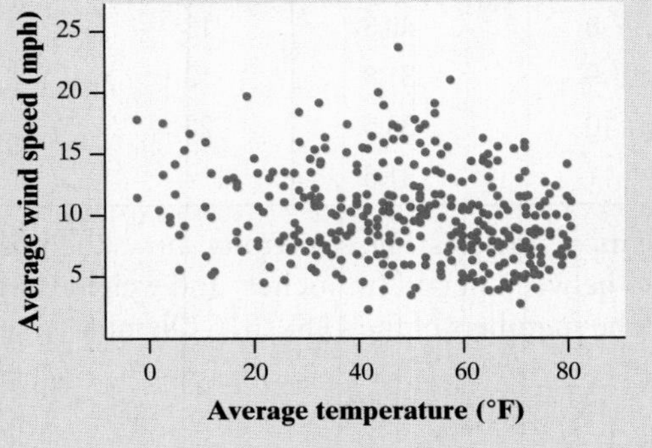

15. **Points and turnovers** Here is a scatterplot showing the relationship between the number of turnovers and the number of points scored for players in a recent NBA season.[15] The correlation for these data is $r = 0.92$. Interpret the correlation.

pg 162

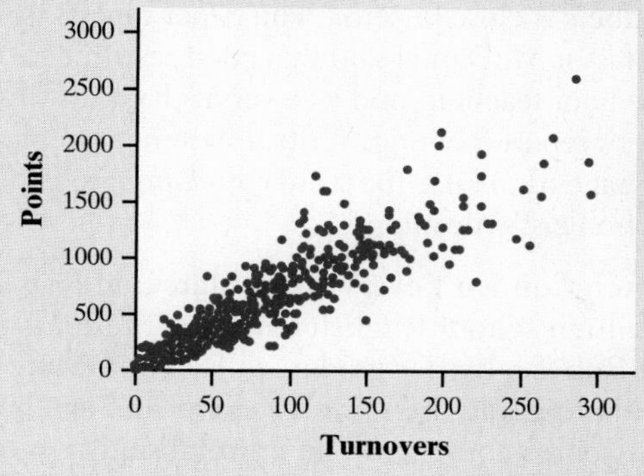


16. **Oh, that smarts!** Infants who cry easily may be more easily stimulated than others. This may be a sign of higher IQ. Child development researchers explored the relationship between the crying of infants 4 to 10 days old and their IQ test scores at age 3 years. A snap of a rubber band on the sole of the foot caused the infants to cry. The researchers recorded the crying and measured its intensity by the number of peaks in the most active 20 seconds. The correlation for these data is $r = 0.45$.[16] Interpret the correlation.

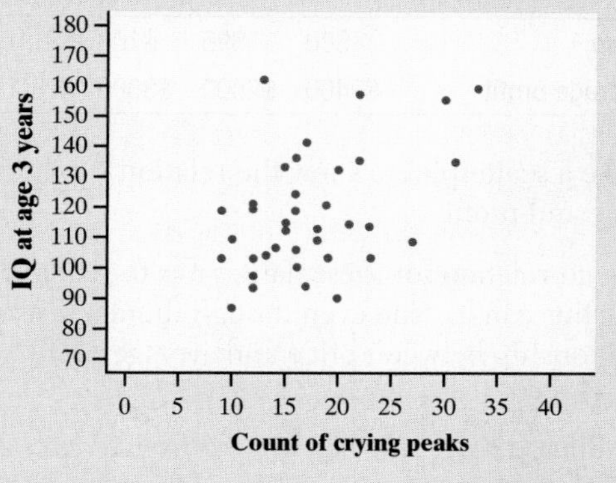


17. **More turnovers?** Refer to Exercise 15. Does the fact that $r = 0.92$ suggest that an increase in turnovers will cause NBA players to score more points? Explain your reasoning.

pg 165

18. **More crying?** Refer to Exercise 16. Does the fact that $r = 0.45$ suggest that making an infant cry will increase his or her IQ later in life? Explain your reasoning.

19. **Hot dogs** Are hot dogs that are high in calories also high in salt? The following scatterplot shows the calories and salt content (measured in milligrams of sodium) in 17 brands of meat hot dogs.[17]

pg 165

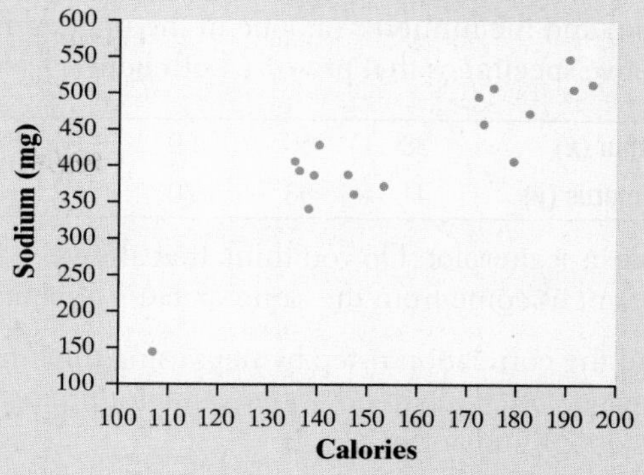


(a) The correlation for these data is $r = 0.87$. Interpret this value.

(b) What effect does the hot dog brand with the smallest calorie content have on the correlation? Justify your answer.

20. **All brawn?** The following scatterplot plots the average brain weight (in grams) versus average body weight (in kilograms) for 96 species of mammals.[18] There are many small mammals whose points overlap at the lower left.

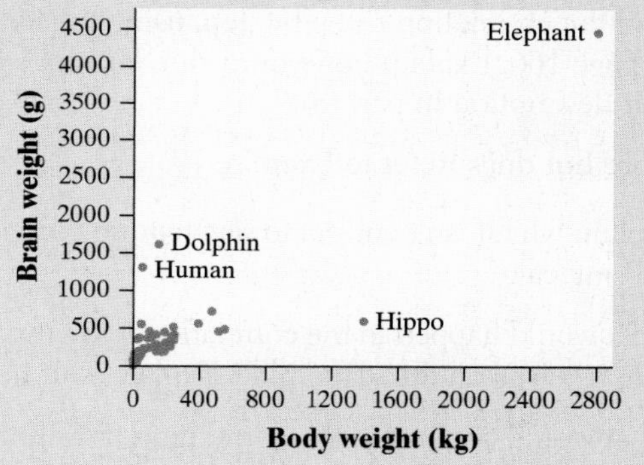


(a) The correlation between body weight and brain weight is $r = 0.86$. Interpret this value.

(b) What effect does the human have on the correlation? Justify your answer.

21. **Dem bones** Archaeopteryx is an extinct beast that had feathers like a bird but teeth and a long bony tail like a reptile. Only six fossil specimens are known to exist today. Because these specimens differ greatly in size, some scientists think they are different species rather than individuals from the same species. If the specimens belong to the same species and differ in size because some are younger than others, there should be a positive linear relationship between the lengths of a pair of bones from all individuals. A point outside the pattern would suggest a different species. Here are data on the lengths (in centimeters) of the femur (a leg

bone) and the humerus (a bone in the upper arm) for the five specimens that preserve both bones:[19]

| | | | | | |
|---|---|---|---|---|---|
| Femur ($x$) | 38 | 56 | 59 | 64 | 74 |
| Humerus ($y$) | 41 | 63 | 70 | 72 | 84 |

(a) Make a scatterplot. Do you think that all five specimens come from the same species? Explain.

(b) Find the correlation $r$ step by step, using the formula on page 166. Explain how your value for $r$ matches your graph in part (a).

22. **Data on dating** A student wonders if tall women tend to date taller men than do short women. She measures herself, her dormitory roommate, and the women in the adjoining dorm rooms. Then she measures the next man each woman dates. Here are the data (heights in inches):

| | | | | | | |
|---|---|---|---|---|---|---|
| Women ($x$) | 66 | 64 | 66 | 65 | 70 | 65 |
| Men ($y$) | 72 | 68 | 70 | 68 | 71 | 65 |

(a) Make a scatterplot of these data. Describe what you see.

(b) Find the correlation $r$ step by step, using the formula on page 166. Explain how your value for $r$ matches your description in part (a).

23. **More hot dogs** Refer to Exercise 19.

pg 169

(a) Explain why it isn't correct to say that the correlation is 0.87 mg/cal.

(b) What would happen to the correlation if the variables were reversed on the scatterplot? Explain your reasoning.

(c) What would happen to the correlation if sodium was measured in grams instead of milligrams? Explain your reasoning.

24. **More brains** Refer to Exercise 20.

(a) Explain why it isn't correct to say that the correlation is 0.86 g/kg.

(b) What would happen to the correlation if the variables were reversed on the scatterplot? Explain your reasoning.

(c) What would happen to the correlation if brain weight was measured in kilograms instead of grams? Explain your reasoning.

25. **Rank the correlations** Consider each of the following relationships: the heights of fathers and the heights of their adult sons, the heights of husbands and the heights of their wives, and the heights of women at age 4 and their heights at age 18. Rank the correlations between these pairs of variables from largest to smallest. Explain your reasoning.

26. **Teaching and research** A college newspaper interviews a psychologist about student ratings of the teaching of faculty members. The psychologist says, "The evidence indicates that the correlation between the research productivity and teaching rating of faculty members is close to zero." The paper reports this as "Professor McDaniel said that good researchers tend to be poor teachers, and vice versa." Explain why the paper's report is wrong. Write a statement in plain language (don't use the word *correlation*) to explain the psychologist's meaning.

27. **Correlation isn't everything** Marc and Rob are both high school English teachers. Students think that Rob is a harder grader, so Rob and Marc decide to grade the same 10 essays and see how their scores compare. The correlation is $r = 0.98$, but Rob's scores are always lower than Marc's. Draw a possible scatterplot that illustrates this situation.

28. **Limitations of correlation** A carpenter sells handmade wooden benches at a craft fair every week. Over the past year, the carpenter has varied the price of the benches from \$80 to \$120 and recorded the average weekly profit he made at each selling price. The prices of the bench and the corresponding average profits are shown in the table.

| | | | | | |
|---|---|---|---|---|---|
| Price | \$80 | \$90 | \$100 | \$110 | \$120 |
| Average profit | \$2400 | \$2800 | \$3000 | \$2800 | \$2400 |

(a) Make a scatterplot to show the relationship between price and profit.

(b) The correlation for these data is $r = 0$. Explain how this can be true even though there is a strong relationship between price and average profit.

**Multiple Choice:** *Select the best answer for Exercises 29–34.*

29. You have data for many years on the average price of a barrel of oil and the average retail price of a gallon of unleaded regular gasoline. If you want to see how well the price of oil predicts the price of gas, then you should make a scatterplot with ____________ as the explanatory variable.

(a) the price of oil

(b) the price of gas

(c) the year

(d) either oil price or gas price

(e) time

**30.** In a scatterplot of the average price of a barrel of oil and the average retail price of a gallon of gas, you expect to see

(a) very little association.

(b) a weak negative association.

(c) a strong negative association.

(d) a weak positive association.

(e) a strong positive association.

**31.** The following graph plots the gas mileage (in miles per gallon) of various cars from the same model year versus the weight of these cars (in thousands of pounds). The points marked with red dots correspond to cars made in Japan. From this plot, we may conclude that

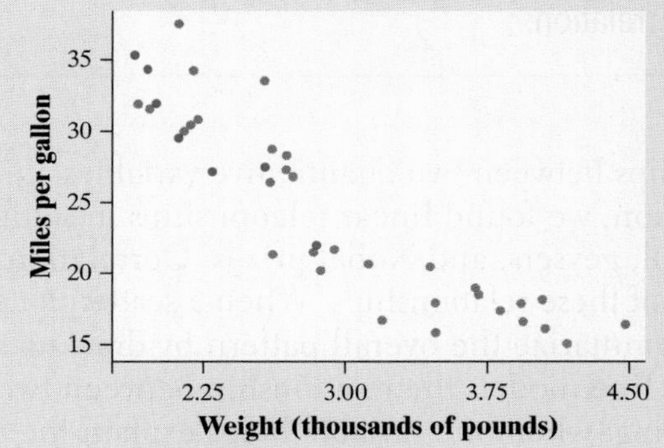


(a) there is a positive association between weight and gas mileage for Japanese cars.

(b) the correlation between weight and gas mileage for all the cars is close to 1.

(c) there is little difference between Japanese cars and cars made in other countries.

(d) Japanese cars tend to be lighter in weight than other cars.

(e) Japanese cars tend to get worse gas mileage than other cars.

**32.** If women always married men who were 2 years older than themselves, what would be the correlation between the ages of husband and wife?

(a) 2 (b) 1 (c) 0.5

(d) 0 (e) Can't tell without seeing the data

**33.** The scatterplot shows reading test scores against IQ test scores for 14 fifth-grade children. What effect does the point at IQ = 124 and reading score = 10 have on the correlation?

(a) It makes the correlation closer to 1.

(b) It makes the correlation closer to 0 but still positive.

(c) It makes the correlation equal to 0.

(d) It makes the correlation negative.

(e) It has no effect on the correlation.

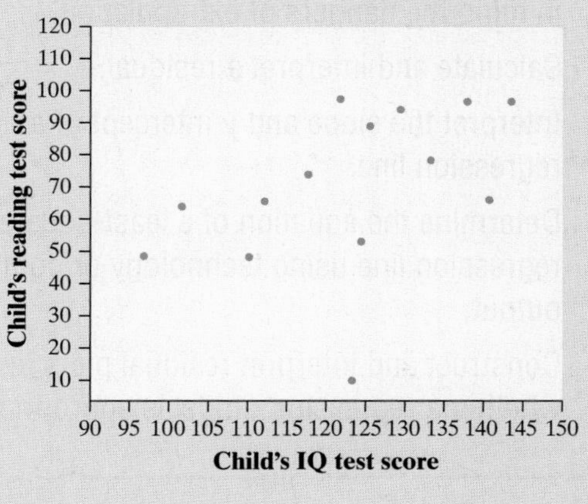


**34.** If we leave out this point, the correlation for the remaining 13 points in the preceding figure is closest to

(a) −0.95. (b) −0.65. (c) 0.

(d) 0.65. (e) 0.95.

## Recycle and Review

**35. Big diamonds** (1.2) Here are the weights (in milligrams) of 58 diamonds from a nodule carried up to the earth's surface in surrounding rock. These data represent a population of diamonds formed in a single event deep in the earth.[20]

| | | | | | | | | | | |
|---|---|---|---|---|---|---|---|---|---|---|
| 13.8 | 3.7 | 33.8 | 11.8 | 27.0 | 18.9 | 19.3 | 20.8 | 25.4 | 23.1 | 7.8 |
| 10.9 | 9.0 | 9.0 | 14.4 | 6.5 | 7.3 | 5.6 | 18.5 | 1.1 | 11.2 | 7.0 |
| 7.6 | 9.0 | 9.5 | 7.7 | 7.6 | 3.2 | 6.5 | 5.4 | 7.2 | 7.8 | 3.5 |
| 5.4 | 5.1 | 5.3 | 3.8 | 2.1 | 2.1 | 4.7 | 3.7 | 3.8 | 4.9 | 2.4 |
| 1.4 | 0.1 | 4.7 | 1.5 | 2.0 | 0.1 | 0.1 | 1.6 | 3.5 | 3.7 | 2.6 |
| 4.0 | 2.3 | 4.5 | | | | | | | | |

Make a histogram to display the distribution of weight. Describe the distribution.

**36. Fruit fly thorax lengths** (2.2) Fruit flies are used frequently in genetic research because of their quick reproductive cycle. The length of the thorax (in millimeters) for male fruit flies is approximately Normally distributed with a mean of 0.80 mm and a standard deviation of 0.08 mm.[21]

(a) What proportion of male fruit flies have a thorax length greater than 1 mm?

(b) What is the 30th percentile for male fruit fly thorax lengths?

# SECTION 3.2 Least-Squares Regression

## LEARNING TARGETS *By the end of the section, you should be able to:*

- Make predictions using regression lines, keeping in mind the dangers of extrapolation.
- Calculate and interpret a residual.
- Interpret the slope and *y* intercept of a regression line.
- Determine the equation of a least-squares regression line using technology or computer output.
- Construct and interpret residual plots to assess whether a regression model is appropriate.
- Interpret the standard deviation of the residuals and $r^2$ and use these values to assess how well a least-squares regression line models the relationship between two variables.
- Describe how the least-squares regression line, standard deviation of the residuals, and $r^2$ are influenced by unusual points.
- Find the slope and *y* intercept of the least-squares regression line from the means and standard deviations of *x* and *y* and their correlation.

Linear (straight-line) relationships between two quantitative variables are fairly common. In the preceding section, we found linear relationships in settings as varied as Major League Baseball, geysers, and Nobel prizes. Correlation measures the strength and direction of these relationships. When a scatterplot shows a linear relationship, we can summarize the overall pattern by drawing a line on the scatterplot. A **regression line** models the relationship between two variables, but only in a specific setting: when one variable helps explain the other. Regression, unlike correlation, requires that we have an explanatory variable and a response variable.

Sometimes regression lines are referred to as *simple linear regression models.* They are called "simple" because they involve only one explanatory variable.

**DEFINITION Regression line**

A **regression line** is a line that models how a response variable *y* changes as an explanatory variable *x* changes. Regression lines are expressed in the form $\hat{y} = a + bx$ where $\hat{y}$ (pronounced "*y*-hat") is the predicted value of *y* for a given value of *x*.

It is common knowledge that cars and trucks lose value the more they are driven. Can we predict the price of a used Ford F-150 SuperCrew 4 × 4 truck if we know how many miles it has on the odometer? A random sample of 16 used Ford F-150 SuperCrew 4 × 4s was selected from among those listed for sale at autotrader.com. The number of miles driven and price (in dollars) were recorded for each of the trucks.[22] Here are the data:

Tim Graham/Alamy

| | | | | | | | | |
|---|---|---|---|---|---|---|---|---|
| **Miles driven** | 70,583 | 129,484 | 29,932 | 29,953 | 24,495 | 75,678 | 8359 | 4447 |
| **Price ($)** | 21,994 | 9500 | 29,875 | 41,995 | 41,995 | 28,986 | 31,891 | 37,991 |
| **Miles driven** | 34,077 | 58,023 | 44,447 | 68,474 | 144,162 | 140,776 | 29,397 | 131,385 |
| **Price ($)** | 34,995 | 29,988 | 22,896 | 33,961 | 16,883 | 20,897 | 27,495 | 13,997 |

Figure 3.6 is a scatterplot of these data. The plot shows a moderately strong, negative linear association between miles driven and price. There are two distinct clusters of trucks: a group of 12 trucks between 0 and 80,000 miles driven and a group of 4 trucks between 120,000 and 160,000 miles driven. The correlation is $r = -0.815$. The line on the plot is a regression line for predicting price from miles driven.

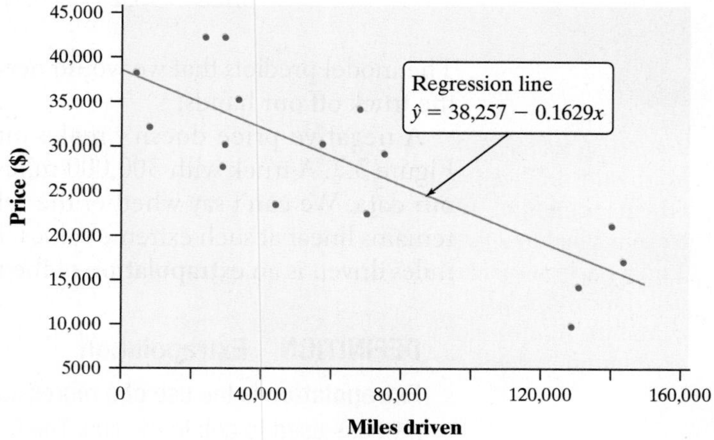


**FIGURE 3.6** Scatterplot showing the price and miles driven of used Ford F-150s, along with a regression line.

# Prediction

We can use a regression line to predict the value of the response variable for a specific value of the explanatory variable. For the Ford F-150 data, the equation of the regression line is

$$\widehat{\text{price}} = 38{,}257 - 0.1629\ (\text{miles driven})$$

When we want to refer to the predicted value of a variable, we add a hat on top. Here, $\widehat{\text{price}}$ refers to the predicted price of a used Ford F-150.

If a used Ford F-150 has 100,000 miles driven, substitute $x = 100{,}000$ in the equation. The predicted price is

$$\widehat{\text{price}} = 38{,}257 - 0.1629(100{,}000) = \$21{,}967$$

This prediction is illustrated in Figure 3.7.

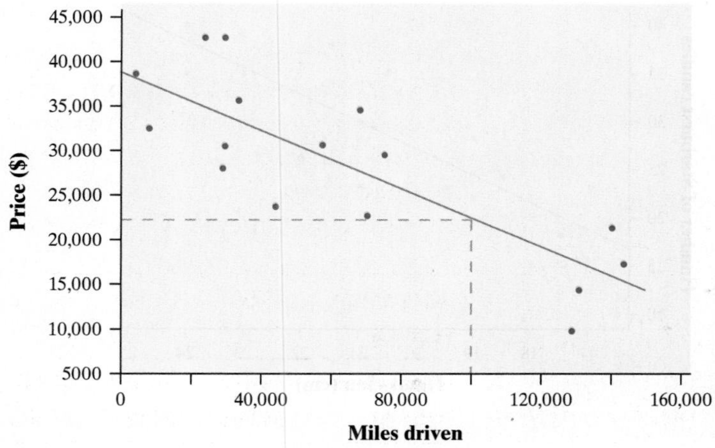


**FIGURE 3.7** Using the regression line to predict price for a Ford F-150 with 100,000 miles driven.

Even though the value $\hat{y} = \$21,967$ is unlikely to be the actual price of a truck that has been driven 100,000 miles, it's our best guess based on the linear model using $x =$ miles driven. We can also think of $\hat{y} = \$21,967$ as the average price for a sample of trucks that have each been driven 100,000 miles.

Can we predict the price of a Ford F-150 with 300,000 miles driven? We can certainly substitute 300,000 into the equation of the line. The prediction is

$$\widehat{\text{price}} = 38,257 - 0.1629(300,000) = -\$10,613$$

The model predicts that we would need to *pay* \$10,613 just to have someone take the truck off our hands!

A negative price doesn't make much sense in this context. Look again at Figure 3.7. A truck with 300,000 miles driven is far outside the set of $x$ values for our data. We can't say whether the relationship between miles driven and price remains linear at such extreme values. Predicting the price for a truck with 300,000 miles driven is an **extrapolation** of the relationship beyond what the data show.

**DEFINITION Extrapolation**

**Extrapolation** is the use of a regression line for prediction outside the interval of $x$ values used to obtain the line. The further we extrapolate, the less reliable the predictions.

Few relationships are linear for all values of the explanatory variable. **Don't make predictions using values of $x$ that are much larger or much smaller than those that actually appear in your data.**

## EXAMPLE How much candy can you grab?

### Prediction

**PROBLEM:** The scatterplot below shows the hand span (in cm) and number of Starburst™ candies grabbed by each student when Mr. Tyson's class did the "Candy grab" activity. The regression line $\hat{y} = -29.8 + 2.83x$ has been added to the scatterplot.

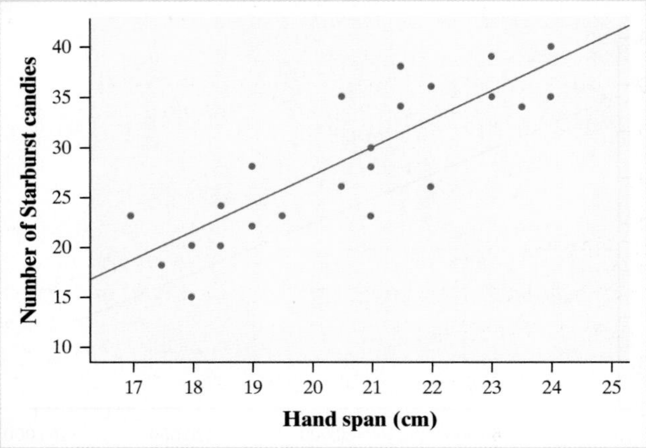

Josh Tabor

(a) Andres has a hand span of 22 cm. Predict the number of Starburst™ candies he can grab.

(b) Mr. Tyson's young daughter McKayla has a hand span of 12 cm. Predict the number of Starburst candies she can grab.

(c) How confident are you in each of these predictions? Explain.

**SOLUTION:**

(a) $\hat{y} = -29.8 + 2.83(22)$

$\hat{y} = 32.46$ Starburst candies

Don't worry that the predicted number of Starburst candies isn't an integer. Think of 32.46 as the average number of Starburst candies that a group of students, each with a hand span of 22 cm, could grab.

(b) $\hat{y} = -29.8 + 2.83(12)$

$\hat{y} = 4.16$ Starburst candies

(c) The prediction for Andres is believable because $x = 22$ is within the interval of x-values used to create the model. However, the prediction for McKayla is not trustworthy because $x = 12$ is far outside of the x-values used to create the regression line. The linear form may not extend to hand spans this small.

**FOR PRACTICE, TRY EXERCISE 37**

## Residuals

In most cases, no line will pass exactly through all the points in a scatterplot. Because we use the line to predict $y$ from $x$, the prediction errors we make are errors in $y$, the vertical direction in the scatterplot.

Figure 3.8 shows a scatterplot of the Ford F-150 data with a regression line added. The prediction errors are marked as bold vertical segments in the graph. These vertical deviations represent "leftover" variation in the response variable after fitting the regression line. For that reason, they are called **residuals**.

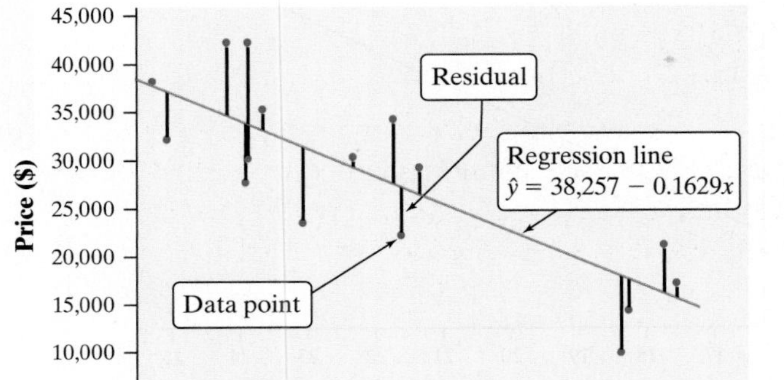


**FIGURE 3.8** Scatterplot of the Ford F-150 data with a regression line added. A good regression line should make the residuals (shown as bold vertical segments) as small as possible.

**DEFINITION Residual**

A **residual** is the difference between the actual value of $y$ and the value of $y$ predicted by the regression line. That is,

$$\text{residual} = \text{actual } y - \text{predicted } y$$
$$= y - \hat{y}$$

In Figure 3.8 above, the highlighted data point represents a Ford F-150 that had 70,583 miles driven and a price of \$21,994. The regression line predicts a price of

$$\widehat{\text{price}} = 38{,}257 - 0.1629(70{,}583) = \$26{,}759$$

for this truck, but its actual price was \$21,994. This truck's residual is

$$\begin{aligned}\text{residual} &= \text{actual } y - \text{predicted } y \\ &= y - \hat{y} \\ &= 21{,}994 - 26{,}759 = -\$4765\end{aligned}$$

The actual price of this truck is \$4765 *less* than the cost predicted by the regression line with $x =$ miles driven. Why is the actual price less than predicted? There are many possible reasons. Perhaps the truck needs body work, has mechanical issues, or has been in an accident.

## EXAMPLE

### Can you grab more than expected?
### Calculating and interpreting a residual

**PROBLEM:** Here again is the scatterplot showing the hand span (in cm) and number of Starburst™ candies grabbed by each student in Mr. Tyson's class. The regression line is $\hat{y} = -29.8 + 2.83x$.

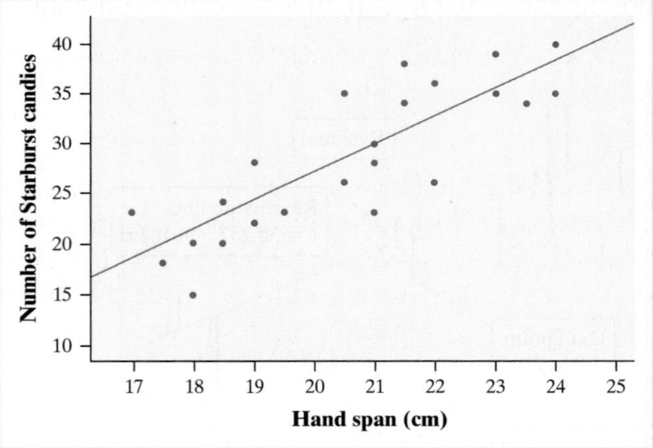

Josh Tabor

Find and interpret the residual for Andres, who has a hand span of 22 cm and grabbed 36 Starburst candies.

**SOLUTION:**

$\hat{y} = -29.8 + 2.83(22) = 32.46$ Starburst candies

Residual $= 36 - 32.46 = 3.54$ Starburst candies

Andres grabbed 3.54 more Starburst candies than the number predicted by the regression line with $x =$ hand span.

> Residual $=$ actual $y$ $-$ predicted $y$
> $= y - \hat{y}$

**FOR PRACTICE, TRY EXERCISE 39**

## CHECK YOUR UNDERSTANDING

Some data were collected on the weight of a male white laboratory rat for the first 25 weeks after its birth. A scatterplot of $y$ = weight (in grams) and $x$ = time since birth (in weeks) shows a fairly strong, positive linear relationship. The regression equation $\hat{y} = 100 + 40x$ models the data fairly well.

1. Predict the rat's weight at 16 weeks old.
2. Calculate and interpret the residual if the rat weighed 700 grams at 16 weeks old.
3. Should you use this line to predict the rat's weight at 2 years old? Use the equation to make the prediction and discuss your confidence in the result. (There are 454 grams in a pound.)

# Interpreting a Regression Line

A regression line is a *model* for the data, much like the density curves of Chapter 2. The **y intercept** and **slope** of the regression line describe what this model tells us about the relationship between the response variable $y$ and the explanatory variable $x$.

The data used to calculate a regression line typically come from a sample. The statistics $a$ and $b$ in the sample regression model estimate the $y$ intercept and slope parameters of the population regression model. You'll learn more about how this works in Chapter 12.

**DEFINITION *y* intercept, Slope**

In the regression equation $\hat{y} = a + bx$:

- $a$ is the ***y* intercept,** the predicted value of $y$ when $x = 0$
- $b$ is the **slope,** the amount by which the predicted value of $y$ changes when $x$ increases by 1 unit

You are probably accustomed to the form $y = mx + b$ for the equation of a line from algebra. Statisticians have adopted a different form for the equation of a regression line. Some use $\hat{y} = b_0 + b_1x$. We prefer the form $\hat{y} = a + bx$ for three reasons: (1) it's simpler, (2) your calculator uses this form, and (3) the formula sheet provided on the AP® exam uses this form. Just remember that the slope is the coefficient of $x$, no matter what form is used.

Let's return to the Ford F-150 data. The equation of the regression line for these data is $\hat{y} = 38{,}257 - 0.1629x$, where $x$ = miles driven and $y$ = price. The slope $b = -0.1629$ tells us that the *predicted* price of a used Ford F-150 goes down by \$0.1629 (16.29 cents) for each additional mile that the truck has been driven. The $y$ intercept $a = 38{,}257$ is the *predicted* price (in dollars) of a used Ford F-150 that has been driven 0 miles.

**AP® EXAM TIP**

When asked to interpret the slope or $y$ intercept, it is very important to include the word *predicted* (or equivalent) in your response. Otherwise, it might appear that you believe the regression equation provides actual values of $y$.

The slope of a regression line is an important numerical description of the relationship between the two variables. Although we need the value of the $y$ intercept to draw the line, it is statistically meaningful only when the explanatory variable can actually take values close to 0, as in the Ford F-150 data. In other cases, using the regression line to make a prediction for $x = 0$ is an extrapolation.

# EXAMPLE

## Grabbing more candy
### Interpreting the slope and *y* intercept

**PROBLEM:** The scatterplot shows the hand span (in cm) and number of Starburst™ candies grabbed by each student in Mr. Tyson's class, along with the regression line $\hat{y} = -29.8 + 2.83x$.

(a) Interpret the slope of the regression line.

(b) Does the value of the *y* intercept have meaning in this context? If so, interpret the *y* intercept. If not, explain why.

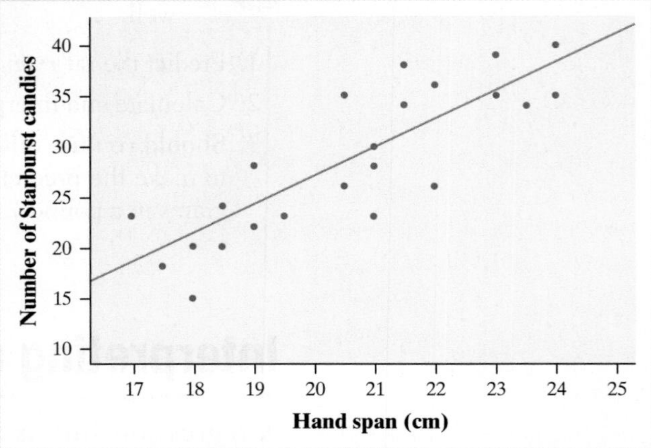


**SOLUTION:**

(a) The predicted number of Starburst candies grabbed goes up by 2.83 for each increase of 1 cm in hand span.

Remember that the slope describes how the *predicted* value of *y* changes, not the actual value of *y*.

(b) The *y* intercept does not have meaning in this case, as it is impossible to have a hand span of 0 cm.

Predicting the number of Starburst candies when $x = 0$ is an extrapolation—and results in an unrealistic prediction of $-29.8$.

**FOR PRACTICE, TRY EXERCISE 41**

For the Ford F-150 data, the slope $b = -0.1629$ is very close to 0. This does *not* mean that change in miles driven has little effect on price. The size of the slope depends on the units in which we measure the two variables. In this setting, the slope is the predicted change in price (in dollars) when the distance driven increases by 1 mile. There are 100 cents in a dollar. If we measured price in cents instead of dollars, the slope would be 100 times steeper, $b = -16.29$. *You can't say how strong a relationship is by looking at the slope of the regression line.*

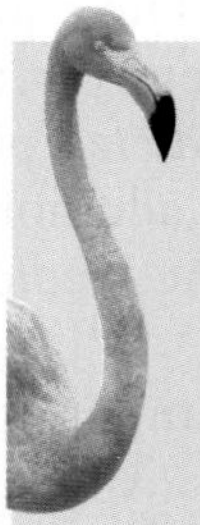

## CHECK YOUR UNDERSTANDING

Some data were collected on the weight of a male white laboratory rat for the first 25 weeks after its birth. A scatterplot of $y$ = weight (in grams) and $x$ = time since birth (in weeks) shows a fairly strong, positive linear relationship. The regression equation $\hat{y} = 100 + 40x$ models the data fairly well.

1. Interpret the slope of the regression line.
2. Does the value of the *y* intercept have meaning in this context? If so, interpret the *y* intercept. If not, explain why.

## The Least-Squares Regression Line

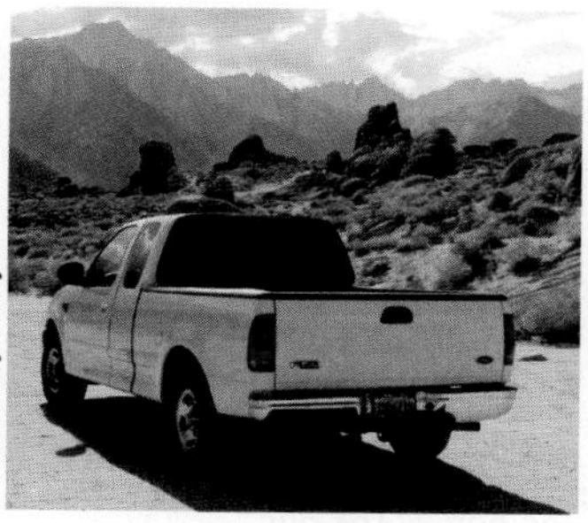

Duncan Selby/Alamy

There are many different lines we could use to model the association in a particular scatterplot. A *good* regression line makes the residuals as small as possible.

In the F-150 example, the regression line we used is $\hat{y} = 38{,}257 - 0.1629x$. How does this line make the residuals "as small as possible"? Maybe this line minimizes the *sum* of the residuals. If we add the prediction errors for all 16 trucks, the positive and negative residuals cancel out, as shown in Figure 3.9(a). That's the same issue we faced when we tried to measure deviation around the mean in Chapter 1. We'll solve the current problem in much the same way—by squaring the residuals.

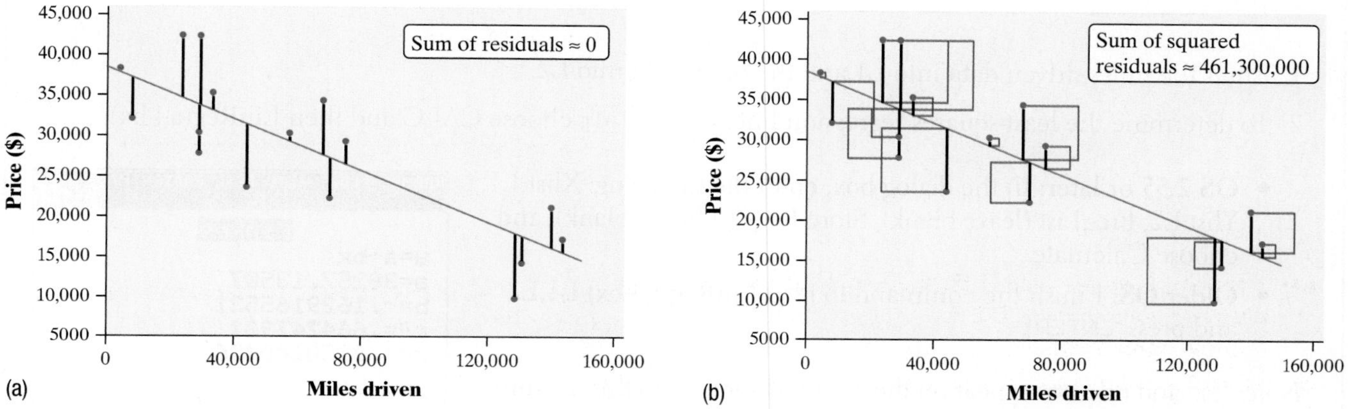


**FIGURE 3.9** Scatterplots of the Ford F-150 data with the regression line added. (a) The residuals will add to approximately 0 when using a good regression line. (b) A good regression line should make the sum of squared residuals as small as possible.

A good regression line will have a sum of residuals near 0. But the regression line we prefer is the one that minimizes the sum of the squared residuals. That's what the line shown in Figure 3.9(b) does for the Ford F-150 data, which is why we call it the **least-squares regression line**. No other regression line would give a smaller sum of squared residuals.

In addition to minimizing the sum of squared residuals, the least-squares regression line always goes through the point $(\bar{x}, \bar{y})$.

**DEFINITION Least-squares regression line**

The **least-squares regression line** is the line that makes the sum of the squared residuals as small as possible.

Your calculator or statistical software will give the equation of the least-squares line from data that you enter. Then you can concentrate on understanding and using the regression line.

**AP® EXAM TIP**

When displaying the equation of a least-squares regression line, the calculator will report the slope and intercept with much more precision than we need. There is no firm rule for how many decimal places to show for answers on the AP® Statistics exam. Our advice: decide how much to round based on the context of the problem you are working on.

## 9. Technology Corner — CALCULATING LEAST-SQUARES REGRESSION LINES

*TI-Nspire and other technology instructions are on the book's website at highschool.bfwpub.com/updatedtps6e.*

Let's use the Ford F-150 data to show how to find the equation of the least-squares regression line on the TI-83/84. Here are the data again:

| | | | | | | | | |
|---|---|---|---|---|---|---|---|---|
| **Miles driven** | 70,583 | 129,484 | 29,932 | 29,953 | 24,495 | 75,678 | 8359 | 4447 |
| **Price ($)** | 21,994 | 9500 | 29,875 | 41,995 | 41,995 | 28,986 | 31,891 | 37,991 |
| **Miles driven** | 34,077 | 58,023 | 44,447 | 68,474 | 144,162 | 140,776 | 29,397 | 131,385 |
| **Price ($)** | 34,995 | 29,988 | 22,896 | 33,961 | 16,883 | 20,897 | 27,495 | 13,997 |

1. Enter the miles driven data into L1 and the price data into L2.
2. To determine the least-squares regression line, press STAT; choose CALC and then LinReg(a+bx).

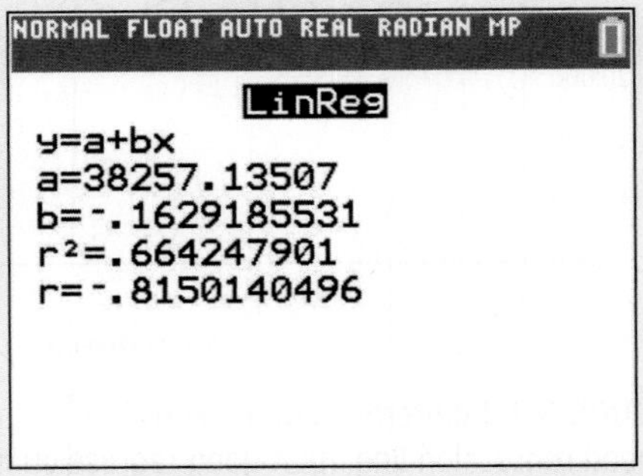


- **OS 2.55 or later:** In the dialog box, enter the following: Xlist:L1, Ylist:L2, FreqList (leave blank), Store RegEQ (leave blank), and choose Calculate.
- **Older OS:** Finish the command to read LinReg(a+bx) L1,L2 and press ENTER.

*Note:* If $r^2$ and $r$ do not appear on the TI-83/84 screen, do this one-time series of keystrokes:

- **OS 2.55 or later:** Press MODE and set STAT DIAGNOSTICS to ON. Then redo Step 2 to calculate the least-squares line. The $r^2$ and $r$ values should now appear.
- **Older OS:** Press 2nd 0 (CATALOG), scroll down to DiagnosticOn, and press ENTER. Press ENTER again to execute the command. The screen should say "Done." Then redo Step 2 to calculate the least-squares line. The $r^2$ and $r$ values should now appear.

To graph the least-squares regression line on the scatterplot:

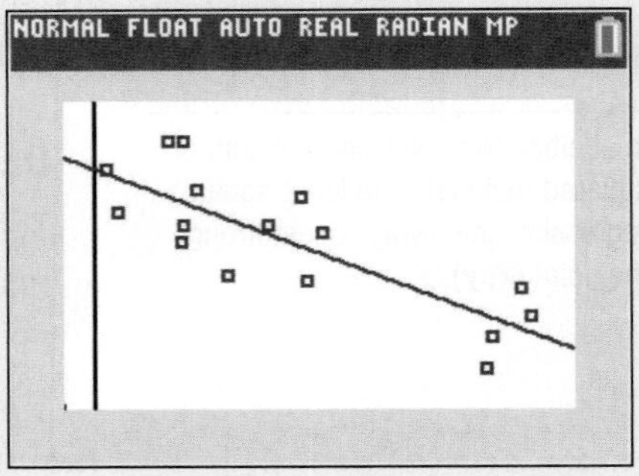


1. Set up a scatterplot (see Technology Corner 8 on page 159).
2. Press Y= and enter the equation of the least-squares regression line in Y1.
3. Press ZOOM and choose ZoomStat to see the scatterplot with the least-squares regression line.

*Note:* When you calculate the equation of the least-squares regression line, you can have the calculator store the equation to Y1. When setting up the calculation, enter Y1 for the StoreRegEq prompt blank (OS 2.55 or later) or use the following command (older OS): LinReg(a+bx) L1,L2,Y1. Y1 is found by pressing VARS and selecting Y-VARS, then Function, then Y1.

# Determining if a Linear Model Is Appropriate: Residual Plots

One of the first principles of data analysis is to look for an overall pattern and for striking departures from the pattern. A regression line describes the overall pattern of a linear relationship between an explanatory variable and a response variable. We see departures from this pattern by looking at a **residual plot**.

Some software packages prefer to plot the residuals against the predicted values $\hat{y}$ instead of against the values of the explanatory variable. The basic shape of the two plots is the same because $\hat{y}$ is linearly related to $x$.

**DEFINITION Residual plot**

A **residual plot** is a scatterplot that displays the residuals on the vertical axis and the explanatory variable on the horizontal axis.

Residual plots help us assess whether or not a linear model is appropriate. In Figure 3.10(a), the scatterplot shows the relationship between the average income (gross domestic product per person, in dollars) and fertility rate (number of children per woman) in 187 countries, along with the least-squares regression line. The residual plot in Figure 3.10(b) shows the average income for each country and the corresponding residual.

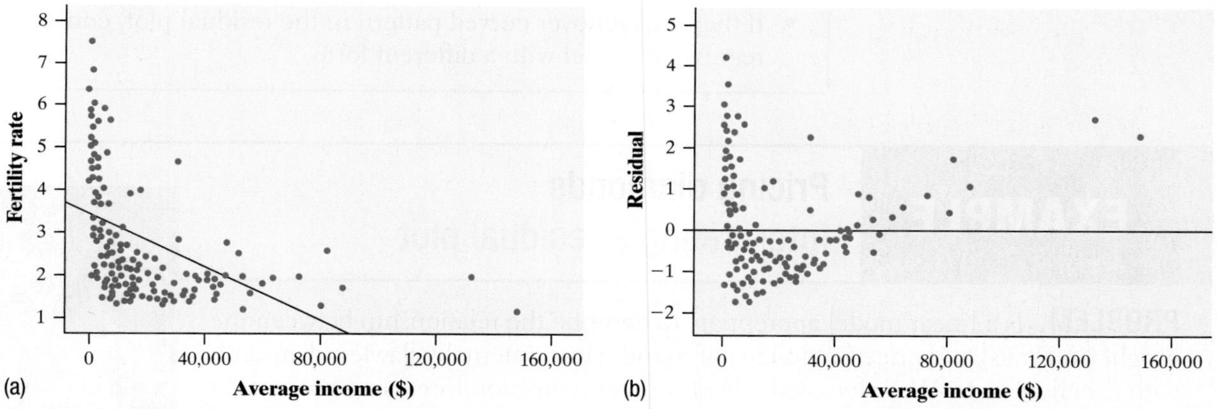


**FIGURE 3.10** The (a) scatterplot and (b) residual plot for the linear model relating fertility rate to average income for a sample of countries.

The least-squares regression line clearly doesn't fit this association very well! For most countries with average incomes under $5000, the actual fertility rates are greater than predicted, resulting in positive residuals. For countries with average incomes between $5000 and $60,000, the actual fertility rates tend to be smaller than predicted, resulting in negative residuals. Countries with average incomes above $60,000 all have fertility rates greater than predicted, again resulting in positive residuals. This U-shaped pattern in the residual plot indicates that the linear form of our model doesn't match the form of the association. A curved model might be better in this case.

In Figure 3.11(a), the scatterplot shows the Ford F-150 data, along with the least-squares regression line. The corresponding residual plot is shown in Figure 3.11(b).

Looking at the scatterplot, the line seems to be a good fit for this relationship. You can "see" that the line is appropriate by the lack of a leftover curved pattern in the residual plot. In fact, the residuals look randomly scattered around the residual = 0 line.

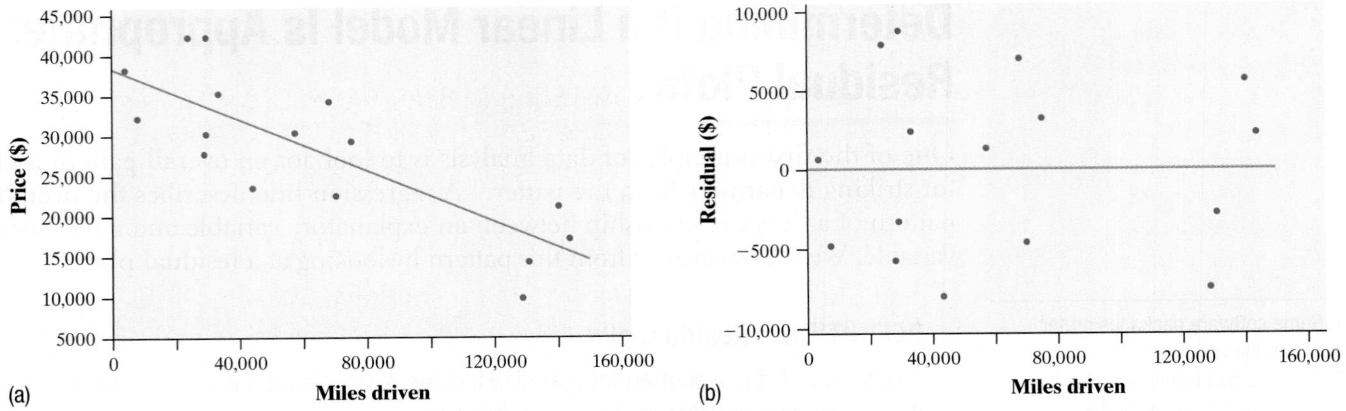


**FIGURE 3.11** The (a) scatterplot and (b) residual plot for the linear model relating price to miles driven for Ford F-150s.

## HOW TO INTERPRET A RESIDUAL PLOT

To determine whether the regression model is appropriate, look at the residual plot.

- If there is no leftover curved pattern in the residual plot, the regression model is appropriate.
- If there is a leftover curved pattern in the residual plot, consider using a regression model with a different form.

## EXAMPLE

### Pricing diamonds
### Interpreting a residual plot

JGI/Getty Images

**PROBLEM:** Is a linear model appropriate to describe the relationship between the weight (in carats) and price (in dollars) of round, clear, internally flawless diamonds with excellent cuts? We calculated a least-squares regression line using $x =$ weight and $y =$ price and made the corresponding residual plot shown.[23] Use the residual plot to determine if the linear model is appropriate.

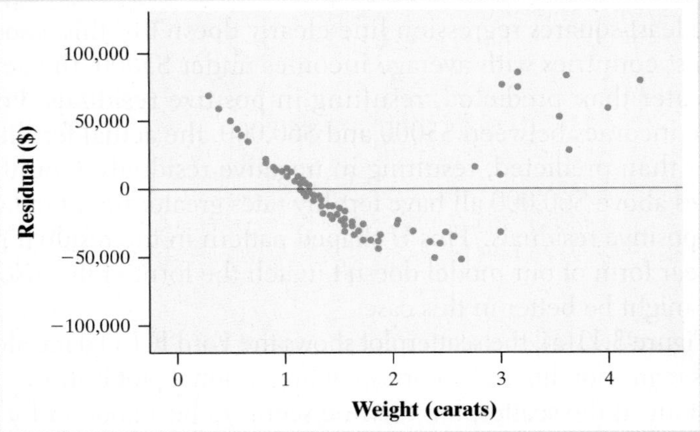


**SOLUTION:**

**The linear model relating price to carat weight is not appropriate because there is a U-shaped pattern left over in the residual plot.**

**FOR PRACTICE, TRY EXERCISE 47**

## Think About It

**WHY DO WE LOOK FOR PATTERNS IN RESIDUAL PLOTS?** The word *residual* comes from the Latin word *residuum*, meaning "left over." When we calculate a residual, we are calculating what is left over after subtracting the predicted value from the actual value:

$$\text{residual} = \text{actual } y - \text{predicted } y$$

Likewise, when we look at the form of a residual plot, we are looking at the form that is left over after subtracting the form of the model from the form of the association:

$$\text{form of residual plot} = \text{form of association} - \text{form of model}$$

When there is a leftover form in the residual plot, the form of the association and form of the model are not the same. However, if the form of the association and form of the model are the *same*, the residual plot should have no form, other than random scatter.

## 10. Technology Corner MAKING RESIDUAL PLOTS

*TI-Nspire and other technology instructions are on the book's website at* highschool.bfwpub.com/updatedtps6e.

Let's continue the analysis of the Ford F-150 miles driven and price data from Technology Corner 9 (page 184). You should have already made a scatterplot, calculated the equation of the least-squares regression line, and graphed the line on the scatterplot. Now, we want to calculate residuals and make a residual plot. Fortunately, your calculator has already done most of the work. Each time the calculator computes a regression line, it computes the residuals and stores them in a list named RESID.

1. Set up a scatterplot in the statistics plots menu.

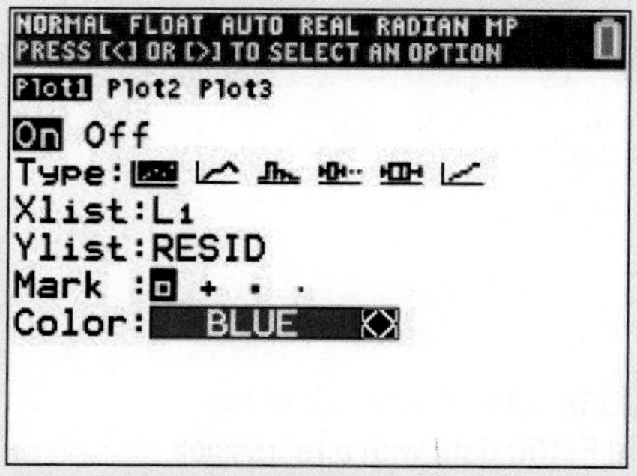


- Press 2nd Y= (STAT PLOT).
- Press ENTER or 1 to go into Plot1.
- Adjust the settings as shown. The RESID list is found in the List menu by pressing 2nd STAT. *Note:* You have to calculate the equation of the least-squares regression line using the calculator *before* making a residual plot. Otherwise, the RESID list will include the residuals from a different least-squares regression line.

2. Use ZoomStat to let the calculator choose an appropriate window.

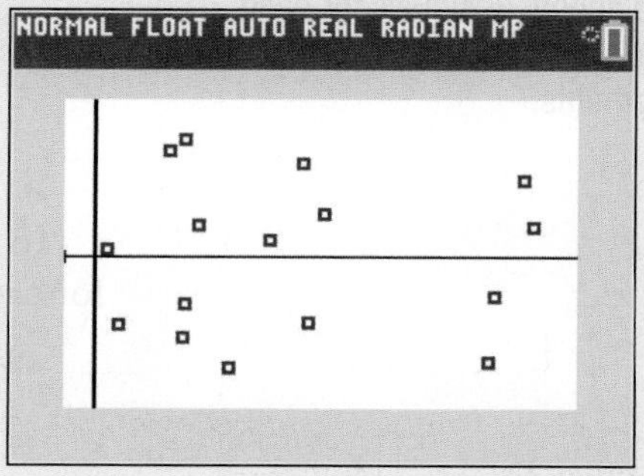


- Press ZOOM and choose 9: ZoomStat.

*Note:* If you want to see the values of the residuals, you can have the calculator put them in L3 (or any list). In the list editor, highlight the heading of L3, choose the RESID list from the LIST menu, and press ENTER.

## CHECK YOUR UNDERSTANDING

In Exercises 3 and 7, we asked you to make and describe a scatterplot for the hiker data shown in the table.

| | | | | | | | | |
|---|---|---|---|---|---|---|---|---|
| **Body weight (lb)** | 120 | 187 | 109 | 103 | 131 | 165 | 158 | 116 |
| **Backpack weight (lb)** | 26 | 30 | 26 | 24 | 29 | 35 | 31 | 28 |

1. Calculate the equation of the least-squares regression line.
2. Make a residual plot for the linear model in Question 1.
3. What does the residual plot indicate about the appropriateness of the linear model? Explain your answer.

# How Well the Line Fits the Data: The Role of $s$ and $r^2$ in Regression

We use a residual plot to determine if a least-squares regression line is an appropriate model for the relationship between two variables. Once we determine that a least-squares regression line is appropriate, it makes sense to ask a follow-up question: How well does the line work? That is, if we use the least-squares regression line to make predictions, how good will these predictions be?

**THE STANDARD DEVIATION OF THE RESIDUALS** We already know that a residual measures how far an actual $y$ value is from its corresponding predicted value $\hat{y}$. Earlier in this section, we calculated the residual for the Ford F-150 with 70,583 miles driven and price \$21,994. As shown in Figure 3.12, the residual was −\$4765, meaning that the actual price was \$4765 *less* than we predicted.

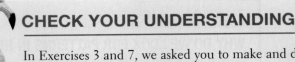

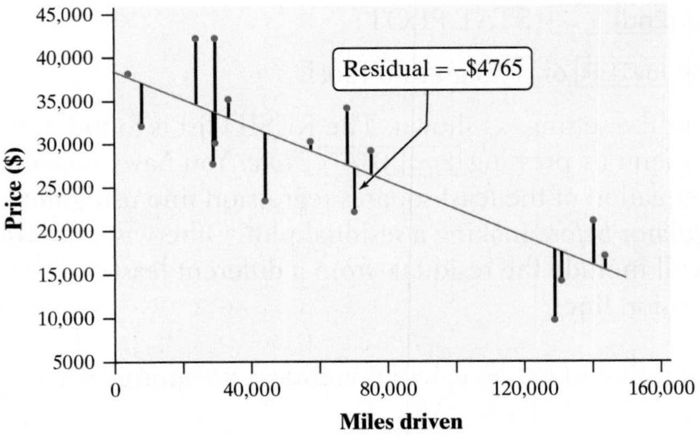


**FIGURE 3.12** Scatterplot of the Ford F-150 data with a regression line added. Residuals for each truck are shown with vertical line segments.

To assess how well the line fits *all* the data, we need to consider the residuals for each of the trucks, not just one. Here are the residuals for all 16 trucks:

| | | | | | | | |
|---|---|---|---|---|---|---|---|
| −4765 | −7664 | −3506 | 8617 | 7728 | 3057 | −5004 | 458 |
| 2289 | 1183 | −8121 | 6858 | 2110 | 5572 | −5973 | −2857 |

Using these residuals, we can estimate the "typical" prediction error when using the least-squares regression line. To do this, we calculate the **standard deviation of the residuals** $s$.

**DEFINITION Standard deviation of residuals $s$**

The **standard deviation of the residuals $s$** measures the size of a typical residual. That is, $s$ measures the typical distance between the actual $y$ values and the predicted $y$ values.

To calculate $s$, use the following formula:

$$s = \sqrt{\frac{\text{sum of squared residuals}}{n-2}} = \sqrt{\frac{\sum(y_i - \hat{y}_i)^2}{n-2}}$$

For the Ford F-150 data, the standard deviation of the residuals is

$$s = \sqrt{\frac{(-4765)^2 + (-7664)^2 + \cdots + (-2857)^2}{16-2}} = \sqrt{\frac{461{,}264{,}136}{14}} = \$5740$$

*Interpretation*: The actual price of a Ford F-150 is typically about \$5740 away from the price predicted by the least-squares regression line with $x$ = miles driven. If we look at the residual plot in Figure 3.11, this seems like a reasonable value. Although some of the residuals are close to 0, others are close to \$10,000 or −\$10,000.

**Think About It**

**DOES THE FORMULA FOR $s$ LOOK SLIGHTLY FAMILIAR?** It should. In Chapter 1, we defined the standard deviation of a set of quantitative data as

$$s_x = \sqrt{\frac{\sum(x_i - \bar{x})^2}{n-1}}$$

We interpreted the resulting value as the "typical" distance of the data points from the mean. In the case of two-variable data, we're interested in the typical (vertical) distance of the data points from the regression line. We find this value in much the same way: first add up the squared deviations, then average them (again, in a funny way), and take the square root to get back to the original units of measurement.

**THE COEFFICIENT OF DETERMINATION $r^2$** There is another numerical quantity that tells us how well the least-squares line predicts values of the response variable $y$. It is $r^2$, the **coefficient of determination**. Some computer packages call it "R-sq." You may have noticed this value in some of the output that we showed earlier. Although it's true that $r^2$ is equal to the square of the correlation $r$, there is much more to this story.

Some people interpret $r^2$ as the proportion of variation in the response variable that is explained by the explanatory variable in the model.

**DEFINITION The coefficient of determination $r^2$**

The **coefficient of determination $r^2$** measures the percent reduction in the sum of squared residuals when using the least-squares regression line to make predictions, rather than the mean value of $y$. In other words, $r^2$ measures the percent of the variability in the response variable that is accounted for by the least-squares regression line.

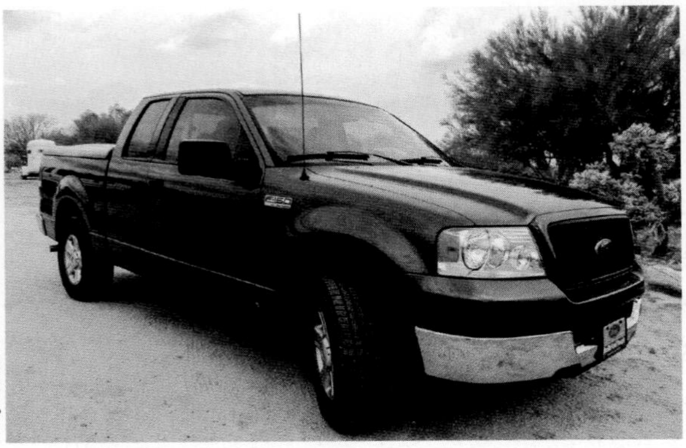

Holly Albrecht

Suppose we wanted to predict the price of a particular used Ford F-150, but we didn't know how many miles it had been driven. Our best guess would be the average cost of a used Ford F-150, $\bar{y} = \$27,834$. Of course, this prediction is unlikely to be very good, as the prices vary quite a bit from the mean ($s_y = \$9570$). If we knew how many miles the truck had been driven, we could use the least-squares regression line to make a better prediction. How much better are predictions that use the least-squares regression line with $x$ = miles driven, rather than predictions that use only the average price? The answer is $r^2$.

The scatterplot in Figure 3.13(a) shows the squared residuals along with the sum of squared residuals (approximately 1,374,000,000) when using the average price as the predicted value. The scatterplot in Figure 3.13(b) shows the squared residuals along with the sum of squared residuals (approximately 461,300,000) when using the least-squares regression line with $x$ = miles driven to predict the price. Notice that the squares in part (b) are quite a bit smaller.

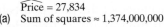

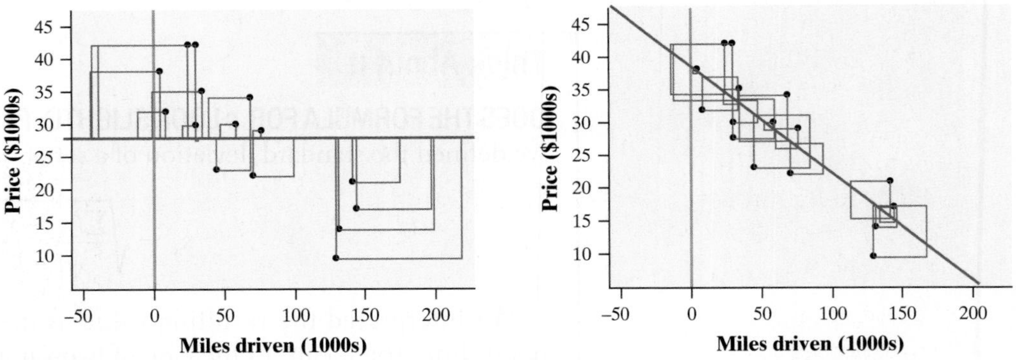


(a) $\widehat{\text{Price}} = 27{,}834$
Sum of squares ≈ 1,374,000,000

(b) $\widehat{\text{Price}} = 38{,}257 - 0.1629$ Miles driven; $r^2 = 0.66$
Sum of squares ≈ 461,300,000

**FIGURE 3.13** (a) The sum of squared residuals is about 1,374,000,000 if we use the mean price as our prediction for all 16 trucks. (b) The sum of squared residuals from the least-squares regression line is about 461,300,000.

To find $r^2$, calculate the percent reduction in the sum of squared residuals:

$$r^2 = \frac{1{,}374{,}000{,}000 - 461{,}300{,}000}{1{,}374{,}000{,}000} = \frac{912{,}700{,}000}{1{,}374{,}000{,}000} = 0.66$$

The sum of squared residuals has been reduced by 66%.

*Interpretation*: About 66% of the variability in the price of a Ford F-150 is accounted for by the least-squares regression line with $x =$ miles driven. The remaining 34% is due to other factors, including age, color, and condition.

If all the points fall directly on the least-squares line, the sum of squared residuals is 0 and $r^2 = 1$. Then all the variation in $y$ is accounted for by the linear relationship with $x$. In the worst-case scenario, the least-squares line does no better at predicting $y$ than $y = \bar{y}$ does. Then the two sums of squared residuals are the same and $r^2 = 0$.

It's fairly remarkable that the coefficient of determination $r^2$ is actually the square of the correlation. This fact provides an important connection between correlation and regression. When you see a linear association, square the correlation to get a better feel for how well the least-squares line fits the data.

### Think About It

**WHAT'S THE RELATIONSHIP BETWEEN *s* AND $r^2$?** Both s and $r^2$ are calculated from the sum of squared residuals. They also both measure how well the line fits the data. The standard deviation of the residuals reports the size of a typical prediction error, in the same units as the response variable. In the truck example, $s = \$5740$. The value of $r^2$, however, does not have units and is usually expressed as a percentage between 0% and 100%, such as $r^2 = 66\%$. Because these values assess how well the line fits the data in different ways, we recommend you follow the example of most statistical software and report both.

Knowing how to interpret s and $r^2$ is much more important than knowing how to calculate them. Consequently, we typically let technology do the calculations.

## EXAMPLE

### Grabbing candy, again
### Interpreting *s* and $r^2$

**PROBLEM:** The scatterplot shows the hand span (in centimeters) and number of Starburst™ candies grabbed by each student in Mr. Tyson's class, along with the regression line $\hat{y} = -29.8 + 2.83x$. For this model, technology gives $s = 4.03$ and $r^2 = 0.697$.

(a) Interpret the value of s.

(b) Interpret the value of $r^2$.

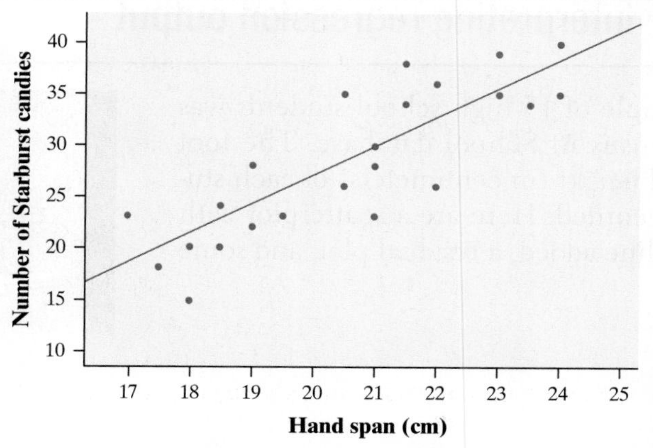

**SOLUTION:**

(a) The actual number of Starburst™ candies grabbed is typically about 4.03 away from the number predicted by the least-squares regression line with x = hand span.

(b) About 69.7% of the variability in number of Starburst candies grabbed is accounted for by the least-squares regression line with x = hand span.

**FOR PRACTICE, TRY EXERCISE 55**

# Interpreting Computer Regression Output

Figure 3.14 displays the basic regression output for the Ford F-150 data from two statistical software packages: Minitab and JMP. Other software produces very similar output. Each output records the slope and *y* intercept of the least-squares line. The software also provides information that we don't yet need, although we will use much of it later. Be sure that you can locate the slope, the *y* intercept, and the values of *s* (called *root mean square error* in JMP) and $r^2$ on both computer outputs. *Once you understand the statistical ideas, you can read and work with almost any software output.*

**Minitab**

Slope — y intercept

```
Predictor        Coef     SE Coef      T       P
Constant        38257        2446  15.64   0.000
Miles Driven  -0.16292    0.03096  -5.26   0.000

S = 5740.13     R-Sq = 66.4%    R-Sq(adj) = 64.0%
```

$r^2$

Standard deviation of the residuals

**JMP**

```
Summary of Fit
RSquare                          0.664248
RSquare Adj                      0.640266
Root Mean Square Error           5740.131
Mean of Response                 27833.69
Observations (or Sum Wgts)             16

Parameter Estimates
Term            Estimate    Std Error   t Ratio   Prob>|t|
Intercept     38257.135     2445.813     15.64     <.0001
Miles Driven  -0.162919     0.030956     -5.26     0.0001
```

$r^2$

Standard deviation of the residuals

y intercept — Slope

**FIGURE 3.14** Least-squares regression results for the Ford F-150 data from Minitab and JMP statistical software. Other software produces similar output.

## EXAMPLE

### Using feet to predict height
### Interpreting regression output

**PROBLEM:** A random sample of 15 high school students was selected from the U.S. Census At School database. The foot length (in centimeters) and height (in centimeters) of each student in the sample were recorded. Here are a scatterplot with the least-squares regression line added, a residual plot, and some computer output:

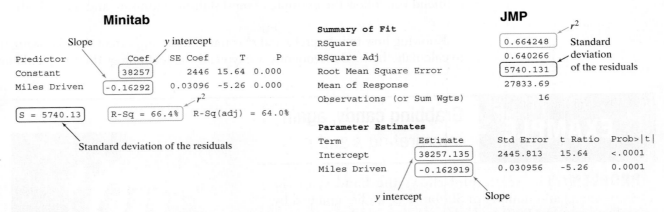

© Fancy/Alamy

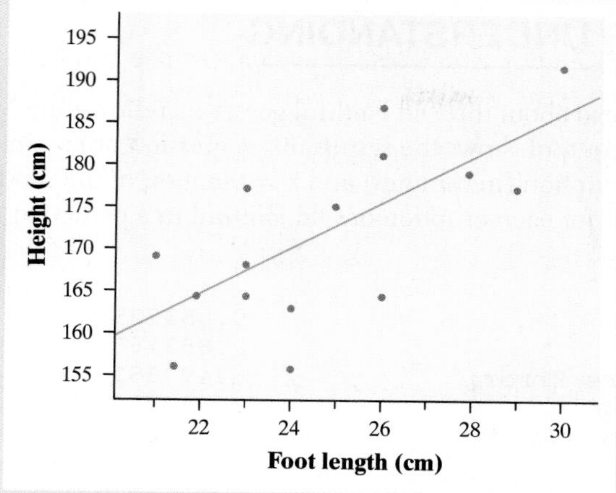


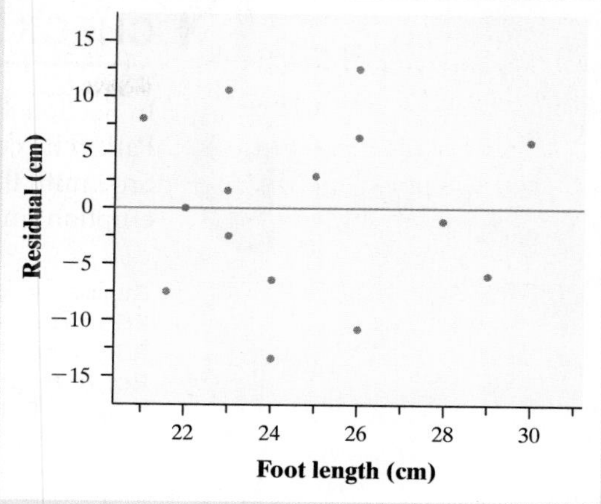


| Predictor | Coef | SE Coef | T | P |
|---|---|---|---|---|
| Constant | 103.4100 | 19.5000 | 5.30 | 0.000 |
| Foot length | 2.7469 | 0.7833 | 3.51 | 0.004 |
| S = 7.95126 | R-Sq = 48.6% | | R-Sq(adj) = 44.7% | |

(a) Is a line an appropriate model to use for these data? Explain how you know.
(b) Find the correlation.
(c) What is the equation of the least-squares regression line that models the relationship between foot length and height? Define any variables that you use.
(d) By about how much do the actual heights typically vary from the values predicted by the least-squares regression line with $x$ = foot length?

## SOLUTION:

(a) Because the scatterplot shows a linear association and the residual plot has no obvious leftover curved patterns, a line is an appropriate model to use for these data.

(b) $r = \sqrt{0.486} = 0.697$

The correlation $r$ is the square root of $r^2$, where $r^2$ is a value between 0 and 1. Because the square root function on your calculator will always give a positive result, make sure to consider whether the correlation is positive or negative. If the slope is negative, so is the correlation.

(c) $\widehat{\text{height}} = 103.41 + 2.7469\,(\text{foot length})$

We could also write the equation as $\hat{y} = 103.41 + 2.7469x$, where $\hat{y}$ = predicted height (cm) and $x$ = foot length (cm).

(d) $s = 7.95$, so the actual heights typically vary by about 7.95 cm from the values predicted by the regression line with $x$ = foot length.

**FOR PRACTICE, TRY EXERCISE 59**

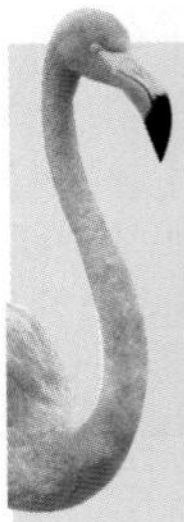

## CHECK YOUR UNDERSTANDING

In Section 3.1, you read about the Old Faithful geyser in Yellowstone National Park. The computer output shows the results of a regression of $y$ = interval of time until the next eruption (in minutes) and $x$ = duration of the most recent eruption (in minutes) for each eruption of Old Faithful in a particular month.

```
Summary of Fit
RSquare                              0.853725
RSquare Adj                          0.853165
Root Mean Square Error               6.493357
Mean of Response                    77.543730
Observations (or Sum Wgts)         263.000000

Parameter Estimates
Term          Estimate    Std Error    t Ratio    Prob>|t|
Intercept    33.347442     1.201081      27.76     <.0001*
Duration     13.285406     0.340393      39.03     <.0001*
```

1. What is the equation of the least-squares regression line that models the relationship between interval and duration? Define any variables that you use.
2. Interpret the slope of the least-squares regression line.
3. Identify and interpret the standard deviation of the residuals.
4. What percent of the variability in interval is accounted for by the least-squares regression line with $x$ = duration?

# Regression to the Mean

Using technology is often the most convenient way to find the equation of a least-squares regression line. It is also possible to calculate the equation of the least-squares regression line using only the means and standard deviations of the two variables and their correlation. Exploring this method will highlight an important relationship between the correlation and the slope of a least-squares regression line—and reveal why we include the word *regression* in the expression *least-squares regression line.*

### HOW TO CALCULATE THE LEAST-SQUARES REGRESSION LINE USING SUMMARY STATISTICS

We have data on an explanatory variable $x$ and a response variable $y$ for $n$ individuals and want to calculate the least-squares regression line $\hat{y} = a + bx$. From the data, calculate the means $\bar{x}$ and $\bar{y}$ and the standard deviations $s_x$ and $s_y$ of the two variables and their correlation $r$. The **slope** is:

$$b = r\frac{s_y}{s_x}$$

Because the least-squares regression line passes through the point $(\bar{x}, \bar{y})$, the **$y$ intercept** is:

$$a = \bar{y} - b\bar{x}$$

The formula for the slope reminds us that the distinction between explanatory and response variables is important in regression. Least-squares regression makes the distances of the data points from the line small only in the $y$ direction. If we reverse the roles of the two variables, the values of $s_x$ and $s_y$ will reverse in the slope formula, resulting in a different least-squares regression line. This is *not* true for correlation: switching $x$ and $y$ does *not* affect the value of $r$.

The formula for the $y$ intercept comes from the fact that the least-squares regression line always passes through the point $(\bar{x}, \bar{y})$. Once we know the slope ($b$) and that the line goes through the point $(\bar{x}, \bar{y})$, we can use algebra to solve for the $y$ intercept. Substituting $(\bar{x}, \bar{y})$ into the equation $\hat{y} = a + bx$ produces the equation $\bar{y} = a + b\bar{x}$. Solving this equation for $a$ gives the equation shown in the definition box, $a = \bar{y} - b\bar{x}$. To see how these formulas work in practice, let's look at an example.

## EXAMPLE

### More about feet and height
### Calculating the least-squares regression line

**PROBLEM:** In the preceding example, we used data from a random sample of 15 high school students to investigate the relationship between foot length (in centimeters) and height (in centimeters). The mean and standard deviation of the foot lengths are $\bar{x} = 24.76$ and $s_x = 2.71$. The mean and standard deviation of the heights are $\bar{y} = 171.43$ and $s_y = 10.69$. The correlation between foot length and height is $r = 0.697$. Find the equation of the least-squares regression line for predicting height from foot length.

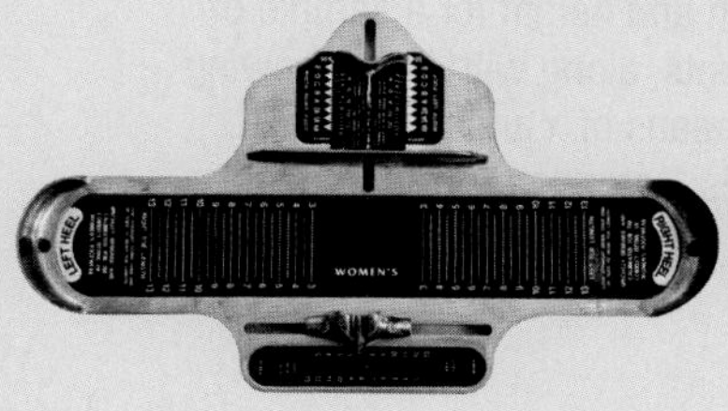


panpote/Getty Images

**SOLUTION:**

$$b = 0.697\frac{10.69}{2.71} = 2.75$$

$$b = r\frac{s_y}{s_x}$$
$$a = \bar{y} - b\bar{x}$$

$$a = 171.43 - 2.75(24.76) = 103.34$$

The equation of the least-squares regression line is $\hat{y} = 103.34 + 2.75x$.

**FOR PRACTICE, TRY EXERCISE 63**

There is a close connection between the correlation and the slope of the least-squares regression line. The slope is

$$b = r\frac{s_y}{s_x} = \frac{r \cdot s_y}{s_x}$$

This equation says that along the regression line, a change of 1 standard deviation in $x$ corresponds to a change of $r$ standard deviations in $y$. When the variables are perfectly correlated ($r = 1$ or $r = -1$), the change in the predicted response $\hat{y}$ is the same (in standard deviation units) as the change in $x$. For example, if $r = 1$ and $x$ is 2 standard deviations above $\bar{x}$, then the corresponding value of $\hat{y}$ will be 2 standard deviations above $\bar{y}$.

However, if the variables are not perfectly correlated ($-1 < r < 1$), the change in $\hat{y}$ is *less than* the change in $x$, when measured in standard deviation units. To illustrate this property, let's return to the foot length and height data from the preceding example.

Figure 3.15 shows the scatterplot of height versus foot length and the regression line $\hat{y} = 103.34 + 2.75x$. We have added four more lines to the graph: a vertical line at the mean foot length $\bar{x}$, a vertical line at $\bar{x} + s_x$ (1 standard deviation above the mean foot length), a horizontal line at the mean height $\bar{y}$, and a horizontal line at $\bar{y} + s_y$ (1 standard deviation above the mean height).

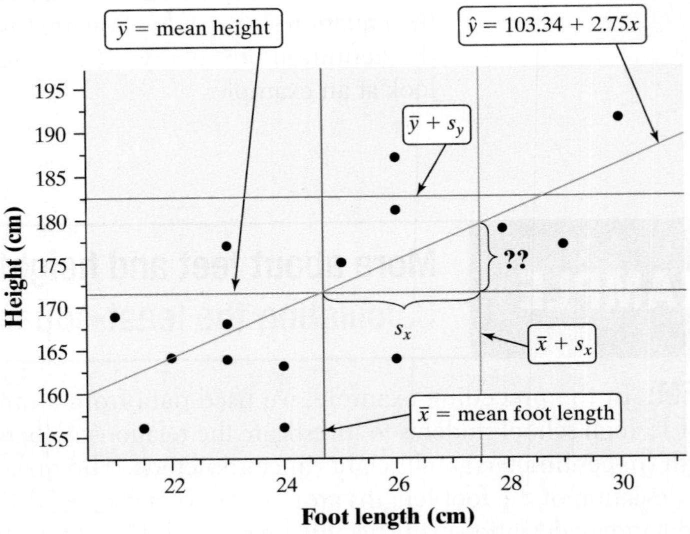


**FIGURE 3.15** Scatterplot showing the relationship between foot length and height for a sample of students, along with lines showing the means of $x$ and $y$ and the values 1 standard deviation above each mean.

When a student's foot length is 1 standard deviation above the mean foot length $\bar{x}$, the predicted height $\hat{y}$ is above the mean height $\bar{y}$—but not an entire standard deviation above the mean. How far above the mean is the value of $\hat{y}$?

From the graph, we can see that

$$b = \text{slope} = \frac{\text{change in } y}{\text{change in } x} = \frac{??}{s_x}$$

From earlier, we know that

$$b = \frac{r \cdot s_y}{s_x}$$

Setting these two equations equal to each other, we have

$$\frac{??}{s_x} = \frac{r \cdot s_y}{s_x}$$

Thus, $\hat{y}$ must be $r \cdot s_y$ above the mean $\bar{y}$.

In other words, for an increase of 1 standard deviation in the value of the explanatory variable $x$, the least-squares regression line predicts an increase of *only* $r$ standard deviations in the response variable $y$. When the correlation isn't $r = 1$ or $r = -1$, the predicted value of $y$ is closer to its mean $\bar{y}$ than the value of $x$ is to its mean $\bar{x}$. *This is called regression to the mean, because the values of y "regress" to their mean.*

Sir Francis Galton (1822–1911) is often credited with discovering the idea of regression to the mean. He looked at data on the heights of children versus the heights of their parents. He found that taller-than-average parents tended to have children who were taller than average but not quite as tall as their parents. Likewise, shorter-than-average parents tended to have children who were shorter than average but not quite as short as their parents. Galton used the symbol $r$ for the correlation because of its important relationship to regression.

# Correlation and Regression Wisdom

Correlation and regression are powerful tools for describing the relationship between two variables. When you use these tools, you should be aware of their limitations.

**CORRELATION AND REGRESSION LINES DESCRIBE ONLY LINEAR RELATIONSHIPS** You can calculate the correlation and the least-squares line for any relationship between two quantitative variables, but the results are useful only if the scatterplot shows a linear pattern. *Always plot your data first!*

The following four scatterplots show very different relationships. Which one do you think shows the greatest correlation?

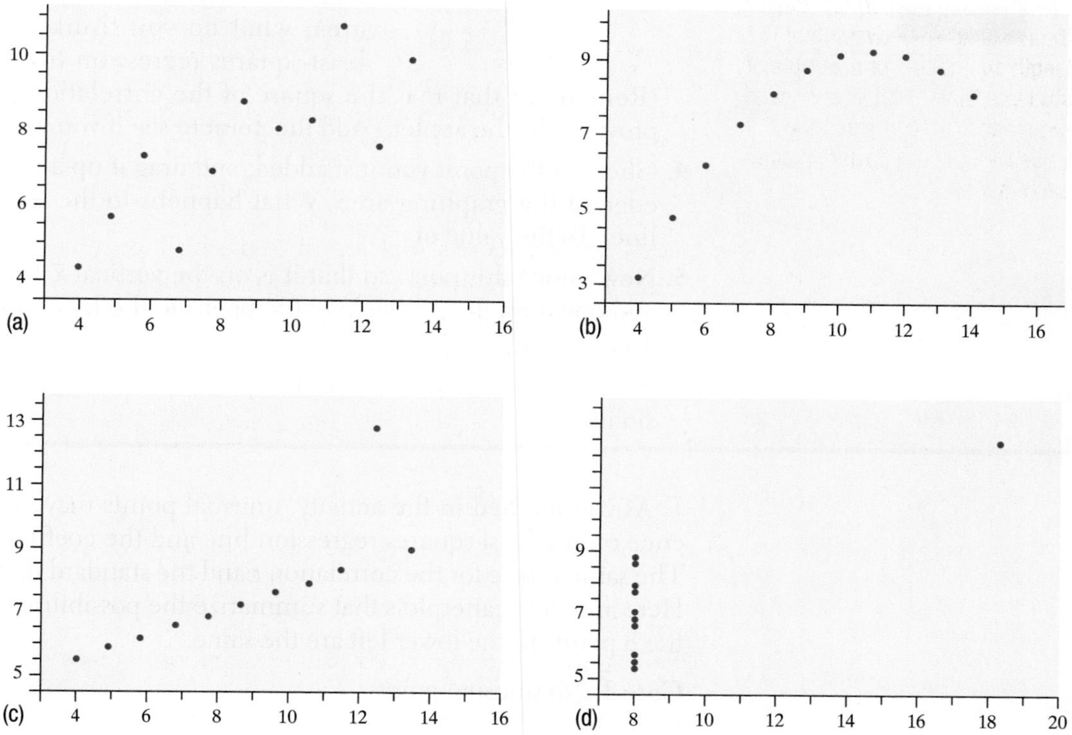


*Answer:* All four have the same correlation, $r = 0.816$. Furthermore, the least-squares regression line for each relationship is exactly the same, $\hat{y} = 3 + 0.5x$. These four data sets, developed by statistician Frank Anscombe, illustrate the importance of graphing data before doing calculations.[24]

**CORRELATION AND LEAST-SQUARES REGRESSION LINES ARE NOT RESISTANT** You already know that the correlation $r$ is not resistant. One unusual point in a scatterplot can greatly change the value of $r$. Is the least-squares line resistant? The following activity will help you answer this question.

**Investigating properties of the least-squares regression line**

In this activity, you will use the *Correlation and Regression* applet to explore some properties of the least-squares regression line.

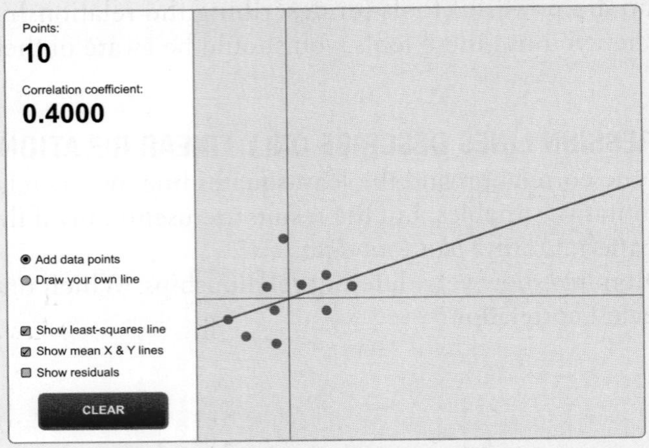

1. Launch the applet at highschool.bfwpub.com/updatedtps6e.

2. Click on the graphing area to add 10 points in the lower-left corner so that the correlation is about $r = 0.40$. Also, check the boxes to show the "Least-Squares Line" and the "Mean X & Y" lines as in the screen shot. Notice that the least-squares regression line goes though the point $(\bar{x}, \bar{y})$.

3. If you were to add a point on the least-squares regression line at the right edge of the graphing area, what do you think would happen to the least-squares regression line? To the value of $r^2$? (Remember that $r^2$ is the square of the correlation coefficient, which is provided by the applet.) Add the point to see if you were correct.

4. Click on the point you just added, and drag it up and down along the right edge of the graphing area. What happens to the least-squares regression line? To the value of $r^2$?

5. Now, move this point so that it is on the vertical $\bar{x}$ line. Drag the point up and down on the $\bar{x}$ line. What happens to the least-squares regression line? To the value of $r^2$?

6. Briefly summarize how unusual points influence the least-squares regression line.

As you learned in the activity, unusual points may or may not have an influence on the least-squares regression line and the coefficient of determination $r^2$. The same is true for the correlation $r$ and the standard deviation of the residuals $s$. Here are four scatterplots that summarize the possibilities. In all four scatterplots, the 8 points in the lower left are the same.

**Case 1:** No unusual points

$\hat{y} = 3.04 + 0.65x$
$r = 0.41$
$r^2 = 0.17$
$s = 2.77$

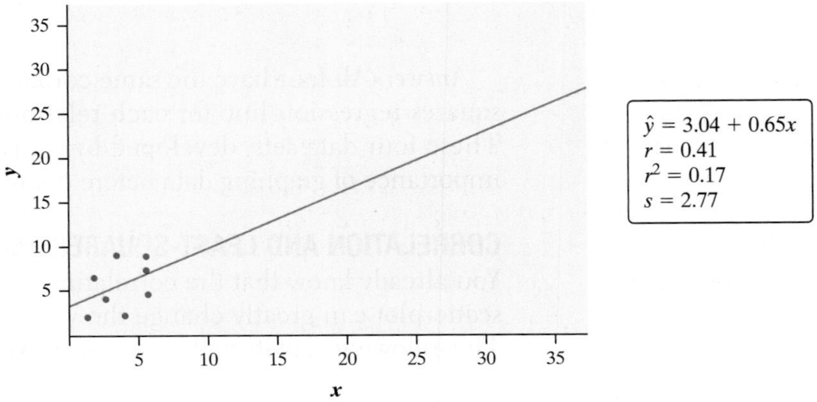

**Case 2:** A point that is far from the other points in the *x* direction, but in the same pattern.

$\hat{y} = 3.04 + 0.65x$
$r = 0.94$
$r^2 = 0.89$
$s = 2.56$

Compared to Case 1, the equation of the least-squares regression line remained the same, but the values of *r* and $r^2$ greatly increased. The standard deviation of the residuals got a bit smaller because the additional point has a very small residual.

**Case 3:** A point that is far from the other points in the *x* direction, and not in the same pattern.

$\hat{y} = 5.80 - 0.09x$
$r = -0.34$
$r^2 = 0.18$
$s = 2.89$

Compared to Case 1, the equation of the least-squares regression line is much different, with the slope going from positive to negative and the *y* intercept increasing. The value of *r* is now negative while the value of $r^2$ stayed almost the same. Even though the new point has a relatively small residual, the standard deviation of the residuals got a bit larger because the line doesn't fit the remaining points nearly as well.

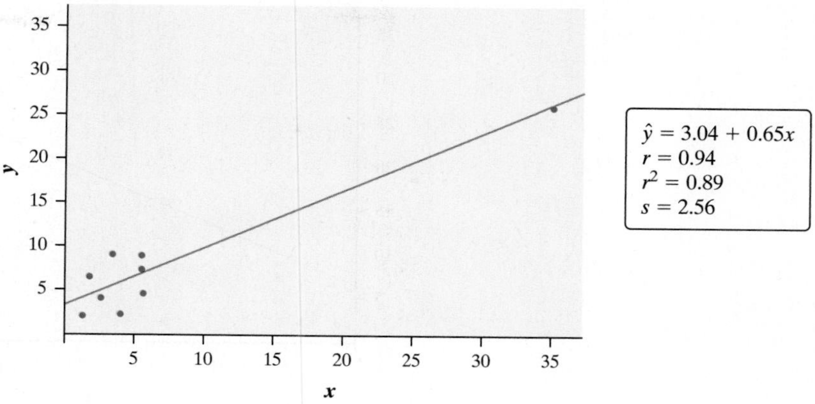

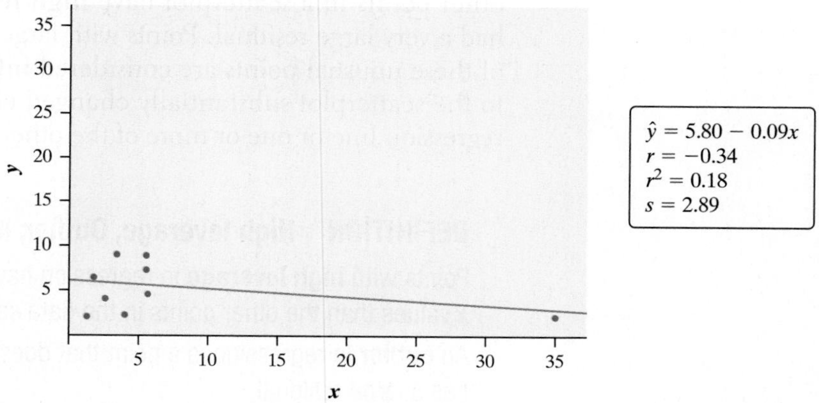

**Case 4:** A point that is far from the other points in the $y$ direction, and not in the same pattern.

$$\hat{y} = 6.31 + 0.65x$$
$$r = 0.11$$
$$r^2 = 0.01$$
$$s = 10.82$$

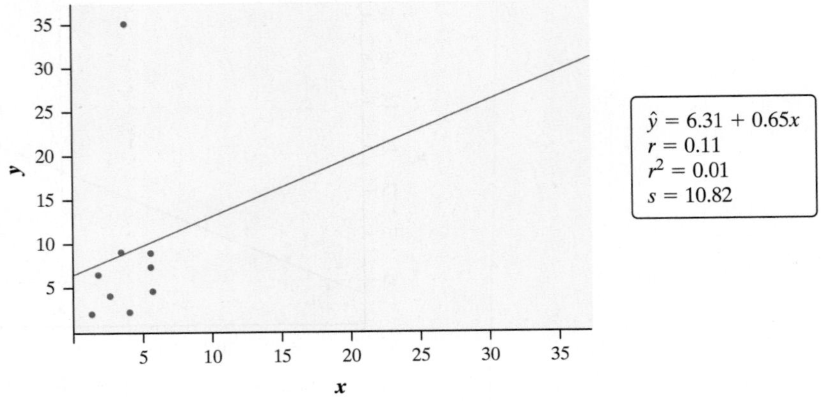

Compared to Case 1, the slope of the least-squares regression line is the same, but the $y$ intercept is a little larger as the line appears to have shifted up slightly. The values of $r$ and $r^2$ are much smaller than before. Because the new point has such a large residual, the standard deviation of the residuals is much larger.

In Cases 2 and 3, the unusual point had a much bigger $x$ value than the other points. Points whose $x$ values are much smaller or much larger than the other points in a scatterplot have **high leverage**. In Case 4, the unusual point had a very large residual. Points with large residuals are called **outliers**. All three of these unusual points are considered **influential points** because adding them to the scatterplot substantially changed either the equation of the least-squares regression line or one or more of the other summary statistics $(r, r^2, s)$.

---

**DEFINITION   High leverage, Outlier, Influential point**

Points with **high leverage** in regression have much larger or much smaller $x$ values than the other points in the data set.

An **outlier** in regression is a point that does not follow the pattern of the data and has a large residual.

An **influential point** in regression is any point that, if removed, substantially changes the slope, $y$ intercept, correlation, coefficient of determination, or standard deviation of the residuals.

---

Outliers and high-leverage points are often influential in regression calculations! The best way to investigate the influence of such points is to do regression calculations with and without them to see how much the results differ. Here is an example that shows what we mean.

Does the age at which a child begins to talk predict a later score on a test of mental ability? A study of the development of young children recorded the age in months at which each of 21 children spoke their first word and their Gesell Adaptive Score, the result of an aptitude test taken much later.[25] A scatterplot of the data appears in Figure 3.16, along with a residual plot, and computer output. Two points, child 18 and child 19, are labeled on each plot.

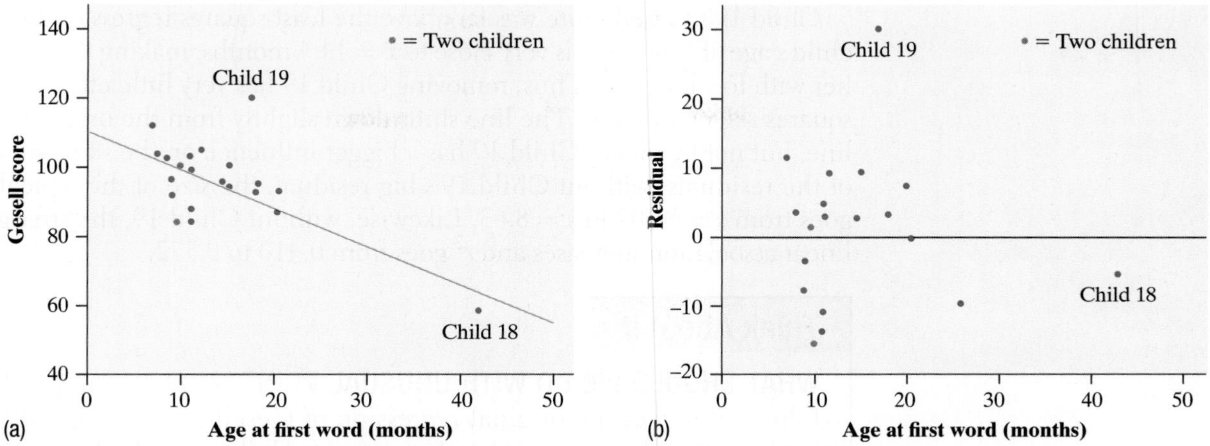

**FIGURE 3.16** (a) Scatterplot of Gesell Adaptive Scores versus the age at first word for 21 children, along with the least-squares regression line. (b) Residual plot for the linear model. The point for Child 18 has high leverage and the point for Child 19 is an outlier. Each purple point in the graphs stands for two individuals.

The point for Child 18 has high leverage because its $x$ value is much larger than the $x$ values of other points. The point for Child 19 is an outlier because it falls outside the pattern of the other points and has a very large residual. How do these two points affect the regression? Figure 3.17 shows the results of removing each of these points on the equation of the least-squares regression line, the standard deviation of the residuals, and $r^2$.

With all 19 points:
$\hat{y} = 109.9 - 1.127x$
$s = 11.02$ and $r^2 = 0.410$

Without Child 18:
$\hat{y} = 105.6 - 0.779x$
$s = 11.11$ and $r^2 = 0.112$

Without Child 19:
$\hat{y} = 109.3 - 1.193x$
$s = 8.63$ and $r^2 = 0.572$

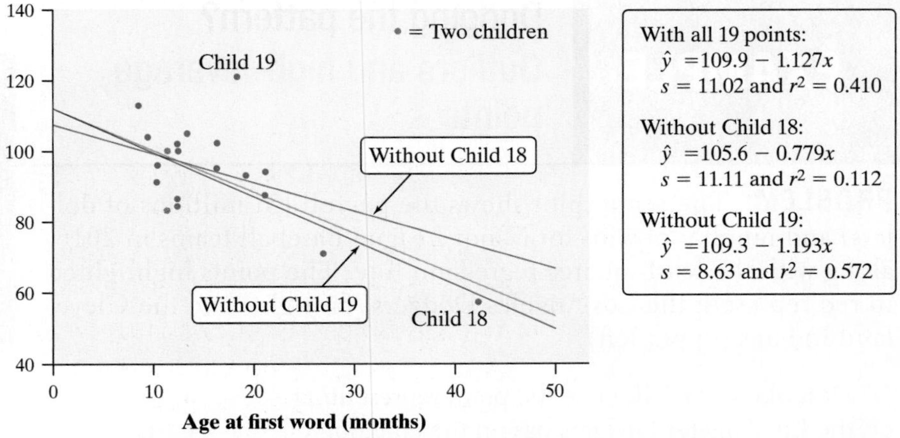

**FIGURE 3.17** Three least-squares regression lines of Gesell score on age at first word. The green line is calculated from all the data. The dark blue line is calculated leaving out only Child 18. The red line is calculated leaving out only Child 19.

You can see that removing the point for Child 18 moves the line quite a bit. Because of Child 18's extreme position on the age ($x$) scale, removing this high-leverage point makes the slope closer to 0 and the $y$ intercept smaller. Removing Child 18 also increases the standard deviation of the residuals because its small residual was making the typical distance from the regression line smaller. Finally, removing Child 18 also decreases $r^2$ (and makes the correlation closer to 0) because the linear association is weaker without this point.

Child 19's Gesell score was far above the least-squares regression line, but this child's age (17 months) is very close to $\bar{x} = 14.4$ months, making this point an outlier with low leverage. Thus, removing Child 19 has very little effect on the least-squares regression line. The line shifts down slightly from the original regression line, but not by much. Child 19 has a bigger influence on the standard deviation of the residuals: without Child 19's big residual, the size of the typical residual goes from $s = 11.02$ to $s = 8.63$. Likewise, without Child 19, the strength of the linear association increases and $r^2$ goes from 0.410 to 0.572.

### Think About It

**WHAT SHOULD WE DO WITH UNUSUAL POINTS?** The strong influence of Child 18 makes the original regression of Gesell score on age at first word misleading. The original data have $r^2 = 0.41$. That is, the least-squares line with $x =$ age at which a child begins to talk accounts for 41% of the variability in Gesell score. This relationship is strong enough to be interesting to parents. If we leave out Child 18, $r^2$ drops to only 11%. The apparent strength of the association was largely due to a single influential observation.

What should the child development researcher do? She must decide whether Child 18 is so slow to speak that this individual should not be allowed to influence the analysis. If she excludes Child 18, much of the evidence for a connection between the age at which a child begins to talk and later ability score vanishes. If she keeps Child 18, she needs data on other children who were also slow to begin talking, so the analysis no longer depends as heavily on just one child.

## EXAMPLE

## Dodging the pattern?

### Outliers and high-leverage points

Robert J Daveant/Shutterstock.com

**PROBLEM:** The scatterplot shows the payroll (in millions of dollars) and number of wins for Major League Baseball teams in 2016, along with the least-squares regression line. The points highlighted in red represent the Los Angeles Dodgers (far right) and the Cleveland Indians (upper left).

(a) Describe what influence the point representing the Los Angeles Dodgers has on the equation of the least-squares regression line. Explain your reasoning.

(b) Describe what influence the point representing the Cleveland Indians has on the standard deviation of the residuals and $r^2$. Explain your reasoning.

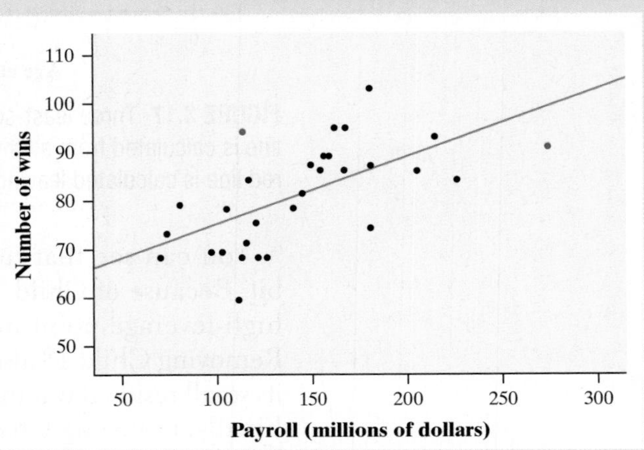

**SOLUTION:**

(a) Because the point for the Los Angeles Dodgers is on the right and below the least-squares regression line, it is making the slope of the line closer to 0 and the *y* intercept greater. If the Dodgers' point was removed, the line would be steeper.

> The point for the Dodgers has high leverage because its *x* value is much larger than the others.

(b) Because the point for the Cleveland Indians has a large residual, it is making the standard deviation of the residuals greater and the value of $r^2$ smaller.

> The point for the Indians is an outlier because it has a large residual.

**FOR PRACTICE, TRY EXERCISE 67**

**ASSOCIATION DOES NOT IMPLY CAUSATION** When we study the relationship between two variables, we often hope to show that changes in the explanatory variable *cause* changes in the response variable. **A strong association between two variables is not enough to draw conclusions about cause and effect.** Sometimes an observed association really does reflect cause and effect. A household that heats with natural gas uses more gas in colder months because cold weather requires burning more gas to stay warm. In other cases, an association is explained by other variables, and the conclusion that $x$ causes $y$ is not valid.

> Remember: It only makes sense to talk about the *correlation* between two *quantitative* variables. If one or both variables are categorical, you should refer to the *association* between the two variables. To be safe, use the more general term *association* when describing the relationship between any two variables.

A study once found that people with two cars live longer than people who own only one car.[26] Owning three cars is even better, and so on. There is a substantial positive association between number of cars $x$ and length of life $y$. Can we lengthen our lives by buying more cars? No. The study used number of cars as a quick indicator of wealth. Well-off people tend to have more cars. They also tend to live longer, probably because they are better educated, take better care of themselves, and get better medical care. The cars have nothing to do with it. There is no cause-and-effect link between number of cars and length of life.

## Section 3.2 Summary

- A **regression line** models how a response variable $y$ changes as an explanatory variable $x$ changes. You can use a regression line to **predict** the value of $y$ for any value of $x$ by substituting this $x$ value into the equation of the line.
- The **slope** $b$ of a regression line $\hat{y} = a + bx$ describes how the predicted value of $y$ changes for each increase of 1 unit in $x$.
- The **$y$ intercept** $a$ of a regression line $\hat{y} = a + bx$ is the predicted value of $y$ when the explanatory variable $x$ equals 0. This prediction does not have a logical interpretation unless $x$ can actually take values near 0.
- Avoid **extrapolation**, using a regression line to make predictions using values of the explanatory variable outside the values of the data from which the line was calculated.
- The most common method of fitting a line to a scatterplot is least squares. The **least-squares regression line** is the line that minimizes the sum of the squares of the vertical distances of the observed points from the line.

- You can examine the fit of a regression line by studying the **residuals,** which are the differences between the actual values of $y$ and predicted values of $y$: Residual $= y - \hat{y}$. Be on the lookout for curved patterns in the **residual plot,** which indicate that a linear model may not be appropriate.
- The **standard deviation of the residuals** $s$ measures the typical size of a residual when using the regression line.
- The **coefficient of determination** $r^2$ is the percent of the variation in the response variable that is accounted for by the least-squares regression line using a particular explanatory variable.
- The least-squares regression line of $y$ on $x$ is the line with slope $b = r\frac{s_y}{s_x}$ and intercept $a = \bar{y} - b\bar{x}$. This line always passes through the point $(\bar{x}, \bar{y})$.
- **Influential points** can greatly affect correlation and regression calculations. Points with $x$ values far from $\bar{x}$ have **high leverage** and can be very influential. Points with large residuals are called **outliers** and can also affect correlation and regression calculations.
- Most of all, be careful not to conclude that there is a cause-and-effect relationship between two variables just because they are strongly associated.

### 3.2 Technology Corners

*TI-Nspire and other technology instructions are on the book's website at highschool.bfwpub.com/updatedtps6e.*



## Section 3.2 Exercises

37. **Predicting wins** Earlier we investigated the relationship between $x$ = payroll (in millions of dollars) and $y$ = number of wins for Major League Baseball teams in 2016. Here is a scatterplot of the data, along with the regression line $\hat{y} = 60.7 + 0.139x$:

pg 178

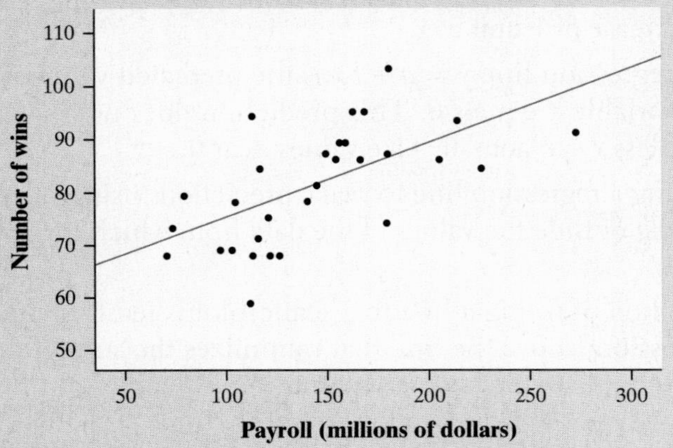


(a) Predict the number of wins for a team that spends \$200 million on payroll.

(b) Predict the number of wins for a team that spends \$400 million on payroll.

(c) How confident are you in each of these predictions? Explain your reasoning.

38. **How much gas?** Joan is concerned about the amount of energy she uses to heat her home. The scatterplot (on page 205) shows the relationship between $x$ = mean temperature in a particular month and $y$ = mean amount of natural gas used per day (in cubic feet) in that month, along with the regression line $\hat{y} = 1425 - 19.87x$.

(a) Predict the mean amount of natural gas Joan will use per day in a month with a mean temperature of 30°F.

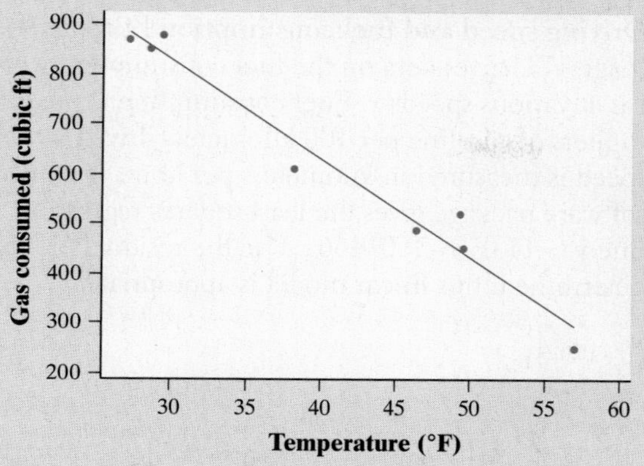


(b) Predict the mean amount of natural gas Joan will use per day in a month with a mean temperature of 65°F.

(c) How confident are you in each of these predictions? Explain your reasoning.

**39. Residual wins** Refer to Exercise 37. The Chicago Cubs won the World Series in 2016. They had 103 wins and spent \$182 million on payroll. Calculate and interpret the residual for the Cubs.
pg 180

**40. Residual gas** Refer to Exercise 38. During March, the average temperature was 46.4°F and Joan used an average of 490 cubic feet of gas per day. Calculate and interpret the residual for this month.

**41. More wins?** Refer to Exercise 37.
pg 182

(a) Interpret the slope of the regression line.

(b) Does the value of the $y$ intercept have meaning in this context? If so, interpret the $y$ intercept. If not, explain why.

**42. Less gas?** Refer to Exercise 38.

(a) Interpret the slope of the regression line.

(b) Does the value of the $y$ intercept have meaning in this context? If so, interpret the $y$ intercept. If not, explain why.

**43. Long strides** The scatterplot shows the relationship between $x$ = height of a student (in inches) and $y$ = number of steps required to walk the length of a school hallway, along with the regression line $\hat{y} = 113.6 - 0.921x$.

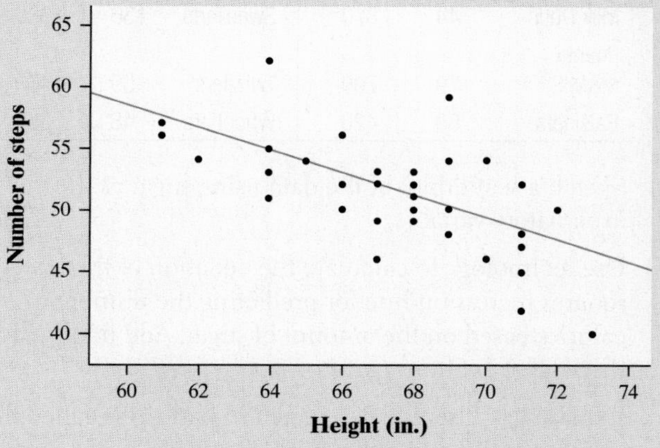


(a) Calculate and interpret the residual for Kiana, who is 67 inches tall and took 49 steps to walk the hallway.

(b) Matthew is 10 inches taller than Samantha. About how many fewer steps do you expect Matthew to take compared to Samantha?

**44. Crickets chirping** The scatterplot shows the relationship between $x$ = temperature in degrees Fahrenheit and $y$ = chirps per minute for the striped ground cricket, along with the regression line $\hat{y} = -0.31 + 0.212x$.[27]

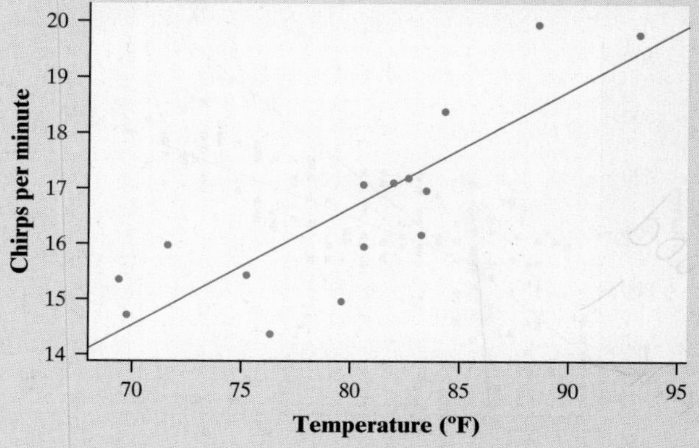


(a) Calculate and interpret the residual for the cricket who chirped 20 times per minute when the temperature was 88.6°F.

(b) About how many additional chirps per minute do you expect a cricket to make if the temperature increases by 10°F?

**45. More Olympic athletes** In Exercises 5 and 11, you described the relationship between height (in inches) and weight (in pounds) for Olympic track and field athletes. The scatterplot shows this relationship, along with two regression lines. The regression line for the shotput, hammer throw, and discus throw athletes (blue squares) is $\hat{y} = -115 + 5.13x$. The regression line for the remaining athletes (black dots) is $\hat{y} = -297 + 6.41x$.

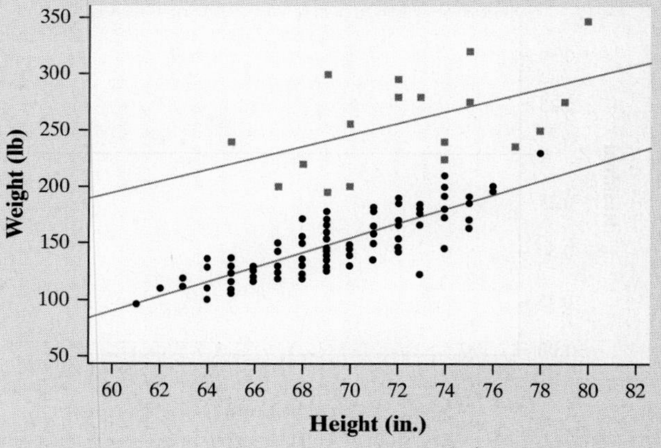

(a) How do the regression lines compare?

(b) How much more do you expect a 72-inch discus thrower to weigh than a 72-inch sprinter?

**46. More Starbucks** In Exercises 6 and 12, you described the relationship between fat (in grams) and the number of calories in products sold at Starbucks. The scatterplot shows this relationship, along with two regression lines. The regression line for the food products (blue squares) is $\hat{y} = 170 + 11.8x$. The regression line for the drink products (black dots) is $\hat{y} = 88 + 24.5x$.

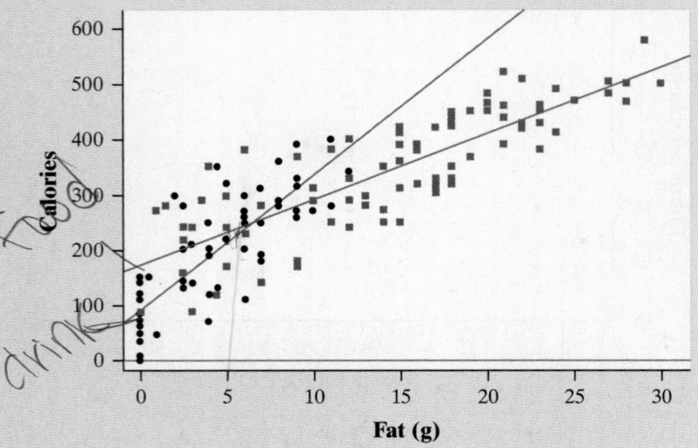


(a) How do the regression lines compare?

(b) How many more calories do you expect to find in a food item with 5 grams of fat compared to a drink item with 5 grams of fat?

**47. Infant weights in Nahya** A study of nutrition in developing countries collected data from the Egyptian village of Nahya. Researchers recorded the mean weight (in kilograms) for 170 infants in Nahya each month during their first year of life. A hasty user of statistics enters the data into software and computes the least-squares line without looking at the scatterplot first. The result is weight = 4.88 + 0.267 (age). Use the residual plot to determine if this linear model is appropriate.

pg 186

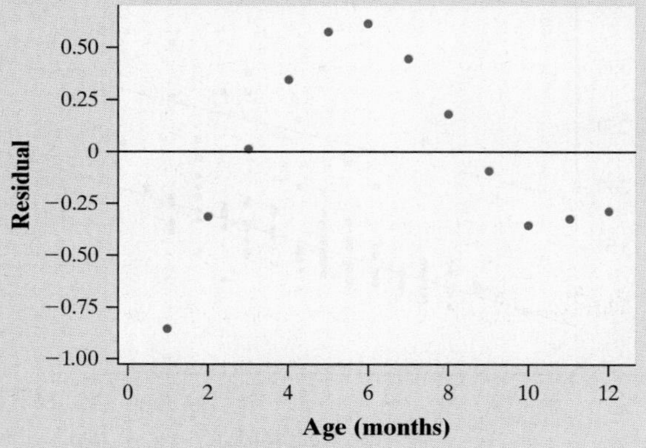


**48. Driving speed and fuel consumption** Exercise 9 (page 171) gives data on the fuel consumption $y$ of a car at various speeds $x$. Fuel consumption is measured in liters of gasoline per 100 kilometers driven, and speed is measured in kilometers per hour. A statistical software package gives the least-squares regression line $\hat{y} = 11.058 - 0.01466x$. Use the residual plot to determine if this linear model is appropriate.

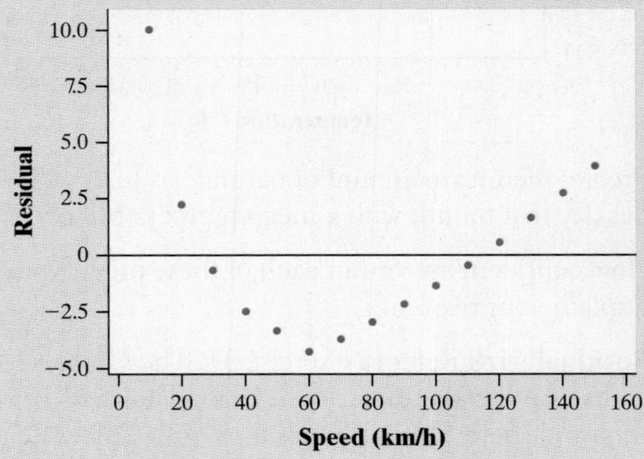


**49. Actual weight** Refer to Exercise 47. Use the equation of the least-squares regression line and the residual plot to estimate the *actual* mean weight of the infants when they were 1 month old.

**50. Actual consumption** Refer to Exercise 48. Use the equation of the least-squares regression line and the residual plot to estimate the *actual* fuel consumption of the car when driving 20 kilometers per hour.

**51. Movie candy** Is there a relationship between the amount of sugar (in grams) and the number of calories in movie-theater candy? Here are the data from a sample of 12 types of candy:

| Name | Sugar (g) | Calories | Name | Sugar (g) | Calories |
|---|---|---|---|---|---|
| Butterfinger Minis | 45 | 450 | Reese's Pieces | 61 | 580 |
| Junior Mints | 107 | 570 | Skittles | 87 | 450 |
| M&M'S® | 62 | 480 | Sour Patch Kids | 92 | 490 |
| Milk Duds | 44 | 370 | SweeTarts | 136 | 680 |
| Peanut M&M'S® | 79 | 790 | Twizzlers | 59 | 460 |
| Raisinets | 60 | 420 | Whoppers | 48 | 350 |

(a) Sketch a scatterplot of the data using sugar as the explanatory variable.

(b) Use technology to calculate the equation of the least-squares regression line for predicting the number of calories based on the amount of sugar. Add the line to the scatterplot from part (a).

(c) Explain why the line calculated in part (b) is called the "least-squares" regression line.

52. **Long jumps** Here are the 40-yard-dash times (in seconds) and long-jump distances (in inches) for a small class of 12 students:

| | | | | | | |
|---|---|---|---|---|---|---|
| Dash time (sec) | 5.41 | 5.05 | 7.01 | 7.17 | 6.73 | 5.68 |
| Long-jump distance (in.) | 171 | 184 | 90 | 65 | 78 | 130 |
| Dash time (sec) | 5.78 | 6.31 | 6.44 | 6.50 | 6.80 | 7.25 |
| Long-jump distance (in.) | 173 | 143 | 92 | 139 | 120 | 110 |

(a) Sketch a scatterplot of the data using dash time as the explanatory variable.

(b) Use technology to calculate the equation of the least-squares regression line for predicting the long-jump distance based on the dash time. Add the line to the scatterplot from part (a).

(c) Explain why the line calculated in part (b) is called the "least-squares" regression line.

53. **More candy** Refer to Exercise 51. Use technology to create a residual plot. Sketch the residual plot and explain what information it provides.

54. **More long jumps** Refer to Exercise 52. Use technology to create a residual plot. Sketch the residual plot and explain what information it provides.

55. **Longer strides** In Exercise 43, we modeled the pg 191 relationship between $x$ = height of a student (in inches) and $y$ = number of steps required to walk the length of a school hallway, with the regression line $\hat{y} = 113.6 - 0.921x$. For this model, technology gives $s = 3.50$ and $r^2 = 0.399$.

(a) Interpret the value of $s$.

(b) Interpret the value of $r^2$.

56. **Crickets keep chirping** In Exercise 44, we modeled the relationship between $x$ = temperature in degrees Fahrenheit and $y$ = chirps per minute for the striped ground cricket, with the regression line $\hat{y} = -0.31 + 0.212x$. For this model, technology gives $s = 0.97$ and $r^2 = 0.697$.

(a) Interpret the value of $s$.

(b) Interpret the value of $r^2$.

57. **Olympic figure skating** For many people, the women's figure skating competition is the highlight of the Olympic Winter Games. Scores in the short program $x$ and scores in the free skate $y$ were recorded for each of the 24 skaters who competed in both rounds during the 2010 Winter Olympics in Vancouver, Canada.[28] Here is a scatterplot with least-squares regression line $\hat{y} = -16.2 + 2.07x$. For this model, $s = 10.2$ and $r^2 = 0.736$.

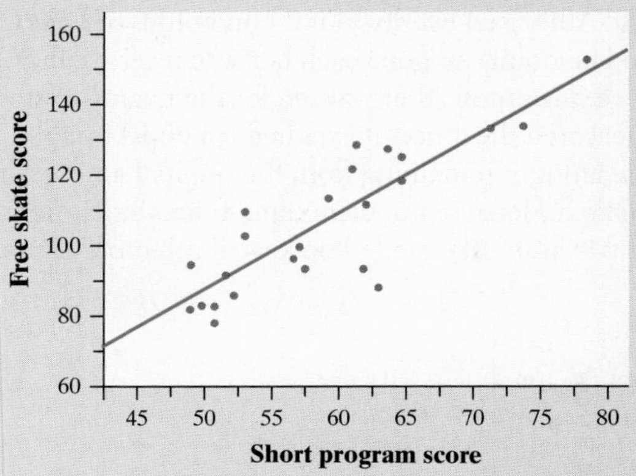


(a) Calculate and interpret the residual for the 2010 gold medal winner Yu-Na Kim, who scored 78.50 in the short program and 150.06 in the free skate.

(b) Interpret the slope of the least-squares regression line.

(c) Interpret the standard deviation of the residuals.

(d) Interpret the coefficient of determination.

58. **Age and height** A random sample of 195 students was selected from the United Kingdom using the Census At School data selector. The age $x$ (in years) and height $y$ (in centimeters) were recorded for each student. Here is a scatterplot with the least-squares regression line $\hat{y} = 106.1 + 4.21x$. For this model, $s = 8.61$ and $r^2 = 0.274$.

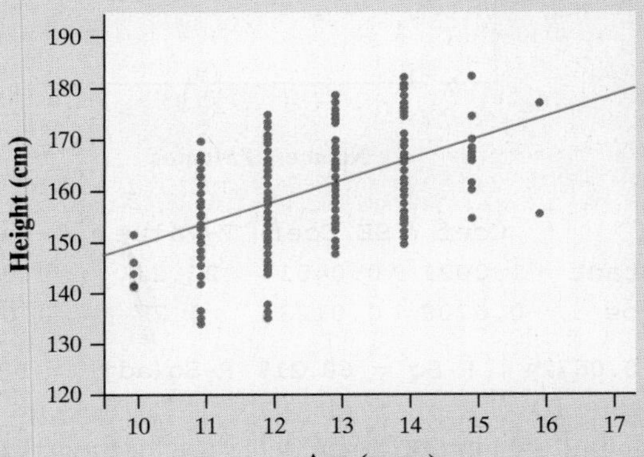


(a) Calculate and interpret the residual for the student who was 141 cm tall at age 10.

(b) Interpret the slope of the least-squares regression line.

(c) Interpret the standard deviation of the residuals.

(d) Interpret the coefficient of determination.

59. **More mess?** pg 192 When Mentos are dropped into a newly opened bottle of Diet Coke, carbon dioxide is released from the Diet Coke very rapidly, causing the Diet Coke to be expelled from the bottle. To see if using more Mentos causes more Diet Coke to be expelled, Brittany and Allie used twenty-four 2-cup bottles of Diet Coke and randomly assigned each bottle to receive either 2, 3, 4, or 5 Mentos. After waiting for the fizzing to stop, they measured the amount expelled (in cups) by subtracting the amount remaining from the original amount in the bottle.[29] Here is computer output from a linear regression of $y$ = amount expelled on $x$ = number of Mentos:

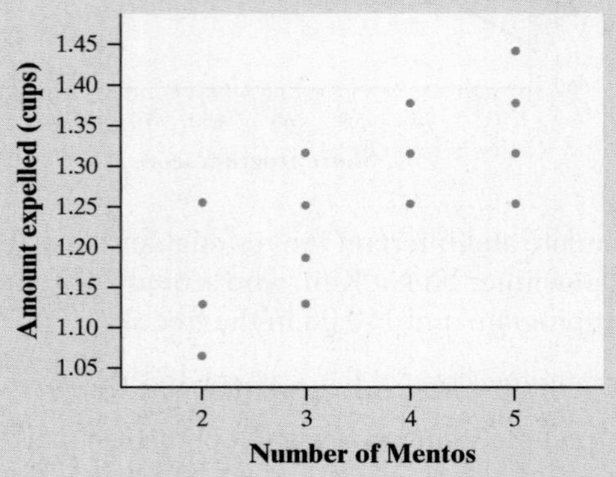


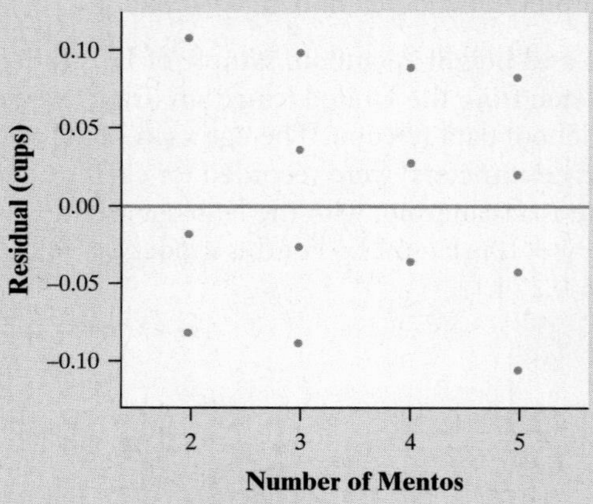


| Term | Coef | SE Coef | T-Value | P-Value |
|---|---|---|---|---|
| Constant | 1.0021 | 0.0451 | 22.21 | 0.000 |
| Mentos | 0.0708 | 0.0123 | 5.77 | 0.000 |

S = 0.06724 R-Sq = 60.21% R-Sq(adj) = 58.40%

(a) Is a line an appropriate model to use for these data? Explain how you know.

(b) Find the correlation.

(c) What is the equation of the least-squares regression line? Define any variables that you use.

(d) Interpret the values of $s$ and $r^2$.

60. **Less mess?** Kerry and Danielle wanted to investigate whether tapping on a can of soda would reduce the amount of soda expelled after the can has been shaken. For their experiment, they vigorously shook 40 cans of soda and randomly assigned each can to be tapped for 0 seconds, 4 seconds, 8 seconds, or 12 seconds. After waiting for the fizzing to stop, they measured the amount expelled (in milliliters) by subtracting the amount remaining from the original amount in the can.[30] Here is computer output from a linear regression of $y$ = amount expelled on $x$ = tapping time:

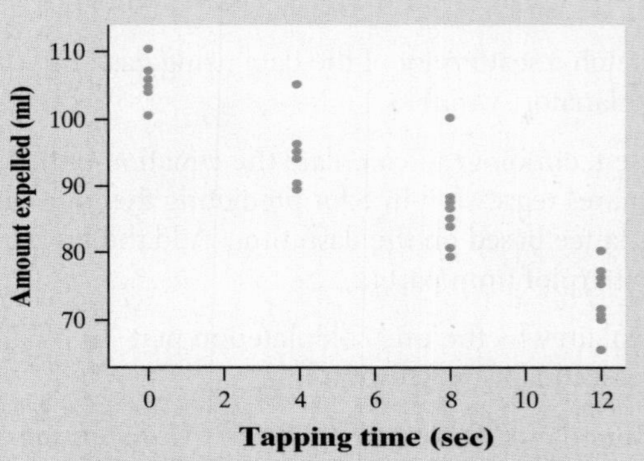


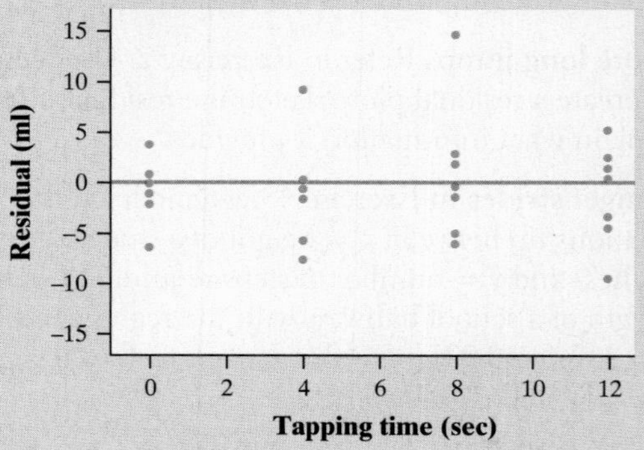


| Term | Coef | SE Coef | T-Value | P-Value |
|---|---|---|---|---|
| Constant | 106.360 | 1.320 | 80.34 | 0.000 |
| Tapping_time | -2.635 | 0.177 | -14.90 | 0.000 |

S = 5.00347 R-Sq = 85.38% R-Sq(adj) = 84.99%

(a) Is a line an appropriate model to use for these data? Explain how you know.

(b) Find the correlation.

(c) What is the equation of the least-squares regression line? Define any variables that you use.

(d) Interpret the values of $s$ and $r^2$.

**61. Temperature and wind** The average temperature (in degrees Fahrenheit) and average wind speed (in miles per hour) were recorded for 365 consecutive days at Chicago's O'Hare International Airport. Here is computer output for a regression of $y$ = average wind speed on $x$ = average temperature:

**Summary of Fit**

| | |
|---|---|
| RSquare | 0.047874 |
| RSquare Adj | 0.045251 |
| Root Mean Square Error | 3.655950 |
| Mean of Response | 9.826027 |
| Observations (or Sum Wgts) | 365 |

**Parameter Estimates**

| Term | Estimate | Std Error | t Ratio | Prob>\|t\| |
|---|---|---|---|---|
| Intercept | 11.897762 | 0.521320 | 22.82 | <.0001* |
| Avg temp | -0.041077 | 0.009615 | -4.27 | <.0001* |

(a) Calculate and interpret the residual for the day where the average temperature was 42°F and the average wind speed was 2.2 mph.

(b) Interpret the slope.

(c) By about how much do the actual average wind speeds typically vary from the values predicted by the least-squares regression line with $x$ = average temperature?

(d) What percent of the variability in average wind speed is accounted for by the least-squares regression line with $x$ = average temperature?

**62. Beetles and beavers** Do beavers benefit beetles? Researchers laid out 23 circular plots, each 4 meters in diameter, in an area where beavers were cutting down cottonwood trees. In each plot, they counted the number of stumps from trees cut by beavers and the number of clusters of beetle larvae. Ecologists believe that the new sprouts from stumps are more tender than other cottonwood growth, so beetles prefer them. If so, more stumps should produce more beetle larvae.[31] Here is computer output for a regression of $y$ = number of beetle larvae on $x$ = number of stumps:

**Summary of Fit**

| | |
|---|---|
| RSquare | 0.839144 |
| RSquare Adj | 0.831484 |
| Root Mean Square Error | 6.419386 |
| Mean of Response | 25.086960 |
| Observations (or Sum Wgts) | 23 |

**Parameter Estimates**

| Term | Estimate | Std Error | t Ratio | Prob>\|t\| |
|---|---|---|---|---|
| Intercept | -1.286104 | 2.853182 | -0.45 | 0.6568 |
| Number of stumps | 11.893733 | 1.136343 | 10.47 | <.0001* |

(a) Calculate and interpret the residual for the plot that had 2 stumps and 30 beetle larvae.

(b) Interpret the slope.

(c) By about how much do the actual number of larvae typically vary from the values predicted by the least-squares regression line with $x$ = number of stumps?

(d) What percent of the variability in number of larvae is accounted for by the least-squares regression line with $x$ = number of stumps?

**63. Husbands and wives** pg 195 The mean height of married American women in their early 20s is 64.5 inches and the standard deviation is 2.5 inches. The mean height of married men the same age is 68.5 inches with standard deviation 2.7 inches. The correlation between the heights of husbands and wives is about $r = 0.5$.

(a) Find the equation of the least-squares regression line for predicting a husband's height from his wife's height for married couples in their early 20s.

(b) Suppose that the height of a randomly selected wife was 1 standard deviation below average. Predict the height of her husband *without* using the least-squares line.

**64. The stock market** Some people think that the behavior of the stock market in January predicts its behavior for the rest of the year. Take the explanatory variable $x$ to be the percent change in a stock market index in January and the response variable $y$ to be the change in the index for the entire year. We expect a positive correlation between $x$ and $y$ because the change during January contributes to the full year's change. Calculation from data for an 18-year period gives

$$\bar{x} = 1.75\% \quad s_x = 5.36\% \quad \bar{y} = 9.07\%$$
$$s_y = 15.35\% \quad r = 0.596$$

(a) Find the equation of the least-squares line for predicting full-year change from January change.

(b) Suppose that the percent change in a particular January was 2 standard deviations above average. Predict the percent change for the entire year *without* using the least-squares line.

**65. Will I bomb the final?** We expect that students who do well on the midterm exam in a course will usually also do well on the final exam. Gary Smith of Pomona College looked at the exam scores of all 346 students who took his statistics class over a 10-year period.[32] Assume that both the midterm and final exam were scored out of 100 points.

(a) State the equation of the least-squares regression line if each student scored the same on the midterm and the final.

(b) The actual least-squares line for predicting final-exam score $y$ from midterm-exam score $x$ was $\hat{y} = 46.6 + 0.41x$. Predict the score of a student who scored 50 on the midterm and a student who scored 100 on the midterm.

(c) Explain how your answers to part (b) illustrate regression to the mean.

66. **It's still early** We expect that a baseball player who has a high batting average in the first month of the season will also have a high batting average the rest of the season. Using 66 Major League Baseball players from a recent season,[33] a least-squares regression line was calculated to predict rest-of-season batting average $y$ from first-month batting average $x$. *Note:* A player's batting average is the proportion of times at bat that he gets a hit. A batting average over 0.300 is considered very good in Major League Baseball.

(a) State the equation of the least-squares regression line if each player had the same batting average the rest of the season as he did in the first month of the season.

(b) The actual equation of the least-squares regression line is $\hat{y} = 0.245 + 0.109x$. Predict the rest-of-season batting average for a player who had a 0.200 batting average the first month of the season and for a player who had a 0.400 batting average the first month of the season.

(c) Explain how your answers to part (b) illustrate regression to the mean.

67. **Who's got hops?** Haley, Jeff, and Nathan measured pg 202 the height (in inches) and vertical jump (in inches) of 74 students at their school.[34] Here is a scatterplot of the data, along with the least-squares regression line. Jacob (highlighted in red) had a vertical jump of nearly 3 feet!

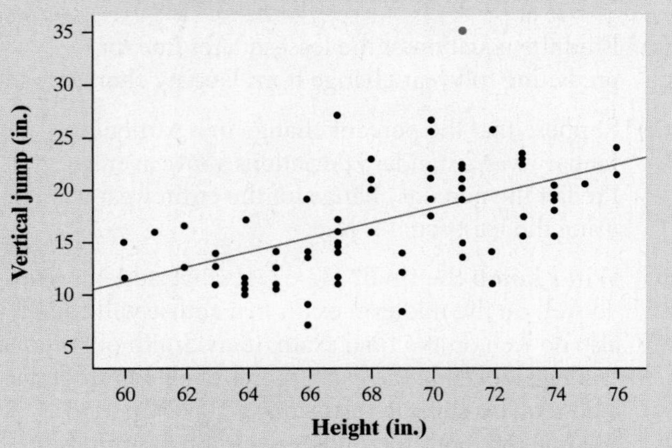


(a) Describe the influence that Jacob's point has on the equation of the least-squares regression line.

(b) Describe the influence that Jacob's point has on the standard deviation of the residuals and $r^2$.

68. **Stand mixers** The scatterplot shows the weight (in pounds) and cost (in dollars) of 11 stand mixers.[35] The mixer from Walmart (highlighted in red) was much lighter—and cheaper—than the other mixers.

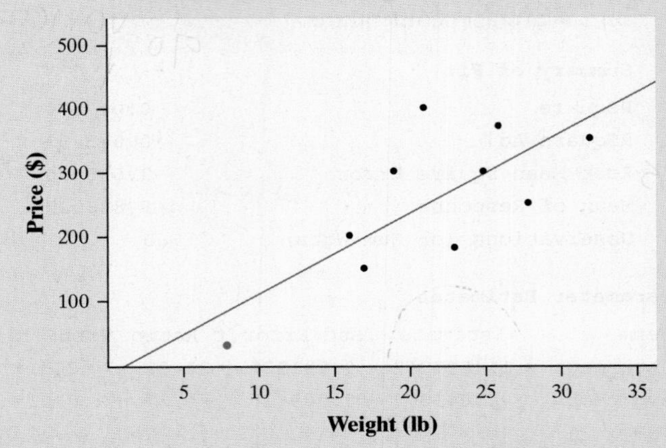


(a) Describe what influence the highlighted point has on the equation of the least-squares regression line.

(b) Describe what influence the highlighted point has on the standard deviation of the residuals and $r^2$.

69. **Managing diabetes** People with diabetes measure their fasting plasma glucose (FPG, measured in milligrams per milliliter) after fasting for at least 8 hours. Another measurement, made at regular medical checkups, is called HbA. This is roughly the percent of red blood cells that have a glucose molecule attached. It measures average exposure to glucose over a period of several months. The table gives data on both HbA and FPG for 18 diabetics five months after they had completed a diabetes education class.[36]

| Subject | HbA (%) | FPG (mg/ml) | Subject | HbA (%) | FPG (mg/ml) |
|---|---|---|---|---|---|
| 1 | 6.1 | 141 | 10 | 8.7 | 172 |
| 2 | 6.3 | 158 | 11 | 9.4 | 200 |
| 3 | 6.4 | 112 | 12 | 10.4 | 271 |
| 4 | 6.8 | 153 | 13 | 10.6 | 103 |
| 5 | 7.0 | 134 | 14 | 10.7 | 172 |
| 6 | 7.1 | 95 | 15 | 10.7 | 359 |
| 7 | 7.5 | 96 | 16 | 11.2 | 145 |
| 8 | 7.7 | 78 | 17 | 13.7 | 147 |
| 9 | 7.9 | 148 | 18 | 19.3 | 255 |

(a) Make a scatterplot with HbA as the explanatory variable. Describe what you see.

(b) Subject 18 has an unusually large $x$ value. What effect do you think this subject has on the correlation? What effect do you think this subject has on the equation of the least-squares regression line? Calculate the correlation and equation of the least-squares regression line with and without this subject to confirm your answer.

(c) Subject 15 has an unusually large $y$ value. What effect do you think this subject has on the correlation? What effect do you think this subject has on the equation of the least-squares regression line? Calculate the correlation and equation of the least-squares regression line with and without this subject to confirm your answer.

70. **Rushing for points** What is the relationship between rushing yards and points scored in the National Football League? The table gives the number of rushing yards and the number of points scored for each of the 16 games played by the Jacksonville Jaguars in a recent season.[37]

| Game | Rushing yards | Points scored | Game | Rushing yards | Points scored |
|---|---|---|---|---|---|
| 1 | 163 | 16 | 9 | 141 | 17 |
| 2 | 112 | 3 | 10 | 108 | 10 |
| 3 | 128 | 10 | 11 | 105 | 13 |
| 4 | 104 | 10 | 12 | 129 | 14 |
| 5 | 96 | 20 | 13 | 116 | 41 |
| 6 | 133 | 13 | 14 | 116 | 14 |
| 7 | 132 | 12 | 15 | 113 | 17 |
| 8 | 84 | 14 | 16 | 190 | 19 |

(a) Make a scatterplot with rushing yards as the explanatory variable. Describe what you see.

(b) Game 16 has an unusually large $x$ value. What effect do you think this game has on the correlation? On the equation of the least-squares regression line? Calculate the correlation and equation of the least-squares regression line with and without this game to confirm your answers.

(c) Game 13 has an unusually large $y$ value. What effect do you think this game has on the correlation? On the equation of the least-squares regression line? Calculate the correlation and equation of the least-squares regression line with and without this game to confirm your answers.

**Multiple Choice:** *Select the best answer for Exercises 71–78.*

71. Which of the following is *not* a characteristic of the least-squares regression line?

(a) The slope of the least-squares regression line is always between –1 and 1.

(b) The least-squares regression line always goes through the point $(\bar{x}, \bar{y})$.

(c) The least-squares regression line minimizes the sum of squared residuals.

(d) The slope of the least-squares regression line will always have the same sign as the correlation.

(e) The least-squares regression line is not resistant to outliers.

72. Each year, students in an elementary school take a standardized math test at the end of the school year. For a class of fourth-graders, the average score was 55.1 with a standard deviation of 12.3. In the third grade, these same students had an average score of 61.7 with a standard deviation of 14.0. The correlation between the two sets of scores is $r = 0.95$. Calculate the equation of the least-squares regression line for predicting a fourth-grade score from a third-grade score.

(a) $\hat{y} = 3.58 + 0.835x$

(b) $\hat{y} = 15.69 + 0.835x$

(c) $\hat{y} = 2.19 + 1.08x$

(d) $\hat{y} = -11.54 + 1.08x$

(e) Cannot be calculated without the data.

73. Using data from the LPGA tour, a regression analysis was performed using $x$ = average driving distance and $y$ = scoring average. Using the output from the regression analysis shown below, determine the equation of the least-squares regression line.

```
Predictor              Coef        SE Coef       T        P
Constant           87.974000    2.391000     36.78    0.000
Driving Distance   -0.060934    0.009536     -6.39    0.000

S = 1.01216       R-Sq = 22.1%       R-Sq(adj) = 21.6%
```

(a) $\hat{y} = 87.974 + 2.391x$

(b) $\hat{y} = 87.974 + 1.01216x$

(c) $\hat{y} = 87.974 - 0.060934x$

(d) $\hat{y} = -0.060934 + 1.01216x$

(e) $\hat{y} = -0.060934 + 87.947x$

*Exercises 74 to 78 refer to the following setting.* Measurements on young children in Mumbai, India, found this least-squares line for predicting $y$ = height (in cm) from $x$ = arm span (in cm):[38]

$$\hat{y} = 6.4 + 0.93x$$

74. By looking at the equation of the least-squares regression line, you can see that the correlation between height and arm span is

(a) greater than zero.

(b) less than zero.

(c) 0.93.

(d) 6.4.

(e) Can't tell without seeing the data.

75. In addition to the regression line, the report on the Mumbai measurements says that $r^2 = 0.95$. This suggests that

(a) although arm span and height are correlated, arm span does not predict height very accurately.

(b) height increases by $\sqrt{0.95} = 0.97$ cm for each additional centimeter of arm span.

(c) 95% of the relationship between height and arm span is accounted for by the regression line.

(d) 95% of the variation in height is accounted for by the regression line with $x$ = arm span.

(e) 95% of the height measurements are accounted for by the regression line with $x$ = arm span.

76. One child in the Mumbai study had height 59 cm and arm span 60 cm. This child's residual is

(a) −3.2 cm. (b) −2.2 cm. (c) −1.3 cm.

(d) 3.2 cm. (e) 62.2 cm.

77. Suppose that a tall child with arm span 120 cm and height 118 cm was added to the sample used in this study. What effect will this addition have on the correlation and the slope of the least-squares regression line?

(a) Correlation will increase, slope will increase.

(b) Correlation will increase, slope will stay the same.

(c) Correlation will increase, slope will decrease.

(d) Correlation will stay the same, slope will stay the same.

(e) Correlation will stay the same, slope will increase.

78. Suppose that the measurements of arm span and height were converted from centimeters to meters by dividing each measurement by 100. How will this conversion affect the values of $r^2$ and $s$?

(a) $r^2$ will increase, $s$ will increase.

(b) $r^2$ will increase, $s$ will stay the same.

(c) $r^2$ will increase, $s$ will decrease.

(d) $r^2$ will stay the same, $s$ will stay the same.

(e) $r^2$ will stay the same, $s$ will decrease.

### Recycle and Review

79. **Fuel economy** (2.2) In its recent *Fuel Economy Guide*, the Environmental Protection Agency (EPA) gives data on 1152 vehicles. There are a number of outliers, mainly vehicles with very poor gas mileage or hybrids with very good gas mileage. If we ignore the outliers, however, the combined city and highway gas mileage of the other 1120 or so vehicles is approximately Normal with mean 18.7 miles per gallon (mpg) and standard deviation 4.3 mpg.

(a) The Chevrolet Malibu with a four-cylinder engine has a combined gas mileage of 25 mpg. What percent of the 1120 vehicles have worse gas mileage than the Malibu?

(b) How high must a vehicle's gas mileage be in order to fall in the top 10% of the 1120 vehicles?

80. **Marijuana and traffic accidents** (1.1) Researchers in New Zealand interviewed 907 drivers at age 21. They had data on traffic accidents and they asked the drivers about marijuana use. Here are data on the numbers of accidents caused by these drivers at age 19, broken down by marijuana use at the same age:[39]

| | Marijuana use per year | | | |
|---|---|---|---|---|
| | Never | 1–10 times | 11–50 times | 51+ times |
| Number of drivers | 452 | 229 | 70 | 156 |
| Accidents caused | 59 | 36 | 15 | 50 |

(a) Make a graph that displays the accident rate for each category of marijuana use. Is there evidence of an association between marijuana use and traffic accidents? Justify your answer.

(b) Explain why we can't conclude that marijuana use *causes* accidents based on this study.

# SECTION 3.3 Transforming to Achieve Linearity

**LEARNING TARGETS** *By the end of the section, you should be able to:*

- Use transformations involving powers, roots, or logarithms to create a linear model that describes the relationship between two quantitative variables, and use the model to make predictions.
- Determine which of several models does a better job of describing the relationship between two quantitative variables.

In Section 3.2, we learned how to analyze relationships between two quantitative variables that showed a linear pattern. When bivariate data show a curved relationship, we must develop new techniques for finding an appropriate model. This section describes several simple *transformations* of data that can straighten a nonlinear pattern. Once the data have been transformed to achieve linearity, we can use least-squares regression to generate a useful model for making predictions.

The Gapminder website (www.gapminder.org) provides loads of data on the health and well-being of the world's inhabitants. Figure 3.18 shows a scatterplot of data from Gapminder.[40] The individuals are all the world's nations for which data were available in 2015. The explanatory variable, income per person, is a measure of how rich a country is. The response variable is life expectancy at birth.

We expect people in richer countries to live longer because they have better access to medical care and typically lead healthier lives. The overall pattern of the scatterplot does show this, but the relationship is not linear. Life expectancy rises very quickly as income per person increases and then levels off. People in

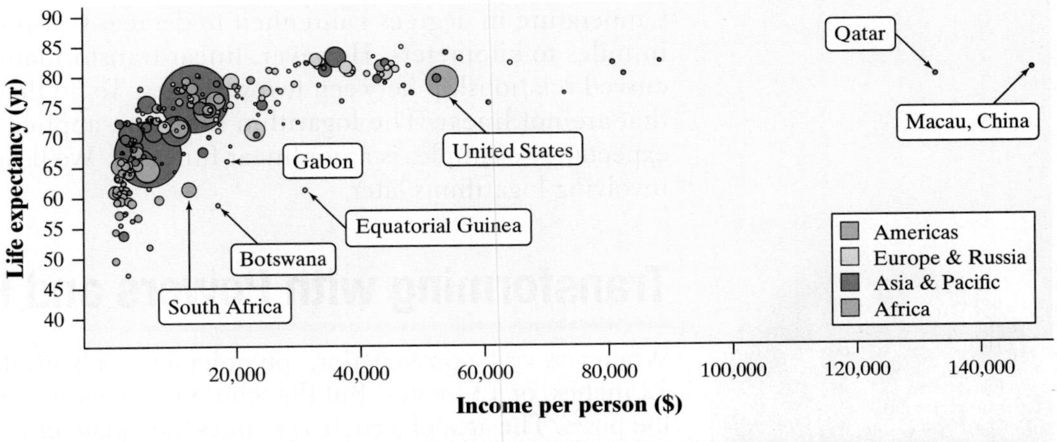


**FIGURE 3.18** Scatterplot of the life expectancy of people in many nations against each nation's income per person. The color of each circle indicates the geographic region in which that country is located. The size of each circle is based on the population of the country—bigger circles indicate larger populations.

very rich countries such as the United States live no longer than people in poorer but not extremely poor nations. In some less wealthy countries, people live longer than in the United States.

Four African nations are outliers. Their life expectancies are much smaller than would be expected based on their income per person. Gabon and Equatorial Guinea produce oil, and South Africa and Botswana produce diamonds. It may be that income from mineral exports goes mainly to a few people and so pulls up income per person without much effect on either the income or the life expectancy of ordinary citizens. That is, income per person is a mean, and we know that mean income can be much higher than median income.

The scatterplot in Figure 3.18 shows a curved pattern. We can straighten things out using logarithms. Figure 3.19 plots the logarithm of income per person against life expectancy for these same countries. The effect is remarkable. This graph has a clear, linear form.

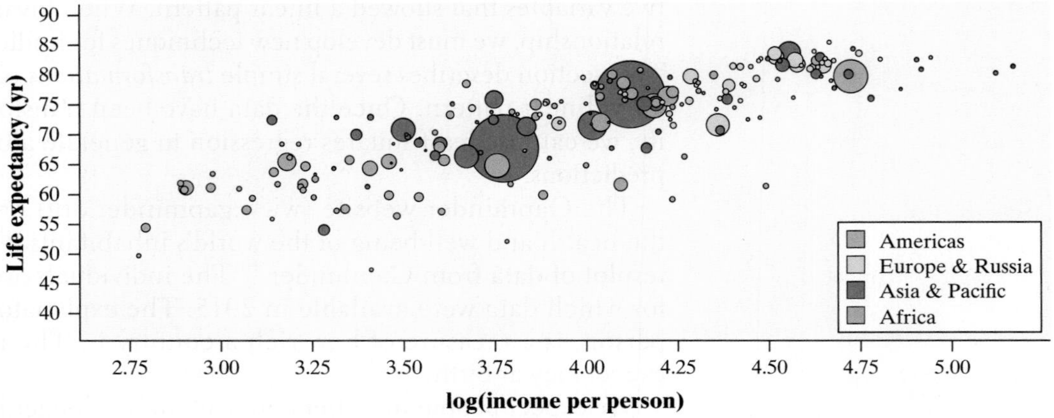


**FIGURE 3.19** Scatterplot of life expectancy against log(income per person) for many nations.

Applying a function such as the logarithm or square root to a quantitative variable is called *transforming the data*. Transforming data amounts to changing the scale of measurement that was used when the data were collected. In Chapter 2, we discussed *linear transformations*, such as converting temperature in degrees Fahrenheit to degrees Celsius or converting distance in miles to kilometers. However, linear transformations cannot straighten a curved relationship between two variables. To do that, we resort to functions that are not linear. The logarithm function, applied in the income and life expectancy example, is a nonlinear function. We'll return to transformations involving logarithms later.

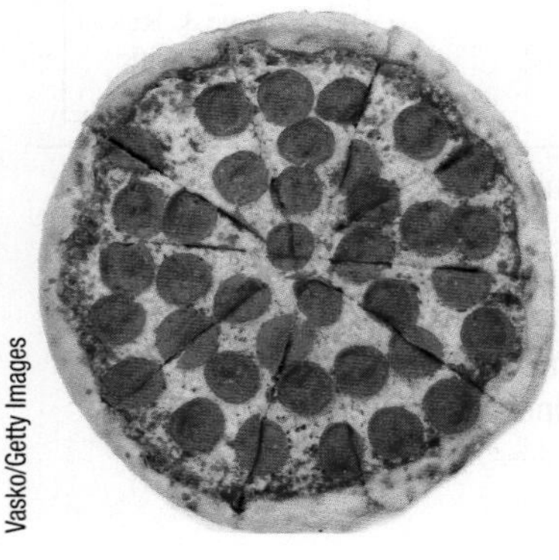

Vasko/Getty Images

# Transforming with Powers and Roots

When you visit a pizza parlor, you order a pizza by its diameter—say, 10 inches, 12 inches, or 14 inches. But the amount you get to eat depends on the area of the pizza. The area of a circle is $\pi$ times the square of its radius $r$. So the area of a round pizza with diameter $x$ is

$$\text{area} = \pi r^2 = \pi\left(\frac{x}{2}\right)^2 = \pi\left(\frac{x^2}{4}\right) = \frac{\pi}{4}x^2$$

This is a *power model* of the form $y = ax^p$ with $a = \pi/4$ and $p = 2$.

When we are dealing with things of the same general form, whether circles or fish or people, we expect area to go up with the square of a dimension such as diameter or height. Volume should go up with the cube of a linear dimension. That is, geometry tells us to expect power models in some settings. There are other physical relationships between two variables that are described by power models. Here are some examples from science.

- The distance that an object dropped from a given height falls is related to time since release by the model

$$\text{distance} = a(\text{time})^2$$

- The time it takes a pendulum to complete one back-and-forth swing (its period) is related to its length by the model

$$\text{period} = a\sqrt{\text{length}} = a(\text{length})^{1/2}$$

- The intensity of a light bulb is related to distance from the bulb by the model

$$\text{intensity} = \frac{a}{\text{distance}^2} = a(\text{distance})^{-2}$$

Although a power model of the form $y = ax^p$ describes the nonlinear relationship between $x$ and $y$ in each of these settings, there is a *linear* relationship between $x^p$ and $y$. If we transform the values of the explanatory variable $x$ by raising them to the $p$ power, and graph the points $(x^p, y)$, the scatterplot should have a linear form. The following example shows what we mean.

## EXAMPLE Go fish!
### Transforming with powers

Les Breault/Alamy

**PROBLEM:** Imagine that you have been put in charge of organizing a fishing tournament in which prizes will be given for the heaviest Atlantic Ocean rockfish caught. You know that many of the fish caught during the tournament will be measured and released. You are also aware that using delicate scales to try to weigh a fish that is flopping around in a moving boat will probably not yield very accurate results. It would be much easier to measure the length of the fish while on the boat. What you need is a way to convert the length of the fish to its weight.

You contact the nearby marine research laboratory, and it provides reference data on the length (in centimeters) and weight (in grams) for Atlantic Ocean rockfish of several sizes.[41] Here is a scatterplot of the data. Note the clear curved form.

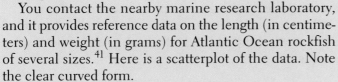

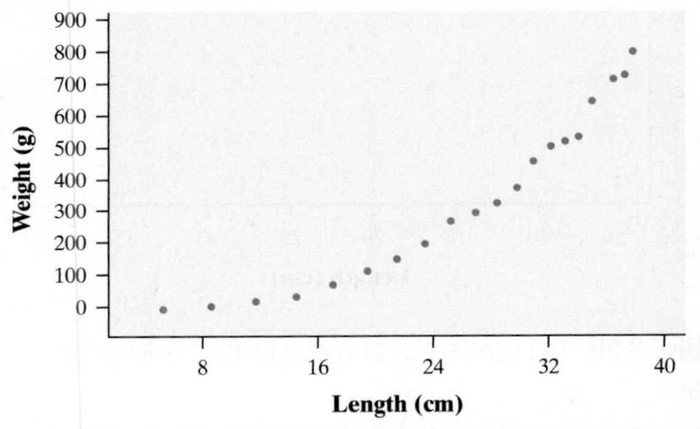


| | | | | | | | | | | |
|---|---|---|---|---|---|---|---|---|---|---|
| **Length** | 5.2 | 8.5 | 11.5 | 14.3 | 16.8 | 19.2 | 21.3 | 23.3 | 25.0 | 26.7 |
| **Weight** | 2 | 8 | 21 | 38 | 69 | 117 | 148 | 190 | 264 | 293 |
| **Length** | 28.2 | 29.6 | 30.8 | 32.0 | 33.0 | 34.0 | 34.9 | 36.4 | 37.1 | 37.7 |
| **Weight** | 318 | 371 | 455 | 504 | 518 | 537 | 651 | 719 | 726 | 810 |

Because length is one-dimensional and weight (like volume) is three-dimensional, a power model of the form weight $= a(\text{length})^3$ should describe the relationship. Here is a scatterplot of weight versus length$^3$:

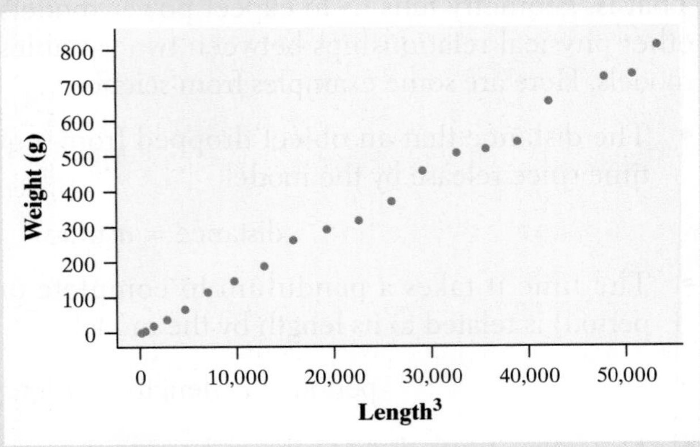


Because the transformation made the association roughly linear, we used computer software to perform a linear regression analysis of $y =$ weight versus $x =$ length$^3$.

```
Regression Analysis: Weight versus Length^3
Predictor   Coef       SE Coef    T     P
Constant    4.066      6.902      0.59  0.563
Length^3    0.0146774  0.0002404  61.07 0.000
S = 18.8412    R-Sq = 99.5%    R-Sq(adj) = 99.5%
```

(a) Give the equation of the least-squares regression line. Define any variables you use.

(b) Suppose a contestant in the fishing tournament catches an Atlantic Ocean rockfish that's 36 centimeters long. Use the model from part (a) to predict the fish's weight.

**SOLUTION:**

(a) $\widehat{\text{weight}} = 4.066 + 0.0146774(\text{length})^3$

(b) $\widehat{\text{weight}} = 4.066 + 0.0146774(36)^3$

$\widehat{\text{weight}} = 688.9$ grams

If you write the equation as $\hat{y} = 4.066 + 0.0146774x^3$ make sure to define $y =$ weight and $x =$ length.

**FOR PRACTICE, TRY EXERCISE 81**

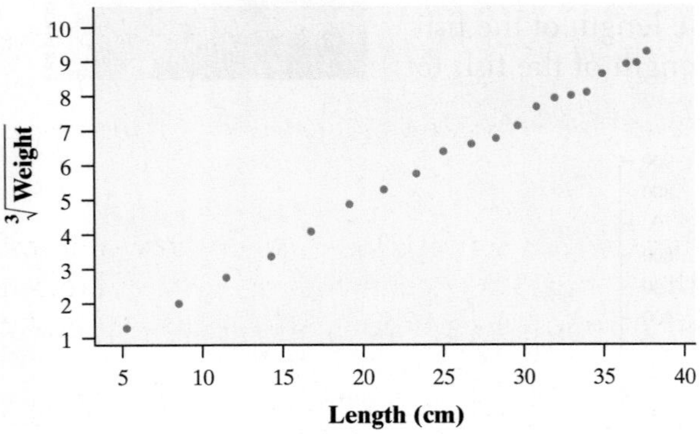


**FIGURE 3.20** The scatterplot of $\sqrt[3]{\text{weight}}$ versus length is linear.

There's another way to transform the data in the "Go fish!" example to achieve linearity. We can take the cube root of the weight values and graph $\sqrt[3]{\text{weight}}$ versus length. Figure 3.20 shows that the resulting scatterplot has a linear form.

Why does this transformation work? Start with weight $= a(\text{length})^3$ and take the cube root of both sides of the equation:

$$\sqrt[3]{\text{weight}} = \sqrt[3]{a(\text{length})^3}$$

$$\sqrt[3]{\text{weight}} = \sqrt[3]{a}(\text{length})$$

That is, there is a linear relationship between length and $\sqrt[3]{\text{weight}}$.

## EXAMPLE

### Go fish!
### Transforming with roots

Doug Wilson/Alamy

**PROBLEM:** Figure 3.20 shows that the relationship between length and $\sqrt[3]{\text{weight}}$ is roughly linear for Atlantic Ocean rockfish. Here is computer output from a linear regression analysis of $y = \sqrt[3]{\text{weight}}$ versus $x$ = length:

**Regression Analysis: $\sqrt[3]{\text{Weight}}$ versus Length**

| Predictor | Coef | SE Coef | T | P |
|---|---|---|---|---|
| Constant | −0.02204 | 0.07762 | −0.28 | 0.780 |
| Length | 0.246616 | 0.002868 | 86.00 | 0.000 |

S = 0.124161 R-Sq = 99.8% R-Sq(adj) = 99.7%

(a) Give the equation of the least-squares regression line. Define any variables you use.

(b) Suppose a contestant in the fishing tournament catches an Atlantic Ocean rockfish that's 36 centimeters long. Use the model from part (a) to predict the fish's weight.

**SOLUTION:**

(a) $\widehat{\sqrt[3]{\text{weight}}} = -0.02204 + 0.246616(\text{length})$

If you write the equation as $\widehat{\sqrt[3]{y}} = -0.02204 + 0.246616x$, make sure to define $y$ = weight and $x$ = length.

(b) $\widehat{\sqrt[3]{\text{weight}}} = -0.02204 + 0.246616(36) = 8.856$

$\widehat{\text{weight}} = 8.856^3 = 694.6$ grams

The least-squares regression line gives the predicted value of the *cube root* of weight. To get the predicted weight, reverse the cube root by raising the result to the third power.

FOR PRACTICE, TRY EXERCISE 83

When experience or theory suggests that a bivariate relationship is described by a power model of the form $y = ax^p$ where $p$ is known, there are two methods for transforming the data to achieve linearity.

1. Raise the values of the explanatory variable $x$ to the $p$ power and plot the points $(x^p, y)$.
2. Take the $p$th root of the values of the response variable $y$ and plot the points $(x, \sqrt[p]{y})$.

What if you have no idea what value of $p$ to use? You could guess and test until you find a transformation that works. Some technology comes with built-in sliders that allow you to dynamically adjust the power and watch the scatterplot change shape as you do.

It turns out that there is a much more efficient method for linearizing a curved pattern in a scatterplot. Instead of transforming with powers and roots, we use logarithms. This more general method works when the data follow an unknown power model or any of several other common mathematical models.

## Transforming with Logarithms: Power Models

To achieve linearity from a power model, we apply the logarithm transformation to *both* variables. Here are the details:

1. A power model has the form $y = ax^p$, where $a$ and $p$ are constants.
2. Take the logarithm of both sides of this equation. Using properties of logarithms, we get

$$\log y = \log(ax^p) = \log a + \log(x^p) = \log a + p\log x$$

The equation

$$\log y = \log a + p\log x$$

shows that taking the logarithm of both variables results in a linear relationship between $\log x$ and $\log y$. *Note:* You can use base-10 logarithms or natural (base-$e$) logarithms to straighten the association.

3. Look carefully: the *power* $p$ in the power model becomes the *slope* of the straight line that links $\log y$ to $\log x$.

If a power model describes the relationship between two variables, a scatterplot of the logarithms of both variables should produce a linear pattern. Then we can fit a least-squares regression line to the transformed data and use the linear model to make predictions. Here's an example.

**EXAMPLE**

### Go fish!

### Transforming with logarithms: Power models

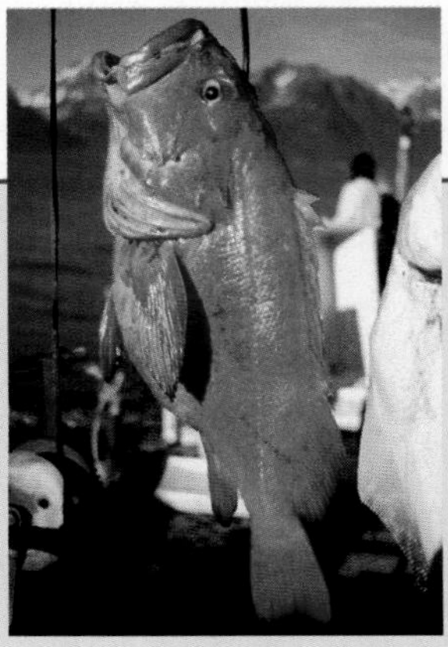

RGB Ventures/SuperStock/Alamy

**PROBLEM:** In the preceding examples, we used powers and roots to find a model for predicting the weight of an Atlantic Ocean rockfish from its length. We still expect a power model of the form weight = $a$(length)$^3$ based on geometry. Here once again is a scatterplot of the data from the local marine research lab:

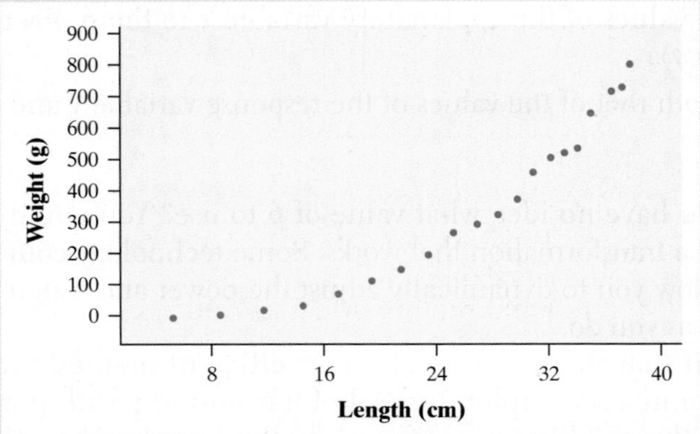


We took the logarithm (base 10) of the values for both variables. Here is some computer output from a linear regression analysis of the transformed data.

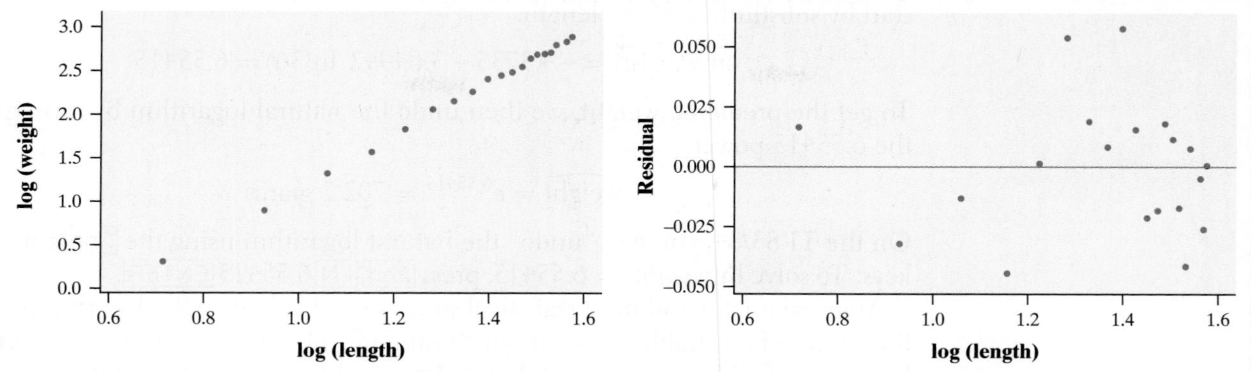


| Regression Analysis: log(Weight) versus log(Length) | | | | |
|---|---|---|---|---|
| Predictor | Coef | SE Coef | T | P |
| Constant | −1.89940 | 0.03799 | −49.99 | 0.000 |
| log(Length) | 3.04942 | 0.02764 | 110.31 | 0.000 |
| S = 0.0281823 | | R-Sq = 99.9% | R-Sq(adj) = 99.8% | |

(a) Based on the output, explain why it would be reasonable to use a power model to describe the relationship between weight and length for Atlantic Ocean rockfish.

(b) Give the equation of the least-squares regression line. Be sure to define any variables you use.

(c) Suppose a contestant in the fishing tournament catches an Atlantic Ocean rockfish that's 36 centimeters long. Use the model from part (b) to predict the fish's weight.

## SOLUTION:

(a) The scatterplot of log(weight) versus log(length) has a linear form, and the residual plot shows a fairly random scatter of points about the residual = 0 line. So a power model seems reasonable here.

> If a power model describes the relationship between two variables *x* and *y*, then a *linear* model should describe the relationship between log *x* and log *y*.

(b) $\widehat{\log(\text{weight})} = -1.89940 + 3.04942 \log(\text{length})$

> If you write the equation as $\widehat{\log(y)} = -1.89940 + 3.04942 \log(x)$, make sure to define $y$ = weight and $x$ = length.

(c) $\widehat{\log(\text{weight})} = -1.89940 + 3.04942 \log(36)$

$\widehat{\log(\text{weight})} = 2.8464$

$\widehat{\text{weight}} = 10^{2.8464} \approx 702.1$ grams

> The least-squares regression line gives the predicted value of the base-10 *logarithm* of weight. To get the predicted weight, undo the logarithm by raising 10 to the 2.8464 power.

**FOR PRACTICE, TRY EXERCISE 85**

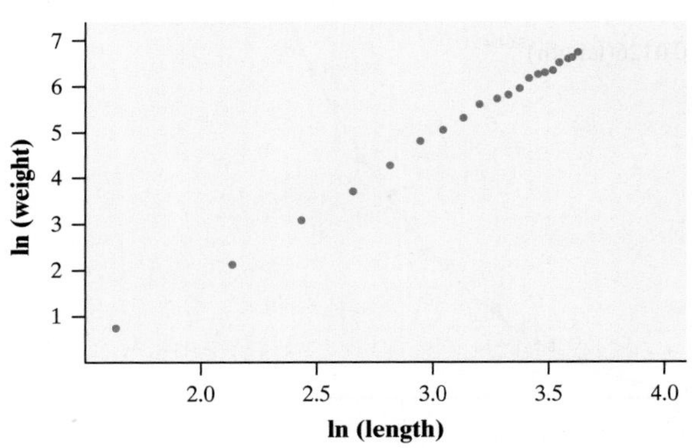


On the TI-83/84, you can "undo" the logarithm using the 2nd function keys. To solve $\widehat{\log(\text{weight})} = 2.8464$, press 2nd LOG 2.8464 ENTER.

In addition to base-10 logarithms, you can also use natural (base-*e*) logarithms to transform the variables. Using the same Atlantic Ocean rockfish data, here is a scatterplot of ln(weight) versus ln(length).

The least-squares regression line for these data is

$$\widehat{\ln(\text{weight})} = -4.3735 + 3.04942 \ln(\text{length})$$

To predict the weight of an Atlantic Ocean rockfish that is 36 centimeters, we start by substituting 36 for length.

$$\widehat{\ln(\text{weight})} = -4.3735 + 3.04942 \ln(36) = 6.55415$$

To get the predicted weight, we then undo the natural logarithm by raising $e$ to the 6.55415 power.

$$\widehat{\text{weight}} = e^{6.55415} = 702.2 \text{ grams}$$

On the TI-83/84, you can "undo" the natural logarithm using the 2nd function keys. To solve ln(weight) = 6.55415, press 2nd LN 6.55415 ENTER.

Your calculator and most statistical software will calculate the logarithms of all the values of a variable with a single command. The important thing to remember is that if a bivariate relationship is described by a power model, then we can linearize the relationship by taking the logarithm of *both* the explanatory and response variables.

## Think About It

**HOW DO WE FIND THE POWER MODEL FOR PREDICTING *Y* FROM *X*?** The least-squares line for the transformed rockfish data is

$$\widehat{\log(\text{weight})} = -1.89940 + 3.04942 \log(\text{length})$$

If we use the definition of the logarithm as an exponent, we can rewrite this equation as

$$\widehat{\text{weight}} = 10^{-1.89940 + 3.04942 \log(\text{length})}$$

Using properties of exponents, we can simplify this as follows:

$$\widehat{\text{weight}} = 10^{-1.89940} \cdot 10^{3.04942 \log(\text{length})} \quad \text{using the fact that } b^m b^n = b^{m+n}$$

$$\widehat{\text{weight}} = 10^{-1.89940} \cdot 10^{\log(\text{length})^{3.04942}} \quad \text{using the fact that } p \log x = \log x^p$$

$$\widehat{\text{weight}} = 0.0126(\text{length})^{3.04942} \quad \text{using the fact that } 10^{\log x} = x$$

This equation is now in the familiar form of a power model $y = ax^p$ with $a = 0.0126$ and $p = 3.04942$. Notice how close the power is to 3, as expected from geometry.

We could use the power model to predict the weight of a 36-centimeter-long Atlantic Ocean rockfish:

$$\widehat{\text{weight}} = 0.0126(36)^{3.04942} \approx 701.76 \text{ grams}$$

This is roughly the same prediction we got earlier. Here is the scatterplot of the original rockfish data with the power model added. Note how well this model fits the association!

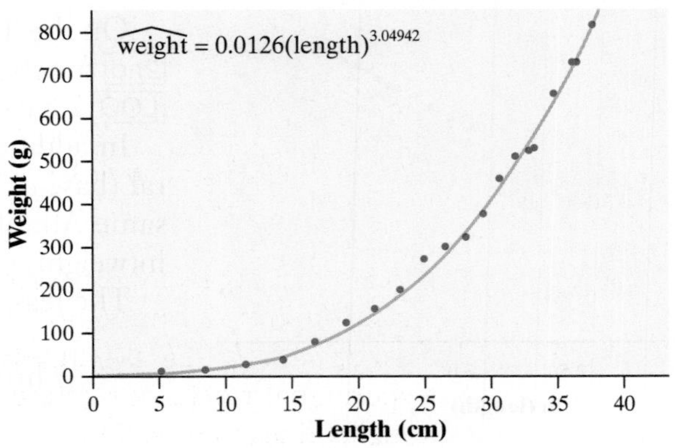

## Transforming with Logarithms: Exponential Models

A linear model has the form $y = a + bx$. The value of $y$ increases (or decreases) at a constant rate as $x$ increases. The slope $b$ describes the constant rate of change of a linear model. That is, for each 1-unit increase in $x$, the model predicts an increase of $b$ units in $y$. You can think of a linear model as describing the repeated addition of a constant amount. Sometimes the relationship between $y$ and $x$ is based on repeated *multiplication* by a constant factor. That is, each time $x$ increases by 1 unit, the value of $y$ is multiplied by $b$. An *exponential model* of the form $y = ab^x$ describes such growth by multiplication.

Populations of living things tend to grow exponentially if not restrained by outside limits such as lack of food or space. More pleasantly (unless we're talking about credit card debt!), money also displays exponential growth when interest is compounded each time period. Compounding means that the last period's income earns income in the next period. Figure 3.21 shows the balance of a savings account where \$100 is invested at 6% interest, compounded annually (assuming no additional deposits or withdrawals). After $x$ years, the account balance $y$ is given by the exponential model $y = 100(1.06)^x$.

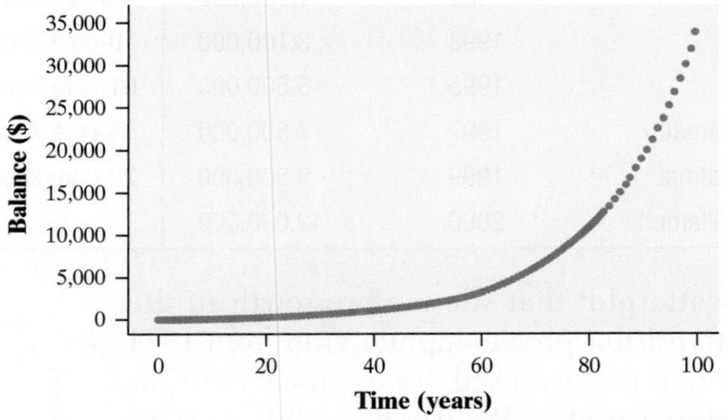


**FIGURE 3.21** Scatterplot of the exponential growth of a \$100 investment in a savings account paying 6% interest, compounded annually.

An exponential model of the form $y = ab^x$ describes the relationship between $x$ and $y$, where $a$ and $b$ are constants. We can use logarithms to produce a linear relationship. Start by taking the logarithm of each side (we'll use base 10, but the natural logarithm ln using base $e$ would work just as well). Then use algebraic properties of logarithms to simplify the resulting expressions. Here are the details:

| | |
|---|---|
| $\log y = \log(ab^x)$ | taking the logarithm of both sides |
| $\log y = \log a + \log(b^x)$ | using the property $\log(mn) = \log m + \log n$ |
| $\log y = \log a + x \log b$ | using the property $\log m^p = p \log m$ |

We can then rearrange the final equation as

$$\log y = \log a + (\log b)x$$

Notice that $\log a$ and $\log b$ are constants because $a$ and $b$ are constants. So the equation gives a linear model relating the explanatory variable $x$ to the transformed variable $\log y$. Thus, if the relationship between two variables follows an exponential model, a scatterplot of the logarithm of $y$ against $x$ should show a roughly linear association.

## EXAMPLE

# Moore's law and computer chips

### Transforming with logarithms: Exponential models

Nandana de silva/Alamy

**PROBLEM:** Gordon Moore, one of the founders of Intel Corporation, predicted in 1965 that the number of transistors on an integrated circuit chip would double every 18 months. This is Moore's law, one way to measure the revolution in computing. Here are data on the dates and number of transistors for Intel microprocessors:[42]

| Processor | Year | Transistors | Processor | Year | Transistors |
|---|---|---|---|---|---|
| Intel 4004 | 1971 | 2,300 | Pentium III Tualatin | 2001 | 45,000,000 |
| Intel 8008 | 1972 | 3,500 | Itanium 2 McKinley | 2002 | 220,000,000 |
| Intel 8080 | 1974 | 4,500 | Itanium 2 Madison 6M | 2003 | 410,000,000 |
| Intel 8086 | 1978 | 29,000 | Itanium 2 with 9 MB cache | 2004 | 592,000,000 |
| Intel 80286 | 1982 | 134,000 | Dual-core Itanium 2 | 2006 | 1,700,000,000 |
| Intel 80386 | 1985 | 275,000 | Six-core Xeon 7400 | 2008 | 1,900,000,000 |
| Intel 80486 | 1989 | 1,180,235 | 8-core Xeon Nehalem-EX | 2010 | 2,300,000,000 |
| Pentium | 1993 | 3,100,000 | 10-core Xeon Westmere-EX | 2011 | 2,600,000,000 |
| Pentium Pro | 1995 | 5,500,000 | 61-core Xeon Phi | 2012 | 5,000,000,000 |
| Pentium II Klamath | 1997 | 7,500,000 | 18-core Xeon Haswell-E5 | 2014 | 5,560,000,000 |
| Pentium III Katmai | 1999 | 9,500,000 | 22-core Xeon Broadwell-E5 | 2016 | 7,200,000,000 |
| Pentium 4 Willamette | 2000 | 42,000,000 | | | |

Here is a scatterplot that shows the growth in the number of transistors on a computer chip from 1971 to 2016. Notice that we used "years since 1970" as the explanatory variable. We'll explain this on page 224. If Moore's law is correct, then an exponential model should describe the relationship between the variables.

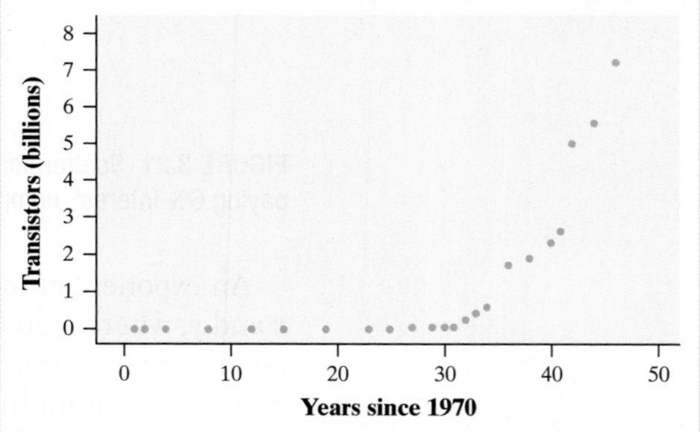


(a) Here is a scatterplot of the natural (base-$e$) logarithm of the number of transistors on a computer chip versus years since 1970. Based on this graph, explain why it would be reasonable to use an exponential model to describe the relationship between number of transistors and years since 1970.

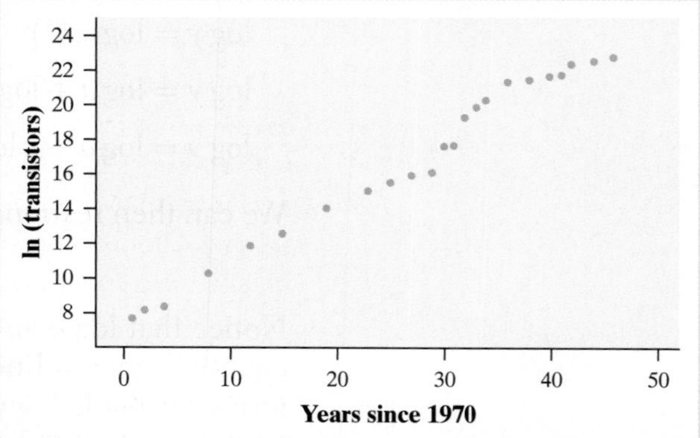

(b) Here is some computer output from a linear regression analysis of the transformed data. Give the equation of the least-squares regression line. Be sure to define any variables you use.

| Predictor | Coef | SE Coef | T | P |
|---|---|---|---|---|
| Constant | 7.2272 | 0.3058 | 23.64 | 0.000 |
| Years since 1970 | 0.3542 | 0.0102 | 34.59 | 0.000 |
| S = 0.6653 | R-Sq = 98.2% | | R-Sq(adj) = 98.2% | |

(c) Use your model from part (b) to predict the number of transistors on an Intel computer chip in 2020.

## SOLUTION:

(a) The scatterplot of ln(transistors) versus years since 1970 has a fairly linear pattern. So an exponential model seems reasonable here.

If an exponential model describes the relationship between two variables *x* and *y*, we expect a scatterplot of (*x*, ln *y*) to be roughly linear.

(b) $\widehat{\ln(\text{transistors})} = 7.2272 + 0.3542(\text{years since } 1970)$

If you write the equation as $\widehat{\ln(y)} = 7.2272 + 0.3542x$, make sure to define $y =$ number of transistors and $x =$ years since 1970.

(c) $\widehat{\ln(\text{transistors})} = 7.2272 + 0.3542(50) = 24.9372$

$\widehat{\text{transistors}} = e^{24.9372} = 67{,}622{,}053{,}360$

2020 is 50 years since 1970.

This model predicts that an Intel chip made in 2020 will have about 68 billion transistors.

The least-squares regression line gives the predicted value of ln(transistors). To get the predicted number of transistors, undo the logarithm by raising *e* to the 24.9372 power.

**FOR PRACTICE, TRY EXERCISE 89**

Here is a residual plot for the linear regression in part (b) of the example:

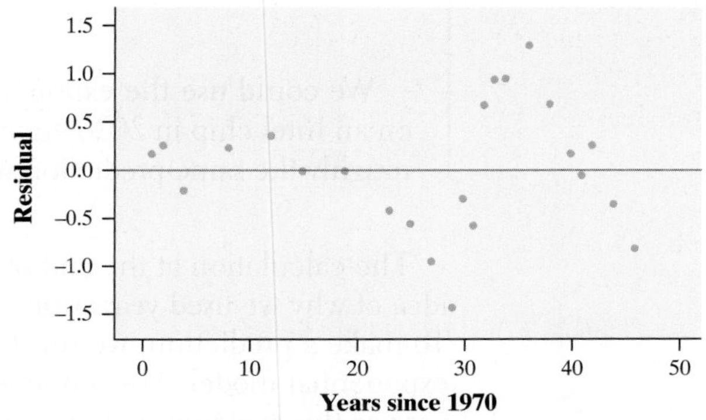


The residual plot shows a leftover pattern, with the residuals going from positive to negative to positive to negative as we move from left to right. However, the residuals are small in size relative to the transformed *y*-values, and the scatterplot of the transformed data is much more linear than the original scatterplot. We feel reasonably comfortable using this model to make predictions about the number of transistors on a computer chip.

Let's recap this big idea: When an association follows an exponential model, the transformation to achieve linearity is carried out by taking the logarithm of the response variable. The crucial property of the logarithm for our purposes is that *if a variable grows exponentially, its logarithm grows linearly*.

**Think About It**

**HOW DO WE FIND THE EXPONENTIAL MODEL FOR PREDICTING *Y* FROM *X*?** The least-squares line for the transformed data in the computer chip example is

$$\widehat{\ln(\text{transistors})} = 7.2272 + 0.3542(\text{years since 1970})$$

If we use the definition of the logarithm as an exponent, we can rewrite this equation as

$$\widehat{\text{transistors}} = e^{7.2272+0.3542(\text{years since 1970})}$$

Using properties of exponents, we can simplify this as follows:

| | |
|---|---|
| $\widehat{\text{transistors}} = e^{7.2272} \cdot e^{0.3542(\text{years since 1970})}$ | using the fact that $b^m b^n = b^{m+n}$ |
| $\widehat{\text{transistors}} = e^{7.2272} \cdot (e^{0.3542})^{(\text{years since 1970})}$ | using the fact that $(b^m)^n = b^{mn}$ |
| $\widehat{\text{transistors}} = 1376.4 \cdot (1.4250)^{(\text{years since 1970})}$ | simplifying |

This equation is now in the familiar form of an exponential model $y = ab^x$ with $a = 1376.4$ and $b = 1.4250$. Here is the scatterplot of the original transistor data with the exponential model added:

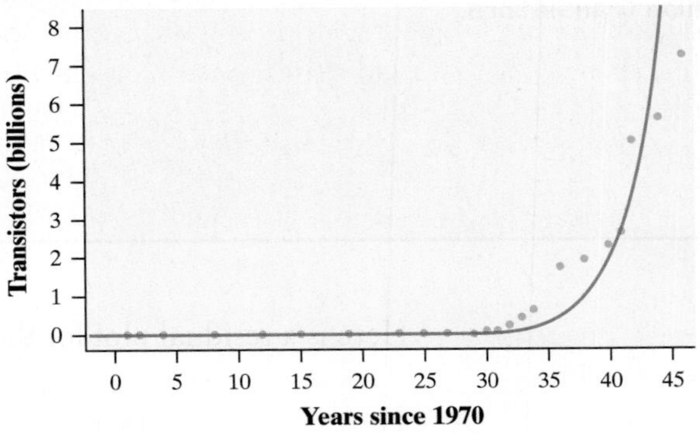


We could use the exponential model to predict the number of transistors on an Intel chip in 2020: $\widehat{\text{transistors}} = 1376.4(1.4250)^{50} \approx 6.7529 \cdot 10^{10}$. This is roughly the same prediction we obtained earlier.

The calculation at the end of the Think About It feature might give you some idea of why we used years since 1970 as the explanatory variable in the example. To make a prediction, we substituted the value $x = 50$ into the equation for the exponential model. This value is the exponent in our calculation. If we had used year as the explanatory variable, our exponent would have been 2020. Such a large exponent can lead to overflow errors on a calculator.

## Putting It All Together: Which Transformation Should We Choose?

Suppose that a scatterplot shows a curved relationship between two quantitative variables $x$ and $y$. How can we decide whether a power model or an exponential model better describes the relationship?

### HOW TO CHOOSE A MODEL

When choosing between different models to describe a relationship between two quantitative variables:

- Choose the model whose residual plot has the most random scatter.
- If there is more than one model with a randomly scattered residual plot, choose the model with the largest coefficient of determination, $r^2$.

It is not advisable to use the standard deviation of the residuals $s$ to help choose a model, as the $y$ values for the different models might be on different scales.

The following example illustrates the process of choosing the most appropriate model for a curved relationship.

## EXAMPLE

### Stop that car!

### Choosing a model

**PROBLEM:** How is the braking distance for a car related to the amount of tread left on the tires? Researchers collected data on the braking distance (measured in car lengths) for a car making a panic stop in standing water, along with the tread depth of the tires (in 1/32 inch).[43]

Here is linear regression output for three different models, along with a residual plot for each model. Model 1 is based on the original data, while Models 2 and 3 involve transformations of the original data.

**Model 1. Braking distance vs. Tread depth**

| Predictor | Coef | SE Coef | T | P |
|---|---|---|---|---|
| Constant | 16.4873 | 0.7648 | 21.557 | 0.0000 |
| Tread depth | −0.7282 | 0.1125 | −6.457 | 0.0001 |

S = 1.1827 R-Sq = 0.822 R-sq(adj) = 0.803

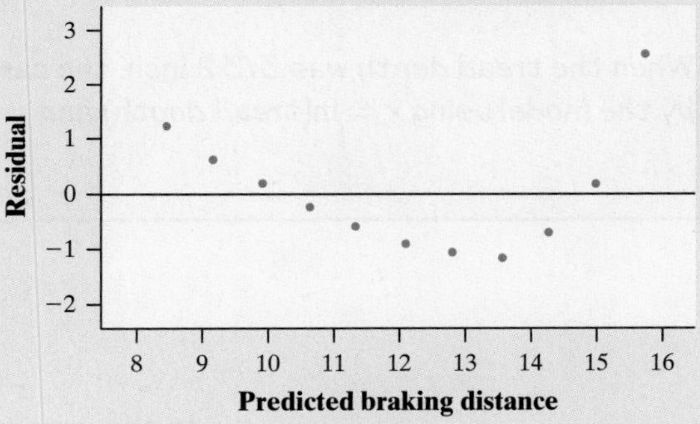


**Model 2. ln(braking distance) vs. ln(tread depth)**

| Predictor | Coef | SE Coef | T | P |
|---|---|---|---|---|
| Constant | 2.9034 | 0.0051 | 566.34 | 0.0000 |
| ln(tread depth) | −0.2690 | 0.0029 | −91.449 | 0.0000 |

S = 0.007 R-sq = 0.999 R-sq(adj) = 0.999

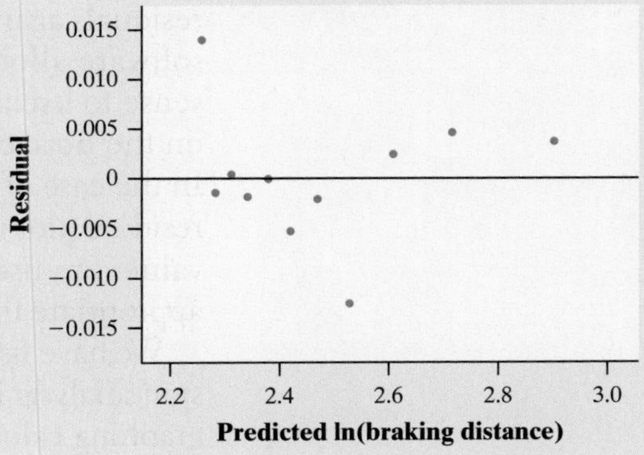

**Model 3. ln(braking distance) vs. Tread depth**

| Predictor | Coef | SE Coef | T | P |
|---|---|---|---|---|
| Constant | 2.8169 | 0.0461 | 61.077 | 0.0000 |
| Tread depth | −0.0569 | 0.0068 | −8.372 | 0.0000 |

S = 0.071 R-sq = 0.886 R-sq(adj) = 0.874

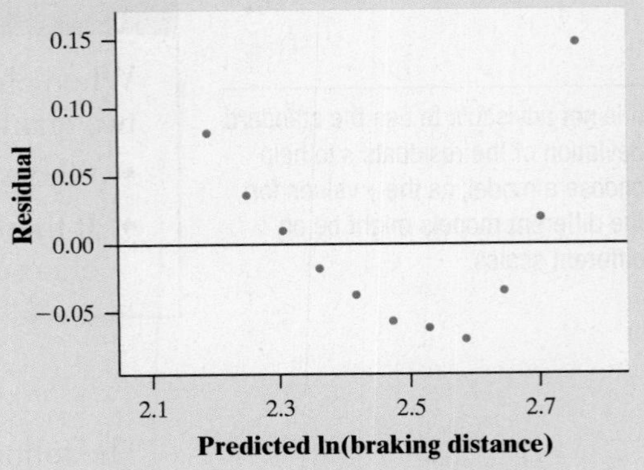


(a) Which model does the best job of summarizing the relationship between tread depth and braking distance? Explain your reasoning.

(b) Use the model chosen in part (a) to calculate and interpret the residual for the trial when the tread depth was 3/32 inch and the stopping distance was 13.6 car lengths.

**SOLUTION:**

(a) Because the model that uses $x = \ln(\text{tread depth})$ and $y = \ln(\text{braking distance})$ produced the most randomly scattered residual plot with no leftover curved pattern, it is the most appropriate model.

Note that the value of $r^2$ is also closest to 1 for the model that uses $x = \ln(\text{tread depth})$ and $y = \ln(\text{braking distance})$.

(b) $\widehat{\ln(\text{braking distance})} = 2.9034 - 0.2690\ln(3) = 2.608$

$\widehat{\text{braking distance}} = e^{2.608} = 13.57$ car lengths

Residual $= 13.6 - 13.57 = 0.03$

The residual calculated here is on the original scale (car lengths), while the residuals shown in the residual plot for this model are on a logarithmic scale.

When the tread depth was 3/32 inch, the car traveled 0.03 car length farther than the distance predicted by the model using $x = \ln(\text{tread depth})$ and $y = \ln(\text{braking distance})$.

**FOR PRACTICE, TRY EXERCISE 91**

In the preceding example, the residual plots used the predicted values on the horizontal axis rather than the values of the explanatory variable. Plotting the residuals against the predicted values is common in statistical software. Because software allows for multiple explanatory variables in a single model, it makes sense to use a combination of the explanatory variables (the predicted values) on the horizontal axis rather than using only one of the explanatory variables. In the case of simple linear regression (one explanatory variable), we interpret a residual plot in the same way, whether the explanatory variable or the predicted values are used on the horizontal axis: the more randomly scattered, the more appropriate the model.

We have used statistical software to do all the transformations and linear regression analysis in this section so far. Now let's look at how the process works on a graphing calculator.

## 11. Technology Corner TRANSFORMING TO ACHIEVE LINEARITY

*TI-Nspire and other technology instructions are on the book's website at highschool.bfwpub.com/updatedtps6e.*

We'll use the Atlantic Ocean rockfish data to illustrate a general strategy for performing transformations with logarithms on the TI-83/84. A similar approach could be used for transforming data with powers and roots.

- Enter the values of the explanatory variable in L1 and the values of the response variable in L2. Make a scatterplot of $y$ versus $x$ and confirm that there is a curved pattern.

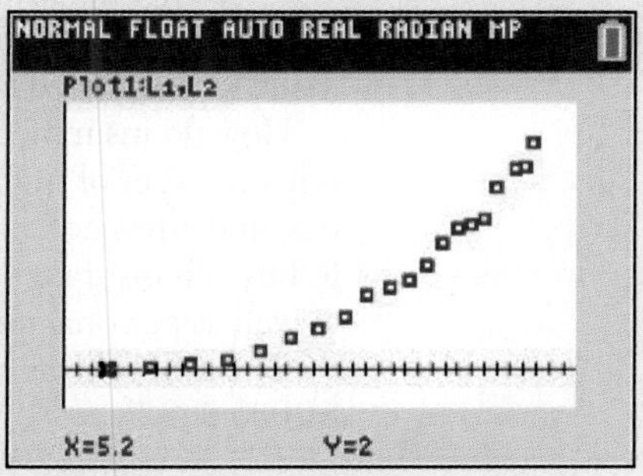


- Define L3 to be the natural logarithm (ln) of L1 and L4 to be the natural logarithm of L2. To see whether a power model fits the original data, make a plot of ln $y$ (L4) versus ln $x$ (L3) and look for linearity. To see whether an exponential model fits the original data, make a plot of ln $y$ (L4) versus $x$ (L1) and look for linearity.

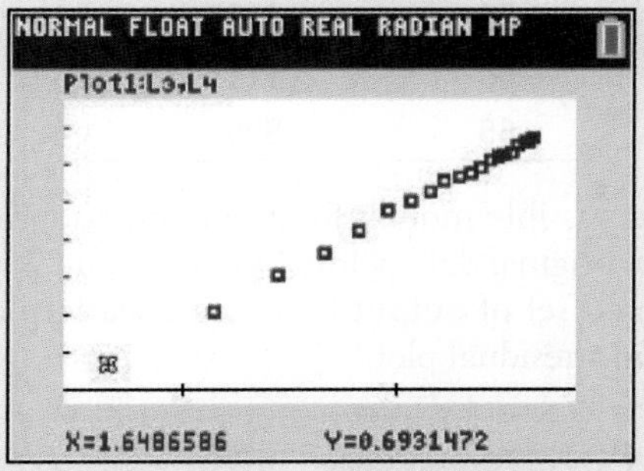


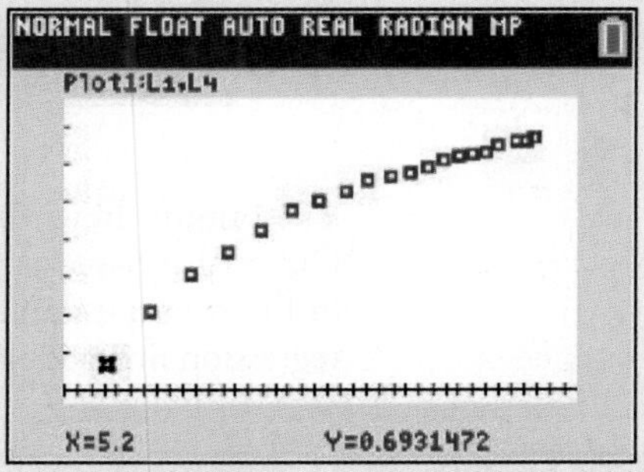


- If a linear pattern is present, calculate the equation of the least-squares regression line. For the Atlantic Ocean rockfish data, we executed the command LinReg(a + bx)L3, L4.

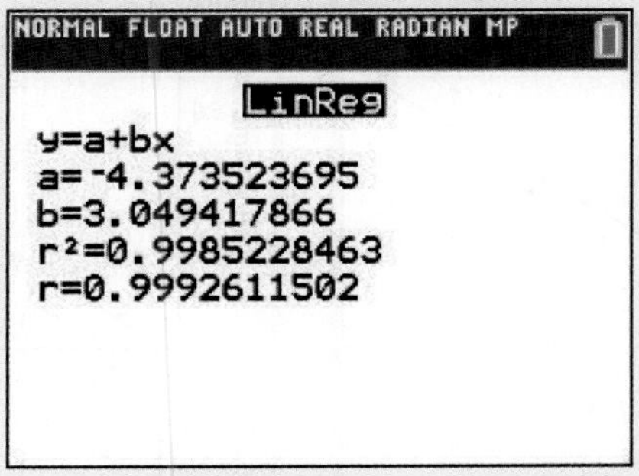


- Construct a residual plot to look for any leftover curved patterns. For Xlist, enter the list you used as the explanatory variable in the linear regression calculation. For Ylist, use the RESID list stored in the calculator. For the Atlantic Ocean rockfish data, we used L3 as the Xlist.

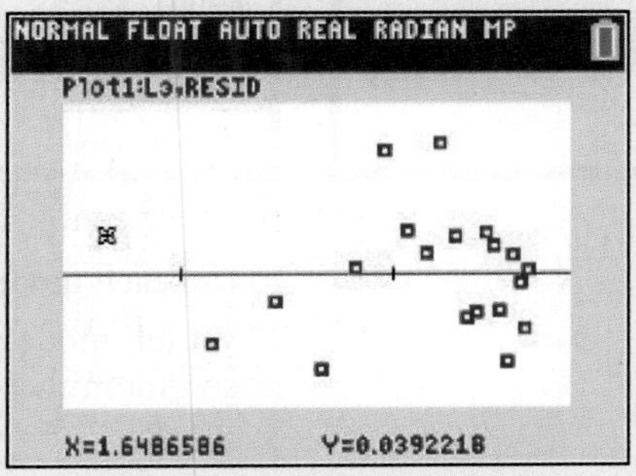

## CHECK YOUR UNDERSTANDING

One sad fact about life is that we'll all die someday. Many adults plan ahead for their eventual passing by purchasing life insurance. Many different types of life insurance policies are available. Some provide coverage throughout an individual's life (whole life), while others last only for a specified number of years (term life). The policyholder makes regular payments (premiums) to the insurance company in return for the coverage. When the insured person dies, a payment is made to designated family members or other beneficiaries.

How do insurance companies decide how much to charge for life insurance? They rely on a staff of highly trained actuaries—people with expertise in probability, statistics, and advanced mathematics—to establish premiums. For an individual who wants to buy life insurance, the premium will depend on the type and amount of the policy as well as personal characteristics like age, sex, and health status.

The table shows monthly premiums for a 10-year term-life insurance policy worth \$1,000,000.[44]

| Age (years) | Monthly premium |
|---|---|
| 40 | \$29 |
| 45 | \$46 |
| 50 | \$68 |
| 55 | \$106 |
| 60 | \$157 |
| 65 | \$257 |

The output shows three possible models for predicting monthly premium from age. Option 1 is based on the original data, while Options 2 and 3 involve transformations of the original data. Each set of output includes a scatterplot with a least-squares regression line added and a residual plot.

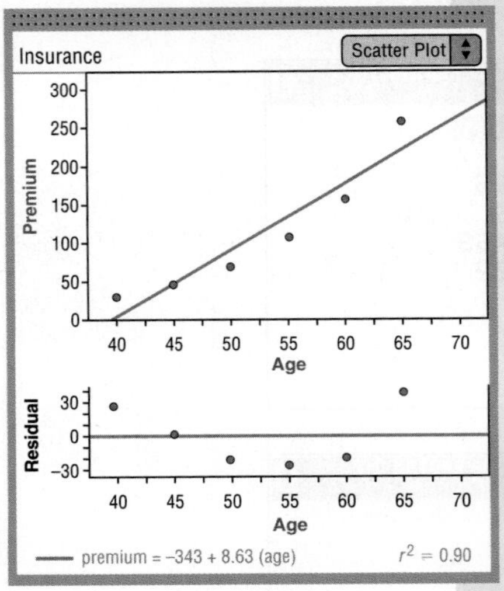


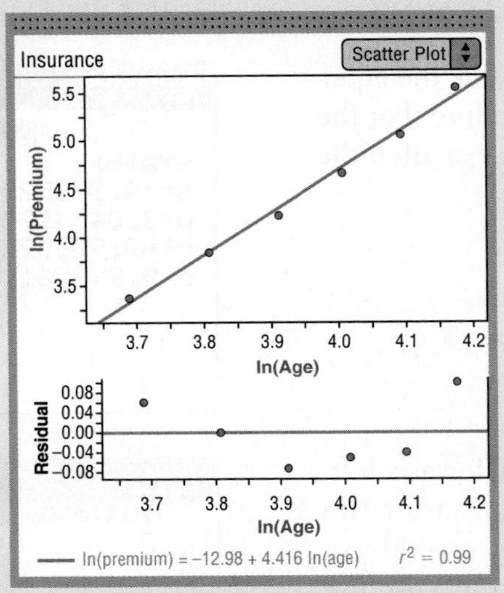


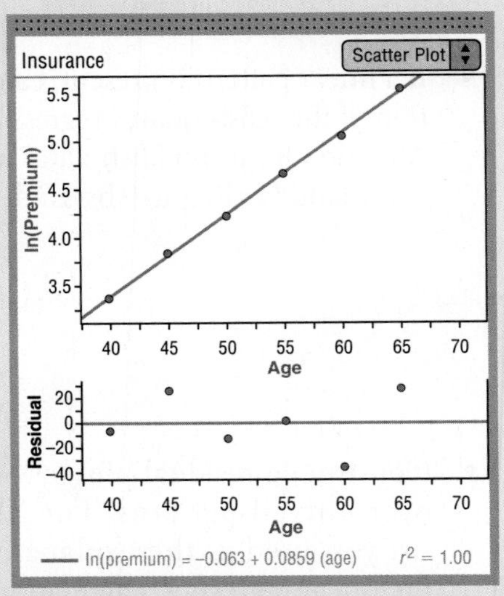


1. Use each model to predict how much a 58-year-old would pay for such a policy.
2. Which model does the best job summarizing the relationship between age and monthly premium? Explain your answer.

## Section 3.3 Summary

- Curved relationships between two quantitative variables can sometimes be changed into linear relationships by **transforming** one or both of the variables. Once we transform the data to achieve linearity, we can fit a least-squares regression line to the transformed data and use this linear model to make predictions.
- When theory or experience suggests that the relationship between two variables follows a **power model** of the form $y = ax^p$, transformations involving powers and roots can linearize a curved pattern in a scatterplot.
  - **Option 1**: Raise the values of the explanatory variable $x$ to the power $p$, then look at a graph of $(x^p, y)$.
  - **Option 2**: Take the $p$th root of the values of the response variable $y$, then look at a graph of $(x, \sqrt[p]{y})$.
- Another useful strategy for straightening a curved pattern in a scatterplot is to take the **logarithm** of one or both variables. When a power model describes the relationship between two variables, a plot of log $y$ versus log $x$ (or ln $y$ versus ln $x$) should be linear.
- For an **exponential model** of the form $y = ab^x$, the predicted values of the response variable are multiplied by a factor of $b$ for each increase of 1 unit in the explanatory variable. When an exponential model describes the relationship between two variables, a plot of log $y$ versus $x$ (or ln $y$ versus $x$) should be linear.
- To decide between competing models, choose the model with the most randomly scattered residual plot. If it is difficult to determine which residual plot is the most randomly scattered, choose the model with the largest value of $r^2$.

### 3.3 Technology Corner

*TI-Nspire and other technology instructions are on the book's website at highschool.bfwpub.com/updatedtps6e.*

**11. Transforming to achieve linearity** **Page 227**

## Section 3.3 Exercises

81. **The swinging pendulum** Mrs. Hanrahan's precalculus pg 215 class collected data on the length (in centimeters) of a pendulum and the time (in seconds) the pendulum took to complete one back-and-forth swing (called its period). The theoretical relationship between a pendulum's length and its period is

$$\text{period} = \frac{2\pi}{\sqrt{g}}\sqrt{\text{length}}$$

where g is a constant representing the acceleration due to gravity (in this case, $g = 980\ \text{cm/s}^2$). Here is a graph of period versus $\sqrt{\text{length}}$, along with output from a linear regression analysis using these variables.

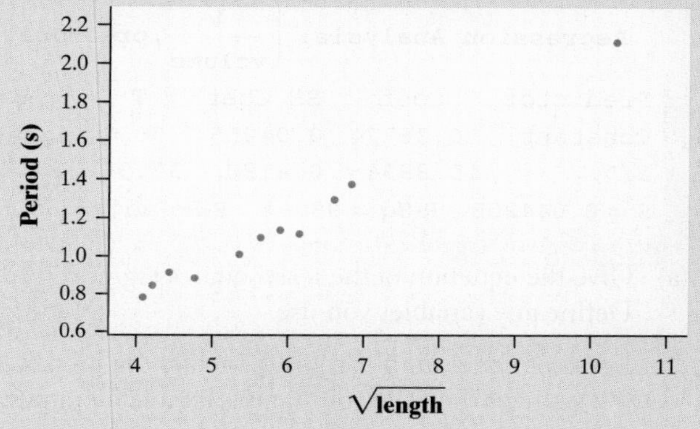

**Regression Analysis: ($\sqrt{\text{length}}$, period)**

| Predictor | Coef | SE Coef | T | P |
|---|---|---|---|---|
| Constant | −0.08594 | 0.05046 | −1.70 | 0.123 |
| sqrt(length) | 0.209999 | 0.008322 | 25.23 | 0.000 |

S = 0.0464223 R-Sq = 98.6% R-Sq(adj) = 98.5%

(a) Give the equation of the least-squares regression line. Define any variables you use.

(b) Use the model from part (a) to predict the period of a pendulum with length 80 cm.

**82. Boyle's law** If you have taken a chemistry or physics class, then you are probably familiar with Boyle's law: for gas in a confined space kept at a constant temperature, pressure times volume is a constant (in symbols, $PV = k$). Students in a chemistry class collected data on pressure and volume using a syringe and a pressure probe. If the true relationship between the pressure and volume of the gas is $PV = k$, then

$$P = k\frac{1}{V}$$

Here is a graph of pressure versus $\frac{1}{\text{volume}}$, along with output from a linear regression analysis using these variables:

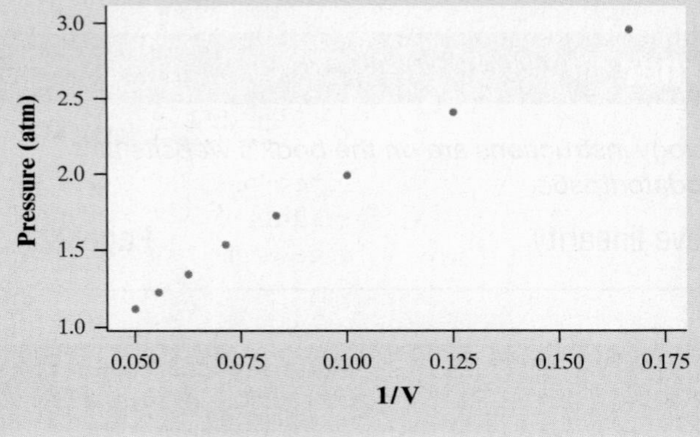


**Regression Analysis: ($\frac{1}{\text{volume}}$, pressure)**

| Predictor | Coef | SE Coef | T | P |
|---|---|---|---|---|
| Constant | 0.36774 | 0.04055 | 9.07 | 0.000 |
| 1/V | 15.8994 | 0.4190 | 37.95 | 0.000 |

S = 0.044205 R-Sq = 99.6% R-Sq(adj) = 99.5%

(a) Give the equation of the least-squares regression line. Define any variables you use.

(b) Use the model from part (a) to predict the pressure in the syringe when the volume is 17 cubic centimeters.

**83. The swinging pendulum** Refer to Exercise 81. pg 217 Here is a graph of period$^2$ versus length, along with output from a linear regression analysis using these variables.

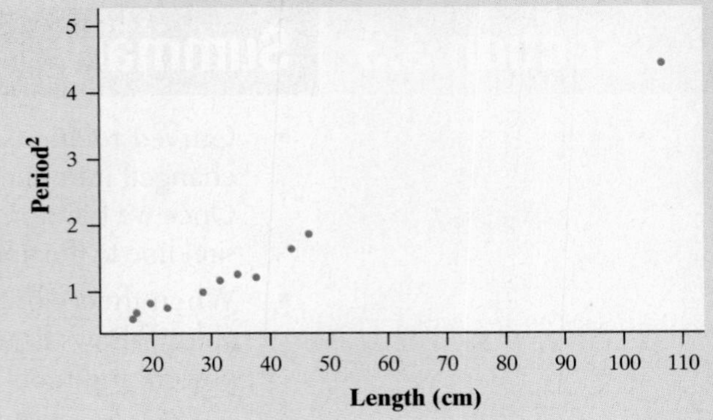


**Regression Analysis: (length, period$^2$)**

| Predictor | Coef | SE Coef | T | P |
|---|---|---|---|---|
| Constant | −0.15465 | 0.05802 | −2.67 | 0.026 |
| Length | 0.042836 | 0.001320 | 32.46 | 0.000 |

S = 0.105469 R-Sq = 99.2% R-Sq(adj) = 99.1%

(a) Give the equation of the least-squares regression line. Define any variables you use.

(b) Use the model from part (a) to predict the period of a pendulum with length 80 centimeters.

**84. Boyle's law** Refer to Exercise 82. Here is a graph of $\frac{1}{\text{pressure}}$ versus volume, along with output from a linear regression analysis using these variables:

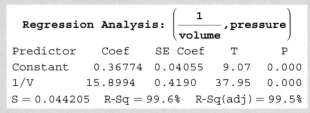

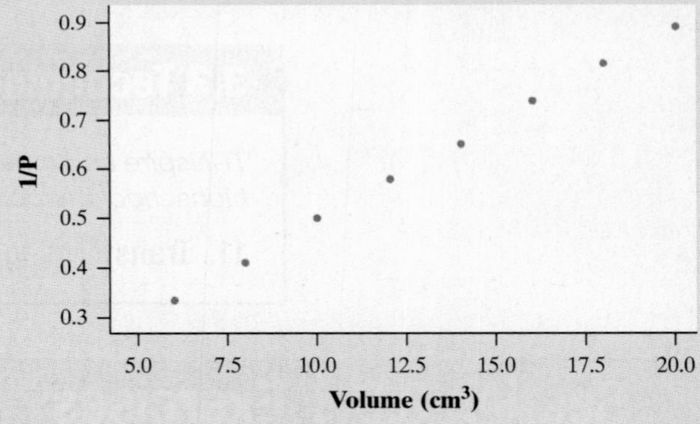


**Regression Analysis: (volume, $\frac{1}{\text{pressure}}$)**

| Predictor | Coef | SE Coef | T | P |
|---|---|---|---|---|
| Constant | 0.100170 | 0.003779 | 26.51 | 0.000 |
| Volume | 0.0398119 | 0.0002741 | 145.23 | 0.000 |

S = 0.003553 R-Sq = 100.0% R-Sq(adj) = 100.0%

(a) Give the equation of the least-squares regression line. Define any variables you use.

(b) Use the model from part (a) to predict the pressure in the syringe when the volume is 17 cubic centimeters.

**85. The swinging pendulum** Refer to Exercise 81. We pg 218 took the logarithm (base 10) of the values for both length and period. Here is some computer output from a linear regression analysis of the transformed data.

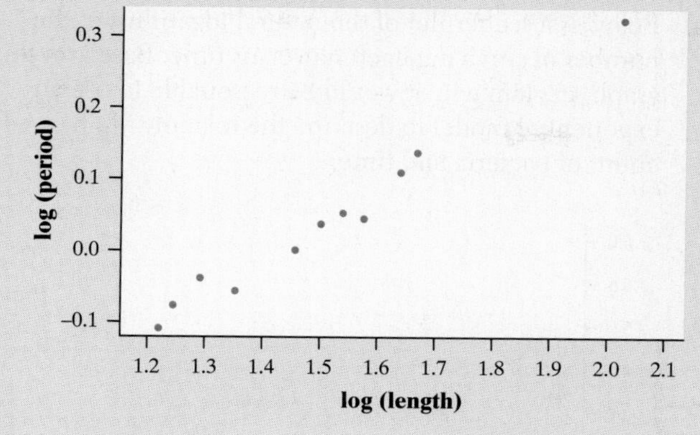


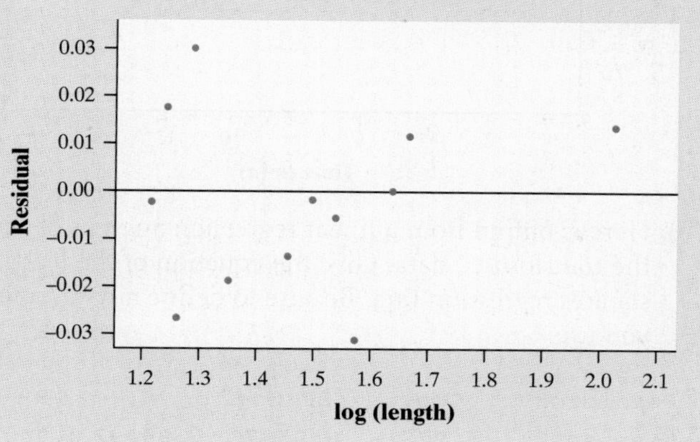


**Regression Analysis: log(Period) versus log(Length)**

| Predictor | Coef | SE Coef | T | P |
|---|---|---|---|---|
| Constant | −0.73675 | 0.03808 | −19.35 | 0.000 |
| log(Length) | 0.51701 | 0.02511 | 20.59 | 0.000 |
| S = 0.0185568 | R-Sq = 97.9% | R-Sq(adj) = 97.7% | | |

(a) Based on the output, explain why it would be reasonable to use a power model to describe the relationship between the length and period of a pendulum.

(b) Give the equation of the least-squares regression line. Be sure to define any variables you use.

(c) Use the model from part (b) to predict the period of a pendulum with length 80 cm.

86. **Boyle's law** Refer to Exercise 82. We took the logarithm (base 10) of the values for both volume and pressure. Here is some computer output from a linear regression analysis of the transformed data.

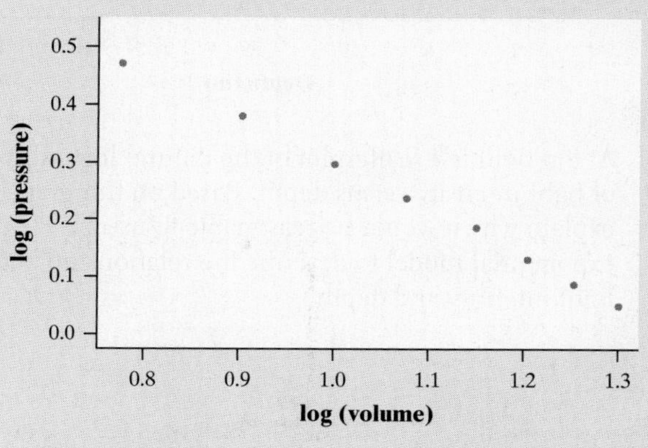


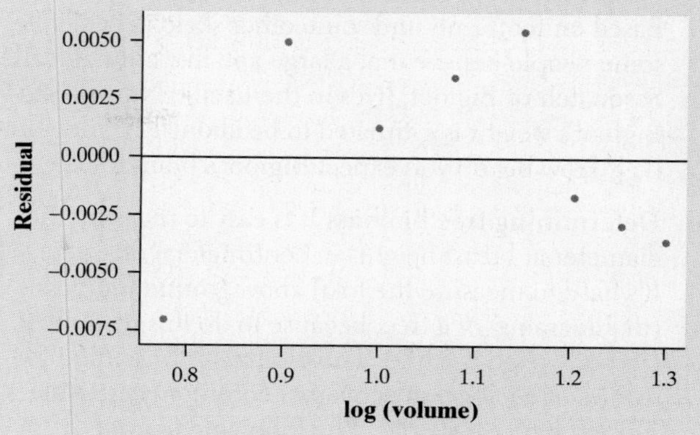


**Regression Analysis: log(Pressure) versus log(Volume)**

| Predictor | Coef | SE Coef | T | P |
|---|---|---|---|---|
| Constant | 1.11116 | 0.01118 | 99.39 | 0.000 |
| log(Volume) | −0.81344 | 0.01020 | −79.78 | 0.000 |
| S = 0.00486926 | R-Sq = 99.9% | R-Sq(adj) = 99.9% | | |

(a) Based on the output, explain why it could be reasonable to use a power model to describe the relationship between pressure and volume.

(b) Give the equation of the least-squares regression line. Be sure to define any variables you use.

(c) Use the model from part (b) to predict the pressure in the syringe when the volume is 17 cubic centimeters.

87. **Brawn versus brain** How is the weight of an animal's brain related to the weight of its body? Researchers collected data on the brain weight (in grams) and body weight (in kilograms) for 96 species of mammals.[45] The following figure is a scatterplot of the logarithm of brain weight against the logarithm of body weight for all 96 species. The least-squares regression line for the transformed data is

$$\widehat{\log y} = 1.01 + 0.72 \log x$$

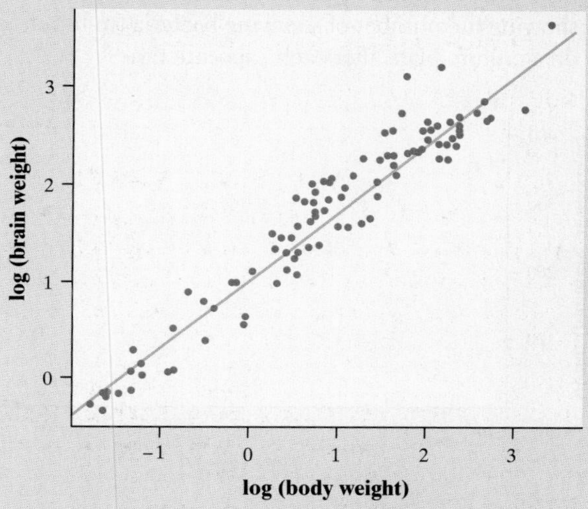

Based on footprints and some other sketchy evidence, some people believe that a large ape-like animal, called Sasquatch or Bigfoot, lives in the Pacific Northwest. Bigfoot's weight is estimated to be about 127 kilograms (kg). How big do you expect Bigfoot's brain to be?

88. **Determining tree biomass** It is easy to measure the diameter at breast height (in centimeters) of a tree. It's hard to measure the total aboveground biomass (in kilograms) of a tree, because to do this, you must cut and weigh the tree. The biomass is important for studies of ecology, so ecologists commonly estimate it using a power model. The following figure is a scatterplot of the natural logarithm of biomass against the natural logarithm of diameter at breast height (DBH) for 378 trees in tropical rain forests.[46] The least-squares regression line for the transformed data is

$$\widehat{\ln y} = -2.00 + 2.42 \ln x$$

Use this model to estimate the biomass of a tropical tree 30 cm in diameter.

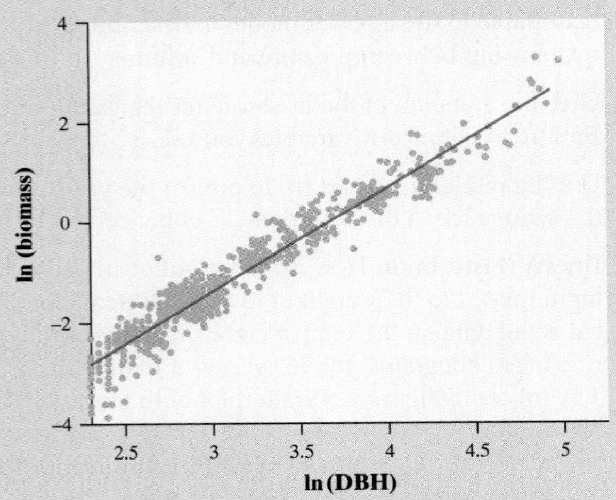


89. **Killing bacteria** Expose marine bacteria to X-rays for pg 222 time periods from 1 to 15 minutes. Here is a scatterplot showing the number of surviving bacteria (in hundreds) on a culture plate after each exposure time:[47]

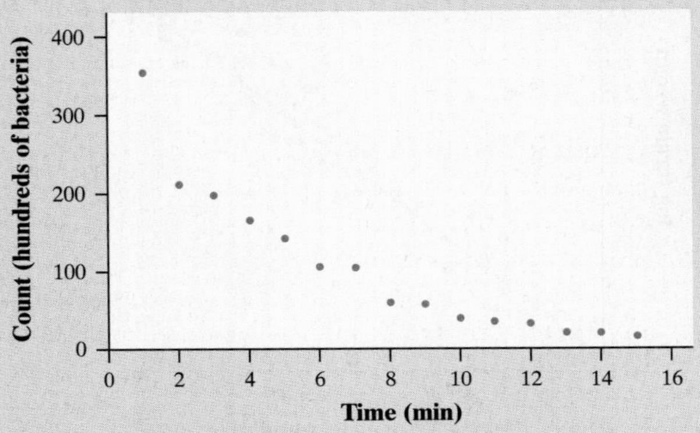


(a) Below is a scatterplot of the natural logarithm of the number of surviving bacteria versus time. Based on this graph, explain why it would be reasonable to use an exponential model to describe the relationship between count of bacteria and time.

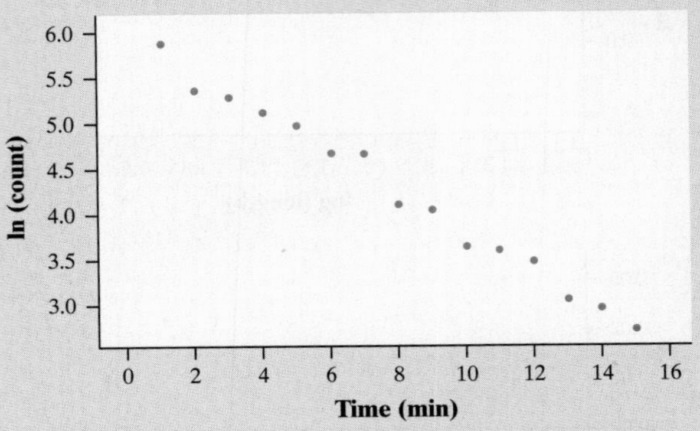


(b) Here is output from a linear regression analysis of the transformed data. Give the equation of the least-squares regression line. Be sure to define any variables you use.

| Predictor | Coef | SE Coef | T | P |
|---|---|---|---|---|
| Constant | 5.97316 | 0.05978 | 99.92 | 0.000 |
| Time | −0.218425 | 0.006575 | −33.22 | 0.000 |
| S = 0.110016 | R-Sq = 98.8% | R-Sq(adj) = 98.7% | | |

(c) Use your model to predict the number of surviving bacteria after 17 minutes.

90. **Light through water** Some college students collected data on the intensity of light at various depths in a lake. Here is a scatterplot of their data:

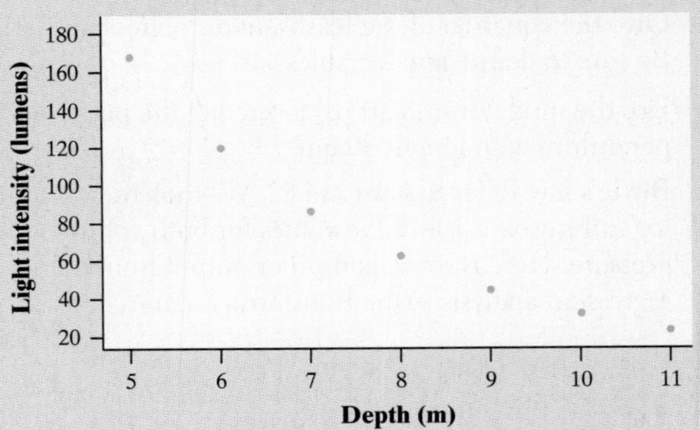


(a) At top right is a scatterplot of the natural logarithm of light intensity versus depth. Based on this graph, explain why it would be reasonable to use an exponential model to describe the relationship between light intensity and depth.

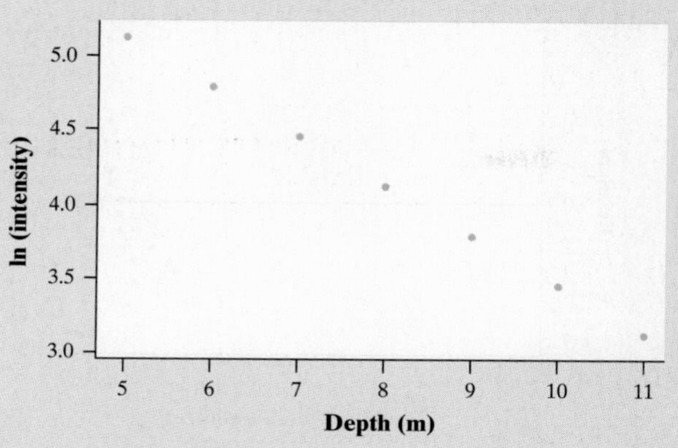


(b) Here is computer output from a linear regression analysis of the transformed data. Give the equation of the least-squares regression line. Be sure to define any variables you use.

| Predictor | Coef | SE Coef | T | P |
|---|---|---|---|---|
| Constant | 6.78910 | 0.00009 | 78575.46 | 0.000 |
| Depth (m) | −0.333021 | 0.000010 | −31783.44 | 0.000 |
| S = 0.000055 | R-Sq = 100.0% | R-Sq(adj) = 100.0% | | |

(c) Use your model to predict the light intensity at a depth of 12 meters.

91. **Putting success** How well do professional golfers putt from different distances? Researchers collected data on the percent of putts made for various distances to the hole (in feet).[48]

pg 225

Here is linear regression output for three different models, along with a residual plot for each model. Model 1 is based on the original data, while Models 2 and 3 involve transformations of the original data.

**Model 1. Percent made vs. Distance**

| Predictor | Coef | SE Coef | T | P |
|---|---|---|---|---|
| Constant | 83.6081 | 4.7206 | 17.711 | 0.0000 |
| Distance | −4.0888 | 0.3842 | −10.64 | 0.0000 |
| S = 9.17 | R-sq = 0.870 | R-Sq(adj) = 0.862 | | |

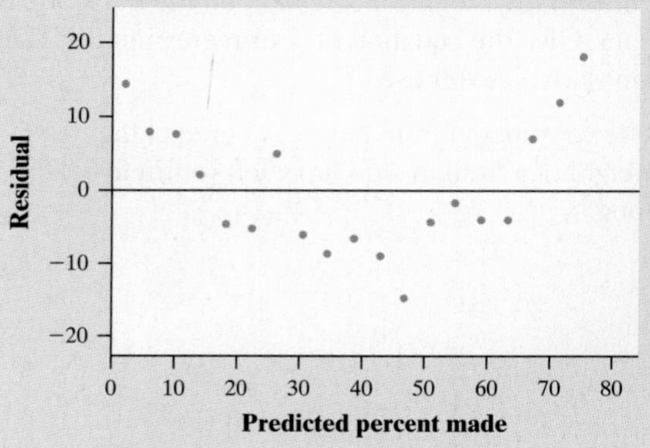


**Model 2. ln(percent made) vs. ln(distance)**

| Predictor | Coef | SE Coef | T | P |
|---|---|---|---|---|
| Constant | 5.5047 | 0.1628 | 33.821 | 0.0000 |
| ln(distance) | −0.9154 | 0.0702 | −13.04 | 0.0000 |
| S = 0.196 | R-sq = 0.909 | R-sq(adj) = 0.904 | | |

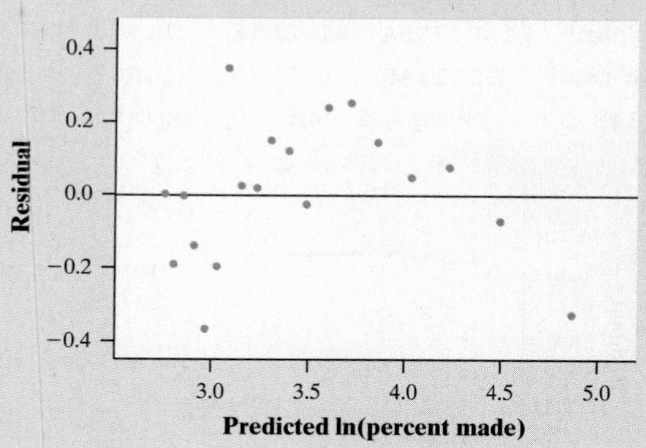


**Model 3. ln(percent made) vs. Distance**

| Predictor | Coef | SE Coef | T | P |
|---|---|---|---|---|
| Constant | 4.6649 | 0.0825 | 56.511 | 0.0000 |
| Distance | −0.1091 | 0.0067 | −16.24 | 0.0000 |
| S = 0.160 | R-sq = 0.939 | R-sq(adj) = 0.936 | | |

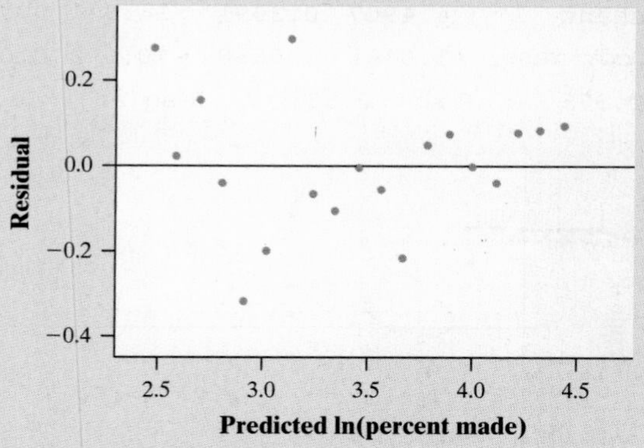


(a) Which model does the best job of summarizing the relationship between distance and percent made? Explain your reasoning.

(b) Using the model chosen in part (a), calculate and interpret the residual for the point where the golfers made 31% of putts from 14 feet away.

92. **Counting carnivores** Ecologists look at data to learn about nature's patterns. One pattern they have identified relates the size of a carnivore (body mass in kilograms) to how many of those carnivores exist in an area. A good measure of "how many" (abundance) is to count carnivores per 10,000 kg of their prey in the area. Researchers collected data on the abundance and body mass for 25 carnivore species.[49]

Here is linear regression output for three different models, along with a residual plot for each model. Model 1 is based on the original data, while Models 2 and 3 involve transformations of the original data.

**Model 1. Abundance vs. Body mass**

| Predictor | Coef | SE Coef | T | P |
|---|---|---|---|---|
| Constant | 158.3094 | 81.2586 | 1.948 | 0.0637 |
| Body mass | −1.1140 | 0.9972 | −1.007 | 0.3245 |
| S = 345.5 | R-sq = 0.042 | R-sq(adj) = 0.001 | | |

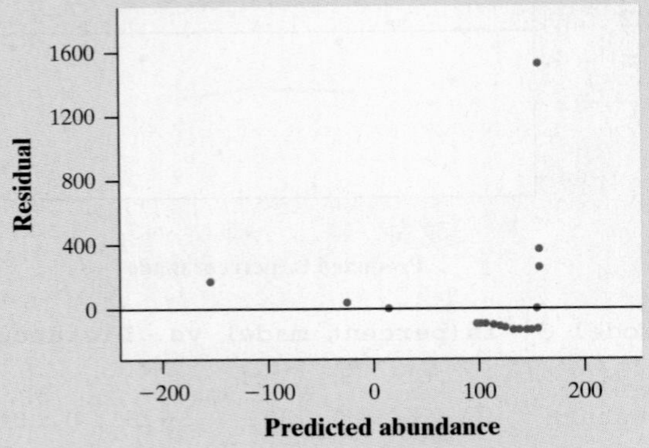


**Model 2. ln(abundance) vs. ln(body mass)**

| Predictor | Coef | SE Coef | T | P |
|---|---|---|---|---|
| Constant | 4.4907 | 0.3091 | 14.531 | 0.0000 |
| ln(body mass) | −1.0481 | 0.0980 | −10.693 | 0.0000 |
| S = 0.975 | R-sq = 0.833 | R-sq(adj) = 0.825 | | |

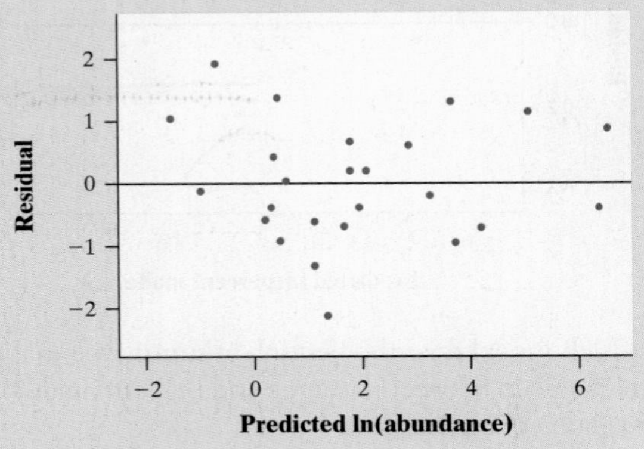


**Model 3. ln(abundance) vs. Body mass**

| Predictor | Coef | SE Coef | T | P |
|---|---|---|---|---|
| Constant | 2.6375 | 0.4843 | 5.447 | 0.0000 |
| Body mass | −0.0166 | 0.0059 | −2.791 | 0.0104 |
| S = 2.059 | R-sq = 0.253 | R-sq(adj) = 0.220 | | |

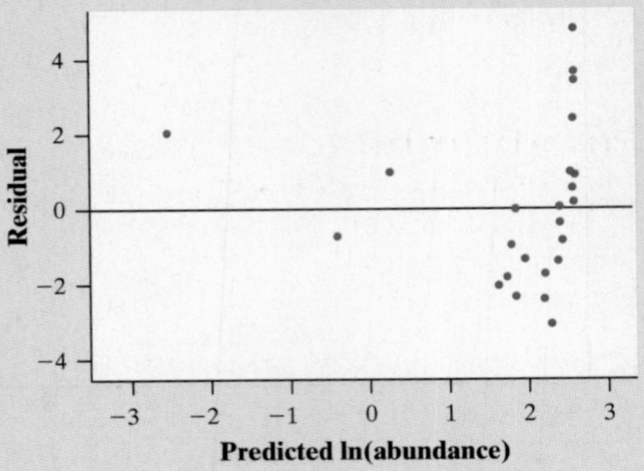


(a) Which model does the best job of summarizing the relationship between body mass and abundance? Explain your reasoning.

(b) Using the model chosen in part (a), calculate and interpret the residual for the coyote, which has a body mass of 13.0 kg and an abundance of 11.65.

93. **Heart weights of mammals** Here are some data on the hearts of various mammals:[50]

| Mammal | Length of cavity of left ventricle (cm) | Heart weight (g) |
|---|---|---|
| Mouse | 0.55 | 0.13 |
| Rat | 1.00 | 0.64 |
| Rabbit | 2.20 | 5.80 |
| Dog | 4.00 | 102.00 |
| Sheep | 6.50 | 210.00 |
| Ox | 12.00 | 2030.00 |
| Horse | 16.00 | 3900.00 |

(a) Make an appropriate scatterplot for predicting heart weight from length. Describe what you see.

(b) Use transformations to linearize the relationship. Does the relationship between heart weight and length seem to follow an exponential model or a power model? Justify your answer.

(c) Perform least-squares regression on the transformed data. Give the equation of your regression line. Define any variables you use.

(d) Use your model from part (c) to predict the heart weight of a human who has a left ventricle 6.8 cm long.

94. **Click-through rates** Companies work hard to have their website listed at the top of an Internet search. Is there a relationship between a website's position in the results of an Internet search (1 = top position, 2 = 2nd position, etc.) and the percentage of people who click on the link for the website? Here are click-through rates for the top 10 positions in searches on a mobile device:[51]

| Position | Click-through rate (%) |
|---|---|
| 1 | 23.53 |
| 2 | 14.94 |
| 3 | 11.19 |
| 4 | 7.47 |
| 5 | 5.29 |
| 6 | 3.80 |
| 7 | 2.79 |
| 8 | 2.11 |
| 9 | 1.57 |
| 10 | 1.18 |

(a) Make an appropriate scatterplot for predicting click-through rate from position. Describe what you see.

(b) Use transformations to linearize the relationship. Does the relationship between click-through rate and position seem to follow an exponential model or a power model? Justify your answer.

(c) Perform least-squares regression on the transformed data. Give the equation of your regression line. Define any variables you use.

(d) Use your model from part (c) to predict the click-through rate for a website in the 11th position.

**Multiple Choice:** *Select the best answer for Exercises 95 and 96.*

95. Students in Mr. Handford's class dropped a kickball beneath a motion detector. The detector recorded the height of the ball (in feet) as it bounced up and down several times. Here is computer output from a linear regression analysis of the transformed data of log(height) versus bounce number. Predict the highest point the ball reaches on its seventh bounce.

| Predictor | Coef | SE Coef | T | P |
|---|---|---|---|---|
| Constant | 0.45374 | 0.01385 | 32.76 | 0.000 |
| Bounce | −0.117160 | 0.004176 | −28.06 | 0.000 |
| S = 0.0132043 | R-Sq = 99.6% | R-Sq(adj) = 99.5% | | |

(a) 0.35 feet (b) 0.37 feet (c) 0.43 feet

(d) 2.26 feet (e) 2.32 feet

96. A scatterplot of $y$ versus $x$ shows a positive, nonlinear association. Two different transformations are attempted to try to linearize the association: using the logarithm of the $y$-values and using the square root of the $y$-values. Two least-squares regression lines are calculated, one that uses $x$ to predict $\log(y)$ and the other that uses $x$ to predict $\sqrt{y}$. Which of the following would be the best reason to prefer the least-squares regression line that uses $x$ to predict $\log(y)$?

(a) The value of $r^2$ is smaller.

(b) The standard deviation of the residuals is smaller.

(c) The slope is greater.

(d) The residual plot has more random scatter.

(e) The distribution of residuals is more Normal.

### Recycle and Review

97. **Shower time** (1.3, 2.2) Marcella takes a shower every morning when she gets up. Her time in the shower varies according to a Normal distribution with mean 4.5 minutes and standard deviation 0.9 minute.

(a) Find the probability that Marcella's shower lasts between 3 and 6 minutes on a randomly selected day.

(b) If Marcella took a 7-minute shower, would it be classified as an outlier by the $1.5IQR$ rule? Justify your answer.

98. **NFL weights** (1.2, 1.3) Players in the National Football League (NFL) are bigger and stronger than ever before. And they are heavier, too.[52]

(a) Here is a boxplot showing the distribution of weight for NFL players in a recent season. Describe the distribution.

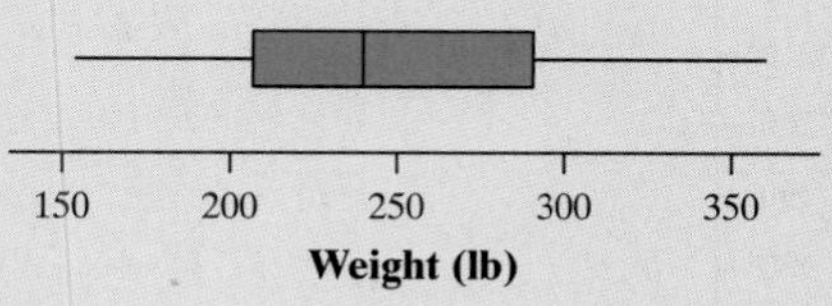


(b) Now, here is a dotplot of the same distribution. What feature of the distribution does the dotplot reveal that wasn't revealed by the boxplot?

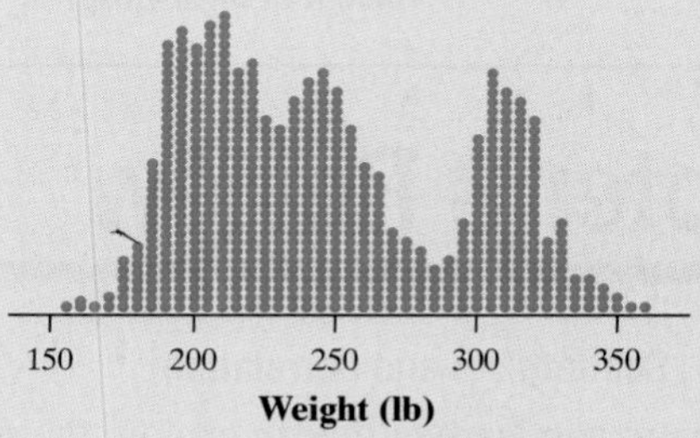

# Chapter 3 Wrap-Up

## FRAPPY! FREE RESPONSE AP® PROBLEM, YAY!

The following problem is modeled after actual AP® Statistics exam free response questions. Your task is to generate a complete, concise response in 15 minutes.

*Directions: Show all your work. Indicate clearly the methods you use, because you will be scored on the correctness of your methods as well as on the accuracy and completeness of your results and explanations.*

Two statistics students went to a flower shop and randomly selected 12 carnations. When they got home, the students prepared 12 identical vases with exactly the same amount of water in each vase. They put one tablespoon of sugar in 3 vases, two tablespoons of sugar in 3 vases, and three tablespoons of sugar in 3 vases. In the remaining 3 vases, they put no sugar. After the vases were prepared, the students randomly assigned 1 carnation to each vase and observed how many hours each flower continued to look fresh. A scatterplot of the data is shown below.

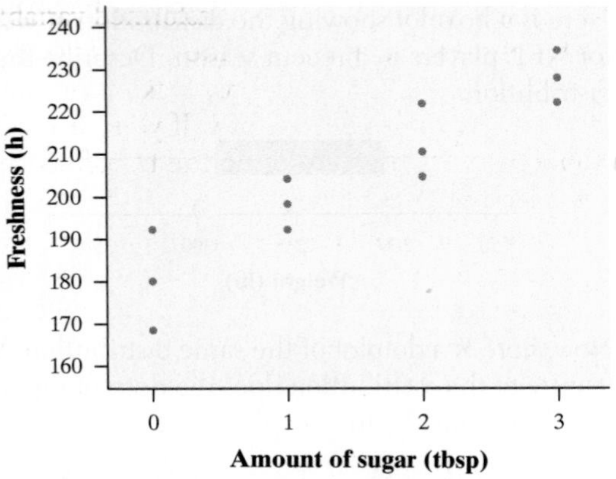


(a) Briefly describe the association shown in the scatterplot.

(b) The equation of the least-squares regression line for these data is $\hat{y} = 180.8 + 15.8x$. Interpret the slope of the line in the context of the study.

(c) Calculate and interpret the residual for the flower that had 2 tablespoons of sugar and looked fresh for 204 hours.

(d) Suppose that another group of students conducted a similar experiment using 12 flowers, but included different varieties in addition to carnations. Would you expect the value of $r^2$ for the second group's data to be greater than, less than, or about the same as the value of $r^2$ for the first group's data? Explain.

After you finish, you can view two example solutions on the book's website (highschool.bfwpub.com/updatedtps6e). Determine whether you think each solution is "complete," "substantial," "developing," or "minimal." If the solution is not complete, what improvements would you suggest to the student who wrote it? Finally, your teacher will provide you with a scoring rubric. Score your response and note what, if anything, you would do differently to improve your own score.

# Chapter 3 Review

### Section 3.1: Scatterplots and Correlation

In this section, you learned how to explore the relationship between two quantitative variables. As with distributions of a single variable, the first step is always to make a graph. A scatterplot is the appropriate type of graph to investigate relationships between two quantitative variables. To describe a scatterplot, be sure to discuss four characteristics: direction, form, strength, and unusual features. The direction of a relationship might be positive, negative, or neither. The form of a relationship can be linear or nonlinear. A relationship is strong if it closely follows a specific form. Finally, unusual features include points that clearly fall outside the pattern of the rest of the data and distinct clusters of points.

The correlation $r$ is a numerical summary for linear relationships that describes the direction and strength of the association. When $r > 0$, the association is positive, and when $r < 0$, the association is negative. The correlation will always take values between $-1$ and 1, with $r = -1$ and $r = 1$ indicating a perfectly linear relationship. Strong linear relationships have correlations near 1 or $-1$, while weak linear relationships have correlations near 0. It isn't possible to determine the form of a relationship from only the correlation. Strong nonlinear relationships can have a correlation close to 1 or a correlation close to 0. You also learned that unusual points can greatly affect the value of the correlation and that correlation does not imply causation. That is, we can't assume that changes in one variable cause changes in the other variable, just because they have a correlation close to 1 or $-1$.

### Section 3.2: Least-Squares Regression

In this section, you learned how to use least-squares regression lines as models for relationships between two quantitative variables that have a linear association. It is important to understand the difference between the actual data and the model used to describe the data. To emphasize that the model only provides predicted values, least-squares regression lines are always expressed in terms of $\hat{y}$ instead of $y$. Likewise, when you are interpreting the slope of a least-squares regression line, describe the change in the *predicted* value of $y$.

The difference between the actual value of $y$ and the predicted value of $y$ is called a residual. Residuals are the key to understanding almost everything in this section. To find the equation of the least-squares regression line, find the line that minimizes the sum of the squared residuals. To see if a linear model is appropriate, make a residual plot. If there is no leftover curved pattern in the residual plot, you know the model is appropriate. To assess how well a line fits the data, calculate the standard deviation of the residuals $s$ to estimate the size of a typical prediction error. You can also calculate $r^2$, which measures the percent of the variation in the $y$ variable that is accounted for by the least-squares regression line.

You also learned how to obtain the equation of a least-squares regression line from computer output and from summary statistics (the means and standard deviations of two variables and their correlation). As with the correlation, the equation of the least-squares regression line and the values of $s$ and $r^2$ can be greatly affected by influential points, such as outliers and points with high leverage. Make sure to plot the data and note any unusual points before making any calculations.

### Section 3.3: Transforming to Achieve Linearity

When the association between two variables is nonlinear, transforming one or both of the variables can result in a linear association.

If the association between two variables follows a power model in the form $y = ax^p$, there are several transformations that will result in a linear association.

- Raise the values of $x$ to the power of $p$ and plot $y$ versus $x^p$.
- Calculate the $p$th root of the $y$-values and plot $\sqrt[p]{y}$ versus $x$.
- Calculate the logarithms of the $x$-values and the $y$-values, and plot log($y$) versus log($x$) or ln($y$) versus ln($x$).

If the association between two variables follows an exponential model in the form $y = ab^x$, transform the data by computing the logarithms of the $y$-values and plot log($y$) versus $x$ or ln($y$) versus $x$.

Once you have achieved linearity, calculate the equation of the least-squares regression line using the transformed data. Remember to include the transformed variables when you are writing the equation of the line. Likewise, when using the line to make predictions, make sure that the prediction is in the original units of $y$. If you transformed the $y$ variable, you will need to undo the transformation after using the least-squares regression line.

To decide which of two or more models is most appropriate, choose the one that produces the most linear association and whose residual plot has the most random scatter. If more than one residual plot is randomly scattered, choose the model with the value of $r^2$ closest to 1.

## What Did You Learn?

| Learning Target | Section | Related Example on Page(s) | Relevant Chapter Review Exercise(s) |
|---|---|---|---|
| Distinguish between explanatory and response variables for quantitative data. | 3.1 | 154 | R3.4 |
| Make a scatterplot to display the relationship between two quantitative variables. | 3.1 | 155 | R3.4 |

| Learning Target | Section | Related Example on Page(s) | Relevant Chapter Review Exercise(s) |
|---|---|---|---|
| Describe the direction, form, and strength of a relationship displayed in a scatterplot and identify unusual features. | 3.1 | 158 | R3.1, R3.2 |
| Interpret the correlation. | 3.1 | 162 | R3.3 |
| Understand the basic properties of correlation, including how the correlation is influenced by unusual points. | 3.1 | 165, 169 | R3.1, R3.2 |
| Distinguish correlation from causation. | 3.1 | 165 | R3.6 |
| Make predictions using regression lines, keeping in mind the dangers of extrapolation. | 3.2 | 178 | R3.4, R3.5 |
| Calculate and interpret a residual. | 3.2 | 180 | R3.3, R3.4 |
| Interpret the slope and *y* intercept of a regression line. | 3.2 | 182 | R3.4 |
| Determine the equation of a least-squares regression line using technology or computer output. | 3.2 | 192 | R3.3, R3.4 |
| Construct and interpret residual plots to assess whether a regression model is appropriate. | 3.2 | 186 | R3.3, R3.4 |
| Interpret the standard deviation of the residuals and $r^2$ and use these values to assess how well a least-squares regression line models the relationship between two variables. | 3.2 | 191 | R3.3, R3.5 |
| Describe how the least-squares regression line, standard deviation of the residuals, and $r^2$ are influenced by unusual points. | 3.2 | 202 | R3.1 |
| Find the slope and *y* intercept of the least-squares regression line from the means and standard deviations of *x* and *y* and their correlation. | 3.2 | 195 | R3.5 |
| Use transformations involving powers, roots, or logarithms to create a linear model that describes the relationship between two quantitative variables, and use the model to make predictions. | 3.3 | 215, 217, 218, 222 | R3.7 |
| Determine which of several models does a better job of describing the relationship between two quantitative variables. | 3.3 | 225 | R3.7 |

# Chapter 3 Review Exercises

*These exercises are designed to help you review the important ideas and methods of the chapter.*

**R3.1** **Born to be old?** Is there a relationship between the gestational period (time from conception to birth) of an animal and its average life span? The figure shows a scatterplot of the gestational period and average life span for 43 species of animals.[53]

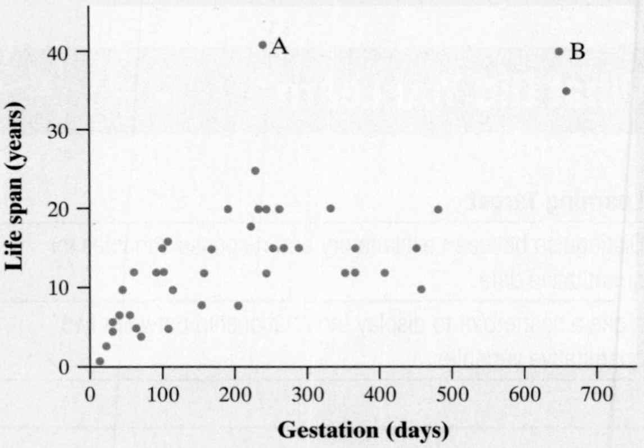

(a) Describe the relationship shown in the scatterplot.

(b) Point A is the hippopotamus. What effect does this point have on the correlation, the equation of the least-squares regression line, and the standard deviation of the residuals?

(c) Point B is the Asian elephant. What effect does this point have on the correlation, the equation of the least-squares regression line, and the standard deviation of the residuals?

**R3.2 Penguins diving** A study of king penguins looked for a relationship between how deep the penguins dive to seek food and how long they stay under water.[54] For all but the shallowest dives, there is an association between $x$ = depth (in meters) and $y$ = dive duration (in minutes) that is different for each penguin. The study gives a scatterplot for one penguin titled "The Relation of Dive Duration ($y$) to Depth ($x$)." The scatterplot shows an association that is positive, linear, and strong.

(a) Explain the meaning of the term *positive association* in this context.

(b) Explain the meaning of the term *linear association* in this context.

(c) Explain the meaning of the term *strong association* in this context.

(d) Suppose the researchers reversed the variables, using $x$ = dive duration and $y$ = depth. Would this change the correlation? The equation of the least-squares regression line?

**R3.3 Stats teachers' cars** A random sample of AP® Statistics teachers was asked to report the age (in years) and mileage of their primary vehicles. Here are a scatterplot, a residual plot, and other computer output:

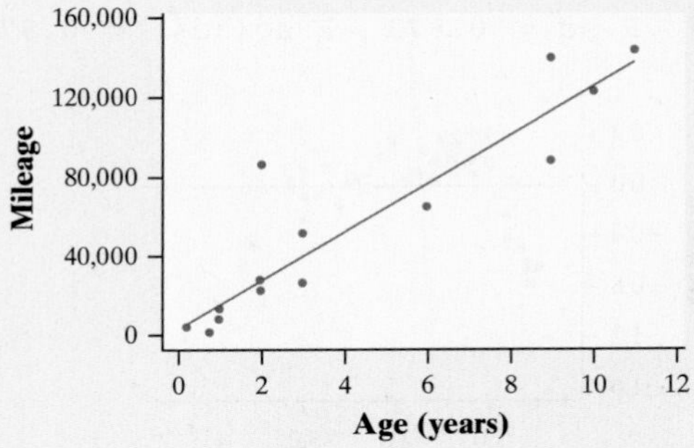


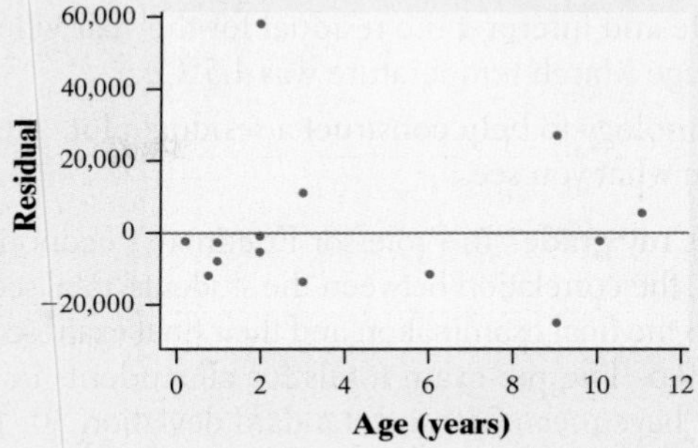


```
Predictor   Coef    SE Coef      T       P
Constant    3704    8268       0.45    0.662
Age        12188    1492       8.17    0.000
S = 20870.5   R-Sq = 83.7%   R-Sq(adj) = 82.4%
```

(a) Is a linear model appropriate for these data? Explain how you know this.

(b) What's the correlation between car age and mileage? Interpret this value in context.

(c) Give the equation of the least-squares regression line for these data. Identify any variables you use.

(d) One teacher reported that her 6-year-old car had 65,000 miles on it. Find and interpret its residual.

(e) Interpret the values of $s$ and $r^2$.

**R3.4 Late bloomers?** Japanese cherry trees tend to blossom early when spring weather is warm and later when spring weather is cool. Here are some data on the average March temperature (in degrees Celsius) and the day in April when the first cherry blossom appeared over a 24-year period:[55]

| | | | | | | | | | | | | |
|---|---|---|---|---|---|---|---|---|---|---|---|---|
| Temperature (°C) | 4.0 | 5.4 | 3.2 | 2.6 | 4.2 | 4.7 | 4.9 | 4.0 | 4.9 | 3.8 | 4.0 | 5.1 |
| Days in April to first blossom | 14 | 8 | 11 | 19 | 14 | 14 | 14 | 21 | 9 | 14 | 13 | 11 |
| Temperature (°C) | 4.3 | 1.5 | 3.7 | 3.8 | 4.5 | 4.1 | 6.1 | 6.2 | 5.1 | 5.0 | 4.6 | 4.0 |
| Days in April to first blossom | 13 | 28 | 17 | 19 | 10 | 17 | 3 | 3 | 11 | 6 | 9 | 11 |

(a) Make a well-labeled scatterplot that's suitable for predicting when the cherry trees will blossom from the temperature. Which variable did you choose as the explanatory variable? Explain your reasoning.

(b) Use technology to calculate the correlation and the equation of the least-squares regression line. Interpret the slope and $y$ intercept of the line in this setting.

(c) Suppose that the average March temperature this year was 8.2°C. Would you be willing to use the equation in part (b) to predict the date of first blossom? Explain your reasoning.

(d) Calculate and interpret the residual for the year when the average March temperature was 4.5°C.

(e) Use technology to help construct a residual plot. Describe what you see.

**R3.5 What's my grade?** In Professor Friedman's economics course, the correlation between the students' total scores prior to the final examination and their final exam scores is $r = 0.6$. The pre-exam totals for all students in the course have mean 280 and standard deviation 30. The final exam scores have mean 75 and standard deviation 8. Professor Friedman has lost Julie's final exam but knows that her total before the exam was 300. He decides to predict her final exam score from her pre-exam total.

(a) Find the equation for the least-squares regression line Professor Friedman should use to make this prediction.

(b) Use the least-squares regression line to predict Julie's final exam score.

(c) Explain the meaning of the phrase "least squares" in the context of this question.

(d) Julie doesn't think this method accurately predicts how well she did on the final exam. Determine $r^2$. Use this result to argue that her actual score could have been much higher (or much lower) than the predicted value.

**R3.6 Calculating achievement** The principal of a high school read a study that reported a high correlation between the number of calculators owned by high school students and their math achievement. Based on this study, he decides to buy each student at his school two calculators, hoping to improve their math achievement. Explain the flaw in the principal's reasoning.

**R3.7 Diamonds!** Diamonds are expensive, especially big ones. To create a model to predict price from size, the weight (in carats) and price (in dollars) was recorded for each of 94 round, clear, internally flawless diamonds with excellent cuts.[56]

Here is linear regression output for three different models, along with a residual plot for each model. Model 1 is based on the original data, while Models 2 and 3 involve transformations of the original data.

**Model 1. Price vs. Weight**

| Predictor | Coef | SE Coef | T | P |
|---|---|---|---|---|
| Constant | −98666 | 7594.1 | −12.992 | 0.0000 |
| Weight | 105932 | 4219.5 | 25.105 | 0.0000 |

S = 34073 R-sq = 0.873 R-sq(adj) = 0.871

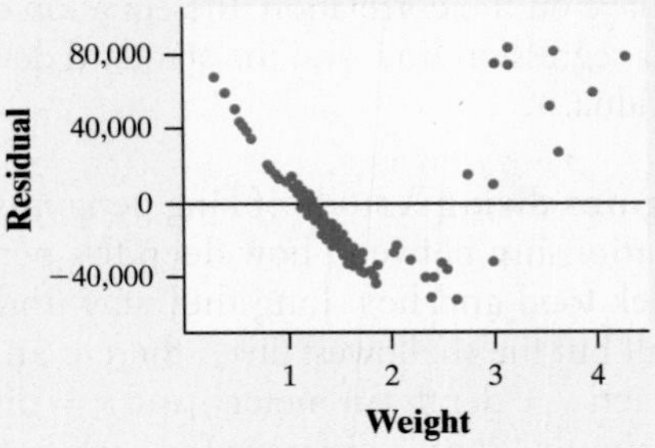


**Model 2. ln(price) vs. ln(weight)**

| Predictor | Coef | SE Coef | T | P |
|---|---|---|---|---|
| Constant | 9.7062 | 0.0209 | 465.102 | 0.0000 |
| ln(weight) | 2.2913 | 0.0332 | 68.915 | 0.0000 |

S = 0.171 R-sq = 0.981 R-sq(adj) = 0.981

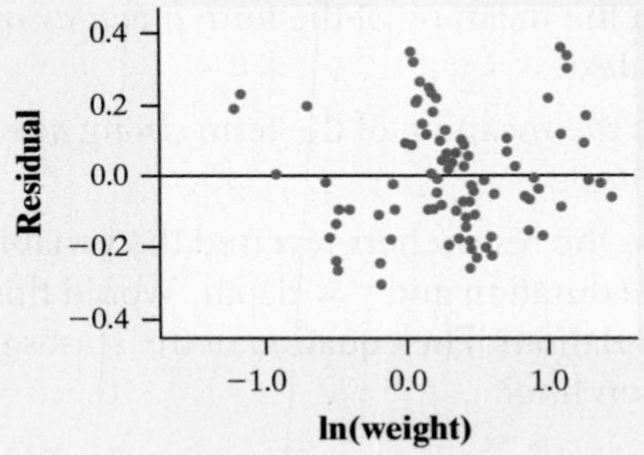


**Model 3. ln(price) vs. Weight**

| Predictor | Coef | SE Coef | T | P |
|---|---|---|---|---|
| Constant | 8.2709 | 0.0988 | 83.716 | 0.0000 |
| Weight | 1.3791 | 0.0549 | 25.123 | 0.0000 |

S = 0.443 R-sq = 0.873 R-sq(adj) = 0.871

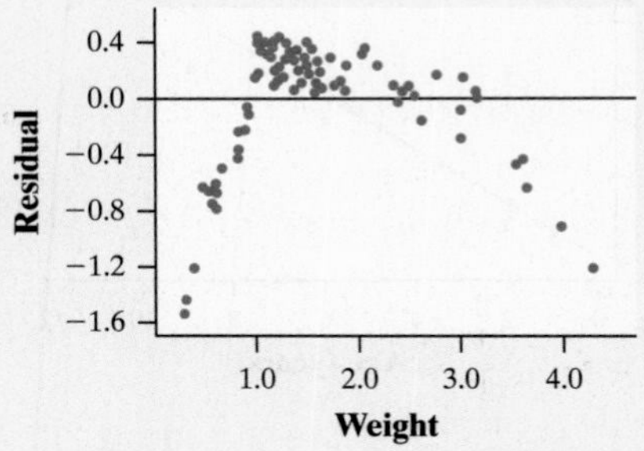


(a) Use each of the three models to predict the price of a diamond of this type that weighs 2 carats.

(b) Which model does the best job of summarizing the relationship between weight and price? Explain your reasoning.

# Chapter 3 AP® Statistics Practice Test

## Section I: Multiple Choice *Select the best answer for each question.*

**T3.1** A school guidance counselor examines how many extracurricular activities students participate in and their grade point average. The guidance counselor says, "The evidence indicates that the correlation between the number of extracurricular activities a student participates in and his or her grade point average is close to 0." Which of the following is the most appropriate conclusion?

(a) Students involved in many extracurricular activities tend to be students with poor grades.

(b) Students with good grades tend to be students who are not involved in many extracurricular activities.

(c) Students involved in many extracurricular activities are just as likely to get good grades as bad grades.

(d) Students with good grades tend to be students who are involved in many extracurricular activities.

(e) No conclusion should be made based on the correlation without looking at a scatterplot of the data.

**T3.2** An AP® Statistics student designs an experiment to see whether today's high school students are becoming too calculator-dependent. She prepares two quizzes, both of which contain 40 questions that are best done using paper-and-pencil methods. A random sample of 30 students participates in the experiment. Each student takes both quizzes—one with a calculator and one without—in a random order. To analyze the data, the student constructs a scatterplot that displays a linear association between the number of correct answers with and without a calculator for the 30 students. A least-squares regression yields the equation

$$\widehat{\text{Calculator}} = -1.2 + 0.865\,(\text{Pencil}) \qquad r = 0.79$$

Which of the following statements is/are true?

I. If the student had used Calculator as the explanatory variable, the correlation would remain the same.

II. If the student had used Calculator as the explanatory variable, the slope of the least-squares line would remain the same.

III. The standard deviation of the number of correct answers on the paper-and-pencil quizzes was smaller than the standard deviation on the calculator quizzes.

(a) I only
(b) II only
(c) III only
(d) I and III only
(e) I, II, and III

*Questions T3.3–T3.5 refer to the following setting.* Scientists examined the activity level of 7 fish at different temperatures. Fish activity was rated on a scale of 0 (no activity) to 100 (maximal activity). The temperature was measured in degrees Celsius. A computer regression printout and a residual plot are provided. Notice that the horizontal axis on the residual plot is labeled "Fitted value," which means the same thing as "predicted value."

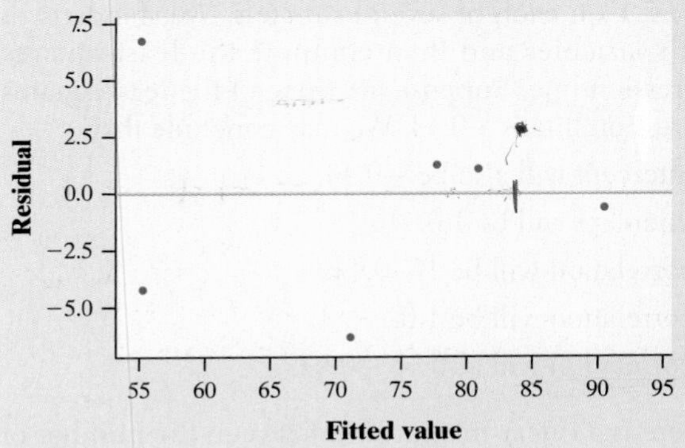


| Predictor | Coef | SE Coef | T | P |
|---|---|---|---|---|
| Constant | 148.62 | 10.71 | 13.88 | 0.000 |
| Temperature | −3.2167 | 0.4533 | −7.10 | 0.001 |
| S = 4.78505 | R-Sq = 91.0% | | R-Sq(adj) = 89.2% | |

**T3.3** What is the correlation between temperature and fish activity?

(a) 0.95 (b) 0.91 (c) 0.45
(d) −0.91 (e) −0.95

**T3.4** What was the actual activity level rating for the fish at a temperature of 20°C?

(a) 87 (b) 84 (c) 81
(d) 66 (e) 3

**T3.5** Which of the following gives a correct interpretation of $s$ in this setting?

(a) For every 1°C increase in temperature, fish activity is predicted to increase by 4.785 units.

(b) The typical distance of the temperature readings from their mean is about 4.785°C.

(c) The typical distance of the activity level ratings from the least-squares line is about 4.785 units.

(d) The typical distance of the activity level readings from their mean is about 4.785 units.

(e) At a temperature of 0°C, this model predicts an activity level of 4.785 units.

**T3.6** Which of the following statements is *not* true of the correlation $r$ between the lengths (in inches) and weights (in pounds) of a sample of brook trout?

(a) $r$ must take a value between $-1$ and 1.

(b) $r$ is measured in inches.

(c) If longer trout tend to also be heavier, then $r > 0$.

(d) $r$ would not change if we measured the lengths of the trout in centimeters instead of inches.

(e) $r$ would not change if we measured the weights of the trout in kilograms instead of pounds.

**T3.7** When we standardize the values of a variable, the distribution of standardized values has mean 0 and standard deviation 1. Suppose we measure two variables X and Y on each of several subjects. We standardize both variables and then compute the least-squares regression line. Suppose the slope of the least-squares regression line is $-0.44$. We may conclude that

(a) the intercept will also be $-0.44$.

(b) the intercept will be 1.0.

(c) the correlation will be $1/-0.44$.

(d) the correlation will be 1.0.

(e) the correlation will also be $-0.44$.

**T3.8** There is a linear relationship between the number of chirps made by the striped ground cricket and the air temperature. A least-squares fit of some data collected by a biologist gives the model $\hat{y} = 25.2 + 3.3x$, where $x$ is the number of chirps per minute and $\hat{y}$ is the estimated temperature in degrees Fahrenheit. What is the predicted increase in temperature for an increase of 5 chirps per minute?

(a) 3.3°F (b) 16.5°F (c) 25.2°F

(d) 28.5°F (e) 41.7°F

**T3.9** The scatterplot shows the relationship between the number of people per television set and the number of people per physician for 40 countries, along with the least-squares regression line. In Ethiopia, there were 503 people per TV and 36,660 people per doctor. Which of the following is correct?

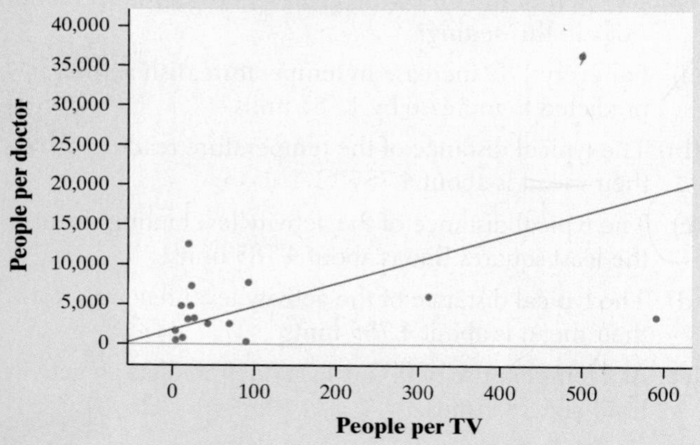


(a) Increasing the number of TVs in a country will attract more doctors.

(b) The slope of the least-squares regression line is less than 1.

(c) The correlation is greater than 1.

(d) The point for Ethiopia is decreasing the slope of the least-squares regression line.

(e) Ethiopia has more people per doctor than expected, based on how many people it has per TV.

**T3.10** The scatterplot shows the lean body mass and metabolic rate for a sample of 5 adults. For each person, the lean body mass is the subject's total weight in kilograms less any weight due to fat. The metabolic rate is the number of calories burned in a 24-hour period.

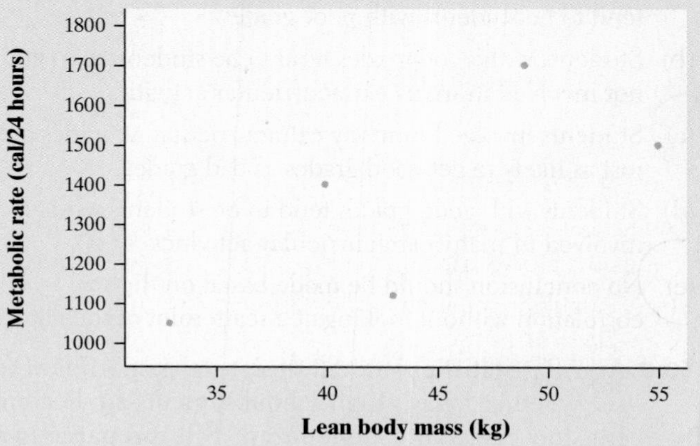


Because a person with no lean body mass should burn no calories, it makes sense to model the relationship with a direct variation function in the form $y = kx$. Models were tried using different values of $k$ ($k = 25$, $k = 26$, etc.) and the sum of squared residuals (SSR) was calculated for each value of $k$. Here is a scatterplot showing the relationship between SSR and $k$:

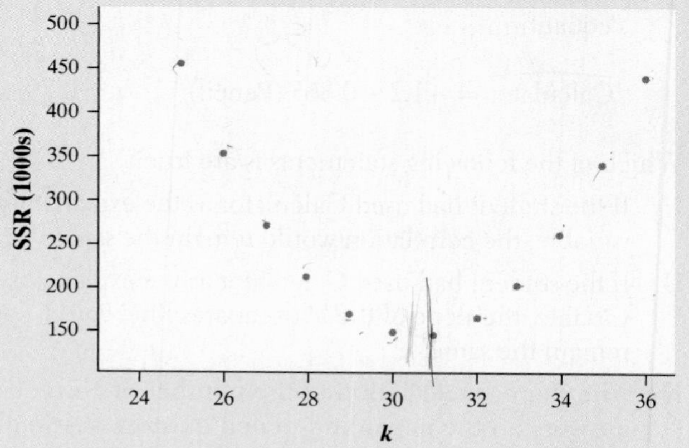


According to the scatterplot, what is the ideal value of $k$ to use for predicting metabolic rate?

(a) 24 (b) 25 (c) 29

(d) 31 (e) 36

**T3.11** We record data on the population of a particular country from 1960 to 2010. A scatterplot reveals a clear curved relationship between population and year. However, a different scatterplot reveals a strong linear relationship between the logarithm (base 10) of the population and the year. The least-squares regression line for the transformed data is

$$\widehat{\log(\text{population})} = -13.5 + 0.01(\text{year})$$

Based on this equation, which of the following is the best estimate for the population of the country in the year 2020?

(a) 6.7
(b) 812
(c) 5,000,000
(d) 6,700,000
(e) 8,120,000

**Section II: Free Response** *Show all your work. Indicate clearly the methods you use, because you will be graded on the correctness of your methods as well as on the accuracy and completeness of your results and explanations.*

**T3.12** Sarah's parents are concerned that she seems short for her age. Their doctor has kept the following record of Sarah's height:

| Age (months) | 36 | 48 | 51 | 54 | 57 | 60 |
|---|---|---|---|---|---|---|
| Height (cm) | 86 | 90 | 91 | 93 | 94 | 95 |

(a) Make a scatterplot of these data using age as the explanatory variable. Describe what you see.

(b) Using your calculator, find the equation of the least-squares regression line.

(c) Calculate and interpret the residual for the point when Sarah was 48 months old.

(d) Would you be confident using the equation from part (b) to predict Sarah's height when she is 40 years old? Explain.

**T3.13** Drilling down beneath a lake in Alaska yields chemical evidence of past changes in climate. Biological silicon, left by the skeletons of single-celled creatures called diatoms, is a measure of the abundance of life in the lake. A rather complex variable based on the ratio of certain isotopes relative to ocean water gives an indirect measure of moisture, mostly from snow. As we drill down, we look further into the past. Here is a scatterplot of data from 2300 to 12,000 years ago:

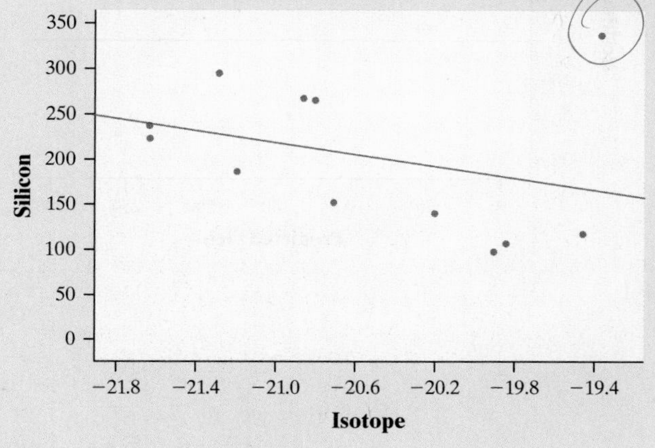


(a) Identify the unusual point in the scatterplot and estimate its $x$ and $y$ coordinates.

(b) Describe the effect this point has on

i. the correlation.

ii. the slope and $y$ intercept of the least-squares line.

iii. the standard deviation of the residuals.

**T3.14** Long-term records from the Serengeti National Park in Tanzania show interesting ecological relationships. When wildebeest are more abundant, they graze the grass more heavily, so there are fewer fires and more trees grow. Lions feed more successfully when there are more trees, so the lion population increases. Researchers collected data on one part of this cycle, wildebeest abundance (in thousands of animals), and the percent of the grass area burned in the same year. The results of a least-squares regression on the data are shown here.[57]

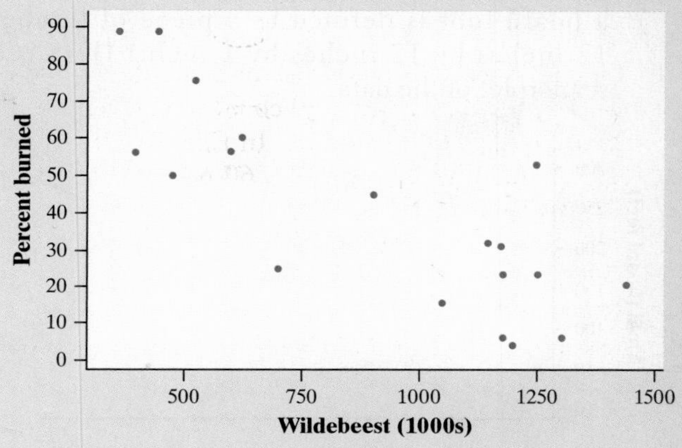

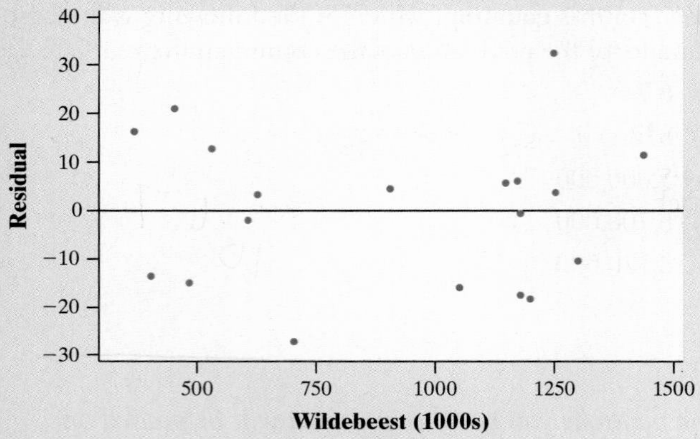


| Predictor | Coef | SE Coef | T | P |
|---|---|---|---|---|
| Constant | 92.29 | 10.06 | 9.17 | 0.000 |
| Wildebeest (1000s) | −0.05762 | 0.01035 | −5.56 | 0.001 |

S = 15.9880 R-Sq = 64.6% R-Sq(adj) = 62.5%

(a) Is a linear model appropriate for describing the relationship between wildebeest abundance and percent of grass area burned? Explain.

(b) Give the equation of the least-squares regression line. Be sure to define any variables you use.

(c) Interpret the slope. Does the value of the *y* intercept have meaning in this context? If so, interpret the *y* intercept. If not, explain why.

(d) Interpret the standard deviation of the residuals and $r^2$.

**T3.15** Foresters are interested in predicting the amount of usable lumber they can harvest from various tree species. They collect data on the diameter at breast height (DBH) in inches and the yield in board feet of a random sample of 20 Ponderosa pine trees that have been harvested. (Note that a board foot is defined as a piece of lumber 12 inches by 12 inches by 1 inch.) Here is a scatterplot of the data.

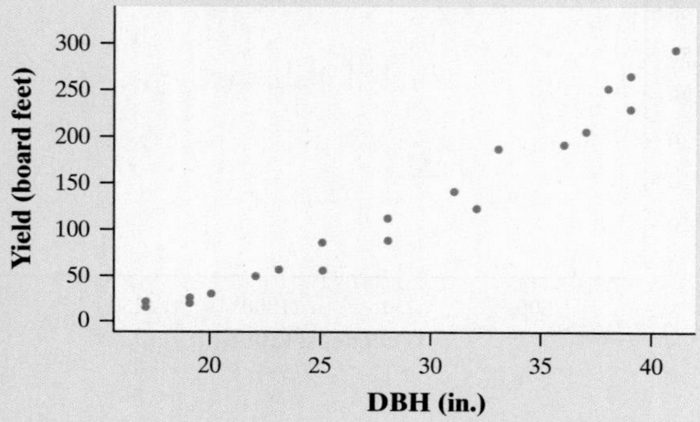


(a) Here is some computer output and a residual plot from a least-squares regression on these data. Explain why a linear model may not be appropriate in this case.

| Predictor | Coef | SE Coef | T | P |
|---|---|---|---|---|
| Constant | −191.12 | 16.98 | −11.25 | 0.000 |
| DBH (inches) | 11.0413 | 0.5752 | 19.19 | 0.000 |

S = 20.3290 R-Sq = 95.3% R-Sq(adj) = 95.1%

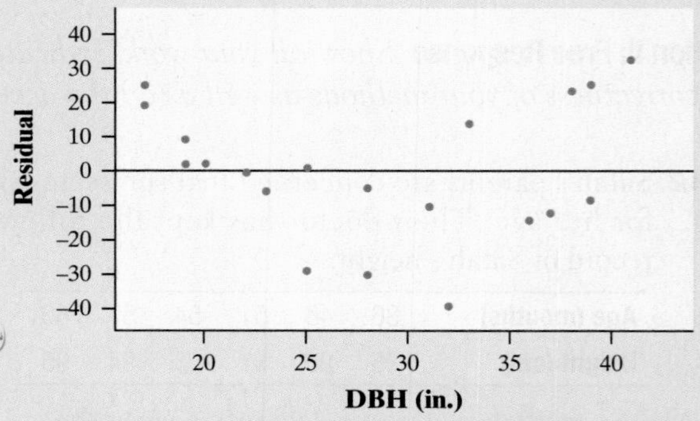


The foresters are considering two possible transformations of the original data: (1) cubing the diameter values or (2) taking the natural logarithm of the yield measurements. After transforming the data, a least-squares regression analysis is performed. Here is some computer output and a residual plot for each of the two possible regression models:

**Option 1: Cubing the diameter values**

| Predictor | Coef | SE Coef | T | P |
|---|---|---|---|---|
| Constant | 2.078 | 5.444 | 0.38 | 0.707 |
| DBH^3 | 0.0042597 | 0.0001549 | 27.50 | 0.000 |

S = 14.3601 R-Sq = 97.7% R-Sq(adj) = 97.5%

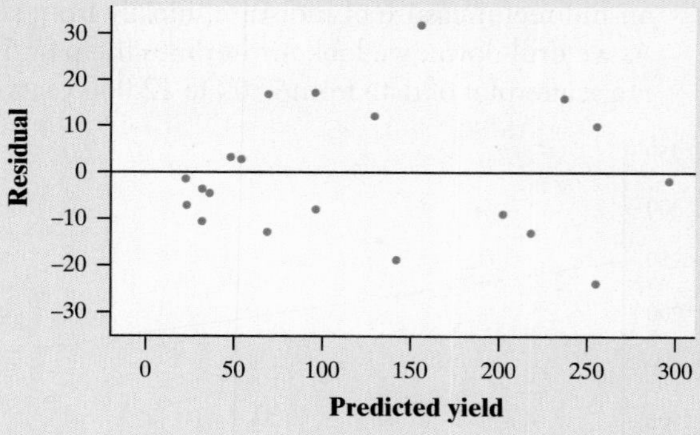

**Option 2: Taking natural logarithm of yield measurements**

| Predictor | Coef | SE Coef | T | P |
|---|---|---|---|---|
| Constant | 1.2319 | 0.1795 | 6.86 | 0.000 |
| DBH | 0.113417 | 0.006081 | 18.65 | 0.000 |

S = 0.214894 R-Sq = 95.1% R-Sq(adj) = 94.8%

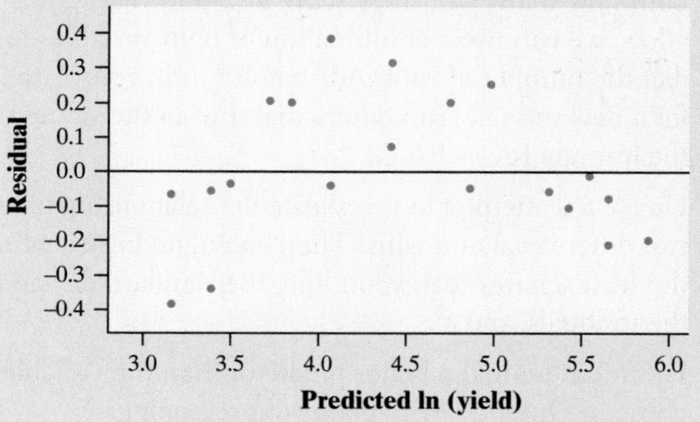


(b) Use both models to predict the amount of usable lumber from a Ponderosa pine with diameter 30 inches.

(c) Which of the predictions in part (b) seems more reliable? Give appropriate evidence to support your choice.

# Chapter 3 Project Investigating Relationships in Baseball

What is a better predictor of the number of wins for a baseball team, the number of runs scored by the team or the number of runs they allow the other team to score? What variables can we use to predict the number of runs a team scores? To predict the number of runs it allows the other team to score? In this project, you will use technology to help answer these questions by exploring a large set of data from Major League Baseball.

## Part 1

1. Download the "MLB Team Data 2012–2016" Excel file from the book's website, along with the "Glossary for MLB Team Data file," which explains each of the variables included in the data set.[58] Import the data into the statistical software package you prefer.
2. Create a scatterplot to investigate the relationship between runs scored per game (R/G) and wins (W). Then calculate the equation of the least-squares regression line, the standard deviation of the residuals, and $r^2$. *Note:* R/G is in the section for hitting statistics and W is in the section for pitching statistics.
3. Create a scatterplot to investigate the relationship between runs *allowed* per game (RA/G) and wins (W). Then calculate the equation of the least-squares regression line, the standard deviation of the residuals, and $r^2$. *Note:* Both of these variables may be found in the section for pitching statistics.
4. Compare the two associations. Is runs scored or runs allowed a better predictor of wins? Explain your reasoning.
5. Because the number of wins a team has is dependent on both how many runs they score and how many runs they allow, we can use a combination of both variables to predict the number of wins. Add a column in your data table for a new variable, run differential. Fill in the values using the formula R/G – RA/G.
6. Create a scatterplot to investigate the relationship between run differential and wins. Then calculate the equation of the least-squares regression line, the standard deviation of the residuals, and $r^2$.
7. Is run differential a better predictor than the variable you chose in Question 4? Explain your reasoning.

## Part 2

It is fairly clear that the number of games a team wins is dependent on both runs scored and runs allowed. But what variables help predict runs scored? Runs allowed?

1. Choose either runs scored (R) or runs allowed (RA) as the response variable you will try to model.
2. Choose at least three different explanatory variables (or combinations of explanatory variables) that might help predict the response variable you chose in Question 1. Create a scatterplot using each explanatory variable. Then calculate the equation of the least-squares regression line, the standard deviation of the residuals, and $r^2$ for each relationship.
3. Which explanatory variable from Question 2 is the best? Explain your reasoning.

# UNIT 3
## Collecting Data

Chapter 4

# Collecting Data

AfriPics.com/Alamy Stock Photo

# INTRODUCTION

You can hardly go a day without hearing the results of a statistical study. Here are some examples:

- The National Highway Traffic Safety Administration (NHTSA) reports that seat belt use in passenger vehicles increased from 88.5% in 2015 to 90.1% in 2016.[1]
- According to a survey, U.S. teens aged 13 to 18 use entertainment media (television, Internet, social media, listening to music, etc.) nearly 9 hours a day, on average.[2]
- A study suggests that lack of sleep increases the risk of catching a cold.[3]
- For their final project, two AP® Statistics students showed that listening to music while studying decreased subjects' performance on a memory task.[4]

Can we trust these results? As you'll learn in this chapter, the answer depends on how the data were produced. Let's take a closer look at where the data came from in each of these studies.

Each year, the NHTSA conducts an *observational study* of seat belt use in vehicles. The NHTSA sends trained observers to record the behavior of people in vehicles at randomly selected locations across the country. The idea of an observational study is simple: you can learn a lot just by watching or by asking a few questions, as in the survey of teens' media habits. Common Sense Media conducted this survey using a random sample of 1399 U.S. 13- to 18-year-olds. Both of these studies use information from a *sample* to draw conclusions about some larger *population*. Section 4.1 examines the issues involved in sampling and surveys.

In the sleep and catching a cold study, 153 volunteers answered questions about their sleep habits over a two-week period. Then researchers gave them a virus and waited to see who developed a cold. This was a complicated observational study. Compare this with the *experiment* performed by the AP® Statistics students. They recruited 30 students and divided them into two groups of 15 by drawing names from a hat. Students in one group tried to memorize a list of words while listening to music. Students in the other group tried to memorize the same list of words while sitting in silence. Section 4.2 focuses on designing experiments.

In Section 4.3, we revisit two key ideas from Sections 4.1 and 4.2: drawing conclusions about a population based on a random sample and drawing conclusions about cause and effect based on a randomized experiment. In both cases, we will focus on the role of randomization in our analysis.

# SECTION 4.1 Sampling and Surveys

**LEARNING TARGETS** *By the end of the section, you should be able to:*

- Identify the population and sample in a statistical study.
- Identify voluntary response sampling and convenience sampling and explain how these sampling methods can lead to bias.
- Describe how to select a simple random sample using slips of paper, technology, or a table of random digits.
- Describe how to select a sample using stratified random sampling, cluster sampling, and systematic random sampling, and explain whether a particular sampling method is appropriate in a given situation.
- Explain how undercoverage, nonresponse, question wording, and other aspects of a sample survey can lead to bias.

Suppose we want to find out what percent of young drivers in the United States text while driving. To answer the question, we will survey 16- to 20-year-olds who live in the United States and drive. Ideally, we would ask them all by conducting a **census**. But contacting every driver in this age group wouldn't be practical: it would take too much time and cost too much money. Instead, we pose the question to a **sample** chosen to represent the entire **population** of young drivers.

**DEFINITION Population, Census, Sample**

The **population** in a statistical study is the entire group of individuals we want information about. A **census** collects data from every individual in the population.

A **sample** is a subset of individuals in the population from which we collect data.

The distinction between population and sample is basic to statistics. To make sense of any sample result, you must know what population the sample represents.

## EXAMPLE Sampling monitors and voters

### Populations and samples

**PROBLEM:** Identify the population and the sample in each of the following settings.

(a) The quality control manager at a factory that produces computer monitors selects 10 monitors from the production line each hour. The manager inspects each monitor for defects in construction and performance.

(b) Prior to an election, a news organization surveys 1000 registered voters to predict which candidate will be elected as president.

Cancan Chu/Getty Images

**SOLUTION:**

(a) The population is all the monitors produced in this factory that hour. The sample is the 10 monitors selected from the production line.

(b) The population is all registered voters. The sample is the 1000 registered voters surveyed.

> Because the sample came from 1 hour's production at this factory, the population is the monitors produced that hour in this factory—not all monitors produced in the world or even all monitors produced by this factory.

**FOR PRACTICE, TRY EXERCISE 1**

## The Idea of a Sample Survey

We often draw conclusions about a population based on a sample. Have you ever tasted a sample of ice cream and ordered a cone because the sample tastes good? Because ice cream is fairly uniform, the single taste represents the whole. Choosing a representative sample from a large and varied population (like all young U.S. drivers) is not so easy. The first step in planning a **sample survey** is to decide *what population* we want to describe. The second step is to decide *what we want to measure.*

**DEFINITION Sample survey**

A **sample survey** is a study that collects data from a sample to learn about the population from which the sample was selected.

By our definition, the individuals in a sample survey can be people, animals, or things. Some people use the terms *survey* or *sample survey* to refer only to studies in which people are asked questions, like the news organization survey in the preceding example. We'll avoid this more restrictive terminology.

The final step in planning a sample survey is to decide how to choose a sample from the population. Here is an activity that illustrates the process of conducting a sample survey.

## ACTIVITY Who wrote the Federalist Papers?

The Federalist Papers are a series of 85 essays supporting the ratification of the U.S. Constitution. At the time they were published, the identity of the authors was a secret known to only a few people. Over time, however, the authors were identified as Alexander Hamilton, James Madison, and John Jay. The authorship of 73 of the essays is fairly certain, leaving 12 in dispute. However, thanks in some part to statistical analysis, most scholars now believe that the 12 disputed essays were written by Madison alone or in collaboration with Hamilton.[5]

Bragin Alexey/Shutterstock.com

There are several ways to use statistics to help determine the authorship of a disputed text. One method is to estimate the average word length in a disputed text and compare it to the average word lengths of works where the authorship is not in dispute.

The following passage is the opening paragraph of Federalist Paper No. 51,[6] one of the disputed essays. The theme of this essay is the separation of powers between the three branches of government.

> To what expedient, then, shall we finally resort, for maintaining in practice the necessary partition of power among the several departments, as laid down in the Constitution? The only answer that can be given is, that as all these exterior provisions are found to be inadequate, the defect must be supplied, by so contriving the interior structure of the government as that its several constituent parts may, by their mutual relations, be the means of keeping each other in their proper places. Without presuming to undertake a full development of this important idea, I will hazard a few general observations, which may perhaps place it in a clearer light, and enable us to form a more correct judgment of the principles and structure of the government planned by the convention.

1. Choose 5 words from this passage. Count the number of letters in each of the words you selected, and find the average word length.
2. Your teacher will draw and label a horizontal axis for a class dotplot. Plot the average word length you obtained in Step 1 on the graph.
3. Your teacher will show you how to use a random number generator to select a random sample of 5 words from the 130 words in the opening passage. Count the number of letters in each of the selected words, and find the average word length.
4. Your teacher will draw and label another horizontal axis with the same scale for a comparative dotplot. Plot the average word length you obtained in Step 3 on the graph.
5. How do the dotplots compare? Can you think of any reasons why they might be different? Discuss with your classmates.

## How to Sample Badly

Suppose we want to know how long students at a large high school spent doing homework last week. We might go to the school library and ask the first 30 students we see about the amount of time they spend on their homework. This method is known as a **convenience sampling**.

**DEFINITION Convenience sampling**

**Convenience sampling** selects individuals from the population who are easy to reach.

**Convenience sampling often produces unrepresentative data.** Consider our sample of 30 students from the school library. It's unlikely that this convenience sample accurately represents the homework habits of all students at the high school. In fact, if we were to repeat this sampling process day after day, we would almost always overestimate the average homework time in the population. Why? Because students who hang out in the library tend to be more studious. This is **bias**: using a method that systematically favors certain outcomes over others.

**DEFINITION Bias**

The design of a statistical study shows **bias** if it is very likely to underestimate or very likely to overestimate the value you want to know.

**AP® EXAM TIP**

If you're asked to describe how the design of a sample survey leads to bias, you're expected to do two things: (1) describe how the members of the sample might respond differently from the rest of the population, and (2) explain how this difference would lead to an underestimate or overestimate. Suppose you were asked to explain how using your statistics class as a sample to estimate the proportion of all high school students who own a graphing calculator could result in bias. You might respond, "This is a convenience sample. It would probably include a much higher proportion of students with a graphing calculator than in the population at large because a graphing calculator is required for the statistics class. So this method would probably lead to an overestimate of the actual population proportion."

caution

**Bias is not just bad luck in one sample.** It's the result of a bad study design that will consistently miss the truth about the population in the same way. Convenience sampling will almost always result in bias. So will **voluntary response sampling**.

**DEFINITION Voluntary response sampling**

**Voluntary response sampling** allows people to choose to be in the sample by responding to a general invitation.

Because voluntary response sampling typically gives a sample that is unrepresentative of the population we want to know about, the bias that results is called *voluntary response bias*.

Most Internet polls, along with call-in, text-in, and write-in polls, rely on voluntary response sampling. *People who self-select to participate in such surveys are usually not representative of the population of interest.* Voluntary response sampling attracts people who feel strongly about an issue, and who often share the same opinion. That leads to bias.

## EXAMPLE Boaty McBoatface
### Biased sampling methods

**PROBLEM:** In 2016, Britain's Natural Environment Research Council invited the public to name its new \$300 million polar research ship. To vote on the name, people simply needed to visit a website and record their choice. Ignoring names suggested by the council, over 124,000 people voted for "Boaty McBoatface," which ended up having more than 3 times as many votes as the second-place finisher.[7]

NERC/SplashNews/Newscom

What type of sampling did the council use in their poll? Explain how bias in this sampling method could have affected the poll results.

**SOLUTION:**

The council used voluntary response sampling: people chose to go online and respond. The people who chose to be in the sample were probably less serious about science than the British population as a whole—and more likely to prefer a funny name. The proportion of people in the sample who prefer the name Boaty McBoatface is likely to be greater than the proportion of all British residents who would choose this name.

Remember to describe how the responses from the members of the sample might differ from the responses from the rest of the population *and* how this difference will affect the estimate.

**FOR PRACTICE, TRY EXERCISE 5**

## CHECK YOUR UNDERSTANDING

For each of the following situations, identify the sampling method used. Then explain how bias in the sampling method could affect the results.

1. A farmer brings a juice company several crates of oranges each week. A company inspector looks at 10 oranges from the top of each crate before deciding whether to buy all the oranges.
2. The ABC program *Nightline* once asked if the United Nations should continue to have its headquarters in the United States. Viewers were invited to call one telephone number to respond "Yes" and another to respond "No." There was a charge for calling either number. More than 186,000 callers responded, and 67% said "No."

## How to Sample Well: Random Sampling

In convenience sampling, the researcher chooses easy-to-reach members of the population. In voluntary response sampling, people decide whether to join the sample. Both sampling methods suffer from bias due to personal choice. As you discovered in The Federalist Papers activity, a good way to avoid bias is to let chance choose the sample. That's the idea of **random sampling**.

In everyday life, some people use the word *random* to mean "haphazard," as in "That's so random." In statistics, random means "using chance." Don't say that a sample was chosen at random if a chance process wasn't used to select the individuals.

**DEFINITION Random sampling**

**Random sampling** involves using a chance process to determine which members of a population are included in the sample.

For example, to choose a random sample of 6 students from a class of 30, start by writing each of the 30 names on a separate slip of paper, making sure the slips are all the same size. Then put the slips in a hat, mix them well, and pull out slips one at a time until you have identified 6 different students. An alternative approach would be to give each member of the population a distinct number and to use the "hat method" with these numbers instead of people's names. Note that this version would work just as well if the population consisted of animals or things. The resulting sample is called a **simple random sample**, or **SRS** for short.

**DEFINITION Simple random sample (SRS)**

A **simple random sample (SRS)** of size *n* is chosen in such a way that every group of *n* individuals in the population has an equal chance to be selected as the sample.

An SRS gives every possible sample of the desired size an equal chance to be chosen. Picture drawing 20 slips of paper (the sample) from a hat containing 200 identical slips (the population). Any set of 20 slips has the same chance as any other set of 20 to be chosen. This also means that each individual has the same chance to be chosen in an SRS. However, giving each individual the same chance to be selected is not enough to guarantee that a sample is an SRS. Some other random sampling methods give each member of the population an equal chance to be selected, but not each possible sample. We'll look at some of these methods later.

**HOW TO CHOOSE A SIMPLE RANDOM SAMPLE** The hat method won't work well if the population is large. Imagine trying to take a simple random sample of 1000 registered voters in the United States using a hat! In practice, most people use random numbers generated by technology to choose samples.

**HOW TO CHOOSE AN SRS WITH TECHNOLOGY**

- **Label.** Give each individual in the population a distinct numerical label from 1 to *N*, where *N* is the number of individuals in the population.
- **Randomize.** Use a random number generator to obtain *n different* integers from 1 to *N*, where *n* is the sample size.
- **Select.** Choose the individuals that correspond to the randomly selected integers.

When choosing an SRS, we **sample without replacement**. That is, once an individual is selected for a sample, that individual cannot be selected again. Many random number generators **sample with replacement**, so it is important to explain that repeated numbers should be ignored when using technology to select an SRS.

**DEFINITION Sampling without replacement, Sampling with replacement**

When **sampling without replacement**, an individual from a population can be selected only once.

When **sampling with replacement**, an individual from a population can be selected more than once.

The following Technology Corner shows you how to select an SRS using a graphing calculator.

## 12. Technology Corner CHOOSING AN SRS

*TI-Nspire and other technology instructions are on the book's website at highschool.bfwpub.com/updatedtps6e.*

Let's use a graphing calculator to select an SRS of 10 students from a population of students numbered 1 to 1750.

1. Check that your calculator's random number generator is working properly.
   - Press MATH, then select PROB (PRB) and choose randInt(.

     **Newer OS:** In the dialogue box, enter these values: lower: 1, upper: 1750, n: 1, choose Paste, and press ENTER.

**Older OS:** Complete the command randInt(1,1750) and press ENTER.

- Compare your results with those of your classmates. If several students got the same number, you'll need to seed your calculator's random integer generator with different numbers before you proceed. Directions for doing this are given in the *Teacher's Edition.*

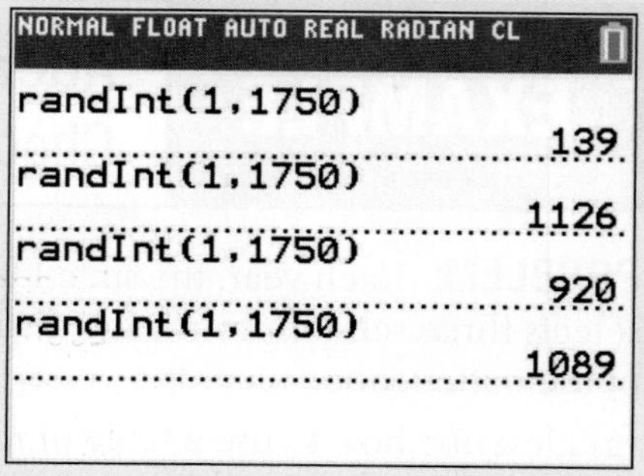


2. Randomly generate 10 distinct numbers from 1 to 1750 by pressing ENTER until you have chosen 10 different labels.

*Note:* If you have a TI-84 Plus CE, use the command RandIntNoRep(1,1750,10) to get 10 distinct integers from 1 to 1750. If you have a TI-84 with OS 2.55 or later, use the command RandIntNoRep(1,1750) to sort the numbers from 1 to 1750 in random order. The first 10 numbers listed give the labels of the chosen students.

There are many random number generators available on the Internet, including those at www.random.org. You can also use the random number generator on your calculator.

If you don't have technology handy, you can use a table of random digits to choose an SRS. We have provided a table of random digits at the back of the book (Table D). Here is an excerpt:

**Table D Random digits**

| LINE | | | | | | | | |
|---|---|---|---|---|---|---|---|---|
| 101 | 19223 | 95034 | 05756 | 28713 | 96409 | 12531 | 42544 | 82853 |
| 102 | 73676 | 47150 | 99400 | 01927 | 27754 | 42648 | 82425 | 36290 |
| 103 | 45467 | 71709 | 77558 | 00095 | 32863 | 29485 | 82226 | 90056 |

You can think of this table as the result of someone putting the digits 0 to 9 in a hat, mixing, drawing one, replacing it, mixing again, drawing another, and so on. The digits have been arranged in groups of five within numbered rows to make the table easier to read. The groups and rows have no special meaning—Table D is just a long list of randomly chosen digits. As with technology, there are three steps in using Table D to choose a random sample.

## HOW TO CHOOSE AN SRS USING TABLE D

- **Label.** Give each member of the population a distinct numerical label with the same number of digits. Use as few digits as possible.
- **Randomize.** Read consecutive groups of digits of the appropriate length from left to right across a line in Table D. Ignore any group of digits that wasn't used as a label or that duplicates a label already in the sample. Stop when you have chosen *n* different labels.
- **Select.** Choose the individuals that correspond to the randomly selected labels.

Always use the shortest labels that will cover your population. For instance, you can label up to 100 individuals with two digits: 01, 02, . . . , 99, 00. As standard practice, we recommend that you begin with label 1 (or 01 or 001 or 0001, as needed). Reading groups of digits from the table gives all individuals the same chance to be chosen because all labels of the same length have the same chance to be found in the table. For example, any pair of digits in the table is equally likely to be any of the 100 possible labels 01, 02, . . . , 99, 00. Here's an example that shows how this process works.

## EXAMPLE

### Attendance audit
### Choosing an SRS with Table D

**PROBLEM:** Each year, the state Department of Education randomly selects three schools from each district and conducts a detailed audit of their attendance records.

Holly Albrecht

(a) Describe how to use a table of random digits to select an SRS of three schools from this list of 19 schools.

| | |
|---|---|
| Amphitheater High School | Keeling Elementary School |
| Amphitheater Middle School | La Cima Middle School |
| Canyon del Oro High School | Mesa Verde Elementary School |
| Copper Creek Elementary School | Nash Elementary School |
| Coronado K-8 School | Painted Sky Elementary School |
| Cross Middle School | Prince Elementary School |
| Donaldson Elementary School | Rio Vista Elementary School |
| Harelson Elementary School | Walker Elementary School |
| Holaway Elementary School | Wilson K-8 School |
| Ironwood Ridge High School | |

(b) Use the random digits here to choose the sample.

62081 64816 87374 09517 84534 06489 87201 97245

**SOLUTION:**

(a) Label the schools from 01 to 19 in alphabetical order. Move along a line of random digits from left to right, reading two-digit numbers, until three different numbers from 01 to 19 have been selected (ignoring repeated numbers and the numbers 20–99, 00). Audit the three schools that correspond with the numbers selected.

Remember to include all three steps:
- Label
- Randomize
- Select

(b) 62-skip, 08-select, 16-select, 48-skip, 16-repeat, 87-skip, 37-skip, 40-skip, 95-skip, 17-select.

The three schools are 08: Harelson Elementary School, 16: Prince Elementary School, and 17: Rio Vista Elementary School.

FOR PRACTICE, TRY EXERCISE 11

## CHECK YOUR UNDERSTANDING

A furniture maker buys hardwood in batches that each contain 1000 pieces. The supplier is supposed to dry the wood before shipping (wood that isn't dry won't hold its size and shape). The furniture maker chooses five pieces of wood from each batch and tests their moisture content. If any piece exceeds 12% moisture content, the entire batch is sent back. Describe how to select a simple random sample of 5 pieces using each of the following:

1. A random number generator
2. A table of random digits

## Other Random Sampling Methods

One of the most common alternatives to simple random sampling is called **stratified random sampling**. This method involves dividing the population into non-overlapping groups (**strata**) of individuals who are expected to have similar responses, sampling from each of these groups, and combining these "subsamples" to form the overall sample.

> The singular form of *strata* is *stratum.*

**DEFINITION Strata, Stratified random sampling**

**Strata** are groups of individuals in a population who share characteristics thought to be associated with the variables being measured in a study. **Stratified random sampling** selects a sample by choosing an SRS from each stratum and combining the SRSs into one overall sample.

Stratified random sampling works best when the individuals within each stratum are similar (homogeneous) with respect to what is being measured and when there are large differences between strata. For example, in a study of sleep habits on school nights, the population of students in a large high school might be divided into freshman, sophomore, junior, and senior strata. After all, it is reasonable to think that freshmen have different sleep habits than seniors. The following activity illustrates the benefit of choosing appropriate strata.

### ACTIVITY Sampling sunflowers

Li Ding/Alamy

A British farmer grows sunflowers for making sunflower oil. Her field is arranged in a grid pattern, with 10 rows and 10 columns as shown in the figure. Irrigation ditches run along the top and bottom of the field. The farmer would like to estimate the number of healthy plants in the field so she can project how much money she'll make from selling them. It would take too much time to count the plants in all 100 squares, so she'll accept an estimate based on a sample of 10 squares.

1. Use Table D or technology to take a simple random sample of 10 grid squares. Record the location (e.g., B6) of each square you select.
2. This time, select a stratified random sample using the *rows* as strata. Use Table D or technology to randomly select one square from each horizontal row. Record the location of each square—for example, Row 1: G, Row 2: B, and so on.
3. Now, take a stratified random sample using the *columns* as strata. Use Table D or technology to randomly select one square from each vertical column. Record the location of each square—for example, Column A: 4, Column B: 1, and so on.
4. Your teacher will provide the actual number of healthy sunflowers in each grid square. Use that information to calculate your estimate of the mean number of healthy sunflowers per square in the entire field for each of your samples in Steps 1, 2, and 3.
5. Make comparative dotplots showing the mean number of healthy sunflowers obtained using the three different sampling methods for all members of the class. Describe any similarities and differences you see.

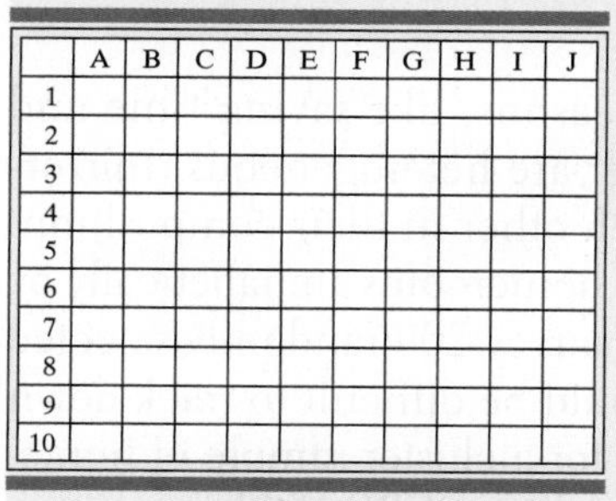

The following dotplots show the mean number of healthy plants in 100 samples using each of the three sampling methods in the activity: simple random sampling, stratified random sampling with rows of the field as strata, and stratified random sampling with columns of the field as strata. Notice that all three distributions are centered at about 102.5, the true mean number of healthy plants in all squares of the field. That makes sense because random sampling tends to yield accurate estimates of unknown population means.

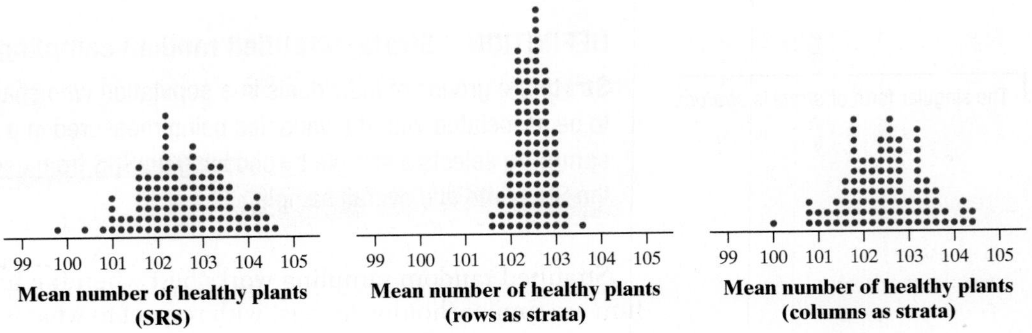


One other detail stands out in the graphs: there is much less variability in the estimates when we use the rows as strata. The table provided by your teacher shows the actual number of healthy sunflowers in each grid square. Notice that the squares within each row are relatively homogeneous because they contain a similar number of healthy plants, but that there are big differences between rows. *When we can choose strata that have similar responses (e.g., number of healthy plants) within strata but different responses between strata, stratified random samples give more precise estimates than simple random samples of the same size.*

Why didn't using the columns as strata reduce the variability of the estimates in a similar way? Because the numbers of healthy plants vary a lot within each column and aren't very different from other columns.

Both simple random sampling and stratified random sampling are hard to use when populations are large and spread out over a wide area. In that situation, we might prefer to use **cluster sampling**. This method involves dividing the population into non-overlapping groups (**clusters**) of individuals that are "near" one another, then randomly selecting whole clusters to form the overall sample.

**DEFINITION Clusters, Cluster sampling**

A **cluster** is a group of individuals in the population that are located near each other. **Cluster sampling** selects a sample by randomly choosing clusters and including each member of the selected clusters in the sample.

Cluster sampling is often used for practical reasons, like saving time and money. Ideally, the individuals within each cluster are heterogeneous (mirroring the population) and clusters are similar to each other in their composition. Imagine a large high school that assigns students to homerooms alphabetically by last name, in groups of 25. Administrators want to survey 200 randomly selected students about a proposed schedule change. It would be difficult to track down an SRS of 200 students, so the administration opts for a cluster sample of homerooms. The principal (who knows some statistics) selects an SRS of 8 homerooms and gives the survey to all 25 students in each of the selected homerooms.

*Be sure you understand the difference between strata and clusters.* We want each stratum to contain similar individuals and for large differences to exist between strata. For a cluster sample, we'd *like* each cluster to look just like the population, but on a smaller scale. Unfortunately, cluster samples don't offer the statistical advantage of better information about the population that stratified random samples do. Here's an example that compares stratified random sampling and cluster sampling.

## EXAMPLE

### Sampling at a school assembly
### Other sampling methods

**PROBLEM:** The student council wants to conduct a survey about use of the school library during the first five minutes of an all-school assembly in the auditorium. There are 800 students present at the assembly. Here is a map of the auditorium. Note that students are seated by grade level and that the seats are numbered from 1 to 800.

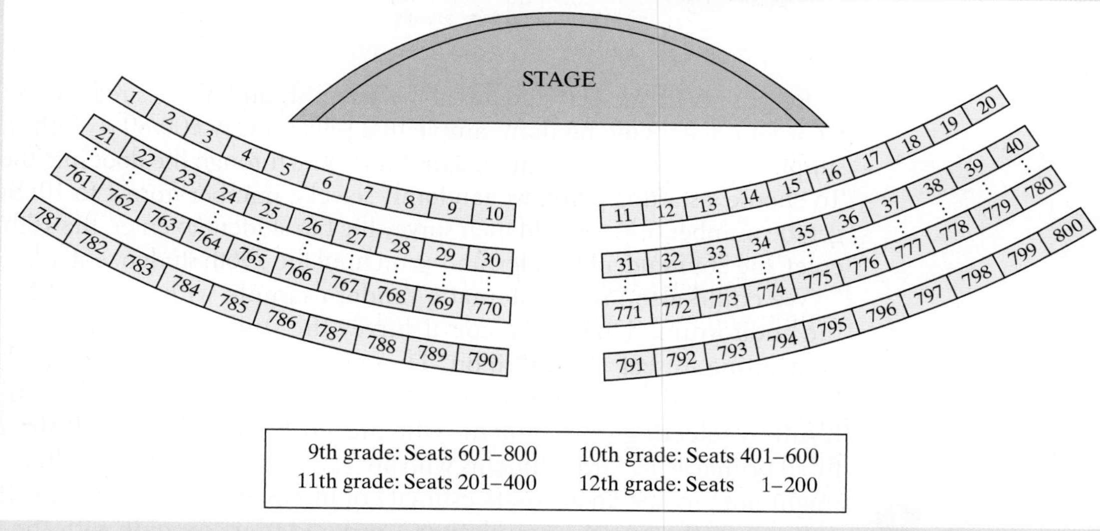


(a) Describe how to obtain a sample of 80 students using stratified random sampling. Explain your choice of strata and why this method might be preferred to simple random sampling.

(b) Describe how to obtain a sample of 80 students using cluster sampling. Explain your choice of clusters and why this method might be preferred to simple random sampling.

**SOLUTION:**

(a) **Because students' library use might be similar within grade levels but different across grade levels, use the grade-level seating areas as strata. For the 9th grade, generate 20 different random integers from 601 to 800 and give the survey to the students in those seats. Do the same for sophomores, juniors, and seniors using their corresponding seat numbers. Stratification by grade level should result in more precise estimates of student opinion than a simple random sample of the same size.**

(b) Each column of seats from the stage to the back of the auditorium could be used as a cluster because it would be relatively easy to hand out the surveys to an entire column. Label the columns from 1 to 20 starting at the left side of the stage, generate 2 different integers from 1 to 20, and give the survey to the 80 students sitting in these two columns. Cluster sampling is much more efficient than finding 80 seats scattered about the auditorium, as required by simple random sampling.

Note that each cluster contains students from all four grade levels, so each should represent the population fairly well. Randomly selecting 4 rows as clusters would also be easy, but this may over- or under-represent one grade level.

**FOR PRACTICE, TRY EXERCISE 21**

Another way to select a random sample of 80 students from the 800 students at the assembly is with **systematic random sampling**.

**DEFINITION Systematic random sampling**

**Systematic random sampling** selects a sample from an ordered arrangement of the population by randomly selecting one of the first *k* individuals and choosing every *k*th individual thereafter.

Because there are 800 students at the assembly and our desired sample size is 80, we want a systematic random sample that selects every 800/80 = 10th student. For simplicity, we could survey students as they come through the door into the assembly. To choose a starting point, we randomly select a number from 1 to 10. Suppose we get the number 6. We would then survey the 6th student to enter, the 16th student to enter, the 26th student to enter, and so on until the 796th student enters the assembly.

We could also use a systematic random sample with every 10th seat inside the auditorium. That is, we could survey the student sitting in seat 6, the student sitting in seat 16, and so on. There is a drawback to this method, however. It is possible that our starting number could have been 1 or 10, which would result in selecting only students who are sitting in aisle seats. If these seats are filled primarily by tardy students who are less responsible, the results of the survey might not provide an accurate estimate of library use for all students. **If there are patterns in the way the population is ordered that coincide with the pattern in a systematic sample, the sample may not be representative of the population.**

Systematic random sampling is particularly useful in certain contexts, such as exit polling at a polling place on Election Day. Because an unknown number of voters will come to the polling place that day, it isn't practical to select a simple random sample. However, it would be quite easy to generate a random number from 1 to 20 and to survey the chosen voter and every 20th voter thereafter.

Most large-scale sample surveys use *multistage sampling*, which combines two or more sampling methods. For example, the U.S. Census Bureau carries out a monthly Current Population Survey (CPS) of about 60,000 households. Researchers start by choosing a stratified random sample of neighborhoods in 756 of the 2007 geographical areas in the United States. Then they divide each neighborhood into clusters of four nearby households and select a cluster sample to interview.

Analyzing data from sampling methods other than simple random sampling takes us beyond basic statistics. But the SRS is the building block of more elaborate methods, and the principles of analysis remain much the same for these other methods.

## CHECK YOUR UNDERSTANDING

A factory runs 24 hours a day, producing 15,000 wood pencils per day over three 8-hour shifts—day, evening, and overnight. In the last stage of manufacturing, the pencils are packaged in boxes of 10 pencils each. Each day a sample of 300 pencils is selected and inspected for quality.

1. Describe how to select a stratified random sample of 300 pencils. Explain your choice of strata.
2. Describe how to select a cluster sample of 300 pencils. Explain your choice of clusters.
3. Describe how to select a systematic random sample of 300 pencils.
4. Explain a benefit of each of these three methods in this context.

# Sample Surveys: What Else Can Go Wrong?

As we have learned, the use of bad sampling methods (convenience or voluntary response) often leads to bias. Researchers can avoid these methods by using random sampling to choose their samples. Other problems in conducting sample surveys are more difficult to avoid.

Sampling is sometimes done using a list of individuals in the population, called a *sampling frame*. Such lists are seldom accurate or complete. The result is **undercoverage**.

Because undercoverage typically results in a sample that is unrepresentative of the population we want to know about, the bias that occurs is sometimes called *undercoverage bias*.

**DEFINITION Undercoverage**

**Undercoverage** occurs when some members of the population are less likely to be chosen or cannot be chosen in a sample.

Most samples suffer from some degree of undercoverage. A sample survey of households, for example, will miss not only homeless people but also prison inmates and students in dormitories. An opinion poll conducted by calling landline telephone numbers will miss households that have only cell phones as well as households without a phone. The results of sample surveys may not be accurate if the people who are undercovered differ from the rest of the population in ways that affect their responses.

Even if every member of the population is equally likely to be selected for a sample, not all members of the population are equally likely to provide a response. Some people are never at home and cannot be reached by pollsters on the phone or in person. Other people see an unfamiliar phone number on their caller ID and never pick up the phone or quickly hang up when they don't recognize the voice of the caller. These are examples of **nonresponse**, another major source of bias in surveys.

Because nonresponse typically results in a sample that is unrepresentative of the population we want to know about, the bias that occurs is sometimes called *nonresponse bias*.

**DEFINITION Nonresponse**

**Nonresponse** occurs when an individual chosen for the sample can't be contacted or refuses to participate.

Nonresponse leads to bias when the individuals who can't be contacted or refuse to participate would respond differently from those who do participate. Consider a landline telephone survey that asks people how many hours of television they watch per day. People who are selected but are out of the house won't be able to respond. Because these people probably watch less television than the people who are at home when the phone call is made, the mean number of hours obtained in the sample is likely to be greater than the mean number of hours of TV watched in the population.

How bad is nonresponse? According to polling guru Nate Silver, "Response rates to political polls are dismal. Even polls that make every effort to contact a representative sample of voters now get no more than 10 percent to complete their surveys—down from about 35 percent in the 1990s."[8] In contrast, the Census Bureau's American Community Survey (ACS) has the lowest nonresponse rate of any poll we know: only about 1% of the households in the sample refuse to respond. The overall nonresponse rate, including "never at home" and other causes, is just 2.5%.[9]

caution

**Some students misuse the term *voluntary response* to explain why certain individuals don't respond in a sample survey.** Their belief is that participation in the survey is optional (voluntary), so anyone can refuse to take part. What the students are describing is *nonresponse.* Think about it this way: nonresponse can occur only after a sample has been selected. In a voluntary response sample, every individual has opted to take part, so there won't be any nonresponse.

The wording of questions has an important influence on the answers given to a sample survey. Confusing or leading questions can introduce *question wording bias.* Even a single word can make a difference. In a recent Quinnipiac University poll, half of the respondents were asked if they support "stronger gun laws" and the other half were asked if they support "stronger gun control laws." In the first group, 52% of respondents supported stronger laws, but when the word *control* was added to the question, only 46% of respondents supported stronger laws.[10]

The gender, age, ethnicity, or behavior of the interviewer can also affect people's responses. People may lie about their age, income, or drug use. They may misremember how many hours they spent on the Internet last week. Or they might make up an answer to a question that they don't understand. All these issues can lead to **response bias**.

**DEFINITION Response bias**

**Response bias** occurs when there is a systematic pattern of inaccurate answers to a survey question.

## EXAMPLE Wash your hands!
### Response bias

**PROBLEM:** What percent of Americans wash their hands after using the bathroom? It depends on how you collect the data. In a telephone survey of 1006 U.S. adults, 96% said they always wash their hands after using a public restroom. An observational study of 6028 adults in public restrooms told a different story: only 85% of those observed washed their hands after using the restroom. Explain why the results of the two studies are so different.[11]

**SOLUTION:**

When asked in person, many people may lie about always washing their hands because they want to appear to have good hygiene. When people are only observed and not asked directly, the percent who wash their hands will be smaller—and much closer to the truth.

FOR PRACTICE, TRY EXERCISE 29

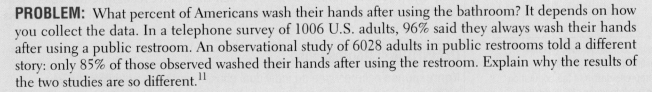

Even the order in which questions are asked is important. For example, ask a sample of college students these two questions:

- "How happy are you with your life in general?" (Answer on a scale of 1 to 5.)
- "How many dates did you have last month?"

There is almost no association between responses to the two questions when asked in this order. It appears that dating has little to do with happiness. Reverse the order of the questions, however, and a much stronger association appears: college students who say they had more dates tend to give higher ratings of happiness about life. The lesson is clear: the order in which questions are asked can influence the results.

## CHECK YOUR UNDERSTANDING

1. Each of the following is a possible source of bias in a sample survey. Name the type of bias that could result.
   (a) The sample is chosen at random from a telephone directory.
   (b) Some people cannot be contacted in five calls.
   (c) Interviewers choose people walking by on the sidewalk to interview.
2. A survey paid for by makers of disposable diapers found that 84% of the sample opposed banning disposable diapers. Here is the actual question:

   > It is estimated that disposable diapers account for less than 2% of the trash in today's landfills. In contrast, beverage containers, third-class mail, and yard wastes are estimated to account for about 21% of the trash in landfills. Given this, in your opinion, would it be fair to ban disposable diapers?[12]

   Do you think the estimate of 84% is less than, greater than, or about equal to the percent of all people in the population who would oppose banning disposable diapers? Explain your reasoning.

# Section 4.1 Summary

- A **census** collects data from every individual in the **population.**
- A **sample survey** selects a **sample** from the population of all individuals about which we desire information. The goal of a sample survey is to draw conclusions about the population based on data from the sample.
- **Convenience sampling** chooses individuals who are easiest to reach. In **voluntary response sampling,** individuals choose to join the sample in response to an open invitation. Both these sampling methods usually lead to **bias:** they will be very likely to underestimate or very likely to overestimate the value you want to know.
- **Random sampling** uses a chance process to select a sample.
- A **simple random sample (SRS)** gives every possible sample of a given size the same chance to be chosen. Choose an SRS by labeling the members of the population and using slips of paper, a table of random digits, or technology to select the sample. Make sure to use **sampling without replacement** when selecting an SRS.
- To use **stratified random sampling,** divide the population into non-overlapping groups of individuals (**strata**) that are similar in some way that might affect their responses. Then choose a separate SRS from each stratum and

combine these SRSs to form the sample. When strata are "similar within (homogeneous) but different between," stratified random samples tend to give more precise estimates of unknown population values than do simple random samples.

- To use **cluster sampling,** divide the population into non-overlapping groups of individuals that are located near each other, called **clusters.** Randomly select some of these clusters. All the individuals in the chosen clusters are included in the sample. Ideally, clusters are "different within (heterogeneous) but similar between." Cluster sampling saves time and money by collecting data from entire groups of individuals that are close together.
- To use **systematic random sampling**, select a value of $k$ based on the population size and desired sample size, randomly select a value from 1 to $k$ to identify the first individual in the sample, and choose every $k$th individual thereafter. If there are patterns in the way the population is ordererd that coincide with the value of $k$, the sample may not be representative of the population. Otherwise, systematic random sampling can be easier to conduct than other sampling methods.
- Random sampling helps avoid bias in choosing a sample. Bias can still occur in the sampling process due to **undercoverage,** which happens when some members of the population are less likely to be chosen or cannot be chosen for the sample.
- Other serious problems in sample surveys can occur after the sample is chosen. The single biggest problem is **nonresponse:** when people can't be contacted or refuse to answer. Untruthful answers by respondents, poorly worded questions, and other problems can lead to **response bias.**

## 4.1 Technology Corner

*TI-Nspire and other technology instructions are on the book's website at highschool.bfwpub.com/updatedtps6e.*

**12. Choosing an SRS** **Page 254**

# Section 4.1 Exercises

1. **Sampling stuffed envelopes** A large retailer prepares its customers' monthly credit card bills using an automatic machine that folds the bills, stuffs them into envelopes, and seals the envelopes for mailing. Are the envelopes completely sealed? Inspectors choose 40 envelopes at random from the 1000 stuffed each hour for visual inspection. Identify the population and the sample. pg 249

2. **Student archaeologists** An archaeological dig turns up large numbers of pottery shards, broken stone tools, and other artifacts. Students working on the project classify each artifact and assign a number to it. The counts in different categories are important for understanding the site, so the project director chooses 2% of the artifacts at random and checks the students' work. Identify the population and the sample.

3. **Students as customers** A high school's student newspaper plans to survey local businesses about the importance of students as customers. From an alphabetical list of all local businesses, the newspaper staff chooses 150 businesses at random. Of these, 73 return the questionnaire mailed by the staff. Identify the population and the sample.

4. **Customer satisfaction** A department store mails a customer satisfaction survey to people who make credit card purchases at the store. This month, 45,000 people made

credit card purchases. Surveys are mailed to 1000 of these people, chosen at random, and 137 people return the survey form. Identify the population and the sample.

5. **Sleepless nights** How much sleep do high school students get on a typical school night? A counselor designed a survey to find out. To make data collection easier, the counselor surveyed the first 100 students to arrive at school on a particular morning. These students reported an average of 7.2 hours of sleep on the previous night.
pg 252

(a) What type of sample did the counselor obtain?

(b) Explain why this sampling method is biased. Is 7.2 hours probably greater than or less than the true average amount of sleep last night for all students at the school? Why?

6. **Online polls** *Parade* magazine posed the following question: "Should drivers be banned from using all cell phones?" Readers were encouraged to vote online at parade.com. The subsequent issue of *Parade* reported the results: 2407 (85%) said "Yes" and 410 (15%) said "No."

(a) What type of sample did the *Parade* survey obtain?

(b) Explain why this sampling method is biased. Is 85% probably greater than or less than the true percent of all adults who believe that all cell phone use while driving should be banned? Why?

7. **Online reviews** Many websites include customer reviews of products, restaurants, hotels, and so on. The manager of a hotel was upset to see that 26% of reviewers on a travel website gave the hotel "1 star"—the lowest possible rating. Explain how bias in the sampling method could affect the estimate.

8. **Funding for fine arts** The band director at a high school wants to estimate the percentage of parents who support a decrease in the budget for fine arts. Because many parents attend the school's annual musical, the director surveys the first 30 parents who arrive at the show. Explain how bias in the sampling method could affect the estimate.

9. **Explain it to the congresswoman** You are on the staff of a member of Congress who is considering a bill that would provide government-sponsored insurance for nursing-home care. You report that 1128 letters have been received on the issue, of which 871 oppose the legislation. "I'm surprised that most of my constituents oppose the bill. I thought it would be quite popular," says the congresswoman. Are you convinced that a majority of the voters oppose the bill? How would you explain the statistical issue to the congresswoman?

10. **Sampling mall shoppers** You may have seen the mall interviewer, clipboard in hand, approaching people passing by. Explain why even a large sample of mall shoppers would not provide a trustworthy estimate of the current unemployment rate in the city where the mall is located.

11. **Do you trust the Internet?** You want to ask a sample of high school students the question "How much do you trust information about health that you find on the Internet—a great deal, somewhat, not much, or not at all?" You try out this and other questions on a pilot group of 5 students chosen from your class.
pg 256

(a) Explain how you would use a line of Table D to choose an SRS of 5 students from the following list.

(b) Use line 107 to select the sample. Show how you use each of the digits.

| | | | |
|---|---|---|---|
| Anderson | Drasin | Kim | Rider |
| Arroyo | Eckstein | Molina | Rodriguez |
| Batista | Fernandez | Morgan | Samuels |
| Bell | Fullmer | Murphy | Shen |
| Burke | Gandhi | Nguyen | Tse |
| Cabrera | Garcia | Palmiero | Velasco |
| Calloway | Glaus | Percival | Wallace |
| Delluci | Helling | Prince | Washburn |
| Deng | Husain | Puri | Zabidi |
| De Ramos | Johnson | Richards | Zhao |

12. **Apartment living** You are planning a report on apartment living in a college town. You decide to select three apartment complexes at random for in-depth interviews with residents.

(a) Explain how you would use a line of Table D to choose an SRS of 3 complexes from the following list.

(b) Use line 117 to select the sample. Show how you use each of the digits.

| | | |
|---|---|---|
| Ashley Oaks | Country View | Mayfair Village |
| Bay Pointe | Country Villa | Nobb Hill |
| Beau Jardin | Crestview | Pemberly Courts |
| Bluffs | Del-Lynn | Peppermill |
| Brandon Place | Fairington | Pheasant Run |
| Briarwood | Fairway Knolls | Richfield |
| Brownstone | Fowler | Sagamore Ridge |
| Burberry | Franklin Park | Salem Courthouse |
| Cambridge | Georgetown | Village Manor |
| Chauncey Village | Greenacres | Waterford Court |
| Country Squire | Lahr House | Williamsburg |

13. **Sampling the forest** To gather data on a 1200-acre pine forest in Louisiana, the U.S. Forest Service laid a grid of 1410 equally spaced circular plots over a map of the forest. A ground survey visited a sample of 10% of the plots.[13]

(a) Explain how you would use a random number generator to choose an SRS of 141 plots. Your description should be clear enough for a classmate to carry out your plan.

(b) Use your method from part (a) to choose the first 3 plots.

14. **Sampling gravestones** The local genealogical society in Coles County, Illinois, has compiled records on all 55,914 gravestones in cemeteries in the county for the years 1825 to 1985. Historians plan to use these records to learn about African Americans in Coles County's history. They first choose an SRS of 395 records to check their accuracy by visiting the actual gravestones.[14]

(a) Explain how you would use a random number generator to choose the SRS. Your description should be clear enough for a classmate to carry out your plan.

(b) Use your method from part (a) to choose the first 3 gravestones.

15. **No tipping** The owner of a large restaurant is considering a new "no tipping" policy and wants to survey a sample of employees. The policy would add 20% to the cost of food and beverages and the additional revenue would be distributed equally among servers and kitchen staff. Describe how to select a stratified random sample of approximately 30 employees. Explain your choice of strata and why stratified random sampling might be preferred in this context.

16. **Parking on campus** The director of student life at a university wants to estimate the proportion of undergraduate students who regularly park a car on campus. Describe how to select a stratified random sample of approximately 100 students. Explain your choice of strata and why stratified random sampling might be preferred in this context.

17. **SRS of engineers?** A corporation employs 2000 male and 500 female engineers. A stratified random sample of 200 male and 50 female engineers gives every individual in the population the same chance to be chosen for the sample. Is it an SRS? Explain your answer.

18. **SRS of students?** At a party, there are 30 students over age 21 and 20 students under age 21. You choose at random 3 of those over 21 and separately choose at random 2 of those under 21 to interview about their attitudes toward alcohol. You have given every student at the party the same chance to be interviewed. Is your sample an SRS? Explain your answer.

19. **High-speed Internet** Laying fiber-optic cable is expensive. Cable companies want to make sure that if they extend their lines to less dense suburban or rural areas, there will be sufficient demand so the work will be cost-effective. They decide to conduct a survey to determine the proportion of households in a rural subdivision that would buy the service. They select a simple random sample of 5 blocks in the subdivision and survey each family that lives on the selected blocks.

(a) What is the name for this kind of sampling method?

(b) Give a possible reason why the cable company chose this method.

20. **Timber!** A lumber company wants to estimate the proportion of trees in a large forest that are ready to be cut down. They use an aerial map to divide the forest into 200 equal-sized rectangles. Then they choose a random sample of 20 rectangles and examine every tree in the selected rectangles.

(a) What is the name for this kind of sampling method?

(b) Give a possible reason why the lumber company chose this method.

21. **How is your room?** A hotel has 30 floors with 40 rooms per floor. The rooms on one side of the hotel face the water, while rooms on the other side face a golf course. There is an extra charge for the rooms with a water view. The hotel manager wants to select 120 rooms and survey the registered guest in each of the selected rooms about his or her overall satisfaction with the property.

pg 259

(a) Describe how to obtain a sample of 120 rooms using stratified random sampling. Explain your choice of strata and why this method might be preferred to simple random sampling.

(b) Describe how to obtain a sample of 120 rooms using cluster sampling. Explain your choice of clusters and why this method might be preferred to simple random sampling.

22. **Go Blue!** Michigan Stadium, also known as "The Big House," seats over 100,000 fans for a football game. The University of Michigan Athletic Department wants to survey fans about concessions that are sold during games. Tickets are most expensive for seats on the sidelines. The cheapest seats are in the end zones (where one of the authors sat as a student). A map of the stadium is shown.

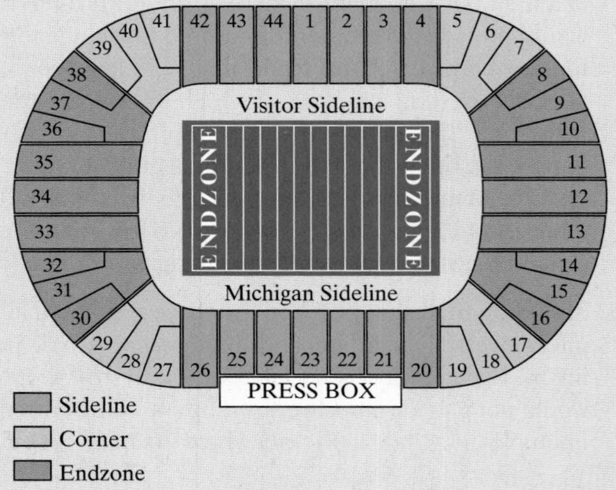

(a) Describe how to obtain a sample using stratified random sampling. Explain your choice of strata and why this method might be preferred to simple random sampling.

(b) Describe how to obtain a sample using cluster sampling. Explain your choice of clusters and why this method might be preferred to simple random sampling.

23. **Dead trees** In Rocky Mountain National Park, many mature pine trees along Highway 34 are dying due to infestation by pine beetles. Scientists would like to use a sample of size 200 to estimate the proportion of the approximately 5000 pine trees along the highway that have been infested.

(a) Explain why it wouldn't be practical for scientists to obtain an SRS in this setting.

(b) A possible alternative would be to use the first 200 pine trees along the highway as you enter the park. Why isn't this a good idea?

(c) Describe how to select a systematic random sample of 200 pine trees along Highway 34.

24. **iPhones** Suppose 1000 iPhones are produced at a factory today. Management would like to ensure that the phones' display screens meet the company's quality standards before shipping them to retail stores. Because it takes about 10 minutes to inspect an individual phone's display screen, managers decide to inspect a sample of 20 phones from the day's production.

(a) Explain why it would be difficult for managers to inspect an SRS of 20 iPhones that are produced today.

(b) An eager employee suggests that it would be easy to inspect the last 20 iPhones that were produced today. Why isn't this a good idea?

(c) Describe how to select a systematic random sample of 20 phones from the day's production.

25. **Eating on campus** The director of student life at a small college wants to know what percent of students eat regularly in the cafeteria. To find out, the director selects an SRS of 300 students who live in the dorms. Describe how undercoverage might lead to bias in this study. Explain the likely direction of the bias.

26. **Immigration reform** A news organization wants to know what percent of U.S. residents support a "pathway to citizenship" for people who live in the United States illegally. The news organization randomly selects registered voters for the survey. Describe how undercoverage might lead to bias in this study. Explain the likely direction of the bias.

27. **Reporting weight loss** A total of 300 people participated in a free 12-week weight-loss course at a community health clinic. After one year, administrators emailed each of the 300 participants to see how much weight they had lost since the end of the course. Only 56 participants responded to the survey. The mean weight loss for this sample was 13.6 pounds. Describe how nonresponse might lead to bias in this study. Explain the likely direction of the bias.

28. **Nonresponse** A survey of drivers began by randomly sampling from all listed residential telephone numbers in the United States. Of 45,956 calls to these numbers, 5029 were completed. The goal of the survey was to estimate how far people drive, on average, per day. Describe how nonresponse might lead to bias in this study. Explain the likely direction of the bias.

29. **Running red lights** An SRS of 880 drivers was asked: pg 262 "Recalling the last ten traffic lights you drove through, how many of them were red when you entered the intersections?" Of the 880 respondents, 171 admitted that at least one light had been red.[15] A practical problem with this survey is that people may not give truthful answers. Explain the likely direction of the bias.

30. **Seat belt use** A study in El Paso, Texas, looked at seat belt use by drivers. Drivers were observed at randomly chosen convenience stores. After they left their cars, they were invited to answer questions that included questions about seat belt use. In all, 75% said they always used seat belts, yet only 61.5% were wearing seat belts when they pulled into the store parking lots.[16] Explain why the two percentages are so different.

31. **Boys don't cry?** Two female statistics students asked a random sample of 60 high school boys if they have ever cried during a movie. Thirty of the boys were asked directly and the other 30 were asked anonymously by means of a "secret ballot." When the responses were anonymous, 63% of the boys said "Yes," whereas only 23% of the other group said "Yes." Explain why the two percentages are so different.

32. **Weight? Wait what?** Marcos asked a random sample of 50 mall shoppers for their weight. Twenty-five of the shoppers were asked directly and the other 25 were asked anonymously by means of a "secret ballot." The mean reported weight was 13 pounds heavier for the anonymous group. Explain why the two means are so different.[17]

33. **Wording bias** Comment on each of the following as a potential sample survey question. Is the question clear? Is it slanted toward a desired response?

(a) "Some cell phone users have developed brain cancer. Should all cell phones come with a warning label explaining the danger of using cell phones?"

(b) "Do you agree that a national system of health insurance should be favored because it would provide health insurance for everyone and would reduce administrative costs?"

(c) "In view of escalating environmental degradation and incipient resource depletion, would you favor economic incentives for recycling of resource–intensive consumer goods?"

**34. Checking for bias** Comment on each of the following as a potential sample survey question. Is the question clear? Is it slanted toward a desired response?

(a) Which of the following best represents your opinion on gun control?

i. The government should confiscate our guns.

ii. We have the right to keep and bear arms.

(b) A freeze in nuclear weapons should be favored because it would begin a much-needed process to stop everyone in the world from building nuclear weapons now and reduce the possibility of nuclear war in the future. Do you agree or disagree?

**Multiple Choice** *Select the best answer for Exercises 35–40.*

35. A popular website places opinion poll questions next to many of its news stories. Simply click your response to join the sample. One of the questions was "Do you plan to diet this year?" More than 30,000 people responded, with 68% saying "Yes." Which of the following is true?

(a) About 68% of Americans planned to diet.

(b) The poll used a convenience sample, so the results tell us little about the population of all adults.

(c) The poll uses voluntary response, so the results tell us little about the population of all adults.

(d) The sample is too small to draw any conclusion.

(e) None of these.

36. To gather information about the validity of a new standardized test for 10th-grade students in a particular state, a random sample of 15 high schools was selected from the state. The new test was administered to every 10th-grade student in the selected high schools. What kind of sample is this?

(a) A simple random sample

(b) A stratified random sample

(c) A cluster sample

(d) A systematic random sample

(e) A voluntary response sample

37. Your statistics class has 30 students. You want to ask an SRS of 5 students from your class whether they use a mobile device for the online quizzes. You label the students 01, 02, . . . , 30. You enter the table of random digits at this line:

14459 26056 31424 80371 65103 62253 22490 61181

Your SRS contains the students labeled

(a) 14, 45, 92, 60, 56.

(b) 14, 31, 03, 10, 22.

(c) 14, 03, 10, 22, 22.

(d) 14, 03, 10, 22, 06.

(e) 14, 03, 10, 22, 11.

38. Suppose that 35% of the voters in a state are registered as Republicans, 40% as Democrats, and 25% as Independents. A newspaper wants to select a sample of 1000 registered voters to predict the outcome of the next election. If it randomly selects 350 Republicans, randomly selects 400 Democrats, and randomly selects 250 Independents, did this sampling procedure result in a simple random sample of registered voters from this state?

(a) Yes, because each registered voter had the same chance of being chosen.

(b) Yes, because random chance was involved.

(c) No, because not all registered voters had the same chance of being chosen.

(d) No, because a different number of registered voters was selected from each party.

(e) No, because not all possible groups of 1000 registered voters had the same chance of being chosen.

39. A local news agency conducted a survey about unemployment by randomly dialing phone numbers during the work day until it gathered responses from 1000 adults in its state. In the survey, 19% of those who responded said they were not currently employed. In reality, only 6% of the adults in the state were not currently employed at the time of the survey. Which of the following best explains the difference in the two percentages?

(a) The difference is due to sampling variability. We shouldn't expect the results of a random sample to match the truth about the population every time.

(b) The difference is due to response bias. Adults who are employed are likely to lie and say that they are unemployed.

(c) The difference is due to undercoverage bias. The survey included only adults and did not include teenagers who are eligible to work.

(d) The difference is due to nonresponse bias. Adults who are employed are less likely to be available for the sample than adults who are unemployed.

(e) The difference is due to voluntary response. Adults are able to volunteer as a member of the sample.

40. A simple random sample of 1200 adult Americans is selected, and each person is asked the following question: "In light of the huge national deficit, should the government at this time spend additional money to send humans to Mars?" Only 39% of those responding answered "Yes." This survey

(a) is reasonably accurate because it used a large simple random sample.

(b) needs to be larger because only about 24 people were drawn from each state.

(c) probably understates the percent of people who favor sending humans to Mars.

(d) is very inaccurate but neither understates nor overstates the percent of people who favor sending humans to Mars. Because simple random sampling was used, it is unbiased.

(e) probably overstates the percent of people who favor sending humans to Mars.

### Recycle and Review

41. **Don't turn it over** (3.2) How many points do turnovers cost teams in the NFL? The scatterplot shows the relationship between $x$ = number of turnovers and $y$ = number of points scored by teams in the NFL during 2015, along with the least-squares regression line $\hat{y} = 460.2 - 4.084x$.

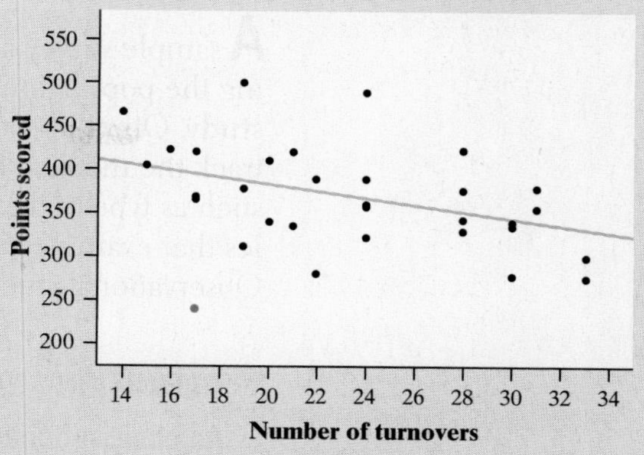


(a) Interpret the slope of the regression line in context.

(b) For this regression line, $s = 57.3$. Interpret this value.

(c) Calculate and interpret the residual for the San Francisco 49ers, who had 17 turnovers and scored 238 points.

(d) How does the point for the 49ers affect the least-squares regression line and standard deviation of the residuals? Explain your answer.

42. **Internet charges** (2.1) Some Internet service providers (ISPs) charge companies based on how much bandwidth they use in a month. One method that ISPs use to calculate bandwidth is to find the 95th percentile of a company's usage based on samples of hundreds of 5-minute intervals during a month.

(a) Explain what "95th percentile" means in this setting.

(b) Is it possible to determine the $z$-score for a usage total that is at the 95th percentile? If so, find the $z$-score. If not, explain why not.

# SECTION 4.2 Experiments

## LEARNING TARGETS *By the end of the section, you should be able to:*

- Explain the concept of confounding and how it limits the ability to make cause-and-effect conclusions.
- Distinguish between an observational study and an experiment, and identify the explanatory and response variables in each type of study.
- Identify the experimental units and treatments in an experiment.
- Describe the placebo effect and the purpose of blinding in an experiment.
- Describe how to randomly assign treatments in an experiment using slips of paper, technology, or a table of random digits.
- Explain the purpose of comparison, random assignment, control, and replication in an experiment.
- Describe a completely randomized design for an experiment.
- Describe a randomized block design and a matched pairs design for an experiment and explain the purpose of blocking in an experiment.

A sample survey aims to gather information about a population without disturbing the population in the process. Sample surveys are one kind of **observational study**. Other observational studies record the behavior of animals in the wild or track the medical history of volunteers to look for associations between variables such as type of diet, amount of exercise, and blood pressure. Observational studies that examine existing data for a sample of individuals are called *retrospective*. Observational studies that track individuals into the future are called *prospective*.

**DEFINITION Observational study**

An **observational study** observes individuals and measures variables of interest but does not attempt to influence the responses.

Section 4.2 is about statistical designs for experiments, a very different way to produce data.

## Observational Studies Versus Experiments

Tatiana Popova/Shutterstock.com

Is taking a vitamin D supplement good for you? Hundreds of observational studies have looked at the relationship between vitamin D concentration in a person's blood and various health outcomes.[18] In one prospective observational study, researchers found that teenage girls with higher vitamin D intakes were less likely to suffer broken bones.[19] Other observational studies have shown that people with higher vitamin D concentration have less cardiovascular disease, better cognitive function, and less risk of diabetes than people with lower concentrations of vitamin D.

In the observational studies involving vitamin D and diabetes, the **explanatory variable** is vitamin D concentration in the blood and the **response variable** is diabetes status—whether or not the person developed diabetes.

**DEFINITION Response variable, Explanatory variable**

A **response variable** measures an outcome of a study. An **explanatory variable** may help explain or predict changes in a response variable.

Unfortunately, it is very difficult to show that taking vitamin D *causes* a lower risk of diabetes using an observational study. As shown in the table, there are many possible differences between the group of people with high vitamin D concentration and the group of people with low vitamin D concentration. Any of these differences could be causing the difference in diabetes risk between the two groups of people.

| Variable | Group 1 | Group 2 |
|---|---|---|
| **Vitamin D concentration (explanatory)** | **High vitamin D concentration** | **Low vitamin D concentration** |
| Quality of diet | Better diet | Worse diet |
| Amount of exercise | More exercise | Less exercise |
| ⋮ | ⋮ | ⋮ |
| Amount of vitamin supplementation | More likely to take other vitamins | Less likely to take other vitamins |
| **Diabetes status (response)** | **Less likely to have diabetes** | **More likely to have diabetes** |

For example, it is possible that people who have healthier diets eat lots of foods that are high in vitamin D. Likewise, it is possible that people with healthier diets are less likely to develop diabetes. Vitamin D concentration may not have anything to do with diabetes status, even though there is an association between the two variables. In this case, we say there is **confounding** between vitamin D concentration and diet because we cannot tell which variable is causing the change in diabetes status.

Some people call a variable that results in confounding, like diet in this case, a *confounding variable.*

**DEFINITION Confounding**

**Confounding** occurs when two variables are associated in such a way that their effects on a response variable cannot be distinguished from each other.

**AP® EXAM TIP**

If you are asked to identify a possible confounding variable in a given setting, you are expected to explain how the variable you choose (1) is associated with the explanatory variable and (2) is associated with the response variable.

Likewise, because sun exposure increases vitamin D concentration, it is possible that people who exercise a lot outside have higher concentrations of vitamin D. If people who exercise a lot are also less likely to get diabetes, then amount of exercise and vitamin D concentration are confounded—we can't say which variable is the cause of the smaller diabetes risk. In other words, exercise is a confounding variable because it is related to both vitamin D concentration and diabetes status.

Here is an example that describes a retrospective observational study.

## EXAMPLE

### Smoking and ADHD
### Confounding

Ai/Getty Images

**PROBLEM:** In a study of more than 4700 children, researchers from Cincinnati Children's Hospital Medical Center found that those children whose mothers smoked during pregnancy were more than twice as likely to develop ADHD as children whose mothers had not smoked.[20] Explain how confounding makes it unreasonable to conclude that a mother's smoking during pregnancy causes an increase in the risk of ADHD in her children based on this study.

**SOLUTION:**

**It is possible that the mothers who smoked during pregnancy were also more likely to have unhealthy diets. If people with unhealthy diets are also more likely to have children with ADHD, then it could be that unhealthy diets caused the increase in ADHD risk, not smoking.**

Notice that the solution describes how diet might be associated with the explanatory variable (smoking status) and with the response variable (ADHD status).

**FOR PRACTICE, TRY EXERCISE 43**

*Observational studies of the effect of an explanatory variable on a response variable often fail because of confounding between the explanatory variable and one or more other variables.* In contrast to observational studies, **experiments** don't just observe individuals or ask them questions. They actively impose some *treatment* to measure the response. Experiments can answer questions like "Does aspirin reduce the chance of a heart attack?" and "Can yoga help dogs live longer?"

**DEFINITION Experiment**

An **experiment** deliberately imposes treatments (conditions) on individuals to measure their responses.

To determine if taking vitamin D actually causes a reduction in diabetes risk, researchers in Norway performed an experiment. The researchers randomly assigned 500 people with pre-diabetes to either take a high dose of vitamin D or to take a **placebo**—a pill that looked exactly like the vitamin D supplement but contained no active ingredient. After 5 years, about 40% of the people in each group were diagnosed with diabetes.[21] In other words, the association between vitamin D concentration and diabetes status disappeared when comparing two groups that were roughly the same to begin with.

**DEFINITION Placebo**

A **placebo** is a treatment that has no active ingredient, but is otherwise like other treatments.

The experiment in Norway avoided confounding by letting chance decide who took vitamin D and who didn't. That way, people with healthier diets were split about evenly between the two groups. So were people who exercise a lot and people who take other vitamins. *When our goal is to understand cause and effect, experiments are the only source of fully convincing data.* For this reason, the distinction between observational study and experiment is one of the most important in statistics.

## EXAMPLE Facebook and financial incentives
### Observational studies and experiments

Rostislav_Sedlacek/Shutterstock.com

**PROBLEM:** In each of the following settings, identify the explanatory and response variables. Then determine if each is an experiment or an observational study. Explain your reasoning.

(a) In a study conducted by researchers at the University of Texas, people were asked about their social media use and satisfaction with their marriage. Of the heavy social media users, 32% had thought seriously about leaving their spouse. Only 16% of non–social media users had thought seriously about leaving their spouse.[22]

(b) In a diet study using 100 overweight volunteers, 50 volunteers were randomly assigned to receive weight-loss counseling, monthly weigh-ins, and a three-month gym pass. The other 50 volunteers were given financial incentives (earning \$20 for losing 4 pounds in a month or paying \$20 otherwise) along with the counseling, weigh-ins, and gym pass. The group with the financial incentives lost 6.7 more pounds, on average.[23]

**SOLUTION:**

(a) Explanatory variable: Frequency of social media use. Response variable: Marital satisfaction. This is an observational study because people weren't assigned to use social media or not.

(b) Explanatory variable: Whether or not financial incentives were given. Response variable: Amount of weight lost. This is an experiment because researchers gave some volunteers financial incentives and did not give financial incentives to the other volunteers.

In part (a), the response variable is not the *percent* who thought about leaving their spouse. This percent is a summary of all the responses. Likewise, in part (b), the response variable is not the *average* weight loss. This average is a summary of all the responses.

**FOR PRACTICE, TRY EXERCISE 45**

In part (a) of the example, it would be incorrect to conclude that using social media causes marital dissatisfaction. It could be that other variables are confounded with social media use—or even that marital dissatisfaction is causing increased social media use. In part (b), it is reasonable to conclude that the financial incentives caused the increase in weight loss because this was a well-designed experiment, and the difference in average weight loss between the two groups was too large to be explained by chance alone.

## CHECK YOUR UNDERSTANDING

1. Does reducing screen brightness increase battery life in laptop computers? To find out, researchers obtained 30 new laptops of the same brand. They chose 15 of the computers at random and adjusted their screens to the brightest setting. The other 15 laptop screens were left at the default setting—moderate brightness. Researchers then measured how long each machine's battery lasted. Was this an observational study or an experiment? Justify your answer.

*Questions 2–4 refer to the following setting.* Does eating dinner with their families improve students' academic performance? According to an ABC News article, "Teenagers who eat with their families at least five times a week are more likely to get better grades in school."[24] This finding was based on a sample survey conducted by researchers at Columbia University.

2. Was this an observational study or an experiment? Justify your answer.
3. What are the explanatory and response variables?
4. Explain clearly why such a study cannot establish a cause-and-effect relationship. Suggest a variable that may be confounded with whether families eat dinner together.

# The Language of Experiments

An experiment is a statistical study in which we actually do something (a **treatment**) to people, animals, or objects (the **experimental units** or **subjects**) to observe the response.

**DEFINITION Treatment, Experimental unit, Subjects**

A specific condition applied to the individuals in an experiment is called a **treatment**. If an experiment has several explanatory variables, a treatment is a combination of specific values of these variables. An **experimental unit** is the object to which a treatment is randomly assigned. When the experimental units are human beings, they are often called **subjects**.

The best way to learn the language of experiments is to practice using it.

## EXAMPLE How can we prevent malaria?
### Vocabulary of experiments

**PROBLEM:** Malaria causes hundreds of thousands of deaths each year, with many of the victims being children. Will regularly screening children for the malaria parasite and treating those who test positive reduce the proportion of children who develop the disease? Researchers worked with children in 101 schools in Kenya, randomly assigning half of the schools to receive regular screenings and follow-up treatments and the remaining schools to receive no regular screening. Children at all 101 schools were tested for malaria at the end of the study.[25] Identify the treatments and the experimental units in this experiment.

**SOLUTION:**

This experiment compares two treatments: (1) regular screenings and follow-up treatments and (2) no regular screening. The experimental units are 101 schools in Kenya.

Note that the experimental units are the schools, not the students. The decision about who to screen was made school by school, not student by student. All students at the same school received the same treatment.

FOR PRACTICE, TRY EXERCISE 49

In the malaria experiment, there was one explanatory variable: screening status. In other experiments, there are multiple explanatory variables. Sometimes, these explanatory variables are called **factors**. In an experiment with multiple factors, the treatments are formed using the various **levels** of each of the factors. When there is only one factor, the levels are equivalent to the treatments.

**DEFINITION Factor, Levels**

In an experiment, a **factor** is an explanatory variable that is manipulated and may cause a change in the response variable. The different values of a factor are called **levels**.

Here's an example of a multifactor experiment.

## EXAMPLE The five-second rule
### Experiments with multiple explanatory variables

**PROBLEM:** Have you ever dropped a tasty piece of food on the ground, then quickly picked it up and eaten it? If so, you probably thought about the "five-second rule," which states that a piece of food is safe to eat if it has been on the floor less than 5 seconds. The rule is based on the belief that bacteria need time to transfer from the floor to the food. But does it work?

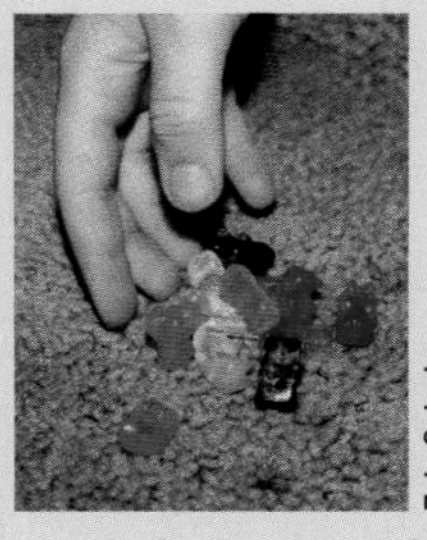
Zaia Snively

Researchers from Rutgers University put the five-second rule to the test. They used four different types of food: watermelon, bread, bread with butter, and gummy candy. They dropped the food onto four different surfaces: stainless steel, ceramic tile, wood, and carpet. And they waited for four different lengths of time: less than 1 second, 5 seconds, 30 seconds, and 300 seconds. Finally, they used bacteria prepared two different ways: in a tryptic soy broth or peptone buffer. Once the bacteria were ready, the researchers spread them out on the different surfaces and started dropping food.[26]

(a) List the factors in this experiment and the number of levels for each factor.

(b) If the researchers used every possible combination of levels to form the treatments, how many treatments were included in the experiment?

(c) List two of the treatments.

**SOLUTION:**

(a) Type of food (4 levels), type of surface (4 levels), amount of time (4 levels), method of bacterial preparation (2 levels)

(b) $4 \times 4 \times 4 \times 2 = 128$ different treatments

(c) Watermelon/stainless steel/less than 1 second/tryptic soy broth; gummy candy/wood/300 seconds/peptone buffer

FOR PRACTICE, TRY EXERCISE 51

What did the researchers discover? The wetter foods had greater bacterial transfer and food dropped on carpet had the least bacterial transfer. There was greater bacterial transfer the longer the food was on the surface, although there was some transfer that happened almost instantaneously. Overall, the researchers concluded that the type of food and type of surface were at least as important as the amount of time the food remained on the surface.

This example shows how experiments allow us to study the combined effect of several factors. The interaction of several factors can produce effects that could not be predicted from looking at the effect of each factor alone. For example, although longer time was associated with more bacterial transfer in general, this relationship might not be true for very moist food.

# Designing Experiments: Comparison

Zaia Snively

Experiments are the preferred method for examining the effect of one variable on another. By imposing the specific treatment of interest and controlling other influences, we can pin down cause and effect. Good designs are essential for effective experiments, just as they are for sampling. To see why, let's start with an example of a bad experimental design.

Does caffeine affect pulse rate? Many students regularly consume caffeine to help them stay alert. So it seems plausible that taking caffeine might increase an individual's pulse rate. Is this true? One way to investigate this claim is to ask volunteers to measure their pulse rates, drink some cola with caffeine, measure their pulse rates again after 10 minutes, and calculate the increase in pulse rate.

This experiment has a very simple design. A group of subjects (the students) were exposed to a treatment (the cola with caffeine), and the outcome (change in pulse rate) was observed. Here is the design:

$$\text{Students} \rightarrow \text{Cola with caffeine} \rightarrow \text{Change in pulse rate}$$

Unfortunately, even if the pulse rate of every student went up, we couldn't attribute the increase to caffeine. Perhaps the excitement of being in an experiment made their pulse rates increase. Maybe it was the sugar in the cola and not the caffeine. Perhaps their teacher told them a funny joke during the 10-minute waiting period and made everyone laugh. In other words, there are many other variables that are potentially confounded with taking caffeine.

Many laboratory experiments use a design like the one in the caffeine example:

$$\text{Experimental units} \rightarrow \text{Treatment} \rightarrow \text{Measure response}$$

In the lab environment, simple designs often work well. Field experiments and experiments with animals or people deal with more varied conditions. *Outside the lab, badly designed experiments often yield worthless results because of confounding.*

The remedy for the confounding in the caffeine example is to do a comparative experiment with two groups: one group that receives caffeine and a **control group** that does not receive caffeine.

**DEFINITION Control group**

In an experiment, a **control group** is used to provide a baseline for comparing the effects of other treatments. Depending on the purpose of the experiment, a control group may be given an inactive treatment (placebo), an active treatment, or no treatment at all.

In all other aspects, these groups should be treated exactly the same so that the only difference is the caffeine. That way, if there is convincing evidence of a difference in the average increase in pulse rates, we can safely conclude it was *caused* by the caffeine. This means that one group could get regular cola with caffeine, while the control group gets caffeine-free cola. Both groups would get the same amount of sugar, so sugar consumption would no longer be confounded with caffeine intake. Likewise, both groups would experience the same events during the experiment, so what happens during the experiment won't be confounded with caffeine intake either.

## EXAMPLE Preventing malaria
### Control groups

Alexander Joe/Getty Images

**PROBLEM:** In an earlier example, we described an experiment in which researchers randomly assigned 101 schools in Kenya to either receive regular malaria screenings and follow-up treatments or to receive no regular screenings. Explain why it was necessary to include a control group of schools that didn't receive regular screenings.

**SOLUTION:**

**The purpose of the control group is to provide a baseline for comparing the effect of the regular screenings and follow-up treatments. Otherwise, researchers wouldn't be able to determine if a decrease in malaria rates was due to the treatment or some other change that occurred during the experiment (like a drought that killed off mosquitos, slowing the spread of malaria).**

FOR PRACTICE, TRY EXERCISE 55

A control group was essential in the malaria experiment to determine if screening was effective. However, *not all experiments include a control group*—as long as comparison takes place. In the experiment about the five-second rule, there were 128 different treatments being compared and no control group. A control group wasn't essential in this experiment because researchers were interested in comparing different amounts of time on the floor, different types of food, and different types of surfaces.

## Designing Experiments: Blinding and the Placebo Effect

In the caffeine experiment, we used comparison to help prevent confounding. But even when there is comparison, confounding is still possible. If the subjects in the experiment know what type of soda they are receiving, the expectations of the two groups will be different. The knowledge that a subject is receiving caffeine may increase his or her pulse rate, apart from the caffeine itself. This is an example of the **placebo effect**.

**DEFINITION Placebo effect**

The **placebo effect** describes the fact that some subjects in an experiment will respond favorably to any treatment, even an inactive treatment.

In one study, researchers zapped the wrists of 24 test subjects with a painful jolt of electricity. Then they rubbed a cream with no active medicine on subjects' wrists and told them the cream should help soothe the pain. When researchers shocked them again, 8 subjects said they experienced significantly less pain.[27] When the ailment is psychological, like depression, some experts think that the placebo effect accounts for about three-quarters of the effect of the most widely used drugs.[28]

Because of the placebo effect, it is important that subjects don't know what treatment they are receiving. It is also better if the people interacting with the subjects and measuring the response variable don't know which subjects are receiving which treatment. When neither group knows who is receiving which treatment, the experiment is **double-blind**. Other experiments are **single-blind**.

**DEFINITION Double-blind, Single-blind**

In a **double-blind** experiment, neither the subjects nor those who interact with them and measure the response variable know which treatment a subject is receiving.

In a **single-blind** experiment, either the subjects or the people who interact with them and measure the response variable don't know which treatment a subject is receiving.

The idea of a double-blind design is simple. Until the experiment ends and the results are in, only the study's statistician knows for sure which treatment a subject is receiving. However, some experiments cannot be carried out in a double-blind manner. For example, if researchers are comparing the effects of exercise and dieting on weight loss, then subjects will know which treatment they are receiving. Such an experiment can still be single-blind if the individuals who are interacting with the subjects and measuring the response variable don't know who is dieting and who is exercising. In other single-blind experiments, the subjects are unaware of which treatment they are receiving, but the people interacting with them and measuring the response variable do know.

## EXAMPLE

### Do magnets repel pain?
### Blinding and the placebo effect

**PROBLEM:** Early research showed that magnetic fields affected living tissue in humans. Some doctors have begun to use magnets to treat patients with chronic pain. Scientists wondered if this type of therapy really worked. They designed a double-blind experiment to find out. A total of 50 patients with chronic pain were recruited for the study. A doctor identified a painful site on each patient and asked him or her to rate the pain on a scale from 0 (mild pain) to 10 (severe pain). Then the doctor selected a sealed envelope containing a magnet at random from a box with a mixture of active and inactive magnets. The chosen magnet was applied to the site of the pain for 45 minutes. After being treated, each patient was again asked to rate the level of pain from 0 to 10.[29]

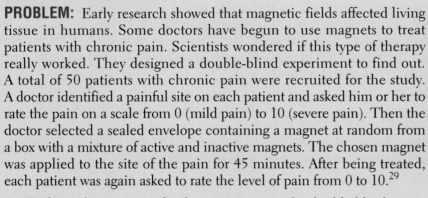

Eric O'Connell/Getty Images

(a) Explain what it means for this experiment to be double-blind.

(b) Why was it important for this experiment to be double-blind?

**SOLUTION:**

(a) Neither the subjects nor the doctors applying the magnets and recording the pain ratings knew which subjects had the active magnets and which had the inactive magnets.

(b) If subjects knew they were receiving an active treatment, researchers wouldn't know if any improvement was due to the magnets or to the expectation of getting better (the placebo effect). If the doctors knew which subjects received which treatments, they might treat one group of subjects differently from the other group. This would make it difficult to know if the magnets were the cause of any improvement.

**FOR PRACTICE, TRY EXERCISE 59**

## CHECK YOUR UNDERSTANDING

A new analysis is casting doubt on a claimed benefit of omega-3 fish oil. For years, doctors have been recommending eating fish and taking fish oil supplements to prevent heart disease. But the new analysis reviewed 20 previous studies and showed that the effects of omega-3 aren't as great as once suspected. One reason is that an early trial of omega-3 supplements was conducted as an open-label study.[30] In this type of study, both patients and researchers know who is receiving which treatment.

1. Describe a potential problem with an open-label study in this context.
2. Describe how you can fix the problem identified in Question 1.

## Designing Experiments: Random Assignment

Comparison alone isn't enough to produce results we can trust. If the treatments are given to groups that differ greatly when the experiment begins, confounding will result. If we allow students to choose what type of cola they will drink in the caffeine experiment, students who consume caffeine on a regular basis might be more likely to choose the regular cola. Due to their caffeine tolerance, these students' pulse rates might not increase as much as other students' pulse rates. In this case, caffeine tolerance would be confounded with the amount of caffeine consumed, making it impossible to conclude cause and effect.

To create roughly equivalent groups at the beginning of an experiment, we use **random assignment** to determine which experimental units get which treatment.

**DEFINITION Random assignment**

In an experiment, **random assignment** means that experimental units are assigned to treatments using a chance process.

Let's look at how random assignment can be used to improve the design of the caffeine experiment.

### EXAMPLE Caffeine and pulse rates
### How random assignment works

**PROBLEM:** A total of 20 students have agreed to participate in an experiment comparing the effects of caffeinated cola and caffeine-free cola on pulse rates. Describe how you would randomly assign 10 students to each of the two treatments:

(a) Using 20 identical slips of paper

(b) Using technology

(c) Using Table D

Zaia Shively

**SOLUTION:**

(a) On 10 slips of paper, write the letter "A"; on the remaining 10 slips, write the letter "B." Shuffle the slips of paper and hand out one slip of paper to each volunteer. Students who get an "A" slip receive the cola with caffeine and students who get a "B" slip receive the cola without caffeine.

When describing a method of random assignment, don't stop after creating the groups. Make sure to identify which group gets which treatment.

(b) Label each student with a different integer from 1 to 20. Then randomly generate 10 different integers from 1 to 20. The students with these labels receive the cola with caffeine. The remaining 10 students receive the cola without caffeine.

When using a random number generator or a table of random digits to assign treatments, make sure to account for the possibility of repeated numbers when describing your method.

(c) Label each student with a different integer from 01 to 20. Go to a line of Table D and read two-digit groups moving from left to right. The first 10 different labels between 01 and 20 identify the 10 students who receive cola with caffeine. The remaining 10 students receive the caffeine-free cola. Ignore groups of digits from 21 to 00.

FOR PRACTICE, TRY EXERCISE 63

Random assignment should distribute the students who regularly consume caffeine in roughly equal numbers to each group. It should also balance out the students with high metabolism and those with larger body sizes in the caffeine and caffeine-free groups. Random assignment helps ensure that the effects of other variables (e.g., caffeine tolerance, metabolism, or body size) are spread evenly among the two groups, making it easier to attribute a difference in mean pulse rate change to the caffeine.

## Designing Experiments: Control

Although random assignment should create two groups of students that are roughly equivalent to begin with, we still have to ensure that the only consistent difference between the groups during the experiment is the type of cola they receive. We can **control** the effects of some variables by keeping them the same for both groups. For example, we should make both treatments contain the same amount of sugar. If one group got regular cola and the other group got caffeine-free *diet* cola, then the amount of sugar would be confounded with the amount of caffeine—we wouldn't know if it was the sugar or the caffeine that was causing a change in pulse rates.

**DEFINITION Control**

In an experiment, **control** means keeping other variables constant for all experimental units.

We also want to control other variables to reduce the variability in the response variable. Suppose we let volunteers in both groups choose how much cola they want to drink. In that case, the changes in pulse rate would be more variable than if we made sure each subject drank the same amount of soda. Letting the amount of cola vary will make it harder to determine if caffeine is really having an effect on pulse rates.

The dotplots on the left show the results of an experiment in which the amount of cola consumed was the same for all participating students. Because there is so little overlap in these graphs, it seems clear that caffeine increases pulse rates. The dotplots on the right show the results of an experiment in which the students were permitted to choose how much or how little cola they consumed. Notice that the centers of the distributions haven't changed, but the distributions are much more variable. The increased overlap in the graphs makes the evidence supporting the effect of caffeine less convincing.

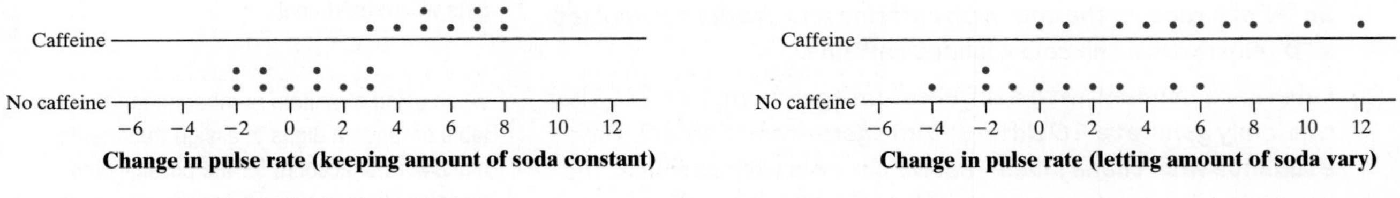


After randomly assigning treatments and controlling other variables, the two groups should be about the same, except for the treatments. Then a difference in the average change in pulse rate must be due either to the treatments themselves—or to the random assignment. We can't say that *any* difference between the average pulse rate changes for students in the two groups must be caused by the difference in caffeine. There would be *some* difference, even if both groups received the same type of cola, because the random assignment is unlikely to produce two groups that are exactly equivalent with respect to every variable that might affect pulse rate.

## Designing Experiments: Replication

Would you trust an experiment with just one student in each group? No, because the results would depend too much on which student was assigned to the caffeinated cola. However, if we randomly assign many subjects to each group, the effects of chance will balance out, and there will be little difference in the average responses in the two groups—unless the treatments themselves cause a difference. This is the idea of **replication**.

There must be at least 2 experimental units receiving each treatment to achieve replication.

**DEFINITION Replication**

In an experiment, **replication** means giving each treatment to enough experimental units so that a difference in the effects of the treatments can be distinguished from chance variation due to the random assignment.

Caution

**In statistics, replication means "use enough subjects." In other fields, the term *replication* has a different meaning.** In these fields, replication means conducting an experiment in one setting and then having other investigators conduct a similar experiment in a different setting. That is, replication means repeatability.

## Experiments: Putting It All Together

The following box summarizes the four key principles of experimental design: comparison, random assignment, control, and replication.

**PRINCIPLES OF EXPERIMENTAL DESIGN**

The basic principles for designing experiments are as follows:

1. **Comparison.** Use a design that compares two or more treatments.
2. **Random assignment.** Use chance to assign experimental units to treatments (or treatments to experimental units). Doing so helps create roughly equivalent groups of experimental units by balancing the effects of other variables among the treatment groups.
3. **Control.** Keep other variables the same for all groups, especially variables that are likely to affect the response variable. Control helps avoid confounding and reduces variability in the response variable.
4. **Replication.** Giving each treatment to enough experimental units so that any differences in the effects of the treatments can be distinguished from chance differences between the groups.

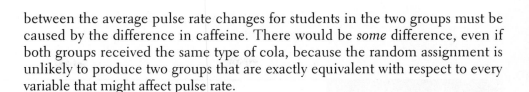

Let's see how these principles were used in designing a famous medical experiment.

## EXAMPLE

### The Physicians' Health Study
### Principles of experimental design

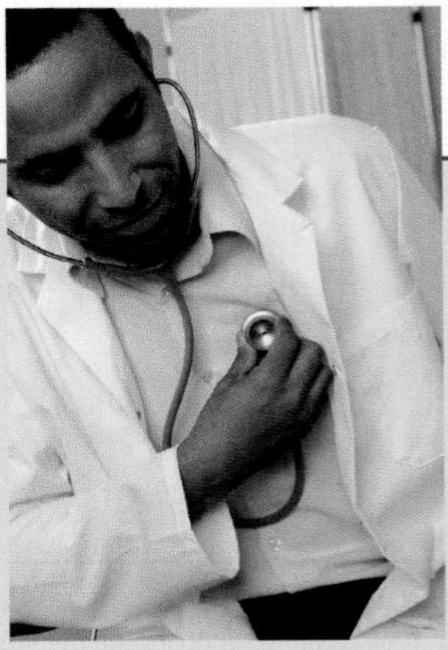

SCIENCE PHOTO LIBRARY/AGE Fotostock

**PROBLEM:** Does regularly taking aspirin help protect people against heart attacks? The Physicians' Health Study was a medical experiment that helped answer this question. In fact, the Physicians' Health Study looked at the effects of two drugs: aspirin and beta-carotene. Researchers wondered if beta-carotene would help prevent some forms of cancer. The subjects in this experiment were 21,996 male physicians. There were two explanatory variables (factors), each having two levels: aspirin (yes or no) and beta-carotene (yes or no). Combinations of the levels of these factors form the four treatments shown in the diagram. One-fourth of the subjects were assigned at random to each of these treatments.

On odd-numbered days, the subjects took either a tablet that contained aspirin or a placebo that looked and tasted like the aspirin but had no active ingredient. On even-numbered days, they took either a capsule containing beta-carotene or a placebo. There were several response variables—the study looked for heart attacks, several kinds of cancer, and other medical outcomes. After several years, 239 of the placebo group but only 139 of the aspirin group had suffered heart attacks. This difference is large enough to give good evidence that taking aspirin does reduce heart attacks.[31] It did not appear, however, that beta-carotene had any effect on preventing cancer.

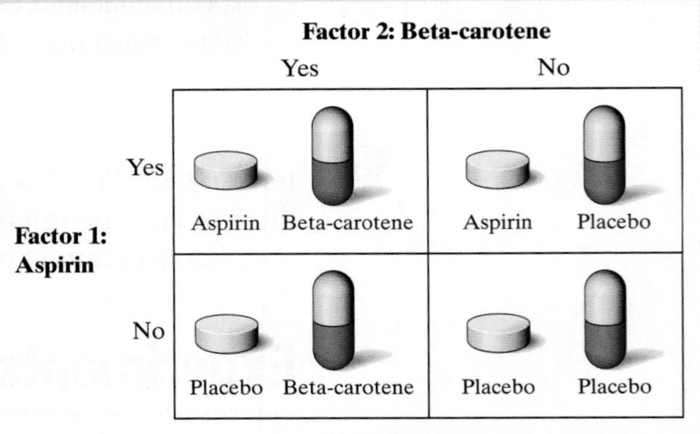


(a) Explain how this experiment used comparison.

(b) Explain the purpose of randomly assigning the physicians to the four treatments.

(c) Name two variables that were controlled in this experiment and why it was beneficial to control these variables.

(d) Explain how this experiment used replication. What is the purpose of replication in this context?

**SOLUTION:**

(a) Researchers used a design that compared each of the active treatments to a placebo.

(b) Random assignment helped ensure that the four groups of physicians were roughly equivalent at the beginning of the experiment.

(c) The experiment used subjects of the same gender and same occupation. Using only male physicians helps to reduce the variability in the response variables.

(d) There were over 5000 subjects in each treatment group. This helped ensure that the difference in heart attacks was due to the aspirin and not to chance variation in the random assignment.

> If women and people with other occupations were included, the results might be more variable, making it harder to determine the effects of aspirin and beta-carotene. However, using only male physicians means we don't know how females or other males would respond to these treatments.

**FOR PRACTICE, TRY EXERCISE 67**

Reports in medical journals regularly begin with words like these from a study of a flu vaccine given as a nose spray: "This study was a randomized, double-blind, placebo-controlled trial. Participants were enrolled from 13 sites across the continental United States between mid-September and mid-November."[32] Doctors are supposed to know what this means. Now you know, too.

The Physicians' Health Study shows how well-designed experiments can yield good evidence that differences in the treatments cause the differences we observe in the response.

### CHECK YOUR UNDERSTANDING

Many utility companies have introduced programs to encourage energy conservation among their customers. An electric company considers placing small digital displays in households to show current electricity use and what the cost would be if this use continued for a month. Will the displays reduce electricity use? One cheaper approach is to give customers a chart and information about monitoring their electricity use from their outside meter. Would this method work almost as well? The company decides to conduct an experiment using 60 households to compare these two approaches (display, chart) with a group of customers who receive information about energy consumption but no help in monitoring electricity use.

1. Explain why it was important to have a control group that didn't get the display or the chart.
2. Describe how to randomly assign the treatments to the 60 households.
3. What is the purpose of randomly assigning treatments in this context?

## Completely Randomized Designs

The diagram in Figure 4.1 presents the details of the caffeine experiment: random assignment, the sizes of the groups and which treatment they receive, and the response variable. This type of design is called a **completely randomized design**.

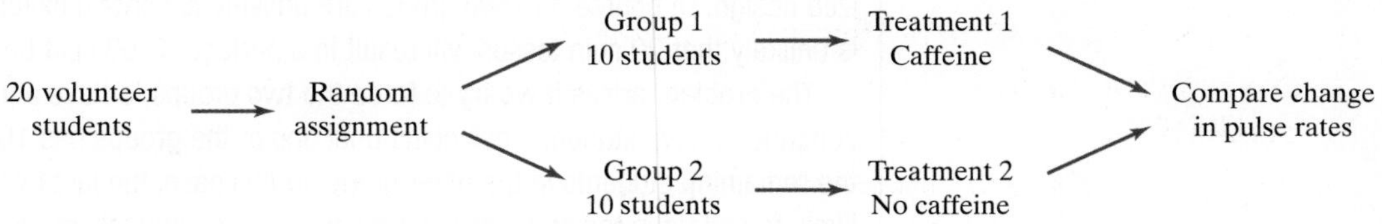


**FIGURE 4.1** Outline of a completely randomized design to compare caffeine and no caffeine.

**DEFINITION Completely randomized design**

In a **completely randomized design**, the experimental units are assigned to the treatments completely at random.

Although there are good statistical reasons for using treatment groups that are about equal in size, the definition of a completely randomized design does not require that each treatment be assigned to an equal number of experimental units. It does specify that the assignment of treatments must occur completely at random.

## EXAMPLE

### Chocolate milk and concussions
### Completely randomized design

Ann Heath

**PROBLEM:** "Concussion-Related Measures Improved in High School Football Players Who Drank New Chocolate Milk" announced a recent headline.[33] In the study, researchers compared a group of concussed football players given a new type of chocolate milk with a group of concussed football players who received no treatment.

(a) Explain why it isn't reasonable to conclude that the new type of chocolate milk is effective for treating high school football players with concussions based on this study.

(b) To test the effectiveness of the new type of chocolate milk, you recruit 50 high school football players who suffered a concussion in the previous 24 hours to participate in an experiment. Write a few sentences describing a completely randomized design for this experiment.

**SOLUTION:**

(a) It is possible that the group who received the new type of chocolate milk improved because they knew they were being treated and expected to get better, not because of the new chocolate milk.

(b) Number the players from 1 to 50. Use a random number generator to produce 25 different integers from 1 to 50 and give the new type of chocolate milk to the players with these numbers. Give regular chocolate milk to the remaining 25 players. Compare the concussion-related measures for the two groups.

FOR PRACTICE, TRY EXERCISE 69

**AP® EXAM TIP**

If you are asked to describe a completely randomized design, stay away from flipping coins. For example, suppose we ask each student in the caffeine experiment to toss a coin. If it's heads, then the student will drink the cola with caffeine. If it's tails, then the student will drink the caffeine-free cola. As long as all 20 students toss a coin, this is still a completely randomized design. Of course, the two groups are unlikely to contain exactly 10 students because it is unlikely that 20 coin tosses will result in a perfect 50-50 split between heads and tails.

The problem arises if we try to force the two groups to have equal sizes. Suppose we continue to have students toss coins until one of the groups has 10 students and then place the remaining students in the other group. In this case, the last two students in line are very likely to end up in the same group. However, in a completely randomized design, the last two subjects should only have a 50% chance of ending up in the same group.

# Randomized Block Designs

Completely randomized designs are the simplest statistical designs for experiments. They illustrate clearly the principles of comparison, random assignment, control, and replication. But just as with sampling, there are times when the simplest method doesn't yield the most precise results. When a population consists of groups of individuals that are "similar within but different between," a stratified random sample gives a better estimate than a simple random sample. This same logic applies in experiments.

Suppose that a mobile phone company is considering two different keyboard designs (A and B) for its new smartphone. The company decides to perform an experiment to compare the two keyboards using a group of 10 volunteers. The response variable is typing speed, measured in words per minute.

Zaia Snively

How should the company address the fact that four of the volunteers already use a smartphone, whereas the remaining six volunteers do not? They could use a completely randomized design and hope that the random assignment distributes the smartphone users and non-smartphone users about evenly between the group using keyboard A and the group using keyboard B. Even so, there might be a lot of variability in typing speed within both treatment groups because some members of each treatment group are more familiar with smartphones than the others. This additional variability might make it difficult to detect a difference in the effectiveness of the two keyboards. What should the researchers do?

Because the company knows that experience with smartphones will affect typing speed, they could start by separating the volunteers into two groups—one with experienced smartphone users and one with inexperienced smartphone users. Each of these groups of similar subjects is known as a **block**. Within each block, the company could then randomly assign half of the subjects to use keyboard A and the other half to use keyboard B. To control other variables, each subject should be given the same passage to type while in a quiet room with no distractions. This **randomized block design** helps account for the variation in typing speed that is due to experience with smartphones.

**DEFINITION Block, Randomized block design**

A **block** is a group of experimental units that are known before the experiment to be similar in some way that is expected to affect the response to the treatments.

In a **randomized block design**, the random assignment of experimental units to treatments is carried out separately within each block.

Figure 4.2 outlines the randomized block design for the smartphone experiment. The subjects are first separated into blocks based on their experience with smartphones. Then the two treatments are randomly assigned within each block.

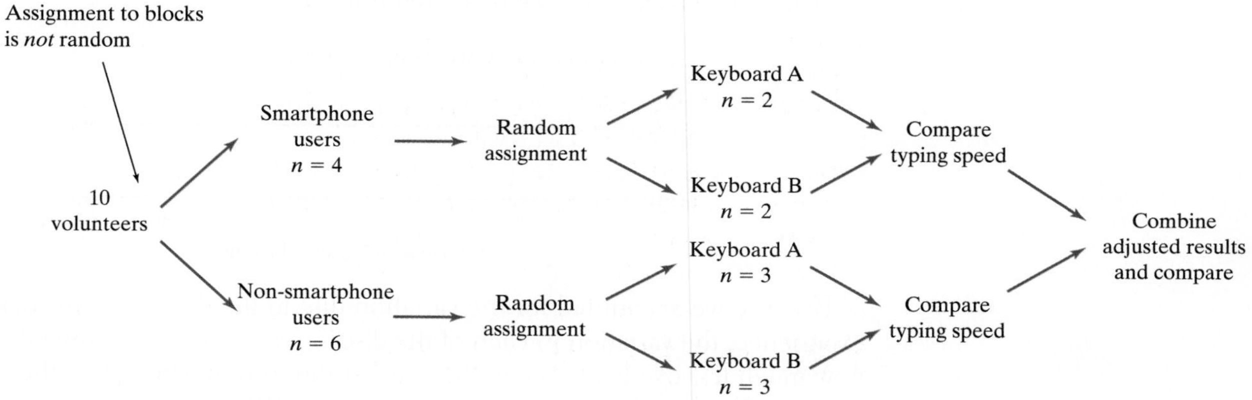


**FIGURE 4.2** Outline of a randomized block design for the smartphone experiment. The blocks consist of volunteers who have used smartphones and volunteers who have not used smartphones. The treatments are keyboard A and keyboard B.

Using a randomized block design allows us to account for the variation in the response that is due to the blocking variable of smartphone experience. This makes it easier to determine if one treatment is really more effective than the other.

To see how blocking helps, let's look at the results of the smartphone experiment. In the block of 4 smartphone users, 2 were randomly assigned to use keyboard A and the other 2 were assigned to use keyboard B. Likewise, in the block of 6 non-smartphone users, 3 were randomly assigned to use keyboard A and the other 3 were assigned to use keyboard B. Each of the 10 volunteers typed the same passage and the typing speed was recorded. Here are the results:

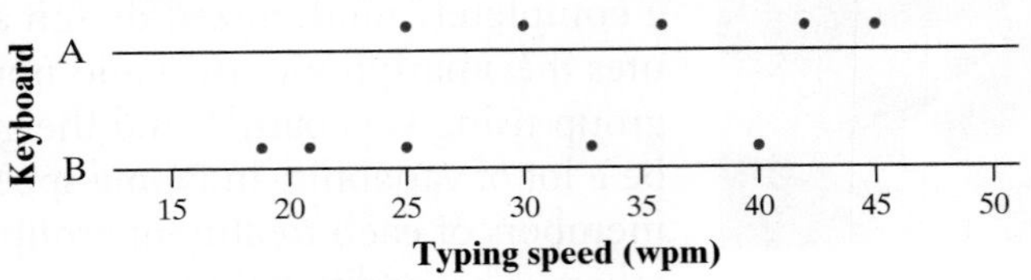


There is some evidence that keyboard A results in higher typing speeds, but the evidence isn't that convincing. Enough overlap occurs in the two distributions that the differences might simply be due to the chance variation in the random assignment.

If we compare the results for the two keyboards within each block, however, a different story emerges. Among the 4 smartphone users (indicated by the red squares), keyboard A was the clear winner. Likewise, among the 6 non-smartphone users (indicated by the black dots), keyboard A was also the clear winner.

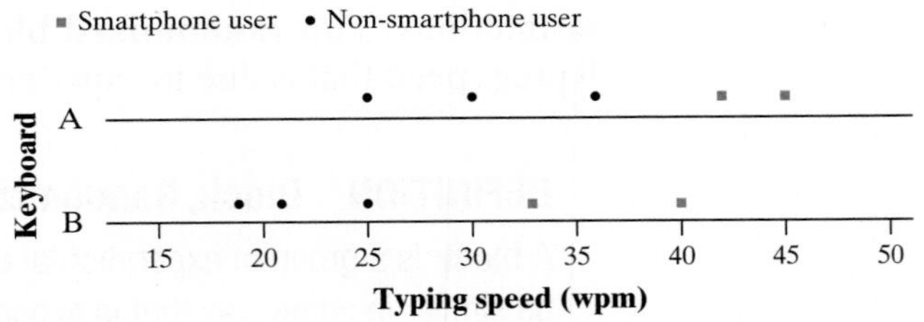


The overlap in the first set of dotplots was due almost entirely to the variation in smartphone experience—smartphone users were generally faster than non-smartphone users, regardless of which keyboard they used. In fact, the average typing speed for the smartphone users was 40, while the average typing speed for non-smartphone users was only 26, a difference of 14 words per minute. To account for the variation created by the difference in smartphone experience, let's subtract 14 from each of the typing speeds in the block of smartphone users to "even the playing field." Here are the results:

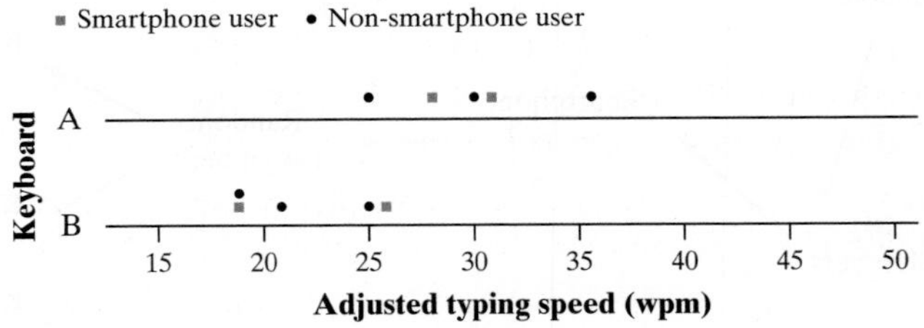


Because we accounted for the variation due to the difference in smartphone experience, the variation in each of the distributions has been reduced. There is now much less overlap between the two distributions, meaning that the evidence in favor of keyboard A is much more convincing. *When blocks are formed wisely, it is easier to find convincing evidence that one treatment is more effective than another.*

**AP® EXAM TIP**

Don't mix the language of experiments and the language of sample surveys or other observational studies. You will lose credit for saying things like "use a randomized block design to select the sample for this survey" or "this experiment suffers from nonresponse because some subjects dropped out during the study."

The idea of blocking is an important additional principle of experimental design. A wise experimenter will form blocks based on the most important unavoidable sources of variation among the experimental units. In other words, the experimenter will form blocks using the variables that are the best predictors of the response variable. Random assignment will then average out the effects of the remaining other variables and allow a fair comparison of the treatments. The moral of the story is: *control what you can, block on what you can't control, and randomize to create comparable groups.*

## EXAMPLE

### Should I use the popcorn button?
### Blocking in an experiment

Ann Health

**PROBLEM:** A popcorn lover wants to determine if it is better to use the "popcorn button" on her microwave oven or use the amount of time recommended on the bag of popcorn. To measure how well each method works, she will count the number of unpopped kernels remaining after popping. To obtain the experimental units, she goes to the store and buys 10 bags each of 4 different varieties of microwave popcorn (butter, cheese, natural, and kettle corn), for a total of 40 bags.

(a) Describe a randomized block design for this experiment. Justify your choice of blocks.

(b) Explain why a randomized block design might be preferable to a completely randomized design for this experiment.

**SOLUTION:**

(a) Form blocks based on variety, because the number of unpopped kernels is likely to differ by variety. Randomly assign 5 bags of each variety to the popcorn button treatment and 5 to the timed treatment by placing all 10 bags of a particular variety in a large box. Shake the box, pick 5 bags without looking, and assign them to be popped using the popcorn button. The remaining 5 bags will be popped using the instructions on the bags. Repeat this process for the remaining 3 varieties. After popping each of the 40 bags in random order, count the number of unpopped kernels in each bag and compare the results within each variety. Then combine the results from the 4 varieties after accounting for the difference in average response for each variety.

> It is important to pop the bags in random order so that changes over time (e.g., temperature, humidity) aren't confounded with the explanatory variable. For example, if the 20 "popcorn button" bags are popped last when the room temperature is greater, we wouldn't know if using the popcorn button or the warmer temperature was the cause of a difference in the number of unpopped kernels.

(b) A randomized block design accounts for the variability in the number of unpopped kernels created by the different varieties of popcorn (butter, cheese, natural, kettle). This makes it easier to determine if using the microwave button is more effective for reducing the number of unpopped kernels.

FOR PRACTICE, TRY EXERCISE 71

Another way to address the variability in unpopped kernels created by the different varieties is to use only one variety of popcorn in the experiment. Because variety of popcorn is no longer a variable, it will not be a source of variability. Of course, this means that the results of the experiment only apply to that one variety of popcorn—not ideal for the popcorn lover in the example!

**MATCHED PAIRS DESIGN** A common type of randomized block design for comparing two treatments is a **matched pairs design**. The idea is to create blocks by matching pairs of similar experimental units. The random assignment of subjects to treatments is done within each matched pair. Just as with other forms of blocking, matching helps account for the variation due to the variable(s) used to form the pairs.

**DEFINITION Matched pairs design**

A **matched pairs design** is a common experimental design for comparing two treatments that uses blocks of size 2. In some matched pairs designs, two very similar experimental units are paired and the two treatments are randomly assigned within each pair. In others, each experimental unit receives both treatments in a random order.

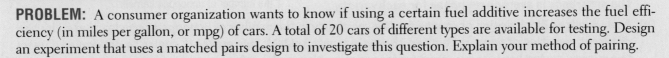

J-Elgaard/Getty Images

Suppose we want to investigate if listening to classical music while taking a math test affects performance. A total of 30 students in a math class volunteer to take part in the experiment. The difference in mathematical ability among the volunteers is likely to create additional variation in the test scores, making it harder to see the effect of classical music. To account for this variation, we could pair the students by their grade in the class—the two students with the highest grades are paired together, the two students with the next highest grades are paired together, and so on. Within each pair, one student is randomly assigned to take a math test while listening to classical music and the other member of the pair is assigned to take the math test in silence.

Sometimes, each "pair" in a matched pairs design consists of just one experimental unit that gets both treatments in random order. In the experiment about the effect of listening to classical music, we could have each student take a math test in both conditions. To decide the order, we might flip a coin for each student. If the coin lands on heads, the student takes a math test with classical music playing today and a similar math test without music playing tomorrow. If it lands on tails, the student does the opposite—no music today and classical music tomorrow.

Randomizing the order of treatments is important to avoid confounding. Suppose everyone did the classical music treatment on the first day and the no-music treatment on the second day, but the air conditioner wasn't working on the second day. We wouldn't know if any difference in mean test score was due to the difference in treatment or the difference in room temperature.

## EXAMPLE

### Will an additive improve my mileage?
### Matched pairs design

**PROBLEM:** A consumer organization wants to know if using a certain fuel additive increases the fuel efficiency (in miles per gallon, or mpg) of cars. A total of 20 cars of different types are available for testing. Design an experiment that uses a matched pairs design to investigate this question. Explain your method of pairing.

**SOLUTION:**

Give each car both treatments. It is reasonable to think that some cars are more fuel efficient than others, so using each car as its own "pair" accounts for the variation in fuel efficiency in the experimental units. For each car, randomly assign the order in which the treatments are assigned by flipping a coin. Heads indicates using the additive first and no additive second. Tails indicates using no additive first and then the additive second. For each car, record the fuel efficiency (mpg) after using each treatment.

FOR PRACTICE, TRY EXERCISE 77

In the preceding example, it is also possible to form pairs of two similar cars. For instance, we could pair together the two most fuel-efficient cars, the next two most fuel-efficient cars, and so on. This is less ideal, however, because there will still be some differences between the members of each pair that may cause additional variation in the results. Using the same car twice creates perfectly matched "pairs," and it also doubles the number of pairs used in the experiment. Both these features make it easier to find convincing evidence that the gas additive is effective, if it really is effective.

## CHECK YOUR UNDERSTANDING

Researchers would like to design an experiment to compare the effectiveness of three different advertisements for a new television series featuring the works of Jane Austen. There are 300 volunteers available for the experiment.

1. Describe a completely randomized design to compare the effectiveness of the three advertisements.
2. Describe a randomized block design for this experiment. Justify your choice of blocks.
3. Why might a randomized block design be preferable in this context?

# Section 4.2 Summary

- Statistical studies often try to show that changing one variable (the **explanatory variable**) causes changes in another variable (the **response variable**). Variables are **confounded** when their effects on a response variable can't be distinguished from each other.
- We can produce data to answer specific questions using **observational studies** or **experiments.** Observational studies that examine existing data for a sample of individuals are called *retrospective.* Observational studies that track individuals into the future are called *prospective.* Experiments actively do something to people, animals, or objects in order to measure their response. Experiments are the best way to show cause and effect.
- In an experiment, we impose one or more **treatments** on a group of **experimental units** (sometimes called **subjects** if they are human). Each treatment is a combination of the **levels** of the explanatory variables (also called **factors**).
- Some experiments give a **placebo** (fake treatment) to a **control group.** That helps prevent confounding due to the **placebo effect,** whereby some patients get better because they expect the treatment to work.
- Many behavioral and medical experiments are **double-blind.** That is, neither the subjects nor those interacting with them and measuring their responses know who is receiving which treatment. If one group knows and the other doesn't, then the experiment is **single-blind.**
- The basic principles of experimental design are:
  - **Comparison:** Use a design that compares two or more treatments.
  - **Random assignment:** Use chance (slips of paper, a random number generator, a table of random digits) to assign experimental units to treatments. This helps create roughly equivalent groups before treatments are imposed.
  - **Control:** Keep other variables the same for all groups. Control helps avoid confounding and reduces the variation in responses, making it easier to decide if a treatment is effective.

- **Replication:** Impose each treatment on enough experimental units so that the effects of the treatments can be distinguished from chance differences between the groups.

- In a **completely randomized design,** the experimental units are assigned to the treatments completely at random.
- A **randomized block design** forms groups (**blocks**) of experimental units that are similar with respect to a variable that is expected to affect the response. Treatments are assigned at random within each block. Responses are then compared within each block and combined with the responses of other blocks after accounting for the differences between the blocks. When blocks are chosen wisely, it is easier to determine if one treatment is more effective than another.
- A **matched pairs design** is a common form of randomized block design for comparing two treatments. In some matched pairs designs, each subject receives both treatments in a random order. In others, two very similar subjects are paired, and the two treatments are randomly assigned within each pair.

## Section 4.2 Exercises

43. **Good for the gut?** Is fish good for the gut? Researchers tracked 22,000 male physicians for 22 years. Those who reported eating seafood of any kind at least 5 times per week had a 40% lower risk of colon cancer than those who said they ate seafood less than once a week. Explain how confounding makes it unreasonable to conclude that eating seafood causes a reduction in the risk of colon cancer, based on this study.[34]
pg 271

44. **Straight A's now, healthy later** A study by Pamela Herd of the University of Wisconsin–Madison found a link between high school grades and health. Analyzing data from the Wisconsin Longitudinal Study, which has tracked the lives of thousands of Wisconsin high school graduates from the class of 1957, Herd found that students with higher grade-point averages were more likely to say they were in excellent or very good health in their early 60s. Explain how confounding makes it unreasonable to conclude that people will live healthier lives if they increase their GPA, based on this study.[35]

45. **Snacking and TV** Does the type of program people watch influence how much they eat? A total of 94 college students were randomly assigned to one of three treatments: watching 20 minutes of a Hollywood action movie (*The Island*), watching the same 20-minute excerpt of the movie with no sound, and watching 20 minutes of an interview program. While watching, participants were given snacks (M&M'S®, cookies, carrots, and grapes) and allowed to eat as much as they wanted. Subjects who watched the highly stimulating excerpt from *The Island* ate 65% more calories than subjects who watched the interview show. Participants who watched the silent version of *The Island* ate 46% more calories than those who watched the interview show.[36] Identify the explanatory and response variables in this study. Then determine if it is an experiment or an observational study. Explain your reasoning.
pg 272

46. **Learning biology with computers** An educator wants to compare the effectiveness of computer software for teaching biology with that of a textbook presentation. She gives a biology pretest to each student in a group of high school juniors, then randomly divides them into two groups. One group uses the computer, and the other studies the text. At the end of the year, she tests all the students again and compares the increase in biology test scores in the two groups. Identify the explanatory and response variables in this study. Then determine if it is an experiment or an observational study. Explain your reasoning.

47. **Child care and aggression** A study of child care enrolled 1364 infants and followed them through their sixth year in school. Later, the researchers published an article in which they stated that "the more time children spent in child care from birth to age 4½, the more adults tended to rate them, both at age 4½ and at kindergarten, as less likely to get along with others, as more assertive, as disobedient, and as aggressive."[37]
    (a) What are the explanatory and response variables?
    (b) Is this a prospective observational study, a retrospective observational study, or an experiment? Justify your answer.
    (c) Does this study show that child care makes children more aggressive? Explain your reasoning.

48. **Chocolate and happy babies** A University of Helsinki (Finland) study wanted to determine if chocolate

consumption during pregnancy had an effect on infant temperament at age 6 months. Researchers began by asking 305 healthy pregnant women to report their chocolate consumption. Six months after birth, the researchers asked mothers to rate their infants' temperament using the traits of smiling, laughter, and fear. The babies born to women who had been eating chocolate daily during pregnancy were found to be more active and "positively reactive"—a measure that the investigators said encompasses traits like smiling and laughter.[38]

(a) What are the explanatory and response variables?

(b) Is this a prospective observational study, a retrospective observational study, or an experiment? Justify your answer.

(c) Does this study show that eating chocolate regularly during pregnancy helps produce infants with good temperament? Explain your reasoning.

49. **Growing in the shade** The ability to grow in shade may pg 274 help pine trees found in the dry forests of Arizona to resist drought. How well do these pines grow in shade? Investigators planted pine seedlings in a greenhouse in either full light, light reduced to 25% of normal by shade cloth, or light reduced to 5% of normal. At the end of the study, they dried the young trees and weighed them. Identify the experimental units and the treatments.

50. **Sealing your teeth** Many children have their molars sealed to help prevent cavities. In an experiment, 120 children aged 6–8 were randomly assigned to a control group, a group in which sealant was applied and reapplied periodically for 36 months, or a group in which fluoride varnish was applied and reapplied periodically for 42 months. After 9 years, the percent of initially healthy molars with cavities was calculated for each group.[39] Identify the experimental units and the treatments.

51. **Improving response rate** How can we reduce the rate of pg 274 refusals in telephone surveys? Most people who answer at all listen to the interviewer's introductory remarks and then decide whether to continue. One study made telephone calls to randomly selected households to ask opinions about the next election. In some calls, the interviewer gave her name; in others, she identified the university she was representing; and in still others, she identified both herself and the university. For each type of call, the interviewer either did or did not offer to send a copy of the final survey results to the person interviewed.

(a) List the factors in this experiment and state how many levels each factor has.

(b) If the researchers used every possible combination of levels to form the treatments, how many treatments were included in the experiment?

(c) List two of the treatments.

52. **Fabric science** A maker of fabric for clothing is setting up a new line to "finish" the raw fabric. The line will use either metal rollers or natural-bristle rollers to raise the surface of the fabric; a dyeing-cycle time of either 30 or 40 minutes; and a temperature of either 150°C or 175°C. Three specimens of fabric will be subjected to each treatment and scored for quality.

(a) List the factors in this experiment and state how many levels each factor has.

(b) If the researchers used every possible combination of levels to form the treatments, how many treatments were included in the experiment?

(c) List two of the treatments.

53. **Want a snack?** Can snacking on fruit rather than candy reduce later food consumption? Researchers randomly assigned 12 women to eat either 65 calories of berries or 65 calories of candy. Two hours later, all 12 women were given an unlimited amount of pasta to eat. The researchers recorded the amount of pasta consumed by each subject. The women who ate the berries consumed 133 fewer calories, on average. Identify the explanatory and response variables, the experimental units, and the treatments.

54. **Pricey pizza?** The cost of a meal might affect how customers evaluate and appreciate food. To investigate, researchers worked with an Italian all-you-can-eat buffet to perform an experiment. A total of 139 subjects were randomly assigned to pay either $4 or $8 for the buffet and then asked to rate the quality of the pizza on a 9-point scale. Subjects who paid $8 rated the pizza 11% higher than those who paid only $4.[40] Identify the explanatory and response variables, the experimental units, and the treatments.

55. **Oils and inflammation** The extracts of avocado and pg 276 soybean oils have been shown to slow cell inflammation in test tubes. Will taking avocado and soybean unsaponifiables (called ASU) help relieve pain for subjects with joint stiffness due to arthritis? In an experiment, 345 men and women were randomly assigned to receive either 300 milligrams of ASU daily for three years or a placebo daily for three years.[41] Explain why it was necessary to include a control group in this experiment.

56. **Supplements for testosterone** As men age, their testosterone levels gradually decrease. This may cause a reduction in energy, an increase in fat, and other undesirable changes. Do testosterone supplements reverse some of these effects? A study in the Netherlands assigned 237 men aged 60 to 80 with low or low-normal testosterone levels to either a testosterone

supplement or a placebo.[42] Explain why it was necessary to include a control group in this experiment.

57. **Cocoa and blood flow** A study of blood flow involved 27 healthy people aged 18 to 72. Each subject consumed a cocoa beverage containing 900 milligrams of flavonols daily for 5 days. Using a finger cuff, blood flow was measured on the first and fifth days of the study. After 5 days, researchers measured what they called "significant improvement" in blood flow and the function of the cells that line the blood vessels.[43] What flaw in the design of this experiment makes it impossible to say if the cocoa really caused the improved blood flow? Explain your answer.

58. **Reducing unemployment** Will cash bonuses speed the return to work of unemployed people? A state department of labor notes that last year 68% of people who filed claims for unemployment insurance found a new job within 15 weeks. As an experiment, this year the state offers $500 to people filing unemployment claims if they find a job within 15 weeks. The percent who do so increases to 77%. What flaw in the design of this experiment makes it impossible to say if the bonus really caused the increase? Explain your answer.

59. **More oil and inflammation** Refer to Exercise 55. pg 278 Could blinding be used in this experiment? Explain your reasoning. Why is blinding an important consideration in this context?

60. **More testosterone** Refer to Exercise 56. Could blinding be used in this experiment? Explain your reasoning. Why is blinding an important consideration in this context?

61. **Meditation for anxiety** An experiment that claimed to show that meditation lowers anxiety proceeded as follows. The experimenter interviewed the subjects and rated their level of anxiety. Then the subjects were randomly assigned to two groups. The experimenter taught one group how to meditate and they meditated daily for a month. The other group was simply told to relax more. At the end of the month, the experimenter interviewed all the subjects again and rated their anxiety level. The meditation group now had less anxiety. Psychologists said that the results were suspect because the ratings were not blind. Explain what this means and how lack of blindness could affect the reported results.

62. **Side effects** Even if an experiment is double-blind, the blinding might be compromised if side effects of the treatments differ. For example, suppose researchers at a skin-care company are comparing their new acne treatment against that of the leading competitor. Fifty subjects are assigned at random to each treatment, and the company's researchers will rate the improvement for each of the 100 subjects. The researchers aren't told which subjects received which treatments, but they know that their new acne treatment causes a slight reddening of the skin. How might this knowledge compromise the blinding? Explain why this is an important consideration in the experiment.

63. **Layoffs and "survivor guilt"** Workers who survive a pg 279 layoff of other employees at their location may suffer from "survivor guilt." A study of survivor guilt and its effects used as subjects 120 students who were offered an opportunity to earn extra course credit by doing proofreading. Each subject worked in the same cubicle as another student, who was an accomplice of the experimenters. At a break midway through the work, one of three things happened:

    *Treatment 1:* The accomplice was told to leave; it was explained that this was because she performed poorly.

    *Treatment 2:* It was explained that unforeseen circumstances meant there was only enough work for one person. By "chance," the accomplice was chosen to be laid off.

    *Treatment 3:* Both students continued to work after the break.

    The subjects' work performance after the break was compared with their performance before the break. Overall, subjects worked harder when told the other student's dismissal was random.[44] Describe how you would randomly assign the subjects to the treatments

    (a) using slips of paper.

    (b) using technology.

    (c) using Table D.

64. **Precise offers** People often use round prices as first offers in a negotiation. But would a more precise number suggest that the offer was more reasoned and informed? In an experiment, 238 adults played the role of a person selling a used car. Each adult received one of three initial offers: $2000, $1865 (a precise under-offer), and $2135 (a precise over-offer). After hearing the initial offer, each subject made a counter-offer. The difference in the initial offer and counter-offer was the largest in the group that received the $2000 offer.[45] Describe how the researchers could have randomly assigned the subjects to the treatments

    (a) using slips of paper.

    (b) using technology.

    (c) using Table D.

65. **Stronger players** A football coach hears that a new exercise program will increase upper-body strength better than lifting weights. He is eager to test this new program in the off-season with the players on his high

school team. The coach decides to let his players choose which of the two treatments they will undergo for 3 weeks—exercise or weight lifting. He will use the number of push-ups a player can do at the end of the experiment as the response variable. Which principle of experimental design does the coach's plan violate? Explain how this violation could lead to confounding.

66. **Killing weeds** A biologist would like to determine which of two brands of weed killer, X or Y, is less likely to harm the plants in a garden at the university. Before spraying near the plants, the biologist decides to conduct an experiment using 24 individual plants. Which of the following two plans for randomly assigning the treatments should the biologist use? Why?

*Plan A:* Choose the 12 healthiest-looking plants. Then flip a coin. If it lands heads, apply Brand X weed killer to these plants and Brand Y weed killer to the remaining 12 plants. If it lands tails, do the opposite.

*Plan B:* Choose 12 of the 24 plants at random. Apply Brand X weed killer to those 12 plants and Brand Y weed killer to the remaining 12 plants.

67. **Boosting preemies** pg 282 Do blood-building drugs help brain development in babies born prematurely? Researchers randomly assigned 53 babies, born more than a month premature and weighing less than 3 pounds, to one of three groups. Babies either received injections of erythropoietin (EPO) three times a week, darbepoetin once a week for several weeks, or no treatment. Results? Babies who got the medicines scored much better by age 4 on measures of intelligence, language, and memory than the babies who received no treatment.[46]

(a) Explain how this experiment used comparison.

(b) Explain the purpose of randomly assigning the babies to the three treatments.

(c) Name two variables that were controlled in this experiment and why it was beneficial to control these variables.

(d) Explain how this experiment used replication. What is the purpose of replication in this context?

68. **The effects of day care** Does preschool help low-income children stay in school and hold good jobs later in life? The Carolina Abecedarian Project (the name suggests the ABCs) has followed a group of 111 children for over 40 years. Back then, these individuals were all healthy but low-income black infants in Chapel Hill, North Carolina. All the infants received nutritional supplements and help from social workers. Half were also assigned at random to an intensive preschool program. Results? Children who were assigned to the preschool program had higher IQ's, higher standardized test scores, and were less likely to repeat a grade in school.[47]

(a) Explain how this experiment used comparison.

(b) Explain the purpose of randomly assigning the infants to the two treatments.

(c) Name two variables that were controlled in this experiment and why it was beneficial to control these variables.

(d) Explain how this experiment used replication. What is the purpose of replication in this context?

69. **Treating prostate disease** pg 284 A large study used records from Canada's national health care system to compare the effectiveness of two ways to treat prostate disease. The two treatments are traditional surgery and a new method that does not require surgery. The records described many patients whose doctors had chosen what method to use. The study found that patients treated by the new method were more likely to die within 8 years.[48]

(a) Further study of the data showed that this conclusion was wrong. The extra deaths among patients who were treated with the new method could be explained by other variables. What other variables might be confounded with a doctor's choice of surgical or nonsurgical treatment?

(b) You have 300 prostate patients who are willing to serve as subjects in an experiment to compare the two methods. Write a few sentences describing a completely randomized design for this experiment.

70. **Diet soda and pregnancy** A large study of 3000 Canadian children and their mothers found that the children of mothers who drank diet soda daily during pregnancy were twice as likely to be overweight at age 1 than children of mothers who avoided diet soda during pregnancy.[49]

(a) A newspaper article about this study had the headline "Diet soda, pregnancy: Mix may fuel childhood obesity." This headline suggests that there is a cause-and-effect relationship between diet soda consumption during pregnancy and the weight of the children 1 year after birth. However, this relationship could be explained by other variables. What other variables might be confounded with a mother's consumption of diet soda during pregnancy?

(b) You have 300 pregnant mothers who are willing to serve as subjects in an experiment that compares three treatments during pregnancy: no diet soda, one diet soda per day, and two diet sodas per day. Write a few sentences describing a completely randomized design for this experiment.

71. **A fruitful experiment** pg 287 A citrus farmer wants to know which of three fertilizers (A, B, and C) is most effective for increasing the number of oranges on his trees. He

is willing to use 30 mature trees of various sizes from his orchard in an experiment with a randomized block design.

(a) Describe a randomized block design for this experiment. Justify your choice of blocks.

(b) Explain why a randomized block design might be preferable to a completely randomized design for this experiment.

72. **In the cornfield** An agriculture researcher wants to compare the yield of 5 corn varieties: A, B, C, D, and E. The field in which the experiment will be carried out increases in fertility from north to south. The researcher therefore divides the field into 25 plots of equal size, arranged in 5 east–west rows of 5 plots each, as shown in the diagram.

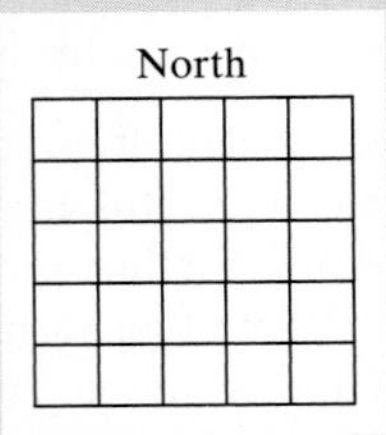


(a) Describe a randomized block design for this experiment. Justify your choice of blocks.

(b) Explain why a randomized block design might be preferable to a completely randomized design for this experiment.

73. **Doctors and nurses** Nurse-practitioners are nurses with advanced qualifications who often act much like primary-care physicians. Are they as effective as doctors at treating patients with chronic conditions? An experiment was conducted with 1316 patients who had been diagnosed with asthma, diabetes, or high blood pressure. Within each condition, patients were randomly assigned to either a doctor or a nurse-practitioner. The response variables included measures of the patients' health and of their satisfaction with their medical care after 6 months.[50]

(a) Which are the blocks in this experiment: the different diagnoses (asthma, diabetes, or high blood pressure) or the type of care (nurse or doctor)? Why?

(b) Explain why a randomized block design is preferable to a completely randomized design in this context.

(c) Suppose the experiment used only diabetes patients, but there were still 1316 subjects willing to participate. What advantage would this offer? What disadvantage?

74. **Comparing cancer treatments** The progress of a type of cancer differs in women and men. Researchers want to design an experiment to compare three therapies for this cancer. They recruit 500 male and 300 female patients who are willing to serve as subjects.

(a) Which are the blocks in this experiment: the three cancer therapies or the two sexes? Why?

(b) What are the advantages of a randomized block design over a completely randomized design using these 800 subjects?

(c) Suppose the researchers had 800 male and no female subjects available for the study. What advantage would this offer? What disadvantage?

75. **Aw, rats!** A nutrition experimenter intends to compare the weight gain of newly weaned male rats fed Diet A with that of rats fed Diet B. To do this, she will feed each diet to 10 rats. She has available 10 rats from one litter and 10 rats from a second litter. Rats in the first litter appear to be slightly healthier.

(a) If the 10 rats from Litter 1 were fed Diet A, then initial health would be a confounding variable. Explain this statement carefully.

(b) Describe how to randomly assign the rats to treatments using a randomized block design with litters as blocks.

(c) Use technology or Table D to carry out the random assignment.

76. **Comparing weight-loss treatments** A total of 20 overweight females have agreed to participate in a study of the effectiveness of four weight-loss treatments: A, B, C, and D. The researcher first calculates how overweight each subject is by comparing the subject's actual weight with her "ideal" weight. The subjects and their excess weights in pounds are as follows:

| | | | | | | | |
|---|---|---|---|---|---|---|---|
| Birnbaum | 35 | Hernandez | 25 | Moses | 25 | Smith | 29 |
| Brown | 34 | Jackson | 33 | Nevesky | 39 | Stall | 33 |
| Brunk | 30 | Kendall | 28 | Obrach | 30 | Tran | 35 |
| Cruz | 34 | Loren | 32 | Rodriguez | 30 | Wilansky | 42 |
| Deng | 24 | Mann | 28 | Santiago | 27 | Williams | 22 |

The response variable is the weight lost after 8 weeks of treatment.

(a) If the 5 most overweight women were assigned Treatment A, the next 5 most overweight women were assigned Treatment B, and so on, then the amount overweight would be a confounding variable. Explain this statement carefully.

(b) Describe how to randomly assign the women to treatments using a randomized block design. Use blocks of size 4 formed by the amount overweight.

(c) Use technology or Table D to carry out the random assignment.

77. **SAT preparation** A school counselor wants to compare the effectiveness of an online SAT preparation program with an in-person SAT preparation class. For an experiment, the counselor recruits 30 students who have already taken the SAT once. The response variable will be the improvement in SAT score.

pg 288

(a) Design an experiment that uses a completely randomized design to investigate this question.

(b) Design an experiment that uses a matched pairs design to investigate this question. Explain your method of pairing.

(c) Which design do you prefer? Explain your answer.

78. **Valve surgery** Medical researchers want to compare the success rate of a new non-invasive method for replacing heart valves using a cardiac catheter with traditional open-heart surgery. They have 40 male patients, ranging in age from 55 to 75, who need valve replacement. One of several response variables will be the percentage of blood that flows backward—in the wrong direction—through the valve on each heartbeat.

(a) Design an experiment that uses a completely randomized design to investigate this question.

(b) Design an experiment that uses a matched pairs design to investigate this question. Explain your method of pairing.

(c) Which design do you prefer? Explain your answer.

79. **Look, Ma, no hands!** Does talking on a hands-free cell phone distract drivers? Researchers recruit 40 student subjects for an experiment to investigate this question. They have a driving simulator equipped with a hands-free phone for use in the study. Each subject will complete two sessions in the simulator: one while talking on the hands-free phone and the other while just driving. The order of the two sessions for each subject will be determined at random. The route, driving conditions, and traffic flow will be the same in both sessions.

(a) What type of design did the researchers use in their study?

(b) Explain why the researchers chose this design instead of a completely randomized design.

(c) Why is it important to randomly assign the order of the treatments?

(d) Explain how and why researchers controlled for other variables in this experiment.

80. **Chocolate gets my heart pumping** Cardiologists at Athens Medical School in Greece wanted to test if chocolate affects blood vessel function. The researchers recruited 17 healthy young volunteers, who were each given a 3.5-ounce bar of dark chocolate, either bittersweet or fake chocolate. On another day, the volunteers received the other treatment. The order in which subjects received the bittersweet and fake chocolate was determined at random. The subjects had no chocolate outside the study, and investigators didn't know if a subject had eaten the real or the fake chocolate. An ultrasound was taken of each volunteer's upper arm to observe the functioning of the cells in the walls of the main artery. The researchers found that blood vessel function was improved when the subjects ate bittersweet chocolate, and that there were no such changes when they ate the placebo (fake chocolate).[51]

(a) What type of design did the researchers use in their study?

(b) Explain why the researchers chose this design instead of a completely randomized design.

(c) Why is it important to randomly assign the order of the treatments for the subjects?

(d) Explain how and why researchers controlled for other variables in this experiment.

81. **Got deodorant?** A group of students wants to perform an experiment to determine whether Brand A or Brand B deodorant lasts longer. One group member suggests the following design: Recruit 40 student volunteers—20 male and 20 female. Separate by gender, because male and female bodies might respond differently to deodorant. Give all the males Brand A deodorant and all the females Brand B. Have the principal judge how well the deodorant is still working at the end of the school day on a 0 to 10 scale. Then compare ratings for the two treatments.

(a) Identify any flaws you see in the proposed design for this experiment.

(b) Describe how you would design the experiment. Explain how your design addresses each of the problems you identified in part (a).

82. **Close shave** Which of two brands (X or Y) of electric razor shaves closer? Researchers want to design and carry out an experiment to answer this question using 50 adult male volunteers. Here's one idea: Have all 50 subjects shave the left sides of their faces with the Brand X razor and shave the right sides of their faces with the Brand Y razor. Then have each man decide which razor gave the closer shave and compile the results.

(a) Identify any flaws you see in the proposed design for this experiment.

(b) Describe how you would design the experiment. Explain how your design addresses each of the problems you identified in part (a).

**Multiple Choice** *Select the best answer for Exercises 83–90.*

83. Can a vegetarian or low-salt diet reduce blood pressure? Men with high blood pressure are assigned at random to one of four diets: (1) normal diet with unrestricted salt; (2) vegetarian with unrestricted salt; (3) normal with restricted salt; and (4) vegetarian with restricted salt. This experiment has

(a) one factor, the type of diet.

(b) two factors, high blood pressure and type of diet.

(c) two factors, normal/vegetarian diet and unrestricted/restricted salt.

(d) three factors, men, high blood pressure, and type of diet.

(e) four factors, the four diets being compared.

84. In the experiment of the preceding exercise, the subjects were randomly assigned to the different treatments. What is the most important reason for this random assignment?

(a) Random assignment eliminates the effects of other variables such as stress and body weight.

(b) Random assignment balances the effects of other variables such as stress and body weight among the four treatment groups.

(c) Random assignment makes it possible to make a conclusion about all men.

(d) Random assignment reduces the amount of variation in blood pressure.

(e) Random assignment prevents the placebo effect from ruining the results of the study.

85. To investigate if standing up while studying affects performance in an algebra class, a teacher assigns half of the 30 students in his class to stand up while studying and assigns the other half to not stand up while studying. To determine who receives which treatment, the teacher identifies the two students who did best on the last exam and randomly assigns one to stand and one to not stand. The teacher does the same for the next two highest-scoring students and continues in this manner until each student is assigned a treatment. Which of the following best describes this plan?

(a) This is an observational study.

(b) This is an experiment with blocking.

(c) This is a completely randomized experiment.

(d) This is a stratified random sample.

(e) This is a systematic random sample.

86. A gardener wants to try different combinations of fertilizer (none, 1 cup, 2 cups) and mulch (none, wood chips, pine needles, plastic) to determine which combination produces the highest yield for a variety of green beans. He has 60 green-bean plants to use in the experiment. If he wants an equal number of plants to be assigned to each treatment, how many plants will be assigned to each treatment?

(a) 1 (b) 3 (c) 4

(d) 5 (e) 12

87. Corn variety 1 yielded 140 bushels per acre last year at a research farm. This year, corn variety 2, planted in the same location, yielded only 110 bushels per acre. Based on these results, is it reasonable to conclude that corn variety 1 is more productive than corn variety 2?

(a) Yes, because 140 bushels per acre is greater than 110 bushels per acre.

(b) Yes, because the study was done at a research farm.

(c) No, because there may be other differences between the two years besides the corn variety.

(d) No, because there was no use of a placebo in the experiment.

(e) No, because the experiment wasn't double-blind.

88. A report in a medical journal notes that the risk of developing Alzheimer's disease among subjects who regularly opted to take the drug ibuprofen was about half the risk of those who did not. Is this good evidence that ibuprofen is effective in preventing Alzheimer's disease?

(a) Yes, because the study was a randomized, comparative experiment.

(b) No, because the effect of ibuprofen is confounded with the placebo effect.

(c) Yes, because the results were published in a reputable professional journal.

(d) No, because this is an observational study. An experiment would be needed to confirm (or not confirm) the observed effect.

(e) Yes, because a 50% reduction can't happen just by chance.

89. A farmer is conducting an experiment to determine which variety of apple tree, Fuji or Gala, will produce more fruit in his orchard. The orchard is divided into 20 equally sized square plots. He has 10 trees of each variety and randomly assigns each tree to a separate plot in the orchard. What are the experimental unit(s) in this study?

(a) The trees (b) The plots (c) The apples

(d) The farmer (e) The orchard

90. Two essential features of all statistically designed experiments are

(a) comparing two or more treatments; using the double-blind method.

(b) comparing two or more treatments; using chance to assign subjects to treatments.

(c) always having a placebo group; using the double-blind method.

(d) using a block design; using chance to assign subjects to treatments.

(e) using enough subjects; always having a control group.

### Recycle and Review

91. **Seed weights** (2.2) Biological measurements on the same species often follow a Normal distribution quite closely. The weights of seeds of a variety of winged bean are approximately Normal with mean 525 milligrams (mg) and standard deviation 110 mg.

(a) What percent of seeds weigh more than 500 mg?

(b) If we discard the lightest 10% of these seeds, what is the smallest weight among the remaining seeds?

92. **Comparing rainfall** (1.3) The boxplots summarize the distributions of average monthly rainfall (in inches) for Tucson, Arizona, and Princeton, New Jersey.[52] Compare these distributions.

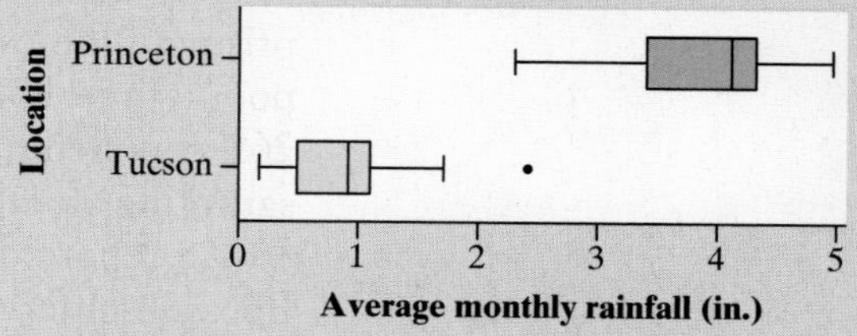


# SECTION 4.3 Using Studies Wisely

## LEARNING TARGETS *By the end of the section, you should be able to:*

- Explain the concept of sampling variability when making an inference about a population and how sample size affects sampling variability.
- Explain the meaning of statistically significant in the context of an experiment and use simulation to determine if the results of an experiment are statistically significant.
- Identify when it is appropriate to make an inference about a population and when it is appropriate to make an inference about cause and effect.
- Evaluate if a statistical study has been carried out in an ethical manner.*

Researchers who conduct statistical studies often want to draw conclusions that go beyond the data they produce. Here are two examples.

- The U.S. Census Bureau carries out a monthly Current Population Survey of about 60,000 households. Their goal is to use data from these randomly selected households to estimate the percent of unemployed individuals in the population.
- Scientists performed an experiment that randomly assigned 21 volunteer subjects to one of two treatments: sleep deprivation for one night or unrestricted sleep. The scientists hoped to show that sleep deprivation causes a decrease in performance two days later.[53]

What conclusions can be drawn from a particular study? The answer depends on how the data were collected.

*This is an important topic, but it is not required for the AP® Statistics exam.

# Inference for Sampling

When the members of a sample are selected at random from a population, we can use the sample results to *infer* things about the population. That is, we can make *inferences* about the population from which the sample was randomly selected. Inference from convenience samples or voluntary response samples would be misleading because these methods of choosing a sample are biased. In these cases, we are almost certain that the sample does *not* fairly represent the population.

Even when making an inference from a random sample, it would be surprising if the estimate from the sample was exactly equal to the truth about the population. For example, in a random sample of 1399 U.S. teens aged 13–18, 26% reported more than 8 hours of entertainment media use per day. Because of **sampling variability**, it would be surprising if exactly 26% of *all* U.S. teens aged 13–18 reported more than 8 hours of entertainment media use per day. Why? Because different samples of 1399 U.S. teens aged 13–18 will include different sets of people and produce different estimates.

**DEFINITION Sampling variability**

**Sampling variability** refers to the fact that different random samples of the same size from the same population produce different estimates.

The following activity explores the idea of sampling variability.

## ACTIVITY Exploring sampling variability

When making an inference about a population from a random sample, we shouldn't expect the estimate to be exactly correct. But how much do sample results vary? Your teacher has prepared a large population of beads, where 30% have a certain color (e.g., red) so you can explore this question.

G. Curt Fiedler/Getty Images

1. In a moment, you will select a random sample of 20 beads. Do you expect that your sample will contain exactly 30% red beads? Explain your reasoning.
2. Mix the beads thoroughly, select a random sample of 20 beads from the population, calculate the percent of red beads in the sample, and replace the beads in the population.
3. After all students have selected a sample, make a class dotplot showing each student's estimate for the percent of red beads. Where is the graph centered? How much does the percent of red beads vary?
4. Imagine that you repeated Steps 2 and 3 with random samples of size 100. How do you expect the dotplots would compare? *If there's time, select random samples of size 100 to confirm your answer.*

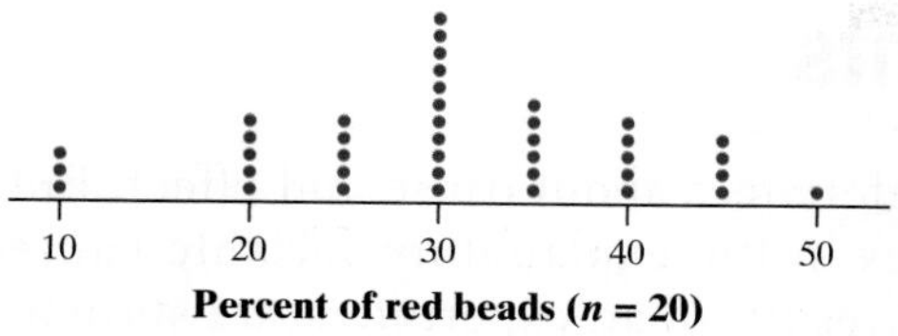


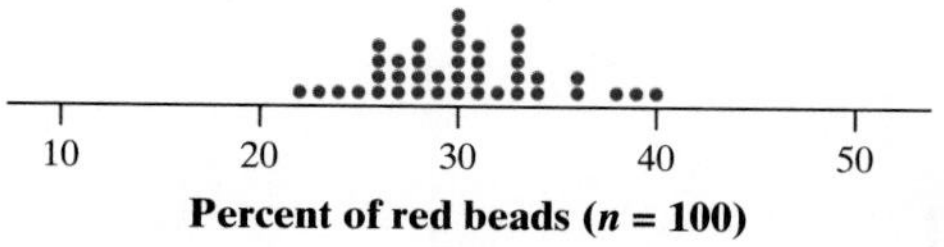


When Mrs. Storrs's class of 40 students did the red bead activity, they produced the dotplot shown at left (top) for samples of size 20. The dotplot is centered around 30%, the true percent of red beads. This shouldn't be surprising because random sampling helps avoid bias. Notice also that the estimates varied from 10% to 50% and that only 11 of the 40 estimates were equal to exactly 30%.

To see the effect of increasing the sample size, we simulated 40 random samples of size 100 from the same population and recorded the percent of red beads in each sample. Notice that the graph is still centered at 30%, but there is much less variability.

**SAMPLING VARIABILITY AND SAMPLE SIZE**

Larger random samples tend to produce estimates that are closer to the true population value than smaller random samples. In other words, estimates from larger samples are more precise.

## EXAMPLE

### Weighing football players
### Inference for sampling

Diamond Images/Getty Images

**PROBLEM:** How much do National Football League (NFL) players weigh, on average? In a random sample of 50 NFL players, the average weight is 244.4 pounds.

(a) Do you think that 244.4 pounds is the true average weight of all NFL players? Explain your answer.

(b) Which would be more likely to give an estimate close to the true average weight of all NFL players: a random sample of 50 players or a random sample of 100 players? Explain your answer.

**SOLUTION:**

(a) No. Different samples of size 50 would produce different average weights. So it would be surprising if this estimate is equal to the true average weight of all NFL players.

(b) A random sample of 100 players, because estimates tend to be closer to the truth when the sample size is larger.

FOR PRACTICE, TRY EXERCISE 93

Estimates from random samples often come with a *margin of error* that allows us to create an interval of plausible values for the true population value. In the preceding example about NFL players, the margin of error for the estimate of 244.4 pounds is 14.2 pounds. Based on this margin of error, it wouldn't be surprising if the true average weight for all NFL players was as small as $244.4 - 14.2 = 230.2$ pounds or as large as $244.4 + 14.2 = 258.6$ pounds.

You will learn how to calculate the margin of error for a population mean in Chapter 10. For now, make sure to remember the effect of sampling variability when using data from a random sample to make an inference about a population.

# Inference for Experiments

Well-designed experiments allow for inferences about cause and effect. But we should only conclude that changes in the explanatory variable cause changes in the response variable if the results of an experiment are **statistically significant**.

> **DEFINITION Statistically significant**
>
> When the observed results of a study are too unusual to be explained by chance alone, the results are called **statistically significant**.

Mr. Wilcox and his students decided to perform the caffeine experiment from the preceding section. In their experiment, 10 student volunteers were randomly assigned to drink cola with caffeine and the remaining 10 students were assigned to drink caffeine-free cola. The table and graph show the change in pulse rate for each student (Final pulse rate – Initial pulse rate), along with the mean change for each group.

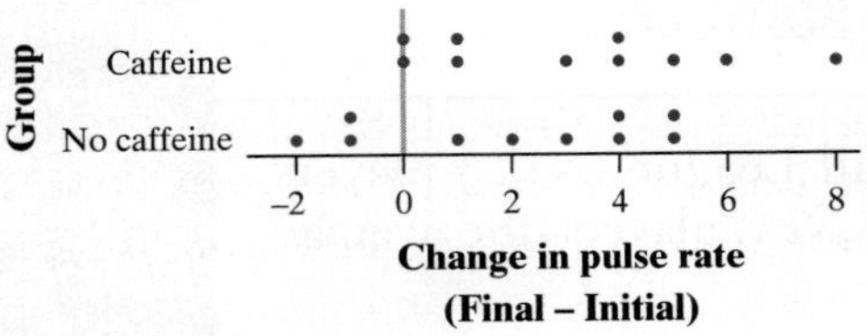


| | Change in pulse rate (Final pulse rate – Initial pulse rate) | | | | | | | | | | Mean change |
|---|---|---|---|---|---|---|---|---|---|---|---|
| Caffeine | 8 | 3 | 5 | 1 | 4 | 0 | 6 | 1 | 4 | 0 | 3.2 |
| No Caffeine | 3 | −2 | 4 | −1 | 5 | 5 | 1 | 2 | −1 | 4 | 2.0 |

The dotplots provide some evidence that caffeine has an effect on pulse rates. The mean change for the 10 students who drank cola with caffeine was 3.2, which is 1.2 greater than for the group who drank caffeine-free cola. But are the results statistically significant?

Recall that the purpose of random assignment in this experiment was to create two groups that were roughly equivalent at the beginning of the experiment. Subjects with high caffeine tolerance should be split up in about equal numbers, subjects with high metabolism should be split up in about equal numbers, and so on.

Of course, the random assignment is unlikely to produce groups that are exactly equivalent. One group might get more "favorable" subjects just by chance. That is, the caffeine group might end up with a few extra subjects who were likely to have a pulse rate increase, just due to chance variation in the random assignment.

There are two ways to explain why the mean change in pulse rate was 1.2 greater for the caffeine group:

1. Caffeine does *not* have an effect on pulse rates, and the difference of 1.2 happened because of chance variation in the random assignment.
2. Caffeine increases pulse rates.

If it is plausible to get a difference of 1.2 or more simply due to the chance variation in random assignment, the results of the experiment are not statistically significant. But if it is very unlikely to get a difference of 1.2 or more by chance alone, we rule out Explanation 1 and say the results are statistically significant—and that caffeine increases pulse rates.

How can we determine if a difference of 1.2 is statistically significant? You'll find out in the following activity.

## ACTIVITY Analyzing the caffeine experiment

Zaia Snively

In the experiment performed by Mr. Wilcox's class, the mean change in pulse rate for the caffeine group was 1.2 greater than the mean change for the no-caffeine group. This provides some evidence that caffeine increases pulse rates. Is this evidence convincing? Or is it plausible that a difference of 1.2 would arise just due to chance variation in the random assignment? In this activity, we'll investigate by seeing what differences typically occur just by chance, assuming caffeine doesn't affect pulse rates. That is, we'll assume that the change in pulse rate for a particular student would be the same regardless of what treatment he or she was assigned.

| | Change in pulse rate (Final pulse rate – Initial pulse rate) | | | | | | | | | | Mean change |
|---|---|---|---|---|---|---|---|---|---|---|---|
| Caffeine | 8 | 3 | 5 | 1 | 4 | 0 | 6 | 1 | 4 | 0 | 3.2 |
| No Caffeine | 3 | –2 | 4 | –1 | 5 | 5 | 1 | 2 | –1 | 4 | 2.0 |

1. Gather 20 index cards to represent the 20 students in this experiment. On each card, write one of the 20 outcomes listed in the table. For example, write "8" on the first card, "3" on the second card, and so on.
2. Shuffle the cards and deal two piles of 10 cards each. This represents randomly assigning the 20 students to the two treatments, *assuming that the treatment received doesn't affect the change in pulse rate.* The first pile of 10 cards represents the caffeine group, and the second pile of 10 cards represents the no-caffeine group.
3. Find the mean change for each group and subtract the means (Caffeine – No caffeine). *Note:* It is possible to get a negative difference.
4. Your teacher will draw and label an axis for a class dotplot. Plot the difference you got in Step 3 on the graph.
5. In Mr. Wilcox's class, the observed difference in means was 1.2. Is a difference of 1.2 statistically significant? Discuss with your classmates.

We used technology to perform 100 trials of the simulation described in the activity. The dotplot in Figure 4.3 shows that getting a difference of 1.2 isn't that unusual. In 19 of the 100 trials, we obtained a difference of 1.2 or more simply due to chance variation in the random assignment.

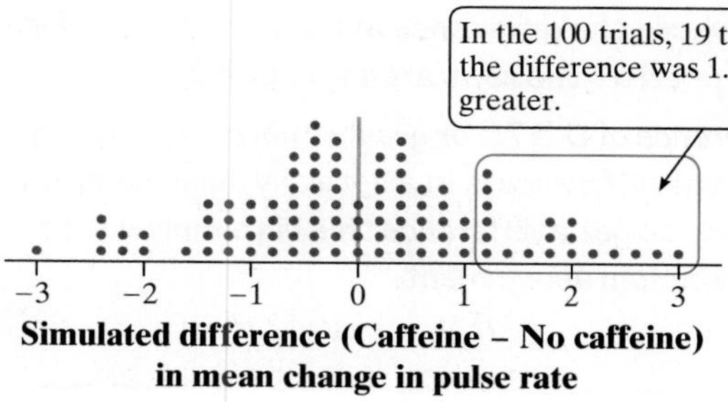


**FIGURE 4.3** Dotplot showing the differences in means that occurred in 100 simulated random assignments, assuming that caffeine has no effect on pulse rates.

Because a difference of 1.2 or greater is somewhat likely to occur by chance alone, the results of Mr. Wilcox's class experiment aren't statistically significant. Based on this experiment, there isn't convincing evidence that caffeine increases pulse rates.

## EXAMPLE

### Distracted driving
### Inference for experiments

**PROBLEM:** Is talking on a cell phone while driving more distracting than talking to a passenger? David Strayer and his colleagues at the University of Utah designed an experiment to help answer this question. They used 48 undergraduate students as subjects. The researchers randomly assigned half of the subjects to drive in a simulator while talking on a cell phone, and the other half to drive in the simulator while talking to a passenger. One response variable was whether or not the driver stopped at a rest area that was specified by researchers before the simulation started. The table shows the results.[54]

Sean Locke Photography/Shutterstock.com

| | | Treatment | | |
|---|---|---|---|---|
| | | Cell phone | Passenger | Total |
| Response | Stopped at rest area | 12 | 21 | 33 |
| | Didn't stop | 12 | 3 | 15 |
| | Total | 24 | 24 | 48 |

(a) Calculate the difference (Passenger – Cell phone) in the proportion of students who stopped at the rest area in the two groups.

One hundred trials of a simulation were performed to see what differences in proportions would occur due only to chance variation in the random assignment, assuming that the type of distraction did not affect whether a subject stopped at the rest area. That is, 33 "stoppers" and 15 "non-stoppers" were randomly assigned to two groups of 24.

(b) There are three dots at 0.29. Explain what these dots mean in this context.

(c) Use the results of the simulation to determine if the difference in proportions from part (a) is statistically significant. Explain your reasoning.

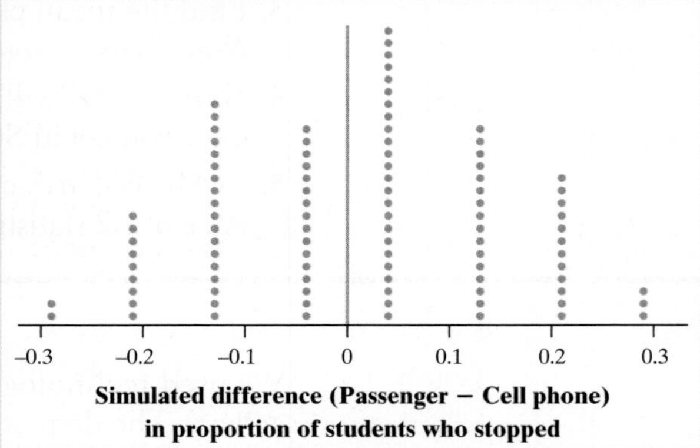


**SOLUTION:**

(a) Difference in proportions $= 21/24 - 12/24 = 0.875 - 0.500 = 0.375$

(b) When we assumed that the type of distraction doesn't matter, there were three simulated random assignments where the difference in the proportion of students who stopped at the rest area was 0.29.

(c) Because a difference of 0.375 or greater never occurred in the simulation, the difference is statistically significant. It is extremely unlikely to get a difference this big simply due to chance variation in the random assignment.

Because the difference is statistically significant, we can make a cause-and-effect conclusion: talking on a cell phone is more distracting than talking with a passenger—at least for subjects like those in the experiment.

**FOR PRACTICE, TRY EXERCISE 99**

In the caffeine example, we said that a difference in means of 1.2 was not unusual because a difference that big or bigger occurred 19% of the time by chance alone. In the distracted drivers example, we said that a difference in proportions of 0.375 was unusual because a difference this big or bigger occurred 0% of the time by chance alone. So the boundary between "not unusual" and "unusual" must be somewhere between 0% and 19%. For now, we recommend using a boundary of 5% so that differences that would occur less than 5% of the time by chance alone are considered statistically significant.

## The Scope of Inference: Putting it All Together

The type of conclusion (inference) that can be drawn from a study depends on how the data in the study were collected.

In the example about average weight in the NFL, the players were *randomly selected* from all NFL players. As you learned in Section 4.1, random sampling helps to avoid bias and produces reliable estimates of the truth about the population. Because the mean weight in the sample of players was 244.4 pounds, our best guess for the mean weight in the population of all NFL players is 244.4 pounds. Even though our estimates are rarely exactly correct, when samples are selected at random, we can make an *inference about the population.*

Both random sampling and random assignment introduce chance variation into a statistical study. When performing inference, statisticians use the laws of probability to describe this chance variation. You'll learn how this works later in the book.

In the distracted driver experiment, subjects were *randomly assigned* to talk on a cell phone or talk to a passenger. As you learned in Section 4.2, random assignment helps ensure that the two groups of subjects are as alike as possible before the treatments are imposed. If the group assigned to talk with a passenger remembers to stop at the rest area more often than the group assigned to talk on a cell phone, and the difference is too large to be explained by chance variation in the random assignment, it must be due to the treatments. In that case, the researchers could safely conclude that talking on a cell phone is more distracting than talking to a passenger. That is, they can make an *inference about cause and effect.* However, because the experiment used volunteer subjects, the scientists can only apply this conclusion to subjects like the ones in their experiment.

Let's recap what we've learned about the scope of inference in a statistical study.

**THE SCOPE OF INFERENCE**

- Random selection of individuals allows inference about the population from which the individuals were chosen.
- Random assignment of individuals to groups allows inference about cause and effect.

The following chart summarizes the possibilities.[55]

| | | Were individuals randomly assigned to groups? | |
|---|---|---|---|
| | | Yes | No |
| Were individuals randomly selected? | Yes | Inference about the population: YES<br>Inference about cause and effect: YES | Inference about the population: YES<br>Inference about cause and effect: NO |
| | No | Inference about the population: NO<br>Inference about cause and effect: YES | Inference about the population: NO<br>Inference about cause and effect: NO |

Well-designed experiments randomly assign individuals to treatment groups. However, most experiments don't select experimental units at random from the larger population. That limits such experiments to inference about cause and effect for individuals like those who received the treatments. Observational studies don't randomly assign individuals to groups, which makes it challenging to make an inference about cause and effect. But an observational study that uses random sampling can make an inference about the population.

## EXAMPLE When will I ever use this stuff?

### The scope of inference

**PROBLEM:** Researchers at the University of North Carolina were concerned about the increasing dropout rate in the state's high schools, especially for low-income students. Surveys of recent dropouts revealed that many of these students had started to lose interest during middle school. They said they saw little connection between what they were studying in school and their future plans. To change this perception, researchers developed a program called CareerStart. The central idea of the program is that teachers show students how the topics they're learning about can be applied to specific careers.

To test the effectiveness of CareerStart, the researchers recruited 14 middle schools in Forsyth County to participate in an experiment. Seven of the schools, determined at random, used CareerStart along with the district's standard curriculum. The other 7 schools just followed the standard curriculum. Researchers followed both groups of students for several years, collecting data on students' attendance, behavior, standardized test scores, level of engagement in school, and whether or not the students graduated from high school.

*Results:* Students at schools that used CareerStart generally had significantly better attendance and fewer discipline problems, earned higher test scores, reported greater engagement in their classes, and were more likely to graduate.[56]

What conclusion can we draw from this study? Explain your reasoning.

**SOLUTION:**

**Because treatments were randomly assigned and the results were significant, we can conclude that using the CareerStart curriculum caused better attendance, fewer discipline problems, higher test scores, greater engagement, and increased graduation rates. However, these results only apply to schools like those in the study because the schools were not randomly selected from any population.**

With no random selection, the results of the study should be applied only to schools like those in the study. With random assignment, it is possible to make an inference about cause and effect.

**FOR PRACTICE, TRY EXERCISE 103**

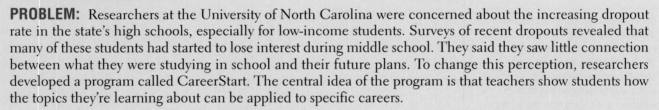

### CHECK YOUR UNDERSTANDING

When an athlete suffers a sports-related concussion, does it help to remove the athlete from play immediately? Researchers recruited 95 athletes seeking care for a sports-related concussion at a medical clinic and followed their progress during recovery. Researchers found statistically significant evidence that athletes who were removed from play immediately recovered more quickly, on average, than athletes who continued to play.[57] What conclusion can we draw from this study? Explain your answer.

## The Challenges of Establishing Causation

A well-designed experiment can tell us that changes in the explanatory variable cause changes in the response variable. More precisely, it tells us that this happened for specific individuals in the specific environment of this specific experiment. In some cases, it isn't practical or even ethical to do an experiment. Consider these important questions:

- Does going to church regularly help people live longer?
- Does smoking cause lung cancer?

To answer these cause-and-effect questions, we just need to perform a randomized comparative experiment. Unfortunately, we can't randomly assign people to attend church or to smoke cigarettes. The best data we have about these and many other cause-and-effect questions come from observational studies.

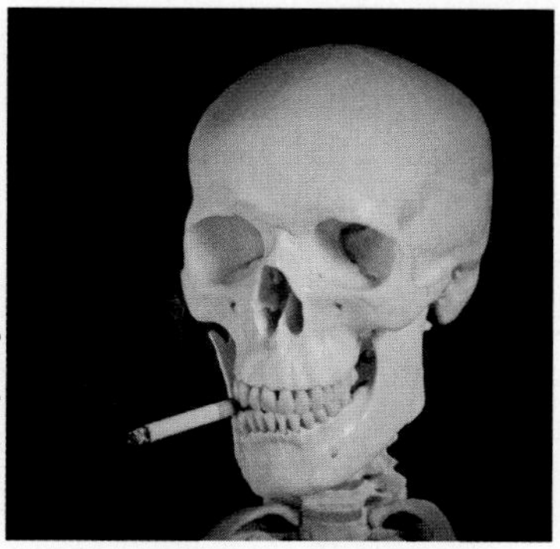

archives/Getty Images

Doctors had long observed that most lung cancer patients were smokers. Comparison of smokers and similar nonsmokers showed a very strong association between smoking and death from lung cancer. Could the association be due to some other variable? Is there some genetic factor that makes people both more likely to become addicted to nicotine and to develop lung cancer? If so, then smoking and lung cancer would be strongly associated even if smoking had no direct effect on the lungs. Or maybe confounding is to blame. It might be that smokers live unhealthy lives in other ways (diet, alcohol, lack of exercise) and that some other habit confounded with smoking is a cause of lung cancer. Still, it is sometimes possible to build a strong case for causation in the absence of experiments. The evidence that smoking causes lung cancer is about as strong as nonexperimental evidence can be.

There are several criteria for establishing causation when we can't do an experiment:

- *The association is strong.* The association between smoking and lung cancer is very strong.
- *The association is consistent.* Many studies of different kinds of people in many countries link smoking to lung cancer. That reduces the chance that some other variable specific to one group or one study explains the association.
- *Larger values of the explanatory variable are associated with stronger responses.* People who smoke more cigarettes per day or who smoke over a longer period get lung cancer more often. People who stop smoking reduce their risk.

- *The alleged cause precedes the effect in time.* Lung cancer develops after years of smoking. The number of men dying of lung cancer rose as smoking became more common, with a lag of about 30 years. Lung cancer kills more men than any other form of cancer. Lung cancer was rare among women until women began to smoke. Lung cancer in women rose along with smoking, again with a lag of about 30 years, and has passed breast cancer as the leading cause of cancer death among women.
- *The alleged cause is plausible.* Experiments with animals show that tars from cigarette smoke do cause cancer.

Medical authorities do not hesitate to say that smoking causes lung cancer. The U.S. Surgeon General states that cigarette smoking is "the largest avoidable cause of death and disability in the United States."[58] The evidence for causation is overwhelming—but it is not as strong as the evidence provided by well-designed experiments. Conducting an experiment in which some subjects were forced to smoke and others were not allowed to would be unethical. In cases like this, observational studies are our best source of reliable information.

## Data Ethics*

There are some potential experiments that are clearly unethical. In other cases, the boundary between "ethical" and "unethical" isn't as clear. Decide if you think each of the following studies is ethical or unethical:

- A promising new drug has been developed for treating cancer in humans. Before giving the drug to human subjects, researchers want to administer the drug to animals to see if there are any potentially serious side effects.
- Are companies discriminating against some individuals in the hiring process? To find out, researchers prepare several equivalent résumés for fictitious job applicants, with the only difference being the gender of the applicant. They send the fake résumés to companies advertising positions and keep track of the number of males and females who are contacted for interviews.
- Will people try to stop someone from driving drunk? A television news program hires an actor to play a drunk driver and uses a hidden camera to record the behavior of individuals who encounter the driver.

The most complex issues of data ethics arise when we collect data from people. The ethical difficulties are more severe for experiments that impose some treatment on people than for sample surveys that simply gather information. Trials of new medical treatments, for example, can do harm as well as good to their subjects.

Here are some basic standards of data ethics that must be obeyed by all studies that gather data from human subjects, both observational studies and experiments. The law requires that studies carried out or funded by the federal government obey these principles.[59] But neither the law nor the consensus of experts is completely clear about the details of their application.

*This is an important topic, but it is not required for the AP® Statistics exam.

**BASIC DATA ETHICS**

- All planned studies must be reviewed in advance by an *institutional review board* charged with protecting the safety and well-being of the subjects.
- All individuals who are subjects in a study must give their *informed consent* before data are collected.
- All individual data must be kept *confidential.* Only statistical summaries for groups of subjects may be made public.

**Institutional Review Boards** The purpose of an *institutional review board* is not to decide whether a proposed study will produce valuable information or if it is statistically sound. The board's purpose is, in the words of one university's board, "to protect the rights and welfare of human subjects (including patients) recruited to participate in research activities." The board reviews the plan of the study and can require changes. It reviews the consent form to be sure that subjects are informed about the nature of the study and about any potential risks. Once research begins, the board monitors its progress at least once a year.

**Informed Consent** Both words in the phrase *informed consent* are important, and both can be controversial. Subjects must be informed in advance about the nature of a study and any risk of harm it may bring. In the case of a questionnaire, physical harm is not possible. But a survey on sensitive issues could result in emotional harm. The participants should be told what kinds of questions the survey will ask and roughly how much of their time it will take. Experimenters must tell subjects the nature and purpose of the study and outline possible risks. Subjects must then consent in writing.

**Confidentiality** It is important to protect individuals' privacy by keeping all data about them *confidential.* The report of an opinion poll may say what percent of the 1200 respondents believed that legal immigration should be reduced. It may not report what *you* said about this or any other issue. Confidentiality is not the same as *anonymity.* Anonymity means that individuals are anonymous—their names are not known even to the director of the study. Anonymity is rare in statistical studies. Even where anonymity is possible (mainly in surveys conducted by mail), it prevents any follow-up to improve nonresponse or inform individuals of results.

## Section 4.3 Summary

- **Sampling variability** refers to the idea that different random samples of the same size from the same population produce different estimates. Reduce sampling variability by increasing the sample size.
- When the observed results of a study are too unusual to be explained by chance alone, we say that the results are **statistically significant.**
- Most studies aim to make inferences that go beyond the data produced.

  - **Inference about a population** requires that the individuals taking part in a study be randomly selected from the population.
  - A well-designed experiment that randomly assigns experimental units to treatments allows **inference about cause and effect.**
- In the absence of an experiment, good evidence of causation requires a strong association that appears consistently in many studies, a clear explanation for the alleged causal link, and careful examination of other variables.
- Studies involving humans must be screened in advance by an **institutional review board.** All participants must give their **informed consent** before taking part. Any information about the individuals in the study must be kept **confidential.**

## Section 4.3 Exercises

93. **Tweet, tweet!** What proportion of students at your school use Twitter? To find out, you survey a simple random sample of students from the school roster.

pg 299

(a) Will your sample result be exactly the same as the true population proportion? Explain your answer.

(b) Which would be more likely to produce a sample result closer to the true population value: an SRS of 50 students or an SRS of 100 students? Explain your answer.

94. **Far from home?** A researcher wants to estimate the average distance that students at a large community college live from campus. To find out, she surveys a simple random sample of students from the registrar's database.

(a) Will the researcher's sample result be exactly the same as the true population mean? Explain your answer.

(b) Which would be more likely to produce a sample result closer to the true population value: an SRS of 100 students or an SRS of 50 students? Explain your answer.

95. **Football on TV** A Gallup poll conducted telephone interviews with a random sample of 1000 adults aged 18 and older. Of these, 37% said that football is their favorite sport to watch on television. The margin of error for this estimate is 3.1 percentage points.

(a) Would you be surprised if a census revealed that 50% of adults in the population would say their favorite sport to watch on TV is football? Explain your answer.

(b) Explain how Gallup could decrease the margin of error.

96. **Car colors in Miami** Using a webcam, a traffic analyst selected a random sample of 800 cars traveling on I-195 in Miami on a weekday morning. Among the 800 cars in the sample, 24% were white. The margin of error for this estimate is 3.0 percentage points.

(a) Would you be surprised if a census revealed that 26% of cars on I-195 in Miami on a weekday morning were white? Explain your answer.

(b) Explain how the traffic analyst could decrease the margin of error.

97. **Kissing the right way** According to a newspaper article, "Most people are kissing the 'right way.'" That is, according to a study, the majority of couples prefer to tilt their heads to the right when kissing. In the study, a researcher observed a random sample of 124 kissing couples and found that 83/124 (66.9%) of the couples tilted to the right.[60] To determine if these data provide convincing evidence that couples are more likely to tilt their heads to the right, 100 simulated SRSs were selected.

Each dot in the graph shows the number of couples that tilt to the right in a simulated SRS of 124 couples, assuming that each couple has a 50% chance of tilting to the right.

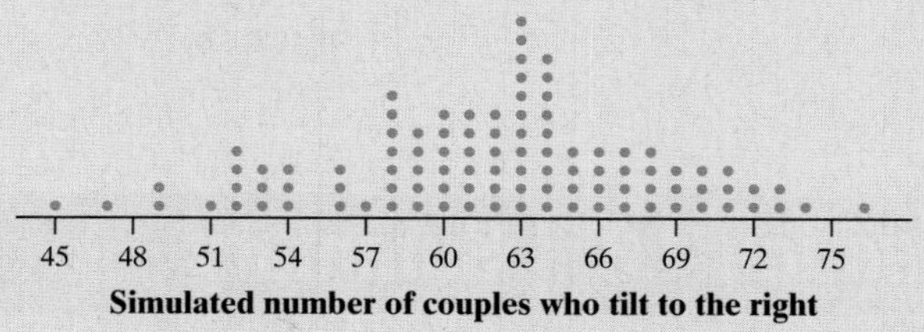


(a) Explain how the graph illustrates the concept of sampling variability.

(b) Based on the data from the study and the results of the simulation, is there convincing evidence that couples prefer to kiss the "right way"? Explain your answer.

98. **Weekend birthdays** Over the years, the percentage of births that are planned caesarean sections has been rising. Because doctors can schedule these deliveries, there might be more children born during the week and fewer born on the weekend than if births were uniformly distributed throughout the week. To investigate, Mrs. McDonald and her class selected an SRS of 73 people born since 1993. Of these people, 24 were born on Friday, Saturday, or Sunday.

To determine if these data provide convincing evidence that fewer than 43% (3/7) of people born since 1993 were born on Friday, Saturday, or Sunday, 100 simulated SRSs were selected. Each dot in the graph shows the number of people that were born on Friday, Saturday, or Sunday in a simulated SRS of 73 people, assuming that each person had a 43% chance of being born on one of these three days.

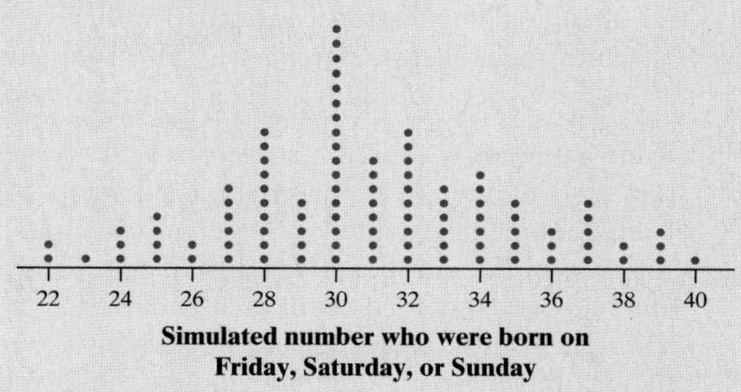


(a) Explain how the dotplot illustrates the concept of sampling variability.

(b) Based on the data from the study and the results of the simulation, is there convincing evidence that fewer than 43% of people born since 1993 were born on Friday, Saturday, or Sunday? Explain your answer.

99. **I work out a lot** Are people influenced by what others say? Michael conducted an experiment in front of a popular gym. As people entered, he asked them how many days they typically work out per week. As he asked the question, he showed the subjects one of two clipboards, determined at random. Clipboard A had the question and many responses written down, where the majority of responses were 6 or 7 days per week. Clipboard B was the same, except most of the responses were 1 or 2 days per week. The mean response for the Clipboard A group was 4.68 and the mean response for the Clipboard B group was 4.21.[61]

pg 302

(a) Calculate the difference (Clipboard A – Clipboard B) in the mean number of days for the two groups.

One hundred trials of a simulation were performed to see what differences in means would occur due only to chance variation in the random assignment, assuming that the responses on the clipboard don't matter. The results are shown in the dotplot.

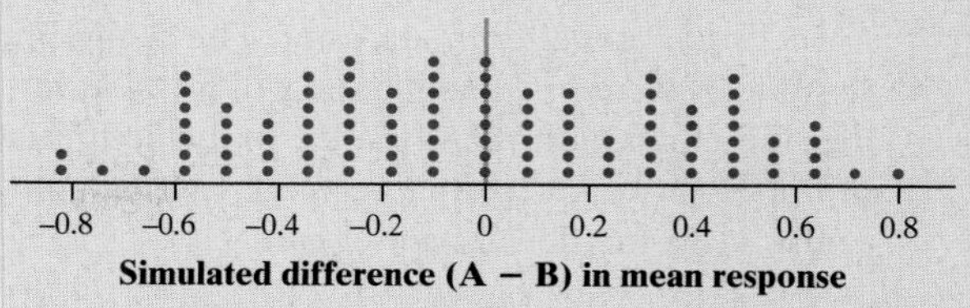


(b) There is one dot at 0.72. Explain what this dot means in this context.

(c) Use the results of the simulation to determine if the difference in means from part (a) is statistically significant. Explain your reasoning.

100. **A louse-y situation** A study published in the *New England Journal of Medicine* compared two medicines to treat head lice: an oral medication called ivermectin and a topical lotion containing malathion. Researchers studied 812 people in 376 households in seven areas around the world. Of the 185 households randomly assigned to ivermectin, 171 were free from head lice after 2 weeks, compared with only 151 of the 191 households randomly assigned to malathion.[62]

(a) Calculate the difference (Ivermectin – Malathion) in the proportion of households that were free from head lice in the two groups.

One hundred trials of a simulation were performed to see what differences in proportions would occur due only to chance variation in the random assignment, assuming that the type of medication doesn't matter. The results are shown in the dotplot.

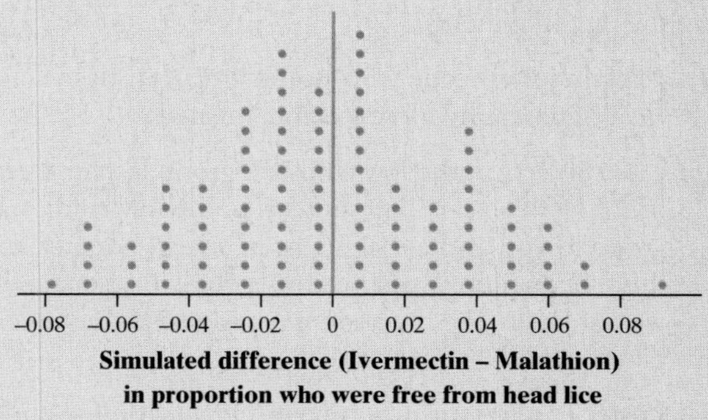


(b) There is one dot at 0.09. Explain what this dot means in this context.

(c) Use the results of the simulation to determine if the difference in proportions from part (a) is statistically significant. Explain your reasoning.

101. **Acupuncture and pregnancy** A study sought to determine if the ancient Chinese art of acupuncture could help infertile women become pregnant.[63] A total of 160 healthy women undergoing assisted reproductive therapy were recruited for the study. Half of the subjects were randomly assigned to receive acupuncture treatment 25 minutes before embryo transfer and

again 25 minutes after the transfer. The remaining 80 subjects were instructed to lie still for 25 minutes after the embryo transfer. *Results:* In the acupuncture group, 34 women became pregnant. In the control group, 21 women became pregnant.

(a) Why did researchers randomly assign the subjects to the two treatments?

(b) The difference in the percent of women who became pregnant in the two groups is statistically significant. Explain what this means to someone who knows little statistics.

(c) Explain why the design of the study prevents us from concluding that acupuncture caused the difference in pregnancy rates.

102. **Do diets work?** Dr. Linda Stern and her colleagues recruited 132 obese adults at the Philadelphia Veterans Affairs Medical Center in Pennsylvania. Half the participants were randomly assigned to a low-carbohydrate diet and the other half to a low-fat diet. Researchers measured each participant's change in weight and cholesterol level after six months and again after one year. Subjects in the low-carb diet group lost significantly more weight than subjects in the low-fat diet group during the first six months. At the end of a year, however, the average weight loss for subjects in the two groups was not significantly different.[64]

(a) Why did researchers randomly assign the subjects to the diet treatments?

(b) Explain to someone who knows little statistics what "lost significantly more weight" means.

(c) The subjects in the low-carb diet group lost an average of 5.1 kg in a year. The subjects in the low-fat diet group lost an average of 3.1 kg in a year. Explain how this information could be consistent with the fact that weight loss in the two groups was not significantly different.

103. **Foster care versus orphanages** Do abandoned children placed in foster homes do better than similar children placed in an institution? The Bucharest Early Intervention Project found statistically significant evidence that they do. The subjects were 136 young children abandoned at birth and living in orphanages in Bucharest, Romania. Half of the children, chosen at random, were placed in foster homes. The other half remained in the orphanages.[65] (Foster care was not easily available in Romania at the time and so was paid for by the study.) What conclusion can we draw from this study? Explain your reasoning.

pg 304

104. **Frozen batteries** Will storing batteries in a freezer make them last longer? To find out, a company that produces batteries takes a random sample of 100 AA batteries from its warehouse. The company statistician randomly assigns 50 batteries to be stored in the freezer and the other 50 to be stored at room temperature for 3 years. At the end of that time period, each battery's charge is tested. *Result:* Batteries stored in the freezer had a significantly higher average charge. What conclusion can we draw from this study? Explain your reasoning.

105. **Attend church, live longer?** One of the better studies of the effect of regular attendance at religious services gathered data from a random sample of 3617 adults. The researchers then measured lots of variables, not just the explanatory variable (religious activities) and the response variable (length of life). A news article said: "Churchgoers were more likely to be nonsmokers, physically active, and at their right weight. But even after health behaviors were taken into account, those not attending religious services regularly still were significantly more likely to have died."[66] What conclusion can we draw from this study? Explain your reasoning.

106. **Rude surgeons** Is a friendly surgeon a better surgeon? In a study of more than 32,000 surgical patients from 7 different medical centers, researchers classified surgeons by the number of unsolicited complaints that had been recorded about their behavior. The researchers found that surgical complications were significantly more common in patients whose surgeons had received the most complaints about their behavior, compared with patients whose surgeons had received the fewest complaints.[67] What conclusion can we draw from this study? Explain your reasoning.

107. **Berry good** Eating blueberries and strawberries might improve heart health, according to a long-term study of 93,600 women who volunteered to take part. These berries are high in anthocyanins due to their pigment. Women who reported consuming the most anthocyanins had a significantly smaller risk of heart attack compared to the women who reported consuming the least. What conclusion can we draw from this study? Explain your reasoning.[68]

108. **Exercise and memory** A study of strength training and memory randomly assigned 46 young adults to two groups. After both groups were shown 90 pictures, one group had to bend and extend one leg against heavy resistance 60 times. The other group stayed relaxed, while the researchers used the same exercise machine to bend and extend their legs with no resistance. Two days later, each subject was shown 180 pictures—the original 90 pictures plus 90 new pictures and asked to identify which pictures were shown two days earlier. The resistance group was significantly more successful in identifying these pictures than was the relax group. What conclusions can we draw from this study? Explain your reasoning.[69]

109.* **Minimal risk?** You have been invited to serve on a college's institutional review board. You must decide whether several research proposals qualify for lighter review because they involve only minimal risk to subjects. Federal regulations say that "minimal risk" means the risks are no greater than "those ordinarily encountered in daily life or during the performance of routine physical or psychological examinations or tests." That's vague. Which of these do you think qualifies as "minimal risk"?

(a) Draw a drop of blood by pricking a finger to measure blood sugar.

(b) Draw blood from the arm for a full set of blood tests.

(c) Insert a tube that remains in the arm so that blood can be drawn regularly.

110.* **Who reviews?** Government regulations require that institutional review boards consist of at least five people, including at least one scientist, one nonscientist, and one person from outside the institution. Most boards are larger, but many contain just one outsider.

(a) Why should review boards contain people who are not scientists?

(b) Do you think that one outside member is enough? How would you choose that member? (For example, would you prefer a medical doctor? A religious leader? An activist for patients' rights?)

111.* **Facebook emotions** In cooperation with researchers from Cornell University, Facebook randomly selected almost 700,000 users for an experiment in "emotional contagion." Users' news feeds were manipulated (without their knowledge) to selectively show postings from their friends that were either more positive or more negative in tone, and the emotional tone of their own subsequent postings was measured. The researchers found evidence that people who read emotionally negative postings were more likely to post messages with a negative tone, whereas those who read positive messages were more likely to post messages with a positive tone. The research was widely criticized for being unethical. Explain why.[70]

112.* **No consent needed?** In which of the circumstances listed here would you allow collecting personal information without the subjects' consent?

(a) A government agency takes a random sample of income tax returns to obtain information on the average income of people in different occupations. Only the incomes and occupations are recorded from the returns, not the names.

(b) A social psychologist attends public meetings of a religious group to study the behavior patterns of its members.

(c) A social psychologist pretends to be converted to membership in a religious group and attends private meetings to study the behavior patterns of its members.

113.* **Anonymous? Confidential?** One of the most important nongovernment surveys in the United States is the National Opinion Research Center's General Social Survey (GSS). The GSS regularly monitors public opinion on a wide variety of political and social issues. Interviews are conducted in person in the subject's home. Are a subject's responses to GSS questions anonymous, confidential, or both? Explain your answer.

114.* **Anonymous? Confidential?** Texas A&M, like many universities, offers screening for HIV, the virus that causes AIDS. Students may choose either anonymous or confidential screening. An announcement says, "Persons who sign up for screening will be assigned a number so that they do not have to give their name." They can learn the results of the test by telephone, still without giving their name. Does this describe anonymous or confidential screening? Why?

115.* **The Willowbrook hepatitis studies** In the 1960s, children entering the Willowbrook State School, an institution for the intellectually disabled on Staten Island in New York, were deliberately infected with hepatitis. The researchers argued that almost all children in the institution quickly became infected anyway. The studies showed for the first time that two strains of hepatitis existed. This finding contributed to the development of effective vaccines. Despite these valuable results, the Willowbrook studies are now considered an example of unethical research. Explain why, according to current ethical standards, useful results are not enough to allow a study.

116.* **Unequal benefits** Researchers on aging proposed to investigate the effect of supplemental health services on the quality of life of older people. Eligible patients on the rolls of a large medical clinic were to be randomly assigned to treatment and control groups. The treatment group would be offered hearing aids, dentures, transportation, and other services not available without charge to the control group. The review board believed that providing these services to some but not other persons in the same institution raised ethical questions. Do you agree?

*Exercises 109–116: This is an important topic, but it is not required for the AP® Statistics exam.

**Multiple Choice** *Select the best answer for Exercises 117 and 118.*

117. Do product labels influence customer perceptions? To find out, researchers recruited more than 500 adults and asked them to estimate the number of calories, amount of added sugar, and amount of fat in a variety of food products. Half of the subjects were randomly assigned to evaluate products with the word "Natural" on the label, while the other half were assigned to evaluate the same products without the "Natural" label. On average, the products with the "Natural" label were judged to have significantly fewer calories. Based on this study, is it reasonable to conclude that including the word "Natural" on the label causes a reduction in estimated calories?

(a) No, because the adults weren't randomly selected from the population of all adults.

(b) No, because there wasn't a control group for comparison.

(c) No, because association doesn't imply causation.

(d) Yes, because the adults were randomly assigned to the treatments.

(e) Yes, because there were a large number of adults involved in the study.

118. Some news organizations maintain a database of customers who have volunteered to share their opinions on a variety of issues. Suppose that one of these databases includes 9000 registered voters in California. To measure the amount of support for a controversial ballot issue, 1000 registered voters in California are randomly selected from the database and asked their opinion. Which of the following is the largest population to which the results of this survey should be generalized?

(a) The 1000 people in the sample

(b) The 9000 registered voters from California in the database

(c) All registered voters in California

(d) All California residents

(e) All registered voters in the United States

**Review and Recycle**

119. **Animal testing** (1.1) "It is right to use animals for medical testing if it might save human lives." The General Social Survey asked 1152 adults to react to this statement. Here is the two-way table of their responses:

| | | Gender | | |
|---|---|---|---|---|
| | | Male | Female | Total |
| Opinion about using animals for medical testing | Strongly agree | 76 | 59 | 135 |
| | Agree | 270 | 247 | 517 |
| | Neither agree nor disagree | 87 | 139 | 226 |
| | Disagree | 61 | 123 | 184 |
| | Strongly disagree | 22 | 68 | 90 |
| | Total | 516 | 636 | 1152 |

(a) Construct segmented bar graphs to display the distribution of opinion for males and for females.

(b) Is there an association between gender and opinion for the members of this sample? Explain your answer.

120. **Initial public offerings** (1.3) The business magazine *Forbes* reports that 4567 companies sold their first stock to the public between 1990 and 2000. The *mean* change in the stock price of these companies since the first stock was issued was +111%. The *median* change was −31%.[71] Explain how this difference could happen.

# Chapter 4 Wrap-Up

## FRAPPY! FREE RESPONSE AP® PROBLEM, YAY!

The following problem is modeled after actual AP® Statistics exam free-response questions. Your task is to generate a complete, concise response in 15 minutes.

*Directions: Show all your work. Indicate clearly the methods you use, because you will be scored on the correctness of your methods as well as on the accuracy and completeness of your results and explanations.*

In a recent study, 166 adults from the St. Louis area were recruited and randomly assigned to receive one of two treatments for a sinus infection. Half of the subjects received an antibiotic (amoxicillin) and the other half received a placebo.[72]

(a) Describe how the researchers could have assigned treatments to subjects if they wanted to use a completely randomized design.

(b) All the subjects in the experiment had moderate, severe, or very severe symptoms at the beginning of the study. Describe one statistical benefit and one statistical drawback for using subjects with moderate, severe, or very severe symptoms instead of just using subjects with very severe symptoms.

(c) At different stages during the next month, all subjects took the sino-nasal outcome test. After 10 days, the difference in average test scores was *not* statistically significant. In this context, explain what it means for the difference to be not statistically significant.

(d) One possible way that researchers could have improved the study is to use a randomized block design. Explain how the researchers could have incorporated blocking in their design.

After you finish, you can view two example solutions on the book's website (highschool.bfwpub.com/updatedtps6e). Determine whether you think each solution is "complete," "substantial," "developing," or "minimal." If the solution is not complete, what improvements would you suggest to the student who wrote it? Finally, your teacher will provide you with a scoring rubric. Score your response and note what, if anything, you would do differently to improve your own score

# Chapter 4 Review

### Section 4.1: Sampling and Surveys

In this section, you learned that a population is the group of all individuals that we want information about. A sample is the subset of the population that we use to gather this information. The goal of most sample surveys is to use information from the sample to draw conclusions about the population. Choosing people for a sample because they are located nearby or letting people choose whether or not to be in the sample are poor ways to select a sample. Because convenience sampling and voluntary response sampling will produce estimates that are likely to underestimate or likely to overestimate the value you want to know, these methods of choosing a sample are biased.

To avoid bias in the way the sample is formed, the members of the sample should be chosen at random. One way to do this is with a simple random sample (SRS), which is equivalent to selecting well-mixed slips of paper from a hat without replacement. It is often more convenient to select an SRS using technology or a table of random digits.

Three other random sampling methods are stratified sampling, cluster sampling, and systematic sampling. To obtain a stratified random sample, divide the population into non-overlapping groups (strata) of individuals that are likely to have similar responses, take an SRS from each stratum, and combine the chosen individuals to form the sample. Stratified random samples can produce estimates with much greater precision than simple random samples. To obtain a cluster sample, divide the population into non-overlapping groups (clusters) of individuals that are in similar locations, randomly select clusters, and use every individual in the chosen clusters. Cluster samples are easier to obtain than simple random samples or stratified random samples, but they may not produce very precise estimates. To obtain a systematic random sample, choose a value of $k$ based on the population size and desired sample size, randomly select a number from 1 to $k$ to determine which member of the population to survey first, and then survey every $k$th member thereafter.

Finally, you learned about other issues in sample surveys that can lead to bias: undercoverage occurs when the sampling method systematically underrepresents one part of the population. Nonresponse describes when answers cannot be obtained from some people that were chosen to be in the sample. Bias can also result when some people in the sample don't give accurate responses due to question wording, interviewer characteristics, or other factors.

### Section 4.2: Experiments

In this section, you learned about the difference between observational studies and experiments. Experiments deliberately impose treatments to see if there is a cause-and-effect relationship between two variables. Observational studies look at relationships between two variables, but make it difficult to show cause and effect because other variables may be confounded with the explanatory variable. Variables are confounded when it is impossible to determine which of the variables is causing a change in the response variable.

A common type of comparative experiment uses a completely randomized design. In this type of design, the experimental units are assigned to the treatments completely at random. With random assignment, the treatment groups should be roughly equivalent at the beginning of the experiment. Replication means giving each treatment to as many experimental units as possible. This makes it easier to see the effects of the treatments because the effects of other variables are more likely to be balanced among the treatment groups.

During an experiment, it is important that other variables be controlled (kept the same) for each experimental unit. Doing so helps avoid confounding and removes a possible source of variation in the response variable. Also, beware of the placebo effect—the tendency for people to improve because they expect to, not because of the treatment they are receiving. One way to make sure that all experimental units have the same expectations is to make them blind—unaware of which treatment they are receiving. When the people interacting with the subjects and measuring the response variable are also blind, the experiment is called double-blind.

Blocking in experiments is similar to stratifying in sampling. To form blocks, group together experimental units that are similar with respect to a variable that is associated with the response. Then randomly assign the treatments within each block. A randomized block design that uses blocks with two experimental units is called a matched pairs design. Blocking helps us estimate the effects of the treatments more precisely because we can account for the variability introduced by the variables used to form the blocks.

### Section 4.3: Using Studies Wisely

In this section, you learned that the types of conclusions we can draw depend on how the data are produced. When samples are selected at random, we can make inferences about the population from which the sample was drawn. However, the estimates we calculate from sample data rarely equal the true population value because of sampling variability. We can reduce sampling variability by increasing the sample size.

When treatments are applied to groups formed at random in an experiment, we can make an inference about cause and effect. Making a cause-and-effect conclusion is often difficult because it is impossible or unethical to perform certain types of experiments. Good data ethics requires that studies should be approved by an institutional review board, subjects should give informed consent, and individual data must be kept confidential.

Finally, the results of a study are statistically significant if they are too unusual to occur by chance alone.

## What Did You Learn?

| Learning Target | Section | Related Example on Page(s) | Relevant Chapter Review Exercise(s) |
|---|---|---|---|
| Identify the population and sample in a statistical study. | 4.1 | 249 | R4.1 |
| Identify voluntary response sampling and convenience sampling and explain how these sampling methods can lead to bias. | 4.1 | 252 | R4.2 |
| Describe how to select a simple random sample using slips of paper, technology, or a table of random digits. | 4.1 | 256 | R4.2 |
| Describe how to select a sample using stratified random sampling, cluster sampling, and systematic random sampling, and explain whether a particular sampling method is appropriate in a given situation. | 4.1 | 259 | R4.2, R4.3 |
| Explain how undercoverage, nonresponse, question wording, and other aspects of a sample survey can lead to bias. | 4.1 | 262 | R4.4 |

| Learning Target | Section | Related Example on Page(s) | Relevant Chapter Review Exercise(s) |
|---|---|---|---|
| Explain the concept of confounding and how it limits the ability to make cause-and-effect conclusions. | 4.2 | 271 | R4.5 |
| Distinguish between an observational study and an experiment, and identify the explanatory and response variables in each type of study. | 4.2 | 272 | R4.5 |
| Identify the experimental units and treatments in an experiment. | 4.2 | 274 | R4.6 |
| Describe the placebo effect and the purpose of blinding in an experiment. | 4.2 | 278 | R4.8 |
| Describe how to randomly assign treatments in an experiment using slips of paper, technology, or a table of random digits. | 4.2 | 279 | R4.9 |
| Explain the purpose of comparison, random assignment, control, and replication in an experiment. | 4.2 | 276, 282 | R4.6, R4.8 |
| Describe a completely randomized design for an experiment. | 4.2 | 284 | R4.6, R4.9 |
| Describe a randomized block design and a matched pairs design for an experiment and explain the purpose of blocking in an experiment. | 4.2 | 287, 288 | R4.6, R4.9 |
| Explain the concept of sampling variability when making an inference about a population and how sample size affects sampling variability. | 4.3 | 299 | R4.1 |
| Explain the meaning of statistically significant in the context of an experiment and use simulation to determine if the results of an experiment are statistically significant. | 4.3 | 302 | R4.8 |
| Identify when it is appropriate to make an inference about a population and when it is appropriate to make an inference about cause and effect. | 4.3 | 304 | R4.7 |
| Evaluate if a statistical study has been carried out in an ethical manner.* | 4.3 | | R4.10 |

*This is an important topic, but it is not required for the AP® Statistics exam.

# Chapter 4 Review Exercises

**R4.1 Nurses are the best** A recent random sample of $n = 805$ adult U.S. residents found that the proportion who rated the honesty and ethical standards of nurses as high or very high is 0.85. This is 0.15 higher than the proportion recorded for doctors, the next highest-ranked profession.[73]

(a) Identify the sample and the population in this setting.

(b) Do you think that the proportion of all U.S. residents who would rate the honesty and ethical standards of nurses as high or very high is exactly 0.85? Explain your answer.

(c) What is the benefit of increasing the sample size in this context?

**R4.2 Parking problems** The administration at a high school with 1800 students wants to gather student opinion about parking for students on campus. It isn't practical to contact all students.

(a) Give an example of a way to obtain a voluntary response sample of students. Explain how this method could lead to bias.

(b) Give an example of a way to obtain a convenience sample of students. Explain how this method could lead to bias.

(c) Describe how to select an SRS of 50 students from the school. Explain how using an SRS helps avoid the biases you described in parts (a) and (b).

(d) Describe how to select a systematic random sample of 50 students from the school. What advantage does this method have over an SRS?

**R4.3 Surveying NBA fans** The manager of a sports arena wants to learn more about the financial status of the people who are attending an NBA basketball game. He would like to give a survey to a representative sample of about 10% of the fans in attendance. Ticket prices for the game vary a great deal: seats near the court cost over $200 each, while seats in the top rows of the arena cost $50 each. The arena is divided into 50 numbered sections, from 101 to 150. Each section has rows of seats labeled with letters from A (nearest the court) to ZZ (top row of the arena).

(a) Explain why it might be difficult to give the survey to an SRS of fans.

(b) Explain why it would be better to select a stratified random sample using the lettered rows rather than the numbered sections as strata. What is the benefit of using a stratified sample in this context?

(c) Explain how to select a cluster sample of fans. What is the benefit of using a cluster sample in this context?

**R4.4 Been to the movies?** An opinion poll calls 2000 randomly chosen residential telephone numbers, then asks to speak with an adult member of the household. The interviewer asks, "Box office revenues are at an all-time high. How many movies have you watched in a movie theater in the past 12 months?" In all, 1131 people responded. The researchers used the responses to estimate the mean number of movies adults had watched in a movie theater over the past 12 months.

(a) Describe a potential source of bias related to the wording of the question. Suggest a change that would help fix this problem.

(b) Describe how using only residential phone numbers might lead to bias and how this will affect the estimate.

(c) Describe how nonresponse might lead to bias and how this will affect the estimate.

**R4.5 Are anesthetics safe?** The National Halothane Study was a major investigation of the safety of anesthetics used in surgery. Records of over 850,000 operations performed in 34 major hospitals showed the following death rates for four common anesthetics:[74]

| Anesthetic | A | B | C | D |
|---|---|---|---|---|
| Death rate | 1.7% | 1.7% | 3.4% | 1.9% |

There seems to be a clear association between the anesthetic used and the death rate of patients. Anesthetic C appears to be more dangerous.

(a) Explain why we call the National Halothane Study an observational study rather than an experiment, even though it compared the results of using different anesthetics in actual surgery.

(b) Identify the explanatory and response variables in this study.

(c) When the study looked at other variables that are related to a doctor's choice of anesthetic, it found that Anesthetic C was not causing extra deaths. Explain the concept of confounding in this context and identify a variable that might be confounded with the doctor's choice of anesthetic.

**R4.6 Ugly fries** Few people want to eat discolored french fries. To prevent spoiling and to preserve flavor, potatoes are kept refrigerated before being cut for french fries. But immediate processing of cold potatoes causes discoloring due to complex chemical reactions. The potatoes must therefore be brought to room temperature before processing. Researchers want to design an experiment in which tasters will rate the color and flavor of french fries prepared from several groups of potatoes. The potatoes will be freshly picked or stored for a month at room temperature or stored for a month refrigerated. They will then be sliced and cooked either immediately or after an hour at room temperature.

(a) Identify the experimental units, the factors, the number of levels for each factor, and the treatments.

(b) Describe a completely randomized design for this experiment using 300 potatoes.

(c) A single supplier has made 300 potatoes available to the researchers. Describe a statistical benefit and a statistical drawback of using potatoes from only one supplier.

(d) The researchers decided to do a follow-up experiment using potatoes from several different suppliers. Describe how they should change the design of the experiment to account for the addition of other suppliers.

**R4.7 Don't catch a cold!** A recent study of 1000 students at the University of Michigan investigated how to prevent catching the common cold. The students were randomly assigned to three different cold prevention methods for 6 weeks. Some wore masks, some wore masks and used hand sanitizer, and others took no precautions. The two groups who used masks reported 10–50% fewer cold symptoms than those who did not wear a mask.[75]

(a) Does this study allow for inference about a population? Explain your answer.

(b) Does this study allow for inference about cause and effect? Explain your answer.

**R4.8 An herb for depression?** Does the herb St. John's wort relieve major depression? Here is an excerpt from the report of one study of this issue: "Design: Randomized, Double-Blind, Placebo-Controlled Clinical Trial."[76] The study concluded that the difference in effectiveness of St. John's wort and a placebo was not statistically significant.

(a) Describe the placebo effect in this context. How did the design of this experiment account for the placebo effect?

(b) Explain the purpose of the random assignment.

(c) Why is a double-blind design a good idea in this setting?

(d) Explain what "not statistically significant" means in this context.

**R4.9 How long did I work?** A psychologist wants to know if the difficulty of a task influences our estimate of how long we spend working at it. She designs two sets of mazes that subjects can work through on a computer. One set has easy mazes and the other has difficult mazes. Subjects work until told to stop (after 6 minutes, but subjects do not know this). They are then asked to estimate how long they worked. The psychologist has 30 students available to serve as subjects.

(a) Describe an experiment using a completely randomized design to learn the effect of difficulty on estimated time. Make sure to carefully explain your method of assigning treatments.

(b) Describe a matched pairs experimental design using the same 30 subjects.

(c) Which design would be more likely to detect a difference in the effects of the treatments? Explain your answer.

**R4.10* Deceiving subjects** Students sign up to be subjects in a psychology experiment. When they arrive, they are told that interviews are running late and are taken to a waiting room. The experimenters then stage the theft of a valuable object left in the waiting room. Some subjects are alone with the thief, and others are present in pairs—these are the treatments being compared. Will the subject report the theft?

(a) The students had agreed to take part in an unspecified study, and the true nature of the experiment is explained to them afterward. Does this meet the requirement of informed consent? Explain your answer.

(b) What two other ethical principles should be followed in this study?

*This is an important topic, but it is not required for the AP® Statistics exam.

# Chapter 4 AP® Statistics Practice Test

**Section I: Multiple Choice** *Select the best answer for each question.*

**T4.1** When we take a census, we attempt to collect data from

(a) a stratified random sample.
(b) every individual chosen in a simple random sample.
(c) every individual in the population.
(d) a voluntary response sample.
(e) a convenience sample.

**T4.2** You want to take a simple random sample (SRS) of 50 of the 816 students who live in a dormitory on campus. You label the students 001 to 816 in alphabetical order. In the table of random digits, you read the entries

| | | | | | | | |
|---|---|---|---|---|---|---|---|
| 95592 | 94007 | 69769 | 33547 | 72450 | 16632 | 81194 | 14873 |

The first three students in your sample have labels

(a) 955, 929, 400.
(b) 400, 769, 769.
(c) 559, 294, 007.
(d) 929, 400, 769.
(e) 400, 769, 335.

**T4.3** A study of treatments for angina (pain due to low blood supply to the heart) compared bypass surgery, angioplasty, and use of drugs. The study looked at the medical records of thousands of angina patients whose doctors had chosen one of these treatments. It found that the average survival time of patients given drugs was the highest. What do you conclude?

(a) This study proves that drugs prolong life and should be the treatment of choice.
(b) We can conclude that drugs prolong life because the study was a comparative experiment.
(c) We can't conclude that drugs prolong life because the patients were volunteers.
(d) We can't conclude that drugs prolong life because the groups might differ in ways besides the treatment.
(e) We can't conclude that drugs prolong life because no placebo was used.

**T4.4** Tonya wanted to estimate the average amount of time that students at her school spend on Facebook each day. She gets an alphabetical roster of students in the school from the registrar's office and numbers the students from 1 to 1137. Then Tonya uses a random number generator to pick 30 distinct labels from 1 to 1137. She surveys those 30 students about their Facebook use. Tonya's sample is a simple random sample because

(a) it was selected using a chance process.
(b) it gave every individual the same chance to be selected.
(c) it gave every possible sample of size 30 an equal chance to be selected.
(d) it doesn't involve strata or clusters.
(e) it is guaranteed to be representative of the population.

**T4.5** Consider an experiment to investigate the effectiveness of different insecticides in controlling pests and their impact on the productivity of tomato plants. What is the best reason for randomly assigning treatment levels (spraying or not spraying) to the experimental units (farms)?

(a) Random assignment eliminates the effects of other variables, like soil fertility.

(b) Random assignment eliminates chance variation in the responses.

(c) Random assignment allows researchers to generalize conclusions about the effectiveness of the insecticides to all farms.

(d) Random assignment will tend to average out all other uncontrolled factors such as soil fertility so that they are not confounded with the treatment effects.

(e) Random assignment helps avoid bias due to the placebo effect.

**T4.6** Researchers randomly selected 1700 people from Canada and rated the happiness of each person. Ten years later, the researchers followed up with each person and found that people who were initially rated as happy were less likely to have a heart problem.[77] Which of the following is the most appropriate conclusion based on this study?

(a) Happiness causes better heart health for all people.

(b) Happiness causes better heart health for Canadians.

(c) Happiness causes better heart health for the 1700 people in the study.

(d) Happier people in Canada are less likely to have heart problems.

(e) Happier people in the study were less likely to have heart problems.

**T4.7** The sales force for a publishing company is constantly on the road trying to sell books. As a result, each salesperson accumulates many travel-related expenses that he or she charges to a company-issued credit card. To prevent fraud, management hires an outside company to audit a sample of these expenses. For each salesperson, the auditor prints out the credit card statements for the entire year, randomly chooses one of the first 20 expenses to examine, and then examines every 20th expense from that point on. Which type of sampling method is the auditor using for each salesperson?

(a) Convenience sampling

(b) Simple random sampling

(c) Stratified random sampling

(d) Cluster sampling

(e) Systematic random sampling

**T4.8** *Bias* in a sampling method is

(a) any difference between the sample result and the truth about the population.

(b) the difference between the sample result and the truth about the population due to using chance to select a sample.

(c) any difference between the sample result and the truth about the population due to practical difficulties such as contacting the subjects selected.

(d) any difference between the sample result and the truth about the population that tends to occur in the same direction whenever you use this sampling method.

(e) racism or sexism on the part of those who take the sample.

**T4.9** You wonder if TV ads are more effective when they are longer or repeated more often or both. So you design an experiment. You prepare 30-second and 60-second ads for a camera. Your subjects all watch the same TV program, but you assign them at random to four groups. One group sees the 30-second ad once during the program; another sees it three times; the third group sees the 60-second ad once; and the last group sees the 60-second ad three times. You ask all subjects how likely they are to buy the camera. Which of the following best describes the design of this experiment?

(a) This is a randomized block design, but not a matched pairs design.

(b) This is a matched pairs design.

(c) This is a completely randomized design with one explanatory variable (factor).

(d) This is a completely randomized design with two explanatory variables (factors).

(e) This is a completely randomized design with four explanatory variables (factors).

**T4.10** Can texting make you healthier? Researchers randomly assigned 700 Australian adults to either receive usual health care or usual heath care plus automated text messages with positive messages, such as "Walking is cheap. It can be done almost anywhere. All you need is comfortable shoes and clothing." The group that received the text messages showed a statistically significant increase in physical activity.[78] What is the meaning of "statistically significant" in this context?

(a) The results of this study are very important.

(b) The results of this study should be generalized to all people.

(c) The difference in physical activity for the two groups is greater than 0.

(d) The difference in physical activity for the two groups is very large.

(e) The difference in physical activity for the two groups is larger than the difference that could be expected to happen by chance alone.

**T4.11** You want to know the opinions of American high school teachers on the issue of establishing a national proficiency test as a prerequisite for graduation from high school. You obtain a list of all high school teachers belonging to the National Education Association (the country's largest teachers' union) and mail a survey to a random sample of 2500 teachers. In all, 1347 of the teachers return the survey. Of those who responded, 32% say that they favor some kind of national proficiency test. Which of the following statements about this situation is true?

(a) Because random sampling was used, we can feel confident that the percent of all American high school teachers who would say they favor a national proficiency test is close to 32%.

(b) We cannot trust these results, because the survey was mailed. Only survey results from face-to-face interviews are considered valid.

(c) Because over half of those who were mailed the survey actually responded, we can feel fairly confident that the actual percent of all American high school teachers who would say they favor a national proficiency test is close to 32%.

(d) The results of this survey may be affected by undercoverage and nonresponse.

(e) The results of this survey cannot be trusted due to voluntary response bias.

**Section II: Free Response** *Show all your work. Indicate clearly the methods you use, because you will be graded on the correctness of your methods as well as on the accuracy and completeness of your results and explanations.*

**T4.12** Elephants sometimes damage trees in Africa. It turns out that elephants dislike bees. They recognize beehives in areas where they are common and avoid them. Can this information be used to keep elephants away from trees? Researchers want to design an experiment to answer these questions using 72 acacia trees and three treatments: active hives, empty hives, and no hives.[79]

(a) Identify the experimental units in this experiment.

(b) Explain why it is beneficial to include some trees that have no hives.

(c) Describe how the researchers could carry out a completely randomized design for this experiment. Include a description of how the treatments should be assigned.

**T4.13** Google and Gallup teamed up to survey a random sample of 1673 U.S. students in grades 7–12. One of the questions was "How confident are you that you could learn computer science if you wanted to?" Overall, 54% of students said they were very confident.[80]

(a) Identify the population and the sample.

(b) Explain why it was better to randomly select the students rather than putting the survey question on a website and inviting students to answer the question.

(c) Do you expect that the percent of all U.S. students in grades 7–12 who would say "very confident" is exactly 54%? Explain your answer.

(d) The report also broke the results down by gender. For this question, 62% of males and 48% of females said they were very confident. Which of the three estimates (54%, 62%, 48%) do you expect is closest to the value it is trying to estimate? Explain your answer.

**T4.14** Many people start their day with a jolt of caffeine from coffee or a soft drink. Most experts agree that people who consume large amounts of caffeine each day may suffer from physical withdrawal symptoms if they stop ingesting their usual amounts of caffeine. Researchers recruited 11 volunteers who were caffeine dependent and who were willing to take part in a caffeine withdrawal experiment. The experiment was conducted on two 2-day periods that occurred one week apart. During one of the 2-day periods, each subject was given a capsule containing the amount of caffeine normally ingested by that subject in one day. During the other study period, the subjects were given placebos. The order in which each subject received the two types of capsules was randomized. The subjects' diets were restricted during each of the study periods. At the end of each 2-day study period, subjects were evaluated using a tapping task in which they were instructed to press a button 200 times as fast as they could.[81]

(a) Identify the explanatory and response variables in this experiment.

(b) How was blocking used in the design of this experiment? What is the benefit of blocking in this context?

(c) Researchers randomized the order of the treatments to avoid confounding. Explain how confounding might occur if the researchers gave all subjects the placebo first and the caffeine second. In this context, what problem does confounding cause?

(d) Could this experiment have been carried out in a double-blind manner? Explain your answer.

# Chapter 4 Project Response Bias

In this project, your team will design and conduct an experiment to investigate the effects of response bias in surveys.[82] You may choose the topic for your surveys, but you must design your experiment so that it can answer at least one of the following questions.

- Does the wording of a question affect the response?
- Do the characteristics of the interviewer affect the response?
- Does anonymity change the responses to sensitive questions?
- Does manipulating the answer choices affect the response?
- Can revealing other peoples' answers to a question change the response?

1. Write a proposal describing the design of your experiment. Be sure to include the following items:

(a) Your chosen topic and which of the above questions you'll try to answer.

(b) A detailed description of how you will obtain your subjects (minimum of 50). Your plan must be practical!

(c) An explanation of the treatments in your experiment and how you will determine which subjects get which treatment.

(d) A clear explanation of how you will implement your design.

(e) Precautions you will take to collect data ethically.

Here are two examples of successful student experiments.

*"Make-Up," by Caryn S. and Trisha T.* (all questions asked to males):

i. "Do you find females who wear makeup attractive?" (Questioner wearing makeup: 75% answered "Yes.")

ii. "Do you find females who wear makeup attractive?" (Questioner not wearing makeup: 30% answered "Yes.")

*"Cartoons" by Sean W. and Brian H.:*

i. "Do you watch cartoons?" (90% answered "Yes.")

ii. "Do you *still* watch cartoons?" (60% answered "Yes.")

2. Once your teacher has approved your design, carry out the experiment. Record your data in a table.

3. Analyze your data. What conclusion do you draw? Provide appropriate graphical and numerical evidence to support your answer.

4. Prepare a report that includes the data you collected, your analysis from Step 3, and a discussion of any problems you encountered and how you dealt with them.

# Cumulative AP® Practice Test 1

**Section I: Multiple Choice** *Choose the best answer for Questions AP1.1–AP1.14.*

**AP1.1** You look at real estate ads for houses in Sarasota, Florida. Many houses have prices from $200,000 to $400,000. The few houses on the water, however, have prices up to $15 million. Which of the following statements best describes the distribution of home prices in Sarasota?

(a) The distribution is most likely skewed to the left, and the mean is greater than the median.

(b) The distribution is most likely skewed to the left, and the mean is less than the median.

(c) The distribution is roughly symmetric with a few high outliers, and the mean is approximately equal to the median.

(d) The distribution is most likely skewed to the right, and the mean is greater than the median.

(e) The distribution is most likely skewed to the right, and the mean is less than the median.

**AP1.2** A child is 40 inches tall, which places her at the 90th percentile of all children of similar age. The heights for children of this age form an approximately Normal distribution with a mean of 38 inches. Based on this information, what is the standard deviation of the heights of all children of this age?

(a) 0.20 inch

(b) 0.31 inch

(c) 0.65 inch

(d) 1.21 inches

(e) 1.56 inches

**AP1.3** A large set of test scores has mean 60 and standard deviation 18. If each score is doubled, and then 5 is subtracted from the result, the mean and standard deviation of the new scores are

(a) mean 115 and standard deviation 31.

(b) mean 115 and standard deviation 36.

(c) mean 120 and standard deviation 6.

(d) mean 120 and standard deviation 31.

(e) mean 120 and standard deviation 36.

**AP1.4** For a certain experiment, the available experimental units are eight rats, of which four are female (F1, F2, F3, F4) and four are male (M1, M2, M3, M4). There are to be four treatment groups, A, B, C, and D. If a randomized block design is used, with the experimental units blocked by gender, which of the following assignments of treatments is impossible?

(a) A → (F1, M1), B → (F2, M2), C → (F3, M3), D → (F4, M4)

(b) A → (F1, M2), B → (F2, M3), C → (F3, M4), D → (F4, M1)

(c) A → (F1, M2), B → (F3, F2), C → (F4, M1), D → (M3, M4)

(d) A → (F4, M1), B → (F2, M3), C → (F3, M2), D → (F1, M4)

(e) A → (F4, M1), B → (F1, M4), C → (F3, M2), D → (F2, M3)

**AP1.5** For a biology project, you measure the weight in grams (g) and the tail length in millimeters (mm) of a group of mice. The equation of the least-squares line for predicting tail length from weight is

$$\text{predicted tail length} = 20 + 3 \times \text{weight}$$

Which of the following is *not* correct?

(a) The slope is 3, which indicates that a mouse's predicted tail length should increase by about 3 mm for each additional gram of weight.

(b) The predicted tail length of a mouse that weighs 38 grams is 134 millimeters.

(c) By looking at the equation of the least-squares line, you can see that the correlation between weight and tail length is positive.

(d) If you had measured the tail length in centimeters instead of millimeters, the slope of the regression line would have been $3/10 = 0.3$.

(e) Mice that have a weight of 0 grams will have a tail of length 20 mm.

**AP1.6** The figure shows a Normal density curve. Which of the following gives the best estimates for the mean and standard deviation of this Normal distribution?

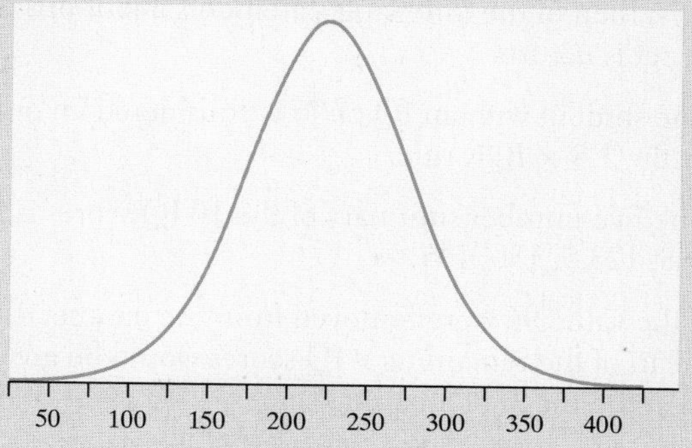


(a) $\mu = 200, \sigma = 50$

(b) $\mu = 200, \sigma = 25$

(c) $\mu = 225, \sigma = 50$

(d) $\mu = 225, \sigma = 25$

(e) $\mu = 225, \sigma = 275$

**AP1.7** The owner of a chain of supermarkets notices that there is a positive correlation between the sales of beer and the sales of ice cream over the course of the previous year. During seasons when sales of beer were above average, sales of ice cream also tended to be above average. Likewise, during seasons when sales of beer were below average, sales of ice cream also tended to be below average. Which of the following would be a valid conclusion from these facts?

(a) Sales records must be in error. There should be no association between beer and ice cream sales.

(b) Evidently, for a significant proportion of customers of these supermarkets, drinking beer causes a desire for ice cream or eating ice cream causes a thirst for beer.

(c) A scatterplot of monthly ice cream sales versus monthly beer sales would show that a straight line describes the pattern in the plot, but it would have to be a horizontal line.

(d) There is a clear negative association between beer sales and ice cream sales.

(e) The positive correlation is most likely a result of the variable temperature; that is, as temperatures increase, so do both beer sales and ice cream sales.

**AP1.8** Here are the IQ scores of 10 randomly chosen fifth-grade students:

| | | | | | | | | | |
|---|---|---|---|---|---|---|---|---|---|
| 145 | 139 | 126 | 122 | 125 | 130 | 96 | 110 | 118 | 118 |

Which of the following statements about this data set is *not* true?

(a) The student with an IQ of 96 is considered an outlier by the $1.5 \times IQR$ rule.

(b) The five-number summary of the 10 IQ scores is 96, 118, 123.5, 130, 145.

(c) If the value 96 were removed from the data set, the mean of the remaining 9 IQ scores would be greater than the mean of all 10 IQ scores.

(d) If the value 96 were removed from the data set, the standard deviation of the remaining 9 IQ scores would be less than the standard deviation of all 10 IQ scores.

(e) If the value 96 were removed from the data set, the *IQR* of the remaining 9 IQ scores would be less than the *IQR* of all 10 IQ scores.

**AP1.9** Before he goes to bed each night, Mr. Kleen pours dishwasher powder into his dishwasher and turns it on. Each morning, Mrs. Kleen weighs the box of dishwasher powder. From an examination of the data, she concludes that Mr. Kleen dispenses a rather consistent amount of powder each night. Which of the following statements is true?

I. There is a high positive correlation between the number of days that have passed since the box of dishwasher powder was opened and the amount of powder left in the box.

II. A scatterplot with days since purchase as the explanatory variable and total amount of dishwasher powder used as the response variable would display a strong positive association.

III. The correlation between the amount of powder left in the box and the amount of powder used should be very close to $-1$.

(a) I only

(b) II only

(c) III only

(d) II and III only

(e) I, II, and III

**AP1.10** The General Social Survey (GSS), conducted by the National Opinion Research Center at the University of Chicago, is a major source of data on social attitudes in the United States. Once each year, 1500 adults are interviewed in their homes all across the country. The subjects are asked their opinions about sex and marriage; attitudes toward women, welfare, foreign policy; and many other issues. The GSS begins by selecting a sample of counties from the 3000 counties in the country. The counties are divided into urban, rural, and suburban; a separate sample of counties is chosen at random from each group. This is a

(a) simple random sample.

(b) systematic random sample.

(c) cluster sample.

(d) stratified random sample.

(e) voluntary response sample.

**AP1.11** You are planning an experiment to determine the effect of the brand of gasoline and the weight of a car on gas mileage measured in miles per gallon. You will use a single test car, adding weights so that its total weight is 3000, 3500, or 4000 pounds. The car will drive on a test track at each weight using each of Amoco, Marathon, and Speedway gasoline. Which is the best way to organize the study?

(a) Start with 3000 pounds and Amoco and run the car on the test track. Then do 3500 and 4000 pounds. Change to Marathon and go through the three weights in order. Then change to Speedway and do the three weights in order once more.

(b) Start with 3000 pounds and Amoco and run the car on the test track. Then change to Marathon and then to Speedway without changing the weight. Then add weights to get 3500 pounds and go through the three gasolines in the same order. Then change to 4000 pounds and do the three gasolines in order again.

(c) Choose a gasoline at random, and run the car with this gasoline at 3000, 3500, and 4000 pounds in order. Choose one of the two remaining gasolines at random and again run the car at 3000, then 3500, then 4000 pounds. Do the same with the last gasoline.

(d) There are nine combinations of weight and gasoline. Run the car several times using each of these combinations. Make all these runs in random order.

(e) Randomly select an amount of weight and a brand of gasoline, and run the car on the test track. Repeat this process a total of 30 times.

**AP1.12** A linear regression was performed using the following five data points: A(2, 22), B(10, 4), C(6, 14), D(14, 2), E(18, −4). The residual for which of the five points has the largest absolute value?

(a) A

(b) B

(c) C

(d) D

(e) E

**AP1.13** The frequency table summarizes the distribution of time that 140 patients at the emergency room of a small-city hospital waited to receive medical attention during the last month.

| Waiting time | Frequency |
|---|---|
| Less than 10 minutes | 5 |
| At least 10 but less than 20 minutes | 24 |
| At least 20 but less than 30 minutes | 45 |
| At least 30 but less than 40 minutes | 38 |
| At least 40 but less than 50 minutes | 19 |
| At least 50 but less than 60 minutes | 7 |
| At least 60 but less than 70 minutes | 2 |

Which of the following represents possible values for the median and *IQR* of waiting times for the emergency room last month?

(a) median = 27 minutes and *IQR* = 15 minutes
(b) median = 28 minutes and *IQR* = 25 minutes
(c) median = 31 minutes and *IQR* = 35 minutes
(d) median = 35 minutes and *IQR* = 45 minutes
(e) median = 45 minutes and *IQR* = 55 minutes

**AP1.14** Boxplots of two data sets are shown.

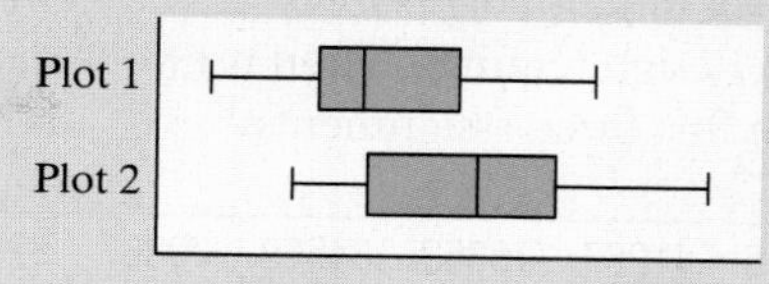


Based on the boxplots, which of the following is true?

(a) The range of both plots is about the same.
(b) The means of both plots are approximately equal.
(c) Plot 2 contains more data points than Plot 1.
(d) The medians are approximately equal.
(e) Plot 1 is more symmetric than Plot 2.

**Section II: Free Response** *Show all your work. Indicate clearly the methods you use, because you will be graded on the correctness of your methods as well as on the accuracy and completeness of your results and explanations.*

**AP1.15** The manufacturer of exercise machines for fitness centers has designed two new elliptical machines that are meant to increase cardiovascular fitness. The two machines are being tested on 30 volunteers at a fitness center near the company's headquarters. The volunteers are randomly assigned to one of the machines and use it daily for two months. A measure of cardiovascular fitness is administered at the start of the experiment and again at the end. The following stemplot contains the differences (After − Before) in the two scores for the two machines. Note that greater differences indicate larger gains in fitness.

| Machine A | | Machine B |
|---|---|---|
| | 0 | 2 |
| 54 | 1 | 0 |
| 876320 | 2 | 159 |
| 97411 | 3 | 2489 |
| 61 | 4 | 257 |
| | 5 | 359 |

Key: 2 | 1 represents a difference (After − Before) of 21 in fitness scores.

(a) Write a few sentences comparing the distributions of cardiovascular fitness gains from the two elliptical machines.

(b) Which machine should be chosen if the company wants to advertise it as achieving the highest overall gain in cardiovascular fitness? Explain your reasoning.

(c) Which machine should be chosen if the company wants to advertise it as achieving the most consistent gain in cardiovascular fitness? Explain your reasoning.

(d) Give one reason why the advertising claims of the company (the scope of inference) for this experiment would be limited. Explain how the company could broaden that scope of inference.

**AP1.16** Those who advocate for monetary incentives in a work environment claim that this type of incentive has the greatest appeal because it allows the winners to do what they want with their winnings. Those in favor of tangible incentives argue that money lacks the emotional appeal of, say, a weekend for two at a romantic country inn or elegant hotel, or a weeklong trip to Europe.

A few years ago a national tire company, in an effort to improve sales of a new line of tires, decided to test which method—offering cash incentives or offering non-cash prizes such as vacations—was more successful in increasing sales. The company had 60 retail sales districts of various sizes across the country and data on the previous sales volume for each district.

(a) Describe a completely randomized design using the 60 retail sales districts that would help answer this question.

(b) Explain how you would use the following excerpt from the table of random digits to do the random assignment that your design requires. Then use your method to make the first three assignments.

| | | | | | | | |
|---|---|---|---|---|---|---|---|
| 07511 | 88915 | 41267 | 16853 | 84569 | 79367 | 32337 | 03316 |
| 81486 | 69487 | 60513 | 09297 | 00412 | 71238 | 27649 | 39950 |

(c) One of the company's officers suggested that it would be better to use a matched pairs design instead of a completely randomized design. Explain how you would change your design to accomplish this.

**AP1.17** In retail stores, there is a lot of competition for shelf space. There are national brands for most products, and many stores carry their own line of in-house brands, too. Because shelf space is not infinite, the question is how many linear feet to allocate to each product and which shelf (top, bottom, or somewhere in the middle) to put it on. The middle shelf is the most popular and lucrative, because many shoppers, if undecided, will simply pick the product that is at eye level.

A local store that sells many upscale goods is trying to determine how much shelf space to allocate to its own brand of men's personal-grooming products. The middle shelf space is randomly varied between 3 and 6 linear feet over the next 12 weeks, and weekly sales revenue (in dollars) from the store's brand of personal-grooming products for men is recorded. Here is some computer output from the study, along with a scatterplot:

| Predictor | Coef | SE Coef | T | P |
|---|---|---|---|---|
| Constant | 317.940 | 31.32 | 10.15 | 0.000 |
| Shelf length | 152.680 | 6.445 | 23.69 | 0.000 |
| S = 22.9212 | R-Sq = 98.2% | R-Sq(adj) = 98.1% | | |

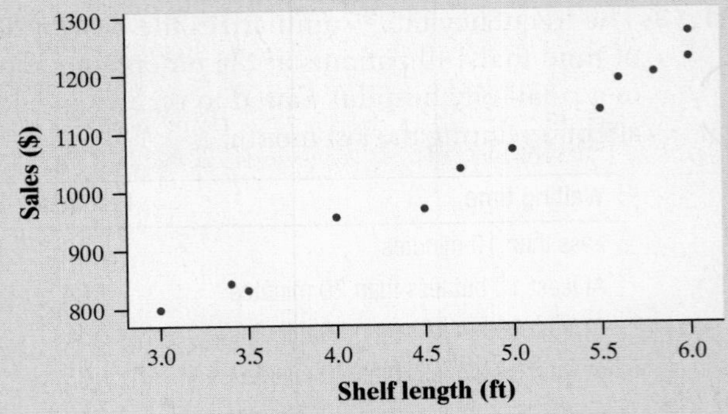


(a) Describe the relationship between shelf length and sales.

(b) Write the equation of the least-squares regression line. Be sure to define any variables you use.

(c) If the store manager were to decide to allocate 5 linear feet of shelf space to the store's brand of men's grooming products, what is the best estimate of the weekly sales revenue?

(d) Interpret the value of $s$.

(e) Identify and interpret the coefficient of determination.

**AP1.18** The manager of the store in the preceding exercise calculated the residual for each point in the scatterplot and made a dotplot of the residuals. The distribution of residuals is roughly Normal with a mean of \$0 and standard deviation of \$22.92.

(a) What percent of the actual sales amounts do you expect to be within \$5 of their expected sales amount?

(b) The middle 95% of residuals should be between which two values? Use this information to give an interval of plausible values for the weekly sales revenue if 5 linear feet are allocated to the store's brand of men's grooming products.

# UNIT 4
## Probability, Random Variables, and Probability Distributions

# Chapter 5

# Probability

Stefano Paterna/Alamy

# INTRODUCTION

Chance is all around us. You and your friend play rock-paper-scissors to determine who gets the last slice of pizza. A coin toss decides which team gets to receive the ball first in a football game. Many adults regularly play the lottery, hoping to win a big jackpot with a few lucky numbers. Others head to casinos or racetracks, hoping that some combination of luck and skill will pay off. People young and old play games of chance involving cards or dice or spinners. The traits that children inherit—hair and eye color, blood type, handedness, dimples, whether they can roll their tongues—are determined by the chance involved in which genes their parents pass along.

The mathematics of chance behavior is called *probability*. Probability is the topic of this chapter. Here is an activity that gives you some idea of what lies ahead.

## ACTIVITY The "1 in 6 wins" game

In this activity, you and your classmates will use simulation to test whether a company's claim is believable.

As a special promotion for its 20-ounce bottles of soda, a soft drink company printed a message on the inside of each bottle cap. Some of the caps said, "Please try again!" while others said, "You're a winner!" The company advertised the promotion with the slogan "1 in 6 wins a prize." The prize is a free 20-ounce bottle of soda.

Grayson's statistics class wonders if the company's claim holds true for the bottles at a nearby convenience store. To find out, all 30 students in the class go to the store and each student buys one 20-ounce bottle of the soda. Two of them get caps that say, "You're a winner!" Does this result give convincing evidence that the company's 1-in-6 claim is inaccurate?

For now, let's assume that the company is telling the truth, and that every 20-ounce bottle of soda it fills has a 1-in-6 chance of getting a cap that says, "You're a winner!" We can model the status of an individual bottle with a six-sided die: let 1 through 5 represent "Please try again!" and 6 represent "You're a winner!"

1. Roll your die 30 times to imitate the process of the students in Grayson's statistics class buying their sodas. How many of them won a prize?
2. Your teacher will draw and label axes for a class dotplot. Plot on the graph the number of prize winners you got in Step 1.
3. Repeat Steps 1 and 2, if needed, to get a total of at least 40 trials of the simulation for your class.
4. Discuss the results with your classmates. What percent of the time did the simulation yield two or fewer prize winners in a class of 30 students just by chance? Does it seem plausible (believable) that the company is telling the truth, but that the class just got unlucky? Or is there convincing evidence that the 1-in-6 claim is wrong? Explain your reasoning.

NIMA Stock/Alamy

As the activity shows, *simulation* is a powerful method for modeling random behavior. Section 5.1 begins by examining the idea of probability and then illustrates how simulation can be used to estimate probabilities. In Sections 5.2 and 5.3, we develop the basic rules and techniques of probability.

Probability calculations are the basis for inference. When we produce data by random sampling or randomized comparative experiments, the laws of probability answer the question "What would happen if we repeated the random sampling or random assignment process many times?" Many of the examples, exercises, and activities in this chapter focus on the connection between probability and inference.

# SECTION 5.1 Randomness, Probability, and Simulation

**LEARNING TARGETS** *By the end of the section, you should be able to:*

- Interpret probability as a long-run relative frequency.
- Use simulation to model a random process.

Imagine tossing a coin 10 times. How likely are you to get a run of 3 or more consecutive heads? An airline knows that a certain percent of customers who purchase tickets will not show up for a flight. If the airline overbooks a particular flight, what are the chances that they'll have enough seats for the passengers who show up? A couple plans to have children until they have at least one boy and one girl. How many children should they expect to have? To answer these questions, you need a better understanding of how random behavior operates.

## The Idea of Probability

In football, a coin toss helps determine which team gets the ball first. Why do the rules of football require a coin toss? Because tossing a coin seems a "fair" way to decide. What does that mean exactly? The following activity should help shed some light on this question.

### ACTIVITY What is probability?

If you toss a fair coin, what's the probability that it shows heads? It's 1/2, or 0.5, right? But what does probability 1/2 really mean? In this activity, you will investigate by flipping a coin several times.

1. Toss your coin once. Record whether you get heads or tails.
2. Toss your coin a second time. Record whether you get heads or tails. What proportion of your first two tosses is heads?

3. Toss your coin 8 more times so that you have 10 tosses in all. Record whether you get heads or tails on each toss in a table like the one that follows.
4. Calculate the overall proportion of heads after each toss and record these values in the bottom row of the table. For instance, suppose you got tails on the first toss and heads on the second toss. Then your overall proportion of heads would be $0/1 = 0.00$ after the first toss and $1/2 = 0.50$ after the second toss.

| Toss | 1 | 2 | 3 | 4 | 5 | 6 | 7 | 8 | 9 | 10 |
|---|---|---|---|---|---|---|---|---|---|---|
| Result (H or T) | | | | | | | | | | |
| Proportion of heads | | | | | | | | | | |

5. Let's use technology to speed things up. Launch the *Idea of Probability* applet at highschool.bfwpub.com/updatedtps6e. Set the number of tosses to 10 and click toss. What proportion of the tosses were heads? Click "Reset" and toss the coin 10 more times. What proportion of heads did you get this time? Repeat this process several more times. What do you notice?

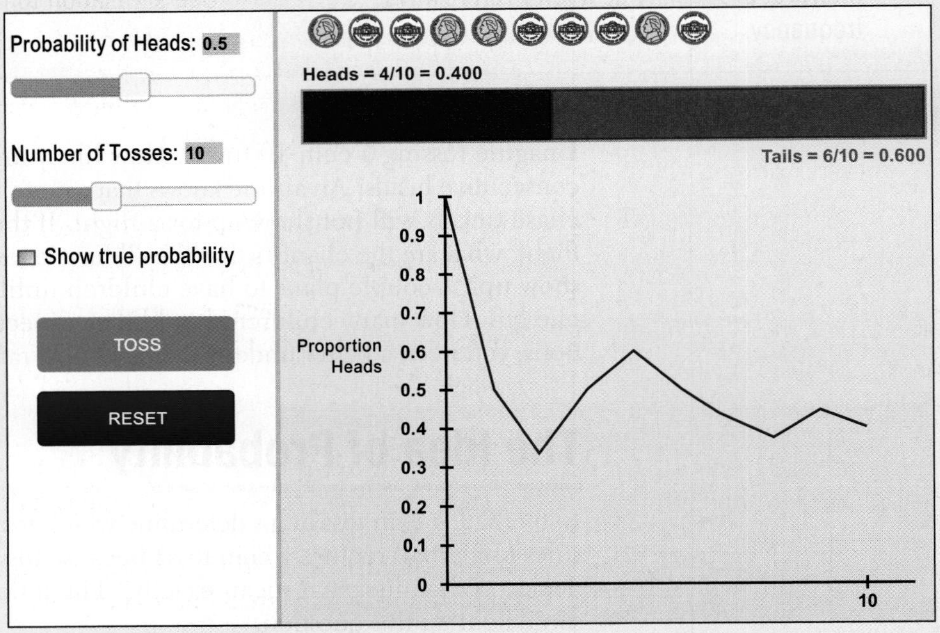


6. What if you toss the coin 100 times? Reset the applet and have it do 100 tosses. Is the proportion of heads exactly equal to 0.5? Close to 0.5?
7. Keep on tossing without hitting "Reset." What happens to the proportion of heads?
8. As a class, discuss what the following statement means: "If you toss a fair coin, the probability of heads is 0.5."
9. If you toss a coin, it can land heads or tails. If you "toss" a thumbtack, it can land with the point sticking up or with the point down. Does that mean the probability of a tossed thumbtack landing point up is 0.5? How can you find out? Discuss with your classmates.

Figure 5.1 shows some results from the preceding activity. The proportion of tosses that land heads varies from 0.30 to 1.00 in the first 10 tosses. As we make more and more tosses, however, the proportion of heads gets closer to 0.5 and stays there.

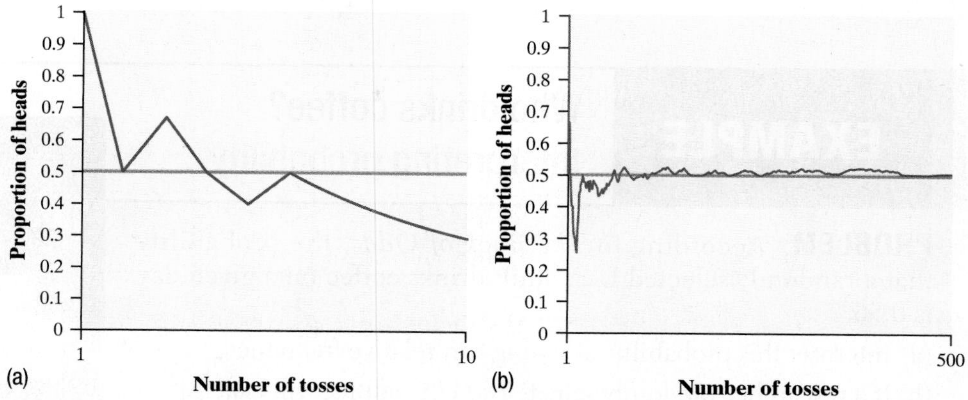


**FIGURE 5.1** (a) The proportion of heads in the first 10 tosses of a coin. (b) The proportion of heads in the first 500 tosses of a coin.

When we watch coin tosses or the results of random sampling and random assignment closely, a remarkable fact emerges: *A* ***random process*** *is unpredictable in the short run but has a regular and predictable pattern in the long run.* This is the basis for the idea of **probability**.

> **DEFINITION Random process, Probability**
>
> A **random process** generates outcomes that are determined purely by chance.
>
> The **probability** of any outcome of a random process is a number between 0 and 1 that describes the proportion of times the outcome would occur in a very long series of trials.

A *trial* is one repetition of a random process.

Outcomes that never occur have probability 0. An outcome that always occurs has probability 1. An outcome that happens half the time in a very long series of trials has probability 0.5.

Some people distinguish between tossing a coin several times to estimate the probability of a head (*empirical probability*) and using a computer applet to simulate this random process.

The fact that the proportion of heads in many tosses eventually closes in on 0.5 is guaranteed by the **law of large numbers**. You can see this in Figure 5.1(b). The horizontal line represents the probability, and the proportion of heads in the simulation approaches this value as the number of trials becomes large.

> **DEFINITION Law of large numbers**
>
> The **law of large numbers** says that if we observe more and more trials of any random process, the proportion of times that a specific outcome occurs approaches its probability.

Probability gives us a language to describe the long-term regularity of a random process that is guaranteed by the law of large numbers. The outcome of a coin toss and the sex of the next baby born in a local hospital are both random. So is the result of a random sample or a random assignment. Even life insurance is based on the fact that deaths occur at random among many individuals. Because men are more likely to die at a younger age than women, insurance companies sometimes charge a man up to 3 times more for a life insurance policy than a woman of the same age.

## EXAMPLE

### Who drinks coffee?
### Interpreting probability

A. and I. Kruk/Shutterstock.com

**PROBLEM:** According to *The Book of Odds*, the probability that a randomly selected U.S. adult drinks coffee on a given day is 0.56.

(a) Interpret this probability as a long-run relative frequency.

(b) If a researcher randomly selects 100 U.S. adults, will exactly 56 of them drink coffee that day? Explain your answer.

**SOLUTION:**

(a) If you take a very large random sample of U.S. adults, about 56% of them will drink coffee that day.

(b) Probably not. With only 100 randomly selected adults, the number who drink coffee that day may not be very close to 56.

The random process consists of randomly selecting a U.S. adult and recording whether or not the person drinks coffee that day.

Probability describes what happens in many, many repetitions (way more than 100) of a random process.

FOR PRACTICE, TRY EXERCISE 1

Life insurance companies, casinos, and others who make important decisions based on probability rely on the long-run predictability of random behavior.

**UNDERSTANDING RANDOMNESS** The idea of probability seems straightforward. It answers the question "What would happen if we did this many times?" Understanding random behavior is important for making decisions, especially when our data collection process includes random sampling or random assignment. But understanding randomness isn't always easy, as the following activity illustrates.

## ACTIVITY

### Investigating randomness

In this activity, you and your classmates will test your ability to imitate a random process.

1. Pretend that you are flipping a fair coin. Without actually flipping a coin, *imagine* the first toss. Write down the result you see in your mind, heads (H) or tails (T).
2. Imagine a second coin flip. Write down the result.

Eyebyte/Alamy

3. Keep doing this until you have recorded the results of 50 imaginary flips. Write your results in groups of 5 to make them easier to read, like this: HTHTH TTHHT, and so on.
4. A *run* is when the same result occurs in consecutive outcomes. In the example in Step 3, there is a run of two tails followed by a run of two heads in the first 10 coin flips. Read through your 50 imagined coin flips and find the longest run.
5. Your teacher will draw and label a number line for a class dotplot. Plot on the graph the length of the longest run you got in Step 4.
6. Use an actual coin, Table D, or technology to generate a similar list of 50 coin flips. Find the longest run that you have.
7. Your teacher will draw and label a number line with the same scale immediately above or below the one in Step 5. Plot on the new dotplot the length of the longest run you got in Step 6.
8. Compare the distributions of longest run from imagined tosses and random tosses. What do you notice?

The idea of probability is that randomness is predictable in the long run. Unfortunately, our intuition about random behavior tries to tell us that randomness should also be predictable in the short run. When it isn't, we look for some explanation other than chance variation.

Suppose you toss a coin 6 times and get TTTTTT. Some people think that the next toss must be more likely to give a head. It's true that in the long run, heads will appear half the time. What is a myth is that future outcomes must make up for an imbalance like six straight tails.

Some people use the phrase *law of averages* to refer to the misguided belief that the results of a random process have to even out in the *short run*.

Coins and dice have no memories. A coin doesn't know that the first 6 outcomes were tails, and it can't try to get a head on the next toss to even things out. Of course, things do even out in the long run. That's the law of large numbers in action. After 10,000 tosses, the results of the first six tosses don't matter. They are overwhelmed by the results of the next 9994 tosses.

When asked to predict the sex—boy (B) or girl (G)—of the next seven babies born in a local hospital, most people will guess something like B-G-B-G-B-G-G. Few people would say G-G-G-B-B-B-G because this sequence of outcomes doesn't "look random." In fact, these two sequences of births are equally likely. "Runs" consisting of several of the same outcome in a row are surprisingly common in a random process. Many students are not aware of this fact when they imagine a sequence of 50 coin tosses in the "Investigating randomness" activity!

Is there such a thing as a "hot hand" in basketball? Belief that runs must result from something other than "just chance" influences behavior. If a basketball player makes several consecutive shots, both the fans and her teammates believe that she has a "hot hand" and is more likely to make the next shot. Several early studies of the hot hand theory showed that runs of baskets made or missed are no more frequent in basketball than would be expected if the result of each shot is unrelated to the outcomes of the player's previous shots. If a player makes half her shots in the long run, her made shots and misses behave just like tosses of a coin—which means that runs of makes and misses are more common than our intuition expects.[1]

Two more recent studies provide some evidence that there is a small hot hand effect for basketball players. These studies also suggest that "hot" shooters take riskier shots, which then masks the hot hand effect.[2]

## CHECK YOUR UNDERSTANDING

1. Pedro drives the same route to work on Monday through Friday. His route includes one traffic light. According to the local traffic department, there is a 55% probability that the light will be red when Pedro reaches the light. Interpret the probability.
2. Probability is a measure of how likely an outcome is to occur. Match one of the probabilities that follow with each statement. Be prepared to defend your answer.

   0 0.001 0.3 0.6 0.99 1

   (a) This outcome is impossible. It can never occur.

   (b) This outcome is certain. It will occur on every trial.

   (c) This outcome is very unlikely, but it will occur once in a while in a long sequence of trials.

   (d) This outcome will occur more often than not.
3. A husband and wife decide to have children until they have at least one child of each sex. The couple has had seven girls in a row. Their doctor assures them that they are much more likely to have a boy next. Explain why the doctor is wrong.

# Simulation

We can model random behavior and estimate probabilities with a **simulation**.

**DEFINITION Simulation**

A **simulation** imitates a random process in such a way that simulated outcomes are consistent with real-world outcomes.

You already have some experience with simulations. In the "Hiring discrimination—it just won't fly!" activity in Chapter 1 (page 6), you drew beads or slips of paper to imitate a random lottery to choose which pilots would become captains. The "Analyzing the caffeine experiment" activity in Chapter 4 (page 301) asked you to shuffle and deal piles of cards to mimic the random assignment of subjects to treatments. The "1 in 6 wins" game that opened this chapter had you roll a die several times to simulate buying 20-ounce sodas and looking under the cap.

The goal in each of these cases was to use simulation to answer a question of interest about some random process. Different random "devices" were used to perform the simulations—beads, slips of paper, cards, or dice. But the same basic strategy was followed each time.

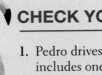

## THE SIMULATION PROCESS

- Describe how to use a random process to perform one trial of the simulation. Tell what you will record at the end of each trial.
- Perform many trials of the simulation.
- Use the results of your simulation to answer the question of interest.

For the 1-in-6 wins game, we wanted to estimate the probability of getting two or fewer prize winners in a class of 30 students if the company's 1-in-6 wins claim is true.

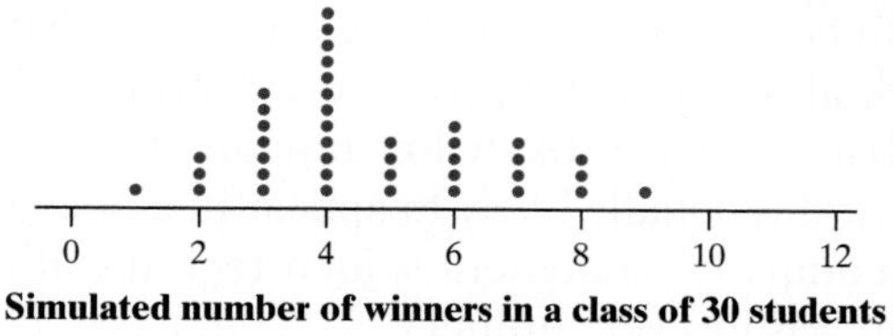


- We rolled a six-sided die 30 times to determine the outcome for each person's bottle of soda: 6 = wins a prize, 1 to 5 = no prize, and recorded the number of winners. The dotplot shows the number of winners in 40 trials of this simulation.
  - In 4 of the 40 trials, two or fewer of the students won a prize. So our estimate of the probability is $4/40 = 0.10 = 10\%$. According to these results, getting 2 winners isn't very likely, but it isn't unusual enough to conclude that the company is lying.

## EXAMPLE NASCAR cards and cereal boxes
### Performing simulations

Daniel Shirey/Getty Images

**PROBLEM:** In an attempt to increase sales, a breakfast cereal company decides to offer a NASCAR promotion. Each box of cereal will contain a collectible card featuring one of the following NASCAR drivers: Joey Lagano, Kevin Harvick, Chase Elliott, Danica Patrick, or Jimmie Johnson. The company claims that each of the 5 cards is equally likely to appear in any box of cereal. A NASCAR fan decides to keep buying boxes of the cereal until she has all 5 drivers' cards. She is surprised when it takes her 23 boxes to get the full set of cards. Does this outcome provide convincing evidence that the 5 cards are not equally likely? To help answer this question, we want to perform a simulation to estimate the probability that it will take 23 or more boxes to get a full set of 5 NASCAR collectible cards.

(a) Describe how to use a random number generator to perform one trial of the simulation.

The dotplot shows the number of cereal boxes it took to get all 5 drivers' cards in 50 trials.

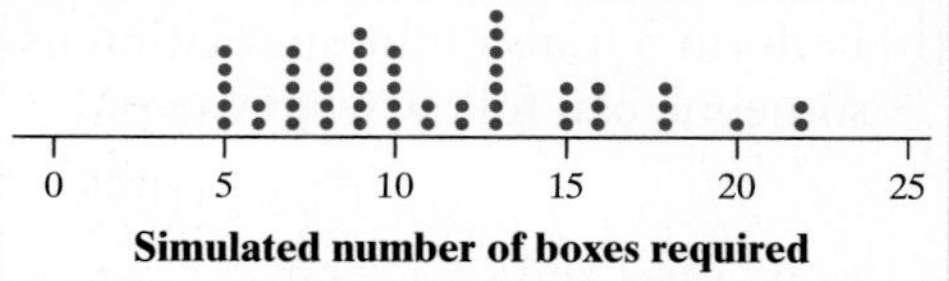


(b) Explain what the dot at 20 represents.

(c) Use the results of the simulation to estimate the probability that it will take 23 or more boxes to get a full set of cards. Does this outcome provide convincing evidence that the 5 cards are not equally likely?

**SOLUTION:**

(a) Let 1 = Joey Lagano, 2 = Kevin Harvick, 3 = Chase Elliott, 4 = Danica Patrick, and 5 = Jimmie Johnson. Generate a random integer from 1 to 5 to simulate buying one box of cereal and looking at which card is inside. Keep generating random integers until all five labels from 1 to 5 appear. Record the number of boxes it takes to get all 5 cards.

> Describe how to use a random process to perform one trial of a simulation. Tell what you will record at the end of each trial.

(b) One trial where it took 20 boxes to get all 5 drivers' cards.

(c) Probability ≈ 0/50 = 0, so there's about a 0% chance it would take 23 or more boxes to get a full set. Because it is so unlikely that it would take 23 or more boxes to get a full set when the drivers' cards are equally likely, this result provides convincing evidence that the 5 NASCAR drivers' cards are not equally likely to appear in each box of cereal.

FOR PRACTICE, TRY EXERCISE 9

It took our NASCAR fan 23 boxes to complete the set of 5 cards. Does that mean the company didn't tell the truth about how the cards were distributed? Not necessarily. Our simulation says that it's very unlikely for someone to have to buy 23 boxes to get a full set *if* each card is equally likely to appear in a box of cereal. The evidence suggests that the company's statement is incorrect. It's still possible, however, that the NASCAR fan was just very unlucky.

Here's one more example that shows the simulation process in action.

## EXAMPLE

### Golden ticket parking lottery
### Performing simulations

Ann Heath

**PROBLEM:** At a local high school, 95 students have permission to park on campus. Each month, the student council holds a "golden ticket parking lottery" at a school assembly. The two lucky winners are given reserved parking spots next to the school's main entrance. Last month, the winning tickets were drawn by a student council member from the AP® Statistics class. When both golden tickets went to members of that same class, some people thought the lottery had been rigged. There are 28 students in the AP® Statistics class, all of whom are eligible to park on campus. We want to perform a simulation to estimate the probability that a fair lottery would result in two winners from the AP® Statistics class.

(a) Describe how you would use a table of random digits to carry out this simulation.

(b) Perform 3 trials of the simulation using the random digits given. Make your procedure clear so that someone can follow what you did.

70708 41098 55181 94904 43563 56934 48394 51719

(c) In 9 of the 100 trials of the simulation, both golden tickets were won by members of the AP® Statistics class. Do these results give convincing evidence that the lottery was not carried out fairly? Explain your reasoning.

**SOLUTION:**

(a) Label the students in the AP® Statistics class from 01 to 28, and label the remaining students from 29 to 95. Numbers from 96 to 00 will be skipped. Moving left to right across a row, look at pairs of digits until we come across two *different* labels from 01 to 95. The two students with these labels win the prime parking spaces. Record whether or not both winners come from the AP® Statistics class. Perform many simulated lotteries. See what percent of the time both winners come from this statistics class.

Describe how to use a random process to perform one trial of the simulation. Tell what you will record at the end of each trial.

(b)

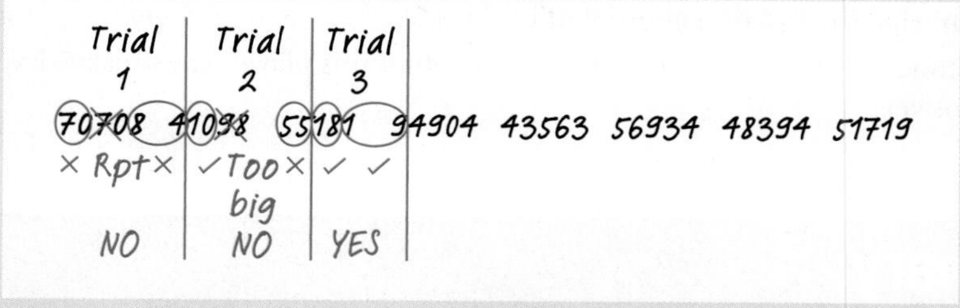


Perform many trials.

There was one trial out of the first 3 in which both golden parking tickets went to members of the AP® Statistics class.

(c) No; there's about a 9% chance of getting both winners from the AP® Statistics class in a fair lottery. Because this probability isn't very small, we don't have convincing evidence that the lottery was unfair. Outcomes like this could occur by chance alone in a fair lottery.

Use the results of your simulation to answer the question of interest.

Does that mean the lottery *was* conducted fairly? Not necessarily.

**FOR PRACTICE, TRY EXERCISE 11**

**AP® EXAM TIP**

On the AP® Statistics exam, you may be asked to describe how to perform a simulation using rows of random digits. If so, provide a clear enough description of your process for the reader to get the same results from *only* your written explanation. Remember that every label needs to be the same length. In the golden ticket lottery example, the labels should be 01 to 95 (all two digits), not 1 to 95. When sampling without replacement, be sure to mention that repeated numbers should be ignored.

In the golden ticket lottery example, we ignored repeated numbers from 01 to 95 within a given trial. That's because the random process involved sampling students *without* replacement. In the NASCAR example, we allowed repeated numbers from 1 to 5 in a given trial. That's because we were selecting a small number of cards from a very large population of cards in thousands of cereal boxes. So the probability of getting, say, a Danica Patrick card in the next box of cereal was still very close to 1/5 even if we had already selected a Danica Patrick card.

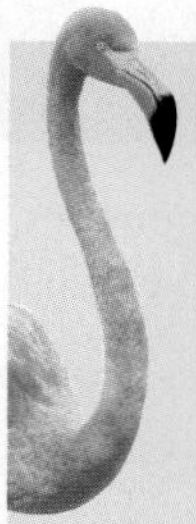

## CHECK YOUR UNDERSTANDING

A basketball announcer suggests that a certain player is a streaky shooter. That is, the announcer believes that if the player makes a shot, the player is more likely to make the next shot. As evidence, the announcer points to a recent game where the player took 30 shots and had a streak of 10 made shots in a row. Is this convincing evidence of streaky shooting by the player? Assume that this player makes 50% of his shots and that the results of a shot don't depend on previous shots.

1. Describe how you would carry out a simulation to estimate the probability that a 50% shooter who takes 30 shots in a game would have a streak of 10 or more made shots.

The dotplot displays the results of 50 simulated games in which this player took 30 shots.

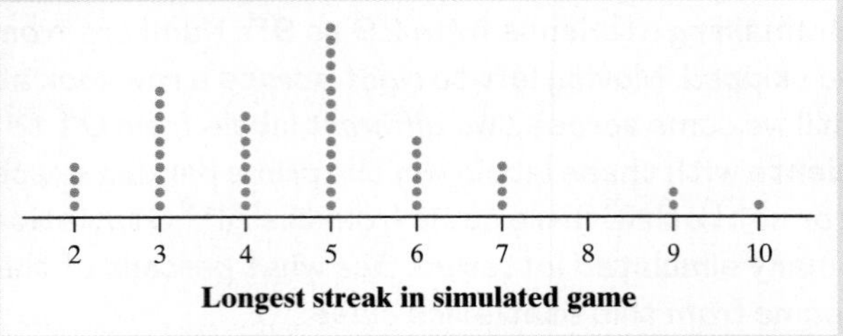


2. Explain what the two dots above 9 indicate.
3. What conclusion would you draw about whether this player was streaky? Explain your answer.

## Section 5.1 Summary

- A **random process** generates outcomes that are determined purely by chance. Random behavior is unpredictable in the short run but has a regular and predictable pattern in the long run.
- The long-run relative frequency of an outcome after many trials of a random process is its **probability.** A probability is a number between 0 (never occurs) and 1 (always occurs).
- The **law of large numbers** says that in many trials of the same random process, the proportion of times that a particular outcome occurs will approach its probability.
- **Simulation** can be used to imitate a random process and to estimate probabilities. To perform a simulation:
  - Describe how to use a random process to perform one trial of the simulation. Tell what you will record at the end of each trial.
  - Perform many trials of the simulation.
  - Use the results of your simulation to answer the question of interest.

## Section 5.1 Exercises

1. **Another commercial** If Aaron tunes into his favorite radio station at a randomly selected time, there is a 0.20 probability that a commercial will be playing.

pg 330

(a) Interpret this probability as a long-run relative frequency.

(b) If Aaron tunes into this station at 5 randomly selected times, will there be exactly 1 time when a commercial is playing? Explain your answer.

2. **Genetics** There are many married couples in which the husband and wife both carry a gene for cystic fibrosis but don't have the disease themselves. Suppose we select one of these couples at random. According to the laws of genetics, the probability that their first child will develop cystic fibrosis is 0.25.

(a) Interpret this probability as a long-run relative frequency.

(b) If researchers randomly select 4 such couples, is one of these couples guaranteed to have a first child who develops cystic fibrosis? Explain your answer.

3. **Mammograms** Many women choose to have annual mammograms to screen for breast cancer after age 40. A mammogram isn't foolproof. Sometimes the test suggests that a woman has breast cancer when she really doesn't (a "false positive"). Other times the test says that

a woman doesn't have breast cancer when she actually does (a "false negative"). Suppose the false negative rate for a mammogram is 0.10.

(a) Explain what this probability means.

(b) Which is a more serious error in this case: a false positive or a false negative? Justify your answer.

4. **Liar, liar!** Sometimes police use a lie detector test to help determine whether a suspect is telling the truth. A lie detector test isn't foolproof—sometimes it suggests that a person is lying when he or she is actually telling the truth (a "false positive"). Other times, the test says that the suspect is being truthful when he or she is actually lying (a "false negative"). For one brand of lie detector, the probability of a false positive is 0.08.

(a) Explain what this probability means.

(b) Which is a more serious error in this case: a false positive or a false negative? Justify your answer.

5. **Three pointers** The figure shows the results of a basketball player attempting many 3-point shots. Explain what this graph tells you about random behavior in the short run and long run.

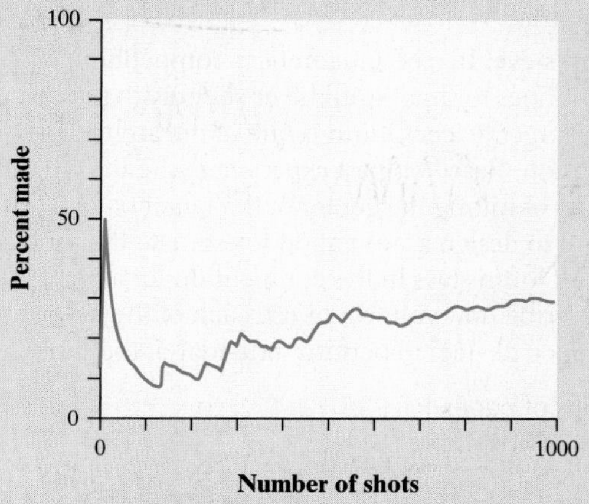


6. **Keep on tossing** The figure shows the results of two different sets of 5000 coin tosses. Explain what this graph tells you about random behavior in the short run and the long run.

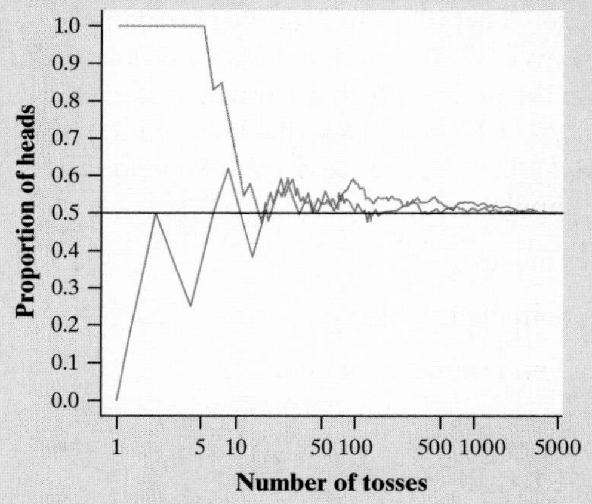


7. **An unenlightened gambler**

(a) A gambler knows that red and black are equally likely to occur on each spin of a roulette wheel. He observes that 5 consecutive reds have occurred and bets heavily on black at the next spin. Asked why, he explains that "black is due." Explain to the gambler what is wrong with this reasoning.

(b) After hearing you explain why red and black are still equally likely after 5 reds on the roulette wheel, the gambler moves to a card game. He is dealt 5 straight red cards from a standard deck with 26 red cards and 26 black cards. He remembers what you said and assumes that the next card dealt in the same hand is equally likely to be red or black. Explain to the gambler what is wrong with this reasoning.

8. **Due for a hit** A very good professional baseball player gets a hit about 35% of the time over an entire season. After the player failed to hit safely in six straight at-bats, a TV commentator said, "He is due for a hit." Explain why the commentator is wrong.

9. **Will Luke pass the quiz?** Luke's teacher has assigned each student in his class an online quiz, which is made up of 10 multiple-choice questions with 4 options each. Luke hasn't been paying attention in class and has to guess on each question. However, his teacher allows each student to take the quiz three times and will record the highest of the three scores. A passing score is 6 or more correct out of 10. We want to perform a simulation to estimate the score that Luke will earn on the quiz if he guesses at random on all the questions.

pg 333

(a) Describe how to use a random number generator to perform one trial of the simulation.

The dotplot shows Luke's simulated quiz score in 50 trials of the simulation.

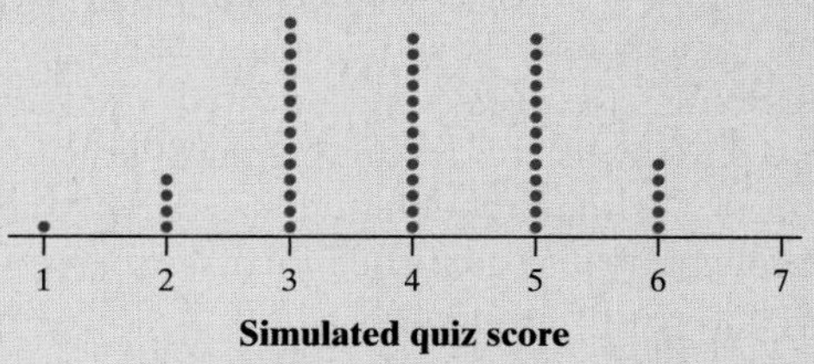


(b) Explain what the dot at 1 represents.

(c) Use the results of the simulation to estimate the probability that Luke passes the quiz.

(d) Doug is in the same class and claims to understand some of the material. If he scored 8 points on the quiz, is there convincing evidence that he understands some of the material? Explain your answer.

10. **Double fault!** A professional tennis player claims to get 90% of her second serves in. In a recent match, the

player missed 5 of her 20 second serves. Is this a surprising result if the player's claim is true? Assume that the player has a 0.10 probability of missing each second serve. We want to carry out a simulation to estimate the probability that she would miss 5 or more of her 20 second serves.

(a) Describe how to use a random number generator to perform one trial of the simulation.

The dotplot displays the number of second serves missed by the player out of 20 second serves in 100 simulated matches.

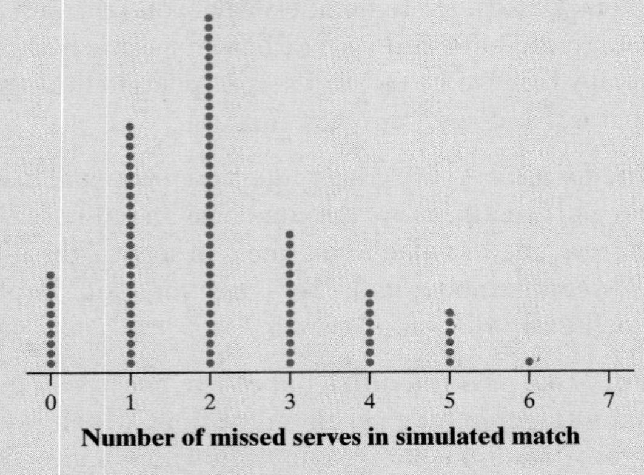


(b) Explain what the dot at 6 represents.

(c) Use the results of the simulation to estimate the probability that the player would miss 5 or more of her 20 second serves in a match.

(d) Is there convincing evidence that the player misses more than 10% of her second serves? Explain your answer.

11. **Airport security** The Transportation Security Administration (TSA) is responsible for airport safety. On some flights, TSA officers randomly select passengers for an extra security check prior to boarding. One such flight had 76 passengers—12 in first class and 64 in coach class. Some passengers were surprised when none of the 10 passengers chosen for screening were seated in first class. We want to perform a simulation to estimate the probability that no first-class passengers would be chosen in a truly random selection.

pg 334

(a) Describe how you would use a table of random digits to carry out this simulation.

(b) Perform one trial of the simulation using the random digits that follow. Copy the digits onto your paper and mark directly on or above them so that someone can follow what you did.

| | | | | | | | |
|---|---|---|---|---|---|---|---|
| 71487 | 09984 | 29077 | 14863 | 61683 | 47052 | 62224 | 51025 |

(c) In 15 of the 100 trials of the simulation, none of the 10 passengers chosen was seated in first class. Does this result provide convincing evidence that the TSA officers did not carry out a truly random selection? Explain your answer.

12. **Scrabble** In the game of Scrabble, each player begins by randomly selecting 7 tiles from a bag containing 100 tiles. There are 42 vowels, 56 consonants, and 2 blank tiles in the bag. Cait chooses her 7 tiles and is surprised to discover that all of them are vowels. We want to perform a simulation to determine the probability that a player will randomly select 7 vowels.

(a) Describe how you would use a table of random digits to carry out this simulation.

(b) Perform one trial of the simulation using the random digits given. Copy the digits onto your paper and mark directly on or above them so that someone can follow what you did.

| | | | | | | | |
|---|---|---|---|---|---|---|---|
| 00694 | 05977 | 19664 | 65441 | 20903 | 62371 | 22725 | 53340 |

(c) In 2 of the 1000 trials of the simulation, all 7 tiles were vowels. Does this result give convincing evidence that the bag of tiles was not well mixed?

13. **Bull's-eye!** In a certain archery competition, each player continues to shoot until he or she misses the center of the target twice. Quinn is one of the archers in this competition. Based on past experience, she has a 0.60 probability of hitting the center of the target on each shot. We want to design a simulation to estimate the probability that Quinn stays in the competition for at least 10 shots. Describe how you would use each of the following chance devices to perform one trial of the simulation.

(a) Slips of paper

(b) Random digits table

(c) Random number generator

14. **Free-throw practice** At the end of basketball practice, each player on the team must shoot free throws until he makes 10 of them. Dwayne is a 70% free-throw shooter. That is, his probability of making any free throw is 0.70. We want to design a simulation to estimate the probability that Dwayne makes 10 free throws in at most 12 shots. Describe how you would use each of the following chance devices to perform one trial of the simulation.

(a) Slips of paper

(b) Random digits table

(c) Random number generator

*In Exercises 15–18, determine whether the simulation design is valid. Justify your answer.*

15. **Smartphone addiction?** A media report claims that 50% of U.S. teens with smartphones feel addicted to their devices.[3] A skeptical researcher believes that this figure is too high. She decides to test the claim by taking a random sample of 100 U.S. teens who have smartphones. Only 40 of the teens in the sample feel addicted to their devices. Does this result give convincing evidence that the media report's 50% claim is too high? To find out, we want to perform a simulation to estimate the probability of getting 40 or fewer teens who feel addicted to their devices in a random sample of size 100 from a very large population of teens with smartphones in which 50% feel addicted to their devices.

    Let 1 = feels addicted and 2 = doesn't feel addicted. Use a random number generator to produce 100 random integers from 1 to 2. Record the number of 1's in the simulated random sample. Repeat this process many, many times. Find the percent of trials on which the number of 1's was 40 or less.

16. **Lefties** A website claims that 10% of U.S. adults are left-handed. A researcher believes that this figure is too low. She decides to test this claim by taking a random sample of 20 U.S. adults and recording how many are left-handed. Four of the adults in the sample are left-handed. Does this result give convincing evidence that the website's 10% claim is too low? To find out, we want to perform a simulation to estimate the probability of getting 4 or more left-handed people in a random sample of size 20 from a very large population in which 10% of the people are left-handed.

    Let 00 to 09 indicate left-handed and 10 to 99 represent right-handed. Move left to right across a row in Table D. Each pair of digits represents one person. Keep going until you get 20 different pairs of digits. Record how many people in the simulated sample are left-handed. Repeat this process many, many times. Find the proportion of trials in which 4 or more people in the simulated sample were left-handed.

17. **Notebook check** Every 9 weeks, Mr. Millar collects students' notebooks and checks their homework. He randomly selects 4 different assignments to inspect for all of the students. Marino is one of the students in Mr. Millar's class. Marino completed 20 homework assignments and did not complete 10 assignments. He is surprised when Mr. Millar only selects 1 assignment that he completed. Should he be surprised? To find out, we want to design a simulation to estimate the probability that Mr. Millar will randomly select 1 or fewer of the homework assignments that Marino completed.

    Get 30 identical slips of paper. Write "N" on 10 of the slips and "C" on the remaining 20 slips. Put the slips into a hat and mix well. Draw 1 slip without looking to represent the first randomly selected homework assignment, and record whether Marino completed it. Put the slip back into the hat, mix again, and draw another slip representing the second randomly selected assignment. Record whether Marino completed this assignment. Repeat this process two more times for the third and fourth randomly selected homework assignments. Record the number out of the 4 randomly selected homework assignments that Marino completed in this trial of the simulation. Perform many trials. Find the proportion of trials in which Mr. Millar randomly selects 1 or fewer of the homework assignments that Marino completed.

18. **Random assignment** Researchers recruited 20 volunteers—8 men and 12 women—to take part in an experiment. They randomly assigned the subjects into two groups of 10 people each. To their surprise, 6 of the 8 men were randomly assigned to the same treatment. Should they be surprised? We want to design a simulation to estimate the probability that a proper random assignment would result in 6 or more of the 8 men ending up in the same group.

    Get 20 identical slips of paper. Write "M" on 8 of the slips and "W" on the remaining 12 slips. Put the slips into a hat and mix well. Draw 10 of the slips without looking and place into one pile representing Group 1. Place the other 10 slips in a pile representing Group 2. Record the largest number of men in either of the two groups from this simulated random assignment. Repeat this process many, many times. Find the percent of trials in which 6 or more men ended up in the same group.

19. **Color-blind men** About 7% of men in the United States have some form of red–green color blindness. Suppose we randomly select one U.S. adult male at a time until we find one who is red–green color-blind. Should we be surprised if it takes us 20 or more men? Describe how you would carry out a simulation to estimate the probability that we would have to randomly select 20 or more U.S. adult males to find one who is red–green color-blind. *Do not perform the simulation.*

20. **Taking the train** According to New Jersey Transit, the 8:00 A.M. weekday train from Princeton to New York City has a 90% chance of arriving on time. To test this claim, an auditor chooses 6 weekdays at random during a month to ride this train. The train arrives late on 2 of those days. Does the auditor have convincing evidence that the company's claim is false? Describe how you would carry out a simulation to estimate the probability that a train with a 90% chance of arriving on time each day would be late on 2 or more of 6 days. *Do not perform the simulation.*

21. **Recycling** Do most teens recycle? To find out, an AP® Statistics class asked an SRS of 100 students at their school whether they regularly recycle. In the sample, 55 students said that they recycle. Is this convincing evidence that more than half of the students at the school would say they regularly recycle? The dotplot shows the results of taking 200 SRSs of 100 students from a population in which the true proportion who recycle is 0.50.

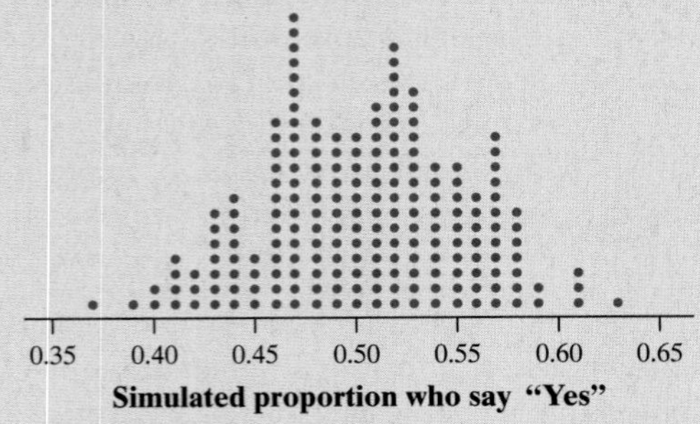


(a) Explain why the sample result (55 out of 100 said "Yes") does not give convincing evidence that more than half of the school's students recycle.

(b) Suppose instead that 63 students in the class's sample had said "Yes." Explain why this result would give convincing evidence that a majority of the school's students recycle.

22. **Brushing teeth, wasting water?** A recent study reported that fewer than half of young adults turn off the water while brushing their teeth. Is the same true for teenagers? To find out, a group of statistics students asked an SRS of 60 students at their school if they usually brush with the water off. In the sample, 27 students said "Yes." The dotplot shows the results of taking 200 SRSs of 60 students from a population in which the true proportion who brush with the water off is 0.50.

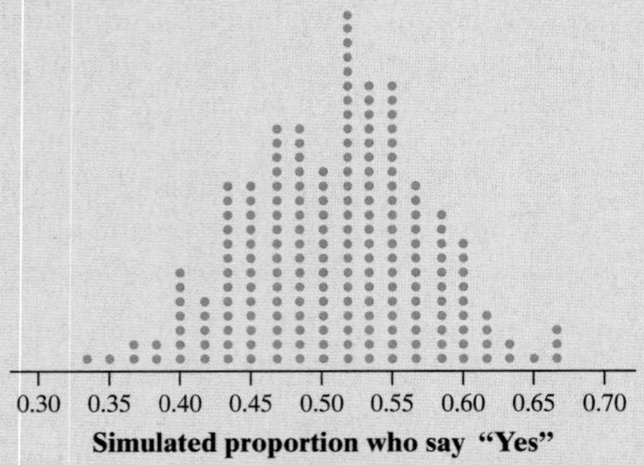


(a) Explain why the sample result (27 of the 60 students said "Yes") does not give convincing evidence that fewer than half of the school's students brush their teeth with the water off.

(b) Suppose instead that 18 of the 60 students in the class's sample had said "Yes." Explain why this result would give convincing evidence that fewer than 50% of the school's students brush their teeth with the water off.

**Multiple Choice:** *Select the best answer for Exercises 23–28.*

23. You read in a book about bridge that the probability that each of the four players is dealt exactly 1 ace is approximately 0.11. This means that

(a) in every 100 bridge deals, each player has 1 ace exactly 11 times.

(b) in 1 million bridge deals, the number of deals on which each player has 1 ace will be exactly 110,000.

(c) in a very large number of bridge deals, the percent of deals on which each player has 1 ace will be very close to 11%.

(d) in a very large number of bridge deals, the average number of aces in a hand will be very close to 0.11.

(e) If each player gets an ace in only 2 of the first 50 deals, then each player should get an ace in more than 11% of the next 50 deals.

24. If I toss a fair coin 5 times and the outcomes are TTTTT, then the probability that tails appears on the next toss is

(a) 0.5. (d) 0.

(b) less than 0.5. (e) 1.

(c) greater than 0.5.

*Exercises 25 to 27 refer to the following setting.* A basketball player claims to make 47% of her shots from the field. We want to simulate the player taking sets of 10 shots, assuming that her claim is true.

25. To simulate the number of makes in 10 shot attempts, you would perform the simulation as follows:

(a) Use 10 random one-digit numbers, where 0–4 are a make and 5–9 are a miss.

(b) Use 10 random two-digit numbers, where 00–46 are a make and 47–99 are a miss.

(c) Use 10 random two-digit numbers, where 00–47 are a make and 48–99 are a miss.

(d) Use 47 random one-digit numbers, where 0 is a make and 1–9 are a miss.

(e) Use 47 random two-digit numbers, where 00–46 are a make and 47–99 are a miss.

26. A total of 25 trials of the simulation were performed. The number of makes in each set of 10 simulated shots was recorded on the dotplot. What is the approximate probability that a 47% shooter makes 5 or more shots in 10 attempts?

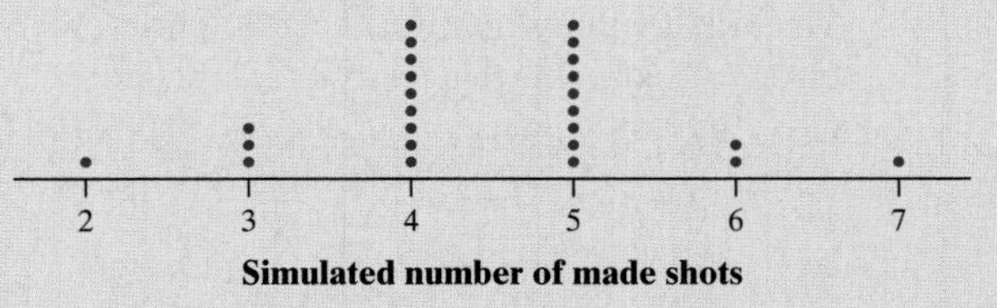

(a) 5/10 (b) 3/10 (c) 12/25

(d) 3/25 (e) 47/100

27. Suppose this player attempts 10 shots in a game and makes only 3 of them. Does this provide convincing evidence that she is less than a 47% shooter?

(a) Yes, because 3/10 (30%) is less than 47%.

(b) Yes, because she never made 47% of her shots in the simulation.

(c) No, because it is plausible (believable) that she would make 3 or fewer shots by chance alone.

(d) No, because the simulation was only repeated 25 times.

(e) No, because more than half of the simulated results were less than 47%.

28. Ten percent of U.S. households contain 5 or more people. You want to simulate choosing a household at random and recording "Yes" if it contains 5 or more people. Which of these is a correct assignment of digits for this simulation?

(a) Odd = Yes; Even = No

(b) 0 = Yes; 1–9 = No

(c) 0–5 = Yes; 6–9 = No

(d) 0–4 = Yes; 5–9 = No

(e) None of these

### Recycle and Review

29. **AARP and Medicare** (4.1) To find out what proportion of Americans support proposed Medicare legislation to help pay medical costs, the AARP conducted a survey of their members (people over age 50 who pay membership dues). One of the questions was: "Even if this plan won't affect you personally either way, do you think it should be passed so that people with low incomes or people with high drug costs can be helped?" Of the respondents, 75% answered "Yes."[4]

(a) Describe how undercoverage might lead to bias in this study. Explain the likely direction of the bias.

(b) Describe how the wording of the question might lead to bias in this study. Explain the likely direction of the bias.

30. **Waiting to park** (1.3, 4.2) Do drivers take longer to leave their parking spaces when someone is waiting? Researchers hung out in a parking lot and collected some data. The graphs and numerical summaries display information about how long it took drivers to exit their spaces.

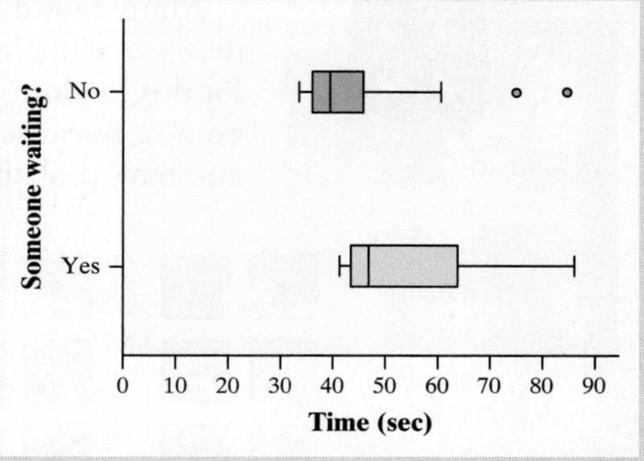


**Descriptive Statistics: Time**

| Waiting | $n$ | Mean | StDev | Min | $Q_1$ | Median | $Q_3$ | Max |
|---|---|---|---|---|---|---|---|---|
| No | 20 | 44.42 | 14.10 | 33.76 | 35.61 | 39.56 | 48.48 | 84.92 |
| Yes | 20 | 54.11 | 14.39 | 41.61 | 43.41 | 47.14 | 66.44 | 85.97 |

(a) Write a few sentences comparing these distributions.

(b) Can we conclude that having someone waiting causes drivers to leave their spaces more slowly? Why or why not?

# SECTION 5.2 Probability Rules

**LEARNING TARGETS** *By the end of the section, you should be able to:*

- Give a probability model for a random process with equally likely outcomes and use it to find the probability of an event.
- Use basic probability rules, including the complement rule and the addition rule for mutually exclusive events.
- Use a two-way table or Venn diagram to model a random process and calculate probabilities involving two events.
- Apply the general addition rule to calculate probabilities.

The idea of probability rests on the fact that random behavior is predictable in the long run. In Section 5.1, we used simulation to imitate a random process. Do we always need to repeat a random process—tossing coins, rolling dice, drawing slips from a hat—many times to determine the probability of a particular outcome? Fortunately, the answer is no.

## Probability Models

Steve Gorton/Getty Images

In Chapter 2, we saw that a Normal density curve could be used to model some distributions of quantitative data. In Chapter 3, we modeled linear relationships between two quantitative variables with a least-squares regression line. Now we're ready to develop a model for random behavior.

Many board games involve rolling dice. Imagine rolling two fair, six-sided dice—one that's red and one that's blue. How do we develop a **probability model** for this random process? Figure 5.2 displays the **sample space** of 36 possible outcomes. Because the dice are fair, each of these outcomes will be equally likely and have probability 1/36.

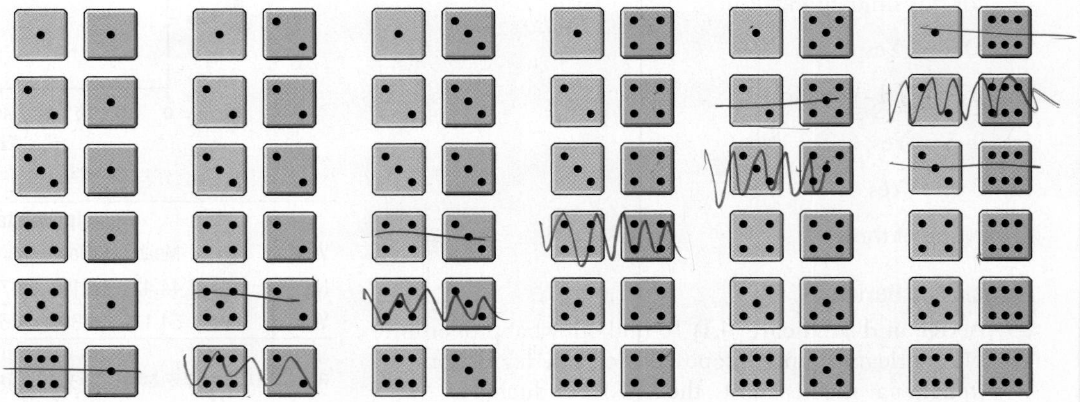

**FIGURE 5.2** The 36 possible outcomes from rolling two six-sided dice, one red and one blue. Each of these equally likely outcomes has probability 1/36.

> **DEFINITION Probability model, Sample space**
>
> A **probability model** is a description of some random process that consists of two parts: a list of all possible outcomes and the probability for each outcome.
>
> The list of all possible outcomes is called the **sample space**.

A sample space can be very simple or very complex. If we toss a coin once, there are only two possible outcomes in the sample space, heads and tails. When Gallup takes a random sample of 1523 U.S. adults and asks a survey question, the sample space consists of all possible sets of responses from 1523 of the over 240 million adults in the country.

A probability model does more than just assign a probability to each outcome. It allows us to find the probability of an **event**.

> **DEFINITION Event**
>
> An **event** is any collection of outcomes from some random process.

Events are usually designated by capital letters, like A, B, C, and so on. For rolling two six-sided dice, we can define event A as getting a sum of 5. We write the probability of event A as $P(\text{A})$ or $P(\text{sum is } 5)$.

It is fairly easy to find the probability of an event in the case of equally likely outcomes. There are 4 outcomes in event A:

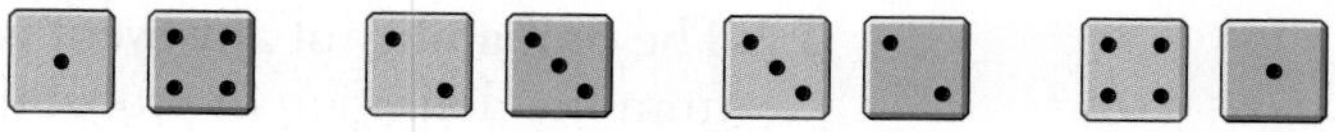

The probability that event A occurs is therefore

$$P(\text{A}) = \frac{\text{number of outcomes with sum of 5}}{\text{total number of outcomes when rolling two dice}} = \frac{4}{36} = 0.111$$

**FINDING PROBABILITIES: EQUALLY LIKELY OUTCOMES**

If all outcomes in the sample space are equally likely, the probability that event A occurs can be found using the formula

$$P(\text{A}) = \frac{\text{number of outcomes in event A}}{\text{total number of outcomes in sample space}}$$

## EXAMPLE

### Spin the spinner

### Probability models: Equally likely outcomes

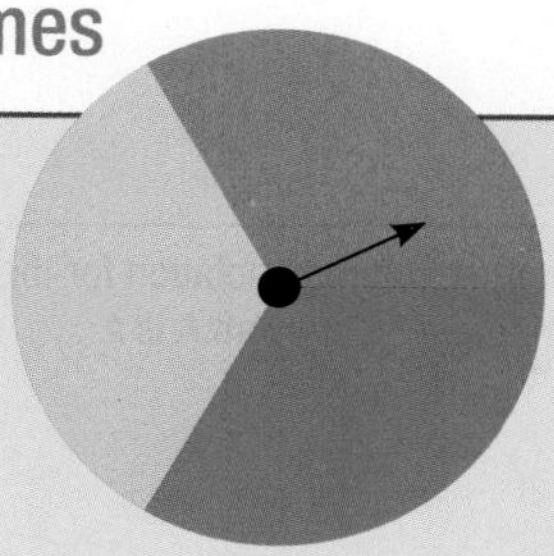

**PROBLEM:** A spinner has three equal sections: red, blue, and yellow. Suppose you spin the spinner two times.

(a) Give a probability model for this random process.

(b) Define event A as spinning blue at least once. Find $P(\text{A})$.

**SOLUTION:**

(a) Sample space: RR RB RY BR BB BY YR YB YY. Because the spinner has equal sections, each of these outcomes will be equally likely and have probability 1/9.

(b) There are 5 outcomes with at least one blue: RB BR BB BY YB. So $P(\text{A}) = \frac{5}{9} = 0.556$.

Remember: A probability model consists of a list of all possible outcomes and the probability of each outcome.

If all outcomes in the sample space are equally likely,

$$P(\text{A}) = \frac{\text{number of outcomes in event A}}{\text{total number of outcomes in sample space}}$$

FOR PRACTICE, TRY EXERCISE 31

## Basic Probability Rules

Our work so far suggests three commonsense rules that a valid probability model must obey:

1. **If all outcomes in the sample space are equally likely, the probability that event A occurs is**

$$P(\text{A}) = \frac{\text{number of outcomes in event A}}{\text{total number of outcomes in sample space}}$$

2. **The probability of any event is a number between 0 and 1.** This rule follows from the definition of probability in Section 5.1: the proportion of times the event would occur in many trials of the random process. A proportion is a number between 0 and 1, so any probability is also a number between 0 and 1.
3. **All possible outcomes together must have probabilities that add up to 1.** Because some outcome must occur on every trial of a random process, the sum of the probabilities for all possible outcomes must be exactly 1.

Here's another rule that follows from the previous two:

4. **The probability that an event does not occur is 1 minus the probability that the event does occur.** If an event occurs in (say) 70% of all trials, it fails to occur in the other 30%. The probability that an event occurs and the probability that it does not occur always add to 100%, or 1. Earlier, we found that the probability of getting a sum of 5 when rolling two fair, six-sided dice is 4/36. What's the probability that the sum is *not* 5?

$$P(\text{sum is not 5}) = 1 - P(\text{sum is 5}) = 1 - \frac{4}{36} = \frac{32}{36} = 0.889$$

We refer to the event "not A" as the **complement** of A and denote it by $\text{A}^C$. For that reason, this handy result is known as the **complement rule**. Using the complement rule in this setting is much easier than counting all 32 possible ways to get a sum that isn't 5.

Another common notation for the complement of event A is A′.

**DEFINITION Complement rule, Complement**

The **complement rule** says that $P(\text{A}^C) = 1 - P(\text{A})$, where $\text{A}^C$ is the **complement** of event A; that is, the event that A does not occur.

Let's consider one more event involving the random process of rolling two fair, six-sided dice: getting a sum of 6. The outcomes in this event are

So $P(\text{sum is 6}) = \frac{5}{36}$. What's the probability that we get a sum of 5 *or* a sum of 6?

$$P(\text{sum is 5 or sum is 6}) = P(\text{sum is 5}) + P(\text{sum is 6}) = \frac{4}{36} + \frac{5}{36} = \frac{9}{36} = 0.25$$

Why does this formula work? Because the events "getting a sum of 5" and "getting a sum of 6" have no outcomes in common—that is, they can't both happen at the same time. We say that these two events are **mutually exclusive (disjoint).** As a result, this intuitive formula is known as the **addition rule for mutually exclusive events.**

**DEFINITION Mutually exclusive (disjoint), Addition rule for mutually exclusive events**

Two events A and B are **mutually exclusive (disjoint)** if they have no outcomes in common and so can never occur together—that is, if $P(\text{A and B}) = 0$.

The **addition rule for mutually exclusive events** A and B says that

$$P(\text{A or B}) = P(\text{A}) + P(\text{B})$$

**Note that this rule works only for mutually exclusive events.** We will soon develop a more general rule for finding $P(\text{A or B})$ that works for any two events.

We can summarize the basic probability rules more concisely in symbolic form.

**BASIC PROBABILITY RULES**

- For any event A, $0 \leq P(\text{A}) \leq 1$.
- If S is the sample space in a probability model, $P(\text{S}) = 1$.
- In the case of *equally likely* outcomes,

$$P(\text{A}) = \frac{\text{number of outcomes in event A}}{\text{total number of outcomes in sample space}}$$

- **Complement rule:** $P(\text{A}^C) = 1 - P(\text{A})$.
- **Addition rule for mutually exclusive events:** If A and B are mutually exclusive, $P(\text{A or B}) = P(\text{A}) + P(\text{B})$.

The earlier dice-rolling and spin the spinner settings involved equally likely outcomes. Here's an example that illustrates use of the basic probability rules when the outcomes of a random process are not equally likely.

## EXAMPLE Avoiding blue M&M'S®
### Basic probability rules

**PROBLEM:** Suppose you tear open the corner of a bag of M&M'S® Milk Chocolate Candies, pour one candy into your hand, and observe the color. According to Mars, Inc., the maker of M&M'S, the probability model for a bag from its Cleveland factory is:

| Color | Blue | Orange | Green | Yellow | Red | Brown |
|---|---|---|---|---|---|---|
| Probability | 0.207 | 0.205 | 0.198 | 0.135 | 0.131 | 0.124 |

Niels Poulsen std/Alamy

(a) Explain why this is a valid probability model.
(b) Find the probability that you don't get a blue M&M.
(c) What's the probability that you get an orange or a brown M&M?

**SOLUTION:**

(a) The probability of each outcome is a number between 0 and 1. The sum of the probabilities is $0.207+0.205+0.198+0.135+0.131+0.124=1$.

(b) $P(\text{not blue})=1-P(\text{blue})=1-0.207=0.793$

Using the complement rule: $P(A^C)=1-P(A)$.

(c) $P(\text{orange or brown})=P(\text{orange})+P(\text{brown})$
$=0.205+0.124=0.329$

Using the addition rule for mutually exclusive events because an M&M® can't be both orange and brown.

**FOR PRACTICE, TRY EXERCISE 35**

For part (b) of the example, we could also use an expanded version of the addition rule for mutually exclusive events:

$$\begin{aligned} P(\text{not blue}) &= P(\text{orange or green or yellow or red or brown}) \\ &= P(\text{orange}) + P(\text{green}) + P(\text{yellow}) + P(\text{red}) + P(\text{brown}) \\ &= 0.205 + 0.198 + 0.135 + 0.131 + 0.124 \\ &= 0.793 \end{aligned}$$

Using the complement rule is much simpler than adding 5 probabilities together!

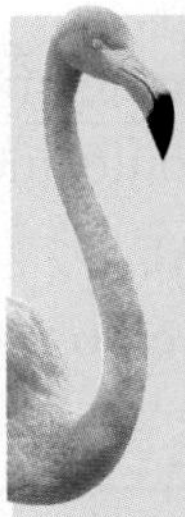

## CHECK YOUR UNDERSTANDING

Suppose we choose an American adult at random. Define two events:

A = the person has a cholesterol level of 240 milligrams per deciliter of blood (mg/dl) or above (high cholesterol)

B = the person has a cholesterol level of 200 to <240 mg/dl (borderline high cholesterol)

According to the American Heart Association, $P(A) = 0.16$ and $P(B) = 0.29$.

1. Explain why events A and B are mutually exclusive.
2. Say in plain language what the event "A or B" is. Then find $P(\text{A or B})$.
3. Let C be the event that the person chosen has a cholesterol level below 200 mg/dl (normal cholesterol). Find $P(C)$.

# Two-Way Tables, Probability, and the General Addition Rule

So far, you have learned how to model random behavior and some basic rules for finding the probability of an event. What if you're interested in finding probabilities involving two events that are not mutually exclusive? For instance, a survey of all residents in a large apartment complex reveals that 68% use Facebook, 28% use Instagram, and 25% do both.[5] Suppose we select a resident at random. What's the probability that the person uses Facebook *or* uses Instagram?

There are two different uses of the word *or* in everyday life. In a restaurant, when you are asked if you want "soup or salad," the waiter wants you to choose one or the other, but not both. However, when you order coffee and are asked if you want "cream or sugar," it's OK to ask for one or the other or both. The same issue arises in statistics.

Mutually exclusive events A and B cannot both happen at the same time. For such events, "A or B" means that *only* event A happens or *only* event B happens. You can find *P*(A or B) with the addition rule for mutually exclusive events:

$$P(\text{A or B}) = P(\text{A}) + P(\text{B})$$

How can we find *P*(A or B) when the two events are *not* mutually exclusive? Now we have to deal with the fact that "A or B" means one or the other *or both*. For instance, "uses Facebook or uses Instagram" in the scenario just described includes U.S. adults who do both.

When you're trying to find probabilities involving two events, like *P*(A or B), a two-way table can display the sample space in a way that makes probability calculations easier.

## EXAMPLE

### Who has pierced ears?
### Two-way tables and probability

**PROBLEM:** Students in a college statistics class wanted to find out how common it is for young adults to have their ears pierced. They recorded data on two variables—gender and whether or not the student had a pierced ear—for all 178 people in the class. The two-way table summarizes the data.

Big Cheese Photo/Superstock

| | | Gender | | |
|---|---|---|---|---|
| | | Male | Female | Total |
| Pierced ear | Yes | 19 | 84 | 103 |
| | No | 71 | 4 | 75 |
| | Total | 90 | 88 | 178 |

Suppose we choose a student from the class at random. Define event A as getting a male student and event B as getting a student with a pierced ear.

(a) Find *P*(B).

(b) Find *P*(A and B). Interpret this value in context.

(c) Find *P*(A or B).

**SOLUTION:**

(a) $P(B) = P(\text{pierced ear}) = \dfrac{103}{178} = 0.579$

(b) $P(\text{A and B}) = P(\text{male and pierced ear}) = \dfrac{19}{178} = 0.107$. There's about an 11% chance that a randomly selected student from this class is male and has a pierced ear.

(c) $P(\text{A or B}) = P(\text{male or pierced ear})$

$$= \frac{71+19+84}{178} = \frac{174}{178} = 0.978$$

In statistics, *or* means "one or the other or both." So "male or pierced ear" includes (i) male but no pierced ear; (ii) pierced ear but not male; and (iii) male and pierced ear.

FOR PRACTICE, TRY EXERCISE 41

When we found *P*(male and pierced ear) in part (b) of the example, we could have described this as either *P*(A and B) or *P*(B and A). Why? Because "male and pierced ear" describes the same event as "pierced ear and male." Likewise, *P*(A or B) is the same as *P*(B or A). Don't get so caught up in the notation that you lose sight of what's really happening!

Part (c) of the example reveals an important fact about finding the probability *P*(A or B): we can't use the addition rule for mutually exclusive events unless events A and B have no outcomes in common. In this case, there are 19 outcomes that are shared by events A and B—the students who are male and have a pierced ear. If we did add the probabilities of A and B, we'd get $90/178 + 103/178 = 193/178$. This is clearly wrong because the probability is bigger than 1! As Figure 5.3 illustrates, outcomes common to both events are counted twice when we add the probabilities of these two events.

The probability *P*(A and B) that events A and B both occur is called a *joint probability.* That's consistent with our description of 19/178 as a *joint relative frequency* in Chapter 1. If the joint probability is 0, the events are disjoint.

$P(\text{A and B}) = \frac{19}{178}$

Outcomes here are double-counted by *P*(A) + *P*(B)

| | | Gender | | |
|---|---|---|---|---|
| | | Male | Female | Total |
| Pierced ear | Yes | 19 | 84 | 103 |
| | No | 71 | 4 | 75 |
| | Total | 90 | 88 | 178 |

$P(\text{B}) = \frac{103}{178}$

$P(\text{A}) = \frac{90}{178}$

**FIGURE 5.3** Two-way table showing events A and B from the pierced-ear example. These events are *not* mutually exclusive, so we can't find *P*(A or B) by just adding the probabilities of the two events.

We can fix the double-counting problem illustrated in the two-way table by subtracting the probability *P*(male and pierced ear) from the sum. That is,

$$\begin{aligned} P(\text{male or pierced ear}) &= P(\text{male}) + P(\text{pierced ear}) - P(\text{male and pierced ear}) \\ &= 90/178 + 103/178 - 19/178 \\ &= 174/178 \end{aligned}$$

This result is known as the **general addition rule**.

**DEFINITION General addition rule**

If A and B are any two events resulting from some random process, the **general addition rule** says that

$$P(\text{A or B}) = P(\text{A}) + P(\text{B}) - P(\text{A and B})$$

Sometimes it's easier to label events with letters that relate to the context, as the following example shows.

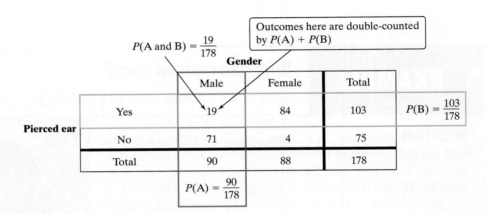

# EXAMPLE

## Facebook versus Instagram
### General addition rule

Ann Heath

**PROBLEM:** A survey of all residents in a large apartment complex reveals that 68% use Facebook, 28% use Instagram, and 25% do both. Suppose we select a resident at random. What's the probability that the person uses Facebook *or* uses Instagram?

**SOLUTION:**

Let event F = uses Facebook and I = uses Instagram.

$$P(\text{F or I}) = P(\text{F}) + P(\text{I}) - P(\text{F and I})$$
$$= 0.68 + 0.28 - 0.25$$
$$= 0.71$$

**FOR PRACTICE, TRY EXERCISE 47**

**AP® EXAM TIP**

Many probability problems involve simple computations that you can do on your calculator. It may be tempting to just write down your final answer without showing the supporting work. Don't do it! A "naked answer," even if it's correct, will usually be penalized on a free-response question.

What happens if we use the general addition rule for two mutually exclusive events A and B? In that case, $P(\text{A and B}) = 0$, and the formula reduces to

$$P(\text{A or B}) = P(\text{A}) + P(\text{B}) - P(\text{A and B}) = P(\text{A}) + P(\text{B}) - 0 = P(\text{A}) + P(\text{B})$$

In other words, the addition rule for mutually exclusive events is just a special case of the general addition rule.

You might be wondering if there is also a rule for finding $P(\text{A and B})$. There is, but it's not quite as intuitive. Stay tuned for that later.

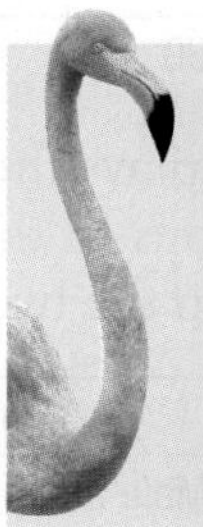

## CHECK YOUR UNDERSTANDING

Yellowstone National Park staff surveyed a random sample of 1526 winter visitors to the park. They asked each person whether they belonged to an environmental club (like the Sierra Club). Respondents were also asked whether they owned, rented, or had never used a snowmobile. The two-way table summarizes the survey responses.

| | | Environmental club | |
|---|---|---|---|
| | | No | Yes |
| **Snowmobile experience** | Never used | 445 | 212 |
| | Renter | 497 | 77 |
| | Owner | 279 | 16 |

Suppose we choose one of the survey respondents at random.

1. What's the probability that the person is an environmental club member?
2. Find *P*(not a snowmobile renter).
3. What's *P*(environmental club member and not a snowmobile renter)?
4. Find the probability that the person is not an environmental club member or is a snowmobile renter.

## Venn Diagrams and Probability

We have seen that two-way tables can be used to illustrate the sample space of a random process involving two events. So can **Venn diagrams**, like the one shown in Figure 5.4.

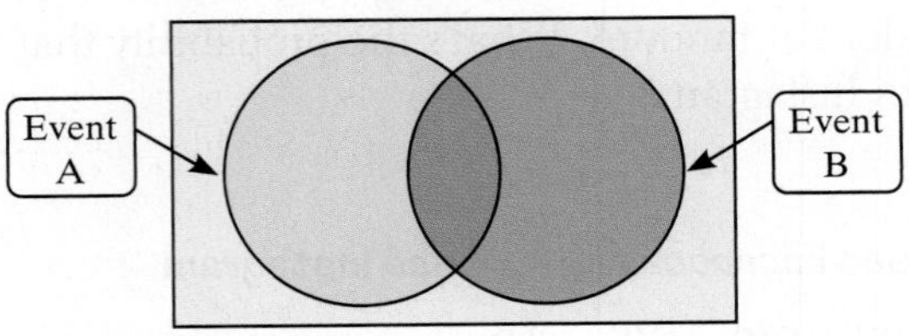


**FIGURE 5.4** A typical Venn diagram that shows the sample space and the relationship between two events A and B.

**DEFINITION Venn diagram**

A **Venn diagram** consists of one or more circles surrounded by a rectangle. Each circle represents an event. The region inside the rectangle represents the sample space of the random process.

In an earlier example, we looked at data from a survey on gender and ear piercings for a large group of college students. The random process was selecting a student in the class at random. Our events of interest were A: is male and B: has a pierced ear. Here is the two-way table that summarizes the data:

| | | Gender | | |
|---|---|---|---|---|
| | | Male | Female | Total |
| Pierced ear | Yes | 19 | 84 | 103 |
| | No | 71 | 4 | 75 |
| | Total | 90 | 88 | 178 |

The Venn diagram in Figure 5.5 displays the sample space in a slightly different way. There are four distinct regions in the Venn diagram. These regions correspond to the four (non-total) cells in the two-way table as follows.

| Region in Venn diagram | In words | Count |
|---|---|---|
| In the intersection of two circles | Male and pierced ear | 19 |
| Inside circle A, outside circle B | Male and no pierced ear | 71 |
| Inside circle B, outside circle A | Female and pierced ear | 84 |
| Outside both circles | Female and no pierced ear | 4 |

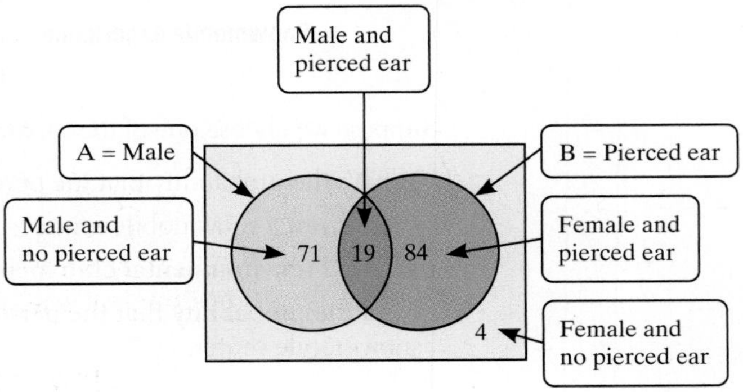


**FIGURE 5.5** The completed Venn diagram for the large group of college students. The circles represent the two events A = male and B = has a pierced ear.

Because Venn diagrams have uses in other branches of mathematics, some standard vocabulary and notation have been developed that will make our work with Venn diagrams a bit easier.

- We introduced the *complement* of an event earlier. In Figure 5.6(a), the complement $A^C$ contains the outcomes that are not in A.
- Figure 5.6(b) shows the event "A and B." You can see why this event is also called the **intersection** of A and B. The corresponding notation is $A \cap B$.
- The event "A or B" is shown in Figure 5.6(c). This event is also known as the **union** of A and B. The corresponding notation is $A \cup B$.

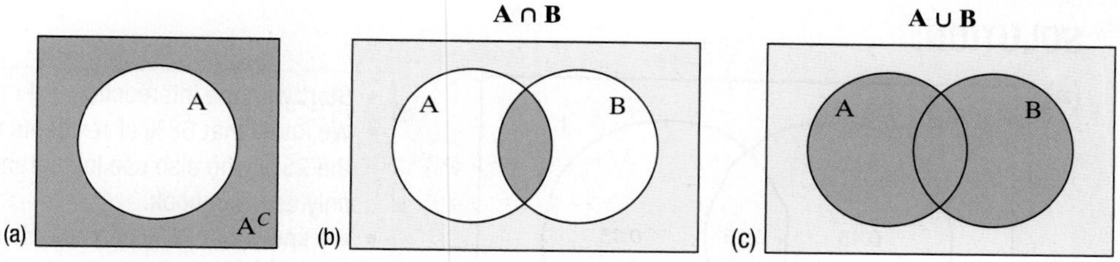


**FIGURE 5.6** The green shaded region in each Venn diagram shows: (a) the *complement* $A^c$ of event A, (b) the *intersection* of events A and B, and (c) the *union* of events A and B.

### DEFINITION Intersection, Union

The event "A and B" is called the **intersection** of events A and B. It consists of all outcomes that are common to both events, and is denoted $A \cap B$.

The event "A or B" is called the **union** of events A and B. It consists of all outcomes that are in event A or event B, or both, and is denoted $A \cup B$.

Here's a way to keep the symbols straight: ∪ for **u**nion; ∩ for **int**ersection.

With this new notation, we can rewrite the general addition rule in symbols as

$$P(A \cup B) = P(A) + P(B) - P(A \cap B)$$

This Venn diagram shows why the formula works in the pierced-ear example.

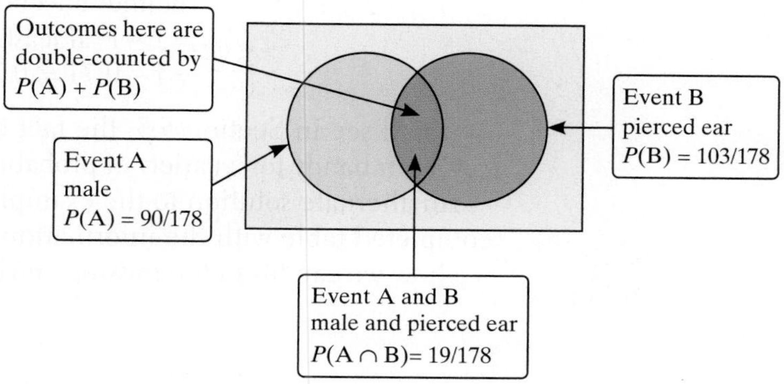


For mutually exclusive (disjoint) events A and B, the *joint probability* $P(A \cap B) = 0$ because the two events have no outcomes in common. So the corresponding Venn diagram consists of two non-overlapping circles. You can see from the figure at left why, in this special case, the general addition rule reduces to

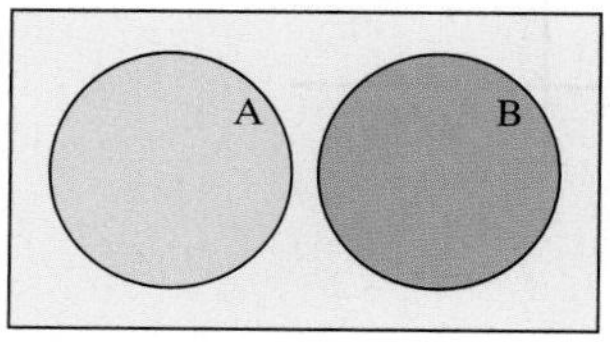


$$P(A \cup B) = P(A) + P(B)$$

## EXAMPLE

### Facebook versus Instagram
### Venn diagrams and probability

Ann Heath

**PROBLEM:** A survey of all residents in a large apartment complex reveals that 68% use Facebook, 28% use Instagram, and 25% do both. Suppose we select a resident at random.

(a) Make a Venn diagram to display the sample space of this random process using the events F: uses Facebook and I: uses Instagram.

(b) Find the probability that the person uses neither Facebook nor Instagram.

**SOLUTION:**

(a)

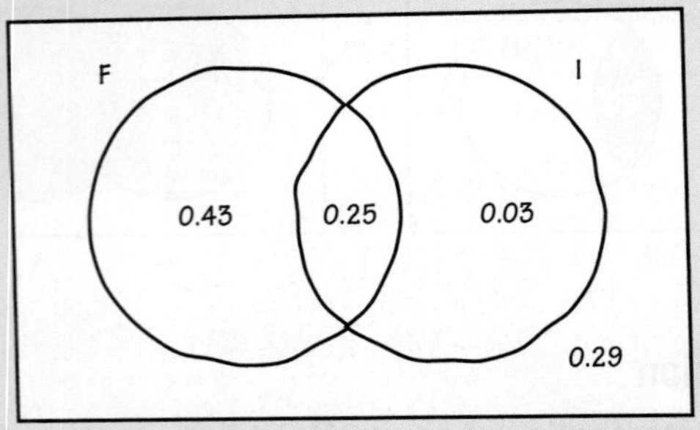


- Start with the intersection: $P(F \cap I) = 0.25$.
- We know that 68% of residents use Facebook. That figure includes the 25% who also use Instagram. So 68% − 25% = 43% = 0.43 only use Facebook.
- We know that 28% of residents use Instagram. That figure includes the 25% who also use Facebook. So 28% − 25% = 3% = 0.03 only use Instagram.
- A total of $0.43 + 0.25 + 0.03 = 0.71 = 71\%$ of residents use at least one of Facebook or Instagram. By the complement rule, $1 - 0.71 = 0.29 = 29\%$ use neither Facebook nor Instagram.

(b) $P(\text{no Facebook and no Instagram}) = 0.29$

**FOR PRACTICE, TRY EXERCISE 51**

In the preceding example, the event "uses neither Facebook nor Instagram" is the complement of the event "uses at least one of Facebook or Instagram." To solve part (b) of the problem, we could have used our answer from the example on page 349 and the complement rule:

$$
\begin{aligned}
&P(\text{neither Facebook nor Instagram})\\
&\quad = 1 - P(\text{at least one of Facebook or Instagram})\\
&\quad = 1 - 0.71 = 0.29
\end{aligned}
$$

As you'll see in Section 5.3, the fact that "none" is the opposite of "at least 1" comes in handy for a variety of probability questions.

An alternate solution to the example uses a two-way table. Here is a partially completed table with the information given in the problem statement. Do you see how we can fill in the missing entries?

| | | Facebook use | | |
|---|---|---|---|---|
| | | Yes | No | Total |
| **Instagram use** | Yes | 25% | | 28% |
| | No | | | |
| | Total | 68% | | 100% |

- 100% – 68% = 32% of residents do not use Facebook.
- 100% – 28% = 72% of residents do not use Instagram.
- 68% – 25% = 43% of residents use Facebook but do not use Instagram.
- 28% – 25% = 3% of residents use Instagram but do not use Facebook.
- 32% – 3% = 72% – 43% = 29% of residents do not use Facebook and do not use Instagram.

The completed table is shown here. We can see the desired probability marked in bold in the table: *P*(neither Facebook nor Instagram) = 0.29.

| | | Facebook use | | |
|---|---|---|---|---|
| | | Yes | No | Total |
| Instagram use | Yes | 25% | 3% | 28% |
| | No | 43% | **29%** | 72% |
| | Total | 68% | 32% | 100% |

## Section 5.2 Summary

- A **probability model** describes a random process by listing all possible outcomes in the **sample space** and giving the probability of each outcome. A valid probability model requires that all possible outcomes have probabilities that add up to 1.
- An **event** is a collection of possible outcomes from the sample space. The probability of any event is a number between 0 and 1.
- The event "A or B" is known as the **union** of A and B, denoted $A \cup B$. It consists of all outcomes in event A, event B, or both.
- The event "A and B" is known as the **intersection** of A and B, denoted $A \cap B$. It consists of all outcomes that are common to both events.
- To find the probability that an event occurs, we use some basic rules:
  - If all outcomes in the sample space are equally likely,

$$P(A) = \frac{\text{number of outcomes in event A}}{\text{total number of outcomes in sample space}}$$

  - **Complement rule:** $P(A^C) = 1 - P(A)$, where $A^C$ is the **complement** of event A; that is, the event that A does not happen.
  - **General addition rule:** For any two events A and B,

$$P(A \cup B) = P(A) + P(B) - P(A \cap B)$$

  - **Addition rule for mutually exclusive events:** Events A and B are **mutually exclusive (disjoint)** if they have no outcomes in common. If A and B are mutually exclusive, $P(\text{A or B}) = P(A) + P(B)$.
- A **two-way table** or **Venn diagram** can be used to display the sample space and to help find probabilities for a random process involving two events.

## Section 5.2 Exercises

31. **Four-sided dice** A four-sided die is a pyramid whose four faces are labeled with the numbers 1, 2, 3, and 4 (see image). Imagine rolling two fair, four-sided dice and recording the number that is showing at the base of each pyramid. For instance, you would record a 4 if the die landed as shown in the figure.
pg 343

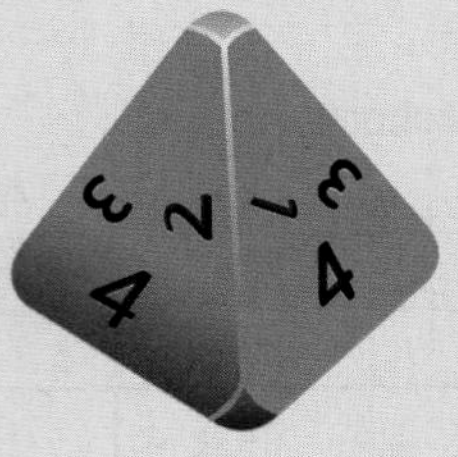

(a) Give a probability model for this random process.

(b) Define event A as getting a sum of 5. Find $P(A)$.

32. **Tossing coins** Imagine tossing a fair coin 3 times.

(a) Give a probability model for this random process.

(b) Define event B as getting more heads than tails. Find $P(B)$.

33. **Grandkids** Mr. Starnes and his wife have 6 grandchildren: Connor, Declan, Lucas, Piper, Sedona, and Zayne. They have 2 extra tickets to a holiday show, and will randomly select which 2 grandkids get to see the show with them.

(a) Give a probability model for this random process.

(b) Find the probability that at least one of the two girls (Piper and Sedona) get to go to the show.

34. **Who's paying?** Abigail, Bobby, Carlos, DeAnna, and Emily go to the bagel shop for lunch every Thursday. Each time, they randomly pick 2 of the group to pay for lunch by drawing names from a hat.

(a) Give a probability model for this random process.

(b) Find the probability that Carlos or DeAnna (or both) ends up paying for lunch.

35. **Mystery box** Ms. Tyson keeps a Mystery Box in her classroom. If a student meets expectations for behavior, she or he is allowed to draw a slip of paper without looking. The slips are all of equal size, are well mixed, and have the name of a prize written on them. One of the "prizes"—extra homework—isn't very desirable! Here is the probability model for the prizes a student can win:
pg 345

| Prize | Pencil | Candy | Stickers | Homework pass | Extra homework |
|---|---|---|---|---|---|
| Probability | 0.40 | 0.25 | 0.15 | 0.15 | 0.05 |

(a) Explain why this is a valid probability model.

(b) Find the probability that a student does not win extra homework.

(c) What's the probability that a student wins candy or a homework pass?

36. **Languages in Canada** Canada has two official languages, English and French. Choose a Canadian at random and ask, "What is your mother tongue?" Here is the distribution of responses, combining many separate languages from the broad Asia/Pacific region:[6]

| Language | English | French | Asian/Pacific | Other |
|---|---|---|---|---|
| Probability | 0.63 | 0.22 | 0.06 | 0.09 |

(a) Explain why this is a valid probability model.

(b) What is the probability that the chosen person's mother tongue is not English?

(c) What is the probability that the chosen person's mother tongue is one of Canada's official languages?

37. **Household size** In government data, a household consists of all occupants of a dwelling unit. Choose an American household at random and count the number of people it contains. Here is the assignment of probabilities for the outcome. The probability of finding 3 people in a household is the same as the probability of finding 4 people.

| Number of people | 1 | 2 | 3 | 4 | 5 | 6 | 7+ |
|---|---|---|---|---|---|---|---|
| Probability | 0.25 | 0.32 | ? | ? | 0.07 | 0.03 | 0.01 |

(a) What probability should replace "?" in the table? Why?

(b) Find the probability that the chosen household contains more than 2 people.

38. **When did you leave?** The National Household Travel Survey gathers data on the time of day when people begin a trip in their car or other vehicle. Choose a trip at random and record the time at which the trip started.[7] Here is an assignment of probabilities for the outcomes:

| Time of day | 10 P.M.–12:59 A.M. | 1 A.M.–5:59 A.M. | 6 A.M.–8:59 A.M. |
|---|---|---|---|
| Probability | 0.040 | 0.033 | 0.144 |
| Time of day | 9 A.M.–12:59 P.M. | 1 P.M.–3:59 P.M. | 4 P.M.–6:59 P.M. |
| Probability | 0.234 | 0.208 | ? |
| Time of day | 7 P.M.–9:59 P.M. | | |
| Probability | 0.123 | | |

(a) What probability should replace "?" in the table? Why?

(b) Find the probability that the chosen trip did not begin between 9 A.M. and 12:59 P.M.

39. **Education among young adults** Choose a young adult (aged 25 to 29) at random. The probability is 0.13 that the person chosen did not complete high school, 0.29 that the person has a high school diploma but no further education, and 0.30 that the person has at least a bachelor's degree.

(a) What must be the probability that a randomly chosen young adult has some education beyond high school but does not have a bachelor's degree? Why?

(b) Find the probability that the young adult completed high school. Which probability rule did you use to find the answer?

(c) Find the probability that the young adult has further education beyond high school. Which probability rule did you use to find the answer?

40. **Preparing for the GMAT** A company that offers courses to prepare students for the Graduate Management Admission Test (GMAT) has collected the following information about its customers: 20% are undergraduate students in business, 15% are undergraduate students in other fields of study, and 60% are college graduates who are currently employed. Choose a customer at random.

(a) What must be the probability that the customer is a college graduate who is not currently employed? Why?

(b) Find the probability that the customer is currently an undergraduate. Which probability rule did you use to find the answer?

(c) Find the probability that the customer is not an undergraduate business student. Which probability rule did you use to find the answer?

41. **Who eats breakfast?** Students in an urban school were curious about how many children regularly eat breakfast. They conducted a survey, asking, "Do you eat breakfast on a regular basis?" All 595 students in the school responded to the survey. The resulting data are shown in the two-way table.[8]

pg 347

| | | Gender | | |
|---|---|---|---|---|
| | | Male | Female | Total |
| Eats breakfast regularly | Yes | 190 | 110 | 300 |
| | No | 130 | 165 | 295 |
| | Total | 320 | 275 | 595 |

Suppose we select a student from the school at random. Define event F as getting a female student and event B as getting a student who eats breakfast regularly.

(a) Find $P(B^C)$.

(b) Find $P(F \text{ and } B^C)$. Interpret this value in context.

(c) Find $P(F \text{ or } B^C)$.

42. **Is this your card?** A standard deck of playing cards (with jokers removed) consists of 52 cards in four suits—clubs, diamonds, hearts, and spades. Each suit has 13 cards, with denominations ace, 2, 3, 4, 5, 6, 7, 8, 9, 10, jack, queen, and king. The jacks, queens, and kings are referred to as "face cards." Imagine that we shuffle the deck thoroughly and deal one card. Define events F: getting a face card, and H: getting a heart. The two-way table summarizes the sample space for this random process.

| | | Card | | |
|---|---|---|---|---|
| | | Face card | Nonface card | Total |
| Suit | Heart | 3 | 10 | 13 |
| | Nonheart | 9 | 30 | 39 |
| | Total | 12 | 40 | 52 |

(a) Find $P(H^C)$.

(b) Find $P(H^C \text{ and } F)$. Interpret this value in context.

(c) Find $P(H^C \text{ or } F)$.

43. **Cell phones** The Pew Research Center asked a random sample of 2024 adult cell-phone owners from the United States their age and which type of cell phone they own: iPhone, Android, or other (including non-smartphones). The two-way table summarizes the data.

| | | Age | | | |
|---|---|---|---|---|---|
| | | 18–34 | 35–54 | 55+ | Total |
| Type of cell phone | iPhone | 169 | 171 | 127 | 467 |
| | Android | 214 | 189 | 100 | 503 |
| | Other | 134 | 277 | 643 | 1054 |
| | Total | 517 | 637 | 870 | 2024 |

Suppose we select one of the survey respondents at random. What's the probability that:

(a) The person is not age 18 to 34 and does not own an iPhone?

(b) The person is age 18 to 34 or owns an iPhone?

44. **Middle school values** Researchers carried out a survey of fourth-, fifth-, and sixth-grade students in Michigan. Students were asked whether good grades, athletic ability, or being popular was most important to them. The two-way table summarizes the survey data.[9]

| | | Grade | | | |
|---|---|---|---|---|---|
| | | 4th grade | 5th grade | 6th grade | Total |
| Most important | Grades | 49 | 50 | 69 | 168 |
| | Athletic | 24 | 36 | 38 | 98 |
| | Popular | 19 | 22 | 28 | 69 |
| | Total | 92 | 108 | 135 | 335 |

Suppose we select one of these students at random. What's the probability of each of the following?

(a) The student is a sixth-grader or rated good grades as important.

(b) The student is not a sixth-grader and did not rate good grades as important.

45. **Roulette** An American roulette wheel has 38 slots with numbers 1 through 36, 0, and 00, as shown in the figure. Of the numbered slots, 18 are red, 18 are black, and 2—the 0 and 00—are green. When the wheel is spun, a metal ball is dropped onto the middle of the wheel. If the wheel is balanced, the ball is equally likely to settle in any of the numbered slots. Imagine spinning a fair wheel once. Define events B: ball lands in a black slot, and E: ball lands in an even-numbered slot. (Treat 0 and 00 as even numbers.)

(a) Make a two-way table that displays the sample space in terms of events B and E.

(b) Find $P(B)$ and $P(E)$.

(c) Describe the event "B and E" in words. Then find the probability of this event.

(d) Explain why $P(B \text{ or } E) \neq P(B) + P(E)$. Then use the general addition rule to compute $P(B \text{ or } E)$.

46. **Colorful disks** A jar contains 36 disks: 9 each of four colors—red, green, blue, and yellow. Each set of disks of the same color is numbered from 1 to 9. Suppose you draw one disk at random from the jar. Define events R: get a red disk, and N: get a disk with the number 9.

(a) Make a two-way table that describes the sample space in terms of events R and N.

(b) Find $P(R)$ and $P(N)$.

(c) Describe the event "R and N" in words. Then find the probability of this event.

(d) Explain why $P(R \text{ or } N) \neq P(R) + P(N)$. Then use the general addition rule to compute $P(R \text{ or } N)$.

47. **Dogs and cats** In one large city, 40% of all households own a dog, 32% own a cat, and 18% own both. Suppose we randomly select a household. What's the probability that the household owns a dog or a cat?
pg 349

48. **Reading the paper** In a large business hotel, 40% of guests read the *Los Angeles Times*. Only 25% read the *Wall Street Journal*. Five percent of guests read both papers. Suppose we select a hotel guest at random and record which of the two papers the person reads, if either. What's the probability that the person reads the *Los Angeles Times* or the *Wall Street Journal?*

49. **Mac or PC?** A recent census at a major university revealed that 60% of its students mainly used Macs. The rest mainly used PCs. At the time of the census, 67% of the school's students were undergraduates. The rest were graduate students. In the census, 23% of respondents were graduate students and used a Mac as their main computer. Suppose we select a student at random from among those who were part of the census. Define events G: is a graduate student, and M: primarily uses a Mac.

(a) Find $P(G \cup M)$. Interpret this value in context.

(b) Consider the event that the randomly selected student is an undergraduate student and primarily uses a PC. Write this event in symbolic form and find its probability.

50. **Gender and political party** In January 2017, 52% of U.S. senators were Republicans and the rest were Democrats or Independents. Twenty-one percent of the senators were females, and 47% of the senators were male Republicans. Suppose we select one of these senators at random. Define events R: is a Republican, and M: is male.

(a) Find $P(R \cup M)$. Interpret this value in context.

(b) Consider the event that the randomly selected senator is a female Democrat or Independent. Write this event in symbolic form and find its probability.

51. **Dogs and cats** Refer to Exercise 47.

(a) Make a Venn diagram to display the outcomes of this random process using events D: owns a dog and C: owns a cat.
pg 352

(b) Find $P(D \cap C^C)$.

52. **Reading the paper** Refer to Exercise 48.

(a) Make a Venn diagram to display the outcomes of this random process using events L: reads the *Los Angeles Times* and W: reads the *Wall Street Journal*.

(b) Find $P(L^C \cap W)$.

53. **Union and intersection** Suppose A and B are two events such that $P(A) = 0.3$, $P(B) = 0.4$, and $P(A \cup B) = 0.58$. Find $P(A \cap B)$.

54. **Union and intersection** Suppose C and D are two events such that $P(C) = 0.6$, $P(D) = 0.45$, and $P(C \cup D) = 0.75$. Find $P(C \cap D)$.

**Multiple Choice:** *Select the best answer for Exercises 55–58.*

55. The partially completed table that follows shows the distribution of scores on the 2016 AP® Statistics exam.

| Score | 1 | 2 | 3 | 4 | 5 |
|---|---|---|---|---|---|
| Probability | 0.235 | 0.155 | 0.249 | 0.217 | ? |

Suppose we randomly select a student who took this exam. What's the probability that he or she earned a score of at least 3?

(a) 0.249 (b) 0.361 (c) 0.390
(d) 0.466 (e) 0.610

56. In a sample of 275 students, 20 say they are vegetarians. Of the vegetarians, 9 eat both fish and eggs, 3 eat eggs but not fish, 1 eats fish but not eggs, and 7 eat neither. Choose one of the vegetarians at random. What is the probability that the chosen student eats fish or eggs?

(a) 9/20 (b) 13/20 (c) 22/20
(d) 9/275 (e) 22/275

*Exercises 57 and 58 refer to the following setting.* The casino game craps is based on rolling two dice. Here is the assignment of probabilities to the sum of the numbers on the up-faces when two dice are rolled:

| Outcome | 2 | 3 | 4 | 5 | 6 | 7 | 8 | 9 | 10 | 11 | 12 |
|---|---|---|---|---|---|---|---|---|---|---|---|
| Probability | 1/36 | 2/36 | 3/36 | 4/36 | 5/36 | 6/36 | 5/36 | 4/36 | 3/36 | 2/36 | 1/36 |

57. The most common bet in craps is the "pass line." A pass line bettor wins immediately if either a 7 or an 11 comes up on the first roll. This is called a *natural*. What is the probability that a natural does *not* occur?

(a) 2/36 (b) 6/36 (c) 8/36
(d) 16/36 (e) 28/36

58. If a player rolls a 2, 3, or 12, it is called *craps*. What is the probability of getting craps or an even sum on one roll of the dice?

(a) 4/36 (b) 18/36 (c) 20/36
(d) 22/36 (e) 32/36

## Recycle and Review

59. **Crawl before you walk** (3.1, 3.2, 4.3) At what age do babies learn to crawl? Does it take longer to learn in the winter, when babies are often bundled in clothes that restrict their movement? Perhaps there might even be an association between babies' crawling age and the average temperature during the month they first try to crawl (around 6 months after birth). Data were collected from parents who brought their babies to the University of Denver Infant Study Center to participate in one of a number of studies. Parents reported the birth month and the age at which their child was first able to creep or crawl a distance of 4 feet within one minute. Information was obtained on 414 infants (208 boys and 206 girls). Crawling age is given in weeks, and average temperature (in degrees Fahrenheit) is given for the month that is 6 months after the birth month.[10]

| Birth month | Average crawling age | Average temperature |
|---|---|---|
| January | 29.84 | 66 |
| February | 30.52 | 73 |
| March | 29.70 | 72 |
| April | 31.84 | 63 |
| May | 28.58 | 52 |
| June | 31.44 | 39 |
| July | 33.64 | 33 |
| August | 32.82 | 30 |
| September | 33.83 | 33 |
| October | 33.35 | 37 |
| November | 33.38 | 48 |
| December | 32.32 | 57 |

(a) Make an appropriate graph to display the relationship between average temperature and average crawling age. Describe what you see.

Some computer output from a linear regression analysis of the data is shown.

| Term | Coef | SE Coef | T-Value | P-Value |
|---|---|---|---|---|
| Constant | 35.68 | 1.32 | 27.08 | 0.000 |
| Average temperature | −0.0777 | 0.0251 | −3.10 | 0.011 |
| S = 1.31920 | R-Sq = 48.96% | | R-Sq(adj) = 43.86% | |

(b) What is the equation of the least-squares regression line that describes the relationship between average temperature and average crawling age? Define any variables that you use.

(c) Interpret the slope of the regression line.

(d) Can we conclude that warmer temperatures 6 months after babies are born causes them to crawl sooner? Justify your answer.

60. **Treating low bone density** (4.2, 4.3) Fractures of the spine are common and serious among women with advanced osteoporosis (low mineral density in the bones). Can taking strontium ranelate help? A large medical trial was conducted to investigate this question. Researchers recruited 1649 women with osteoporosis who had previously had at least one fracture for an experiment. The women were assigned to take either strontium ranelate or a placebo each day. All the women were taking calcium supplements and receiving standard medical care. One response variable was the number of new fractures over 3 years.[11]

(a) Describe a completely randomized design for this experiment.

(b) Explain why it is important to keep the calcium supplements and medical care the same for all the women in the experiment.

(c) The women who took strontium ranelate had statistically significantly fewer new fractures, on average, than the women who took a placebo over a 3-year period. Explain what this means to someone who knows little statistics.

# SECTION 5.3 Conditional Probability and Independence

**LEARNING TARGETS** *By the end of the section, you should be able to:*

- Calculate and interpret conditional probabilities.
- Determine if two events are independent.
- Use the general multiplication rule to calculate probabilities.
- Use a tree diagram to model a random process involving a sequence of outcomes and to calculate probabilities.
- When appropriate, use the multiplication rule for independent events to calculate probabilities.

The probability of an event can change if we know that some other event has occurred. For instance, suppose you toss a fair coin twice. The probability of getting two heads is 1/4 because the sample space consists of the 4 equally likely outcomes

HH HT TH TT

Suppose that the first toss lands tails. Now what's the probability of getting two heads? It's 0. Knowing that the first toss is a tail changes the probability that you get two heads.

This idea is the key to many applications of probability.

## What Is Conditional Probability?

Let's return to the college statistics class from Section 5.2. Earlier, we used the two-way table shown on the next page to find probabilities involving events A: is male and B: has a pierced ear for a randomly selected student.

| | | Gender | | |
|---|---|---|---|---|
| | | Male | Female | Total |
| Pierced ear | Yes | 19 | 84 | 103 |
| | No | 71 | 4 | 75 |
| | Total | 90 | 88 | 178 |

Here is a summary of our previous results:

$P(\text{A}) = P(\text{male}) = 90/178$ $\qquad$ $P(\text{A} \cap \text{B}) = P(\text{male and pierced ear}) = 19/178$

$P(\text{B}) = P(\text{pierced ear}) = 103/178$ $\qquad$ $P(\text{A} \cup \text{B}) = P(\text{male or pierced ear}) = 174/178$

Now let's turn our attention to some other interesting probability questions.

| | | Gender | | |
|---|---|---|---|---|
| | | Male | Female | Total |
| Pierced ear | Yes | 19 | 84 | 103 |
| | No | 71 | 4 | 75 |
| | Total | 90 | 88 | 178 |

1. **If we know that a randomly selected student has a pierced ear, what is the probability that the student is male?** There are 103 students in the class with a pierced ear. We can restrict our attention to this group, since we are told that the chosen student has a pierced ear. Because there are 19 males among the 103 students with a pierced ear, the desired probability is

$$P(\text{male } \textit{given} \text{ pierced ear}) = 19/103 = 0.184$$

| | | Gender | | |
|---|---|---|---|---|
| | | Male | Female | Total |
| Pierced ear | Yes | 19 | 84 | 103 |
| | No | 71 | 4 | 75 |
| | Total | 90 | 88 | 178 |

2. **If we know that a randomly selected student is male, what's the probability that the student has a pierced ear?** This time, our attention is focused on the males in the class. Because 19 of the 90 males in the class have a pierced ear,

$$P(\text{pierced ear } \textit{given} \text{ male}) = 19/90 = 0.211$$

These two questions sound alike, but they actually ask two very different things. Each of these probabilities is an example of a **conditional probability**. The name comes from the fact that we are trying to find the probability that one event will happen under the *condition* that some other event is already known to have occurred. We often use the phrase "given that" to signal the condition.

**DEFINITION Conditional probability**

The probability that one event happens given that another event is known to have happened is called a **conditional probability**. The conditional probability that event A happens given that event B has happened is denoted by $P(\text{A}|\text{B})$.

With this new notation available, we can restate the answers to the two questions just posed as

$$P(\text{male}|\text{pierced ear}) = P(\text{A}|\text{B}) = 19/103$$

and

$$P(\text{pierced ear}|\text{male}) = P(\text{B}|\text{A}) = 19/90$$

Here's an example that illustrates how conditional probability works in a familiar setting.

## EXAMPLE

### A *Titanic* disaster
### Two-way tables and conditional probabilities

**PROBLEM:** In 1912, the luxury liner *Titanic*, on its first voyage across the Atlantic, struck an iceberg and sank. Some passengers got off the ship in lifeboats, but many died. The two-way table gives information about adult passengers who survived and who died, by class of travel.

Images Group/REX/Shutterstock

| | | Class of travel | | | |
|---|---|---|---|---|---|
| | | First | Second | Third | Total |
| Survival status | Survived | 197 | 94 | 151 | 442 |
| | Died | 122 | 167 | 476 | 765 |
| | Total | 319 | 261 | 627 | 1207 |

Suppose we randomly select one of the adult passengers from the *Titanic*. Define events F: first-class passenger, S: survived, and T: third-class passenger.

(a) Find $P(T \mid S)$. Interpret this value in context.

(b) Given that the chosen person is not a first-class passenger, what's the probability that she or he survived? Write your answer as a probability statement using correct symbols for the events.

**SOLUTION:**

(a) $P(T|S) = P(\text{third-class passenger}|\text{survived}) = 151/442 = 0.342.$ Given that the randomly chosen person survived, there is about a 34.2% chance that she or he was a third-class passenger.

To answer part (a), only consider values in the "Survived" row.

(b) $P(\text{survived}|\text{not first-class passenger})$

$$= P(S|F^C) = \frac{94+151}{261+627} = \frac{245}{888} = 0.276$$

To answer part (b), only consider values in the "Second class" and "Third class" columns.

**FOR PRACTICE, TRY EXERCISE 61**

Is there is a connection between conditional probability and conditional relative frequency from Chapter 1? Yes! In part (a) of the example, we found the conditional probability $P(\text{third-class passenger} \mid \text{survived}) = 151/442 = 0.342$. In Chapter 1, we asked, "What proportion of survivors were third-class passengers?" Our answer was also $151/442 = 0.342$. This is a conditional relative frequency because we are finding the percent or proportion of third-class passengers among those who survived. So a conditional probability is just a conditional relative frequency that comes from a random process—in this case, randomly selecting an adult passenger.

Let's look more closely at how conditional probabilities are calculated using the data from the college statistics class. From the two-way table that follows, we see that

$$P(\text{male} \mid \text{pierced ear}) = \frac{19}{103} = \frac{\text{number of students who are male and have a pierced ear}}{\text{number of students with a pierced ear}}$$

| | | Gender | | |
|---|---|---|---|---|
| | | Male | Female | Total |
| Pierced ear | Yes | 19 | 84 | 103 |
| | No | 71 | 4 | 75 |
| | Total | 90 | 88 | 178 |

What if we focus on probabilities instead of numbers of students? Notice that

$$\frac{P(\text{male and pierced ear})}{P(\text{pierced ear})} = \frac{\frac{19}{178}}{\frac{103}{178}} = \frac{19}{103} = P(\text{male} \mid \text{pierced ear})$$

This observation leads to a general formula for computing a conditional probability.

**CALCULATING CONDITIONAL PROBABILITIES**

To find the conditional probability $P(A \mid B)$, use the formula

$$P(A \mid B) = \frac{P(A \text{ and } B)}{P(B)} = \frac{P(A \cap B)}{P(B)} = \frac{P(\text{both events occur})}{P(\text{given event occurs})}$$

By the same reasoning,

$$P(B \mid A) = \frac{P(B \text{ and } A)}{P(A)} = \frac{P(B \cap A)}{P(A)}$$

## EXAMPLE

### Facebook or Instagram?
### Calculating conditional probability

**PROBLEM:** A survey of all residents in a large apartment complex reveals that 68% use Facebook, 28% use Instagram, and 25% do both. Suppose we select a resident at random. Given that the person uses Facebook, what's the probability that she or he uses Instagram?

**SOLUTION:**

$$P(\text{Instagram} \mid \text{Facebook}) = P(I \mid F) = \frac{P(I \cap F)}{P(F)} = \frac{0.25}{0.68} = 0.368$$

FOR PRACTICE, TRY EXERCISE 69

**AP® EXAM TIP**

You can write statements like $P(A \mid B)$ if events A and B are clearly defined in a problem. Otherwise, it's probably easier to use contextual labels, like $P(I \mid F)$ in the preceding example. Or you can just use words: $P(\text{Instagram} \mid \text{Facebook})$.

Refer back to the example. If the person chosen is an Instagram user, what is the probability that he or she uses Facebook? By the conditional probability formula, it's

$$P(\text{Facebook} \mid \text{Instagram}) = P(F \mid I) = \frac{P(F \cap I)}{P(I)} = \frac{0.25}{0.28} = 0.893$$

If the chosen resident uses Instagram, it is extremely likely that he or she uses Facebook. However, if the chosen resident uses Facebook, he or she is not nearly so likely to use Instagram.

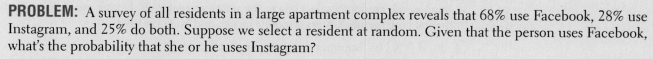

You could also use a two-way table to help you find these conditional probabilities. Here's the table that we made for this setting in Section 5.2 (page 353). It is easy to see that

$$P(\text{Instagram} \mid \text{Facebook}) = \frac{0.25}{0.68} = 0.368 \quad \text{and} \quad P(\text{Facebook} \mid \text{Instagram}) = \frac{0.25}{0.28} = 0.893$$

| | | Facebook use | | |
|---|---|---|---|---|
| | | Yes | No | Total |
| Instagram use | Yes | 25% | 3% | 28% |
| | No | 43% | 29% | 72% |
| | Total | 68% | 32% | 100% |

## CHECK YOUR UNDERSTANDING

Yellowstone National Park surveyed a random sample of 1526 winter visitors to the park. They asked each person whether he or she owned, rented, or had never used a snowmobile. Respondents were also asked whether they belonged to an environmental organization (like the Sierra Club). The two-way table summarizes the survey responses.

| | | Environmental club | | |
|---|---|---|---|---|
| | | No | Yes | Total |
| Snowmobile experience | Never used | 445 | 212 | 657 |
| | Renter | 497 | 77 | 574 |
| | Owner | 279 | 16 | 295 |
| | Total | 1221 | 305 | 1526 |

Suppose we randomly select one of the survey respondents. Define events E: environmental club member, S: snowmobile owner, and N: never used.

1. Find $P(\text{N} \mid \text{E})$. Interpret this value in context.
2. Given that the chosen person is not a snowmobile owner, what's the probability that she or he is an environmental club member? Write your answer as a probability statement using correct symbols for the events.
3. Is the chosen person more likely to not be an environmental club member if he or she has never used a snowmobile, is a snowmobile owner, or a snowmobile renter? Justify your answer.

# Conditional Probability and Independence

Suppose you toss a fair coin twice. Define events A: first toss is a head, and B: second toss is a head. We know that $P(\text{A}) = 1/2$ and $P(\text{B}) = 1/2$.

- What's $P(\text{B} \mid \text{A})$? It's the conditional probability that the second toss is a head given that the first toss was a head. The coin has no memory, so $P(\text{B} \mid \text{A}) = 1/2$.
- What's $P(\text{B} \mid \text{A}^C)$? It's the conditional probability that the second toss is a head given that the first toss was not a head. Getting a tail on the first toss

does not change the probability of getting a head on the second toss, so $P(B \mid A^C) = 1/2$.

In this case, $P(B \mid A) = P(B \mid A^C) = P(B)$. Knowing the outcome of the first toss does not change the probability that the second toss is a head. We say that A and B are **independent events**.

**DEFINITION Independent events**

A and B are **independent events** if knowing whether or not one event has occurred does not change the probability that the other event will happen. In other words, events A and B are independent if

$$P(A|B) = P(A|B^C) = P(A)$$

Alternatively, events A and B are independent if

$$P(B|A) = P(B|A^C) = P(B)$$

Let's contrast the coin-toss scenario with our earlier pierced-ear example. In that case, the random process involved randomly selecting a student from a college statistics class. The events of interest were A: is male, and B: has a pierced ear. Are these two events independent?

| | | Gender | | |
|---|---|---|---|---|
| | | Male | Female | Total |
| Pierced ear | Yes | 19 | 84 | 103 |
| | No | 71 | 4 | 75 |
| | Total | 90 | 88 | 178 |

- Suppose that the chosen student is male. We can see from the two-way table that $P(\text{pierced ear} \mid \text{male}) = P(B \mid A) = 19/90 = 0.211$.
- Suppose that the chosen student is female. From the two-way table, we see that $P(\text{pierced ear} \mid \text{female}) = P(B \mid A^C) = 84/88 = 0.955$.

Knowing that the chosen student is a male changes (greatly reduces) the probability that the student has a pierced ear. So these two events are not independent.

Another way to determine if two events A and B are independent is to compare $P(A \mid B)$ to $P(A)$ or $P(B \mid A)$ to $P(B)$. For the pierced-ear setting,

$$P(\text{pierced ear}|\text{male}) = P(B|A) = 19/90 = 0.211$$

The unconditional probability that the chosen student has a pierced ear is

$$P(\text{pierced ear}) = P(B) = 103/178 = 0.579$$

Again, knowing that the chosen student is male changes (reduces) the probability that the individual has a pierced ear. So these two events are not independent.

## EXAMPLE

### Gender and handedness

### Checking for independence

Graeme Harris/Getty Images

**PROBLEM:** Is there a relationship between gender and handedness? To find out, we used Census At School's Random Data Selector to choose an SRS of 100 Australian high school students who completed a survey. The two-way table summarizes the relationship between gender and dominant hand for these students.

| | | Gender | | |
|---|---|---|---|---|
| | | Male | Female | Total |
| Dominant hand | Right | 39 | 51 | 90 |
| | Left | 7 | 3 | 10 |
| | Total | 46 | 54 | 100 |

Suppose we choose one of the students in the sample at random. Are the events "male" and "left-handed" independent? Justify your answer.

**SOLUTION:**

$P(\text{left-handed} \mid \text{male}) = 7/46 = 0.152$

$P(\text{left-handed} \mid \text{female}) = 3/54 = 0.056$

**Because these probabilities are not equal, the events "male" and "left-handed" are not independent. Knowing that the student is male increases the probability that the student is left-handed.**

**FOR PRACTICE, TRY EXERCISE 71**

In the example, we could have also determined that the two events are not independent by showing that

$$P(\text{left-handed} \mid \text{male}) = 7/46 = 0.152 \neq P(\text{left-handed}) = 10/100 = 0.100$$

Or we could have focused on whether knowing that the chosen student is left-handed changes the probability that the person is male. Because

$$P(\text{male} \mid \text{left-handed}) = 7/10 = 0.70 \neq P(\text{male} \mid \text{right-handed}) = 39/90 = 0.433$$

the events "male" and "left-handed" are not independent.

You might have thought, "Surely there's no connection between gender and handedness. The events 'male' and 'left-handed' are bound to be independent." As the example shows, you can't use your intuition to check whether events are independent. To be sure, you have to calculate some probabilities.

*Is there a connection between independence of events and association between two variables?* Yes! In the preceding example, we found that the events "male" and "left-handed" were not independent for the sample of 100 Australian high school students. Knowing a student's gender helped us predict his or her dominant hand. By what you learned in Chapter 1, there is an association between gender and handedness *for the students in the sample.* The segmented bar graph shows the association in picture form.

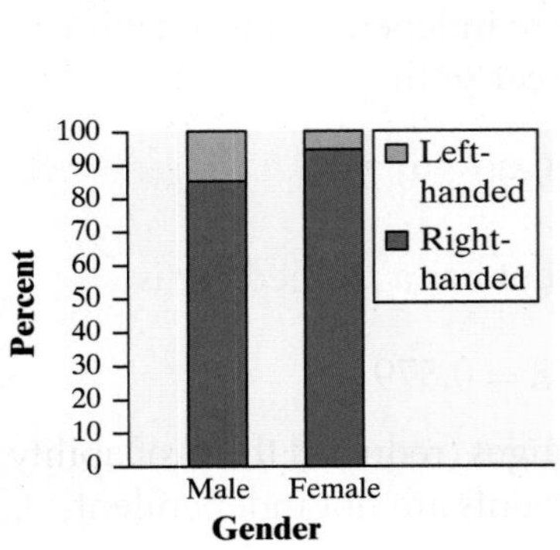

Does that mean an association exists between gender and handedness in the larger population? Maybe. If there is no association between the variables in the population, it would be surprising to choose a random sample of 100 students for which $P$(left-handed | male), $P$(left-handed | female), and $P$(left-handed) were *exactly* equal. But these probabilities should be close to equal if there's no association between the variables in the population. How close is close? We'll discuss this issue further in Chapter 12.

**CHECK YOUR UNDERSTANDING**

For each random process given, determine whether the events are independent. Justify your answer.

1. Shuffle a standard deck of cards, and turn over the top card. Put it back in the deck, shuffle again, and turn over the top card. Define events A: first card is a heart, and B: second card is a heart.
2. Shuffle a standard deck of cards, and turn over the top two cards, one at a time. Define events A: first card is a heart, and B: second card is a heart.
3. The 28 students in Mr. Tabor's AP® Statistics class took a quiz on conditional probability and independence. The two-way table summarizes the class's quiz results based on gender and whether the student got an A. Choose a student from the class at random. The events of interest are "female" and "got an A."

| | | Gender | |
|---|---|---|---|
| | | Female | Male |
| Got an A | Yes | 3 | 1 |
| | No | 18 | 6 |

## The General Multiplication Rule

Suppose that A and B are two events resulting from the same random process. We can find the probability $P$(A or B) with the general addition rule:

$$P(\text{A or B}) = P(\text{A}) + P(\text{B}) - P(\text{A and B})$$

How do we find the probability that both events happen, $P$(A and B)?

Consider this situation: about 55% of high school students participate in a school athletic team at some level. Roughly 6% of these athletes go on to play on a college team in the NCAA.[12] What percent of high school students play a sport in high school *and* go on to play on an NCAA team? About 6% of 55%, or roughly 3.3%.

Let's restate the situation in probability language. Suppose we select a high school student at random. What's the probability that the student plays a sport in high school and goes on to play on an NCAA team? The given information suggests that

$$P(\text{high school sport}) = 0.55 \text{ and } P(\text{NCAA team} \mid \text{high school sport}) = 0.06$$

By the logic just stated,

$$\begin{aligned} &P(\text{high school sport and NCAA team}) \\ &\quad = P(\text{high school sport}) \cdot P(\text{NCAA team} \mid \text{high school sport}) \\ &\quad = (0.55)(0.06) = 0.033 \end{aligned}$$

This is an example of the **general multiplication rule**.

**DEFINITION General multiplication rule**

For any random process, the probability that events A and B both occur can be found using the **general multiplication rule**:

$$P(\text{A and B}) = P(\text{A} \cap \text{B}) = P(\text{A}) \cdot P(\text{B} \mid \text{A})$$

The general multiplication rule says that for both of two events to occur, first one must occur. Then, given that the first event has occurred, the second must occur. To confirm that this result is correct, start with the conditional probability formula

$$P(\text{B} \mid \text{A}) = \frac{P(\text{B} \cap \text{A})}{P(\text{A})}$$

The numerator gives the probability we want because $P(\text{B} \cap \text{A})$ is the same as $P(\text{A} \cap \text{B})$. Multiply both sides of the previous equation by $P(\text{A})$ to get

$$P(\text{A}) \cdot P(\text{B} \mid \text{A}) = P(\text{A} \cap \text{B})$$

**EXAMPLE**

### Teens and social media
### The general multiplication rule

**PROBLEM:** The Pew Internet and American Life Project reported that 79% of teenagers (ages 13 to 17) use social media, and that 39% of teens who use social media feel pressure to post content that will be popular and get lots of comments or likes.[13] Find the probability that a randomly selected teen uses social media and feels pressure to post content that will be popular and get lots of comments or likes.

**SOLUTION:**

$$P(\text{use social media and feel pressure}) = P(\text{use social media}) \cdot P(\text{feel pressure} \mid \text{use social media})$$

$$= (0.79)(0.39) = 0.308$$

**FOR PRACTICE, TRY EXERCISE 77**

## Tree Diagrams and Conditional Probability

Shannon hits the snooze button on her alarm on 60% of school days. If she hits snooze, there is a 0.70 probability that she makes it to her first class on time. If she doesn't hit snooze and gets up right away, there is a 0.90 probability that she makes it to class on time. Suppose we select a school day at random and record whether Shannon hits the snooze button and whether she arrives in class on time. Figure 5.7 shows a **tree diagram** for this random process.

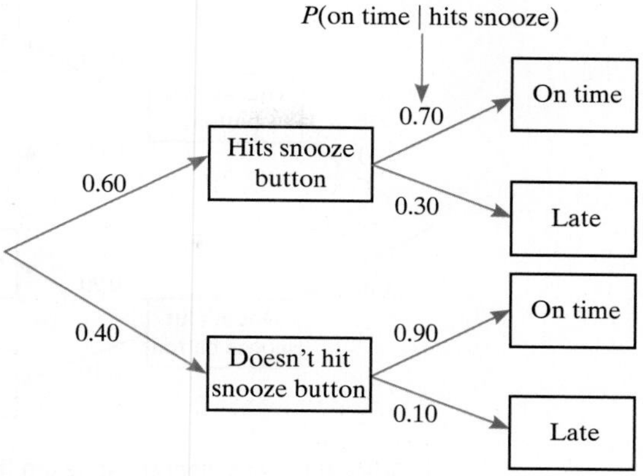


**FIGURE 5.7** A tree diagram displaying the sample space of randomly choosing a school day and noting if Shannon hits the snooze button or not and whether she gets to her first class on time.

There are only two possible outcomes at the first "stage" of this random process: Shannon hits the snooze button or she doesn't. The first set of branches in the tree diagram displays these outcomes with their probabilities. The second set of branches shows the two possible results at the next "stage" of the process—Shannon gets to her first class on time or arrives late—and the probability of each result based on whether or not she hit the snooze button. Note that the probabilities on the second set of branches are *conditional* probabilities, like $P(\text{on time} \mid \text{hits snooze}) = 0.70$.

**DEFINITION Tree diagram**

A **tree diagram** shows the sample space of a random process involving multiple stages. The probability of each outcome is shown on the corresponding branch of the tree. All probabilities after the first stage are conditional probabilities.

We can ask some interesting questions related to the tree diagram:

- **What is the probability that Shannon hits the snooze button and is late for class on a randomly selected school day?** The general multiplication rule provides the answer:

$$\begin{aligned} P(\text{hits snooze and late}) &= P(\text{hits snooze}) \cdot P(\text{late} \mid \text{hits snooze}) \\ &= (0.60)(0.30) \\ &= 0.18 \end{aligned}$$

  There is an 18% chance that Shannon hits the snooze button and is late for class. Note that the previous calculation amounts to multiplying probabilities along the branches of the tree diagram.

- **What's the probability that Shannon is late to class on a randomly selected school day?** Figure 5.8 on the next page illustrates two ways this can happen: Shannon hits the snooze button and is late or she doesn't hit snooze and is late. Because these outcomes are mutually exclusive,

$$P(\text{late}) = P(\text{hits snooze and late}) + P(\text{doesn't hits snooze and late})$$

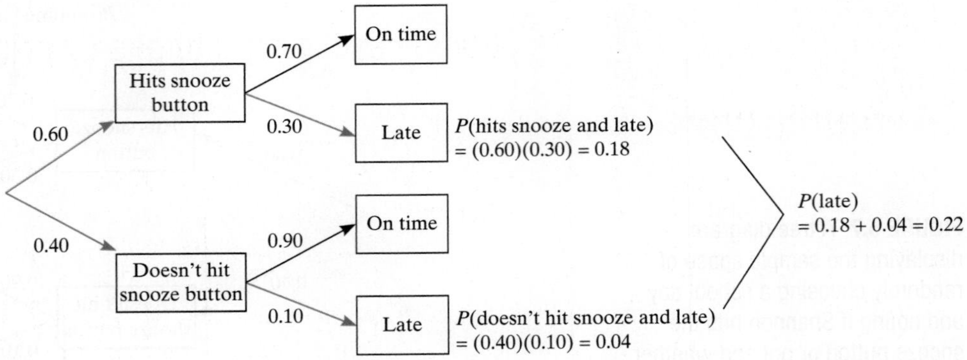


**FIGURE 5.8** Tree diagram showing the two possible ways that Shannon can be late to class on a randomly selected day.

The general multiplication rule tells us that

$$
\begin{aligned}
P(\text{doesn't hit snooze and late}) &= P(\text{doesn't hit snooze}) \cdot P(\text{late} \mid \text{doesn't hit snooze}) \\
&= (0.40)(0.10) \\
&= 0.04
\end{aligned}
$$

So $P(\text{late}) = 0.18 + 0.04 = 0.22$. There is a 22% chance that Shannon will be late to class.

- **Suppose that Shannon is late for class on a randomly chosen school day. What is the probability that she hit the snooze button that morning?** To find this probability, we start with the given information that Shannon is late, which is displayed on the second set of branches in the tree diagram, and ask whether she hit the snooze button, which is shown on the first set of branches. We can use the information from the tree diagram and the conditional probability formula to do the required calculation:

$$
\begin{aligned}
P(\text{hit snooze button} \mid \text{late}) &= \frac{P(\text{hit snooze button and late})}{P(\text{late})} \\
&= \frac{0.18}{0.22} \\
&= 0.818
\end{aligned}
$$

Given that Shannon is late for school on a randomly selected day, there is a 0.818 probability that she hit the snooze button.

This method for solving conditional probability problems that involve "going backward" in a tree diagram is sometimes referred to as *Bayes's theorem.* It was developed by the Reverend Thomas Bayes in the 1700s.

Some interesting conditional probability questions—like this one about P(hit snooze button | late)—involve "going in reverse" on a tree diagram. Note that we just use the conditional probability formula and plug in the appropriate values to answer such questions.

## EXAMPLE

### Do people read more ebooks or print books?
Tree diagrams and probability

Burlingham/Shutterstock.com

**PROBLEM:** Recently, Harris Interactive reported that 20% of millennials, 25% of Gen Xers, 21% of baby boomers, and 17% of matures (age 68 and older) read more ebooks than print books. According to the U.S. Census Bureau, 34% of those 18 and over are millennials, 22% are Gen Xers, 30% are baby boomers, and 14% are matures. Suppose we select one U.S. adult at random and record which generation the person is from and whether she or he reads more ebooks or print books.

(a) Draw a tree diagram to model this random process.

(b) Find the probability that the person reads more ebooks than print books.

(c) Suppose the chosen person reads more ebooks than print books. What's the probability that she or he is a millennial?

**SOLUTION:**

(a)

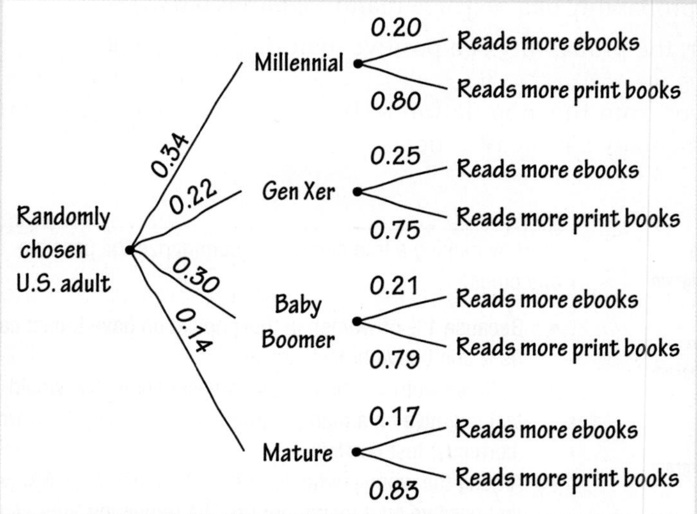


(b) $P(\text{reads more ebooks}) = (0.34)(0.20) + (0.22)(0.25) + (0.30)(0.21) + (0.14)(0.17)$
$= 0.0680 + 0.0550 + 0.0630 + 0.0238$
$= 0.2098$

(c) $$P(\text{millennial} \mid \text{reads more ebooks}) = \frac{P(\text{millennial and reads more ebooks})}{P(\text{reads more ebooks})} = \frac{0.068}{0.2098} = 0.3241$$

$$P(A \mid B) = \frac{P(A \cap B)}{P(B)}$$

**FOR PRACTICE, TRY EXERCISE 81**

One of the most important applications of conditional probability is in the area of drug and disease testing.

## EXAMPLE

### Mammograms
### Tree diagrams and conditional probability

**PROBLEM:** Many women choose to have annual mammograms to screen for breast cancer after age 40. A mammogram isn't foolproof. Sometimes the test suggests that a woman has breast cancer when she really doesn't (a "false positive"). Other times, the test says that a woman doesn't have breast cancer when she actually does (a "false negative").

Sam Edwards/Getty Images

Suppose that we know the following information about breast cancer and mammograms in a particular population:

- One percent of the women aged 40 or over in this population have breast cancer.
- For women who have breast cancer, the probability of a negative mammogram is 0.03.
- For women who don't have breast cancer, the probability of a positive mammogram is 0.06.

A randomly selected woman aged 40 or over from this population tests positive for breast cancer in a mammogram. Find the probability that she actually has breast cancer.

**SOLUTION:**

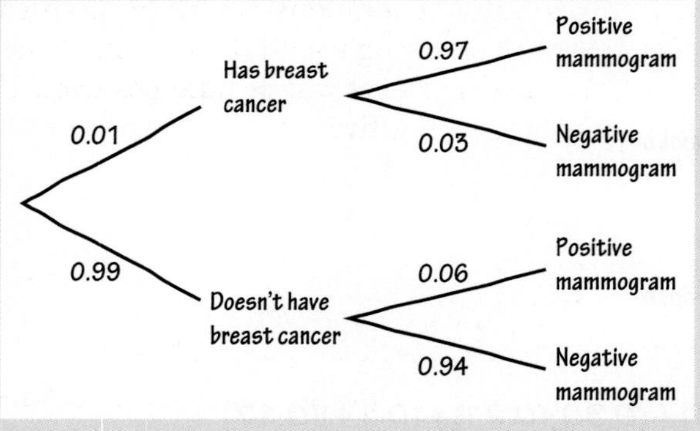


Start by making a tree diagram to summarize the possible outcomes.

- Because 1% of women in this population have breast cancer, 99% don't have breast cancer.
- Of those women who do have breast cancer, 3% would test negative on a mammogram. The remaining 97% would (correctly) test positive.
- Among the women who don't have breast cancer, 6% would test positive on a mammogram. The remaining 94% would (correctly) test negative.

$$
\begin{aligned}
P(\text{breast cancer} \mid \text{positive mammogram}) &= \frac{P(\text{breast cancer and positive mammogram})}{P(\text{positive mammogram})} \\
&= \frac{(0.01)(0.97)}{(0.01)(0.97)+(0.99)(0.06)} \\
&= \frac{0.0097}{0.0691} \\
&= 0.14
\end{aligned}
$$

**FOR PRACTICE, TRY EXERCISE 83**

Are you surprised by the final result of the example—given that a randomly selected woman from the population in question has a positive mammogram, there is only about a 14% chance that she has breast cancer? Most people are. Sometimes a two-way table that includes counts is more convincing.

To make calculations simple, we'll suppose that there are exactly 10,000 women aged 40 or over in this population, and that exactly 100 have breast cancer (that's 1% of the women).

- How many of those 100 would have a positive mammogram? It would be 97% of 100, or 97 of them. That leaves 3 who would test negative.
- How many of the 9900 women who don't have breast cancer would get a positive mammogram? Six percent of them, or $(9900)(0.06) = 594$ women. The remaining $9900 - 594 = 9306$ would test negative.
- In total, $97 + 594 = 691$ women would have positive mammograms and $3 + 9306 = 9309$ women would have negative mammograms.

This information is summarized in the two-way table.

| | | Has breast cancer? | | |
|---|---|---|---|---|
| | | Yes | No | Total |
| **Mammogram result** | Positive | 97 | 594 | 691 |
| | Negative | 3 | 9306 | 9309 |
| | Total | 100 | 9900 | 10,000 |

Given that a randomly selected woman has a positive mammogram, the two-way table shows that the conditional probability is

$$P(\text{breast cancer} \mid \text{positive mammogram}) = 97/691 = 0.14$$

This example illustrates an important fact when considering proposals for widespread testing for serious diseases or illegal drug use: if the condition being tested is uncommon in the population, many positives will be false positives. The best remedy is to retest any individual who tests positive.

## CHECK YOUR UNDERSTANDING

A computer company makes desktop, laptop, and tablet computers at factories in two states: California and Texas. The California factory produces 40% of the company's computers and the Texas factory makes the rest. Of the computers made in California, 25% are desktops, 30% are laptops, and the rest are tablets. Of those made in Texas, 10% are desktops, 20% are laptops, and the rest are tablets. All computers are first shipped to a distribution center in Missouri before being sent out to stores. Suppose we select a computer at random from the distribution center and observe where it was made and whether it is a desktop, laptop, or tablet.[14]

1. Construct a tree diagram to model this random process.
2. Find the probability that the computer is a tablet.
3. Given that a tablet computer is selected, what is the probability that it was made in California?

# The Multiplication Rule for Independent Events

What happens to the general multiplication rule in the special case when events A and B are independent? In that case, $P(B \mid A) = P(B)$. We can simplify the general multiplication rule as follows:

$$\begin{aligned} P(\text{A and B}) = P(A \cap B) &= P(A) \cdot P(B \mid A) \\ &= P(A) \cdot P(B) \end{aligned}$$

This result is known as the **multiplication rule for independent events**.

This rule gives us another way to determine whether two events are independent. If $P(A \cap B) = P(A) \cdot P(B)$, then A and B are independent events.

**DEFINITION Multiplication rule for independent events**

If A and B are independent events, the probability that A and B both occur is

$$P(\text{A and B}) = P(A \cap B) = P(A) \cdot P(B)$$

caution !

**Note that this rule applies only to independent events.**

Suppose that Pedro drives the same route to work on Monday through Friday. His route includes one traffic light. The probability that the light will be green when Pedro arrives is 0.42, yellow is 0.03, and red is 0.55.

1. **What's the probability that the light is green on Monday and red on Tuesday?** Let event A be green light on Monday and event B be red light on Tuesday. These two events are independent because knowing that the light was green on Monday doesn't help us predict the color of the light on Tuesday. By the multiplication rule for independent events,

$$\begin{aligned} P(\text{green on Monday and red on Tuesday}) &= P(\text{A and B}) \\ &= P(A) \cdot P(B) \\ &= (0.42)(0.55) \\ &= 0.231 \end{aligned}$$

There's about a 23% chance that the light will be green on Monday and red on Tuesday.

2. **What's the probability that Pedro finds the light red on Monday through Friday?** We can extend the multiplication rule for independent events to more than two events:

$$\begin{aligned} &P(\text{red Monday } \textit{and} \text{ red Tuesday } \textit{and} \text{ red Wednesday } \textit{and} \text{ red Thursday } \textit{and} \text{ red Friday}) \\ &= P(\text{red Monday}) \cdot P(\text{red Tuesday}) \cdot P(\text{red Wednesday}) \cdot P(\text{red Thursday}) \cdot P(\text{red Friday}) \\ &= (0.55)(0.55)(0.55)(0.55)(0.55) \\ &= (0.55)^5 \\ &= 0.0503 \end{aligned}$$

There is about a 5% chance that Pedro will encounter a red light on all five days in a work week.

## EXAMPLE The *Challenger* disaster
### Multiplication rule for independent events

**PROBLEM:** On January 28, 1986, the space shuttle *Challenger* exploded on takeoff. All seven crew members were killed. Following the disaster, scientists and statisticians helped analyze what went wrong. They determined that the failure of O-ring joints in the shuttle's booster rockets was to blame. Under the cold conditions that day, experts estimated that the probability that an individual O-ring joint would function properly was 0.977. But there were six of these O-ring joints, and all six had to function properly for the shuttle to launch safely. Assuming that O-ring joints succeed or fail independently, find the probability that the shuttle would launch safely under similar conditions.

Thom Baur/AP Images

**SOLUTION:**

$P(\text{O-ring 1 OK and O-ring 2 OK and O-ring 3 OK and O-ring 4 OK and O-ring 5 OK and O-ring 6 OK})$

$= P(\text{O-ring 1 OK}) \cdot P(\text{O-ring 2 OK}) \cdot P(\text{O-ring 3 OK}) \cdot P(\text{O-ring 4 OK}) \cdot P(\text{O-ring 5 OK}) \cdot P(\text{O-ring 6 OK})$

$= (0.977)(0.977)(0.977)(0.977)(0.977)(0.977)$

$= (0.977)^6$

$= 0.870$

**FOR PRACTICE, TRY EXERCISE 89**

The multiplication rule for independent events can also be used to help find *P*(at least one). In the preceding example, the shuttle would *not* launch safely under similar conditions if 1 or 2 or 3 or 4 or 5 or all 6 O-ring joints fail—that is, if *at least one* O-ring fails. The only possible number of O-ring failures excluded is 0. So the events "at least one O-ring joint fails" and "no O-ring joints fail" are complementary events. By the complement rule,

$$\begin{aligned} P(\text{at least one O-ring fails}) &= 1 - P(\text{no O-ring fails}) \\ &= 1 - 0.87 \\ &= 0.13 \end{aligned}$$

That's a very high chance of failure! As a result of this analysis following the *Challenger* disaster, NASA made important safety changes to the design of the shuttle's booster rockets.

## EXAMPLE Rapid HIV testing
### Finding the probability of "at least one"

**PROBLEM:** Many people who visit clinics to be tested for HIV, the virus that causes AIDS, don't come back to learn their test results. Clinics now use "rapid HIV tests" that give a result while the client waits. In a clinic in Malawi, for example, use of rapid tests increased the percentage of clients who learned their test results from 69% to over 99%.

The trade-off for fast results is that rapid tests are less accurate than slower laboratory tests. Applied to people who have no HIV antibodies, one rapid test has a probability of about 0.004 of producing a false positive (i.e., of falsely indicating that antibodies are present).[15]

If a clinic tests 200 randomly selected people who are free of HIV antibodies, what is the probability that at least one false positive will occur? Assume that test results for different individuals are independent.

**SOLUTION:**

$$P(\text{no false positives}) = P(\text{all 200 tests negative})$$
$$= (0.996)(0.996)\cdots(0.996)$$
$$= 0.996^{200}$$
$$= 0.4486$$
$$P(\text{at least one false positive}) = 1 - 0.4486 = 0.5514$$

Start by finding *P*(no false positives).

The probability that any individual test result is negative is $1 - 0.004 = 0.996$.

FOR PRACTICE, TRY EXERCISE 91

**USING THE MULTIPLICATION RULE FOR INDEPENDENT EVENTS WISELY**

caution !

The multiplication rule $P(\text{A and B}) = P(\text{A}) \cdot P(\text{B})$ holds if A and B are *independent* but not otherwise. The addition rule $P(\text{A or B}) = P(\text{A}) + P(\text{B})$ holds if A and B are *mutually exclusive* but not otherwise. **Resist the temptation to use these simple rules when the conditions that justify them are not met.**

Hagar the Horrible
World rights reserved.
©2008 by King Features Syndicate, Inc.
VERY INTERESTING!
WHAT ARE THE ODDS OF TWO ROPES BREAKING AT THE EXACT SAME TIME?!
TWANG!
TWANG!
CHRIS BROWNE
12-22

## EXAMPLE

### Watch the weather!
### Beware lack of independence!

**PROBLEM:** Hacienda Heights and La Puente are two neighboring suburbs in the Los Angeles area. According to the local newspaper, there is a 50% chance of rain tomorrow in Hacienda Heights and a 50% chance of rain in La Puente. Does this mean that there is a $(0.5)(0.5) = 0.25$ probability that it will rain in both cities tomorrow?

**SOLUTION:** No; it is not appropriate to multiply the two probabilities, because "raining tomorrow in Hacienda Heights" and "raining tomorrow in La Puente" are not independent events. If it is raining in one of these locations, there is a high probability that it is raining in the other location because they are geographically close to each other.

**FOR PRACTICE, TRY EXERCISE 93**

*Is there a connection between mutually exclusive and independent?* Let's start with a new random process. Choose a U.S. adult at random. Define event A: the person is male, and event B: the person is pregnant. It's fairly clear that these two events are mutually exclusive (can't happen together)! Are they also independent?

If you know that event A has occurred, does this change the probability that event B happens? Of course! If we know the person is male, then the chance that the person is pregnant is 0. But the probability of selecting *someone* who is pregnant is greater than 0. Because $P(B \mid A) \neq P(B)$, the two events are not independent. Two mutually exclusive events (with nonzero probabilities) can *never* be independent, because if one event happens, the other event is guaranteed not to happen.

### CHECK YOUR UNDERSTANDING

*Questions 1 and 2 refer to the following setting.* New Jersey Transit claims that its 8:00 A.M. train from Princeton to New York has probability 0.9 of arriving on time. Assume that this claim is true.

1. Find the probability that the train arrives late on Monday but on time on Tuesday.
2. What's the probability that the train arrives late at least once in a 5-day week?
3. Government data show that 8% of adults are full-time college students and that 15% of adults are age 65 or older. If we randomly select an adult, is $P(\text{full-time college student and age 65 or older}) = (0.08)(0.15)$? Why or why not?

## Section 5.3 Summary

- A **conditional probability** describes the probability that one event happens given that another event is already known to have happened.
- One way to calculate a conditional probability is to use the formula

$$P(A \mid B) = \frac{P(A \text{ and } B)}{P(B)} = \frac{P(A \cap B)}{P(B)} = \frac{P(\text{both events occur})}{P(\text{given event occurs})}$$

- When knowing whether or not one event has occurred does not change the probability that another event happens, we say that the two events are independent. Events A and B are independent if

$$P(A \mid B) = P(A \mid B^C) = P(A)$$

  or, alternatively, if

$$P(B \mid A) = P(B \mid A^C) = P(B)$$

- Use the **general multiplication rule** to calculate the probability that events A and B both occur:

$$P(A \text{ and } B) = P(A \cap B) = P(A) \cdot P(B \mid A)$$

- When a random process involves multiple stages, a **tree diagram** can be used to display the sample space and to help answer questions involving conditional probability.
- In the special case of independent events, the multiplication rule becomes

$$P(\text{A and B}) = P(\text{A} \cap \text{B}) = P(\text{A}) \cdot P(\text{B})$$

## Section 5.3 Exercises

61. **Superpowers** A random sample of 415 children from England and the United States who completed a survey in a recent year was selected. Each student's country of origin was recorded along with which superpower they would most like to have: the ability to fly, ability to freeze time, invisibility, superstrength, or telepathy (ability to read minds). The data are summarized in the two-way table.

pg 360

| | | Country | | |
|---|---|---|---|---|
| | | England | U.S. | Total |
| Superpower | Fly | 54 | 45 | 99 |
| | Freeze time | 52 | 44 | 96 |
| | Invisibility | 30 | 37 | 67 |
| | Superstrength | 20 | 23 | 43 |
| | Telepathy | 44 | 66 | 110 |
| | Total | 200 | 215 | 415 |

Suppose we randomly select one of these students. Define events E: England, T: telepathy, and S: superstrength.

(a) Find $P(T \mid E)$. Interpret this value in context.

(b) Given that the student did not choose superstrength, what's the probability that this child is from England? Write your answer as a probability statement using correct symbols for the events.

62. **Get rich** A survey of 4826 randomly selected young adults (aged 19 to 25) asked, "What do you think are the chances you will have much more than a middle-class income at age 30?" The two-way table summarizes the responses.[16]

| | | Gender | | |
|---|---|---|---|---|
| | | Female | Male | Total |
| Opinion | Almost no chance | 96 | 98 | 194 |
| | Some chance but probably not | 426 | 286 | 712 |
| | A 50-50 chance | 696 | 720 | 1416 |
| | A good chance | 663 | 758 | 1421 |
| | Almost certain | 486 | 597 | 1083 |
| | Total | 2367 | 2459 | 4826 |

Choose a survey respondent at random. Define events G: a good chance, M: male, and N: almost no chance.

(a) Find $P(G \mid M)$. Interpret this value in context.

(b) Given that the chosen survey respondent didn't say "almost no chance," what's the probability that this person is female? Write your answer as a probability statement using correct symbols for the events.

63. **Body image** A random sample of 1200 U.S. college students was asked, "What is your perception of your own body? Do you feel that you are overweight, underweight, or about right?" The two-way table below summarizes the data on perceived body image by gender.[17]

| | | Gender | |
|---|---|---|---|
| | | Female | Male |
| Body image | About right | 560 | 295 |
| | Overweight | 163 | 72 |
| | Underweight | 37 | 73 |

Suppose we randomly select one of the survey respondents.

(a) Given that the person perceived his or her body image as about right, what's the probability that the person is female?

(b) If the person selected is female, what's the probability that she did not perceive her body image as overweight?

64. **Temperature and hatching** How is the hatching of water python eggs influenced by the temperature of a snake's nest? Researchers randomly assigned newly laid eggs to one of three water temperatures: cold, neutral, or hot. Hot duplicates the extra warmth provided by the mother python, and cold duplicates the absence of the mother.

| | | Nest temperature | | |
|---|---|---|---|---|
| | | Cold | Neutral | Hot |
| Hatching status | Hatched | 16 | 38 | 75 |
| | Didn't hatch | 11 | 18 | 29 |

Suppose we select one of the eggs at random.

(a) Given that the chosen egg was assigned to hot water, what is the probability that it hatched?

(b) If the chosen egg hatched, what is the probability that it was not assigned to hot water?

**65. Foreign-language study** Choose a student in grades 9 to 12 at random and ask if he or she is studying a language other than English. Here is the distribution of results:

| Language | Spanish | French | German | All others | None |
|---|---|---|---|---|---|
| Probability | 0.26 | 0.09 | 0.03 | 0.03 | 0.59 |

(a) What's the probability that the student is studying a language other than English?

(b) What is the probability that a student is studying Spanish given that he or she is studying some language other than English?

**66. Income tax returns** Here is the distribution of the adjusted gross income (in thousands of dollars) reported on individual federal income tax returns in a recent year:

| Income | $< 25$ | 25–49 | 50–99 | 100–499 | $\geq 500$ |
|---|---|---|---|---|---|
| Probability | 0.431 | 0.248 | 0.215 | 0.100 | 0.006 |

(a) What is the probability that a randomly chosen return shows an adjusted gross income of $50,000 or more?

(b) Given that a randomly chosen return shows an income of at least $50,000, what is the conditional probability that the income is at least $100,000?

**67. Tall people and basketball players** Select an adult at random. Define events T: person is over 6 feet tall, and B: person is a professional basketball player. Rank the following probabilities from smallest to largest. Justify your answer.

$P(T)$ $P(B)$ $P(T|B)$ $P(B|T)$

**68. Teachers and college degrees** Select an adult at random. Define events D: person has earned a college degree, and T: person's career is teaching. Rank the following probabilities from smallest to largest. Justify your answer.

$P(D)$ $P(T)$ $P(D|T)$ $P(T|D)$

**69. Dogs and cats** In one large city, 40% of all households own a dog, 32% own a cat, and 18% own both. Suppose we randomly select a household and learn that the household owns a cat. Find the probability that the household owns a dog.
pg 361

**70. Mac or PC?** A recent census at a major university revealed that 60% of its students mainly used Macs. The rest mainly used PCs. At the time of the census, 67% of the school's students were undergraduates. The rest were graduate students. In the census, 23% of respondents were graduate students and used a Mac as their main computer. Suppose we select a student at random from among those who were part of the census and learn that the person mainly uses a Mac. Find the probability that the person is a graduate student.

**71. Who owns a home?** What is the relationship between educational achievement and home ownership? A random sample of 500 U.S. adults was selected. Each member of the sample was identified as a high school graduate (or not) and as a homeowner (or not). The two-way table summarizes the data.
pg 364

| | | High school graduate | | |
|---|---|---|---|---|
| | | Yes | No | Total |
| Homeowner | Yes | 221 | 119 | 340 |
| | No | 89 | 71 | 160 |
| | Total | 310 | 190 | 500 |

Suppose we choose 1 member of the sample at random. Are the events "homeowner" and "high school graduate" independent? Justify your answer.

**72. Is this your card?** A standard deck of playing cards (with jokers removed) consists of 52 cards in four suits—clubs, diamonds, hearts, and spades. Each suit has 13 cards, with denominations ace, 2, 3, 4, 5, 6, 7, 8, 9, 10, jack, queen, and king. The jacks, queens, and kings are referred to as "face cards." Imagine that we shuffle the deck thoroughly and deal one card. The two-way table summarizes the sample space for this random process based on whether or not the card is a face card and whether or not the card is a heart.

| | | Type of card | | |
|---|---|---|---|---|
| | | Face card | Nonface card | Total |
| Suit | Heart | 3 | 10 | 13 |
| | Nonheart | 9 | 30 | 39 |
| | Total | 12 | 40 | 52 |

Are the events "heart" and "face card" independent? Justify your answer.

**73. Cell phones** The Pew Research Center asked a random sample of 2024 adult cell-phone owners from the United States their age and which type of cell phone they own: iPhone, Android, or other (including non-smartphones). The two-way table summarizes the data.

| | | Age | | | |
|---|---|---|---|---|---|
| | | 18–34 | 35–54 | 55+ | Total |
| Type of cell phone | iPhone | 169 | 171 | 127 | 467 |
| | Android | 214 | 189 | 100 | 503 |
| | Other | 134 | 277 | 643 | 1054 |
| | Total | 517 | 637 | 870 | 2024 |

Suppose we select one of the survey respondents at random.

(a) Find $P(\text{iPhone} \mid 18\text{–}34)$.

(b) Use your answer from part (a) to help determine if the events "iPhone" and "18–34" are independent.

**74. Middle school values** Researchers carried out a survey of fourth-, fifth-, and sixth-grade students in Michigan. Students were asked whether good grades, athletic ability, or being popular was most important to them. The two-way table summarizes the survey data.[18]

| | | Grade | | | |
|---|---|---|---|---|---|
| | | 4th | 5th | 6th | Total |
| **Most important** | Grades | 49 | 50 | 69 | 168 |
| | Athletic | 24 | 36 | 38 | 98 |
| | Popular | 19 | 22 | 28 | 69 |
| | Total | 92 | 108 | 135 | 335 |

Suppose we select one of these students at random.

(a) Find $P(\text{athletic} \mid \text{5th grade})$.

(b) Use your answer from part (a) to help determine if the events "5th grade" and "athletic" are independent.

**75. Rolling dice** Suppose you roll two fair, six-sided dice—one red and one green. Are the events "sum is 7" and "green die shows a 4" independent? Justify your answer. (See Figure 5.2 on page 342 for the sample space of this random process.)

**76. Rolling dice** Suppose you roll two fair, six-sided dice—one red and one green. Are the events "sum is 8" and "green die shows a 4" independent? Justify your answer. (See Figure 5.2 on page 342 for the sample space of this random process.)

**77. Free downloads?** pg 366 Illegal music downloading is a big problem: 29% of Internet users download music files, and 67% of downloaders say they don't care if the music is copyrighted.[19] Find the probability that a randomly selected Internet user downloads music and doesn't care if it's copyrighted.

**78. At the gym** Suppose that 10% of adults belong to health clubs, and 40% of these health club members go to the club at least twice a week. Find the probability that a randomly selected adult belongs to a health club and goes there at least twice a week.

**79. Box of chocolates** According to Forrest Gump, "Life is like a box of chocolates. You never know what you're gonna get." Suppose a candymaker offers a special "Gump box" with 20 chocolate candies that look alike. In fact, 14 of the candies have soft centers and 6 have hard centers. Suppose you choose 3 of the candies from a Gump box at random. Find the probability that all three candies have soft centers.

**80. Sampling students** A statistics class with 30 students has 10 males and 20 females. Suppose you choose 3 of the students in the class at random. Find the probability that all three are female.

**81. Fill 'er up!** pg 369 In a certain month, 88% of automobile drivers filled their vehicles with regular gasoline, 2% purchased midgrade gas, and 10% bought premium gas.[20] Of those who bought regular gas, 28% paid with a credit card; of customers who bought midgrade and premium gas, 34% and 42%, respectively, paid with a credit card. Suppose we select a customer at random.

(a) Draw a tree diagram to model this random process.

(b) Find the probability that the customer paid with a credit card.

(c) Suppose the chosen customer paid with a credit card. What's the probability that the customer bought premium gas?

**82. Media usage and good grades** The Kaiser Family Foundation released a study about the influence of media in the lives of young people aged 8–18.[21] In the study, 17% of the youth were classified as light media users, 62% were classified as moderate media users, and 21% were classified as heavy media users. Of the light users who responded, 74% described their grades as good (A's and B's), while only 68% of the moderate users and 52% of the heavy users described their grades as good. Suppose that we select one young person from the study at random.

(a) Draw a tree diagram to model this random process.

(b) Find the probability that this person describes his or her grades as good.

(c) Suppose the chosen person describes his or her grades as good. What's the probability that he or she is a heavy user of media?

**83. First serve** pg 370 Tennis great Andy Murray made 60% of his first serves in a recent season. When Murray made his first serve, he won 76% of the points. When Murray missed his first serve and had to serve again, he won only 54% of the points.[22] Suppose you randomly choose a point on which Murray served. You get distracted before seeing his first serve but look up in time to see Murray win the point. What's the probability that he missed his first serve?

**84. Lactose intolerance** Lactose intolerance causes difficulty in digesting dairy products that contain lactose (milk sugar). It is particularly common among people of African and Asian ancestry. In the United States (not including other groups and people who consider themselves to belong to more than one race), 82% of the population is White, 14% is Black, and 4% is Asian. Moreover, 15% of Whites, 70% of Blacks, and 90% of Asians are lactose intolerant.[23] Suppose we select a U.S. person at random and find that the person is lactose intolerant. What's the probability that she or he is Asian?

**85. HIV testing** Enzyme immunoassay (EIA) tests are used to screen blood specimens for the presence of antibodies to HIV, the virus that causes AIDS. Antibodies indicate the presence of the virus. The test is quite accurate but is not always correct. A false positive occurs when the test gives a positive result but no HIV antibodies are actually present in the blood. A false negative occurs when the test gives a negative result but HIV antibodies are present in the blood. Here are approximate probabilities of positive and negative EIA outcomes when the blood tested does and does not actually contain antibodies to HIV: [24]

| | | Test result | |
|---|---|---|---|
| | | + | − |
| Truth | Antibodies present | 0.9985 | 0.0015 |
| | Antibodies absent | 0.0060 | 0.9940 |

Suppose that 1% of a large population carries antibodies to HIV in their blood. Imagine choosing a person from this population at random. If the person's EIA test is positive, what's the probability that the person has the HIV antibody?

86. **Metal detector** A boy uses a homemade metal detector to look for valuable metal objects on a beach. The machine isn't perfect—it beeps for only 98% of the metal objects over which it passes, and it beeps for 4% of the nonmetallic objects over which it passes. Suppose that 25% of the objects that the machine passes over are metal. Choose an object from this beach at random. If the machine beeps when it passes over this object, find the probability that the boy has found a metal object.

87. **Fundraising by telephone** Tree diagrams can organize problems having more than two stages. The figure shows probabilities for a charity calling potential donors by telephone.[25] Each person called is either a recent donor, a past donor, or a new prospect. At the next stage, the person called either does or does not pledge to contribute, with conditional probabilities that depend on the donor class to which the person belongs. Finally, those who make a pledge either do or don't actually make a contribution. Suppose we randomly select a person who is called by the charity.

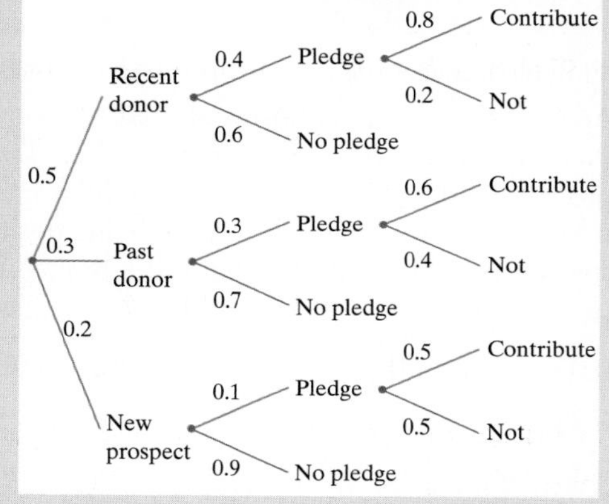


(a) What is the probability that the person contributed to the charity?

(b) Given that the person contributed, find the probability that he or she is a recent donor.

88. **HIV and confirmation testing** Refer to Exercise 85. Many of the positive results from EIA tests are false positives. It is therefore common practice to perform a second EIA test on another blood sample from a person whose initial specimen tests positive. Assume that the false positive and false negative rates remain the same for a person's second test. Find the probability that a person who gets a positive result on both EIA tests has HIV antibodies.

89. **Merry and bright?** A string of Christmas lights contains 20 lights. The lights are wired in series so that if any light fails, the whole string will go dark. Each light has probability 0.98 of working for a 3-year period. The lights fail independently of each other. Find the probability that the string of lights will remain bright for 3 years. pg 373

90. **Get rid of the penny** Harris Interactive reported that 29% of all U.S. adults favor abolishing the penny. Assuming that responses from different individuals are independent, what is the probability of randomly selecting 3 U.S. adults who all say that they favor abolishing the penny?

91. **Is the package late?** A shipping company claims that 90% of its shipments arrive on time. Suppose this claim is true. If we take a random sample of 20 shipments made by the company, what's the probability that at least 1 of them arrives late? pg 373

92. **On a roll** Suppose that you roll a fair, six-sided die 10 times. What's the probability that you get at least one 6?

93. **Who's pregnant?** According to the Current Population Survey (CPS), 27% of U.S. females are older than 55. The Centers for Disease Control and Prevention (CDC) report that 6% of all U.S. females are pregnant. Suppose that these results are accurate. If we randomly select a U.S. female, is $P(\text{pregnant and over } 55) = (0.06)(0.27) = 0.0162$? Why or why not? pg 374

94. **Late flights** An airline reports that 85% of its flights arrive on time. To find the probability that a random sample of 4 of this airline's flights into LaGuardia Airport in New York City on the same night all arrive on time, can we multiply $(0.85)(0.85)(0.85)(0.85)$? Why or why not?

95. **Fire or medical?** Many fire stations handle more emergency calls for medical help than for fires. At one fire station, 81% of incoming calls are for medical help. Suppose we choose 4 incoming calls to the station at random.

(a) Find the probability that all 4 calls are for medical help.

(b) What's the probability that at least 1 of the calls is not for medical help?

(c) Explain why the calculation in part (a) may not be valid if we choose 4 consecutive calls to the station.

96. **Broken links** Internet sites often vanish or move so that references to them can't be followed. In fact, 87% of Internet sites referred to in major scientific journals still work within two years of publication.[26] Suppose we randomly select 7 Internet references from scientific journals.

(a) Find the probability that all 7 references still work two years later.

(b) What's the probability that at least 1 of them doesn't work two years later?

(c) Explain why the calculation in part (a) may not be valid if we choose 7 Internet references from one issue of the same journal.

97. **Mutually exclusive versus independent** The two-way table summarizes data on the gender and eye color of students in a college statistics class. Imagine choosing a student from the class at random. Define event A: student is male, and event B: student has blue eyes.[27]

| | | Gender | | |
|---|---|---|---|---|
| | | Male | Female | Total |
| Eye color | Blue | | | 10 |
| | Brown | | | 40 |
| | Total | 20 | 30 | 50 |

(a) Copy and complete the two-way table so that events A and B are mutually exclusive.

(b) Copy and complete the two-way table so that events A and B are independent.

(c) Copy and complete the two-way table so that events A and B are not mutually exclusive and not independent.

98. **Independence and association** The two-way table summarizes data from an experiment comparing the effectiveness of three different diets (A, B, and C) on weight loss. Researchers randomly assigned 300 volunteer subjects to the three diets. The response variable was whether each subject lost weight over a 1-year period.

| | | Diet | | | |
|---|---|---|---|---|---|
| | | A | B | C | Total |
| Lost weight? | Yes | | 60 | | 180 |
| | No | | 40 | | 120 |
| | Total | 90 | 100 | 110 | 300 |

(a) Suppose we randomly select one of the subjects from the experiment. Show that the events "Diet B" and "Lost weight" are independent.

(b) Copy and complete the table so that there is no association between type of diet and whether a subject lost weight.

(c) Copy and complete the table so that there is an association between type of diet and whether a subject lost weight.

99. **Checking independence** Suppose A and B are two events such that $P(A) = 0.3$, $P(B) = 0.4$, and $P(A \cap B) = 0.12$. Are events A and B independent? Justify your answer.

100. **Checking independence** Suppose C and D are two events such that $P(C) = 0.6$, $P(D) = 0.45$, and $P(C \cap D) = 0.3$. Are events C and D independent? Justify your answer.

101. **The geometric distributions** You are tossing a pair of fair, six-sided dice in a board game. Tosses are independent. You land in a danger zone that requires you to roll doubles (both faces showing the same number of spots) before you are allowed to play again.

(a) What is the probability of rolling doubles on a single toss of the dice?

(b) What is the probability that you do not roll doubles on the first toss, but you do on the second toss?

(c) What is the probability that the first two tosses are not doubles and the third toss is doubles? This is the probability that the first doubles occurs on the third toss.

(d) Do you see the pattern? What is the probability that the first doubles occurs on the $k$th toss?

102. **Matching suits** A standard deck of playing cards consists of 52 cards with 13 cards in each of four suits: spades, diamonds, clubs, and hearts. Suppose you shuffle the deck thoroughly and deal 5 cards face-up onto a table.

(a) What is the probability of dealing five spades in a row?

(b) Find the probability that all 5 cards on the table have the same suit.

**Multiple Choice:** *Select the best answer for Exercises 103–106.*

103. An athlete suspected of using steroids is given two tests that operate independently of each other. Test A has probability 0.9 of being positive if steroids have been used. Test B has probability 0.8 of being positive if steroids have been used. What is the probability that neither test is positive if the athlete has used steroids?

(a) 0.08 (b) 0.28 (c) 0.02
(d) 0.38 (e) 0.72

104. In an effort to find the source of an outbreak of food poisoning at a conference, a team of medical detectives carried out a study. They examined all 50 people who had food poisoning and a random sample of 200 people attending the conference who didn't get food poisoning. The detectives found that 40% of the people with food poisoning went to a cocktail party on the second night of the conference, while only 10% of the people in the random sample attended the same party. Which of the following statements is appropriate for describing the 40% of people who went to the party? (Let F = got food poisoning and A = attended party.)

(a) $P(F \mid A) = 0.40$
(b) $P(A \mid F^C) = 0.40$
(c) $P(F \mid A^C) = 0.40$
(d) $P(A^C \mid F) = 0.40$
(e) $P(A \mid F) = 0.40$

105. Suppose a loaded die has the following probability model:

| Outcome | 1 | 2 | 3 | 4 | 5 | 6 |
|---|---|---|---|---|---|---|
| Probability | 0.3 | 0.1 | 0.1 | 0.1 | 0.1 | 0.3 |

If this die is thrown and the top face shows an odd number, what is the probability that the die shows a 1?

(a) 0.10 (b) 0.17 (c) 0.30
(d) 0.50 (e) 0.60

106. If $P(A) = 0.24$, $P(B) = 0.52$, and A and B are independent events, what is $P(A \text{ or } B)$?

(a) 0.1248
(b) 0.28
(c) 0.6352
(d) 0.76
(e) The answer cannot be determined from the information given.

## Recycle and Review

107. **BMI** (2.2, 5.2, 5.3) Your body mass index (BMI) is your weight in kilograms divided by the square of your height in meters. Online BMI calculators allow you to enter weight in pounds and height in inches. High BMI is a common but controversial indicator of being overweight or obese. A study by the National Center for Health Statistics found that the BMI of American young women (ages 20 to 29) is approximately Normally distributed with mean 26.8 and standard deviation 7.4.[28]

(a) People with BMI less than 18.5 are often classed as "underweight." What percent of young women are underweight by this criterion?

(b) Suppose we select two American young women in this age group at random. Find the probability that at least one of them is classified as underweight.

108. **Snappy dressers** (4.2, 4.3) Matt and Diego suspect that people are more likely to agree to participate in a survey if the interviewers are dressed up. To test this idea, they went to the local grocery store to survey customers on two consecutive Saturday mornings at 10 A.M. On the first Saturday, they wore casual clothing (tank tops and jeans). On the second Saturday, they dressed in button-down shirts and nicer slacks. Each day, they asked every fifth person who walked into the store to participate in a survey. Their response variable was whether or not the person agreed to participate. Here are their results:

| | | Clothing | |
|---|---|---|---|
| | | Casual | Nice |
| Participation | Agreed | 14 | 27 |
| | Declined | 36 | 23 |

(a) Calculate the difference (Casual – Nice) in the proportion of subjects that agreed to participate in the survey in the two groups.

(b) Assume the study design is equivalent to randomly assigning shoppers to the "casual" or "nice" groups. A total of 100 trials of a simulation were performed to see what differences in proportions would occur due only to chance variation in this random assignment. Use the results of the simulation in the following dotplot to determine if the difference in proportions from part (a) is statistically significant. Explain your reasoning.

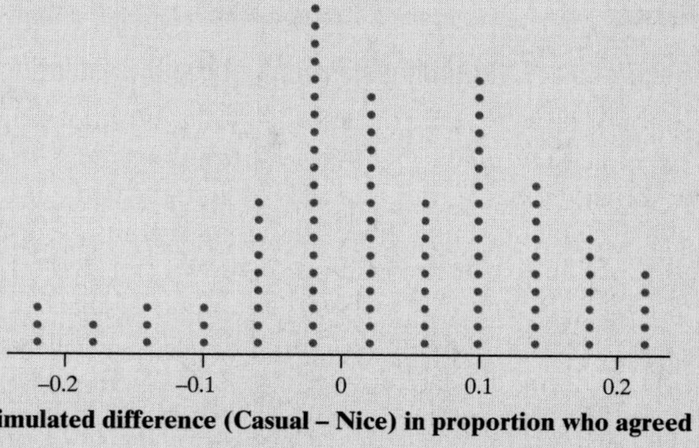


(c) What flaw in the design of this experiment would prevent Matt and Diego from drawing a cause-and-effect conclusion about the impact of an interviewer's attire on nonresponse in a survey?

# Chapter 5 Wrap-Up

## FRAPPY! FREE RESPONSE AP® PROBLEM, YAY!

The following problem is modeled after actual AP® Statistics exam free response questions. Your task is to generate a complete, concise response in 15 minutes.

*Directions: Show all your work. Indicate clearly the methods you use, because you will be scored on the correctness of your methods as well as on the accuracy and completeness of your results and explanations.*

A statistics teacher has 40 students in his class, 23 females and 17 males. At the beginning of class on a Monday, the teacher planned to spend time reviewing an assignment due that day. Unknown to the teacher, only 19 of the females and 11 of the males had completed the assignment. The teacher plans to randomly select students to do problems from the assignment on the whiteboard.

(a) What is the probability that a randomly selected student has completed the assignment?

(b) Are the events "selecting a female" and "selecting a student who completed the assignment" independent? Justify your answer.

Suppose that the teacher randomly selects 4 students to do a problem on the whiteboard and only 2 of the students had completed the assignment.

(c) Describe how to use a table of random digits to estimate the probability that 2 or fewer of the 4 randomly selected students completed the assignment.

(d) Complete three trials of your simulation using the random digits below and use the results to estimate the probability described in part (c).

| | | | | | | | |
|---|---|---|---|---|---|---|---|
| 12975 | 13258 | 13048 | 45144 | 72321 | 81940 | 00360 | 02428 |
| 96767 | 35964 | 23822 | 96012 | 94951 | 65194 | 50842 | 55372 |
| 37609 | 59057 | 66967 | 83401 | 60705 | 02384 | 90597 | 93600 |

After you finish, you can view two example solutions on the book's website (highschool.bfwpub.com/updatedtps6e). Determine whether you think each solution is "complete," "substantial," "developing," or "minimal." If the solution is not complete, what improvements would you suggest to the student who wrote it? Finally, your teacher will provide you with a scoring rubric. Score your response and note what, if anything, you would do differently to improve your own score.

# Chapter 5 Review

### Section 5.1: Randomness, Probability, and Simulation

In this section, you learned about the idea of probability. The law of large numbers says that when you repeat a random process many, many times, the relative frequency of an outcome will approach a single number. This single number is called the probability of the outcome—how often we expect the outcome to occur in a very large number of trials of the random process. Be sure to remember the "large" part of the law of large numbers. Although clear patterns emerge in a large number of trials, we shouldn't expect such regularity in a small number of trials.

Simulation is a powerful tool that we can use to imitate a random process and estimate a probability. To conduct a simulation, describe how to use a random process to perform one trial of the simulation. Tell what you will record at the end of each trial. Then perform many trials, and use the results of your simulation to answer the question of interest. If you are using random digits to perform your simulation, be sure to consider whether digits can be repeated within each trial.

### Section 5.2: Probability Rules

In this section, you learned that random behavior can be described by a probability model. Probability models have two parts, a list of possible outcomes (the sample space) and a probability for each outcome. The probability of each outcome in a probability model must be between 0 and 1, and the probabilities of all the outcomes in the sample space must add to 1.

An event is a collection of possible outcomes from the sample space. The complement rule says the probability that an event occurs is 1 minus the probability that the event doesn't occur. In symbols, the complement rule says that $P(A) = 1 - P(A^C)$. Given two events A and B from some random process, use the general addition rule to find the probability that event A or event B occurs:

$$P(\text{A or B}) = P(A \cup B) = P(A) + P(B) - P(A \cap B)$$

If the events A and B have no outcomes in common, use the addition rule for mutually exclusive events: $P(A \cup B) = P(A) + P(B)$.

Finally, you learned how to use two-way tables and Venn diagrams to display the sample space for a random process involving two events. Using a two-way table or a Venn diagram is a helpful way to organize information and calculate probabilities involving the union $(A \cup B)$ and the intersection $(A \cap B)$ of two events.

### Section 5.3: Conditional Probability and Independence

In this section, you learned that a conditional probability describes the probability of an event occurring given that another event is known to have already occurred. To calculate the probability that event A occurs given that event B has occurred, use the formula

$$P(A \mid B) = \frac{P(A \cap B)}{P(B)} = \frac{P(A \text{ and } B)}{P(B)}$$

Two-way tables and tree diagrams are useful ways to organize the information provided in a conditional probability problem. Two-way tables are best when the problem describes the number or proportion of cases with certain characteristics. Tree diagrams are best when the problem provides the conditional probabilities of different events or describes a sequence of events.

Use the general multiplication rule for calculating the probability that event A and event B both occur:

$$P(A \text{ and } B) = P(A \cap B) = P(A) \cdot P(B \mid A)$$

If knowing whether or not event B occurs doesn't change the probability that event A occurs, then events A and B are independent. That is, events A and B are independent if $P(A \mid B) = P(A \mid B^C) = P(A)$. If events A and B are independent, use the multiplication rule for independent events to find the probability that events A and B both occur: $P(A \cap B) = P(A) \cdot P(B)$.

## What Did You Learn?

| Learning Target | Section | Related Example on Page(s) | Relevant Chapter Review Exercise(s) |
|---|---|---|---|
| Interpret probability as a long-run relative frequency. | 5.1 | 330 | R5.1 |
| Use simulation to model a random process. | 5.1 | 333, 334 | R5.2 |
| Give a probability model for a random process with equally likely outcomes and use it to find the probability of an event. | 5.2 | 343 | R5.3 |
| Use basic probability rules, including the complement rule and the addition rule for mutually exclusive events. | 5.2 | 345 | R5.4 |
| Use a two-way table or Venn diagram to model a random process and calculate probabilities involving two events. | 5.2 | 347, 352 | R5.5 |
| Apply the general addition rule to calculate probabilities. | 5.2 | 349 | R5.5 |
| Calculate and interpret conditional probabilities. | 5.3 | 360, 361 | R5.4, R5.5, R5.7 |
| Determine if two events are independent. | 5.3 | 364 | R5.6 |
| Use the general multiplication rule to calculate probabilities. | 5.3 | 366 | R5.6, R5.7 |
| Use a tree diagram to model a random process involving a sequence of outcomes and to calculate probabilities. | 5.3 | 369, 370 | R5.7 |
| When appropriate, use the multiplication rule for independent events to calculate probabilities. | 5.3 | 373, 374, 375 | R5.8 |

# Chapter 5 Review Exercises

*These exercises are designed to help you review the important ideas and methods of the chapter.*

**R5.1** **Butter side down** Researchers at Manchester Metropolitan University in England determined that if a piece of toast is dropped from a 2.5-foot-high table, the probability that it lands butter side down is 0.81.

(a) Explain what this probability means.

(b) Suppose that the researchers dropped 4 pieces of toast, and all of them landed butter side down. Does that make it more likely that the next piece of toast will land with the butter side up? Explain your answer.

**R5.2** **Butter side down** Refer to the preceding exercise. Maria decides to test this probability and drops 10 pieces of toast from a 2.5-foot table. Only 4 of them land butter side down. Maria wants to perform a simulation to estimate the probability that 4 or fewer pieces of toast out of 10 would land butter side down if the researchers' 0.81 probability value is correct.

(a) Describe how you would use a table of random digits to perform the simulation.

(b) Perform 3 trials of the simulation using the random digits given. Copy the digits onto your paper and mark directly on or above them so that someone can follow what you did.

| | | | | | |
|---|---|---|---|---|---|
| 29077 | 14863 | 61683 | 47052 | 62224 | 51025 |
| 95052 | 90908 | 73592 | 75186 | 87136 | 95761 |
| 27102 | 56027 | 55892 | 33063 | 41842 | 81868 |

(c) The dotplot displays the results of 50 simulated trials of dropping 10 pieces of toast. Is there convincing evidence that the researchers' 0.81 probability value is incorrect? Explain your answer.

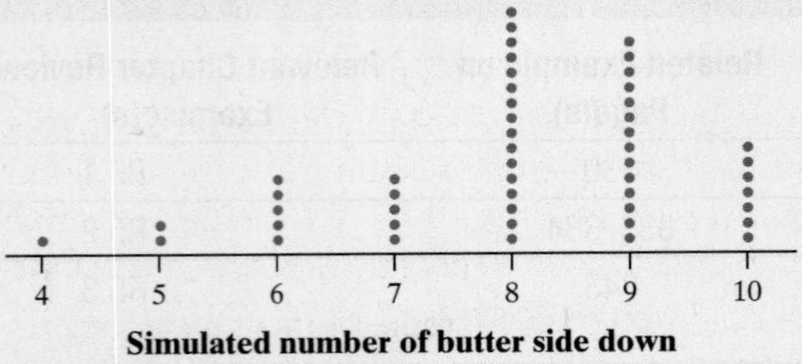


**R5.3 Rock smashes scissors** Almost everyone has played the game rock-paper-scissors at some point. Two players face each other and, at the count of 3, make a fist (rock), an extended hand, palm side down (paper), or a "V" with the index and middle fingers (scissors). The winner is determined by these rules: rock smashes scissors; paper covers rock; and scissors cut paper. If both players choose the same object, then the game is a tie. Suppose that Player 1 and Player 2 are both equally likely to choose rock, paper, or scissors.

(a) Give a probability model for this random process.

(b) Find the probability that Player 1 wins the game on the first throw.

**R5.4 What kind of vehicle?** Randomly select a new vehicle sold in the United States in a certain month.[29] The probability model for the type of vehicle chosen is given here.

| Vehicle type | Passenger car | Pickup truck | SUV | Crossover | Minivan |
|---|---|---|---|---|---|
| Probability | 0.46 | 0.15 | 0.10 | ? | 0.05 |

(a) What is the probability that the vehicle is a crossover? How do you know?

(b) Find the probability that the vehicle is not an SUV or a minivan.

(c) Given that the vehicle is not a passenger car, what is the probability that it is a pickup truck?

**R5.5 Drive to exercise?** The two-way table summarizes the responses of 120 people to a survey in which they were asked, "Do you exercise for at least 30 minutes 4 or more times per week?" and "What kind of vehicle do you drive?"

| | | Car type | | |
|---|---|---|---|---|
| | | Sedan | SUV | Truck |
| Exercise? | Yes | 25 | 15 | 12 |
| | No | 20 | 24 | 24 |

Suppose one person from this sample is randomly selected.

(a) Find the probability that the person drives an SUV.

(b) Find the probability that the person drives a sedan or exercises for at least 30 minutes 4 or more times per week.

(c) Find the probability that the person does not drive a truck, given that she or he exercises for at least 30 minutes 4 or more times per week.

**R5.6 Mike's pizza** You work at Mike's pizza shop. You have the following information about the 9 pizzas in the oven: 3 of the 9 have thick crust and 2 of the 3 thick-crust pizzas have mushrooms. Of the remaining 6 pizzas, 4 have mushrooms. Suppose you randomly select one of the pizzas in the oven.

(a) Are the events "thick-crust pizza" and "pizza with mushrooms" mutually exclusive? Justify your answer.

(b) Are the events "thick-crust pizza" and "pizza with mushrooms" independent? Justify your answer.

Now suppose you randomly select 2 of the pizzas in the oven.

(c) Find the probability that both have mushrooms.

**R5.7 Does the new hire use drugs?** Many employers require prospective employees to take a drug test. A positive result on this test suggests that the prospective employee uses illegal drugs. However, not all people who test positive use illegal drugs. The test result could be a false positive. A negative test result could be a false negative if the person really does use illegal drugs. Suppose that 4% of prospective employees use drugs, and that the drug test has a false positive rate of 5%, and a false negative rate of 10%.[30] Imagine choosing a prospective employee at random.

(a) Draw a tree diagram to model this random process.

(b) Find the probability that the drug test result is positive.

(c) If the prospective employee's drug test result is positive, find the probability that she or he uses illegal drugs.

**R5.8 Lucky penny?** Harris Interactive reported that 33% of U.S. adults believe that finding and picking up a penny is good luck. Assuming that responses from different individuals are independent, what is the probability of randomly selecting 10 U.S. adults and finding at least 1 person who believes that finding and picking up a penny is good luck?

# Chapter 5  AP® Statistics Practice Test

**Section I: Multiple Choice** *Select the best answer for each question.*

*Questions T5.1 to T5.3 refer to the following setting.* A group of 125 truck owners were asked what brand of truck they owned and whether or not the truck has four-wheel drive. The results are summarized in the two-way table below. Suppose we randomly select one of these truck owners.

|  |  | Four-wheel drive? | |
|---|---|:---:|:---:|
|  |  | Yes | No |
|  | Ford | 28 | 17 |
| **Brand of truck** | Chevy | 32 | 18 |
|  | Dodge | 20 | 10 |

*30/125*
*+*
*80/125*
*(20/125)*

**T5.1** What is the probability that the person owns a Dodge or has four-wheel drive?

(a) 20/80 (b) 20/125 (c) 80/125
(d) 90/125 (e) 110/125

**T5.2** What is the probability that the person owns a Chevy, given that the truck has four-wheel drive?

(a) 32/50 (b) 32/80 (c) 32/125
(d) 50/125 (e) 80/125

**T5.3** Which one of the following is true about the events "Owner's truck is a Chevy" and "Owner's truck has four-wheel drive"?

(a) These two events are mutually exclusive and independent.

(b) These two events are mutually exclusive, but not independent. *can't occur together*

(c) These two events are not mutually exclusive, but they are independent. *1 event occuring doesn't change other*

(d) These two events are neither mutually exclusive nor independent.

(e) These two events are mutually exclusive, but we do not have enough information to determine if they are independent.

*1/3 · 297*

**T5.4** A spinner has three equally sized regions: blue, red, and green. Jonny spins the spinner 3 times and gets 3 blues in a row. If he spins the spinner 297 more times, how many more blues is he most likely to get?

(a) 97 (b) 99 (c) 100 (d) 101 (e) 103

*Questions T5.5 and T5.6 refer to the following setting.* Wilt is a fine basketball player, but his free-throw shooting could use some work. For the past three seasons, he has made only 56% of his free throws. His coach sends him to a summer clinic to work on his shot, and when he returns, his coach has him step to the free-throw line and take 50 shots. He makes 34 shots. Is this result convincing evidence that Wilt's free-throw shooting has improved? We want to perform a simulation to

estimate the probability that a 56% free-throw shooter would make 34 or more in a sample of 50 shots.

**T5.5** Which of the following is a correct way to perform the simulation?

(a) Let integers from 1 to 34 represent making a free throw and 35 to 50 represent missing a free throw. Generate 50 random integers from 1 to 50. Count the number of made free throws. Repeat this process many times.

(b) Let integers from 1 to 34 represent making a free throw and 35 to 50 represent missing a free throw. Generate 50 random integers from 1 to 50 with no repeats allowed. Count the number of made free throws. Repeat this process many times.

(c) Let integers from 1 to 56 represent making a free throw and 57 to 100 represent missing a free throw. Generate 50 random integers from 1 to 100. Count the number of made free throws. Repeat this process many times.

(d) Let integers from 1 to 56 represent making a free throw and 57 to 100 represent missing a free throw. Generate 50 random integers from 1 to 100 with no repeats allowed. Count the number of made free throws. Repeat this process many times.

(e) None of the above is correct.

**T5.6** The dotplot displays the number of made shots in 100 simulated sets of 50 free throws by someone with probability 0.56 of making a free throw.

**Simulated number of made shots**

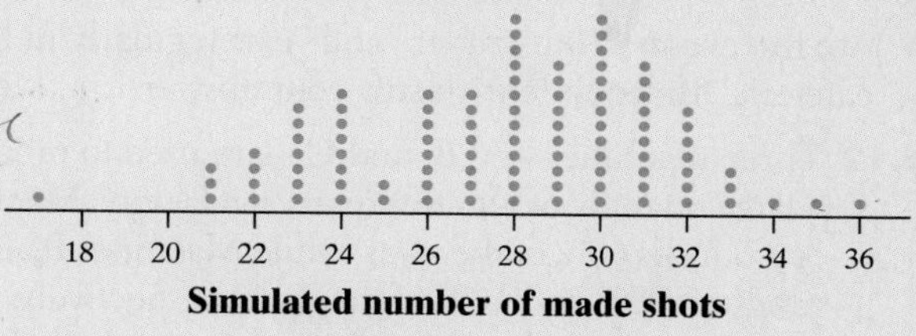

Which of the following is an appropriate statement about Wilt's free-throw shooting based on this dotplot?

(a) If Wilt were still only a 56% shooter, the probability that he would make at least 34 of his shots is about 0.03.

(b) If Wilt were still only a 56% shooter, the probability that he would make at least 34 of his shots is about 0.97.

(c) If Wilt is now shooting better than 56%, the probability that he would make at least 34 of his shots is about 0.03.

(d) If Wilt is now shooting better than 56%, the probability that he would make at least 34 of his shots is about 0.97.

(e) If Wilt were still only a 56% shooter, the probability that he would make at least 34 of his shots is about 0.01.

**T5.7** The partially complete table that follows shows the distribution of scores on the AP® Statistics exam for a class of students.

| Score | 1 | 2 | 3 | 4 | 5 |
|---|---|---|---|---|---|
| Probability | 0.10 | 0.20 | ??? | 0.25 | 0.15 |

Select a student from this class at random. If the student earned a score of 3 or higher on the AP® Statistics exam, what is the probability that the student scored a 5?

(a) 0.150 (b) 0.214 (c) 0.300 (d) 0.428 (e) 0.700

**T5.8** In a class, there are 18 girls and 14 boys. If the teacher selects two students at random to attend a party with the principal, what is the probability that the two students are the same sex?

(a) 0.49 (b) 0.50 (c) 0.51 (d) 0.52 (e) 0.53

**T5.9** Suppose that a student is randomly selected from a large high school. The probability that the student is a senior is 0.22. The probability that the student has a driver's license is 0.30. If the probability that the student is a senior or has a driver's license is 0.36, what is the probability that the student is a senior and has a driver's license?

(a) 0.060 (b) 0.066 (c) 0.080 (d) 0.140 (e) 0.160

**T5.10** The security system in a house has two units that set off an alarm when motion is detected. Neither one is entirely reliable, but one or both *always* go off when there is motion anywhere in the house. Suppose that for motion in a certain location, the probability that detector A goes off and detector B does not go off is 0.25, and the probability that detector A does not go off is 0.35. What is the probability that detector B goes off?

(a) 0.1 (b) 0.35 (c) 0.4 (d) 0.65 (e) 0.75

**Section II: Free Response** *Show all your work. Indicate clearly the methods you use, because you will be graded on the correctness of your methods as well as on the accuracy and completeness of your results and explanations.*

**T5.11** The two-way table summarizes data on whether students at a certain high school eat regularly in the school cafeteria by grade level.

| | | Grade | | | | |
|---|---|---|---|---|---|---|
| | | 9th | 10th | 11th | 12th | Total |
| Eat in cafeteria? | Yes | 130 | 175 | 122 | 68 | 495 |
| | No | 18 | 34 | 88 | 170 | 310 |
| | Total | 148 | 209 | 210 | 238 | 805 |

(a) If you choose a student at random, what is the probability that the student eats regularly in the cafeteria and is not a 10th-grader?

(b) If you choose a student at random who eats regularly in the cafeteria, what is the probability that the student is a 10th-grader?

(c) Are the events "10th-grader" and "eats regularly in the cafeteria" independent? Justify your answer.

**T5.12** Three machines—A, B, and C—are used to produce a large quantity of identical parts at a factory. Machine A produces 60% of the parts, while Machines B and C produce 30% and 10% of the parts, respectively. Historical records indicate that 10% of the parts produced by Machine A are defective, compared with 30% for Machine B and 40% for Machine C. Suppose we randomly select a part produced at the factory.

(a) Find the probability that the part is defective.

(b) If the part is inspected and found to be defective, what's the probability that it was produced by Machine B?

**T5.13** At Dicey Dave's Diner, the dinner buffet usually costs $12.99. Once a month, Dave sponsors "lucky buffet" night. On that night, each patron can either pay the usual price or roll two fair, six-sided dice and pay a number of dollars equal to the product of the numbers showing on the two faces. The table shows the sample space of this random process.

| | | First die | | | | | |
|---|---|---|---|---|---|---|---|
| | | 1 | 2 | 3 | 4 | 5 | 6 |
| Second die | 1 | 1 | 2 | 3 | 4 | 5 | 6 |
| | 2 | 2 | 4 | 6 | 8 | 10 | 12 |
| | 3 | 3 | 6 | 9 | 12 | 15 | 18 |
| | 4 | 4 | 8 | 12 | 16 | 20 | 24 |
| | 5 | 5 | 10 | 15 | 20 | 25 | 30 |
| | 6 | 6 | 12 | 18 | 24 | 30 | 36 |

(a) A customer decides to play Dave's "lucky buffet" game. Find the probability that the customer will pay less than the usual cost of the buffet.

(b) A group of 4 friends comes to Dicey Dave's Diner to play the "lucky buffet" game. Find the probability that all 4 of these friends end up paying less than the usual cost of the buffet.

(c) Find the probability that at least 1 of the 4 friends ends up paying *more* than the usual cost of the buffet.

**T5.14** Based on previous records, 17% of the vehicles passing through a tollbooth have out-of-state plates. A bored tollbooth worker decides to pass the time by counting how many vehicles pass through until he sees two with out-of-state plates. We would like to perform a simulation to estimate the average number of vehicles it takes to find two with out-of-state plates.[31]

(a) Describe how you would use a table of random digits to perform the simulation.

(b) Perform 3 trials of the simulation using the random digits given here. Copy the digits onto your paper and mark directly on or above them so that someone can follow what you did.

| | | | | | |
|---|---|---|---|---|---|
| 41050 | 92031 | 06449 | 05059 | 59884 | 31880 |
| 53115 | 84469 | 94868 | 57967 | 05811 | 84514 |
| 84177 | 06757 | 17613 | 15582 | 51506 | 81435 |

# UNIT 4

## Probability, Random Variables, and Probability Distributions

Chapter 6

# Random Variables and Probability Distributions

Stefano Paterna/Alamy

## INTRODUCTION

Do you drink bottled water or tap water? According to an online news report, about 75% of people drink bottled water regularly. Some people do so because they believe bottled water is safer than tap water. (There's little evidence to support this belief.) Others say they prefer the taste of bottled water. Can people really tell the difference?

## ACTIVITY Bottled water versus tap water

This activity will give you and your classmates a chance to discover whether or not you can taste the difference between bottled water and tap water.

magnetcreative/E+/Getty Images

1. Before class begins, your teacher will prepare numbered stations with cups of water. You will be given an index card with a station number on it.
2. Go to the corresponding station. Pick up three cups (labeled A, B, and C) and take them back to your seat.
3. Your task is to determine which one of the three cups contains the bottled water. Drink all the water in Cup A first, then the water in Cup B, and finally the water in Cup C. Write down the letter of the cup that you think held the bottled water. Do not discuss your results with any of your classmates yet!
4. While you're tasting, your teacher will make a chart on the board like this one:

| Station number | Bottled water cup? | Truth |
|---|---|---|

5. When you are told to do so, go to the board and record your station number and the letter of the cup you identified as containing bottled water.
6. Your teacher will now reveal the truth about the cups of drinking water. How many students in the class identified the bottled water correctly? What percent of the class is this?
7. Let's assume that no one in your class can distinguish tap water from bottled water. In that case, students would just be guessing which cup of water tastes different. If so, what's the probability that an individual student would guess correctly?
8. How many correct identifications would you need to provide convincing evidence that the students in your class aren't just guessing? With your classmates, design and carry out a simulation to answer this question. What do you conclude about your class's ability to distinguish tap water from bottled water?

When Mr. Hogarth's class did the preceding activity, 13 out of 21 students made correct identifications. If we assume that the students in his class can't tell tap water from bottled water, then each one is basically guessing, with a 1/3 chance of being correct. So we'd expect about one-third of Mr. Hogarth's 21

The ABC News program *20/20* set up a blind taste test in which people were asked to rate four different brands of bottled water and New York City tap water without knowing which they were drinking. Can you guess the result? Tap water came out the clear winner in terms of taste.

students (i.e., about 7 students) to guess correctly. How likely is it that 13 or more of the 21 students would guess correctly? To answer this question without a simulation, we need a different kind of probability model from the ones we saw in Chapter 5.

Section 6.1 introduces the concept of a *random variable,* a numerical outcome of some random process (like the number of students who correctly guess the type of water). Each random variable has a *probability distribution* that gives us information about the likelihood that a specific event happens (like 13 or more correct guesses out of 21) and about what's expected to happen if the random process is repeated many times. Section 6.2 examines the effect of transforming and combining random variables on the shape, center, and variability of their probability distributions. In Section 6.3, we'll look at two random variables with probability distributions that are used enough to have their own names—*binomial* and *geometric.*

# SECTION 6.1 Discrete and Continuous Random Variables

**LEARNING TARGETS** *By the end of the section, you should be able to:*

- Use the probability distribution of a discrete random variable to calculate the probability of an event.
- Make a histogram to display the probability distribution of a discrete random variable and describe its shape.
- Calculate and interpret the mean (expected value) of a discrete random variable.
- Calculate and interpret the standard deviation of a discrete random variable.
- Use the probability distribution of a continuous random variable (uniform or Normal) to calculate the probability of an event.

A probability model describes the possible outcomes of a random process and the likelihood that those outcomes will occur. For example, suppose you toss a fair coin 3 times. The sample space for this random process is

HHH HHT HTH THH HTT THT TTH TTT

Because there are 8 equally likely outcomes, the probability is 1/8 for each possible outcome.

Define the **random variable** $X$ = the number of heads obtained in 3 tosses. The value of $X$ will vary from one set of tosses to another, but it will always be one of the numbers 0, 1, 2, or 3. How likely is $X$ to take each of those values? It will be easier to answer this question if we group the possible outcomes by the number of heads obtained:

$X = 0$: TTT $X = 1$: HTT THT TTH $X = 2$: HHT HTH THH $X = 3$: HHH

We can summarize the **probability distribution** of X in a table:

| Value | 0 | 1 | 2 | 3 |
|---|---|---|---|---|
| Probability | 1/8 | 3/8 | 3/8 | 1/8 |

**DEFINITION Random variable, Probability distribution**

A **random variable** takes numerical values that describe the outcomes of a random process.

The **probability distribution** of a random variable gives its possible values and their probabilities.

Recall the two types of quantitative variables from Chapter 1. *Discrete variables* typically result from counting something, while *continuous variables* typically result from measuring something.

We use capital, italic letters (like *X* or *Y*) to designate random variables and lowercase, italic letters (like *x* or *y*) to designate specific values of those variables. There are two main types of probability distributions, corresponding to two types of random variables: *discrete* and *continuous.*

# Discrete Random Variables

The random variable X in the coin-tossing setting is a **discrete random variable.**

**DEFINITION Discrete random variable**

A **discrete random variable** *X* takes a fixed set of possible values with gaps between them.

We can list the possible values of $X =$ the number of heads in 3 tosses of a coin as 0, 1, 2, 3. Note that there are gaps between these values on a number line. For instance, a gap exists between $X = 1$ and $X = 2$ because X cannot take values such as 1.2 or 1.84. The probability distribution of X is

| Value | 0 | 1 | 2 | 3 |
|---|---|---|---|---|
| Probability | 1/8 | 3/8 | 3/8 | 1/8 |

This probability distribution is valid because all the probabilities are between 0 and 1, and their sum is

$$1/8 + 3/8 + 3/8 + 1/8 = 1$$

**PROBABILITY DISTRIBUTION FOR A DISCRETE RANDOM VARIABLE**

The probability distribution of a discrete random variable X lists the values $x_i$ and their probabilities $p_i$:

| Value | $x_1$ | $x_2$ | $x_3$ | ... |
|---|---|---|---|---|
| Probability | $p_1$ | $p_2$ | $p_3$ | ... |

For the probability distribution to be valid, the probabilities $p_i$ must satisfy two requirements:

1. Every probability $p_i$ is a number between 0 and 1, inclusive.
2. The sum of the probabilities is 1: $p_1 + p_2 + p_3 + \cdots = 1$.

The AP® Statistics exam formula sheet uses the notation $P(x_i)$ for the probability that $X = x_i$. We prefer the simpler notation $p_i$ to represent $P(X = x_i)$.

We can use the probability distribution of a discrete random variable to find the probability of an event. For instance, what's the probability that we get at least one head in three tosses of the coin? In symbols, we want to find $P(X \geq 1)$. We know that

$$P(X \geq 1) = P(X = 1 \text{ or } X = 2 \text{ or } X = 3)$$

Because the events $X = 1$, $X = 2$, and $X = 3$ are mutually exclusive, we can add their probabilities to get the answer:

$$P(X \geq 1) = P(X = 1) + P(X = 2) + P(X = 3)$$
$$= 3/8 + 3/8 + 1/8 = 7/8$$

Or we could use the complement rule from Chapter 5:

$$P(X \geq 1) = 1 - P(X < 1) = 1 - P(X = 0)$$
$$= 1 - 1/8 = 7/8$$

## EXAMPLE

### Apgar scores: Babies' health at birth
### Discrete random variables

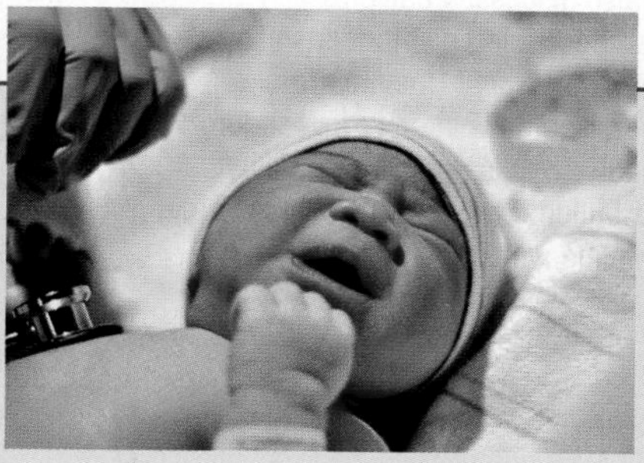
susaro/iStock/Getty Images

**PROBLEM:** In 1952, Dr. Virginia Apgar suggested five criteria for measuring a baby's health at birth: skin color, heart rate, muscle tone, breathing, and response when stimulated. She developed a 0-1-2 scale to rate a newborn on each of the five criteria. A baby's Apgar score is the sum of the ratings on each of the five scales, which gives a whole-number value from 0 to 10. Apgar scores are still used today to evaluate the health of newborns. Although this procedure was later named for Dr. Apgar, the acronym APGAR also represents the five scales: Appearance, Pulse, Grimace, Activity, and Respiration.

What Apgar scores are typical? To find out, researchers recorded the Apgar scores of over 2 million newborn babies in a single year.[1] Imagine selecting a newborn baby at random. (That's our random process.) Define the random variable $X$ = Apgar score of a randomly selected newborn baby. The table gives the probability distribution of $X$.

| Value $x_i$ | 0 | 1 | 2 | 3 | 4 | 5 | 6 | 7 | 8 | 9 | 10 |
|---|---|---|---|---|---|---|---|---|---|---|---|
| Probability $p_i$ | ??? | 0.006 | 0.007 | 0.008 | 0.012 | 0.020 | 0.038 | 0.099 | 0.319 | 0.437 | 0.053 |

(a) Write the event "the baby has an Apgar score of 0" in terms of $X$. Then find its probability.

(b) Doctors decided that Apgar scores of 7 or higher indicate a healthy newborn baby. What's the probability that a randomly selected newborn baby is healthy?

**SOLUTION:**

(a) $P(X=0) = 1 - (0.006 + 0.007 + \cdots + 0.053)$
$= 1 - 0.999$
$= 0.001$

Use the complement rule: $P(X = 0) = 1 - P(X \neq 0)$

(b) $P(X \geq 7) = 0.099 + 0.319 + 0.437 + 0.053$
$= 0.908$

The probability of choosing a healthy baby is $P(X \geq 7) = P(X = 7) + P(X = 8) + P(X = 9) + P(X = 10)$

**FOR PRACTICE, TRY EXERCISE 1**

Note that the probability of randomly selecting a newborn whose Apgar score is *at least* 7 is not the same as the probability that the baby's Apgar score is *greater than* 7. The latter probability is

$$P(X > 7) = P(X = 8) + P(X = 9) + P(X = 10)$$
$$= 0.319 + 0.437 + 0.053$$
$$= 0.809$$

The outcome X = 7 is included in "at least 7" but is not included in "greater than 7." **Be sure to consider whether to include the boundary value in your calculations when dealing with discrete random variables.**

## Analyzing Discrete Random Variables: Describing Shape

We also discussed outliers when describing distributions of quantitative data. Outliers are generally of less interest for random variables because their probability distributions specify which values are likely and which are unlikely.

When we analyzed distributions of quantitative data in Chapter 1, we made it a point to discuss their shape, center, and variability. We'll do the same with probability distributions of random variables.

For the discrete random variable X = Apgar score of a randomly selected newborn baby, the probability distribution is

| **Value $x_i$** | 0 | 1 | 2 | 3 | 4 | 5 | 6 | 7 | 8 | 9 | 10 |
|---|---|---|---|---|---|---|---|---|---|---|---|
| **Probability $p_i$** | 0.001 | 0.006 | 0.007 | 0.008 | 0.012 | 0.020 | 0.038 | 0.099 | 0.319 | 0.437 | 0.053 |

We can display the probability distribution with a histogram. Values of the variable go on the horizontal axis and probabilities go on the vertical axis. There is one bar in the histogram for each value of X. The height of each bar gives the probability for the corresponding value of the variable.

Figure 6.1 shows a histogram of the probability distribution of X. This distribution is skewed to the left and unimodal, with a single peak at an Apgar score of 9.

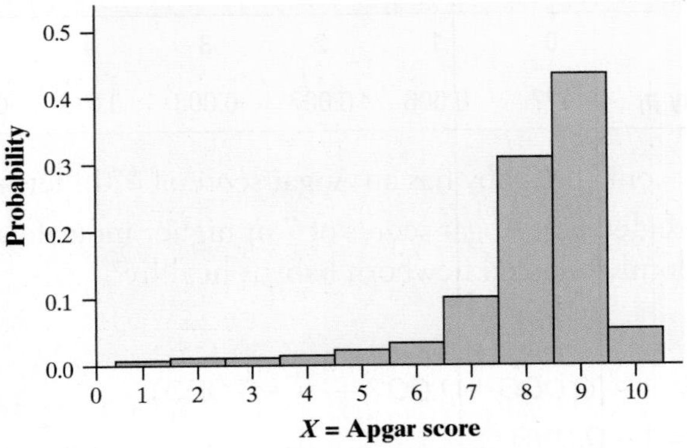


**FIGURE 6.1** Histogram of the probability distribution for the random variable $X$ = Apgar score of a randomly selected newborn baby.

There's another way to think about the graph displayed in Figure 6.1. The probability distribution of the random variable X models the *population distribution* of the quantitative variable "Apgar score of a newborn baby." So we can interpret our earlier result, $P(X > 7) = 0.809$, as saying that about 81% of all newborn babies have Apgar scores greater than 7. We also know that the shape of the population distribution is left-skewed with a single peak at 9.

## EXAMPLE

### Pete's Jeep Tours
### Displaying a probability distribution

**PROBLEM:** Pete's Jeep Tours offers a popular day trip in a tourist area. There must be at least 2 passengers for the trip to run, and the vehicle will hold up to 6 passengers. Pete charges $150 per passenger. Let C = the total amount of money that Pete collects on a randomly selected trip. The probability distribution of C is given in the table.

Judy M. Starnes

| **Total collected $c_i$** | 300 | 450 | 600 | 750 | 900 |
|---|---|---|---|---|---|
| **Probability $p_i$** | 0.15 | 0.25 | 0.35 | 0.20 | 0.05 |

Make a histogram of the probability distribution. Describe its shape.

**SOLUTION:**

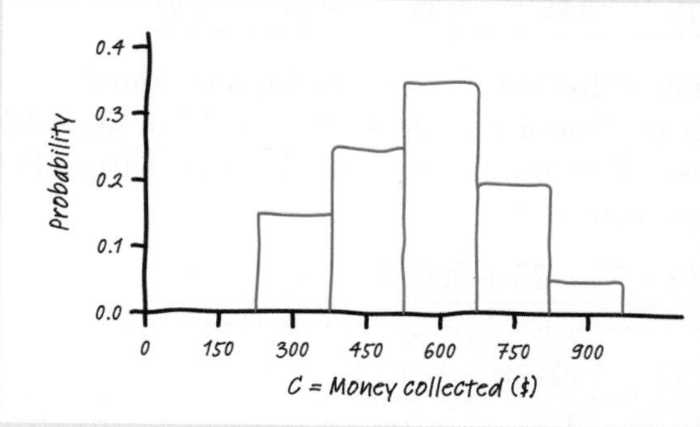


*Remember:* Values of the variable go on the horizontal axis and probabilities go on the vertical axis. Don't forget to properly label and scale each axis!

**The graph is roughly symmetric and has a single peak at $600.**

**FOR PRACTICE, TRY EXERCISE 5**

Notice the use of the label C (collects) for the random variable in the example. Sometimes we prefer contextual labels like this to the more generic X and Y.

## CHECK YOUR UNDERSTANDING

Indiana University Bloomington posts the grade distributions for its courses online.[2] Suppose we choose a student at random from a recent semester of this university's Business Statistics course. The student's grade on a 4-point scale (with A = 4) is a random variable X with this probability distribution:

| **Value** | 0 | 1 | 2 | 3 | 4 |
|---|---|---|---|---|---|
| **Probability** | 0.011 | 0.032 | ??? | 0.362 | 0.457 |

1. Write the event "the student got a C" using probability notation. Then find this probability.
2. Explain in words what $P(X \geq 3)$ means. What is this probability?
3. Make a histogram of the probability distribution. Describe its shape.

# Measuring Center: The Mean (Expected Value) of a Discrete Random Variable

In Chapter 1, you learned how to summarize the center of a distribution of quantitative data with either the median or the mean. For random variables, the mean is typically used to summarize the center of a probability distribution. Because a probability distribution can model the population distribution of a quantitative variable, we label the mean of a random variable X as $\mu_X$. Like any parameter, $\mu_X$ is a single, fixed value.

To find the mean of a quantitative data set, we compute the sum of the individual observations and divide by the total number of data values. How do we find the *mean of a discrete random variable?*

Consider the random variable C = the total amount of money that Pete collects on a randomly selected jeep tour from the previous example. The probability distribution of C is given in the table.

| **Total collected $c_i$** | 300 | 450 | 600 | 750 | 900 |
|---|---|---|---|---|---|
| **Probability $p_i$** | 0.15 | 0.25 | 0.35 | 0.20 | 0.05 |

What's the average amount of money that Pete collects on his jeep tours?

Imagine a hypothetical 100 trips. Pete will collect \$300 on 15 of these trips, \$450 on 25 trips, \$600 on 35 trips, \$750 on 20 trips, and \$900 on 5 trips. Pete's average amount collected for these trips is

$$\frac{300 \cdot 15 + 450 \cdot 25 + 600 \cdot 35 + 750 \cdot 20 + 900 \cdot 5}{100}$$
$$= \frac{300 \cdot 15}{100} + \frac{450 \cdot 25}{100} + \frac{600 \cdot 35}{100} + \frac{750 \cdot 20}{100} + \frac{900 \cdot 5}{100}$$
$$= 300(0.15) + 450(0.25) + 600(0.35) + 750(0.20) + 900(0.05)$$
$$= 562.50$$

The third line of the calculation is just the values of the random variable *C* times their corresponding probabilities.

That is, the **mean of the discrete random variable** C is $\mu_C = \$562.50$. This is also known as the **expected value** of C, denoted by $E(C)$.

The mean (expected value) of any discrete random variable is found in a similar way. It is an average of the possible outcomes, but a weighted average in which each outcome is weighted by its probability.

**DEFINITION Mean (expected value) of a discrete random variable**

The **mean (expected value) of a discrete random variable** is its average value over many, many trials of the same random process.

Suppose that *X* is a discrete random variable with probability distribution

| **Value** | $x_1$ | $x_2$ | $x_3$ | ... |
|---|---|---|---|---|
| **Probability** | $p_1$ | $p_2$ | $p_3$ | ... |

To find the mean (expected value) of *X*, multiply each possible value of *X* by its probability, then add all the products:

$$\mu_X = E(X) = x_1p_1 + x_2p_2 + x_3p_3 + \cdots = \sum x_ip_i$$

The AP® Statistics exam formula sheet gives the formula for the mean (expected value) of a discrete random variable as $\sum x_i \cdot P(x_i)$.

An alternate interpretation of $\mu_C = \$562.50$ comes from thinking of the probability distribution of $C$ as a model for the *population distribution* of money collected: For all of his jeep tours, the average amount that Pete collects on a trip is about \$562.50.

Recall that the mean is the balance point of a distribution. For Pete's distribution of money collected on a randomly selected jeep tour, the histogram balances at $\mu_C = 562.50$. How do we interpret this parameter? If we randomly select many, many jeep tours, Pete will make an average of about \$562.50 per trip.

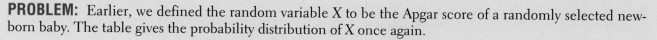

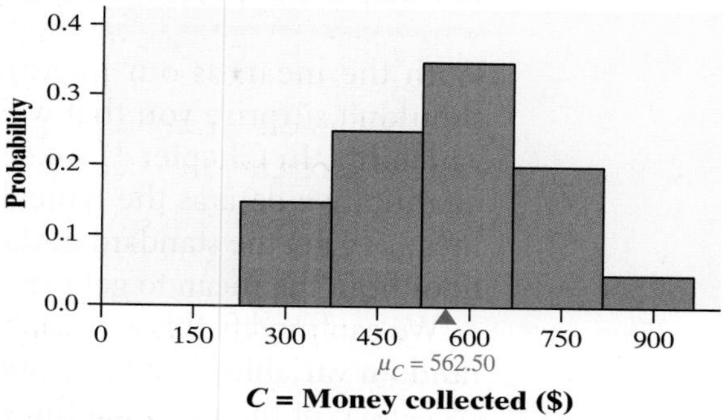


## EXAMPLE

## Apgar scores: What's typical?
### Finding and interpreting the mean

**PROBLEM:** Earlier, we defined the random variable X to be the Apgar score of a randomly selected newborn baby. The table gives the probability distribution of X once again.

| Value $x_i$ | 0 | 1 | 2 | 3 | 4 | 5 | 6 | 7 | 8 | 9 | 10 |
|---|---|---|---|---|---|---|---|---|---|---|---|
| Probability $p_i$ | 0.001 | 0.006 | 0.007 | 0.008 | 0.012 | 0.020 | 0.038 | 0.099 | 0.319 | 0.437 | 0.053 |

Calculate and interpret the expected value of X.

**SOLUTION:**

$E(X) = \mu_X = (0)(0.001) + (1)(0.006) + \cdots + (10)(0.053) = 8.128$

$E(X) = \mu_X = \sum x_i p_i$

If many, many newborns are randomly selected, their average Apgar score will be about 8.128.

**FOR PRACTICE, TRY EXERCISE 7**

**AP® EXAM TIP**

If the mean of a random variable has a non-integer value but you report it as an integer, your answer will not get full credit.

Notice that the mean Apgar score, 8.128, is not a possible value of the random variable X because it is not a whole number between 0 and 10. The non-integer value of the mean shouldn't bother you if you think of the mean (expected value) as a long-run average over many trials of the random process.

How can we find the *median* of a discrete random variable? In Chapter 1, we defined the median as "the midpoint of a distribution, the number such that about half the observations are smaller and about half are larger." The median of a discrete random variable is the 50th percentile of its probability distribution. We can find the median from a *cumulative probability distribution*, like the one shown here for the random variable $C$ = total amount of money Pete collects on a randomly selected jeep tour. We see from the table that $P(C \leq 300) = 0.15$ and that $P(C \leq 450) = 0.40$. So \$300 is the 15th percentile and \$450 is the 40th percentile of the probability distribution of C. The median of a discrete random variable is the smallest value for which the cumulative probability equals or exceeds 0.5. So the median amount of money Pete collects on a randomly selected jeep tour is \$600.

| Total collected $c_i$ | 300 | 450 | 600 | 750 | 900 |
|---|---|---|---|---|---|
| Probability $p_i$ | 0.15 | 0.25 | 0.35 | 0.20 | 0.05 |
| Cumulative probability | 0.15 | 0.40 | 0.75 | 0.95 | 1.00 |

# Measuring Variability: The Standard Deviation (and Variance) of a Discrete Random Variable

With the mean as our measure of center for a discrete random variable, it shouldn't surprise you that we'll use the standard deviation as our measure of variability. In Chapter 1, we defined the standard deviation of a distribution of quantitative data as the typical distance of the values in the data set from the mean. To get the standard deviation, we started by "averaging" the squared deviations from the mean to get the variance and then took the square root.

We can modify this approach to calculate the **standard deviation of a discrete random variable** $X$. Start by finding a weighted average of the squared deviations $(x_i - \mu_X)^2$ of the values of the variable $X$ from its mean $\mu_X$. The probability distribution gives the appropriate weight for each squared deviation. We call this weighted average of squared deviations the **variance** of $X$. Then take the square root to get the standard deviation. Because a probability distribution can model the population distribution of a quantitative variable, we label the variance of a random variable $X$ as $\sigma_X^2$ and the standard deviation as $\sigma_X$. Like any parameter, $\sigma_X$ is a single, fixed value.

**DEFINITION Standard deviation of a discrete random variable, Variance**

The **standard deviation of a discrete random variable** measures how much the values of the variable typically vary from the mean in many, many trials of the random process.

Suppose that $X$ is a discrete random variable with probability distribution

| **Value** | $x_1$ | $x_2$ | $x_3$ | ... |
|---|---|---|---|---|
| **Probability** | $p_1$ | $p_2$ | $p_3$ | ... |

and that $\mu_X$ is the mean of $X$. The **variance** of $X$ is

$$\sigma_X^2 = (x_1 - \mu_X)^2 p_1 + (x_2 - \mu_X)^2 p_2 + (x_3 - \mu_X)^2 p_3 + \cdots$$
$$= \sum (x_i - \mu_X)^2 p_i$$

The standard deviation of $X$ is the square root of the variance:

$$\sigma_X = \sqrt{\sigma_X^2} = \sqrt{(x_1 - \mu_X)^2 p_1 + (x_2 - \mu_X)^2 p_2 + (x_3 - \mu_X)^2 p_3 + \cdots}$$
$$= \sqrt{\sum (x_i - \mu_X)^2 p_i}$$

The AP® Statistics exam formula sheet gives the formula for the standard deviation of a discrete random variable as $\sqrt{\sum (x_i - \mu_X)^2 \cdot P(x_i)}$.

Let's return to the random variable $C$ = the total amount of money that Pete collects on a randomly selected jeep tour. The left two columns of the following table give the probability distribution. Recall that the mean of $C$ is $\mu_C = 562.50$. The third column of the table shows the squared deviation of each value from the mean. The fourth column gives the weighted squared deviations.

| Total collected $c_i$ | Probability $p_i$ | Squared deviation from the mean $(c_i - \mu_C)^2$ | Weighted squared deviation $(c_i - \mu_C)^2 p_i$ |
|---|---|---|---|
| 300 | 0.15 | $(300 - 562.50)^2$ | $(300 - 562.50)^2 (0.15) =$ 10335.94 |
| 450 | 0.25 | $(450 - 562.50)^2$ | $(450 - 562.50)^2 (0.25) =$ 3164.06 |
| 600 | 0.35 | $(600 - 562.50)^2$ | $(600 - 562.50)^2 (0.35) =$ 492.19 |
| 750 | 0.20 | $(750 - 562.50)^2$ | $(750 - 562.50)^2 (0.20) =$ 7031.25 |
| 900 | 0.05 | $(900 - 562.50)^2$ | $(900 - 562.50)^2 (0.05) =$ 5695.31 |
| | | | Sum = 26,718.75 |

Adding the weighted average of the squared deviations in the fourth column gives the variance of $C$:

$$\begin{aligned}\sigma_C^2 &= \sum (c_i - \mu_C)^2 p_i \\ &= (300 - 562.50)^2(0.15) + (450 - 562.50)^2(0.25) + \cdots + (900 - 562.50)^2(0.05) \\ &= 10335.94 + 3164.06 + \cdots + 5695.31 \\ &= 26{,}718.75 \text{ (squared dollars)}\end{aligned}$$

Because the probability distribution of $C$ is a model for the *population distribution* of money collected, we can also interpret the parameter $\sigma_C$ in this way: "For all of his jeep tours, the amount of money that Pete collects on a trip typically varies by about \$163.46 from the mean of \$562.50."

The standard deviation of $C$ is the square root of the variance:

$$\sigma_C = \sqrt{26{,}718.75} = \$163.46$$

How do we interpret this parameter? If many, many jeep tours are randomly selected, the amount of money that Pete collects typically varies by about \$163.46 from the mean of \$562.50.

## EXAMPLE

### How much do Apgar scores vary?

### Finding and interpreting the standard deviation

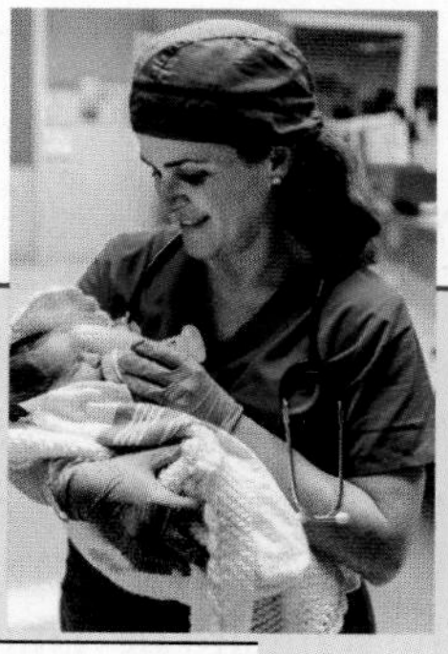

Juanmonino/Getty Images

**PROBLEM:** Earlier, we defined the random variable $X$ to be the Apgar score of a randomly selected newborn baby. The table gives the probability distribution of $X$ once again. In the last example, we calculated the mean Apgar score of a randomly chosen newborn to be $\mu_x = 8.128$.

| **Value $x_i$** | 0 | 1 | 2 | 3 | 4 | 5 | 6 | 7 | 8 | 9 | 10 |
|---|---|---|---|---|---|---|---|---|---|---|---|
| **Probability $p_i$** | 0.001 | 0.006 | 0.007 | 0.008 | 0.012 | 0.020 | 0.038 | 0.099 | 0.319 | 0.437 | 0.053 |

Calculate and interpret the standard deviation of $X$.

**SOLUTION:**

$$\sigma_X^2 = (0 - 8.128)^2(0.001) + (1 - 8.128)^2(0.006) + \cdots + (10 - 8.128)^2(0.053) = 2.066$$

$$\sigma_X = \sqrt{2.066} = 1.437$$

If many, many newborns are randomly selected, the babies' Apgar scores will typically vary by about 1.437 units from the mean of 8.128.

Start by calculating the variance $\sigma_X^2 = \sum (x_i - \mu_X)^2 p_i$.

**FOR PRACTICE, TRY EXERCISE 13**

You can use your calculator to graph the probability distribution of a discrete random variable and to calculate measures of center and variability, as the following Technology Corner illustrates.

## 13. Technology Corner ANALYZING DISCRETE RANDOM VARIABLES

*TI-Nspire and other technology instructions are on the book's website at highschool.bfwpub.com/updatedtps6e.*

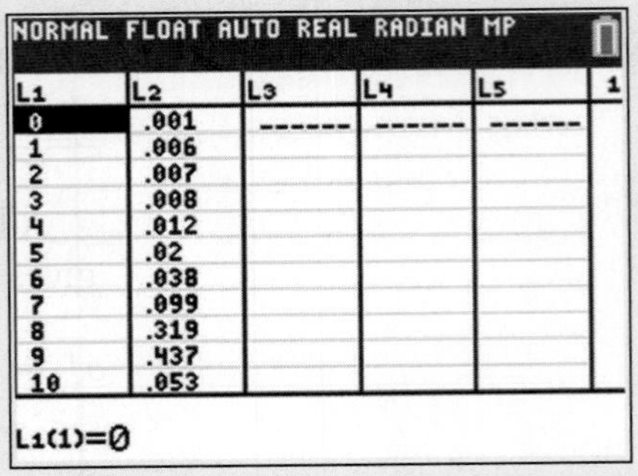


Let's explore what the TI-83/84 can do using the random variable X = Apgar score of a randomly selected newborn.

1. Enter the values of the random variable in list L1 and the corresponding probabilities in list L2.
2. To graph a histogram of the probability distribution:
   - In the statistics plot menu, define Plot 1 to be a histogram with Xlist: L1 and Freq: L2.
   - Adjust your window settings as follows: Xmin = −1, Xmax = 11, Xscl = 1, Ymin = −0.1, Ymax = 0.5, Yscl = 0.1.
   - Press GRAPH.

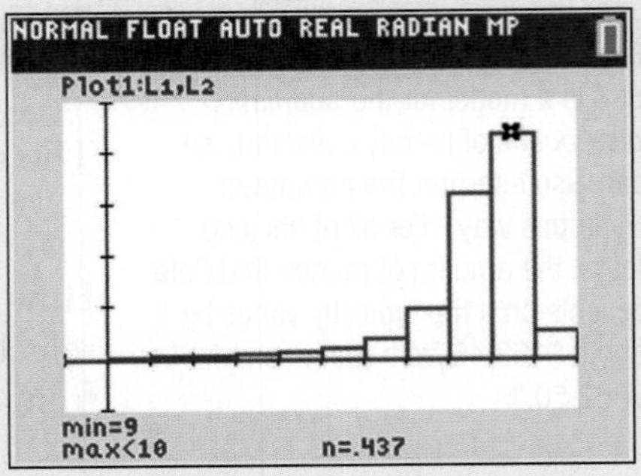


3. To calculate the mean and standard deviation of the random variable, use one-variable statistics with the values in L1 and the probabilities (relative frequencies) in L2. Press STAT, arrow to the CALC menu, and choose 1-Var Stats.

   **OS 2.55 or later:** In the dialog box, specify List: L1 and FreqList: L2. Then choose Calculate.

   **Older OS:** Execute the command 1-Var Stats L1,L2.

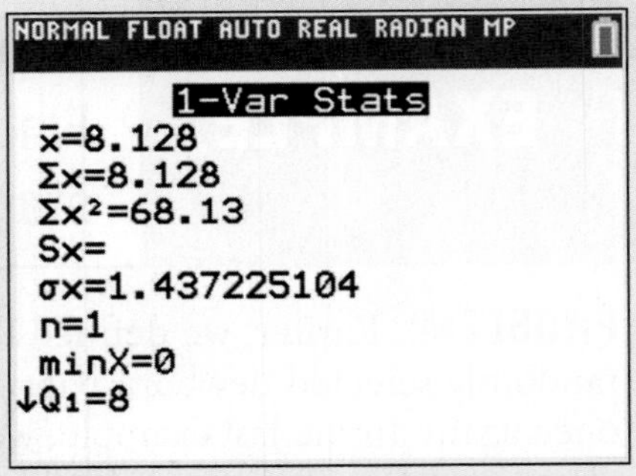


*Note:* If you leave Freq: L2 and try to calculate summary statistics for a quantitative data set that does not include frequencies, you will likely get an error message. Be sure to clear Freq when you are done with calculations to avoid this issue.

**The calculator's notation for the mean of the random variable X is incorrect.** We should write $\mu_X = 8.128$. Fortunately, the notation for the standard deviation is correct: $\sigma_X = 1.437$.

### AP® EXAM TIP

If you are asked to calculate the mean or standard deviation of a discrete random variable on a free response question, you must show numerical values substituted into the appropriate formula, as in the previous two examples. Feel free to use ellipses (. . .) if there are many terms in the summation, as we did. You may then use the method described in Technology Corner 13 to perform the calculation with 1-Var Stats. Writing only 1-Var Stats L1, L2 and then giving the correct values of the mean and standard deviation will *not* earn credit for showing work. Also, be sure to avoid incorrect notation when labeling these parameters.

## CHECK YOUR UNDERSTANDING

Indiana University Bloomington posts the grade distributions for its courses online.[2] Suppose we choose a student at random from a recent semester of this university's Business Statistics course. The student's grade on a 4-point scale (with A = 4) is a random variable X with this probability distribution:

| **Value** | 0 | 1 | 2 | 3 | 4 |
|---|---|---|---|---|---|
| **Probability** | 0.011 | 0.032 | 0.138 | 0.362 | 0.457 |

1. Find the mean of X. Interpret this parameter.
2. Find the standard deviation of X. Interpret this parameter.

# Continuous Random Variables

When we use Table D of random digits to select an integer from 0 to 9, the result is a discrete random variable (call it X). The probability distribution assigns probability 1/10 to each of the 10 equally likely values of X.

Suppose we want to choose a number at random between 0 and 9, allowing *any* number between 0 and 9 as the outcome (like 0.84522 or 7.1111119). Calculator and computer random number generators will do this. The sample space of this random process is the entire interval of values between 0 and 9 on the number line. If we define Y = randomly generated number between 0 and 9, then Y is a **continuous random variable**.

**DEFINITION** **Continuous random variable**

A **continuous random variable** can take any value in an interval on the number line.

Most discrete random variables result from counting something, like the number of siblings that a randomly selected student has. Continuous random variables typically result from measuring something, like the height of a randomly selected student or the time it takes that student to run a mile.

How can we find the probability $P(3 \leq Y \leq 7)$ that the random number generator produces a value between 3 and 7? As in the case of selecting a random digit, we would like all possible outcomes to be equally likely. But we cannot assign probabilities to each individual value of Y and then add them, because there are infinitely many possible values.

*The probability distribution of a continuous random variable is described by a density curve.* Recall from Chapter 2 that any density curve has area exactly 1 underneath it, corresponding to a total probability of 1. We use areas under the density curve to assign probabilities to events.

For the continuous random variable Y = randomly generated number between 0 and 9, its probability distribution is a uniform density curve with constant height 1/9 on the interval from 0 to 9. Note that this probability distribution is valid because the total area under the density curve is

$$\text{Area} = \text{base} \times \text{height} = 9 \times 1/9 = 1$$

Figure 6.2 shows the probability distribution of Y with the area of interest shaded. The area under the density curve between 3 and 7 is

$$\text{Area} = \text{base} \times \text{height} = 4 \times 1/9 = 4/9$$

So $P(3 \le Y \le 7) = 4/9 = 0.444$.

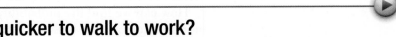

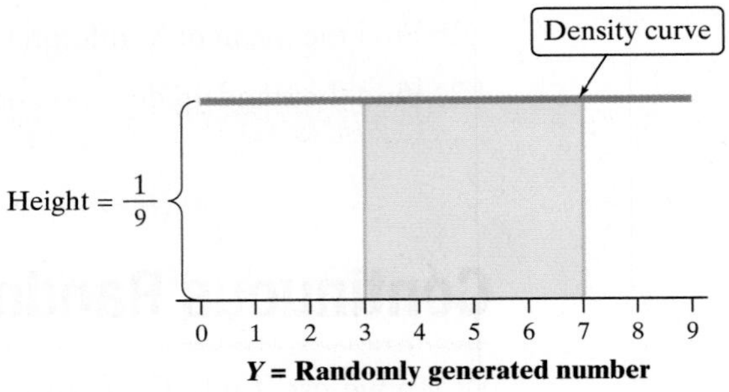


**FIGURE 6.2** The probability distribution of the continuous random variable $Y$ = randomly generated number between 0 and 9. The shaded area represents $P(3 \le Y \le 7)$.

**HOW TO FIND PROBABILITIES FOR A CONTINUOUS RANDOM VARIABLE**

The probability of any event involving a continuous random variable is the area under the density curve and directly above the values on the horizontal axis that make up the event.

Consider a specific outcome from the random number generator setting, such as $P(Y = 7)$. The probability of this event is the area under the density curve that's above the point 7.0000 . . . on the horizontal axis. But this vertical line segment has no width, so the area is 0. In fact, all continuous probability distributions assign probability 0 to every individual outcome. For that reason,

$$P(3 \le Y \le 7) = P(3 \le Y < 7) = P(3 < Y \le 7) = P(3 < Y < 7) = 0.444$$

*Remember:* The probability distribution for a continuous random variable assigns probabilities to *intervals* of outcomes rather than to individual outcomes.

## EXAMPLE

### Will it be quicker to walk to work?

### Continuous random variables

**PROBLEM:** Selena works at a bookstore in the Denver International Airport. She takes the airport train from the main terminal to get to work each day. The airport just opened a new walkway that would allow Selena to get from the main terminal to the bookstore in 4 minutes. She wonders if it will be faster to walk or take the train to work. Let Y = Selena's journey time to work (in minutes) by train on a randomly selected day. The probability distribution of Y can be modeled by a uniform density curve on the interval from 2 to 5 minutes. Find the probability that it will be quicker for Selena to take the train than to walk that day.

**SOLUTION:**

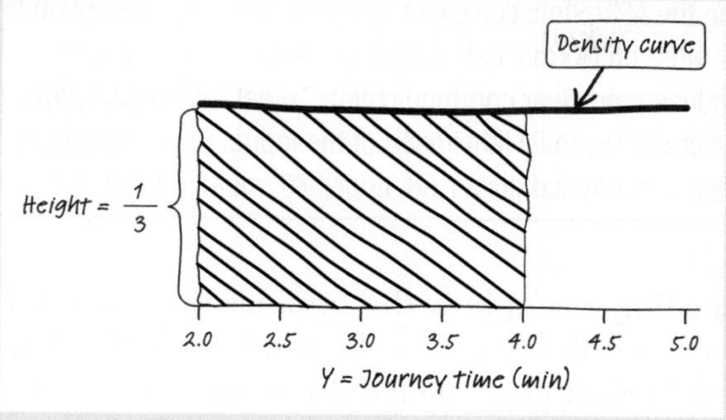


Start by drawing the density curve with the area of interest shaded. The height of the curve needs to be 1/3 so that

Area = base × height = 3 × 1/3 = 1

Shaded area = base × height = 2 × 1/3 = 2/3

$P(Y < 4) = 2/3 = 0.667$

There is a 66.7% chance that it will be quicker for Selena to take the train to work on a randomly selected day.

**FOR PRACTICE, TRY EXERCISE 23**

Density curves are probability distributions. They can also model populations of quantitative data, like Selena's train journey times in the preceding example.

## EXAMPLE Young women's heights
### Normal probability distributions

**PROBLEM:** The heights of young women can be modeled by a Normal distribution with mean $\mu = 64$ inches and standard deviation $\sigma = 2.7$ inches. Suppose we choose a young woman at random and let Y = her height (in inches). Find $P(68 \le Y \le 70)$. Interpret this value.

**SOLUTION:**

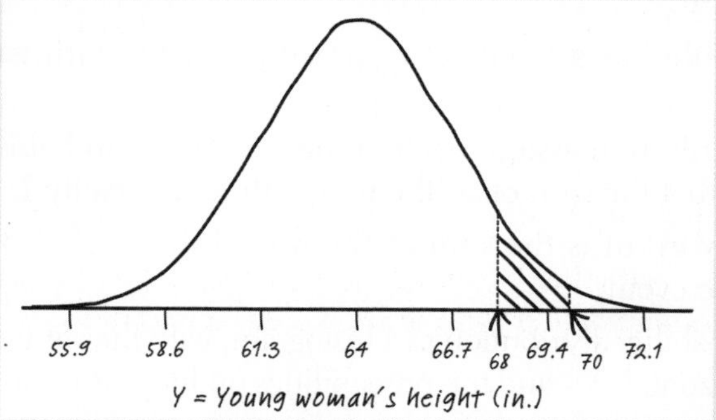


**1. Draw a Normal distribution.**
**2. Perform calculations—show your work!**
(i) Standardize and use Table A or technology; or
(ii) Use technology without standardizing.
Be sure to answer the question that was asked.

(i) $z = \dfrac{68 - 64}{2.7} = 1.48$

$z = \dfrac{70 - 64}{2.7} = 2.22$

*Using Table A:* $0.9868 - 0.9306 = 0.0562$

$P(68 \le Y \le 70) = P(1.48 \le Z \le 2.22)$

*Using technology:* normalcdf(lower:1.48, upper:2.22, mean:0, SD:1) = 0.0562

(ii) normalcdf(lower:68, upper:70, mean:64, SD:2.7) = 0.0561

The probability that a randomly selected young woman has a height between 68 and 70 inches is about 0.06.

**FOR PRACTICE, TRY EXERCISE 27**

**AP® EXAM TIP**

Students often do not get full credit on the AP® Statistics exam because they only use option (ii) with "calculator-speak" to show their work on Normal calculation questions—for example, normalcdf(68,70,64,2.7). This is not considered clear communication. To get full credit, follow the two-step process above, making sure to carefully label each of the inputs in the calculator command if you use technology in Step 2: normalcdf(lower:68, upper:70, mean:64, SD: 2.7).

The calculation in the preceding example is the same as those we did in Chapter 2. Only the language of probability is new. By thinking of the density curve as a model for the population distribution of a quantitative variable, we can interpret the result as: "About 6% of all young women have heights between 68 and 70 inches."

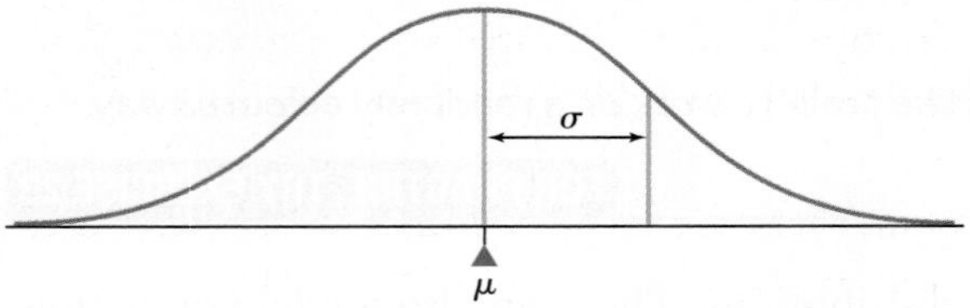


What about the mean $\mu$ and standard deviation $\sigma$ for continuous random variables? We interpret these parameters in the same way as we did for discrete random variables. Chapter 2 showed us how to find the mean of the distribution: it is the point at which the area under the density curve would balance if it were made out of solid material. The mean lies at the center of symmetric density curves such as Normal curves. We can locate the standard deviation of a Normal distribution from its inflection points. Exact calculation of the mean and standard deviation for most continuous random variables requires advanced mathematics.[3]

## Section 6.1 Summary

- A **random variable** takes numerical values determined by the outcome of a random process. The **probability distribution** of a random variable gives its possible values and their probabilities. There are two types of random variables: *discrete* and *continuous.*
- A **discrete random variable** has a fixed set of possible values with gaps between them.
  - A valid probability distribution assigns each of these values a probability between 0 and 1 such that the sum of all the probabilities is exactly 1.
  - The probability of any event is the sum of the probabilities of all the values that make up the event.
  - We can display the probability distribution as a histogram, with the values of the variable on the horizontal axis and the probabilities on the vertical axis.
- A **continuous random variable** can take any value in an interval on the number line.
  - A valid probability distribution for a continuous random variable is described by a density curve with area 1 underneath.
  - The probability of any event is the area under the density curve directly above the values on the horizontal axis that make up the event.
- A probability distribution can also be used to model the population distribution of a quantitative variable.
- We can describe the *shape* of a probability distribution histogram or density curve in the same way as we did a distribution of quantitative data—by identifying symmetry or skewness and any major peaks.

- Use the mean to summarize the *center* of a probability distribution. The **mean of a random variable** $\mu_X$ is the balance point of the probability distribution histogram or density curve.
  - The mean is the long-run average value of the variable after many, many trials of the random process. It is also known as the **expected value** of the random variable, $E(X)$.
  - If X is a discrete random variable, the mean is the average of the values of X, each weighted by its probability:

$$\mu_X = E(X) = \sum x_i p_i = x_1 p_1 + x_2 p_2 + x_3 p_3 + \cdots$$

- Use the standard deviation to summarize the variability of a probability distribution. The **standard deviation of a random variable** $\sigma_X$ measures how much the values of the variable typically vary from the mean in many, many trials of the random process.
  - If X is a discrete random variable, the **variance** of X is the "average" squared deviation of the values of the variable from their mean:

$$\sigma_X^2 = \sum (x_i - \mu_X)^2 p_i = (x_1 - \mu_X)^2 p_1 + (x_2 - \mu_X)^2 p_2 + (x_3 - \mu_X)^2 p_3 + \cdots$$

  The standard deviation $\sigma_X$ is the square root of the variance.

### 6.1 Technology Corner

*TI-Nspire and other technology instructions are on the book's website at highschool.bfwpub.com/updatedtps6e.*

**13. Analyzing discrete random variables** **Page 398**

## Section 6.1 Exercises

1. **Kids and toys** In an experiment on the behavior of young children, each subject is placed in an area with five toys. Past experiments have shown that the probability distribution of the number X of toys played with by a randomly selected subject is as follows: pg 391

| Number of toys $x_i$ | 0 | 1 | 2 | 3 | 4 | 5 |
|---|---|---|---|---|---|---|
| Probability $p_i$ | 0.03 | 0.16 | 0.30 | 0.23 | 0.17 | ??? |

(a) Write the event "child plays with 5 toys" in terms of X. Then find its probability.

(b) What's the probability that a randomly selected subject plays with at most 3 toys?

2. **Spell-checking** Spell-checking software catches "nonword errors," which result in a string of letters that is not a word, as when "the" is typed as "teh." When undergraduates are asked to write a 250-word essay (without spell-checking), the number Y of nonword errors in a randomly selected essay has the following probability distribution.

| Value $y_i$ | 0 | 1 | 2 | 3 | 4 |
|---|---|---|---|---|---|
| Probability $p_i$ | 0.1 | ??? | 0.3 | 0.3 | 0.1 |

(a) Write the event "one nonword error" in terms of Y. Then find its probability.

(b) What's the probability that a randomly selected essay has at least two nonword errors?

3. **Get on the boat!** A small ferry runs every half hour from one side of a large river to the other. The probability distribution for the random variable Y = money collected (in dollars) on a randomly selected ferry trip is shown here.

| Money collected | 0 | 5 | 10 | 15 | 20 | 25 |
|---|---|---|---|---|---|---|
| Probability | 0.02 | 0.05 | 0.08 | 0.16 | 0.27 | 0.42 |

(a) Find $P(X < 20)$. Interpret this result.

(b) Express the event "at least \$20 is collected" in terms of Y. What is the probability of this event?

4. **Skee Ball** Ana is a dedicated Skee Ball player (see photo) who always rolls for the 50-point slot. The probability distribution of Ana's score X on a randomly selected roll of the ball is shown here.

| Score | 10 | 20 | 30 | 40 | 50 |
|---|---|---|---|---|---|
| Probability | 0.32 | 0.27 | 0.19 | 0.15 | 0.07 |

Stan Rohrer/Getty Images

(a) Find $P(X > 20)$. Interpret this result.

(b) Express the event "Ana scores at most 20" in terms of X. What is the probability of this event?

pg 393 5. **Get on the boat!** Refer to Exercise 3. Make a histogram of the probability distribution. Describe its shape.

6. **Skee Ball** Refer to Exercise 4. Make a histogram of the probability distribution. Describe its shape.

7. **Get on the boat!** Refer to Exercise 3. Find the mean of Y. Interpret this parameter. pg 395

| Money collected | 0 | 5 | 10 | 15 | 20 | 25 |
|---|---|---|---|---|---|---|
| Probability | 0.02 | 0.05 | 0.08 | 0.16 | 0.27 | 0.42 |

8. **Skee Ball** Refer to Exercise 4. Find the mean of X. Interpret this parameter.

| Score | 10 | 20 | 30 | 40 | 50 |
|---|---|---|---|---|---|
| Probability | 0.32 | 0.27 | 0.19 | 0.15 | 0.07 |

9. **Benford's law** Faked numbers in tax returns, invoices, or expense account claims often display patterns that aren't present in legitimate records. Some patterns, like too many round numbers, are obvious and easily avoided by a clever crook. Others are more subtle. It is a striking fact that the first digits of numbers in legitimate records often follow a model known as Benford's law.[4] Call the first digit of a randomly chosen legitimate record X for short. The probability distribution for X is shown here (note that a first digit cannot be 0).

| First digit $x_i$ | 1 | 2 | 3 | 4 | 5 | 6 | 7 | 8 | 9 |
|---|---|---|---|---|---|---|---|---|---|
| Probability $p_i$ | 0.301 | 0.176 | 0.125 | 0.097 | 0.079 | 0.067 | 0.058 | 0.051 | 0.046 |

(a) A histogram of the probability distribution is shown. Describe its shape.

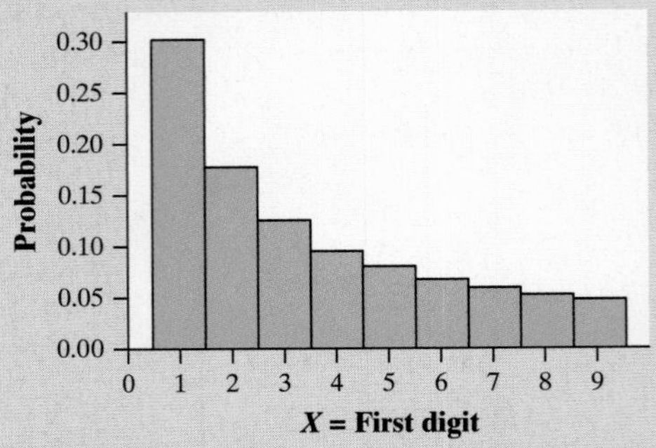


(b) Calculate and interpret the expected value of X.

10. **Working out** Choose a person aged 19 to 25 years at random and ask, "In the past seven days, how many times did you go to an exercise or fitness center or work out?" Call the response Y for short. Based on a large sample survey, here is the probability distribution of Y.[5]

| Days $y_i$ | 0 | 1 | 2 | 3 | 4 | 5 | 6 | 7 |
|---|---|---|---|---|---|---|---|---|
| Probability $p_i$ | 0.68 | 0.05 | 0.07 | 0.08 | 0.05 | 0.04 | 0.01 | 0.02 |

(a) A histogram of the probability distribution is shown. Describe its shape.

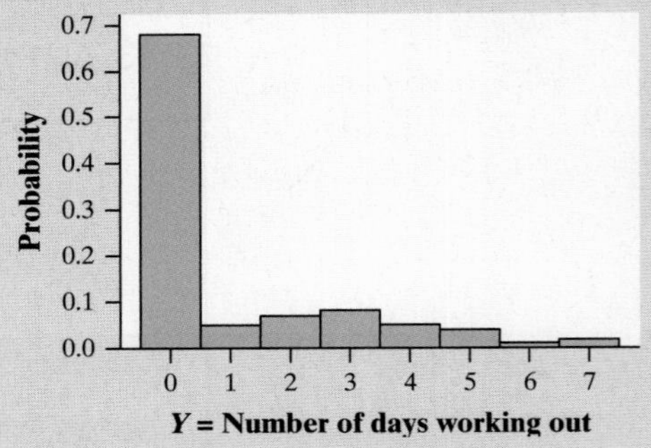


(b) Calculate and interpret the expected value of Y.

11. **Get on the boat!** A small ferry runs every half hour from one side of a large river to the other. The probability distribution for the random variable Y = money collected on a randomly selected ferry trip is shown here. From Exercise 7, $\mu_Y = \$19.35$.

| Money collected | 0 | 5 | 10 | 15 | 20 | 25 |
|---|---|---|---|---|---|---|
| Probability | 0.02 | 0.05 | 0.08 | 0.16 | 0.27 | 0.42 |

(a) Construct the cumulative probability distribution for Y.

(b) Use the cumulative probability distribution to find the median of Y.

(c) Compare the mean and median. Explain why this relationship makes sense based on the probability distribution.

12. **Skee Ball** Ana is a dedicated Skee Ball player (see photo in Exercise 4) who always rolls for the 50-point slot. The probability distribution of Ana's score X on a randomly selected roll of the ball is shown here. From Exercise 8, $\mu_X = 23.8$.

| Score | 10 | 20 | 30 | 40 | 50 |
|---|---|---|---|---|---|
| Probability | 0.32 | 0.27 | 0.19 | 0.15 | 0.07 |

(a) Construct the cumulative probability distribution for X.

(b) Use the cumulative probability distribution to find the median of X.

(c) Compare the mean and median. Explain why this relationship makes sense based on the probability distribution.

13. **Get on the boat!** A small ferry runs every half hour pg 397 from one side of a large river to the other. The probability distribution for the random variable $Y$ = money collected on a randomly selected ferry trip is shown here. From Exercise 7, $\mu_Y = \$19.35$. Calculate and interpret the standard deviation of Y.

| Money collected | 0 | 5 | 10 | 15 | 20 | 25 |
|---|---|---|---|---|---|---|
| Probability | 0.02 | 0.05 | 0.08 | 0.16 | 0.27 | 0.42 |

14. **Skee Ball** Ana is a dedicated Skee Ball player (see photo in Exercise 4) who always rolls for the 50-point slot. The probability distribution of Ana's score X on a randomly selected roll of the ball is shown here. From Exercise 8, $\mu_X = 23.8$. Calculate and interpret the standard deviation of X.

| Score | 10 | 20 | 30 | 40 | 50 |
|---|---|---|---|---|---|
| Probability | 0.32 | 0.27 | 0.19 | 0.15 | 0.07 |

15. **Benford's law** Exercise 9 described how the first digits of numbers in legitimate records often follow a model known as Benford's law. Call the first digit of a randomly chosen legitimate record X for short. The probability distribution for X is shown here (note that a first digit can't be 0). From Exercise 9, $E(X) = 3.441$. Find the standard deviation of X. Interpret this parameter.

| First digit $x_i$ | 1 | 2 | 3 | 4 | 5 | 6 | 7 | 8 | 9 |
|---|---|---|---|---|---|---|---|---|---|
| Probability $p_i$ | 0.301 | 0.176 | 0.125 | 0.097 | 0.079 | 0.067 | 0.058 | 0.051 | 0.046 |

16. **Working out** Exercise 10 described a large sample survey that asked a sample of people aged 19 to 25 years, "In the past seven days, how many times did you go to an exercise or fitness center or work out?" The response Y for a randomly selected survey respondent has the probability distribution shown here. From Exercise 10, $E(Y) = 1.03$. Find the standard deviation of Y. Interpret this parameter.

| Days $y_i$ | 0 | 1 | 2 | 3 | 4 | 5 | 6 | 7 |
|---|---|---|---|---|---|---|---|---|
| Probability $p_i$ | 0.68 | 0.05 | 0.07 | 0.08 | 0.05 | 0.04 | 0.01 | 0.02 |

17. **Life insurance** A life insurance company sells a term insurance policy to 21-year-old males that pays \$100,000 if the insured dies within the next 5 years. The probability that a randomly chosen male will die each year can be found in mortality tables. The company collects a premium of \$250 each year as payment for the insurance. The amount Y that the company profits on a randomly selected policy of this type is \$250 per year, less the \$100,000 that it must pay if the insured dies. Here is the probability distribution of Y:

| Age at death | 21 | 22 | 23 |
|---|---|---|---|
| Profit $y_i$ | −\$99,750 | −\$99,500 | −\$99,250 |
| Probability $p_i$ | 0.00183 | 0.00186 | 0.00189 |
| Age at death | 24 | 25 | 26 or more |
| Profit $y_i$ | −\$99,000 | −\$98,750 | \$1250 |
| Probability $p_i$ | 0.00191 | 0.00193 | 0.99058 |

(a) Explain why the company suffers a loss of \$98,750 on such a policy if a client dies at age 25.

(b) Calculate the expected value of Y. Explain what this result means for the insurance company.

(c) Calculate the standard deviation of Y. Explain what this result means for the insurance company.

18. **Fire insurance** Suppose a homeowner spends \$300 for a home insurance policy that will pay out \$200,000 if the home is destroyed by fire in a given year. Let $P$ = the profit made by the company on a single policy. From previous data, the probability that a home in this area will be destroyed by fire is 0.0002.

(a) Make a table that shows the probability distribution of $P$.

(b) Calculate the expected value of $P$. Explain what this result means for the insurance company.

(c) Calculate the standard deviation of $P$. Explain what this result means for the insurance company.

19. **Size of American households** In government data, a household consists of all occupants of a dwelling unit, while a family consists of two or more persons who live together and are related by blood or marriage. So all families form households, but some households are not families. Here are the distributions of household size and family size in the United States:

| | Number of persons | | | | | | |
|---|---|---|---|---|---|---|---|
| | 1 | 2 | 3 | 4 | 5 | 6 | 7 |
| Household probability | 0.25 | 0.32 | 0.17 | 0.15 | 0.07 | 0.03 | 0.01 |
| Family probability | 0 | 0.42 | 0.23 | 0.21 | 0.09 | 0.03 | 0.02 |

Let $H$ = the number of people in a randomly selected U.S. household and $F$ = the number of people in a randomly selected U.S. family.

(a) Here are histograms comparing the probability distributions of $H$ and $F$. Describe any differences that you observe.

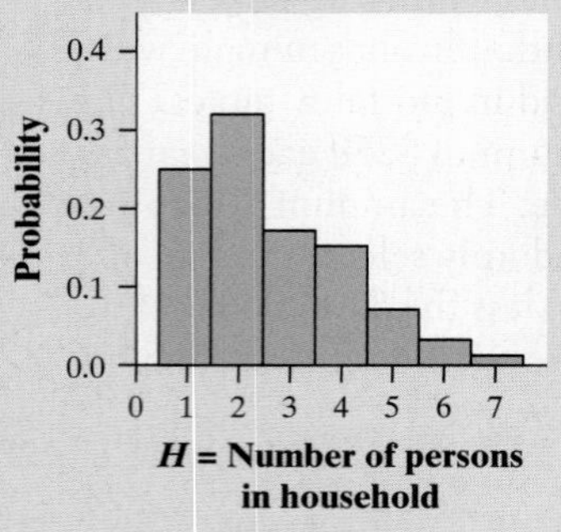


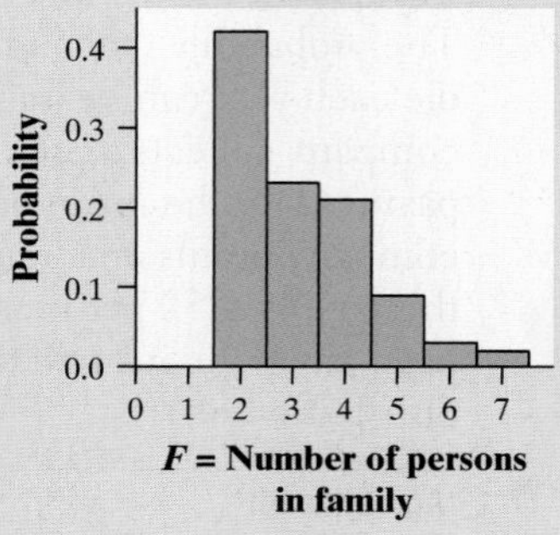


(b) Find the expected value of each random variable. Explain why this difference makes sense.

(c) The standard deviations of the two random variables are $\sigma_H = 1.421$ and $\sigma_F = 1.249$. Explain why this difference makes sense.

**20. Housing in San José** How do rented housing units differ from units occupied by their owners? Here are the distributions of the number of rooms for owner-occupied units and renter-occupied units in San José, California:[6]

| | Number of rooms | | | | | | | | | |
|---|---|---|---|---|---|---|---|---|---|---|
| | 1 | 2 | 3 | 4 | 5 | 6 | 7 | 8 | 9 | 10 |
| **Owned** | 0.003 | 0.002 | 0.023 | 0.104 | 0.210 | 0.224 | 0.197 | 0.149 | 0.053 | 0.035 |
| **Rented** | 0.008 | 0.027 | 0.287 | 0.363 | 0.164 | 0.093 | 0.039 | 0.013 | 0.003 | 0.003 |

Let $X$ = the number of rooms in a randomly selected owner-occupied unit and $Y$ = the number of rooms in a randomly chosen renter-occupied unit.

(a) Here are histograms comparing the probability distributions of $X$ and $Y$. Describe any differences you observe.

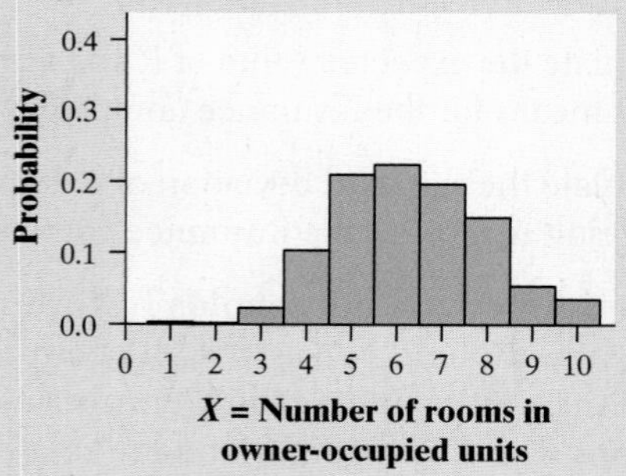


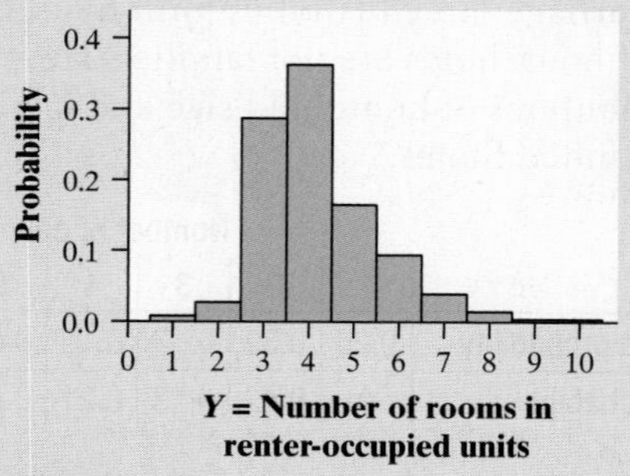


(b) Find the expected number of rooms for both types of housing unit. Explain why this difference makes sense.

(c) The standard deviations of the two random variables are $\sigma_X = 1.640$ and $\sigma_Y = 1.308$. Explain why this difference makes sense.

*Exercises 21 and 22 examine how Benford's law (Exercise 9) can be used to detect fraud.*

**21. Benford's law and fraud** A not-so-clever employee decided to fake his monthly expense report. He believed that the first digits of his expense amounts should be equally likely to be any of the numbers from 1 to 9. In that case, the first digit $Y$ of a randomly selected expense amount would have the probability distribution shown in the histogram.

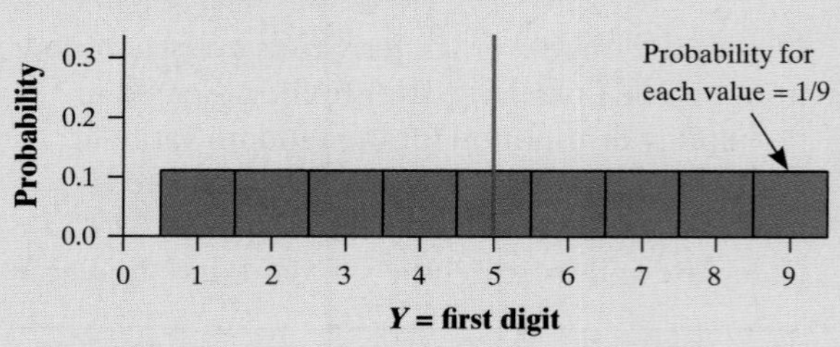


(a) What's $P(Y > 6)$? According to Benford's law (see Exercise 9), what proportion of first digits in the employee's expense amounts should be greater than 6? How could this information be used to detect a fake expense report?

(b) Explain why the mean of the random variable $Y$ is located at the solid red line in the figure.

(c) According to Benford's law, the expected value of the first digit is $\mu_X = 3.441$. Explain how this information could be used to detect a fake expense report.

**22. Benford's law and fraud**

(a) Using the graph from Exercise 21, calculate the standard deviation $\sigma_Y$. This gives us an idea of how much variation we'd expect in the employee's expense records if he assumed that first digits from 1 to 9 were equally likely.

(b) The standard deviation of the first digits of randomly selected expense amounts that follow Benford's law is $\sigma_X = 2.46$. Would using standard deviations be a good way to detect fraud? Explain your answer.

**23. Still waiting for the server?** pg 400 How does your web browser get a file from the Internet? Your computer sends a request for the file to a web server, and the web server sends back a response. Let $Y$ = the amount of time (in seconds) after the start of an hour at which a randomly selected request is received by a particular web server. The probability distribution of $Y$ can be modeled by a uniform density curve on the interval from 0 to 3600 seconds. Find the probability that the request is received by this server within the first 5 minutes (300 seconds) after the hour.

24. **Where's the bus?** Sally takes the same bus to work every morning. Let $X$ = the amount of time (in minutes) that she has to wait for the bus on a randomly selected day. The probability distribution of $X$ can be modeled by a uniform density curve on the interval from 0 minutes to 8 minutes. Find the probability that Sally has to wait between 2 and 5 minutes for the bus.

25. **Class is over!** Mr. Shrager does not always let his statistics class out on time. In fact, he seems to end class according to his own "internal clock." The density curve here models the distribution of $Y$, the amount of time after class ends (in minutes) when Mr. Shrager dismisses the class on a randomly selected day. (A negative value indicates he ended class early.)

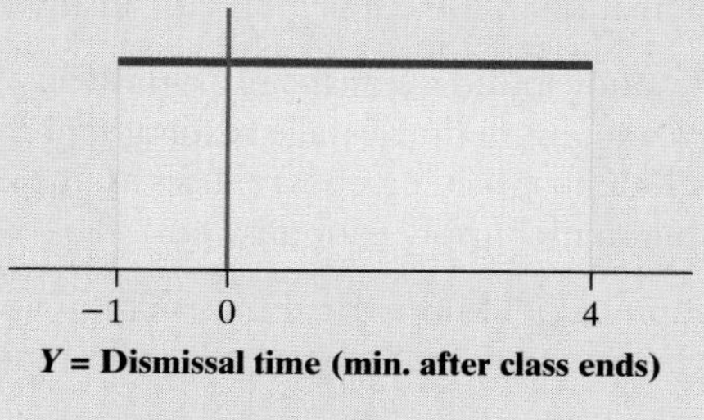


(a) Find and interpret $P(-1 \leq Y \leq 1)$.

(b) What is $\mu_Y$? Explain your answer.

(c) Find the value of $x$ that makes this statement true: $P(Y \geq x) = 0.25$.

26. **Quick, click!** An Internet reaction time test asks subjects to click their mouse button as soon as a light flashes on the screen. The light is programmed to go on at a randomly selected time after the subject clicks "Start." The density curve models the amount of time $Y$ (in seconds) that the subject has to wait for the light to flash.

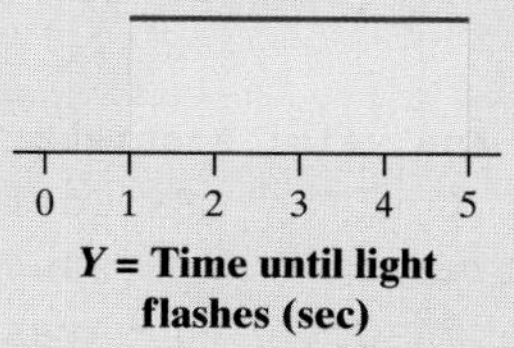


(a) Find and interpret $P(Y > 3.75)$.

(b) What is $\mu_Y$? Explain your answer.

(c) Find the value of $x$ that makes this statement true: $P(Y \geq x) = 0.62$.

27. **Running a mile** A study of 12,000 able-bodied male students at the University of Illinois found that their times for the mile run were approximately Normal with mean 7.11 minutes and standard deviation 0.74 minute.[7] Choose a student at random from this group and call his time for the mile $Y$. Find $P(Y < 6)$. Interpret this value.

pg 401

28. **Give me some sugar!** Machines that fill bags with powdered sugar are supposed to dispense 32 ounces of powdered sugar into each bag. Let $X$ = the weight (in ounces) of the powdered sugar dispensed into a randomly selected bag. Suppose that $X$ can be modeled by a Normal distribution with mean 32 ounces and standard deviation 0.6 ounce. Find $P(X \leq 31)$. Interpret this value.

29. **Horse pregnancies** Bigger animals tend to carry their young longer before birth. The length of horse pregnancies from conception to birth varies according to a roughly Normal distribution with mean 336 days and standard deviation 6 days. Let $X$ = the length of a randomly selected horse pregnancy.

(a) Write the event "pregnancy lasts between 325 and 345 days" in terms of $X$. Then find its probability.

(b) Find the 80th percentile of the distribution.

30. **Ace!** Professional tennis player Novak Djokovic hits the ball extremely hard. His first-serve speeds follow an approximately Normal distribution with mean 115 miles per hour (mph) and standard deviation 6 mph. Choose one of Djokovic's first serves at random. Let $Y$ = its speed, measured in miles per hour.

(a) Write the event "speed is between 100 and 120 miles per hour" in terms of $Y$. Then find its probability.

(b) Find the 15th percentile of the distribution.

**Multiple Choice** *Select the best answer for Exercises 31–34.*

*Exercises 31–33 refer to the following setting.* Choose an American household at random and let the random variable $X$ be the number of cars (including SUVs and light trucks) they own. Here is the probability distribution if we ignore the few households that own more than 5 cars:

| Number of cars | 0 | 1 | 2 | 3 | 4 | 5 |
|---|---|---|---|---|---|---|
| Probability | 0.09 | 0.36 | 0.35 | 0.13 | 0.05 | 0.02 |

31. What's the expected number of cars in a randomly selected American household?

(a) 1.00
(b) 1.75
(c) 1.84
(d) 2.00
(e) 2.50

32. The standard deviation of X is $\sigma_X = 1.08$. If many households were selected at random, which of the following would be the best interpretation of the value 1.08?

(a) The mean number of cars would be about 1.08.

(b) The number of cars would typically be about 1.08 from the mean.

(c) The number of cars would be at most 1.08 from the mean.

(d) The number of cars would be within 1.08 from the mean about 68% of the time.

(e) The mean number of cars would be about 1.08 from the expected value.

33. About what percentage of households have a number of cars within 2 standard deviations of the mean?

(a) 68%
(b) 71%
(c) 93%
(d) 95%
(e) 98%

34. A deck of cards contains 52 cards, of which 4 are aces. You are offered the following wager: Draw one card at random from the deck. You win \$10 if the card drawn is an ace. Otherwise, you lose \$1. If you make this wager very many times, what will be the mean amount you win?

(a) About −\$1, because you will lose most of the time.

(b) About \$9, because you win \$10 but lose only \$1.

(c) About −\$0.15; that is, on average, you lose about 15 cents.

(d) About \$0.77; that is, on average, you win about 77 cents.

(e) About \$0, because the random draw gives you a fair bet.

## Recycle and Review

*Exercises 35 and 36 refer to the following setting.* Many chess masters and chess advocates believe that chess play develops general intelligence, analytical skill, and the ability to concentrate. According to such beliefs, improved reading skills should result from study to improve chess-playing skills. To investigate this belief, researchers conducted a study. All the subjects in the study participated in a comprehensive chess program, and their reading performances were measured before and after the program. The graphs and numerical summaries that follow provide information on the subjects' pretest scores, posttest scores, and the difference (Post − Pre) between these two scores.

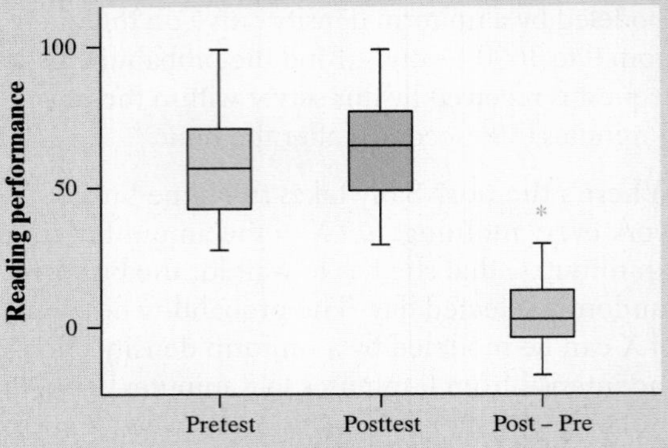


**Descriptive Statistics: Pretest, Posttest, Post − Pre**

| Variable | *N* | Mean | Median | StDev | Min | Max | $Q_1$ | $Q_3$ |
|---|---|---|---|---|---|---|---|---|
| Pretest | 53 | 57.70 | 58.00 | 17.84 | 23.00 | 99.00 | 44.50 | 70.50 |
| Posttest | 53 | 63.08 | 64.00 | 18.70 | 28.00 | 99.00 | 48.00 | 76.00 |
| Post − Pre | 53 | 5.38 | 3.00 | 13.02 | −19.00 | 42.00 | −3.50 | 14.00 |

35. **Better readers?** (1.3, 4.3)

(a) Did students tend to have higher reading scores after participating in the chess program? Justify your answer.

(b) If the study found a statistically significant improvement in the average reading score, could you conclude that playing chess causes an increase in reading skills? Justify your answer.

Some graphical and numerical information about the relationship between pretest and posttest scores is provided here.

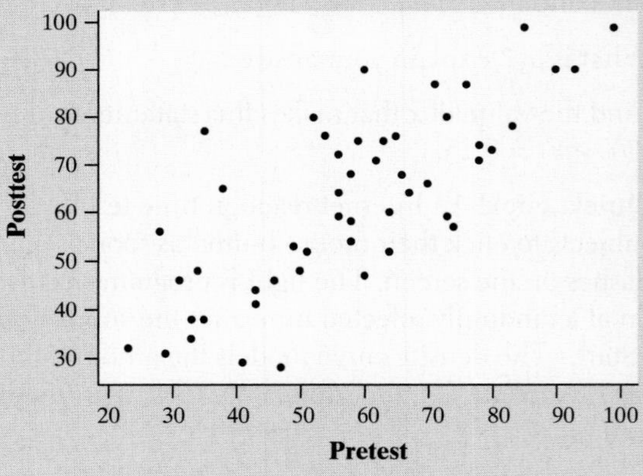


**Regression Analysis: Posttest Versus Pretest**

| Predictor | Coef | SE Coef | T | P |
|---|---|---|---|---|
| Constant | 17.897 | 5.889 | 3.04 | 0.004 |
| Pretest | 0.78301 | 0.09758 | 8.02 | 0.000 |
| S = 12.55 | R-Sq = 55.8% | R-Sq(adj) = 54.9% | | |

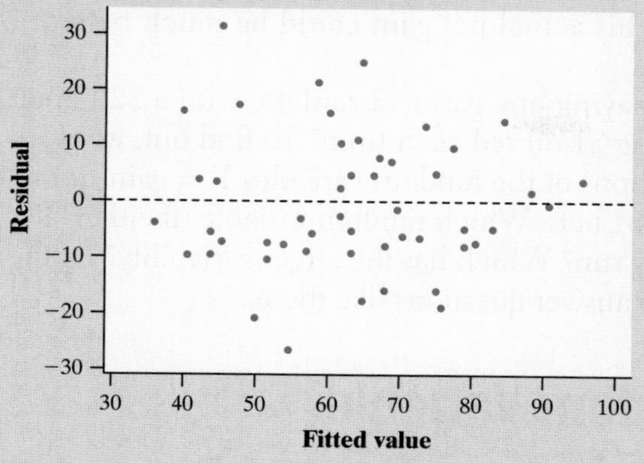


36. **Predicting posttest scores** (3.2)

(a) What is the equation of the least-squares regression line relating posttest and pretest scores? Define any variables used.

(b) Is a linear model appropriate for describing this relationship? Justify your answer.

(c) If we use the least-squares regression line to predict students' posttest scores from their pretest scores, how far off will our predictions typically be?

# SECTION 6.2 Transforming and Combining Random Variables

**LEARNING TARGETS** *By the end of the section, you should be able to:*

- Describe the effect of adding or subtracting a constant or multiplying or dividing by a constant on the probability distribution of a random variable.
- Calculate the mean and standard deviation of the sum or difference of random variables.
- Find probabilities involving the sum or difference of independent Normal random variables.

sjharmon/Getty Images

In Section 6.1, we looked at several examples of random variables and their probability distributions. We also saw that the parameters $\mu_X$ and $\sigma_X$ give us important information about a random variable.

Consider this new setting. An American roulette wheel has 38 slots numbered 1 through 36, plus 0 and 00. Half of the slots from 1 to 36 are red; the other half are black. Both the 0 and 00 slots are green. Suppose that a player places a \$1 bet on red. If the ball lands in a red slot, the player gets the original dollar back, plus an extra dollar for winning the bet. If the ball lands in a different-colored slot, the player loses the \$1 bet. Let X = the net gain on a single \$1 bet on red. Because there is an 18/38 chance that the ball lands in a red slot, the probability distribution of X is as shown in the table.

| **Value $x_i$** | −\$1 | \$1 |
|---|---|---|
| **Probability $p_i$** | 20/38 | 18/38 |

The mean of X is

$$\mu_X = (-1)\left(\frac{20}{38}\right) + (1)\left(\frac{18}{38}\right) = -\$0.05$$

That is, a player can expect to lose an average of 5 cents per \$1 bet if he plays many, many games. You can verify that the standard deviation is $\sigma_X = \$1.00$. If the player only plays a few games, his actual net gain could be much better or worse than this expected value.

Would the player be better off playing one game of roulette with a \$2 bet on red or playing two games and betting \$1 on red each time? To find out, we need to compare the probability distributions of the random variables $Y =$ gain from a \$2 bet and $T =$ total gain from two \$1 bets. Which random variable (if either) has the higher expected gain in the long run? Which has the larger variability? By the end of this section, you'll be able to answer questions like these.

## Transforming a Random Variable

In Chapter 2, we studied the effects of transformations on the shape, center, and variability of a distribution of quantitative data. Here's what we discovered:

1. *Adding (or subtracting) a constant:* Adding the same positive number $a$ to (subtracting $a$ from) each observation:
   - Adds $a$ to (subtracts $a$ from) measures of center and location (mean, median, quartiles, percentiles).
   - Does not change measures of variability (range, *IQR*, standard deviation).
   - Does not change the shape of the distribution.
2. *Multiplying or dividing by a constant:* Multiplying (or dividing) each observation by the same positive number $b$:
   - Multiplies (divides) measures of center and location (mean, median, quartiles, percentiles) by $b$.
   - Multiplies (divides) measures of variability (range, *IQR*, standard deviation) by $b$.
   - Does not change the shape of the distribution.

How are the probability distributions of random variables affected by similar transformations?

Volodymyr Burdiak/Shutterstock.com

**EFFECT OF ADDING OR SUBTRACTING A CONSTANT** Let's return to a familiar setting from Section 6.1. Pete's Jeep Tours offers a popular day trip in a tourist area. There must be at least 2 passengers for the trip to run, and the vehicle will hold up to 6 passengers. Pete charges \$150 per passenger. Let $C =$ the total amount of money that Pete collects on a randomly selected trip. The probability distribution of $C$ is shown in the table and the histogram.

| **Total collected $c_i$** | 300 | 450 | 600 | 750 | 900 |
|---|---|---|---|---|---|
| **Probability $p_i$** | 0.15 | 0.25 | 0.35 | 0.20 | 0.05 |

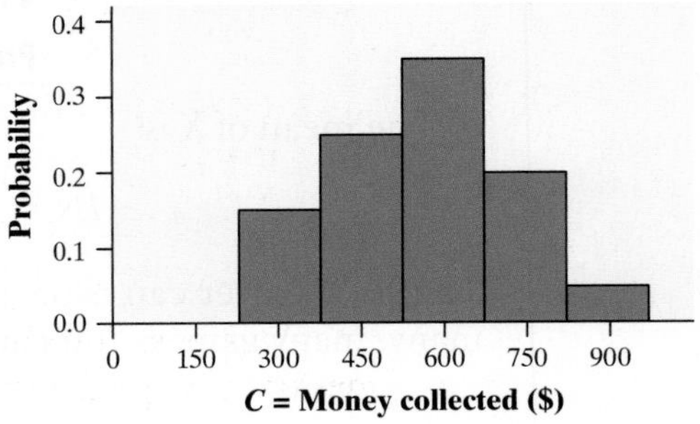

Earlier, we calculated the mean of C as $\mu_C = \$562.50$ and the standard deviation of C as $\sigma_C = \$163.46$. We can describe the probability distribution of C as follows:

**Shape:** Roughly symmetric with a single peak

**Center:** $\mu_C = \$562.50$

**Variability:** $\sigma_C = \$163.46$

It costs Pete \$100 to buy permits, gas, and a ferry pass for each day trip. The amount of profit V that Pete makes on a randomly selected trip is the total amount of money C that he collects from passengers minus \$100. That is, $V = C - 100$. The probability distribution of V is

| Profit $v_i$ | 200 | 350 | 500 | 650 | 800 |
|---|---|---|---|---|---|
| Probability $p_i$ | 0.15 | 0.25 | 0.35 | 0.20 | 0.05 |

A histogram of this probability distribution is shown here.

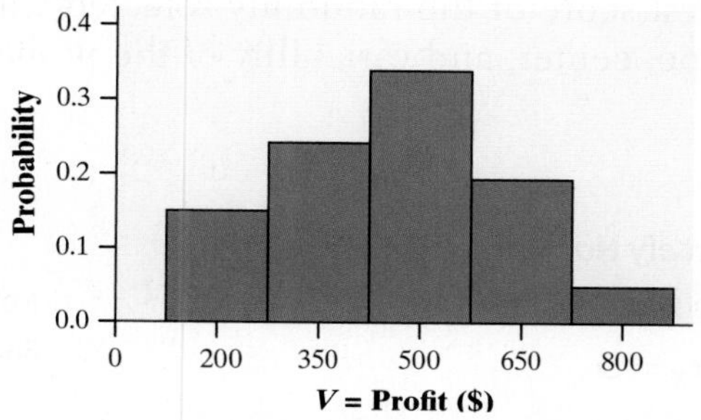


We can see that the probability distribution of V has the same shape as the probability distribution of C. The mean of V is

$$\mu_V = (200)(0.15) + (350)(0.25) + (500)(0.35) + (650)(0.20) + (800)(0.05)$$
$$= \$462.50$$

On average, Pete will make a profit of \$462.50 from the trip. That's \$100 less than $\mu_C$, his mean amount of money collected per trip. The standard deviation of V is

$$\sigma_V = \sqrt{(200 - 462.50)^2(0.15) + (350 - 462.50)^2(0.25) + \cdots + (800 - 462.50)^2(0.05)}$$
$$= \$163.46$$

That's the same as the standard deviation of C.

It's fairly clear that subtracting 100 from the values of the random variable C just shifts the probability distribution to the left by 100. This transformation decreases the mean by 100 (from \$562.50 to \$462.50), but it doesn't change the standard deviation (\$163.46) or the shape. These results can be generalized for any random variable.

## THE EFFECT OF ADDING OR SUBTRACTING A CONSTANT ON A PROBABILITY DISTRIBUTION

Adding the same positive number *a* to (subtracting *a* from) each value of a random variable:

- Adds *a* to (subtracts *a* from) measures of center and location (mean, median, quartiles, percentiles).
- Does not change measures of variability (range, *IQR*, standard deviation).
- Does not change the shape of the probability distribution.

Note that adding or subtracting a constant affects the distribution of a quantitative variable and the probability distribution of a random variable in exactly the same way.

## EXAMPLE

### Scaling test scores
### Effect of adding/subtracting a constant

**PROBLEM:** In a large introductory statistics class, the score X of a randomly selected student on a test worth 50 points can be modeled by a Normal distribution with mean 35 and standard deviation 5. Due to a difficult question on the test, the professor decides to add 5 points to each student's score. Let Y be the scaled test score of the randomly selected student. Describe the shape, center, and variability of the probability distribution of Y.

monkeybusinessimages/Getty Images

**SOLUTION:**

Shape: Approximately Normal

Center: $\mu_Y = \mu_X + 5 = 35 + 5 = 40$

Variability: $\sigma_Y = \sigma_X = 5$

Notice that $Y = X + 5$. Adding a constant doesn't affect the shape or the standard deviation of the probability distribution.

**FOR PRACTICE, TRY EXERCISE 37**

**EFFECT OF MULTIPLYING OR DIVIDING BY A CONSTANT** The professor in the preceding example decides to convert his students' scaled test scores Y to percentages. Because the test was scored out of 50 points, the professor multiplies each student's scaled score by 2 to convert to a percent score W. That is, $W = 2Y$. Figure 6.3 displays the probability distributions of the random variables Y and W. From the graphs, we can see that the measures of center, location, and variability have all doubled—just like the individual students' scores. But the shape of the two distributions is the same.

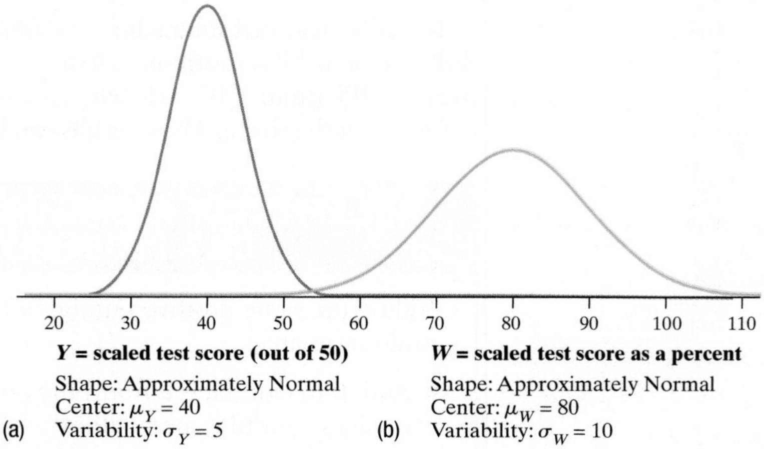


**FIGURE 6.3** Probability distribution of (a) $Y$ = a randomly selected statistics student's scaled test score out of 50 and (b) $W$ = the student's scaled test score as a percent.

It is not common to multiply (or divide) a random variable by a negative number $b$. Doing so would multiply (or divide) the measures of variability by $|b|$. Multiplying or dividing by a negative number would also affect the shape of the probability distribution, as all values would be reflected over the $y$ axis.

## THE EFFECT OF MULTIPLYING OR DIVIDING BY A CONSTANT ON A PROBABILITY DISTRIBUTION

Multiplying (or dividing) each value of a random variable by the same positive number $b$:

- Multiplies (divides) measures of center and location (mean, median, quartiles, percentiles) by $b$.
- Multiplies (divides) measures of variability (range, *IQR*, standard deviation) by $b$.
- Does not change the shape of the distribution.

Once again, multiplying or dividing by a constant has the same effect on the probability distribution of a random variable as it does on a distribution of quantitative data.

## EXAMPLE

### How much does college cost?
### Effect of multiplying/dividing by a constant

**PROBLEM:** El Dorado Community College considers a student to be full-time if he or she is taking between 12 and 18 units. The number of units X that a randomly selected El Dorado Community College full-time student is taking in the fall semester has the following distribution.

| **Number of units** | 12 | 13 | 14 | 15 | 16 | 17 | 18 |
|---|---|---|---|---|---|---|---|
| **Probability** | 0.25 | 0.10 | 0.05 | 0.30 | 0.10 | 0.05 | 0.15 |

At right is a histogram of the probability distribution. The mean is $\mu_X = 14.65$ and the standard deviation is $\sigma_X = 2.056$.

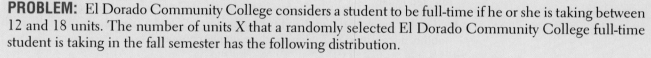

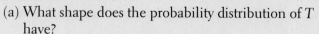

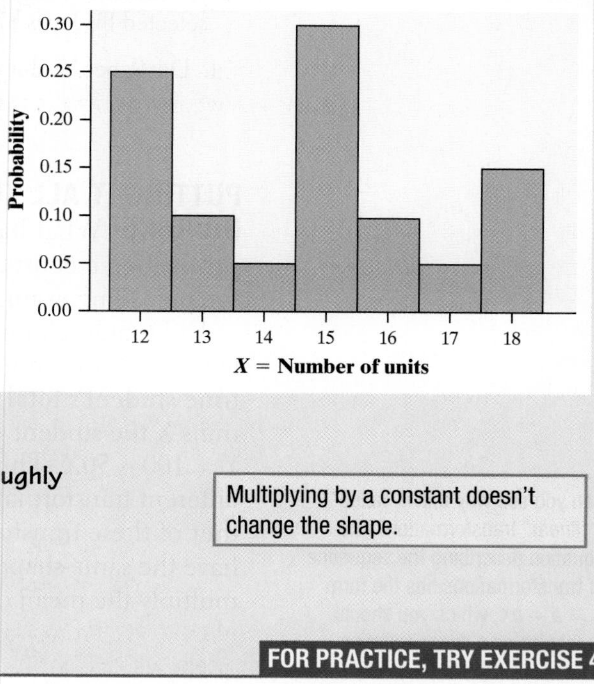


At El Dorado Community College, the tuition for full-time students is \$50 per unit. That is, if $T$ = tuition charge for a randomly selected full-time student, $T = 50X$.

(a) What shape does the probability distribution of $T$ have?

(b) Find the mean of $T$.

(c) Calculate the standard deviation of $T$.

**SOLUTION:**

(a) The same shape as the probability distribution of X: roughly symmetric with three peaks.

Multiplying by a constant doesn't change the shape.

(b) $\mu_T = 50\mu_X = 50(14.65) = \$732.50$

(c) $\sigma_T = 50\sigma_X = 50(2.056) = \$102.80$

FOR PRACTICE, TRY EXERCISE 41

### Think About It

**HOW DOES MULTIPLYING BY A CONSTANT AFFECT THE VARIANCE?** For El Dorado Community College, the variance of the number of units that a randomly selected full-time student takes is $\sigma_X^2 = 4.2275$. The variance of the tuition charge for such a student is $\sigma_T^2 = 10{,}568.75$. That's $(2500)(4.2275)$. So $\sigma_T^2 = 2500\sigma_X^2$. Where did 2500 come from? It's just $(50)^2$. In other words, $\sigma_T^2 = (50)^2\sigma_X^2$. Multiplying a random variable by a constant $b$ multiplies the variance by $b^2$.

## CHECK YOUR UNDERSTANDING

A large auto dealership keeps track of sales made during each hour of the day. Let X = the number of cars sold during the first hour of business on a randomly selected Friday. Based on previous records, the probability distribution of X is as follows:

| Cars sold | 0 | 1 | 2 | 3 |
|---|---|---|---|---|
| Probability | 0.3 | 0.4 | 0.2 | 0.1 |

The random variable X has mean $\mu_X = 1.1$ and standard deviation $\sigma_X = 0.943$. Suppose the dealership's manager receives a \$500 bonus from the company for each car sold. Let Y = the bonus received from car sales during the first hour on a randomly selected Friday.

1. Sketch a graph of the probability distribution of X and a separate graph of the probability distribution of Y. How do their shapes compare?
2. Find the mean of Y.
3. Calculate and interpret the standard deviation of Y.

The manager spends \$75 to provide coffee and doughnuts to prospective customers each morning. So the manager's net profit $T$ during the first hour on a randomly selected Friday is \$75 less than the bonus earned.

4. Describe the shape, center, and variability of the probability distribution of $T$.

**PUTTING IT ALL TOGETHER: ADDING/SUBTRACTING AND MULTIPLYING/DIVIDING** What happens if we transform a random variable by both adding or subtracting a constant and multiplying or dividing by a constant? Let's return to the preceding example.

El Dorado Community College charges each student a \$100 fee per semester in addition to tuition charges. We can calculate a randomly selected full-time student's total charges Y for the fall semester directly from the number of units X the student is taking, using the equation $Y = 50X + 100$ or, equivalently, $Y = 100 + 50X$. This *linear transformation* of the random variable X includes two different transformations: (1) multiplying by 50 and (2) adding 100. Because neither of these transformations affects shape, the probability distribution of Y will have the same shape as the probability distribution of X. To get the mean of Y, we multiply the mean of X by 50, then add 100:

> Can you see why this is called a "linear" transformation? The equation describing the sequence of transformations has the form $Y = a + bX$, which you should recognize as a linear equation.

$$\mu_Y = 100 + 50\mu_X = 100 + 50(14.65) = \$832.50$$

To get the standard deviation of Y, we multiply the standard deviation of X by 50 (adding 100 doesn't affect SD):

$$\sigma_Y = 50\sigma_X = 50(2.056) = \$102.80$$

This logic generalizes to any linear transformation.

**THE EFFECT OF A LINEAR TRANSFORMATION ON A RANDOM VARIABLE**

If $Y = a + bX$ is a linear transformation of the random variable X,

- The probability distribution of Y has the same shape as the probability distribution of X if $b > 0$.
- $\mu_Y = a + b\mu_X$.
- $\sigma_Y = |b|\sigma_X$ (because $b$ could be a negative number).

Note that these results apply to both discrete and continuous random variables.

## EXAMPLE

### The baby and the bathwater
### Analyzing the effect of transformations

**PROBLEM:** One brand of baby bathtub comes with a dial to set the water temperature. When the "babysafe" setting is selected and the tub is filled, the temperature X of the water in a randomly selected bath follows a Normal distribution with a mean of 34°C and a standard deviation of 2°C. Let Y be the water temperature in degrees Fahrenheit for the randomly selected bath. Recall that $F = 32 + \frac{9}{5}C$.

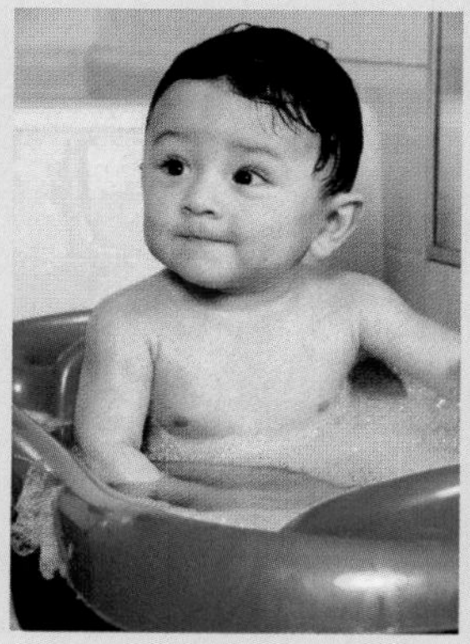
Michael DeLeon/Getty Images

(a) Find the mean of Y.

(b) Calculate and interpret the standard deviation of Y.

**SOLUTION:**

(a) $\mu_Y = 32 + \frac{9}{5}(34) = 93.2°F$

Note that $Y = 32 + \frac{9}{5}X$.

$\mu_Y = a + b\mu_X$

(b) $\sigma_Y = \frac{9}{5}(2) = 3.6°F$

$\sigma_Y = |b|\sigma_X$

If we randomly select many days when the dial is set on "babysafe," the temperature of the bath typically varies about 3.6°F from the mean of 93.2°F.

FOR PRACTICE, TRY EXERCISE 47

The probability distribution of Y = water temperature on a randomly selected day when the "babysafe" setting is used is Normal because the original distribution is Normal, and adding a constant and multiplying by a constant don't affect shape. We can use this Normal distribution to find probabilities as we did in Section 6.1. For instance, according to one source, the temperature of a baby's bathwater should be between 90°F and 100°F. What's $P(90 \le Y \le 100)$? Figure 6.4 shows the desired probability as an area under a Normal curve.

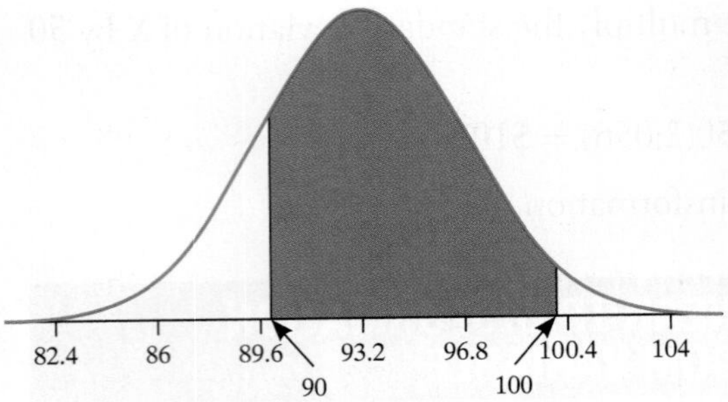


**FIGURE 6.4** The Normal probability distribution of the random variable $Y$ = the temperature (in degrees Fahrenheit) of the bathwater when the dial is set on "babysafe." The shaded area is the probability that the water temperature is between 90°F and 100°F.

To find the probability, we can either (i) standardize the boundary values and use Table A or technology; or (ii) use technology without standardizing.

(i) $z = \dfrac{90 - 93.2}{3.6} = -0.89$ $\qquad z = \dfrac{100 - 93.2}{3.6} = 1.89$

*Using Table* A: $0.9706 - 0.1867 = 0.7839$

*Using technology:* normalcdf(lower: – 0.89, upper:1.89, mean:0, SD:1) = 0.7839

(ii) normalcdf(lower:90, upper:100, mean:93.2, SD:3.6) = 0.7835

When set on "babysafe" mode, there's about a 78% probability that the water temperature meets the recommendation for a randomly selected bath.

# Combining Random Variables

So far, we have looked at settings that involved a single random variable. Many interesting statistics problems require us to combine two or more random variables.

Let's return to the familiar setting of Pete's Jeep Tours. Earlier, we focused on the amount of money C that Pete collects on a randomly selected day trip. This time we'll consider a different but related random variable: X = the number of passengers on a randomly selected trip. Here is its probability distribution:

| **Number of passengers $x_i$** | 2 | 3 | 4 | 5 | 6 |
|---|---|---|---|---|---|
| **Probability $p_i$** | 0.15 | 0.25 | 0.35 | 0.20 | 0.05 |

You can use what you learned earlier to confirm that $\mu_X = 3.75$ passengers and $\sigma_X = 1.0897$ passengers.

Pete's sister Erin runs jeep tours in another part of the country on the same days as Pete in her slightly smaller vehicle, under the name Erin's Adventures. The number of passengers Y on a randomly selected trip has the following probability distribution. You can confirm that $\mu_Y = 3.10$ passengers and $\sigma_Y = 0.943$ passengers.

| **Number of passengers $y_i$** | 2 | 3 | 4 | 5 |
|---|---|---|---|---|
| **Probability $p_i$** | 0.3 | 0.4 | 0.2 | 0.1 |

Here are two questions that we would like to answer based on this scenario:

- What is the distribution of the sum $S = X + Y$ of the number of passengers Pete and Erin will have on their tours on a randomly selected day?
- What is the distribution of the difference $D = X - Y$ in the number of passengers Pete and Erin will have on a randomly selected day?

As this setting suggests, we want to investigate what happens when we add or subtract random variables.

**MEAN (EXPECTED VALUE) OF THE SUM OR DIFFERENCE OF TWO RANDOM VARIABLES** How many total passengers S can Pete and Erin expect to have on their tours on a randomly selected day? Because Pete averages $\mu_X = 3.75$ passengers per trip and Erin averages $\mu_Y = 3.10$ passengers per trip, they will average a total of $\mu_S = 3.75 + 3.10 = 6.85$ passengers per day. We can generalize this result for any two random variables.

## MEAN (EXPECTED VALUE) OF A SUM OF RANDOM VARIABLES

For any two random variables X and Y, if $S = X + Y$, the mean (expected value) of $S$ is

$$\mu_S = \mu_{X+Y} = \mu_X + \mu_Y$$

In other words, the mean of the sum of two random variables is equal to the sum of their means.

What's the mean of the difference $D = X - Y$ in the number of passengers that Pete and Erin have on their tours on a randomly selected day? Because Pete averages $\mu_X = 3.75$ passengers per trip and Erin averages $\mu_Y = 3.10$ passengers per trip, the mean difference is $\mu_D = 3.75 - 3.10 = 0.65$ passengers. That is, Pete averages 0.65 more passengers per day than Erin does. Once again, we can generalize this result for any two random variables.

## MEAN (EXPECTED VALUE) OF A DIFFERENCE OF RANDOM VARIABLES

For any two random variables X and Y, if $D = X - Y$, the mean (expected value) of $D$ is

$$\mu_D = \mu_{X-Y} = \mu_X - \mu_Y$$

In other words, the mean of the difference of two random variables is equal to the difference of their means.

caution !

**The order of subtraction is important.** If we had defined $D = Y - X$, then $\mu_D = \mu_Y - \mu_X = 3.10 - 3.75 = -0.65$. In other words, Erin averages 0.65 fewer passengers than Pete does on a randomly chosen day.

## EXAMPLE

### How much do Pete and Erin make?

### Mean of a sum or difference of random variables

**PROBLEM:** Pete charges \$150 per passenger and Erin charges \$175 per passenger for a jeep tour. Let $C =$ the amount of money that Pete collects and $E =$ the amount of money that Erin collects on a randomly selected day. From our earlier work, we know that $\mu_C = 562.50$ and it is easy to show that $\mu_E = 542.50$. Define $S = C + E$. Calculate and interpret the mean of $S$.

**SOLUTION:**

$\mu_S = \mu_C + \mu_E = 562.50 + 542.50 = \$1105.00$

Pete and Erin expect to collect a total of \$1105 per day, on average, over many randomly selected days.

FOR PRACTICE, TRY EXERCISE 49

How did we calculate $\mu_C$ and $\mu_E$ in the example? Earlier, we defined $X =$ the number of passengers that Pete has and $Y =$ the number of passengers that Erin has on a randomly selected day trip. Recall that $\mu_X = 3.75$ and $\mu_Y = 3.10$. Because Pete charges \$150 per passenger, the amount of money that he collects on a randomly selected day is $C = 150X$. Multiplying a random variable by a constant multiplies the value of the mean by the same constant:

$$\mu_C = 150\mu_X = 150(3.75) = \$562.50$$

Because Erin charges \$175 per passenger, $E = 175Y$ and

$$\mu_E = 175\mu_Y = 175(3.10) = \$542.50$$

The expression $aX + bY$ is called a *linear combination* of the random variables $X$ and $Y$.

We can also write the total amount of money collected as $S = 150X + 175Y$. The discussion here shows that $\mu_S = 150\mu_X + 175\mu_Y$. More generally, if $S = aX + bY$, then $\mu_S = a\mu_X + b\mu_Y$.

What's the mean of the difference $D = C - E$ in the amounts that Pete and Erin collect on a randomly chosen day? It's

$$\mu_D = \mu_C - \mu_E = 562.50 - 542.50 = \$20.00$$

On average, Pete collects \$20 more per day than Erin does.

## Standard Deviation of the Sum or Difference of Two Random Variables

How much variation is there in the total number of passengers $S = X + Y$ who go on Pete's and Erin's tours on a randomly chosen day? Here are the probability distributions of X and Y once again. Let's think about the possible values of S. The number of passengers X on Pete's tour is between 2 and 6, and the number of passengers Y on Erin's tour is between 2 and 5. So the total number of passengers S is between 4 and 11. That is, there's more variability in the values of S than in the values of X or Y alone. This makes sense, because the variation in X and the variation in Y both contribute to the variation in S.

**Pete's Jeeps**

| **Number of passengers $x_i$** | 2 | 3 | 4 | 5 | 6 |
|---|---|---|---|---|---|
| **Probability $p_i$** | 0.15 | 0.25 | 0.35 | 0.20 | 0.05 |

$\mu_X = 3.75 \quad \sigma_X = 1.0897$

**Erin's Adventures**

| **Number of passengers $y_i$** | 2 | 3 | 4 | 5 |
|---|---|---|---|---|
| **Probability $p_i$** | 0.3 | 0.4 | 0.2 | 0.1 |

$\mu_Y = 3.10 \quad \sigma_Y = 0.943$

What's the standard deviation of $S = X + Y$? If we had the probability distribution of S, then we could calculate $\sigma_S$. Let's try to construct this probability distribution starting with the smallest possible value, $S = 4$. The only way to get a total of 4 passengers is if Pete has $X = 2$ passengers and Erin has $Y = 2$ passengers. We know that $P(X = 2) = 0.15$ and that $P(Y = 2) = 0.3$. If the events $X = 2$ and $Y = 2$ are *independent*, we can use the multiplication rule for independent events to find $P(X = 2 \text{ and } Y = 2)$. Otherwise, we're stuck. In fact, we can't calculate the probability for any value of S unless X and Y are **independent random variables**.

**DEFINITION Independent random variables**

If knowing the value of $X$ does not help us predict the value of $Y$, then $X$ and $Y$ are **independent random variables**. In other words, two random variables are independent if knowing the value of one variable does not change the probability distribution of the other variable.

It's reasonable to treat the random variables $X$ = number of passengers on Pete's trip and $Y$ = number of passengers on Erin's trip on a randomly chosen day as independent, because the siblings operate their trips in different parts of the country. Because X and Y are independent,

$$\begin{aligned} P(S=4) &= P(X=2 \text{ and } Y=2) \\ &= (0.15)(0.3) = 0.045 \end{aligned}$$

There are two ways to get a total of $S = 5$ passengers on a randomly selected day: $X = 3$, $Y = 2$ or $X = 2$, $Y = 3$. So

$$\begin{aligned} P(S=5) &= P(X=2 \text{ and } Y=3) + P(X=3 \text{ and } Y=2) \\ &= (0.15)(0.4) + (0.25)(0.3) \\ &= 0.06 + 0.075 = 0.135 \end{aligned}$$

We can construct the probability distribution by listing all combinations of X and Y that yield each possible value of S and adding the corresponding probabilities. Here is the result:

| **Sum $s_i$** | 4 | 5 | 6 | 7 | 8 | 9 | 10 | 11 |
|---|---|---|---|---|---|---|---|---|
| **Probability $p_i$** | 0.045 | 0.135 | 0.235 | 0.265 | 0.190 | 0.095 | 0.030 | 0.005 |

The mean of S is

$$\mu_S = \sum s_i p_i = (4)(0.045) + (5)(0.135) + \cdots + (11)(0.005) = 6.85$$

Our calculation confirms that

$$\mu_S = \mu_X + \mu_Y = 3.75 + 3.10 = 6.85$$

The variance of S is

$$\begin{aligned} \sigma_S^2 &= \sum (s_i - \mu_S)^2 p_i \\ &= (4-6.85)^2(0.045) + (5-6.85)^2(0.135) + \cdots + (11-6.85)^2(0.005) \\ &= 2.0775 \end{aligned}$$

The variances of X and Y are $\sigma_X^2 = (1.0897)^2 = 1.1875$ and $\sigma_Y^2 = (0.943)^2 = 0.89$. Notice that

$$\sigma_X^2 + \sigma_Y^2 = 1.1875 + 0.89 = 2.0775 = \sigma_S^2$$

In other words, the variance of a sum of two independent random variables is the sum of their variances. To find the standard deviation of S, take the square root of the variance:

$$\sigma_S = \sqrt{2.0775} = 1.441$$

Over many randomly selected days, the total number of passengers on Pete's and Erin's trips typically varies by about 1.441 passengers from the mean of 6.85 passengers.

The formula $\sigma_S^2 = \sigma_X^2 + \sigma_Y^2$ is sometimes referred to as the "Pythagorean theorem of statistics." It certainly looks similar to $c^2 = a^2 + b^2$! Just as the real Pythagorean theorem only applies to right triangles, the formula $\sigma_S^2 = \sigma_X^2 + \sigma_Y^2$ only applies if $X$ and $Y$ are independent random variables.

## STANDARD DEVIATION OF THE SUM OF TWO INDEPENDENT RANDOM VARIABLES

For any two *independent* random variables X and Y, if $S = X + Y$, the variance of S is

$$\sigma_S^2 = \sigma_{X+Y}^2 = \sigma_X^2 + \sigma_Y^2$$

To get the standard deviation of S, take the square root of the variance:

$$\sigma_S = \sigma_{X+Y} = \sqrt{\sigma_X^2 + \sigma_Y^2}$$

You might be wondering whether there's a formula for computing the variance or standard deviation of the sum of two random variables that are *not* independent. There is, but it's beyond the scope of this course.

When we add two independent random variables, their variances add. **Standard deviations do not add.** For Pete's and Erin's passenger totals,

$$\sigma_X + \sigma_Y = 1.0897 + 0.943 = 2.0327$$

This is very different from $\sigma_S = 1.441$.

Can you guess what the variance of the *difference* of two independent random variables will be? If you were thinking something like "the difference of their variances," think again! Here are the probability distributions of X and Y from the jeep tours scenario once again:

**Pete's Jeeps**

| Number of passengers $x_i$ | 2 | 3 | 4 | 5 | 6 |
|---|---|---|---|---|---|
| Probability $p_i$ | 0.15 | 0.25 | 0.35 | 0.20 | 0.05 |

$\mu_X = 3.75$ $\sigma_X = 1.0897$

**Erin's Adventures**

| Number of passengers $y_i$ | 2 | 3 | 4 | 5 |
|---|---|---|---|---|
| Probability $p_i$ | 0.3 | 0.4 | 0.2 | 0.1 |

$\mu_Y = 3.10$ $\sigma_Y = 0.943$

By following the process we used earlier with the random variable $S = X + Y$, you can build the probability distribution of $D = X - Y$.

| Value $d_i$ | −3 | −2 | −1 | 0 | 1 | 2 | 3 | 4 |
|---|---|---|---|---|---|---|---|---|
| Probability $p_i$ | 0.015 | 0.055 | 0.145 | 0.235 | 0.260 | 0.195 | 0.080 | 0.015 |

You can use the probability distribution to confirm that:

1. $\mu_D = 0.65 = 3.75 - 3.10 = \mu_X - \mu_Y$
2. $\sigma_D^2 = 2.0775 = 1.1875 + 0.89 = \sigma_X^2 + \sigma_Y^2$
3. $\sigma_D = \sqrt{2.0775} = 1.441$

Result 2 shows that, just as with addition, when we subtract two independent random variables, variances add. There's more variability in the values of the difference D than in the values of X or Y alone. This should make sense, because the variation in X and the variation in Y both contribute to the variation in D.

## STANDARD DEVIATION OF THE DIFFERENCE OF TWO INDEPENDENT RANDOM VARIABLES

For any two *independent* random variables X and Y, if $D = X - Y$, the variance of D is

$$\sigma_D^2 = \sigma_{X-Y}^2 = \sigma_X^2 + \sigma_Y^2$$

To get the standard deviation of D, take the square root of the variance:

$$\sigma_D = \sigma_{X-Y} = \sqrt{\sigma_X^2 + \sigma_Y^2}$$

Let's put this new rule to use in a familiar setting.

## EXAMPLE

### How much do Pete's and Erin's earnings vary?
### SD of a sum or difference of random variables

**PROBLEM:** Pete charges \$150 per passenger and Erin charges \$175 per passenger for a jeep tour. Let $C$ = the amount of money that Pete collects and $E$ = the amount of money that Erin collects on a randomly selected day. From our earlier work, it is easy to show that $\sigma_C = \$163.46$ and $\sigma_E = \$165.03$. You may assume that these two random variables are independent. Define $D = C - E$. Earlier, we found that $\mu_D = \$20$. Calculate and interpret the standard deviation of $D$.

**SOLUTION:**

$D = C - E$. Because $C$ and $E$ are independent random variables,

$\sigma_D^2 = \sigma_C^2 + \sigma_E^2 = (163.46)^2 + (165.03)^2 = 53{,}954.07$

$\sigma_D = \sqrt{53{,}954.07} = \$232.28$

Note that variances add when you are dealing with the sum *or* difference of independent random variables.

Over many randomly selected days, the difference (Pete − Erin) in the amount collected on their jeep tours typically varies by about \$232.28 from the mean difference of \$20.

FOR PRACTICE, TRY EXERCISE 57

How did we calculate $\sigma_C$ and $\sigma_E$ in the example? Earlier, we defined $X$ = the number of passengers that Pete has and $Y$ = the number of passengers that Erin has on a randomly selected day trip. Recall that $\sigma_X = 1.0897$ and $\sigma_Y = 0.943$. Because Pete charges \$150 per passenger, the amount of money that he collects on a randomly selected day is $C = 150X$. Multiplying a random variable by a constant multiplies the value of the standard deviation by the same constant:

$$\sigma_C = 150\sigma_X = 150(1.0897) = \$163.46$$

Because Erin charges \$175 per passenger, $E = 175Y$ and

$$\sigma_E = 175\sigma_Y = 175(0.943) = \$165.03$$

We can write the difference in the amount of money collected as $D = 150X - 175Y$. The discussion here shows that

$$\sigma_D^2 = (150\sigma_X)^2 + (175\sigma_Y)^2 = 150^2\sigma_X^2 + 175^2\sigma_Y^2$$

More generally, if $S = aX + bY$, then $\sigma_S^2 = a^2\sigma_X^2 + b^2\sigma_Y^2$ for independent random variables $X$ and $Y$.

Recall that $aX + bY$ is a *linear combination* of the random variables $X$ and $Y$.

### MEAN AND STANDARD DEVIATION OF A LINEAR COMBINATION OF RANDOM VARIABLES

If $aX + bY$ is a linear combination of the random variables $X$ and $Y$,

- Its mean is $a\mu_X + b\mu_Y$.
- Its standard deviation is $\sqrt{a^2\sigma_X^2 + b^2\sigma_Y^2}$ if $X$ and $Y$ are independent.

Note that these results apply to both discrete and continuous random variables.

**COMBINING VERSUS TRANSFORMING RANDOM VARIABLES** We can extend our rules for combining random variables to situations involving repeated observations of the same random process. Let's return to the gambler we met at the beginning of this section. Suppose he plays two games of roulette, each time placing a \$1 bet on red. What can we say about his total gain (or loss) from playing two games? Earlier, we showed that if X = the amount gained on a single \$1 bet on red, then $\mu_X = -\$0.05$ and $\sigma_X = \$1.00$. Because we're interested in the player's total gain over two games, we'll define $X_1$ as the amount he gains from the first game and $X_2$ as the amount he gains from the second game. Then his total gain $T = X_1 + X_2$. Both $X_1$ and $X_2$ have the same probability distribution as X and, therefore, the same mean (−\$0.05) and standard deviation (\$1.00). The player's expected gain in two games is

$$\mu_T = \mu_{X_1} + \mu_{X_2} = (-\$0.05) + (-\$0.05) = -\$0.10$$

Because knowing the result of one game tells the player nothing about the result of the other game, $X_1$ and $X_2$ are independent random variables. As a result,

$$\sigma_T^2 = \sigma_{X_1}^2 + \sigma_{X_2}^2 = (1.00)^2 + (1.00)^2 = 2.00$$

and the standard deviation of the player's total gain is

$$\sigma_T = \sqrt{2.00} = \$1.41$$

At the beginning of the section, we asked whether a roulette player would be better off placing two separate \$1 bets on red or a single \$2 bet on red. We just showed that the expected total gain from two \$1 bets is $\mu_T = -\$0.10$ with a standard deviation of $\sigma_T = \$1.41$. Now think about what happens if the gambler places a \$2 bet on red in a single game of roulette. Because the random variable X represents a player's gain from a \$1 bet, the random variable Y = 2X represents his gain from a \$2 bet.

What's the player's expected gain from a single \$2 bet on red? It's

$$\mu_Y = 2\mu_X = 2(-\$0.05) = -\$0.10$$

That's the same as his expected gain from playing two games of roulette with a \$1 bet each time. But the standard deviation of the player's gain from a single \$2 bet is

$$\sigma_Y = 2\sigma_X = 2(\$1.00) = \$2.00$$

caution

Compare this result to $\sigma_T = \$1.41$. There's more variability in the gain from a single \$2 bet than in the total gain from two \$1 bets. **Bottom line: $X_1 + X_2$ is not the same as 2X.**

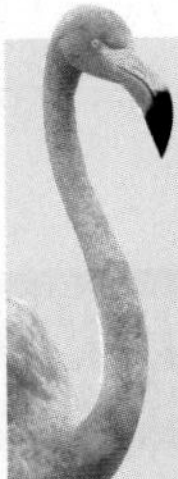

## CHECK YOUR UNDERSTANDING

A large auto dealership keeps track of sales and lease agreements made during each hour of the day. Let X = the number of cars sold and Y = the number of cars leased during the first hour of business on a randomly selected Friday. Based on previous records, the probability distributions of X and Y are as follows:

| **Cars sold $x_i$** | 0 | 1 | 2 | 3 |
|---|---|---|---|---|
| **Probability $p_i$** | 0.3 | 0.4 | 0.2 | 0.1 |

$\mu_X = 1.1 \quad \sigma_X = 0.943$

| **Cars leased $y_i$** | 0 | 1 | 2 |
|---|---|---|---|
| **Probability $p_i$** | 0.4 | 0.5 | 0.1 |

$\mu_Y = 0.7 \quad \sigma_Y = 0.64$

Define $T = X + Y$. Assume that X and Y are independent.

1. Find and interpret $\mu_T$.
2. Calculate and interpret $\sigma_T$.
3. The dealership's manager receives a \$500 bonus for each car sold and a \$300 bonus for each car leased. Find the mean and standard deviation of the manager's total bonus $B$.

## Combining Normal Random Variables

So far, we have concentrated on developing rules for means and variances of random variables. If a random variable is Normally distributed, we can use its mean and standard deviation to compute probabilities. What happens if we combine two *independent* Normal random variables?

We used software to simulate separate random samples of size 1000 for each of two independent, Normally distributed random variables, X and Y. Their means and standard deviations are as follows:

$$\mu_X = 3, \sigma_X = 0.9 \qquad \mu_Y = 1, \sigma_Y = 1.2$$

Figure 6.5(a) shows the results. What do we know about the sum and difference of these two random variables? The histograms in Figure 6.5(b) came from adding and subtracting the corresponding values of X and Y for the 1000 randomly generated observations from each probability distribution.

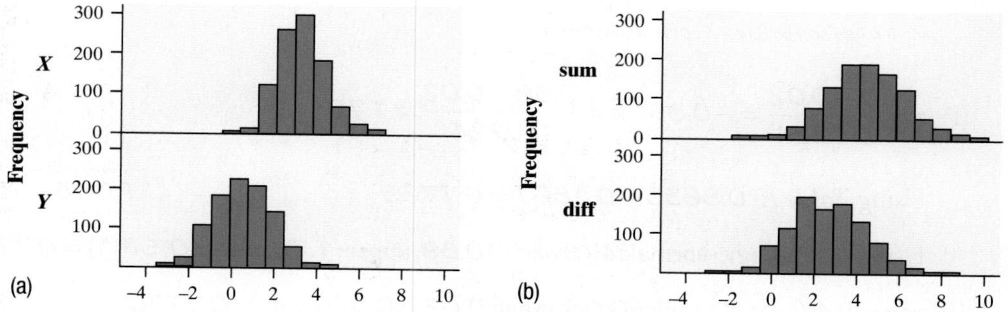


**FIGURE 6.5** (a) Histograms showing the results of randomly selecting 1000 values of two independent, Normal random variables *X* and *Y*. (b) Histograms of the sum and difference of the 1000 randomly selected values of *X* and *Y*.

In fact, any *linear combination* of independent Normal random variables is Normally distributed.

As the simulation illustrates, *any sum or difference of independent Normal random variables is also Normally distributed.* The mean and standard deviation of the resulting Normal distribution can be found using the appropriate rules for means and standard deviations:

| | **Sum X + Y** | **Difference X − Y** |
|---|---|---|
| Mean | $\mu_{X+Y} = \mu_X + \mu_Y = 3 + 1 = 4$ | $\mu_{X-Y} = \mu_X - \mu_Y = 3 - 1 = 2$ |
| SD | $\sigma^2_{X+Y} = \sigma^2_X + \sigma^2_Y = 0.9^2 + 1.2^2 = 2.25$<br>$\sigma_{X+Y} = \sqrt{2.25} = 1.5$ | $\sigma^2_{X-Y} = \sigma^2_X + \sigma^2_Y = 0.9^2 + 1.2^2 = 2.25$<br>$\sigma_{X-Y} = \sqrt{2.25} = 1.5$ |

### EXAMPLE Will the lid fit?
#### Combining Normal random variables

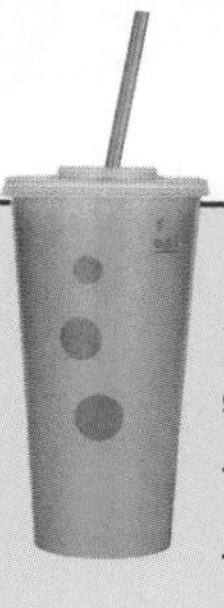

karandaev/Getty Images

**PROBLEM:** The diameter *C* of the top of a randomly selected large drink cup at a fast-food restaurant follows a Normal distribution with a mean of 3.96 inches and a standard deviation of 0.01 inch. The diameter *L* of a randomly selected large lid at this restaurant follows a Normal distribution with mean 3.98 inches and standard deviation 0.02 inch. Assume that *L* and *C* are independent random variables. Let the random variable $D = L - C$ be the difference between the lid's diameter and the cup's diameter.

(a) Describe the distribution of *D*.

(b) For a lid to fit on a cup, the value of *L* has to be bigger than the value of *C*, but not by more than 0.06 inch. Find the probability that a randomly selected lid will fit on a randomly selected cup. Interpret this value.

**SOLUTION:**

(a) Shape: Normal

Center: $\mu_D = 3.98 - 3.96 = 0.02$ inch

Variability: $\sigma_D = \sqrt{(0.02)^2 + (0.01)^2} = 0.0224$ inch

(b) The lid will fit if $0 < L - C \leq 0.06$, that is, if $0 < D \leq 0.06$.

*D* is the difference of two independent Normal random variables.

$\mu_D = \mu_L - \mu_C$

$\sigma_D = \sqrt{\sigma_L^2 + \sigma_C^2}$

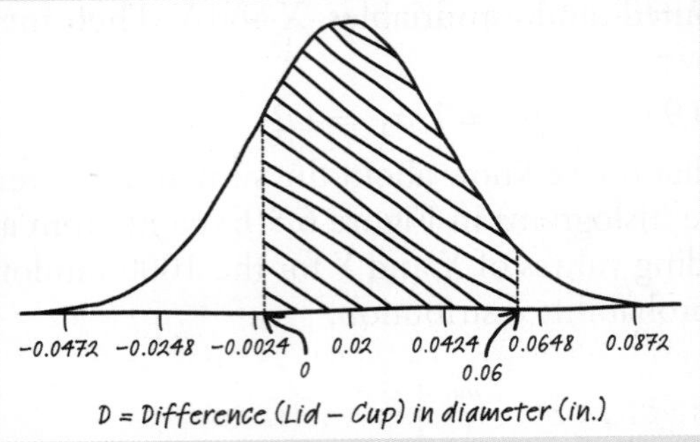


**1. Draw a Normal distribution.**
**2. Perform calculations—show your work!**
(i) Standardize and use Table A or technology; or
(ii) Use technology without standardizing.
Be sure to answer the question that was asked.

(i) $z = \dfrac{0 - 0.02}{0.0224} = -0.89 \quad z = \dfrac{0.06 - 0.02}{0.0224} = 1.79$

*Using Table A:* $0.9633 - 0.1867 = 0.7766$

$P(0 < D \leq 0.06) = P(-0.89 < Z \leq 1.79)$

*Using technology:* normalcdf(lower: −0.89, upper: 1.79, mean: 0, SD: 1) = 0.7765

(ii) normalcdf(lower:0, upper:0.06, mean:0.02, SD:0.0224) = 0.7770

There's about a 78% chance that a randomly selected lid will fit on a randomly selected cup.

**FOR PRACTICE, TRY EXERCISE 65**

domin_domin/Getty Images

We can extend what we have learned about combining independent Normal random variables to settings that involve repeated observations from the same probability distribution. Consider this scenario. Mr. Starnes likes sugar in his hot tea. From experience, he needs between 8.5 and 9 grams of sugar in a cup of tea for the drink to taste right. While making his tea one morning, Mr. Starnes adds four randomly selected packets of sugar. Suppose the amount of sugar in these packets follows a Normal distribution with mean 2.17 grams and standard deviation 0.08 gram. What's the probability that Mr. Starnes's tea tastes right?

Let $X$ = the amount of sugar in a randomly selected packet. Then $X_1$ = amount of sugar in Packet 1, $X_2$ = amount of sugar in Packet 2, $X_3$ = amount of sugar in Packet 3, and $X_4$ = amount of sugar in Packet 4. Each of these random variables has a Normal distribution with mean 2.17 grams and standard deviation 0.08 grams. We're interested in the total amount of sugar that Mr. Starnes puts in his tea: $T = X_1 + X_2 + X_3 + X_4$.

The random variable $T$ is a sum of four independent Normal random variables. So $T$ follows a Normal distribution with mean

$$\mu_T = \mu_{X_1} + \mu_{X_2} + \mu_{X_3} + \mu_{X_4} = 2.17 + 2.17 + 2.17 + 2.17 = 8.68 \text{ grams}$$

and variance

$$\sigma_T^2 = \sigma_{X_1}^2 + \sigma_{X_2}^2 + \sigma_{X_3}^2 + \sigma_{X_4}^2 = (0.08)^2 + (0.08)^2 + (0.08)^2 + (0.08)^2 = 0.0256$$

The standard deviation of $T$ is

$$\sigma_T = \sqrt{0.0256} = 0.16 \text{ gram}$$

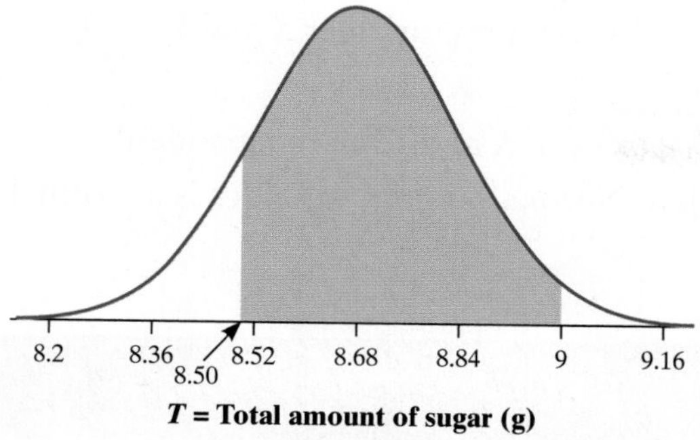


**FIGURE 6.6** Normal distribution of the total amount of sugar in Mr. Starnes's tea.

We want to find the probability that the total amount of sugar in Mr. Starnes's tea is between 8.5 and 9 grams. Figure 6.6 shows this probability as the area under a Normal curve.

To find this area, we can use either of our two familiar methods:

(i) Standardize the boundary values and use Table A or technology:

$$z = \frac{8.5 - 8.68}{0.16} = -1.13 \quad \text{and} \quad z = \frac{9 - 8.68}{0.16} = 2.00$$

*Using Table A:* $P(-1.13 \leq Z \leq 2.00) = 0.9772 - 0.1292 = 0.8480$

*Using technology:* normalcdf(lower: – 1.13, upper:2.00, mean:0, SD:1) = 0.8480

(ii) Use technology to find the desired area without standardizing.

normalcdf(lower:8.5, upper:9, mean:8.68, SD:0.16) = 0.8470

There's about an 85% probability that Mr. Starnes's tea will taste right.

## Section 6.2 Summary

- Adding a positive constant $a$ to (subtracting $a$ from) a random variable increases (decreases) measures of center and location by $a$, but does not affect measures of variability (range, *IQR*, standard deviation) or the shape of its probability distribution.
- Multiplying (dividing) a random variable by a positive constant $b$ multiplies (divides) measures of center and location by $b$ and multiplies (divides) measures of variability (range, *IQR*, standard deviation) by $b$, but does not change the shape of its probability distribution.
- If $Y = a + bX$ is a linear transformation of the random variable $X$ with $b > 0$,
  - The probability distribution of $Y$ has the same shape as the probability distribution of $X$.
  - $\mu_Y = a + b\mu_X$
  - $\sigma_Y = b\sigma_X$
- If $X$ and $Y$ are *any* two random variables,

  $\mu_{X+Y} = \mu_X + \mu_Y$: The mean of the sum of two random variables is the sum of their means.

  $\mu_{X-Y} = \mu_X - \mu_Y$: The mean of the difference of two random variables is the difference of their means.
- If $X$ and $Y$ are **independent random variables**, then knowing the value of one variable does not change the probability distribution of the other variable. In that case, variances add:

  $\sigma^2_{X+Y} = \sigma^2_X + \sigma^2_Y$: The variance of the sum of two independent random variables is the sum of their variances.

  $\sigma^2_{X-Y} = \sigma^2_X + \sigma^2_Y$: The variance of the difference of two independent random variables is the sum of their variances.
- To get the standard deviation of the sum or difference of two independent random variables, calculate the variance and then take the square root:

$$\sigma_{X+Y} = \sigma_{X-Y} = \sqrt{\sigma^2_X + \sigma^2_Y}$$

- If $aX + bY$ is a linear combination of the random variables X and Y,
  - Its mean is $a\mu_X + b\mu_Y$.
  - Its standard deviation is $\sqrt{a^2\sigma_X^2 + b^2\sigma_Y^2}$ if X and Y are independent.
- A linear combination of independent Normal random variables is a Normal random variable.

## Section 6.2 Exercises

37. **Driving to work** The time X it takes Hattan to drive to work on a randomly selected day follows a distribution that is approximately Normal with mean 15 minutes and standard deviation 6.5 minutes. Once he parks his car in his reserved space, it takes 5 more minutes for him to walk to his office. Let $T$ = the total time it takes Hattan to reach his office on a randomly selected day, so $T = X + 5$. Describe the shape, center, and variability of the probability distribution of $T$.
pg 412

38. **Toy shop sales** Total gross profits G on a randomly selected day at Tim's Toys follow a distribution that is approximately Normal with mean \$560 and standard deviation \$185. The cost of renting and maintaining the shop is \$65 per day. Let $P$ = profit on a randomly selected day, so $P = G - 65$. Describe the shape, center, and variability of the probability distribution of $P$.

39. **Airline overbooking** Airlines typically accept more reservations for a flight than the number of seats on the plane. Suppose that for a certain route, an airline accepts 40 reservations on a plane that carries 38 passengers. Based on experience, the probability distribution of Y = the number of passengers who actually show up for a randomly selected flight is given in the following table. You can check that $\mu_Y = 37.4$ and $\sigma_Y = 1.24$.

| Number of passengers $y_i$ | 35 | 36 | 37 | 38 | 39 | 40 |
|---|---|---|---|---|---|---|
| Probability $p_i$ | 0.10 | 0.10 | 0.30 | 0.35 | 0.10 | 0.05 |

There is also a crew of two flight attendants and two pilots on each flight. Let X = the total number of people (passengers plus crew) on a randomly selected flight.

(a) Make a graph of the probability distribution of X. Describe its shape.

(b) Find and interpret $\mu_X$.

(c) Calculate and interpret $\sigma_X$.

40. **City parking** Victoria parks her car at the same garage every time she goes to work. Because she stays at work for different lengths of time each day, the fee the parking garage charges on a randomly selected day is a random variable, G. The table gives the probability distribution of G. You can check that $\mu_G$ = \$14 and $\sigma_G$ = \$2.74.

| Garage fee $g_i$ | \$10 | \$13 | \$15 | \$20 |
|---|---|---|---|---|
| Probability $p_i$ | 0.20 | 0.25 | 0.45 | 0.10 |

In addition to the garage's fee, the city charges a \$3 use tax each time Victoria parks her car. Let $T$ = the total amount of money she pays on a randomly selected day.

(a) Make a graph of the probability distribution of $T$. Describe its shape.

(b) Find and interpret $\mu_T$.

(c) Calculate and interpret $\sigma_T$.

41. **Get on the boat!** A small ferry runs every half hour from one side of a large river to the other. The number of cars X on a randomly chosen ferry trip has the probability distribution shown here with mean $\mu_X = 3.87$ and standard deviation $\sigma_X = 1.29$.
pg 413

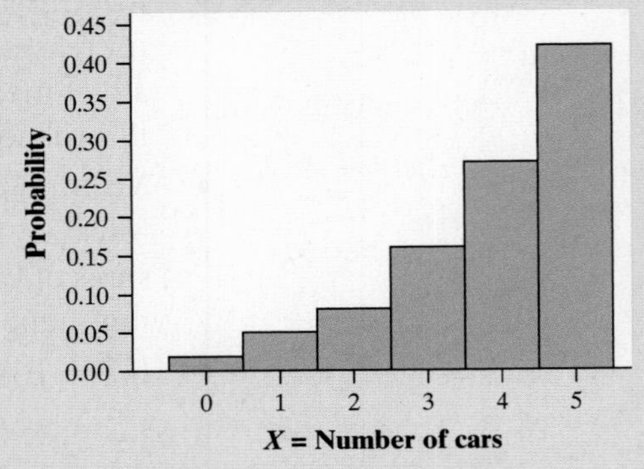


The cost for the ferry trip is \$5. Define $M$ = money collected on a randomly selected ferry trip.

(a) What shape does the probability distribution of $M$ have?

(b) Find the mean of $M$.

(c) Calculate the standard deviation of $M$.

42. **Skee Ball** Ana is a dedicated Skee Ball player who always rolls for the 50-point slot. Ana's score X on a randomly selected roll of the ball has the probability distribution shown here with mean $\mu_X = 23.8$ and standard deviation $\sigma_X = 12.63$.

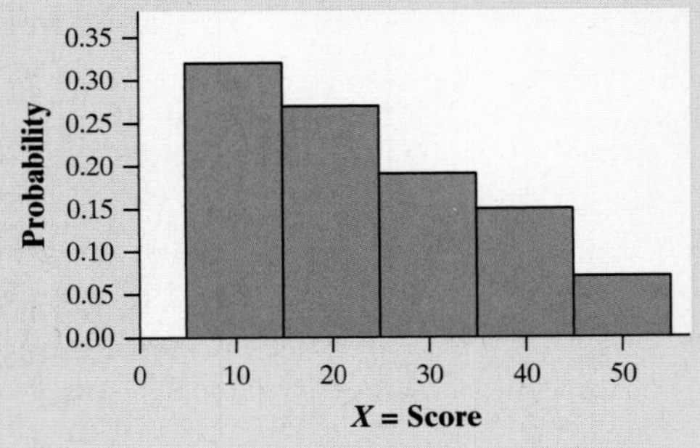

A player receives one ticket from the game for every 10 points scored. Define $T$ = number of tickets Ana gets on a randomly selected roll.

(a) What shape does the probability distribution of $T$ have?

(b) Find the mean of $T$.

(c) Calculate the standard deviation of $T$.

43. **Still waiting for the server?** How does your web browser get a file from the Internet? Your computer sends a request for the file to a web server, and the web server sends back a response. Let $Y$ = the amount of time (in seconds) after the start of an hour at which a randomly selected request is received by a particular web server. The probability distribution of $Y$ can be modeled by a uniform density curve on the interval from 0 to 3600 seconds. Define the random variable $W = Y/60$.

(a) Explain what $W$ represents.

(b) What probability distribution does $W$ have?

44. **Where's the bus?** Sally takes the same bus to work every morning. Let $X$ = the amount of time (in minutes) that she has to wait for the bus on a randomly selected day. The probability distribution of $X$ can be modeled by a uniform density curve on the interval from 0 minutes to 8 minutes. Define the random variable $V = 60X$.

(a) Explain what $V$ represents.

(b) What probability distribution does $V$ have?

*Exercises 45 and 46 refer to the following setting.* Ms. Hall gave her class a 10-question multiple-choice quiz. Let $X$ = the number of questions that a randomly selected student in the class answered correctly. The computer output gives information about the probability distribution of $X$. To determine each student's grade on the quiz (out of 100), Ms. Hall will multiply his or her number of correct answers by 5 and then add 50. Let $G$ = the grade of a randomly chosen student in the class.

| Mean | Median | StDev | Min | Max | $Q_1$ | $Q_3$ |
|---|---|---|---|---|---|---|
| 7.6 | 8.5 | 1.32 | 4 | 10 | 8 | 9 |

45. **Easy quiz**

(a) Find the median of $G$.

(b) Find the interquartile range ($IQR$) of $G$.

46. **More easy quiz**

(a) Find the mean of $G$.

(b) Find the range of $G$.

47. **Too cool at the cabin?** During the winter months, the temperatures at the Starneses' Colorado cabin can stay well below freezing (32°F or 0°C) for weeks at a time. To prevent the pipes from freezing, Mrs. Starnes sets the thermostat at 50°F. She also buys a digital thermometer that records the indoor temperature each night at midnight. Unfortunately, the thermometer is programmed to measure the temperature in degrees Celsius. Based on several years' worth of data, the temperature $T$ in the cabin at midnight on a randomly selected night can be modeled by a Normal distribution with mean 8.5°C and standard deviation 2.25°C. Let $Y$ = the temperature in the cabin at midnight on a randomly selected night in degrees Fahrenheit (recall that $F = (9/5)C + 32$).

pg 415

(a) Find the mean of $Y$.

(b) Calculate and interpret the standard deviation of $Y$.

(c) Find the probability that the midnight temperature in the cabin is less than 40°F.

48. **How much cereal?** A company's single-serving cereal boxes advertise 1.63 ounces of cereal. In fact, the amount of cereal $X$ in a randomly selected box can be modeled by a Normal distribution with a mean of 1.70 ounces and a standard deviation of 0.03 ounce. Let $Y$ = the *excess* amount of cereal beyond what's advertised in a randomly selected box, measured in grams (1 ounce = 28.35 grams).

(a) Find the mean of $Y$.

(b) Calculate and interpret the standard deviation of $Y$.

(c) Find the probability of getting at least 1 gram more cereal than advertised.

49. **Community college costs** El Dorado Community College has a main campus in the suburbs and a downtown campus. The amount $X$ spent on tuition by a randomly selected student at the main campus has mean \$732.50 and standard deviation \$103. The amount $Y$ spent on tuition by a randomly selected student at the downtown campus has mean \$825 and standard deviation \$126.50. Suppose we randomly select one full-time student from each of the two campuses. Calculate and interpret the mean of the sum $S = X + Y$.

pg 417

50. **Essay errors** Typographical and spelling errors can be either "nonword errors" or "word errors." A nonword error is not a real word, as when "the" is typed as "teh." A word error is a real word, but not the right word, as when "lose" is typed as "loose." When students are asked to write a 250-word essay (without spell-checking), the number of nonword errors $X$ in a randomly selected essay has mean 2.1 and standard deviation 1.136. The number of word errors $Y$ in the essay has mean 1.0 and standard deviation 1.0. Calculate and interpret the mean of the sum $S = X + Y$.

51. **Study habits** The Survey of Study Habits and Attitudes (SSHA) is a psychological test that measures academic motivation and study habits. The SSHA score $F$ of a randomly selected female student at a large university has mean 120 and standard deviation 28, and the SSHA score $M$ of a randomly selected male student at the university has mean 105 and standard deviation 35. Suppose we select one male student and one female student at random from this university and give them the SSHA test. Calculate and interpret the mean of the difference $D = F - M$ in their scores.

52. **Commuting to work** Sulé's job is just a few bus stops away from his house. While it can be faster to take the bus to work than to walk, the travel time is more variable due to traffic. The commute time *B* if Sulé takes the bus to work on a randomly selected day has mean 12 minutes and standard deviation 4 minutes. The commute time *W* if Sulé walks to work on a randomly selected day has mean 16 minutes and standard deviation 1 minute. Calculate and interpret the mean of the difference $D = B - W$ in the time it would take Sulé to get to work on a randomly selected day.

53. **Community college costs** Refer to Exercise 49. At the main campus, full-time students pay \$50 per unit. At the downtown campus, full-time students pay \$55 per unit. Find the mean of the difference *D* (Main – Downtown) in the number of units that the two randomly selected students take.

54. **Essay scores** Refer to Exercise 50. An English professor deducts 3 points from a student's essay score for each nonword error and 2 points for each word error. Find the mean of the total score deductions *T* for a randomly selected essay.

55. **Rainy days** Imagine that we randomly select a day from the past 10 years. Let X be the recorded rainfall on this date at the airport in Orlando, Florida, and Y be the recorded rainfall on this date at Disney World just outside Orlando. Suppose that you know the means $\mu_X$ and $\mu_Y$ and the variances $\sigma_X^2$ and $\sigma_Y^2$ of both variables.

(a) Can we calculate the mean of the total rainfall $X + Y$ to be $\mu_X + \mu_Y$? Explain your answer.

(b) Can we calculate the variance of the total rainfall to be $\sigma_X^2 + \sigma_Y^2$? Explain your answer.

56. **His and her earnings** Researchers randomly select a married couple in which both spouses are employed. Let X be the income of the husband and Y be the income of the wife. Suppose that you know the means $\mu_X$ and $\mu_Y$ and the variances $\sigma_X^2$ and $\sigma_Y^2$ of both variables.

(a) Can we calculate the mean of the total income $X + Y$ to be $\mu_X + \mu_Y$? Explain your answer.

(b) Can we calculate the variance of the total income to be $\sigma_X^2 + \sigma_Y^2$? Explain your answer.

57. **Community college costs** Refer to Exercise 49. pg 421 Note that X and Y are independent random variables because the two students are randomly selected from each of the campuses. Calculate and interpret the standard deviation of the sum $S = X + Y$.

58. **Essay errors** Refer to Exercise 50. Assume that the number of non-word errors X and word errors Y in a randomly selected essay are independent random variables. Calculate and interpret the standard deviation of the sum $S = X + Y$.

59. **Study habits** Refer to Exercise 51.

(a) Assume that *F* and *M* are independent random variables. Explain what this means in context.

(b) Calculate and interpret the standard deviation of the difference $D = F - M$ in their scores.

(c) From the information given, can you find the probability that the randomly selected female student has a higher SSHA score than the randomly selected male student? Explain why or why not.

60. **Commuting to work** Refer to Exercise 52.

(a) Assume that *B* and *W* are independent random variables. Explain what this means in context.

(b) Calculate and interpret the standard deviation of the difference *D* (Bus – Walk) in the time it would take Sulé to get to work on a randomly selected day.

(c) From the information given, can you find the probability that it will take Sulé longer to get to work on the bus than if he walks on a randomly selected day? Explain why or why not.

61. **Community college costs** Refer to Exercise 49. Note that X and Y are independent random variables because the two students are randomly selected from each of the campuses. At the main campus, full-time students pay \$50 per unit. At the downtown campus, full-time students pay \$55 per unit. Suppose we randomly select one full-time student from each of the two campuses. Find the standard deviation of the difference *D* (Main – Downtown) in the number of units that the two randomly selected students take.

62. **Essay scores** Refer to Exercise 50. Assume that the number of nonword errors X and word errors Y in a randomly selected essay are independent random variables. An English professor deducts 3 points from a student's essay score for each nonword error and 2 points for each word error. Find the standard deviation of the total score deductions *T* for a randomly selected essay.

*Exercises 63 and 64 refer to the following setting.* In Exercise 17 of Section 6.1, we examined the probability distribution of the random variable X = the amount a life insurance company earns on a randomly chosen 5-year term life policy. Calculations reveal that $\mu_X = \$303.35$ and $\sigma_X = \$9707.57$.

63. **Life insurance** The risk of insuring one person's life is reduced if we insure many people. Suppose that we randomly select two insured 21-year-old males, and that their ages at death are independent. If $X_1$ and $X_2$ are the insurer's income from the two insurance policies, the insurer's average income W on the two policies is

$$W = \frac{X_1 + X_2}{2}$$

Find the mean and standard deviation of W. (You see that the mean income is the same as for a single policy, but the standard deviation is less.)

64. **Life insurance** If we randomly select four insured 21-year-old men, the insurer's average income is

$$V = \frac{X_1 + X_2 + X_3 + X_4}{4}$$

where $X_i$ is the income from insuring one man. Assuming that the amount of income earned on individual policies is independent, find the mean and standard deviation of $V$. (If you compare with the results of Exercise 63, you should see that averaging over more insured individuals reduces risk.)

65. **Time and motion** A time-and-motion study measures the time required for an assembly-line worker to perform a repetitive task. The data show that the time $X$ required to bring a part from a bin to its position on an automobile chassis follows a Normal distribution with mean 11 seconds and standard deviation 2 seconds. The time $Y$ required to attach the part to the chassis follows a Normal distribution with mean 20 seconds and standard deviation 4 seconds. The study finds that the times required for the two steps are independent.

pg 423

(a) Describe the distribution of the total time required for the entire operation of positioning and attaching a randomly selected part.

(b) Management's goal is for the entire process to take less than 30 seconds. Find the probability that this goal will be met for a randomly selected part.

66. **Ohm-my!** The design of an electronic circuit for a toaster calls for a 100-ohm resistor and a 250-ohm resistor connected in series so that their resistances add. The resistance $X$ of a 100-ohm resistor in a randomly selected toaster follows a Normal distribution with mean 100 ohms and standard deviation 2.5 ohms. The resistance $Y$ of a 250-ohm resistor in a randomly selected toaster follows a Normal distribution with mean 250 ohms and standard deviation 2.8 ohms. The resistances $X$ and $Y$ are independent.

(a) Describe the distribution of the total resistance of the two components in series for a randomly selected toaster.

(b) Find the probability that the total resistance for a randomly selected toaster lies between 345 and 355 ohms.

67. **Yard work** Lamar and Hareesh run a two-person lawn-care service. They have been caring for Mr. Johnson's very large lawn for several years, and they have found that the time $L$ it takes Lamar to mow the lawn on a randomly selected day is approximately Normally distributed with a mean of 105 minutes and a standard deviation of 10 minutes. The time $H$ it takes Hareesh to use the edger and string trimmer on a randomly selected day is approximately Normally distributed with a mean of 98 minutes and a standard deviation of 15 minutes. Assume that $L$ and $H$ are independent random variables. Find the probability that Lamar and Hareesh finish their jobs within 5 minutes of each other on a randomly selected day.

68. **Hit the track** Andrea and Barsha are middle-distance runners for their school's track team. Andrea's time $A$ in the 400-meter race on a randomly selected day is approximately Normally distributed with a mean of 62 seconds and a standard deviation of 0.8 second. Barsha's time $B$ in the 400-meter race on a randomly selected day is approximately Normally distributed with a mean of 62.8 seconds and a standard deviation of 1 second. Assume that $A$ and $B$ are independent random variables. Find the probability that Barsha beats Andrea in the 400-meter race on a randomly selected day.

69. **Swim team** Hanover High School has the best women's swimming team in the region. The 400-meter freestyle relay team is undefeated this year. In the 400-meter freestyle relay, each swimmer swims 100 meters. The times, in seconds, for the four swimmers this season are approximately Normally distributed with means and standard deviations as shown. Assume that the swimmer's individual times are independent. Find the probability that the total team time in the 400-meter freestyle relay for a randomly selected race is less than 220 seconds.

| Swimmer | Mean | StDev |
|---|---|---|
| Wendy | 55.2 | 2.8 |
| Jill | 58.0 | 3.0 |
| Carmen | 56.3 | 2.6 |
| Latrice | 54.7 | 2.7 |

70. **Toothpaste** Ken is traveling for his business. He has a new 0.85-ounce tube of toothpaste that's supposed to last him the whole trip. The amount of toothpaste Ken squeezes out of the tube each time he brushes can be modeled by a Normal distribution with mean 0.13 ounce and standard deviation 0.02 ounce. If Ken brushes his teeth six times on a randomly selected trip and the amounts are independent, what's the probability that he'll use all the toothpaste in the tube?

71. **Auto emissions** The amount of nitrogen oxides (NOX) present in the exhaust of a particular model of old car varies from car to car according to a Normal distribution with mean 1.4 grams per mile (g/mi) and standard deviation 0.3 g/mi. Two randomly selected cars of this model are tested. One has 1.1 g/mi of NOX; the other has 1.9 g/mi. The test station attendant finds this difference in emissions between two similar cars surprising. If the NOX levels for two randomly chosen cars of this type are independent, find the probability that the difference is greater than 0.8 or less than −0.8.

72. **Loser buys the pizza** Leona and Fred are friendly competitors in high school. Both are about to take the ACT college entrance examination. They agree that if one of them scores 5 or more points better than the other, the loser will buy the winner a pizza. Suppose that, in fact, Fred and Leona have equal ability so that each score on a randomly selected test varies Normally with mean 24 and standard deviation 2. (The variation is due to luck in guessing and the accident of the specific questions being familiar to the student.) The two scores are independent. What is the probability that the scores differ by 5 or more points in either direction?

**Multiple Choice** *Select the best answer for Exercises 73 and 74, which refer to the following setting.*

The number of calories in a 1-ounce serving of a certain breakfast cereal is a random variable with mean 110 and standard deviation 10. The number of calories in a cup of whole milk is a random variable with mean 140 and standard deviation 12. For breakfast, you eat 1 ounce of the cereal with 1/2 cup of whole milk. Let $T$ be the random variable that represents the total number of calories in this breakfast.

73. The mean of $T$ is

(a) 110. (b) 140. (c) 180. (d) 195. (e) 250.

74. The standard deviation of $T$ is

(a) 22. (b) 16. (c) 15.62. (d) 11.66. (e) 4.

**Recycle and Review**

75. **Fluoride varnish** (4.2) In an experiment to measure the effect of fluoride "varnish" on the incidence of tooth cavities, thirty-four 10-year-old girls whose parents volunteered them for the study were randomly assigned to two groups. One group was given fluoride varnish annually for 4 years, along with standard dental hygiene; the other group followed only the standard dental hygiene regimen. The mean number of cavities in the two groups was compared at the end of the 4 years.

(a) Are the participants in this experiment subject to the placebo effect? Explain.

(b) Describe how you could alter this experiment to make it double-blind.

(c) Explain the purpose of the random assignment in this experiment.

76. **Buying stock** (5.3, 6.1) You purchase a hot stock for \$1000. The stock either gains 30% or loses 25% each day, each with probability 0.5. Its returns on consecutive days are independent of each other. You plan to sell the stock after two days.

(a) What are the possible values of the stock after two days, and what is the probability for each value? What is the probability that the stock is worth more after two days than the \$1000 you paid for it?

(b) What is the mean value of the stock after two days?

*Comment:* You see that these two criteria give different answers to the question "Should I invest?"

# SECTION 6.3 Binomial and Geometric Random Variables

**LEARNING TARGETS** *By the end of the section, you should be able to:*

- Determine whether the conditions for a binomial setting are met.
- Calculate and interpret probabilities involving binomial random variables.
- Calculate the mean and standard deviation of a binomial distribution. Interpret these values.
- When appropriate, use the Normal approximation to the binomial distribution to calculate probabilities.
- Calculate and interpret probabilities involving geometric random variables.
- Calculate the mean and standard deviation of a geometric distribution. Interpret these values.

When the same random process is repeated several times, we are often interested in whether a particular outcome does or doesn't happen on each trial. Here are some examples:

- To test whether someone has extrasensory perception (ESP), choose one of four cards at random—a star, wave, cross, or circle. Ask the person to identify the card without seeing it. Do this a total of 50 times and see how many cards the person identifies correctly. *Random process*: choose a card at random. *Outcome of interest*: person identifies card correctly. *Random variable*: $X$ = number of correct identifications.
- A shipping company claims that 90% of its shipments arrive on time. To test this claim, take a random sample of 100 shipments made by the company last month and see how many arrived on time. *Random process*: randomly select a shipment and check when it arrived. *Outcome of interest*: arrived on time. *Random variable*: $Y$ = number of on-time shipments.

Alex Clark/Alamy

- In the game of Pass the Pigs, a player rolls a pair of pig-shaped dice. On each roll, the player earns points according to how the pigs land. If the player gets a "pig out," in which the two pigs land on opposite sides, she loses all points earned in that round and must pass the pigs to the next player. A player can choose to stop rolling at any point during her turn and to keep the points that she has earned before passing the pigs. *Random process*: roll the pig dice. *Outcome of interest*: pig out. *Random variable*: $T$ = number of rolls it takes the player to pig out.

Some random variables, like $X$ and $Y$ in the first two bullets, count the number of times the outcome of interest occurs in a fixed number of trials. They are called *binomial random variables*. Other random variables, like $T$ in the Pass the Pigs setting, count the number of trials of the random process it takes for the outcome of interest to occur. They are known as *geometric random variables*. These two special types of discrete random variables are the focus of this section.

## Binomial Settings and Binomial Random Variables

Let's start with an activity that involves repeating a random process several times.

### ACTIVITY Pop quiz!

It's time for a pop quiz! We hope you are ready. The quiz consists of 10 multiple-choice questions. Each question has five answer choices, labeled A through E. Now for the bad news: you will not get to see the questions. You just have to guess the answer for each one!

1. Get out a blank sheet of paper. Write your name at the top. Number your paper from 1 to 10. Then guess the answer to each question: A, B, C, D, or E. Do not look at anyone else's paper! You have 2 minutes.
2. Now it's time to grade the quizzes. Exchange papers with a classmate. Your teacher will display the answer key. The correct answer for each of the 10 questions was determined randomly so that A, B, C, D, or E was equally likely to be chosen.
3. How did you do on your quiz? Make a class dotplot that shows the number of correct answers for each student in your class. As a class, describe what you see.

In the "Pop quiz" activity, each student is performing repeated *trials* of the same random process: guessing the answer to a multiple-choice question. We're interested in the number of times that a specific event occurs: getting a correct answer (which we'll call a "success"). Knowing the outcome of one question (right or wrong guess) tells us nothing about the outcome of any other question. That is, the trials are independent. The number of trials is fixed in advance: $n = 10$. And a student's probability of getting a "success" is the same on each trial: $p = 1/5 = 0.2$. When these conditions are met, we have a **binomial setting**.

**DEFINITION Binomial setting**

A **binomial setting** arises when we perform *n* independent trials of the same random process and count the number of times that a particular outcome (called a "success") occurs.

The four conditions for a binomial setting are:

- **B**inary? The possible outcomes of each trial can be classified as "success" or "failure."
- **I**ndependent? Trials must be independent. That is, knowing the outcome of one trial must not tell us anything about the outcome of any other trial.
- **N**umber? The number of trials *n* of the random process must be fixed in advance.
- **S**ame probability? There is the same probability of success *p* on each trial.

The boldface letters in the definition box give you a helpful way to remember the conditions for a binomial setting: just check the BINS!

When checking the binary condition, note that there can be more than two possible outcomes per trial—in the "Pop Quiz" Activity, each question (trial) had five possible answer choices: A, B, C, D, or E. If we define "success" as guessing the correct answer to a question, then "failure" occurs when the student guesses any of the four incorrect answer choices.

**EXAMPLE**

## From blood types to aces
### Identifying binomial settings

tam_odin/Shutterstock.com

**PROBLEM:** Determine whether the given scenario describes a binomial setting. Justify your answer.

(a) Genetics says that the genes children receive from their parents are independent from one child to another. Each child of a particular set of parents has probability 0.25 of having type O blood. Suppose these parents have 5 children. Count the number of children with type O blood.

(b) Shuffle a standard deck of 52 playing cards. Turn over the first 10 cards, one at a time. Record the number of aces you observe.

(c) Shuffle a deck of cards. Turn over the top card. Put the card back in the deck, and shuffle again. Repeat this process until you get an ace. Count the number of cards you had to turn over.

**SOLUTION:**

(a)
- Binary? "Success" = has type O blood. "Failure" = doesn't have type O blood.
- Independent? Knowing one child's blood type tells you nothing about another child's because they inherit genes independently from their parents.
- Number? $n = 5$
- Same probability? $p = 0.25$

This is a binomial setting.

> Check the BINS! A trial consists of observing the blood type for one of these parents' children.

> All the conditions are met and we are counting the number of successes (children with type O blood).

(b)
- Binary? "Success" = get an ace. "Failure" = don't get an ace.
- Independent? No. If the first card you turn over is an ace, then the next card is less likely to be an ace because you're not replacing the top card in the deck. If the first card isn't an ace, the second card is more likely to be an ace.

This is not a binomial setting because the independent condition is not met.

> Check the BINS! A trial consists of turning over a card from the deck and observing what's on the card.

> To check for independence, you could also write
> $P(\text{2nd card ace} \mid \text{1st card ace}) = 3/51$ and
> $P(\text{2nd card ace} \mid \text{1st card not ace}) = 4/51$
> Because the two probabilities are not equal, the trials are not independent.

(c)
- Binary? "Success" = get an ace. "Failure" = don't get an ace.
- Independent? Yes. Because you are replacing the card in the deck and shuffling each time, the result of one trial doesn't tell you anything about the outcome of any other trial.
- Number? No. The number of trials is not fixed in advance.

Because there is no fixed number of trials, this is not a binomial setting.

> Check the BINS! A trial consists of turning over a card from the shuffled deck of 52 cards and observing what's on the card.

> There's another clue that this is not a binomial setting: you're counting the number of trials to get a success and not the number of successes in a fixed number of trials.

**FOR PRACTICE, TRY EXERCISE 77**

The Independent condition involves *conditional* probabilities. In part (b) of the example,

$$P(\text{2nd card ace} \mid \text{1st card ace}) = 3/51 \neq P(\text{2nd card ace} \mid \text{1st card not ace}) = 4/51$$

so the trials are not independent. The Same probability of success condition is about *unconditional* probabilities. Because

$$P(x\text{th card in a shuffled deck is an ace}) = 4/52$$

this condition is met in part (b) of the example. Be sure you understand the difference between these two conditions. When sampling is done without replacement, the Independent condition is violated.

The blood type scenario in part (a) of the example is a binomial setting. If we let X = the number of children with type O blood, then X is a **binomial random variable**. The probability distribution of X is called a **binomial distribution**.

**DEFINITION Binomial random variable, Binomial distribution**

The count of successes *X* in a binomial setting is a **binomial random variable.** The possible values of *X* are 0, 1, 2, ..., *n*.

The probability distribution of *X* is a **binomial distribution.** Any binomial distribution is completely specified by two numbers: the number of trials *n* of the random process and the probability *p* of success on each trial.

In the "Pop quiz" activity at the beginning of the lesson, X = the number of correct answers is a binomial random variable with $n = 10$ and $p = 0.2$.

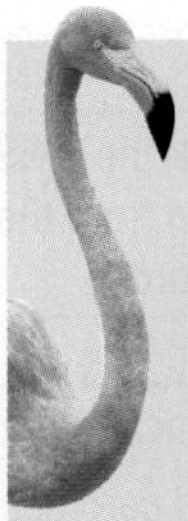

**CHECK YOUR UNDERSTANDING**

For each of the following situations, determine whether or not the given random variable has a binomial distribution. Justify your answer.

1. Shuffle a deck of cards. Turn over the top card. Put the card back in the deck, and shuffle again. Repeat this process 10 times. Let X = the number of aces you observe.
2. Choose 5 students at random from your class. Let Y = the number who are over 6 feet tall.
3. Roll a fair die 100 times. Sometime during the 100 rolls, one corner of the die chips off. Let W = the number of 5s you roll.

# Calculating Binomial Probabilities

How can we calculate probabilities involving binomial random variables? Let's return to the scenario from part (a) of the preceding example:

> Genetics says that the genes children receive from their parents are independent from one child to another. Each child of a particular set of parents has probability 0.25 of having type O blood. Suppose these parents have 5 children. Count the number of children with type O blood.

In this binomial setting, a child with type O blood is a "success" (S) and a child with another blood type is a "failure" (F). The count X of children with type O blood is a binomial random variable with $n = 5$ trials and probability $p = 0.25$ of success on each trial.

- What's $P(X = 0)$? That is, what's the probability that *none* of the 5 children has type O blood? The probability that any one of this couple's children doesn't have type O blood is $1 - 0.25 = 0.75$ (complement rule). By the multiplication rule for independent events (Section 5.3),

$$P(X=0)=P(FFFFF) = (0.75)(0.75)(0.75)(0.75)(0.75)=(0.75)^5=0.2373$$

- How about $P(X = 1)$? There are several different ways in which exactly 1 of the 5 children could have type O blood. For instance, the first child born might have type O blood, while the remaining 4 children don't have type O blood. The probability that this happens is

$$P(SFFFF) = (0.25)(0.75)(0.75)(0.75)(0.75) = (0.25)^1(0.75)^4$$

Alternatively, Child 2 could be the one that has type O blood. The corresponding probability is

$$P(FSFFF) = (0.75)(0.25)(0.75)(0.75)(0.75) = (0.25)^1(0.75)^4$$

There are three more possibilities to consider—those in which Child 3, Child 4, and Child 5 are the only ones to inherit type O blood. Of course, the probability will be the same for each of those cases. In all, there are five different ways in which exactly 1 child would have type O blood, each with the same probability of occurring. As a result,

$$P(X = 1) = P(\text{exactly 1 child with type O blood})$$
$$= 5(0.25)^1(0.75)^4 = 0.3955$$

Number of ways to get 1 child out of 5 with type O blood | 1 child with type O blood | 4 children don't have type O blood

The pattern of this calculation works for any binomial probability:

$$P(X = x) = (\#\text{ of ways to get } x \text{ successes in } n \text{ trials})\,(\text{success probability})^x\,(\text{failure probability})^{n-x}$$

To use this formula, we must count the number of arrangements of $x$ successes in $n$ trials. This number is called the **binomial coefficient**. We use the following fact to do the counting without actually listing all the arrangements.

**DEFINITION Binomial coefficient**

The number of ways to arrange $x$ successes among $n$ trials is given by the **binomial coefficient**

$$\binom{n}{x} = \frac{n!}{x!(n-x)!}$$

for $x = 0, 1, 2, \ldots, n$ where $n!$ (read as "$n$ factorial") is given by

$$n! = n(n-1)(n-2)\cdot\cdots\cdot(3)(2)(1)$$

and $0! = 1$.

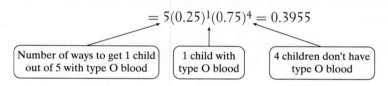

The larger of the two factorials in the denominator of a binomial coefficient will cancel much of the $n!$ in the numerator. For example, the binomial coefficient we need to find the probability that exactly 2 of the couple's 5 children inherit type O blood is

$$\binom{5}{2} = \frac{5!}{2!3!} = \frac{(5)(4)(\not{3})(\not{2})(\not{1})}{(2)(1)(\not{3})(\not{2})(\not{1})} = \frac{(5)(4)}{(2)(1)} = 10$$

**The binomial coefficient $\binom{5}{2}$ is not related to the fraction $\frac{5}{2}$.** A helpful way to remember its meaning is to read it as "5 choose 2"—as in, how many ways are there to choose which 2 children have type O blood in a family with 5 children? Binomial coefficients have many uses, but we are interested in them only as an aid to finding binomial probabilities. If you need to compute a binomial coefficient, use your calculator.

Some people prefer the notation ${}_5C_2$ instead of $\binom{5}{2}$ for the binomial coefficient.

## 14. Technology Corner CALCULATING BINOMIAL COEFFICIENTS

*TI-Nspire and other technology instructions are on the book's website at highschool.bfwpub.com/updatedtps6e.*

To calculate a binomial coefficient like $\binom{5}{2}$ on the TI-83/84, proceed as follows:

- Type 5, press MATH, arrow over to PROB, choose nCr, and press ENTER. Then type 2 and press ENTER again to execute the command 5 nCr 2 (which displays as ${}_5C_2$ on devices with pretty print).

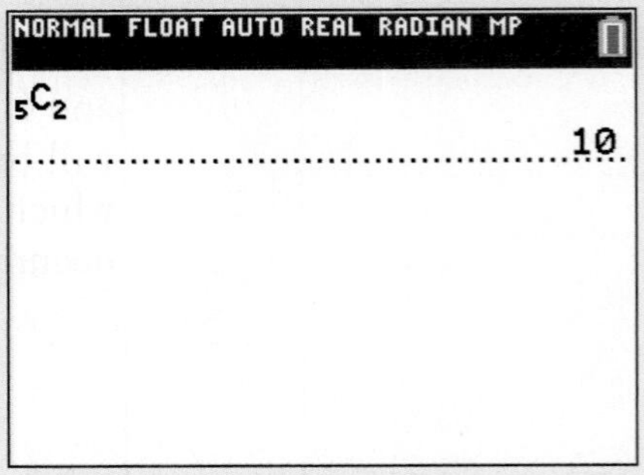


The binomial coefficient $\binom{n}{x}$ counts the number of different ways in which $x$ successes can be arranged among $n$ trials. The binomial probability $P(X = x)$ is this count multiplied by the probability of any one specific arrangement of the $x$ successes.

### BINOMIAL PROBABILITY FORMULA

Suppose that X is a binomial random variable with $n$ trials and probability $p$ of success on each trial. The probability of getting exactly $x$ successes in $n$ trials ($x = 0, 1, 2, \ldots, n$) is

$$P(X = x) = \binom{n}{x} p^x (1-p)^{n-x}$$

where

$$\binom{n}{x} = \frac{n!}{x!(n-x)!}$$

A function (like the binomial probability formula) can be used to specify the probability distribution of a random variable, in addition to a table or a graph.

With our formula in hand, we can now calculate any binomial probability.

## EXAMPLE

### Inheriting blood type
### Calculating a binomial probability

**PROBLEM:** Genetics says that the genes children receive from their parents are independent from one child to another. Each child of a particular set of parents has probability 0.25 of having type O blood. Suppose these parents have 5 children. Let $X$ = the number of children with type O blood. Find $P(X = 3)$. Interpret this value.

**SOLUTION:**

$$P(X = x) = \binom{n}{x} p^x (1-p)^{n-x}$$

$X$ is a binomial random variable with $n = 5$ and $p = 0.25$.

$$P(X=3) = \binom{5}{3}(0.25)^3(0.75)^2$$

$$= 10(0.25)^3(0.75)^2$$

$$= 0.08789$$

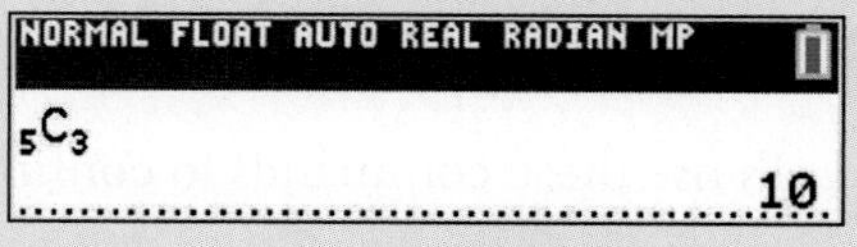


There is about a 9% probability that exactly 3 of the 5 children have type O blood.

**FOR PRACTICE, TRY EXERCISE 81**

There are times when we want to calculate a probability involving more than one value of a binomial random variable. As the following example illustrates, we can just use the binomial probability formula for each value of interest.

## EXAMPLE

### Inheriting blood type
### Calculating binomial probabilities

**PROBLEM:** The preceding example tells us that each child of a particular set of parents has probability 0.25 of having type O blood. Suppose these parents have 5 children. Should the parents be surprised if more than 3 of their children have type O blood? Calculate an appropriate probability to support your answer.

**SOLUTION:**

Let $X$ = the number of children with type O blood. $X$ has a binomial distribution with $n = 5$ and $p = 0.25$.

$$P(X > 3) = P(X = 4) + P(X = 5)$$

$$= \binom{5}{4}(0.25)^4(0.75)^1 + \binom{5}{5}(0.25)^5(0.75)^0$$

$$= 5(0.25)^4(0.75)^1 + 1(0.25)^5(0.75)^0$$

$$= 0.01465 + 0.00098$$

$$= 0.01563$$

Because there's only about a 1.5% probability of having more than 3 children with type O blood, the parents should definitely be surprised if this happens.

**FOR PRACTICE, TRY EXERCISE 85**

We can also use the calculator's binompdf and binomcdf commands to perform the calculations in the previous two examples. The following Technology Corner shows how to do it.

## 15. Technology Corner CALCULATING BINOMIAL PROBABILITIES

*TI-Nspire and other technology instructions are on the book's website at highschool.bfwpub.com/updatedtps6e.*

There are two handy commands on the TI-83/84 for finding binomial probabilities: binompdf and binomcdf. The inputs for both commands are the number of trials $n$, the success probability $p$, and the values of interest for the binomial random variable X.

binompdf(n,p,x) computes $P(X = x)$

binomcdf(n,p,x) computes $P(X \leq x)$

Let's use these commands to confirm our answers in the previous two examples.

1. Find $P(X = 3)$.

- Press 2nd VARS (DISTR) and choose binompdf(.

  **OS 2.55 or later:** In the dialog box, enter these values: trials:5, p:0.25, $x$ value:3, choose Paste, and then press ENTER.

  **Older OS:** Complete the command binompdf (5,0.25,3) and press ENTER.

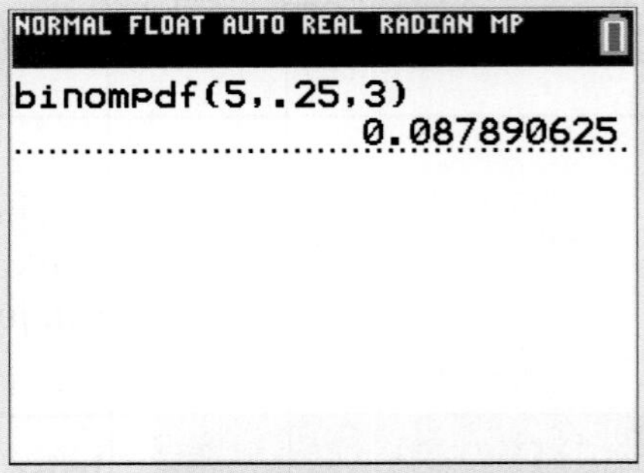


These results agree with our previous answer using the binomial probability formula: 0.08789.

2. Should the parents be surprised if more than 3 of their children have type O blood? To find $P(X > 3)$, use the complement rule:

$$P(X > 3) = 1 - P(X \leq 3) = 1 - \text{binomcdf}(5,0.25,3)$$

- Press 2nd VARS (DISTR) and choose binomcdf(.

  **OS 2.55 or later:** In the dialog box, enter these values: trials:5, p:0.25, $x$ value:3, choose Paste, and then press ENTER. Subtract this result from 1 to get the answer.

  **Older OS:** Complete the command binomcdf(5,0.25,3) and press ENTER. Subtract this result from 1 to get the answer.

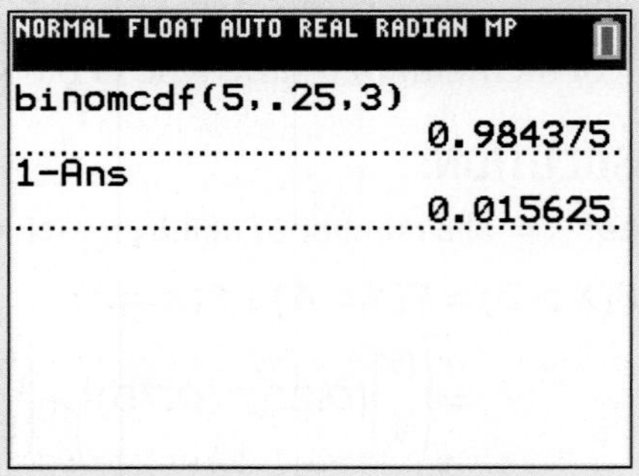


This result agrees with our previous answer using the binomial probability formula: 0.01563.

We could also have done the calculation for part (b) as $P(X > 3) = P(X = 4) + P(X = 5)$ $= \text{binompdf}(5, 0.25, 4) + \text{binompdf}(5, 0.25, 5) = 0.01465 + 0.00098 = 0.01563$.

Note the use of the complement rule to find $P(X > 3)$ in the Technology Corner: $P(X > 3) = 1 - P(X \leq 3)$. This is necessary because the calculator's

> **AP® EXAM TIP**
>
> Don't rely on "calculator speak" when showing your work on free response questions. Writing binompdf(5,0.25,3) = 0.08789 will *not* earn you full credit for a binomial probability calculation. At the very least, you must indicate what each of those calculator inputs represents. For example, "binompdf(trials:5,p:0.25,x value:3) = 0.08789."

binomcdf(n,p,x) command computes the probability of getting *x or fewer* successes in *n* trials. Remember that a *cumulative probability distribution* always gives $P(X \le x)$.

Students often have trouble identifying the correct third input for the binomcdf command when a question asks them to find the probability of getting less than, more than, or at least so many successes. Here's a helpful tip to avoid making such a mistake: write out the possible values of the variable, circle the ones you want to find the probability of, and cross out the rest. In the preceding example, X can take values from 0 to 5 and we want to find $P(X > 3)$:

~~0~~ ~~1~~ ~~2~~ ~~3~~ (4 5)

Crossing out the values from 0 to 3 shows why the correct calculation is $1 - P(X \le 3)$.

Take another look at the solutions in the two blood-type examples. The structure is much like the one we used when doing Normal calculations. Here is a summary box that describes the process.

### HOW TO FIND BINOMIAL PROBABILITIES

**Step 1: State the distribution and the values of interest.** Specify a binomial distribution with the number of trials *n*, success probability *p*, and the values of the variable clearly identified.

**Step 2: Perform calculations—show your work!** Do one of the following:

(i) Use the binomial probability formula to find the desired probability; or

(ii) Use the binompdf or binomcdf command and label each of the inputs.

Be sure to answer the question that was asked.

Here's an example that shows the method at work.

## EXAMPLE

### Free lunch?
### Calculating a cumulative binomial probability

**PROBLEM:** A local fast-food restaurant is running a "Draw a three, get it free" lunch promotion. After each customer orders, a touchscreen display shows the message "Press here to win a free lunch." A computer program then simulates one card being drawn from a standard deck. If the chosen card is a 3, the customer's order is free. Otherwise, the customer must pay the bill.

(a) On the first day of the promotion, 250 customers place lunch orders. Find the probability that fewer than 10 of them win a free lunch.

(b) In fact, only 9 customers won a free lunch. Does this result give convincing evidence that the computer program is flawed?

eldadcarin/Getty Images

**SOLUTION:**

(a) Let $Y =$ the number of customers who win a free lunch. $Y$ has a binomial distribution with $n = 250$ and $p = 4/52$.

$P(Y < 10) = P(Y \le 9)$

$= \text{binomcdf(trials: } 250, p\text{: } 4/52, x \text{ value: } 9)$

$= 0.00613$

(b) There is only a 0.006 probability that fewer than 10 customers would win a free lunch if the computer program is working properly. Because only 9 customers won a free lunch on this day, we have convincing evidence that the computer program is flawed.

**Step 1: State the distribution and the values of interest.**

The values of $Y$ that interest us are

(0 1 2 3 4 5 6 7 8 9) 10 11 12 ... 250

**Step 2: Perform calculations—show your work!**

(i) Use the binomial probability formula to find the desired probability; or

(ii) use the binompdf or binomcdf command and label each of the inputs.

To use the binomial formula, you would have to add the probabilities for $Y = 0, 1, \ldots, 9$. That's too much work!

**FOR PRACTICE, TRY EXERCISE 89**

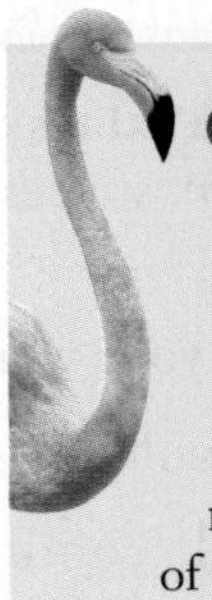

## CHECK YOUR UNDERSTANDING

To introduce his class to binomial distributions, Mr. Miller does the "Pop quiz" activity at the beginning of this section (page 431). Each student in the class guesses an answer from A through E on each of the 10 multiple-choice questions. Mr. Miller determines the "correct" answer for each of the 10 questions randomly so that A, B, C, D, or E was equally likely to be chosen. Hannah is one of the students in this class. Let $X =$ the number of questions that Hannah answers correctly.

1. What probability distribution does $X$ have? Justify your answer.
2. Use the binomial probability formula to find $P(X = 3)$. Interpret this result.
3. To get a passing score on the quiz, a student must answer at least 6 questions correctly. Would you be surprised if Hannah earned a passing score? Calculate an appropriate probability to support your answer.

# Describing a Binomial Distribution: Shape, Center, and Variability

What does the probability distribution of a binomial random variable look like? The table shows the possible values and corresponding probabilities for $X =$ the number of children with type O blood from two previous examples. This is a binomial random variable with $n = 5$ and $p = 0.25$.

| **Value $x_i$** | 0 | 1 | 2 | 3 | 4 | 5 |
|---|---|---|---|---|---|---|
| **Probability $p_i$** | 0.23730 | 0.39551 | 0.26367 | 0.08789 | 0.01465 | 0.00098 |

Figure 6.7 shows a histogram of the probability distribution. This binomial distribution with $n = 5$ and $p = 0.25$ has a clear right-skewed shape. Why? Because

the probability that any one of the couple's children inherits type O blood is 0.25, it's quite likely that 0, 1, or 2 of the children will have type O blood. Larger values of $X$ are much less likely.

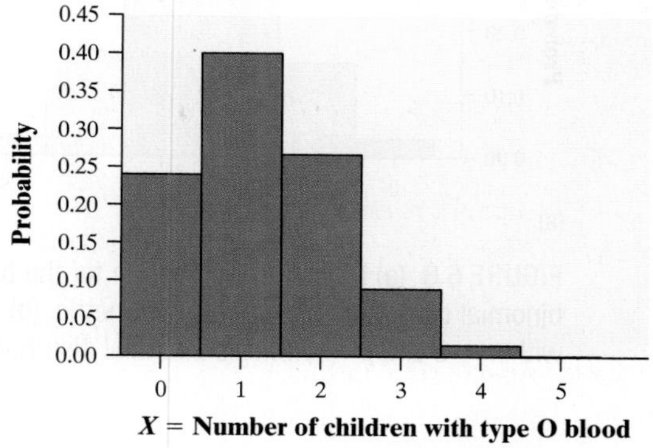


**FIGURE 6.7** Histogram showing the probability distribution of the binomial random variable $X$ = number of children with type O blood in a family with 5 children.

You can use technology to graph a binomial probability distribution like the one shown in Figure 6.7.

## 16. Technology Corner

### GRAPHING BINOMIAL PROBABILITY DISTRIBUTIONS

*TI-Nspire and other technology instructions are on the book's website at highschool.bfwpub.com/updatedtps6e.*

To graph the binomial probability distribution for $n = 5$ and $p = 0.25$:

- Type the possible values of the random variable X into list $L_1$: 0, 1, 2, 3, 4, and 5.
- Highlight $L_2$ with your cursor. Enter the command binompdf(5,0.25) and press ENTER.
- Make a histogram of the probability distribution using the method shown in Technology Corner 13 (page 398).

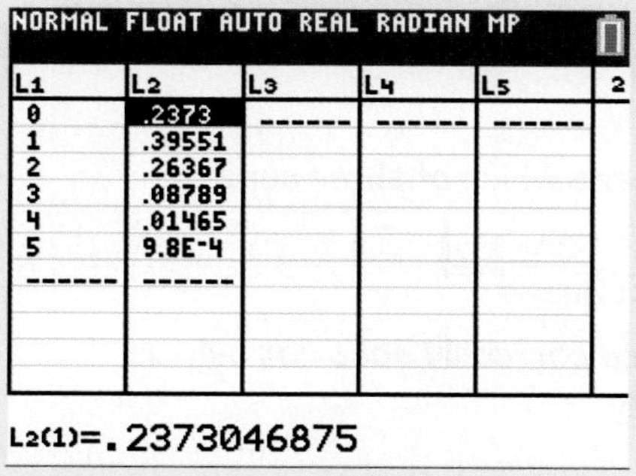


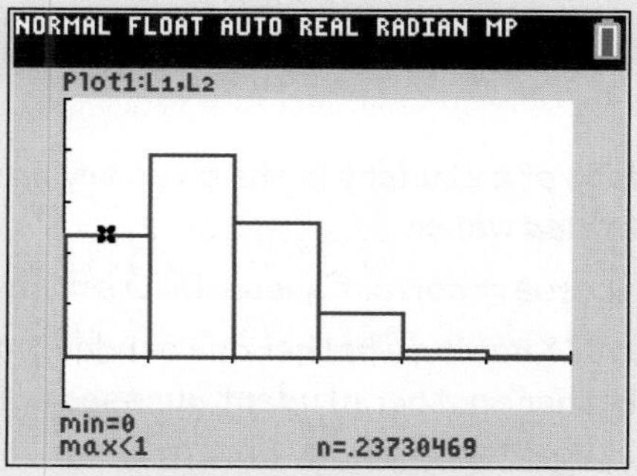


The binomial distribution with $n = 5$ and $p = 0.25$ is skewed to the right. Figure 6.8 on the next page shows two more binomial distributions with different shapes. The binomial distribution with $n = 5$ and $p = 0.51$ in Figure 6.8(a) is roughly symmetric. The binomial distribution with $n = 5$ and $p = 0.8$ in Figure 6.8(b) is skewed to the left. In general, when $n$ is small, the probability distribution of a binomial random variable will be roughly symmetric if $p$ is close to 0.5, right-skewed if $p$ is much less than 0.5, and left-skewed if $p$ is much greater than 0.5.

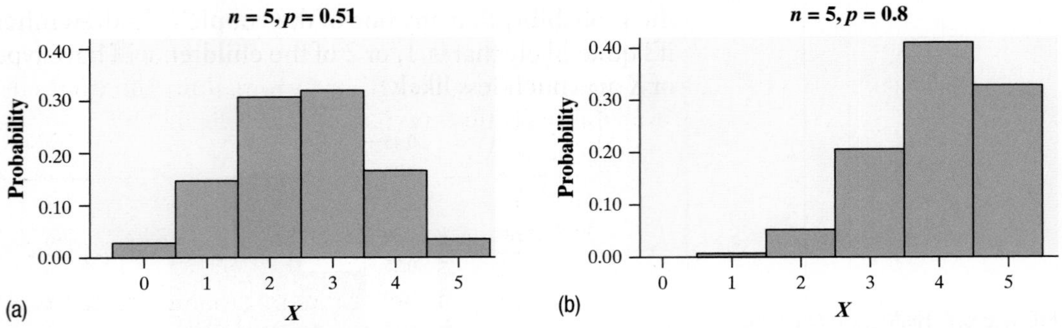


**FIGURE 6.8** (a) Probability histogram for the binomial random variable $X$ with $n = 5$ and $p = 0.51$. This binomial distribution is roughly symmetric. (b) Probability histogram for the binomial random variable $X$ with $n = 5$ and $p = 0.8$. This binomial distribution has a left-skewed shape.

## EXAMPLE

### Bottled water versus tap water
### Describing a binomial distribution

**PROBLEM:** Mr. Hogarth's AP® Statistics class did the activity on page 388. There were 21 students in the class. If we assume that the students in his class could *not* tell tap water from bottled water, then each one is guessing, with a 1/3 probability of being correct. Let X = the number of students who correctly identify the cup containing bottled water. Here is a histogram of the probability distribution of X:

(a) What probability distribution does X have? Justify your answer.

(b) Describe the shape of the probability distribution.

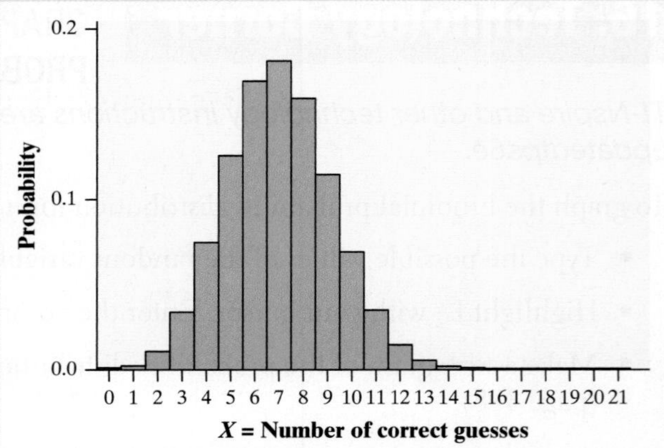


**SOLUTION:**

(a) A trial consists of a student in the class trying to guess which of three cups contained bottled water.

- Binary? Success = correct guess; failure = incorrect guess
- Independent? Knowing whether one student guessed correctly does not help us predict whether another student guessed correctly.
- Number? $n = 21$
- Same probability? $p = 1/3$

Check the BINS!

X has a binomial distribution with $n = 21$ and $p = 1/3$.

(b) The probability distribution of $X$ looks roughly symmetric with a single peak at $X = 7$.

You could also say that the graph is slightly right-skewed due to the long tail that extends out to $X = 21$.

FOR PRACTICE, TRY EXERCISE 91

### MEAN AND STANDARD DEVIATION OF A BINOMIAL RANDOM VARIABLE

The random variable $X =$ the number of children with type O blood from the previous two examples has a binomial distribution with $n = 5$ and $p = 0.25$. Its probability distribution is shown in the table.

| Value $x_i$ | 0 | 1 | 2 | 3 | 4 | 5 |
|---|---|---|---|---|---|---|
| Probability $p_i$ | 0.23730 | 0.39551 | 0.26367 | 0.08789 | 0.01465 | 0.00098 |

Because X is a discrete random variable, we can calculate its mean using the formula

$$\mu_X = E(X) = \sum x_i p_i = x_1p_1 + x_2p_2 + x_3p_3 + \cdots$$

from Section 6.1. We get

$$\mu_X = (0)(0.23730) + (1)(0.39551) + \cdots + (5)(0.00098) = 1.25$$

So the expected number of children with type O blood in families like this one with 5 children is 1.25.

Did you think about why the mean is $\mu_X = 1.25$? Because each child has a 0.25 chance of inheriting type O blood, we'd expect one-fourth of the 5 children to have this blood type. In other words,

$$\mu_X = 5(0.25) = 1.25$$

This method can be used to find the mean of any binomial random variable.

### MEAN (EXPECTED VALUE) OF A BINOMIAL RANDOM VARIABLE

If a count X of successes has a binomial distribution with number of trials $n$ and probability of success $p$, the mean (expected value) of X is

$$\mu_X = E(X) = np$$

To calculate the standard deviation of X, we start by finding the variance.

$$\begin{aligned}\sigma_X^2 &= \sum (x_i - \mu_X)^2 p_i \\ &= (0-1.25)^2(0.23730) + (1-1.25)^2(0.39551) + \cdots + (5-1.25)^2(0.00098) \\ &= 0.9375\end{aligned}$$

So the standard deviation of X is

$$\sigma_X = \sqrt{0.9375} = 0.968$$

The number of children with type O blood will typically vary by about 0.968 from the mean of 1.25 in families like this one with 5 children.

There is a simple formula for the standard deviation of a binomial random variable, but it isn't easy to explain (see the Think About It on page 445). For our family with $n = 5$ children and $p = 0.25$ of type O blood, the *variance* of X is

$$5(0.25)(0.75) = 0.9375$$

To get the standard deviation, we just take the square root:

$$\sigma_X = \sqrt{5(0.25)(0.75)} = \sqrt{0.9375} = 0.968$$

This method works for any binomial random variable.

**STANDARD DEVIATION OF A BINOMIAL RANDOM VARIABLE**

If a count X of successes has a binomial distribution with number of trials $n$ and probability of success $p$, the standard deviation of X is

$$\sigma_X = \sqrt{np(1-p)}$$

caution

**Remember that these formulas for the mean and standard deviation work only for binomial distributions.** The interpretation of the parameters $\mu$ and $\sigma$ is the same as for any discrete random variable.

## EXAMPLE

### Bottled water versus tap water
### Describing a binomial distribution

Gerald Bernard/ Shutterstock

**PROBLEM:** Assume that each of the 21 students in Mr. Hogarth's AP® Statistics class who did the bottled water versus tap water activity was just guessing, so there was a 1/3 chance of each student identifying the cup containing bottled water correctly. Let X = the number of students who make a correct identification. At right is a histogram of the probability distribution of X.

(a) Calculate and interpret the mean of X.

(b) Calculate and interpret the standard deviation of X.

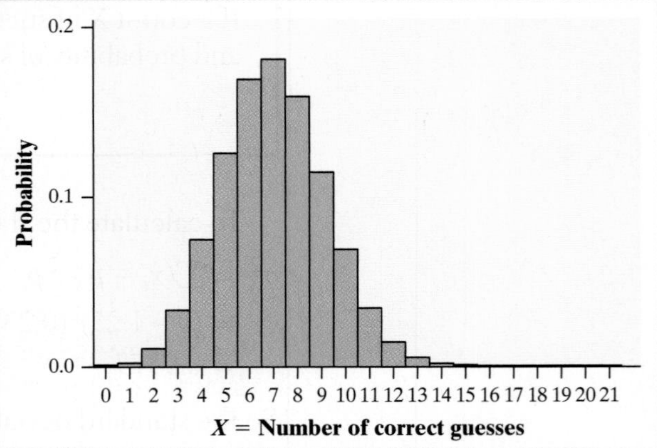


**SOLUTION:**

(a) $\mu_X = np = 21(1/3) = 7$

If all the students in Mr. Hogarth's class were just guessing and repeated the activity many times, the average number of students who guess correctly would be about 7.

(b) $\sigma_X = \sqrt{np(1-p)} = \sqrt{21(1/3)(2/3)} = 2.16$

If all the students in Mr. Hogarth's class were just guessing and repeated the activity many times, the number of students who guess correctly would typically vary by about 2.16 from the mean of 7.

**FOR PRACTICE, TRY EXERCISE 95**

Of the 21 students in Mr. Hogarth's class, 13 made correct identifications. Are you convinced that some of Mr. Hogarth's students could tell bottled water from tap water? The class's result corresponds to X = 13, a value that's nearly 3 standard deviations above the mean. How likely is it that 13 or more of Mr. Hogarth's students would guess correctly? It's

$$\begin{aligned} P(X \geq 13) &= 1 - P(X \leq 12) \\ &= 1 - \text{binomcdf(trials:21, p:1/3, xvalue:12)} \\ &= 1 - 0.9932 \\ &= 0.0068 \end{aligned}$$

The students had less than a 1% chance of getting so many correct identifications if they were all just guessing. This result gives convincing evidence that some of the students in the class could tell bottled water from tap water.

## Think About It

**WHERE DO THE BINOMIAL MEAN AND VARIANCE FORMULAS COME FROM?** We can derive the formulas for the mean and variance of a binomial distribution using what we learned about combining random variables in Section 6.2. Let's start with the random variable $B$ that's described by the following probability distribution.

| **Value $b_i$** | 0 | 1 |
|---|---|---|
| **Probability $p_i$** | $1-p$ | $p$ |

You can think of $B$ as representing the result of a single trial of some random process. If a success occurs (probability $p$), then $B = 1$. If a failure occurs, then $B = 0$. Notice that the mean of $B$ is

$$\mu_B = \sum b_i p_i = (0)(1-p) + (1)(p) = p$$

and that the variance of $B$ is

$$\begin{aligned} \sigma_B^2 &= \sum (b_i - \mu_B)^2 p_i = (0-p)^2(1-p) + (1-p)^2 p \\ &= p^2(1-p) + (1-p)^2 p \\ &= p(1-p)[p + (1-p)] \\ &= p(1-p) \end{aligned}$$

Now consider the random variable $X = B_1 + B_2 + \cdots + B_n$. We can think of $X$ as counting the number of successes in $n$ independent trials of this random process, with each trial having success probability $p$. In other words, $X$ is a binomial random variable. By the rules from Section 6.2, the mean of $X$ is

$$\mu_X = \mu_{B_1} + \mu_{B_2} + \cdots + \mu_{B_n} = p + p + \cdots + p = np$$

and the variance of $X$ is

$$\begin{aligned} \sigma_X^2 &= \sigma_{B_1}^2 + \sigma_{B_2}^2 + \cdots + \sigma_{B_n}^2 \\ &= p(1-p) + p(1-p) + \cdots + p(1-p) \\ &= np(1-p) \end{aligned}$$

The standard deviation of $X$ is therefore

$$\sigma_X = \sqrt{np(1-p)}$$

### CHECK YOUR UNDERSTANDING

To introduce his class to binomial distributions, Mr. Miller does the "Pop quiz" activity at the beginning of this section (page 431). Each student in the class guesses an answer from A through E on each of the 10 multiple-choice questions. Mr. Miller determines the "correct" answer for each of the 10 questions randomly so that A, B, C, D, or E was equally likely to be chosen. Hannah is one of the students in this class. Let $Y$ = the number of questions that Hannah answers *incorrectly*.

1. Use technology to make a histogram of the probability distribution of $Y$. Describe its shape.
2. Calculate and interpret the mean of $Y$.
3. Calculate and interpret the standard deviation of $Y$.
4. On page 440, we defined $X$ = the number of *correct* answers that Hannah got on the quiz. How do the shape, center, and variability of the probability distribution of $X$ compare to your answers for Questions 1 to 3?

## Binomial Distributions in Statistical Sampling

fatamorgana-999/Shutterstock.com

The binomial distributions are important in statistics when we wish to make inferences about the proportion $p$ of successes in a population. For instance, suppose that a supplier inspects a random sample of 10 flash drives from a shipment of 10,000 flash drives in which 200 are defective (bad). Let $X$ = the number of bad flash drives in the sample.

This is not quite a binomial setting. Because we are sampling without replacement, the independence condition is violated. The conditional probability that the second flash drive chosen is bad changes when we know whether the first is good or bad: $P(\text{second is bad} \mid \text{first is good}) = 200/9999 = 0.0200$ but $P(\text{second is bad} \mid \text{first is bad}) = 199/9999 = 0.0199$. These probabilities are very close because removing 1 flash drive from a shipment of 10,000 changes the makeup of the remaining 9999 flash drives very little. The distribution of $X$ is very close to the binomial distribution with $n = 10$ and $p = 0.02$.

To illustrate this, let's compute the probability that none of the 10 flash drives is defective. Using the binomial distribution, it's

$$P(X = 0) = \binom{10}{0}(0.02)^0(0.98)^{10} = 0.8171$$

The actual probability of getting no defective flash drives is

$$P(\text{no defectives}) = \frac{9800}{10{,}000} \times \frac{9799}{9999} \times \frac{9798}{9798} \times \cdots \times \frac{9791}{9991} = 0.8170$$

Those two probabilities are quite close!

Almost all real-world sampling, such as taking an SRS from a population of interest, is done without replacement. As the preceding example illustrates, sampling without replacement leads to a violation of the Independent condition.

However, the flash drives context shows how we can use binomial distributions in the statistical setting of selecting a random sample. When the population is much larger than the sample, a count of successes in an SRS of size $n$ has approximately the binomial distribution with $n$ equal to the sample size and $p$ equal to the proportion of successes in the population. What counts as "much larger"? In practice, the binomial distribution gives a good approximation as long as we sample less than 10% of the population. We refer to this as the **10% condition**.

**DEFINITION 10% condition**

When taking a random sample of size $n$ from a population of size $N$, we can treat individual observations as independent when performing calculations as long as $n < 0.10N$.

Here's a scenario that shows why it's important to check the 10% condition before calculating a binomial probability. You might recognize the setting from the first activity in the book (page 6).

> An airline has just finished training 25 pilots—15 male and 10 female—to become captains. Unfortunately, only 8 captain positions are available right now. Airline managers announce that they will use a lottery to determine which pilots will fill the available positions. One day later, managers reveal the results of the lottery: Of the 8 captains chosen, 5 are female and 3 are male. Some of the male pilots who weren't selected suspect that the lottery was not carried out fairly.

What's the probability of choosing 5 female pilots in a fair lottery? Let X = the number of female pilots selected in a random sample of size $n = 8$ from the population of $N = 25$ pilots. Notice that the sample size is almost 1/3 of the population size. If we ignore this fact and use a binomial probability calculation, we get

$$P(X = 5) = \binom{8}{5}(0.40)^5(0.60)^3 = 0.124$$

The correct probability, however, is 0.106. You can see that the binomial probability is off by about 17% (0.018/0.106) from the correct answer.

## EXAMPLE Teens and debit cards

### Binomial distributions and sampling

**PROBLEM:** In a survey of 500 U.S. teenagers aged 14 to 18, subjects were asked a variety of questions about personal finance.[8] One question asked whether teens had a debit card. Suppose that exactly 12% of teens aged 14 to 18 have debit cards. Let X = the number of teens in a random sample of size 500 who have a debit card.

(a) Explain why X can be modeled by a binomial distribution even though the sample was selected without replacement.

(b) Use a binomial distribution to estimate the probability that 50 or fewer teens in the sample have debit cards.

**SOLUTION:**

(a) 500 is less than 10% of all U.S. teenagers aged 14 to 18.

Check the 10% condition: $n < 0.10N$

(b) $X$ is approximately binomial with $n = 500$ and $p = 0.12$.

$P(X \leq 50) = $ binomcdf(trials: 500, p: 0.12, x value: 50)

$= 0.0932$

**FOR PRACTICE, TRY EXERCISE 99**

# The Normal Approximation to Binomial Distributions

As you saw earlier, the shape of a binomial distribution can be skewed to the right, skewed to the left, or roughly symmetric. Something interesting happens to the shape as the number of trials $n$ increases. You can investigate the relationship between $n$ and $p$ yourself using the *Normal Approximation to Binomial Distributions* applet at the book's website, highschool.bfwpub.com/updatedtps6e.

APPLET

Figure 6.9 shows histograms of binomial distributions for different values of $n$ and $p$. As the number of observations $n$ becomes larger, the binomial distribution gets close to a Normal distribution.

**FIGURE 6.9** Histograms of binomial distributions with (a) $n = 10$ and $p = 0.8$, (b) $n = 20$ and $p = 0.8$, and (c) $n = 50$ and $p = 0.8$. As $n$ increases, the shape of the probability distribution gets closer and closer to Normal.

When $n$ is large, we can use Normal probability calculations to approximate binomial probabilities. To see if $n$ is large enough, check the **Large Counts condition**.

**DEFINITION Large Counts condition**

Suppose that a count $X$ of successes has the binomial distribution with $n$ trials and success probability $p$. The **Large Counts condition** says that the probability distribution of $X$ is approximately Normal if

$$np \geq 10 \quad \text{and} \quad n(1-p) \geq 10$$

That is, the expected numbers (counts) of successes and failures are both at least 10.

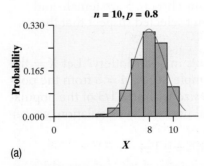

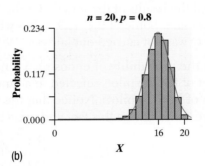

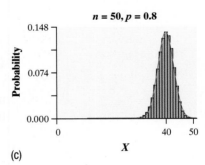

This condition is called "large counts" because $np$ is the expected (mean) count of successes and $n(1-p)$ is the expected (mean) count of failures in a binomial setting. Why do we require that both these values be at least 10? Look back at Figure 6.9. It is clear that a Normal distribution does not approximate the probability distributions in parts (a) or (b) very well. This isn't surprising because the Large Counts condition is not met in either case. For graph (a), both $np = 10(0.8) = 8$ and $n(1-p) = 10(0.2) = 2$ are less than 10. For graph (b), $np = 20(0.8) = 16 \geq 10$, but $n(1-p) = 20(0.2) = 4$ is less than 10. The Normal curve in graph (c) appears to model the binomial probability distribution well. This time, the Large Counts condition is met: $np = 50(0.8) = 40 \geq 10$ and $n(1-p) = 50(0.2) = 10 \geq 10$.

The accuracy of the Normal approximation improves as the sample size $n$ increases. It is most accurate for any fixed $n$ when $p$ is close to 1/2 and least accurate when $p$ is near 0 or 1. This is why the Large Counts condition depends on $p$ as well as $n$.

## EXAMPLE

### Teens and debit cards
### Normal approximation to a binomial distribution

**PROBLEM:** In a survey of 500 U.S. teenagers aged 14 to 18, subjects were asked a variety of questions about personal finance. One question asked whether teens had a debit card. Suppose that exactly 12% of teens aged 14 to 18 have debit cards. Let X = the number of teens in a random sample of size 500 who have a debit card.

(a) Justify why X can be approximated by a Normal distribution.

(b) Use a Normal distribution to estimate the probability that 50 or fewer teens in the sample have debit cards.

**SOLUTION:**

(a) $X$ is approximately binomial with $n = 500$ and $p = 0.12$. Because $np = 500(0.12) = 60 \geq 10$ and $n(1-p) = 500(0.88) = 440 \geq 10$, we can approximate $X$ with a Normal distribution.

(b) $\mu_X = np = 500(0.12) = 60$ and

$\sigma_X = \sqrt{np(1-p)} = \sqrt{500(0.12)(0.88)} = 7.266$

Start by calculating the mean and standard deviation of the binomial random variable *X*.

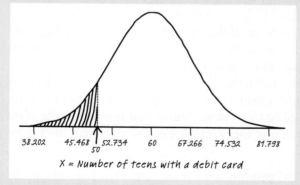

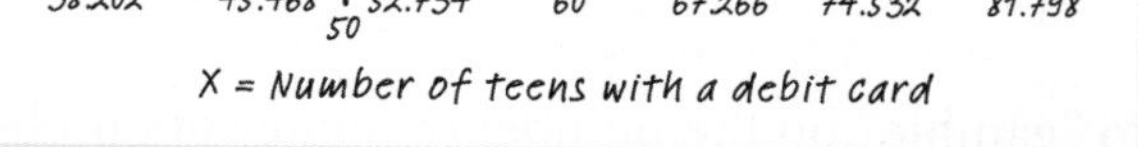


1. **Draw a Normal distribution.**
2. **Perform calculations—show your work!**
   (i) Standardize and use Table A or technology; or
   (ii) Use technology without standardizing.

Be sure to answer the question that was asked.

(i) $z = \frac{50-60}{7.266} = -1.38$

*Using Table A:* 0.0838

*Using technology:* normalcdf(lower: −1000, upper: −1.38, mean:0, SD:1) = 0.0838

$P(X \leq 50) = P(Z \leq -1.38)$

(ii) normalcdf(lower:0, upper:50, mean:60, SD:7.266) = 0.0844

**FOR PRACTICE, TRY EXERCISE 103**

The probability from the Normal approximation in this example, 0.0844, misses the exact binomial probability of 0.0932 from the preceding example by about 0.0088.

## Geometric Random Variables

In a binomial setting, the number of trials $n$ is fixed in advance, and the binomial random variable X counts the number of successes. The possible values of X are 0, 1, 2, . . . , $n$. In other situations, the goal is to repeat a random process *until a success occurs:*

- Roll a pair of dice until you get doubles.
- In basketball, attempt a 3-point shot until you make one.
- Keep placing a $1 bet on the number 15 in roulette until you win.

These are all examples of a **geometric setting**.

**DEFINITION Geometric setting**

A **geometric setting** arises when we perform independent trials of the same random process and record the number of trials it takes to get one success. On each trial, the probability $p$ of success must be the same.

Here's an activity your class can try that involves a geometric setting.

### ACTIVITY Is this your lucky day?

Your teacher is planning to give you 10 problems for homework. As an alternative, you can agree to play the Lucky Day game. Here's how it works. A student will be selected at random from your class and asked to pick a day of the week (e.g., Thursday). Then your teacher will use technology to randomly choose a day of the week as the "lucky day." If the student picks the correct day, the class will have only one homework problem. If the student picks the wrong day, your teacher will select another student from the class at random. The chosen student will pick a day of the week and your teacher will use technology to choose a "lucky day." If this student gets it right, the class will have two homework problems. The game continues until a student correctly picks the lucky day. Your teacher will assign a number of homework problems that is equal to the total number of picks made by members of your class. Are you ready to play the Lucky Day game?

1. Decide as a class whether to "gamble" on the number of homework problems you will receive. You have 30 seconds.
2. Play the Lucky Day game and see what happens!

In a geometric setting, if we define the random variable X to be the number of trials needed to get the first success, then X is called a **geometric random variable**. The probability distribution of X is a **geometric distribution**.

**DEFINITION Geometric random variable, Geometric distribution**

The number of trials *X* that it takes to get a success in a geometric setting is a **geometric random variable.** The probability distribution of *X* is a **geometric distribution** with probability *p* of success on any trial. The possible values of *X* are 1, 2, 3, . . . .

As with binomial random variables, it's important to be able to distinguish situations in which a geometric distribution does and doesn't apply. Let's consider the Lucky Day game. The random variable of interest in this game is $X =$ the number of picks it takes to correctly match the lucky day. Each pick is one trial of the random process. Knowing the result of one student's pick tells us nothing about the result of any other pick. On each trial, the probability of a correct pick is 1/7. This is a geometric setting. Because X counts the number of trials to get the first success, it is a geometric random variable with $p = 1/7$.

What is the probability that the first student picks correctly and wins the Lucky Day game? It's $P(X = 1) = 1/7$. That's also the class's chance of having only one homework problem assigned. For the class to have two homework problems assigned, the first student selected must pick an incorrect day of the week and the second student must pick the lucky day correctly. The probability that this happens is

$$P(X = 2) = (6/7)(1/7) = 0.1224$$

Likewise,

$$P(X = 3) = (6/7)(6/7)(1/7) = 0.1050$$

In general, the probability that the first correct pick comes on the *x*th trial is

$$P(X = x) = (6/7)^{x-1}(1/7)$$

Let's summarize what we've learned about calculating a geometric probability.

**GEOMETRIC PROBABILITY FORMULA**

If X has the geometric distribution with probability $p$ of success on each trial, the possible values of X are 1, 2, 3, . . . . If $x$ is any one of these values,

$$P(X = x) = (1 - p)^{x-1}p$$

With the geometric probability formula in hand, we can now compute any geometric probability.

**EXAMPLE**

## The Lucky Day game
### Calculating geometric probabilities

**PROBLEM:** Mr. Lochel's class decides to play the Lucky Day game. Let $X =$ the number of homework problems that the class receives.

(a) Find the probability that the class receives exactly 10 homework problems as a result of playing the Lucky Day game.

(b) Find $P(X < 10)$ and interpret this value.

**SOLUTION:**

*X* has a geometric distribution with $p = 1/7$.

(a) $P(X = 10) = (6/7)^9(1/7) = 0.0357$

(b) $P(X < 10) = P(X = 1) + P(X = 2) + P(X = 3) + \cdots + P(X = 9)$

$$= 1/7 + (6/7)(1/7) + (6/7)^2(1/7) + \cdots + (6/7)^8(1/7)$$
$$= 0.7503$$

There's about a 75% probability that the class will get fewer than 10 homework problems by playing the Lucky Day game.

**FOR PRACTICE, TRY EXERCISE 107**

There's a clever alternative approach to finding the probability in part (b) of the example. By the complement rule, $P(X < 10) = 1 - P(X \geq 10)$. What's the probability that it will take at least 10 picks for Mr. Lochel's class to win the Lucky Day game? It's the chance that the first 9 picks are all incorrect: $\left(\frac{6}{7}\right)^9 = 0.250$. So the probability that the class will win the Lucky Day game in fewer than 10 picks (and therefore have fewer than 10 homework problems assigned) is

$$P(X < 10) = 1 - P(X \geq 10) = 1 - 0.250 = 0.750$$

As you probably guessed, we can use technology to calculate geometric probabilities. The following Technology Corner shows how to do it.

## 17. Technology Corner — CALCULATING GEOMETRIC PROBABILITIES

*TI-Nspire and other technology instructions are on the book's website at highschool.bfwpub.com/updatedtps6e.*

There are two handy commands on the TI-83/84 for finding geometric probabilities: geometpdf and geometcdf. The inputs for both commands are the success probability $p$ and the value(s) of interest for the geometric random variable X.

geometpdf (p,x) computes $P(X = x)$

geometcdf (p,x) computes $P(X \leq x)$

Let's use these commands to confirm our answers in the previous example.

(a) Find the probability that the class receives exactly 10 homework problems as a result of playing the Lucky Day game.

- Press 2nd VARS (DISTR) and choose geometpdf(.

  **OS 2.55 or later:** In the dialog box, enter these values: p:1/7, x value:10, choose Paste, and then press ENTER.

  **Older OS:** Complete the command geometpdf(1/7,10) and press ENTER.

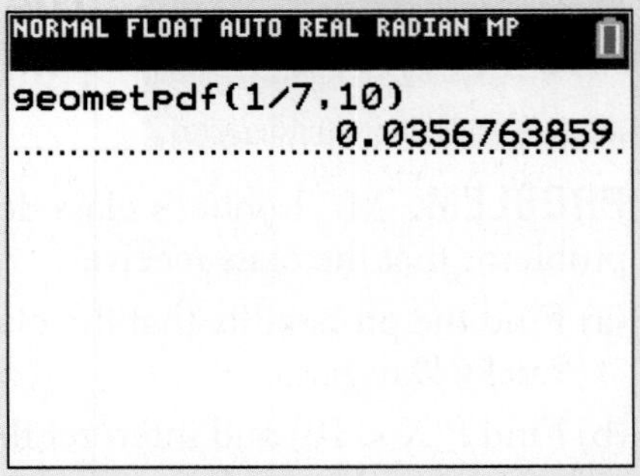


These results agree with our previous answer using the geometric probability formula: 0.0357.

(b) Find $P(X < 10)$ and interpret this value. To find $P(X < 10)$, use the geometcdf command:

$$P(X < 10) = P(X \leq 9) = \text{geometcdf}(1/7, 9)$$

- Press [2nd] [VARS] (DISTR) and choose geometcdf.

  **OS 2.55 or later:** In the dialog box, enter these values: p:1/7, x value:9, choose Paste, and then press [ENTER].

  **Older OS:** Complete the command geometcdf(1/7,9) and press [ENTER].

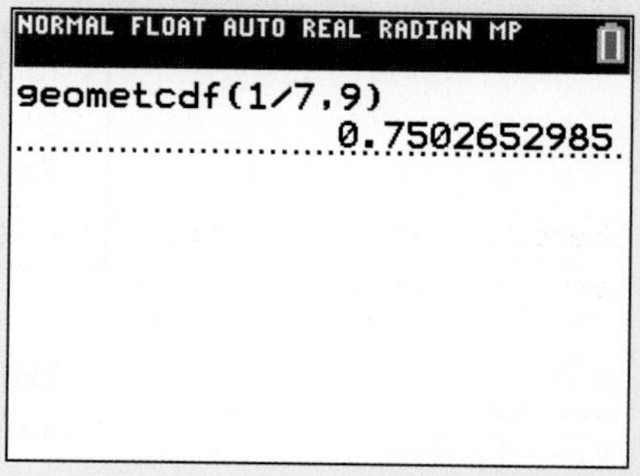


These results agree with our previous answer using the geometric probability formula: 0.7503.

**DESCRIBING A GEOMETRIC DISTRIBUTION: SHAPE, CENTER, AND VARIABILITY** The table shows part of the probability distribution of X = the number of picks it takes to match the lucky day. We can't show the entire distribution because the number of trials it takes to get the first success could be a very large number.

| **Value $x_i$** | 1 | 2 | 3 | 4 | 5 | 6 | 7 | 8 | 9 | ... |
|---|---|---|---|---|---|---|---|---|---|---|
| **Probability $p_i$** | 0.143 | 0.122 | 0.105 | 0.090 | 0.077 | 0.066 | 0.057 | 0.049 | 0.042 | |

Figure 6.10 is a histogram of the probability distribution for values of X from 1 to 26. Let's describe what we see.

***Shape:*** Skewed to the right. Every geometric distribution has this shape. That's because the most likely value of a geometric random variable is 1. The probability of each successive value decreases by a factor of $(1 - p)$.

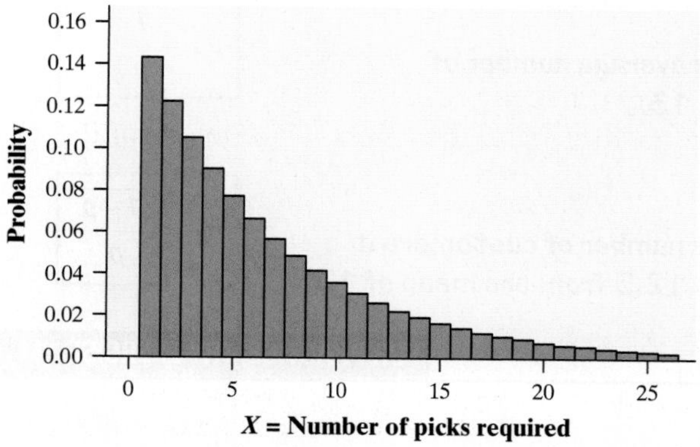


**FIGURE 6.10** Histogram showing the probability distribution of the geometric random variable $X$ = number of trials needed for students to pick correctly in the Lucky Day game.

***Center:*** The mean (expected value) of X is $\mu_X = 7$. If the class played the Lucky Day game many times, they would receive an average of 7 homework problems. It's no coincidence that $p = 1/7$ and $\mu_X = 7$. With probability of success 1/7 on each trial, we'd expect it to take an average of 7 trials to get the first success. That is, $\mu_X = 1/(1/7) = 7$.

***Variability:*** The standard deviation of X is $\sigma_X = 6.48$. If the class played the Lucky Day game many times, the number of homework problems they receive would typically vary by about 6.5 problems from the mean of 7. That could mean a lot of homework! There is a simple formula for the standard deviation of a geometric random variable, but it isn't easy to explain. For the Lucky Day game,

$$\sigma_X = \frac{\sqrt{1 - 1/7}}{1/7} = 6.48$$

We can generalize these results for the mean and standard deviation of a geometric random variable.

## MEAN (EXPECTED VALUE) AND STANDARD DEVIATION OF A GEOMETRIC RANDOM VARIABLE

If X is a geometric random variable with probability of success $p$ on each trial, then its mean (expected value) is $\mu_X = E(X) = \frac{1}{p}$ and its standard deviation is $\sigma_X = \frac{\sqrt{1-p}}{p}$.

We interpret the parameters $\mu$ and $\sigma$ in the same way as for any discrete random variable.

## EXAMPLE

### Waiting for a free lunch
### Mean and SD of a geometric distribution

**PROBLEM:** A local fast-food restaurant is running a "Draw a three, get it free" lunch promotion. After each customer orders, a touchscreen display shows the message "Press here to win a free lunch." A computer program then simulates one card being drawn from a standard deck. If the chosen card is a 3, the customer's order is free. Otherwise, the customer must pay the bill. Let X = the number of customers it takes to get the first free order on a given day.

(a) Calculate and interpret the mean of X.

(b) Calculate and interpret the standard deviation of X.

**SOLUTION:**

*X* has a geometric distribution with $p = 4/52$.

(a) $\mu_X = \frac{1}{(4/52)} = 13$

If the restaurant runs this lunch promotion on many days, the average number of customers it takes to get the first free order would be about 13.

$\mu_X = \frac{1}{p}$

(b) $\sigma_X = \frac{\sqrt{1-4/52}}{4/52} = 12.49$

If the restaurant runs this lunch promotion on many days, the number of customers it takes to get the first free order would typically vary by about 12.5 from the mean of 13.

$\sigma_X = \frac{\sqrt{1-p}}{p}$

FOR PRACTICE, TRY EXERCISE 111

## CHECK YOUR UNDERSTANDING

Suppose you roll a pair of fair, six-sided dice until you get doubles. Let $T$ = the number of rolls it takes. Note that the probability of getting doubles on any roll is $6/36 = 1/6$.

1. Show that $T$ is a geometric random variable.
2. Find $P(T = 3)$. Interpret this result.
3. In the game of Monopoly, a player can get out of jail free by rolling doubles within 3 turns. Find the probability that this happens.
4. Calculate the mean and standard deviation of $T$. Interpret these parameters.

# Section 6.3 Summary

- A **binomial setting** arises when we perform $n$ independent trials of the same random process and count the number of times that a particular outcome (a "success") occurs. The conditions for a binomial setting are:
  - Binary? The possible outcomes of each trial can be classified as "success" or "failure."
  - Independent? Trials must be independent. That is, knowing the result of one trial must not tell us anything about the result of any other trial.
  - Number? The number of trials $n$ of the random process must be fixed in advance.
  - Same probability? There is the same probability of success $p$ on each trial.

  Remember to check the BINS!
- The count of successes X in a binomial setting is a special type of discrete random variable known as a **binomial random variable.** Its probability distribution is a **binomial distribution.** Any binomial distribution is completely specified by two numbers: the number of trials $n$ of the random process and the probability of success $p$ on any trial. The possible values of $X$ are the whole numbers 0, 1, 2, . . . , $n$.
- Use the binomial probability formula to calculate the probability of getting exactly $x$ successes in $n$ trials:

$$P(X = x) = \binom{n}{x} p^x (1 - p)^{n-x}$$

  - The **binomial coefficient**

$$\binom{n}{x} = \frac{n!}{x!(n-x)!}$$

    counts the number of ways $x$ successes can be arranged among $n$ trials.
  - The factorial of $n$ is

$$n! = n(n-1)(n-2) \cdot \cdots \cdot (3)(2)(1)$$

    for positive whole numbers $n$, and $0! = 1$.
- You can also use technology to calculate binomial probabilities. The TI-83/84 command binompdf(n,p,x) computes $P(X = x)$. The TI-83/84 command binomcdf(n,p,x) computes the cumulative probability $P(X \leq x)$.
- A binomial distribution can have a shape that is roughly symmetric, skewed to the right, or skewed to the left.
- The mean and standard deviation of a binomial random variable X are

$$\mu_X = np \text{ and } \sigma_X = \sqrt{np(1-p)}$$

- The binomial distribution with $n$ trials and probability $p$ of success gives a good approximation to the count of successes in a random sample of size $n$ from a large population containing proportion $p$ of successes. This is true as long as the sample size $n$ is less than 10% of the population size N (the **10% condition**). When the 10% condition is met, we can view individual observations as independent.

- The Normal approximation to the binomial distribution says that if X is a count of successes having the binomial distribution with $n$ trials and success probability $p$, then when $n$ is large, X is approximately Normally distributed. You can use this approximation when $np \geq 10$ and $n(1-p) \geq 10$ (the **Large Counts condition**).
- A **geometric setting** consists of repeated trials of the same random process in which the probability $p$ of success is the same on each trial, and the goal is to count the number of trials it takes to get one success. If X = the number of trials required to obtain the first success, then X is a **geometric random variable.** Its probability distribution is called a **geometric distribution.**
- If X has the geometric distribution with probability of success $p$, the possible values of X are the positive integers 1, 2, 3, . . . . The probability that it takes exactly $x$ trials to get the first success is given by

$$P(X = x) = (1-p)^{x-1}p$$

- You can also use technology to calculate geometric probabilities. The TI-83/84 command geometpdf(p,x) computes $P(X = x)$. The TI-83/84 command geometcdf(p,x) computes the cumulative probability $P(X \leq x)$.
- The mean and standard deviation of a geometric random variable X are

$$\mu_X = \frac{1}{p} \text{ and } \sigma_X = \frac{\sqrt{1-p}}{p}$$

## 6.3 Technology Corners

*TI-Nspire and other technology instructions are on the book's website at highschool.bfwpub.com/updatedtps6e.*



## Section 6.3 Exercises

*In Exercises 77–80, determine whether the given scenario describes a binomial setting. Justify your answer.*

77. **Baby elk** Biologists estimate that a randomly selected baby elk has a 44% chance of surviving to adulthood. Assume this estimate is correct. Suppose researchers choose 7 baby elk at random to monitor. Let X = the number that survive to adulthood. pg 433

78. **Long or short?** Put the names of all the students in your statistics class in a hat. Mix up the names, and draw 4 without looking. Let X = the number whose last names have more than six letters.

79. **Bull's-eye!** Lawrence likes to shoot a bow and arrow in his free time. On any shot, he has about a 10% chance of hitting the bull's-eye. As a challenge one day, Lawrence decides to keep shooting until he gets a bull's-eye. Let Y = the number of shots he takes.

80. **Taking the train** According to New Jersey Transit, the 8:00 A.M. weekday train from Princeton to New York City has a 90% chance of arriving on time on a randomly selected day. Suppose this claim is true. Choose 6 days at random. Let Y = the number of days on which the train arrives on time.

81. **Baby elk** Refer to Exercise 77. Use the binomial probability formula to find $P(X = 4)$. Interpret this value.
pg 437

82. **Taking the train** Refer to Exercise 80. Use the binomial probability formula to find $P(Y = 4)$. Interpret this value.

83. **Take a spin** An online spinner has two colored regions—blue and yellow. According to the website, the probability that the spinner lands in the blue region on any spin is 0.80. Assume for now that this claim is correct. Suppose we spin the spinner 12 times and let $X =$ the number of times it lands in the blue region.

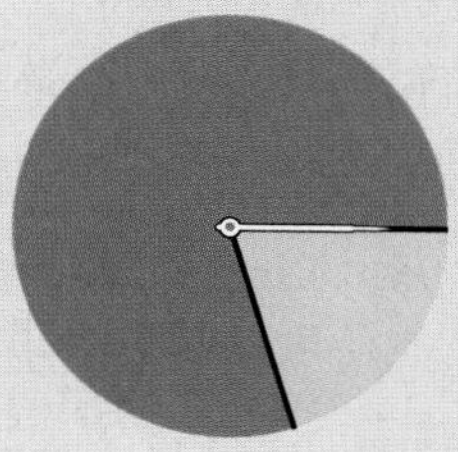

(a) Explain why $X$ is a binomial random variable.

(b) Find the probability that exactly 8 spins land in the blue region.

84. **Red light!** Pedro drives the same route to work on Monday through Friday. His route includes one traffic light. According to the local traffic department, there is a 55% chance that the light will be red on a randomly selected work day. Suppose we choose 10 of Pedro's work days at random and let $Y =$ the number of times that the light is red.

(a) Explain why $Y$ is a binomial random variable.

(b) Find the probability that the light is red on exactly 7 days.

85. **Baby elk** Refer to Exercise 77. How surprising would it be for more than 4 elk in the sample to survive to adulthood? Calculate an appropriate probability to support your answer.
pg 437

86. **Taking the train** Refer to Exercise 80. Would you be surprised if the train arrived on time on fewer than 4 days? Calculate an appropriate probability to support your answer.

87. **Take a spin** Refer to Exercise 83. Calculate and interpret $P(X \leq 7)$.

88. **Red light!** Refer to Exercise 84. Calculate and interpret $P(Y \geq 7)$.

89. **The last kiss** Do people have a preference for the last thing they taste? Researchers at the University of Michigan designed a study to find out. The researchers gave 22 students five different Hershey's Kisses (milk chocolate, dark chocolate, crème, caramel, and almond) in random order and asked the student to rate each one. Participants were not told how many Kisses they would be tasting. However, when the 5th and final Kiss was presented, participants were told that it would be their last one.[9] Assume that the participants in the study don't have a special preference for the last thing they taste. That is, assume that the probability a person would prefer the last Kiss tasted is $p = 0.20$.
pg 439

(a) Find the probability that 14 or more students would prefer the last Kiss tasted.

(b) Of the 22 students, 14 gave the final Kiss the highest rating. Does this give convincing evidence that the participants have a preference for the last thing they taste?

90. **1 in 6 wins** As a special promotion for its 20-ounce bottles of soda, a soft drink company printed a message on the inside of each bottle cap. Some of the caps said, "Please try again!" while others said, "You're a winner!" The company advertised the promotion with the slogan "1 in 6 wins a prize." Grayson's statistics class wonders if the company's claim holds true at a nearby convenience store. To find out, all 30 students in the class go to the store and each buys one 20-ounce bottle of the soda.

(a) Find the probability that two or fewer students would win a prize if the company's claim is true.

(b) Two of the students in Grayson's class got caps that say, "You're a winner!" Does this result give convincing evidence that the company's 1-in-6 claim is false?

91. **Bag check** Thousands of travelers pass through the airport in Guadalajara, Mexico, each day. Before leaving the airport, each passenger must go through the customs inspection area. Customs agents want to be sure that passengers do not bring illegal items into the country. But they do not have time to search every traveler's luggage. Instead, they require each person to press a button. Either a red or a green bulb lights up. If the red light flashes, the passenger will be searched by customs agents. A green light means "go ahead." Customs agents claim that the light has probability 0.30 of showing red on any push of the button. Assume for now that this claim is true. Suppose we watch 20 passengers press the button. Let $R =$ the number who get a red light. Here is a histogram of the probability distribution of $R$:
pg 442

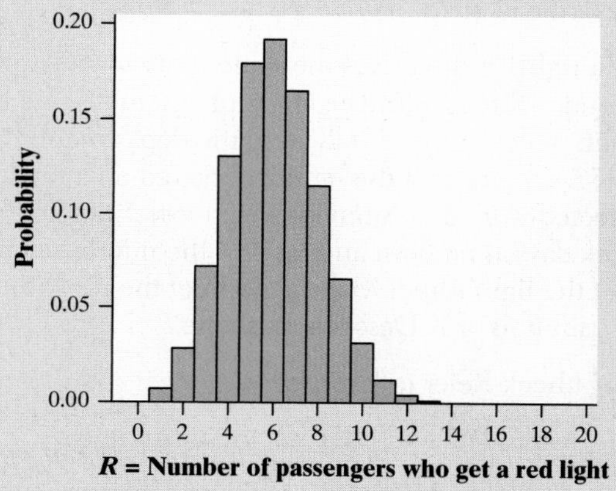

(a) What probability distribution does $R$ have? Justify your answer.

(b) Describe the shape of the probability distribution.

92. **Easy-start mower?** A company has developed an "easy-start" mower that cranks the engine with the push of a button. The company claims that the probability the mower will start on any push of the button is 0.9. Assume for now that this claim is true. On the next 30 uses of the mower, let $T$ = the number of times it starts on the first push of the button. Here is a histogram of the probability distribution of $T$:

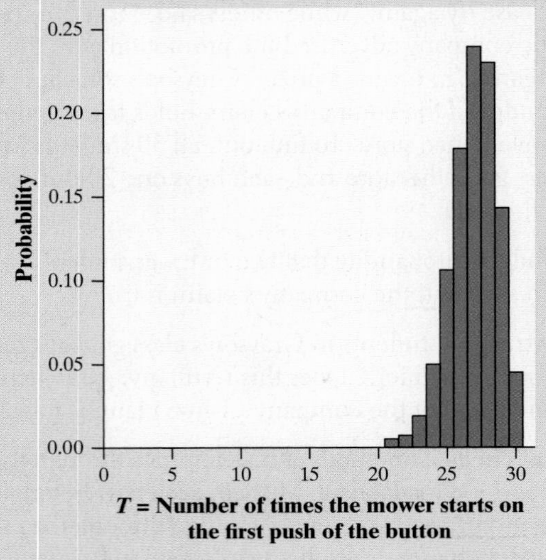


(a) What probability distribution does $T$ have? Justify your answer.

(b) Describe the shape of the probability distribution.

93. **Take a spin** An online spinner has two colored regions—blue and yellow. According to the website, the probability that the spinner lands in the blue region on any spin is 0.80. Assume for now that this claim is correct. Suppose we spin the spinner 12 times and let $X$ = the number of times it lands in the blue region. Make a graph of the probability distribution of $X$. Describe its shape.

94. **Red light!** Pedro drives the same route to work on Monday through Friday. His route includes one traffic light. According to the local traffic department, there is a 55% chance that the light will be red on a randomly selected work day. Suppose we choose 10 of Pedro's work days at random and let $Y$ = the number of times that the light is red. Make a graph of the probability distribution of $Y$. Describe its shape.

95. **Bag check** Refer to Exercise 91.

pg 444

(a) Calculate and interpret the mean of $R$.

(b) Calculate and interpret the standard deviation of $R$.

96. **Easy-start mower** Refer to Exercise 92.

(a) Calculate and interpret the mean of $T$.

(b) Calculate and interpret the standard deviation of $T$.

97. **Random digit dialing** When a polling company calls a telephone number at random, there is only a 9% chance that the call reaches a live person and the survey is successfully completed.[10] Suppose the random digit dialing machine makes 15 calls. Let $X$ = the number of calls that result in a completed survey.

(a) Find the probability that more than 12 calls are *not* completed.

(b) Calculate and interpret $\mu_X$.

(c) Calculate and interpret $\sigma_X$.

98. **Lie detectors** A federal report finds that lie detector tests given to truthful persons have probability 0.2 of suggesting that the person is deceptive.[11] A company asks 12 job applicants about thefts from previous employers, using a lie detector to assess their truthfulness. Suppose that all 12 answer truthfully. Let $Y$ = the number of people whom the lie detector indicates are being deceptive.

(a) Find the probability that the lie detector indicates that at least 10 of the people are being *honest*.

(b) Calculate and interpret $\mu_Y$.

(c) Calculate and interpret $\sigma_Y$.

99. **Lefties** A total of 11% of students at a large high school are left-handed. A statistics teacher selects a random sample of 100 students and records $L$ = the number of left-handed students in the sample.

pg 447

(a) Explain why $L$ can be modeled by a binomial distribution even though the sample was selected without replacement.

(b) Use a binomial distribution to estimate the probability that 15 or more students in the sample are left-handed.

100. **In debt?** According to financial records, 24% of U.S. adults have more debt on their credit cards than they have money in their savings accounts. Suppose that we take a random sample of 100 U.S. adults. Let $D$ = the number of adults in the sample with more debt than savings.

(a) Explain why $D$ can be modeled by a binomial distribution even though the sample was selected without replacement.

(b) Use a binomial distribution to estimate the probability that 30 or more adults in the sample have more debt than savings.

101. **Airport security** The Transportation Security Administration (TSA) is responsible for airport safety. On some flights, TSA officers randomly select passengers for an extra security check before boarding. One such flight had 76 passengers—12 in first class and 64 in coach class. Some passengers were surprised when none of the 10 passengers chosen for screening were seated in first class. Should we use a binomial distribution to approximate this probability? Justify your answer.

102. **Scrabble** In the game of Scrabble, each player begins by drawing 7 tiles from a bag containing 100 tiles. There are 42 vowels, 56 consonants, and 2 blank tiles in the bag. Cait chooses her 7 tiles and is surprised to discover that all of them are vowels. Should we use a binomial distribution to approximate this probability? Justify your answer.

103. **Lefties** Refer to Exercise 99.

pg 449

(a) Justify why $L$ can be approximated by a Normal distribution.

(b) Use a Normal distribution to estimate the probability that 15 or more students in the sample are left-handed.

104. **In debt?** Refer to Exercise 100.

(a) Justify why $D$ can be approximated by a Normal distribution.

(b) Use a Normal distribution to estimate the probability that 30 or more adults in the sample have more debt than savings.

105. **10% condition** To use a binomial distribution to approximate the count of successes in an SRS, why do we require that the sample size $n$ be less than 10% of the population size $N$?

106. **Large Counts condition** To use a Normal distribution to approximate binomial probabilities, why do we require that both $np$ and $n(1-p)$ be at least 10?

107. **Cranky mower** To start her old lawn mower, Rita has to pull a cord and hope for some luck. On any particular pull, the mower has a 20% chance of starting.

pg 451

(a) Find the probability that it takes her exactly 3 pulls to start the mower.

(b) Find the probability that it takes her more than 6 pulls to start the mower.

108. **1-in-6 wins** Alan decides to use a different strategy for the 1-in-6 wins game of Exercise 90. He keeps buying one 20-ounce bottle of the soda at a time until he gets a winner.

(a) Find the probability that he buys exactly 5 bottles.

(b) Find the probability that he buys at most 6 bottles. Show your work.

109. **Geometric or not?** Determine whether each of the following scenarios describes a geometric setting. If so, define an appropriate geometric random variable.

(a) A popular brand of cereal puts a card bearing the image of 1 of 5 famous NASCAR drivers in each box. There is a 1/5 chance that any particular driver's card ends up in any box of cereal. Buy boxes of the cereal until you have all 5 drivers' cards.

(b) In a game of 4-Spot Keno, Lola picks 4 numbers from 1 to 80. The casino randomly selects 20 winning numbers from 1 to 80. Lola wins money if she picks 2 or more of the winning numbers. The probability that this happens is 0.259. Lola decides to keep playing games of 4-Spot Keno until she wins some money.

110. **Geometric or not?** Determine whether each of the following scenarios describes a geometric setting. If so, define an appropriate geometric random variable.

(a) Shuffle a standard deck of playing cards well. Then turn over one card at a time from the top of the deck until you get an ace.

(b) Billy likes to play cornhole in his free time. On any toss, he has about a 20% chance of getting a bag into the hole. As a challenge one day, Billy decides to keep tossing bags until he gets one in the hole.

111. **Using Benford's law** According to Benford's law (Exercise 9, page 404), the probability that the first digit of the amount of a randomly chosen invoice is an 8 or a 9 is 0.097. Suppose you examine randomly selected invoices from a vendor until you find one whose amount begins with an 8 or a 9. Let $X$ = the number of invoices examined.

pg 454

(a) Calculate and interpret the mean of $X$.

(b) Calculate and interpret the standard deviation of $X$.

112. **Roulette** Marti decides to keep placing a $1 bet on number 15 in consecutive spins of a roulette wheel until she wins. On any spin, there's a 1-in-38 chance that the ball will land in the 15 slot. Let $Y$ = the number of spins it takes for Marti to win.

(a) Calculate and interpret the mean of $Y$.

(b) Calculate and interpret the standard deviation of $Y$.

113. **Using Benford's Law** Refer to Exercise 111. Would you be surprised if it took 40 or more invoices to find the first one with an amount that starts with an 8 or 9? Calculate an appropriate probability to support your answer.

114. **Roulette** Refer to Exercise 112. Would you be surprised if Marti won in 3 or fewer spins? Compute an appropriate probability to support your answer.

**Multiple Choice:** *Select the best answer for Exercises 115–119.*

**115.** Joe reads that 1 out of 4 eggs contains salmonella bacteria. So he never uses more than 3 eggs in cooking. If eggs do or don't contain salmonella independently of each other, the number of contaminated eggs when Joe uses 3 eggs chosen at random has the following distribution:

(a) binomial; $n = 4$ and $p = 1/4$

(b) binomial; $n = 3$ and $p = 1/4$

(c) binomial; $n = 3$ and $p = 1/3$

(d) geometric; $p = 1/4$

(e) geometric; $p = 1/3$

*Exercises 116 and 117 refer to the following setting.* A fast-food restaurant runs a promotion in which certain food items come with game pieces. According to the restaurant, 1 in 4 game pieces is a winner.

**116.** If Jeff gets 4 game pieces, what is the probability that he wins exactly 1 prize?

(a) 0.25

(b) 1.00

(c) $\binom{4}{1}(0.25)^1(0.75)^3$

(d) $\binom{4}{1}(0.25)^3(0.75)^1$

(e) $(0.75)^3(0.25)^1$

**117.** If Jeff keeps playing until he wins a prize, what is the probability that he has to play the game exactly 5 times?

(a) $(0.25)^5$

(b) $(0.75)^4$

(c) $(0.75)^5$

(d) $(0.75)^4(0.25)$

(e) $\binom{5}{1}(0.75)^4(0.25)$

**118.** Each entry in a table of random digits like Table D has probability 0.1 of being a 0, and the digits are independent of one another. Each line of Table D contains 40 random digits. The mean and standard deviation of the number of 0s in a randomly selected line will be approximately

(a) mean = 0.1, standard deviation = 0.05.

(b) mean = 0.1, standard deviation = 0.1.

(c) mean = 4, standard deviation = 0.05.

(d) mean = 4, standard deviation = 1.90.

(e) mean = 4, standard deviation = 3.60.

**119.** In which of the following situations would it be appropriate to use a Normal distribution to approximate probabilities for a binomial distribution with the given values of $n$ and $p$?

(a) $n = 10, p = 0.5$

(b) $n = 40, p = 0.88$

(c) $n = 100, p = 0.2$

(d) $n = 100, p = 0.99$

(e) $n = 1000, p = 0.003$

## Recycle and Review

**120. Spoofing** (4.2) To collect information such as passwords, online criminals use "spoofing" to direct Internet users to fraudulent websites. In one study of Internet fraud, students were warned about spoofing and then asked to log into their university account starting from the university's home page. In some cases, the log-in link led to the genuine dialog box. In others, the box looked genuine but, in fact, was linked to a different site that recorded the ID and password the student entered. The box that appeared for each student was determined at random. An alert student could detect the fraud by looking at the true Internet address displayed in the browser status bar, but most just entered their ID and password.

(a) Is this an observational study or an experiment? Justify your answer.

(b) What are the explanatory and response variables? Identify each variable as categorical or quantitative.

**121. Standard deviations** (6.1) Continuous random variables A, B, and C all take values between 0 and 10. Their density curves, drawn on the same horizontal scales, are shown here. Rank the standard deviations of the three random variables from smallest to largest. Justify your answer.

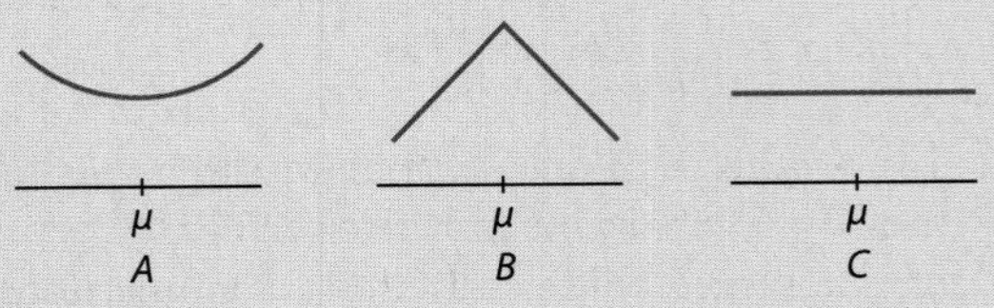

# Chapter 6 Wrap-Up

## FRAPPY! FREE RESPONSE AP® PROBLEM, YAY!

The following problem is modeled after actual AP® Statistics exam free response questions. Your task is to generate a complete, concise response in 15 minutes.

*Directions: Show all your work. Indicate clearly the methods you use, because you will be scored on the correctness of your methods as well as on the accuracy and completeness of your results and explanations.*

Buckley Farms produces homemade potato chips that it sells in bags labeled "16 ounces." The total weight of each bag follows an approximately Normal distribution with a mean of 16.15 ounces and a standard deviation of 0.12 ounce.

(a) If you randomly selected 1 bag of these chips, what is the probability that the total weight is less than 16 ounces?

(b) If you randomly selected 10 bags of these chips, what is the probability that exactly 2 of the bags will have a total weight less than 16 ounces?

(c) Buckley Farms ships its chips in boxes that contain 6 bags. The empty boxes have a mean weight of 10 ounces and a standard deviation of 0.05 ounce. Calculate the mean and standard deviation of the total weight of a box containing 6 bags of chips.

(d) Buckley Farms decides to increase the mean weight of each bag of chips so that only 5% of the bags have weights that are less than 16 ounces. Assuming that the standard deviation remains 0.12 ounce, what mean weight should Buckley Farms use?

After you finish, you can view two example solutions on the book's website (highschool.bfwpub.com/updatedtps6e). Determine whether you think each solution is "complete," "substantial," "developing," or "minimal." If the solution is not complete, what improvements would you suggest to the student who wrote it? Finally, your teacher will provide you with a scoring rubric. Score your response and note what, if anything, you would do differently to improve your own score.

# Chapter 6 Review

### Section 6.1: Discrete and Continuous Random Variables

A random variable assigns numerical values to the outcomes of a random process. The probability distribution of a random variable describes its possible values and their probabilities. There are two types of random variables: discrete and continuous. Discrete random variables take on a fixed set of values with gaps in between. Continuous random variables can take on any value in an interval of numbers.

As in Chapter 1, we are often interested in the shape, center, and variability of a probability distribution. The shape of a discrete probability distribution can be identified by graphing a probability histogram, with the height of each bar representing the probability of a single value. The center is usually identified by the mean (expected value) of the random variable. The mean (expected value) is the average value of the random variable if the random process is repeated many times. The variability of a probability distribution is usually identified by the standard deviation, which describes how much the values of a random variable typically vary from the mean value, in many trials of the random process.

The probability distribution of a continuous random variable is described by a density curve. Probabilities for continuous random variables are determined by finding the area under the density curve and above the values of interest.

### Section 6.2: Transforming and Combining Random Variables

In this section, you learned how linear transformations of a random variable affect the shape, center, and variability of its probability distribution. Similar to what you learned in Chapter 2, adding a positive constant to (or subtracting it from) each value of a random variable changes the measures of center and location, but not the shape or variability of the probability distribution. Multiplying or dividing each value of a random variable by a positive constant changes the measures of center and location and measures of variability, but not the shape of the probability distribution. A linear transformation $Y = a + bX$ (with $b > 0$) does not change the shape of the probability distribution. The mean and standard deviation of Y are:

$$\mu_Y = a + b\mu_X \quad \text{and} \quad \sigma_Y = b\sigma_X$$

You also learned how to calculate the mean and standard deviation for a combination of two or more random variables. The mean of a sum or difference of any two random variables X and Y is given by

$$\mu_{X+Y} = \mu_X + \mu_Y \quad \text{and} \quad \mu_{X-Y} = \mu_X - \mu_Y$$

If X and Y are any two *independent* random variables, variances add:

$$\sigma^2_{X+Y} = \sigma^2_X + \sigma^2_Y \quad \text{and} \quad \sigma^2_{X-Y} = \sigma^2_X + \sigma^2_Y$$

To find the standard deviation, just take the square root of the variance. Recall that X and Y are independent if knowing the value of one variable does not change the probability distribution of the other variable. Also, if independent random variables X and Y are both Normally distributed, then their sum $X + Y$ and difference $X - Y$ are both Normally distributed as well.

The linear combination $aX + bY$ has mean $a\mu_X + b\mu_Y$. Its standard deviation is $\sqrt{a^2\sigma^2_X + b^2\sigma^2_Y}$ if X and Y are independent.

### Section 6.3: Binomial and Geometric Random Variables

In this section, you learned about two common types of discrete random variables, binomial random variables and geometric random variables. Binomial random variables count the number of successes in a fixed number of trials ($n$) of the same random process, whereas geometric random variables count the number of trials needed to get one success. Otherwise, the binomial and geometric settings have the same conditions: there must be two possible outcomes for each trial (success or failure), the trials must be independent, and the probability of success $p$ must stay the same throughout all trials.

To calculate probabilities for a binomial distribution with $n$ trials and probability of success $p$ on each trial, use technology or the binomial probability formula

$$P(X = x) = \binom{n}{x} p^x (1-p)^{n-x}$$

The mean and standard deviation of a binomial random variable X are

$$\mu_X = np \quad \text{and} \quad \sigma_X = \sqrt{np(1-p)}$$

The shape of a binomial distribution depends on both the number of trials $n$ and the probability of success $p$. When the number of trials is large enough that both $np$ and $n(1-p)$ are at least 10, the probability distribution of the binomial random variable X can be modeled with a Normal density curve. Be sure to check the Large Counts condition before using a Normal approximation to a binomial distribution.

A common application of the binomial distribution is when we count the number of times a particular outcome occurs in a random sample from some population. Because sampling is almost always done without replacement, the Independent condition is violated. However, if the sample size is a small fraction of the population size (less than 10%), we can view individual observations as independent. Be sure to check the 10% condition when sampling is done without replacement before using a binomial distribution.

Finally, to calculate probabilities for a geometric distribution with probability of success $p$ on each trial, use technology or the geometric probability formula

$$P(X = x) = (1-p)^{x-1} p$$

A geometric distribution is always skewed to the right and unimodal, with a single peak at 1. The mean and standard deviation of a geometric random variable X are

$$\mu_X = 1/p \quad \text{and} \quad \sigma_X = \frac{\sqrt{1-p}}{p}$$

## What Did You Learn?

| Learning Target | Section | Related Example on Page(s) | Relevant Chapter Review Exercise(s) |
|---|---|---|---|
| Use the probability distribution of a discrete random variable to calculate the probability of an event. | 6.1 | 391 | R6.1 |
| Make a histogram to display the probability distribution of a discrete random variable and describe its shape. | 6.1 | 393 | R6.3 |
| Calculate and interpret the mean (expected value) of a discrete random variable. | 6.1 | 395 | R6.1, R6.3 |
| Calculate and interpret the standard deviation of a discrete random variable. | 6.1 | 397 | R6.1, R6.3 |

| Learning Target | Section | Related Example on Page(s) | Relevant Chapter Review Exercise(s) |
|---|---|---|---|
| Use the probability distribution of a continuous random variable (uniform or Normal) to calculate the probability of an event. | 6.1 | 400, 401 | R6.4 |
| Describe the effect of adding or subtracting a constant or multiplying or dividing by a constant on the probability distribution of a random variable. | 6.2 | 412, 413, 415 | R6.2, R6.3 |
| Calculate the mean and standard deviation of the sum or difference of random variables. | 6.2 | 417, 421 | R6.3, R6.4 |
| Find probabilities involving the sum or difference of independent Normal random variables. | 6.2 | 423 | R6.4 |
| Determine whether the conditions for a binomial setting are met. | 6.3 | 433, 442 | R6.5 |
| Calculate and interpret probabilities involving binomial random variables. | 6.3 | 437, 439, 447 | R6.5 |
| Calculate the mean and standard deviation of a binomial distribution. Interpret these values. | 6.3 | 444 | R6.6 |
| When appropriate, use the Normal approximation to the binomial distribution to calculate probabilities. | 6.3 | 449 | R6.8 |
| Calculate and interpret probabilities involving geometric random variables. | 6.3 | 451 | R6.7 |
| Calculate the mean and standard deviation of a geometric random variable. Interpret these values. | 6.3 | 454 | R6.7 |

# Chapter 6 Review Exercises

*These exercises are designed to help you review the important ideas and methods of the chapter.*

**R6.1 Knees** Patients receiving artificial knees often experience pain after surgery. The pain is measured on a subjective scale with possible values of 1 (low) to 5 (high). Let Y be the pain score for a randomly selected patient. The following table gives the probability distribution for Y.

| Value | 1 | 2 | 3 | 4 | 5 |
|---|---|---|---|---|---|
| Probability | 0.1 | 0.2 | 0.3 | 0.3 | ?? |

(a) Find $P(Y = 5)$. Interpret this value.

(b) Find the probability that a randomly selected patient has a pain score of at most 2.

(c) Calculate the expected pain score and the standard deviation of the pain score.

**R6.2 A glass act** In a process for manufacturing glassware, glass stems are sealed by heating them in a flame. Let X be the temperature (in degrees Celsius) for a randomly chosen glass. The mean and standard deviation of X are $\mu_X = 550°\text{C}$ and $\sigma_X = 5.7°\text{C}$.

(a) Is temperature a discrete or continuous random variable? Explain your answer.

(b) The target temperature is 550°C. What are the mean and standard deviation of the number of degrees off target, $D = X - 550$?

(c) A manager asks for results in degrees Fahrenheit. The conversion of X into degrees Fahrenheit is given by $Y = \frac{9}{5}X + 32$. What are the mean $\mu_Y$ and the standard deviation $\sigma_Y$ of the temperature of the flame in the Fahrenheit scale?

**R6.3 Keno** In a game of 4-Spot Keno, the player picks 4 numbers from 1 to 80. The casino randomly selects 20 winning numbers from 1 to 80. The table shows the possible outcomes of the game and their probabilities, along with the amount of money (Payout) that the player wins for a \$1 bet. If X = the payout for a single \$1 bet, you can check that $\mu_X = \$0.70$ and $\sigma_X = \$6.58$.

| Matches | 0 | 1 | 2 | 3 | 4 |
|---|---|---|---|---|---|
| Payout $x_i$ | \$0 | \$0 | \$1 | \$3 | \$120 |
| Probability $p_i$ | 0.308 | 0.433 | 0.213 | 0.043 | 0.003 |

(a) Make a graph of the probability distribution. Describe what you see.

(b) Interpret the values of $\mu_X$ and $\sigma_X$.

(c) Jerry places a single \$5 bet on 4-Spot Keno. Find the expected value and the standard deviation of his winnings.

(d) Marla plays five games of 4-Spot Keno, betting \$1 each time. Find the expected value and the standard deviation of her total winnings.

**R6.4 Applying torque** A machine fastens plastic screw-on caps onto containers of motor oil. If the machine applies more torque than the cap can withstand, the cap will break. Both the torque applied and the strength of the caps vary. The capping-machine torque $T$ follows a Normal distribution with mean 7 inch-pounds and standard deviation 0.9 inch-pound. The cap strength $C$ (the torque that would break the cap) follows a Normal distribution with mean 10 inch-pounds and standard deviation 1.2 inch-pounds.

(a) Find the probability that a randomly selected cap has a strength greater than 11 inch-pounds.

(b) Explain why it is reasonable to assume that the cap strength and the torque applied by the machine are independent.

(c) Let the random variable $D = C - T$. Find its mean and standard deviation.

(d) What is the probability that a randomly selected cap will break while being fastened by the machine?

*Exercises R6.5 and R6.6 refer to the following setting.* According to Mars, Incorporated, 20.5% of the M&M'S® Milk Chocolate Candies made at its Cleveland factory are orange. Assume that the company's claim is true. Suppose you take a random sample of 8 candies from a large bag of M&M'S. Let $X$ = the number of orange candies you get.

**R6.5 Orange M&M'S**

(a) Explain why it is reasonable to use the binomial distribution for probability calculations involving X.

(b) What's the probability that you get 3 orange M&M'S?

(c) Calculate $P(X \geq 4)$. Interpret this result.

(d) Suppose that you get 4 orange M&M'S in your sample. Does this result provide convincing evidence that Mars's claim about its M&M'S is false? Justify your answer.

**R6.6 Orange M&M'S**

(a) Find and interpret the expected value of X.

(b) Find and interpret the standard deviation of X.

**R6.7 Sushi Roulette** In the Japanese game show *Sushi Roulette*, the contestant spins a large wheel that's divided into 12 equal sections. Nine of the sections show a sushi roll, and three have a "wasabi bomb." When the wheel stops, the contestant must eat whatever food applies to that section. Then the game show host replaces the item of food on the wheel. To win the game, the contestant must eat one wasabi bomb. Let $Y$ = the number of spins required to get a wasabi bomb.

(a) Find the probability that it takes 3 or fewer spins for the contestant to get a wasabi bomb.

(b) Find the mean of Y. Interpret this parameter.

(c) Show that $\sigma_Y = 3.464$. Interpret this parameter.

**R6.8 Public transportation** In a large city, 34% of residents use public transportation at least once per week. Suppose the city's mayor selects a random sample of 200 residents. Let $T$ = the number in the sample who use public transportation at least once per week.

(a) What type of probability distribution does $T$ have? Justify your answer.

(b) Explain why $T$ can be approximated by a Normal distribution.

(c) Calculate the probability that at most 60 residents in the sample use public transportation at least once per week.

# Chapter 6 AP® Statistics Practice Test

**Section I: Multiple Choice** *Select the best answer for each question.*

*Questions T6.1–T6.3 refer to the following setting.* A psychologist studied the number of puzzles that subjects were able to solve in a 5-minute period while listening to soothing music. Let X be the number of puzzles completed successfully by a randomly chosen subject. The psychologist found that X had the following probability distribution.

| Value | 1 | 2 | 3 | 4 |
|---|---|---|---|---|
| Probability | 0.2 | 0.4 | 0.3 | 0.1 |

**T6.1** What is the probability that a randomly chosen subject completes more than the expected number of puzzles in the 5-minute period while listening to soothing music?

(a) 0.1 (b) 0.4 (c) 0.8
(d) 1 (e) Cannot be determined

**T6.2** The standard deviation of X is 0.9. Which of the following is the best interpretation of this value?

(a) About 90% of subjects solved 3 or fewer puzzles.

(b) About 68% of subjects solved between 0.9 puzzles less and 0.9 puzzles more than the mean.

(c) The typical subject solved an average of 0.9 puzzles.

(d) The number of puzzles solved by subjects typically differed from the mean by about 0.9 puzzles.

(e) The number of puzzles solved by subjects typically differed from one another by about 0.9 puzzles.

**T6.3** Let D be the difference in the number of puzzles solved by two randomly selected subjects in a 5-minute period. What is the standard deviation of D?

(a) 0

(b) 0.81

(c) 0.9

(d) 1.27

(e) 1.8

$\sqrt{\sigma_X^2 + \sigma_Y^2}$  n·p

$\sqrt{0.9^2 + 0.9^2} = .9$

**T6.4** Suppose a student is randomly selected from your school. Which of the following pairs of random variables are most likely independent?

(a) X = student's height; Y = student's weight

(b) X = student's IQ; Y = student's GPA

(c) X = student's PSAT Math score; Y = student's PSAT Verbal score

(d) X = average amount of homework the student does per night; Y = student's GPA

(e) X = average amount of homework the student does per night; Y = student's height

**T6.5** A certain vending machine offers 20-ounce bottles of soda for $1.50. The number of bottles X bought from the machine on any day is a random variable with mean 50 and standard deviation 15. Let the random variable Y equal the total revenue from this machine on a randomly selected day. Assume that the machine works properly and that no sodas are stolen from the machine. What are the mean and standard deviation of Y?

(a) $\mu_Y = \$1.50, \sigma_Y = \$22.50$

(b) $\mu_Y = \$1.50, \sigma_Y = \$33.75$

(c) $\mu_Y = \$75, \sigma_Y = \$18.37$

(d) $\mu_Y = \$75, \sigma_Y = \$22.50$

(e) $\mu_Y = \$75, \sigma_Y = \$33.75$

1.5 · 50

15 · 1.5

**T6.6** The weight of tomatoes chosen at random from a bin at the farmer's market follows a Normal distribution with mean $\mu = 10$ ounces and standard deviation $\sigma = 1$ ounce. Suppose we pick four tomatoes at random from the bin and find their total weight T. The random variable T is

$M_T = 10 + 10 + 10 + 10 = 400$

$\sigma_T = \sqrt{1^2 + 1^2 + 1^2 + 1^2} = 2$

(a) Normal, with mean 10 ounces and standard deviation 1 ounce.

(b) Normal, with mean 40 ounces and standard deviation 2 ounces.

(c) Normal, with mean 40 ounces and standard deviation 4 ounces.

(d) binomial, with mean 40 ounces and standard deviation 2 ounces.

(e) binomial, with mean 40 ounces and standard deviation 4 ounces.

**T6.7** Which of the following random variables is geometric?

(a) The number of times I have to roll a single die to get two 6s

(b) The number of cards I deal from a well-shuffled deck of 52 cards to get a heart

(c) The number of digits I read in a randomly selected row of the random digits table to get a 7

(d) The number of 7s in a row of 40 random digits  bin

(e) The number of 6s I get if I roll a die 10 times

**T6.8** Seventeen people have been exposed to a particular disease. Each one independently has a 40% chance of contracting the disease. A hospital has the capacity to handle 10 cases of the disease. What is the probability that the hospital's capacity will be exceeded?

(a) 0.011

(b) 0.035

(c) 0.092

(d) 0.965

(e) 0.989

binomcdf

(17, .40, 10)

X ≥ 10

**T6.9** The figure shows the probability distribution of a discrete random variable X. Which of the following best describes this random variable?

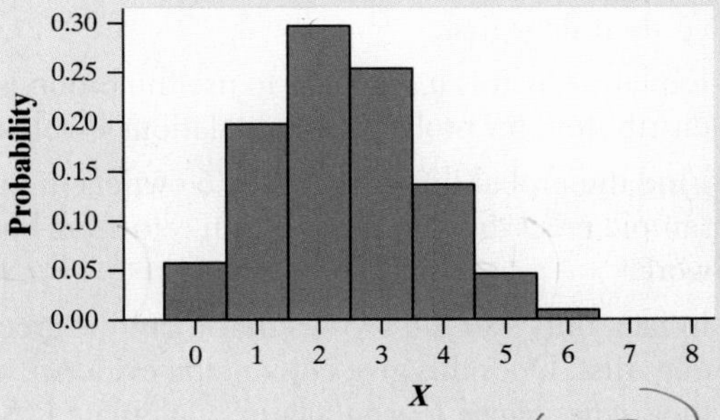

(a) Binomial with $n = 8$, $p = 0.1$

(b) Binomial with $n = 8$, $p = 0.3$

(c) Binomial with $n = 8$, $p = 0.8$

(d) Geometric with $p = 0.1$

(e) Geometric with $p = 0.2$

8(.3)

n·p = 2.4

center ~ 2

**T6.10** A test for extrasensory perception (ESP) involves asking a person to tell which of 5 shapes—a circle, star, triangle, diamond, or heart—appears on a hidden computer screen. On each trial, the computer is equally likely to select any of the 5 shapes. Suppose researchers are testing a person who does not have ESP and so is just guessing on each trial. What is the probability that the person guesses the first 4 shapes incorrectly but gets the fifth one correct?

(a) $1/5$

(b) $\left(\frac{4}{5}\right)^4$

(c) $\left(\frac{4}{5}\right)^4 \cdot \left(\frac{1}{5}\right)$

(d) $\binom{5}{1} \cdot \left(\frac{4}{5}\right)^4 \cdot \left(\frac{1}{5}\right)$

(e) $4/5$

**Section II: Free Response** *Show all your work. Indicate clearly the methods you use, because you will be graded on the correctness of your methods as well as on the accuracy and completeness of your results and explanations.*

**T6.11** Let $Y$ denote the number of broken eggs in a randomly selected carton of one dozen "store brand" eggs at a local supermarket. Suppose that the probability distribution of $Y$ is as follows.

| Value $y_i$ | 0 | 1 | 2 | 3 | 4 |
|---|---|---|---|---|---|
| Probability $p_i$ | 0.78 | 0.11 | 0.07 | 0.03 | 0.01 |

(a) What is the probability that at least 10 eggs in a randomly selected carton are *unbroken*?   0.96

(b) Calculate and interpret $\mu_Y$.   .38

(c) Calculate and interpret $\sigma_Y$.   .82

(d) A quality control inspector at the store keeps looking at randomly selected cartons of eggs until he finds one with at least 2 broken eggs. Find the probability that this happens in one of the first three cartons he inspects.   $(1-P)^{x-1}p$   $0.89^2(.11) = .29$

**T6.12** *Ladies Home Journal* magazine reported that 66% of all dog owners greet their dog before greeting their spouse or children when they return home at the end of the workday. Assume that this claim is true. Suppose 12 dog owners are selected at random. Let $X$ = the number of owners who greet their dogs first.

(a) Explain why it is reasonable to use the binomial distribution for probability calculations involving $X$.

(b) Find the probability that exactly 6 owners in the sample greet their dogs first when returning home from work.   $\binom{12}{6}(.66)^6(.34)^6 = .1179$

(c) In fact, only 4 of the owners in the sample greeted their dogs first. Does this give convincing evidence against the *Ladies Home Journal* claim? Calculate $P(X \le 4)$ and use the result to support your answer.

$\binom{12}{0}(.66)^0(.34)^{12} + \dots + \binom{12}{4}(.66)^4(.34)^8$
$= .0125$ gives sufficient against claim

**T6.13** Ed and Adelaide attend the same high school but are in different math classes. The time $E$ that it takes Ed to do his math homework follows a Normal distribution with mean 25 minutes and standard deviation 5 minutes. Adelaide's math homework time $A$ follows a Normal distribution with mean 50 minutes and standard deviation 10 minutes. Assume that $E$ and $A$ are independent random variables.

(a) Randomly select one math assignment of Ed's and one math assignment of Adelaide's. Let the random variable $D$ be the difference in the amount of time each student spent on their assignments: $D = A - E$. Find the mean and the standard deviation of $D$.   $\mu = 25$

(b) Find the probability that Ed spent longer on his assignment than Adelaide did on hers.

**T6.14** According to the Census Bureau, 13% of American adults (aged 18 and over) are Hispanic. An opinion poll plans to contact an SRS of 1200 adults.

(a) What is the mean number of Hispanics in such samples? What is the standard deviation?   $n \cdot p$   $1200 \cdot .13$   $156$

(b) Should we be suspicious if the sample selected for the opinion poll contains 10% or less Hispanic people? Calculate an appropriate probability to support your answer.

$\sigma = \sqrt{1200(.13)(.8)}$

11. b) $\mu_Y = 0(.78) + 1(.11) + 2(.07) + 3(.03) + 4(.01)$

☆ $\sqrt{(0-.38)^2(.78) + \ldots + (4-.38)^2(.01)}$

12 a It is reasonable bc

B → greet dog 1st or don't N → out of 12 owners

I → Independent S → success rate 66%

13 a) $50 - 25 = 25$ min

$\sqrt{10^2 - 5^2} = 11.18$ min

☆ b) $P(A-E) > 0$

$z \; \frac{0 - 25}{11.18} = -2.24$ (μ over 25, σ under 11.18)

table 0.0125

14 b) 10% or less is 120 people or less

Distribution is approx normal as long as success/fail $\geq 10$

$np = 156$ ? $n(1-p) = 1044$

$\frac{120 - 156}{11.65} = -3.09$ table .001

Only 0.1% chance, very unlikely

suspicious

# Solutions

## Chapter 1

### Introduction

### Answers to Check Your Understanding

*page 5:* **1.** The individuals are the cars in the student parking lot.
**2.** He recorded the car's license plate to identify the individuals. The variables recorded for each individual are the model (categorical), number of cylinders (quantitative), color (categorical), highway gas mileage (quantitative), weight (quantitative), and whether or not it has a navigation system (categorical).
**3.** Number of cylinders (quantitative/discrete), highway gas mileage and weight (quantitative/continuous).

### Answers to Odd-Numbered Introduction Exercises

I.1 **(a)** The AP® Statistics students who completed a questionnaire on the first day of class. **(b)** gender (categorical), grade level (categorical), GPA (quantitative), children in family (quantitative), homework last night (min) (categorical), and type of phone (categorical).
I.3 The individuals are movies. The variables are year (quantitative), rating (categorical), time (min) (quantitative), genre (categorical), and box office ($) (quantitative). *Note:* Year might be considered categorical if we want to know how many of these movies were made each year, rather than the average year.
I.5 The categorical variables are type of wood, type of water repellent, and paint color. The quantitative variables are paint thickness and weathering time.
I.7 The discrete variables are number of siblings and how many books they have read in the past month. The continuous variables are the distance from their home to campus and how long it took them to complete an online survey.
I.9 b

### Section 1.1

### Answers to Check Your Understanding

*page 12:*

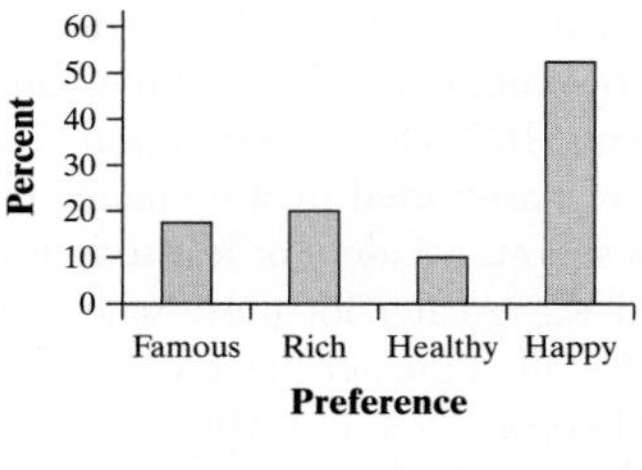


A slight majority (52.5%) of students in the sample said they would prefer to be Happy. Rich and Famous were preferred about equally (20% and 17.5%). The least popular choice was Healthy (10%).
*page 16:* **1.** 113/338 = 0.334. This value makes sense because there were three treatments, so we would expect about one-third of the subjects to be assigned to this treatment.
**2.** The distribution of change in depression is: Full response: 91/338 = 0.269, Partial response: 55/338 = 0.163, and No response: 192/338 = 0.568.
**3.** 27/338 = 0.08 = 8%

*page 22:* **1.** 27/91 = 0.297
**2.** 70/113 = 0.619 = 61.9%
**3.** The distribution of change in depression for the subjects receiving each of the three treatments is:
*St. John's wort* Full response: 27/113 = 0.239; Partial response: 16/113 = 0.142; No response: 70/113 = 0.619
*Zoloft* Full response: 27/109 = 0.248; Partial response: 26/109 = 0.239; No response: 56/109 = 0.514
*Placebo* Full response: 37/116 = 0.319; Partial response: 13/116 = 0.112; No response: 66/116 = 0.569

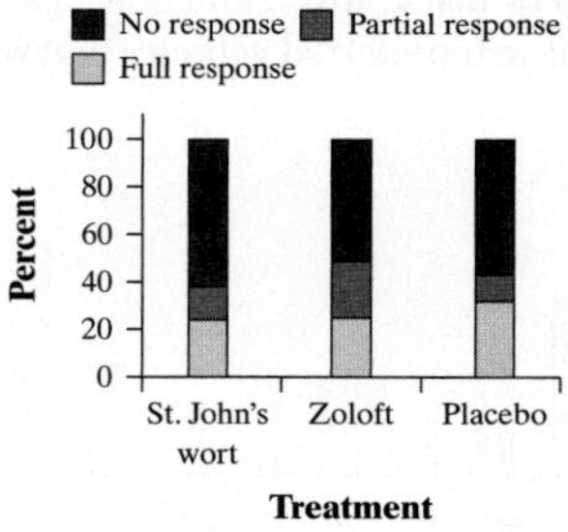


**4.** There does not appear to be a strong association between treatment and change in depression for these subjects because the distribution of response status is very similar for the three different treatments. Also, note that the treatment with the highest rate of "full response" was the placebo.

### Answers to Odd-Numbered Section 1.1 Exercises

1.11 **(a)** The individuals are the babies. **(b)** It appears that births occurred with similar frequencies on weekdays (Monday through Friday), but with noticeably smaller frequencies on the weekend days (Saturday and Sunday).

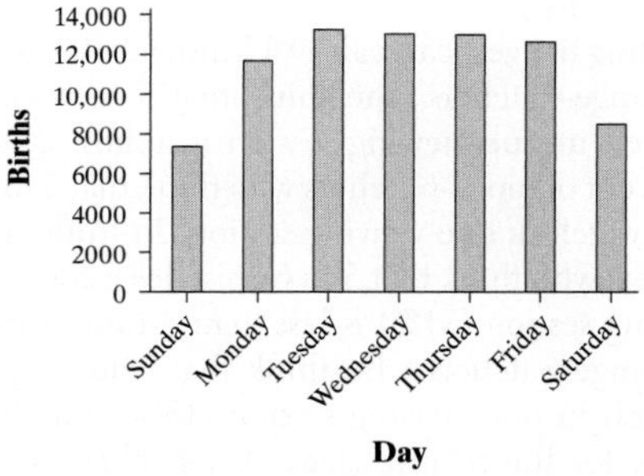


1.13 First, a relative frequency table must be constructed.

| Camera brand | Relative frequency |
|---|---|
| Canon | 23/45 = 51.1% |
| Sony | 6/45 = 13.3% |
| Nikon | 11/45 = 24.4% |
| Fujifilm | 3/45 = 6.7% |
| Olympus | 2/45 = 4.4% |

The relative frequency bar graph is given here.

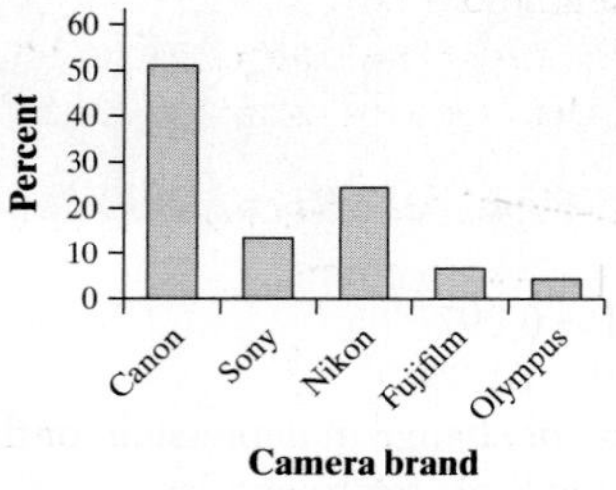


The most popular brand of camera among the 45 most recent purchases on the Internet auction site is Canon, followed by Nikon, Sony, Fujifilm, and Olympus. Canon is the overwhelming favorite with over 50% of the customers purchasing this brand. Also noteworthy is that almost 25% of the customers purchased a Nikon camera.

1.15 **(a)** 2% **(b)** The most popular color of vehicles sold that year was white, followed by black, silver, and gray. It appears that a majority of car buyers that year preferred vehicles that were shades of black and white.

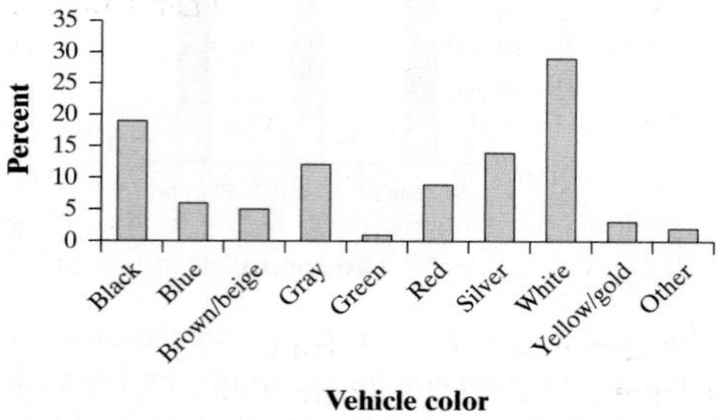


**(c)** It would be appropriate to make a pie chart of these data (including the other category) because the numbers in the table refer to parts of a single whole.

1.17 Estimates will vary, but should be close to 63% Mexican and 9% Puerto Rican.

1.19 The areas of the pictures should be proportional to the numbers of students they represent. As drawn, it appears that most of the students arrived by car, but in reality most came by bus (14 took the bus, 9 came in cars).

1.21 By starting the vertical scale at 12 instead of 0, it looks like the percent of binge-watchers who think that 5 to 6 episodes is too many to watch in one viewing session is almost 20 times higher than the percent of binge-watchers who think that 3 to 4 episodes is too many to watch in one viewing session. In truth, the percent of binge-watchers who think that 5 to 6 episodes is too many to watch in one viewing session (31%) is less than 3 times higher than the percent of binge-watchers who think that 3 to 4 episodes is too many to watch in one viewing session (13%). Similar arguments can be made for the relative sizes of the other categories represented in the bar graph.

1.23 **(a)** $50/150 = 0.333$ **(b)** $29/150 = 19.3\%$ said they saw broken glass at the accident; $121/150 = 80.7\%$ said they did not. **(c)** 10.67%

1.25 **(a)** 71.25% **(b)** 0.633 **(c)** 12.1%

1.27 **(a)** 0.241 **(b)** 68%

1.29 **(a)** The distributions of responses for the three treatment groups are:

| "Smashed into" | "Hit" | Control |
|---|---|---|
| Yes: 16/50 = 32% | Yes: 7/50 = 14% | Yes: 6/50 = 12% |
| No: 34/50 = 68% | No: 43/50 = 86% | No: 44/50 = 88% |

The segmented bar graph is shown here.

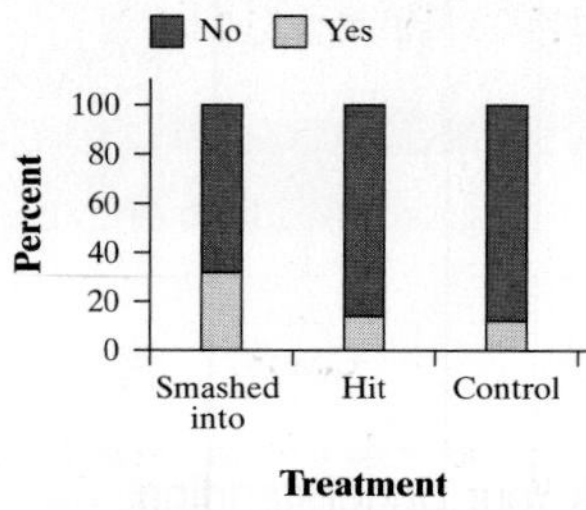


**(b)** The segmented bar graph reveals that there is an association between opinion about broken glass at the accident and treatment received for subjects in the study. Knowing which treatment a subject received helps us predict whether or not that person will respond that he or she saw broken glass at the accident.

1.31 **(a)** 0.655 **(b)** 73.7% **(c)** Based on the mosiac plot, there is an association between perceived body image and gender. Males are about 4 times as likely as females to perceive that they are underweight. Males are also less likely to perceive that they are overweight. However, the overwhelming majority of both genders perceive that their body image is about right.

1.33 Answers may vary. Regardless of whether a student went to a private or public college, most students chose a school that was at least 11 miles from home. Those who went to a public university were most likely to choose a school that was 11 to 50 miles from home (about 30%), while those who went to a private university were most likely to choose a school that was 101 to 500 miles from home (about 29%).

1.35 **(a)** The graph reveals that as age increases, the percent that use smartphones for navigation decreases. **(b)** It would not be appropriate to make a pie chart of these data because the category percentages are not parts of the same whole.

1.37 Answers will vary. Two possible tables are given here.

| | |
|---|---|
| 10 | 40 |
| 50 | 0 |

| | |
|---|---|
| 30 | 20 |
| 30 | 20 |

1.39 **(a)** 64 of the 200 patients transported by helicopter, or 32% of patients died. 260 of the 1100 patients transported by ambulance, or 23.6% of patients died. **(b)** Of the patients who were in serious accidents, 48% who were transported by helicopter died and 60% who were transported by ambulance died. Of the patients who were in less serious accidents, 16% who were transported by helicopter died and 20% who were transported by ambulance died. Whether the patients were in a serious accident or less serious accident, the percentage who died was greater for those who were transported by ambulance. **(c)** Overall, a greater percentage of patients who were transported by helicopter died, but when broken down by seriousness of the accident, in both instances, a greater percentage of patients who were transported by ambulance died. This is because people in more serious accidents were also more likely to be transported by helicopter and were more likely to die. People who were involved in less serious accidents were less likely to be transported by helicopter and were less likely to die. Overall, this makes it appear that those who are transported by helicopters are more likely to die; but when the patients are broken out by seriousness of the accident, we can see that the driving factor for the overall death rates is the seriousness of the accident, not the method of transportation.

1.41 d

1.43 c

## Section 1.2

### Answers to Check Your Understanding

*page 34:* **1.**

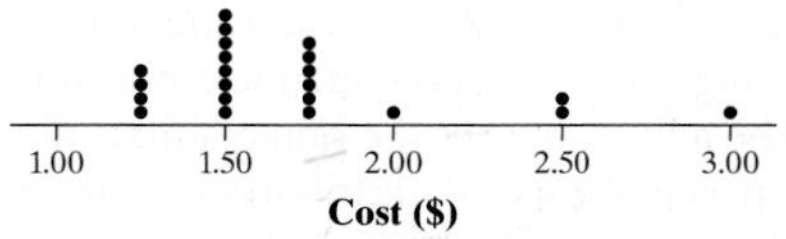


**2.** The distribution of cost is skewed to the right and unimodal, with a single peak at $1.50. There are two small gaps at $2.25 and $2.75.

*page 37:* **1.** *Shape:* The distribution of time to eat ice cream (in seconds) for the female sample is skewed to the right with no clear peak. The distribution of time to eat ice cream (in seconds) for the male sample is skewed to the right, with a single peak at about 17.5 seconds. *Outliers:* The female distribution appears to have one outlier: the female who took approximately 105 seconds to eat the ice cream. The male distribution does not appear to contain any outliers. *Center:* The time it took female students to eat ice cream was generally longer (median $\approx$ 45 seconds) than for male students (median $\approx$ 20 seconds). *Variability:* The ice cream eating times for female students varied more (from about 13 seconds to about 107 seconds) than for the male students (from about 5 seconds to about 50 seconds).

*page 40:* **1.** *Shape:* The distribution of resting pulse rates is skewed to the right, with a single peak on the 70s stem. The distribution of after-exercise pulse rates is also skewed to the right, with a single peak on the 90s stem. *Outliers:* The distribution of resting pulse rates appears to have one outlier: the student whose resting pulse rate was 120 bpm. The distribution of after-exercise pulse rates appears to have one outlier as well: the student whose after-exercise pulse rate was 146 bpm. *Center:* The students' resting pulse rates tended to be lower (median = 76 bpm) than their "after exercise" pulse rates (median = 98 bpm). *Variability:* The "after exercise" pulse rates vary more (from 86 bpm to 146 bpm) than the resting pulse rates (which vary from 68 bpm to 120 bpm).

**2.** b

**3.** e

**4.** c

*page 44:* **1.** One possible histogram is displayed here.

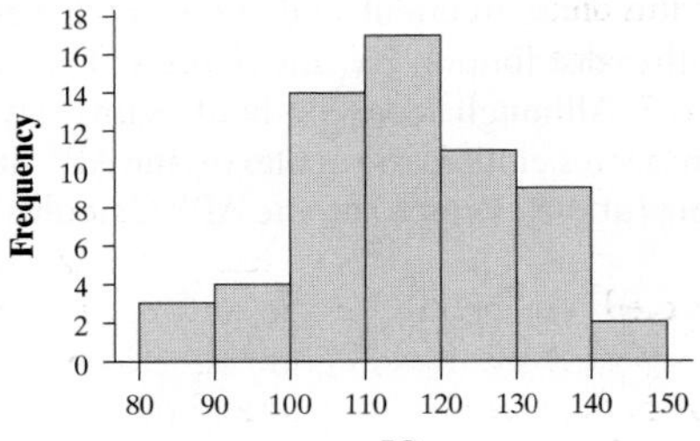


**2.** The distribution of IQ scores is roughly symmetric and bell shaped with a single peak (unimodal). There are no obvious outliers. The typical IQ score appears to be between 110 and 120 (median = 114). The IQ scores vary from about 80 to 150.

*page 46:* **1.** The distribution of word length for both the journal and the magazine have shapes that are skewed to the right and single-peaked. Neither distribution appears to have any outliers. The centers for both distributions are about the same at approximately 5–6 letters per word. Both distributions have similar variability as the length of words in the journal varies from 1 letter to 14 letters and the length of words in the magazine varies from 2 letters to 14 letters.

**2.** This is a bar graph. It displays categorical data about first-year students' planned field of study.

**3.** No, because the variable is categorical and the categories could be listed in any order on the horizontal axis.

### Answers to Odd-Numbered Section 1.2 Exercises

1.45 **(a)** The graph is shown here.

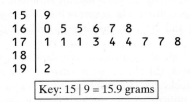

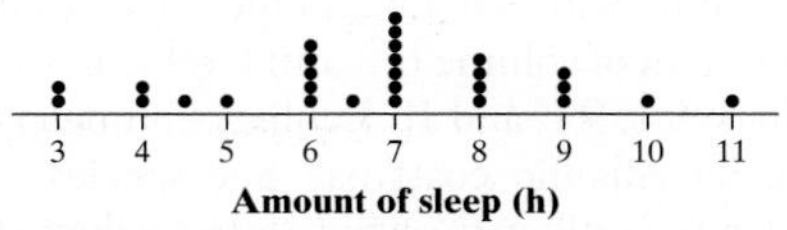


**(b)** 0.179

1.47 **(a)** The dot above 3 indicates that there was one game in which the U.S women's soccer team scored 3 more goals than their opponent. **(b)** All 20 of the values are zero or more, which indicates that the U.S. women's soccer team had a very good season. They won 17/20 = 85% of their games, tied the other team in 3/20 = 15% of their games, and never lost.

1.49 The shape of the distribution is left-skewed with a peak between 90 and 100 years. There is a small gap around 70 years.

1.51 The shape of the distribution is roughly symmetric and unimodal, with a single peak at 7.

1.53 The distribution of difference (U.S. – Opponent) in goals scored is skewed to the right, with two potential outliers: when the U.S. team outscored their opponents by 9 and 10 goals. The median difference was 2 goals and the differences varied from 0 to 10 goals.

1.55 The distribution of total family income for Indiana is roughly symmetric, while the distribution of total family income for New Jersey is slightly skewed right. The value of $125,000 may be an outlier in the Indiana distribution. There are no obvious outliers in the New Jersey distribution. The median for both distributions is about the same, approximately $49,000. The distribution of total family income in Indiana is less variable than the New Jersey distribution. The incomes in Indiana vary from $0 to about $125,000. The incomes in New Jersey vary from $0 to about $170,000.

1.57 **(a)** Both distributions have about the same amount of variability. The "external reward" distribution varies from 5 to about 24. And the "internal reward" distribution varies from about 12 to 30. **(b)** The center of the internal distribution is greater than the center of the external distribution, indicating that external rewards do not promote creativity.

1.59 **(a)** The stemplot is shown here.

| | |
|---|---|
| 15 | 9 |
| 16 | 0 5 5 6 7 8 |
| 17 | 1 1 1 3 4 4 7 7 8 |
| 18 | |
| 19 | 2 |

Key: 15 | 9 = 15.9 grams

**(b)** The graph reveals that there was one Fun Size Snickers® bar that is "gigantic"! It weighs 19.2 grams. **(c)** 0.412

1.61 **(a)** The area of the largest South Carolina county is 1,220 square miles (rounded to the nearest 10 $mi^2$). **(b)** The distribution of the area for the 46 South Carolina counties is right-skewed with distinct peaks on the 500 $mi^2$ and 700 $mi^2$ stems. There are no clear outliers. A typical South Carolina county has an area of about 655 square miles. The area of the counties varies from about 390 square miles to 1,220 square miles.

1.63 **(a)** If we had not split the stems, most of the data would appear on just a few stems, making it hard to identify the shape of the distribution. **(b)** Key: 16 | 0 means that 16.0% of that state's residents are aged 25 to 34. **(c)** The distribution of percent of residents aged 25–34 is roughly symmetric with a possible outlier at 16.0%.

1.65 The distribution of acorn volume for the Atlantic coast is skewed to the right. The distribution of acorn volume for California is roughly symmetric with one high outlier of 17.1 cubic centimeters. The distribution of volume of acorn for the Atlantic coast has 3 potential outliers: 8.1, 9.1, and 10.5 cubic centimeters. The typical acorn volume for Atlantic coast oak tree species (median = 1.7 cubic centimeters) is less than the typical acorn volume for California oak tree species (median = 4.1 cubic centimeters). The Atlantic coast distribution (with acorn volumes from 0.3 to 10.5 cubic centimeters) varies less than the California distribution (with acorn volumes from 0.4 to 17.1 cubic centimeters).

1.67 **(a)** The histogram is shown here.

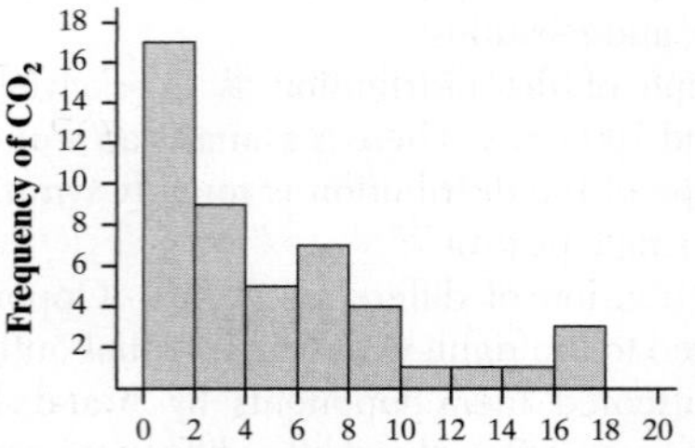


**(b)** The distribution of amount of $CO_2$ emissions per person in these 48 countries is right-skewed. Visually, none of the countries appear to be outliers.

1.69 The data vary from 14 to 54. We chose intervals of width 6, beginning at 12.

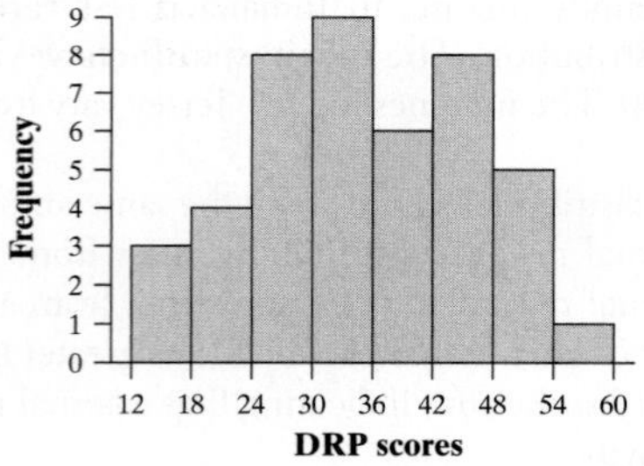


The distribution of DRP scores is roughly symmetric. There do not appear to be any outliers. The center of the DRP score distribution is between 30 and 36 (with median = 35). The DRP scores vary from 14 to 54.

1.71 **(a)** The shape of the distribution is slightly skewed to the left with a single peak. **(b)** The center is between 0% and 2.5% return on common stocks. **(c)** The exact value for the minimum return cannot be identified because we only have a histogram of the returns, not the actual data. The lowest return was in the interval −22.5% to −25%. **(d)** About 37% of these months (102 out of 273) had negative returns.

1.73 It is difficult to effectively compare the salaries of the two teams with these two histograms because the scale on the horizontal axis is very different from one graph to the other. It also does not help that the scales on the *y*-axis differ as well.

1.75 **(a)** No, it would not be appropriate to use frequency histograms instead of relative frequency histograms in this setting because there were many more graduates surveyed (314) than non-graduates (57). **(b)** The distribution of total personal income for each group is skewed to the right and single peaked. There are some possible high outliers in the graduate distribution. There do not appear to be any outliers in the non-graduate distribution. The center of the personal income distribution is larger for graduates than non-graduates, indicating that graduates typically have higher incomes than non-graduates in this sample. The incomes for graduates vary a lot more (from \$0 to \$150,000) than non-graduates (from \$0 to \$60,000).

1.77 A bar graph should be used because birth month is a categorical variable. A possible bar graph is given here.

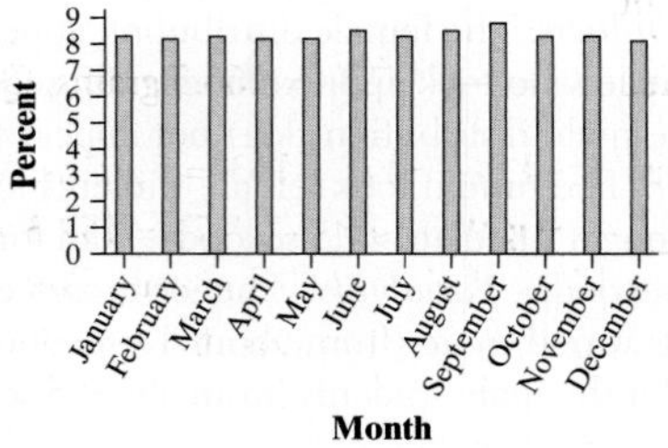


1.79 The histograms shown here use relative frequencies because there are many more students who take the AP® Calculus AB exam than take the AP® Statistics exam.

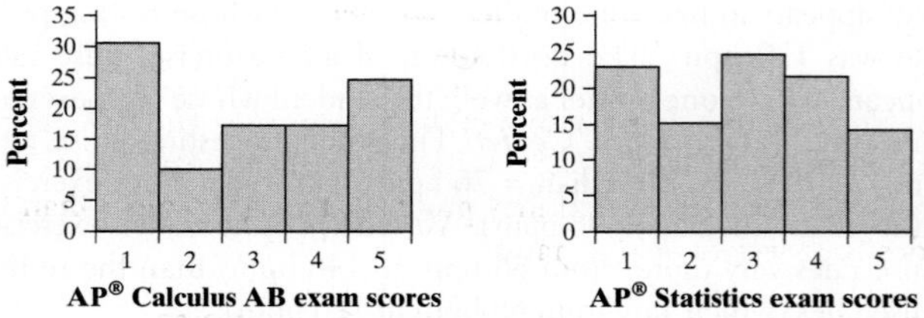


The shapes of the two distributions are very different. The distribution of scores on the AP® Calculus AB exam has a peak at 1 and another slightly lower peak at 5. The distribution of scores on the AP® Statistics exam, however, is more uniform with scores of 1, 3, and 4 being the most frequent and scores of 5 being the least frequent. Neither distribution has any outliers. The center of both distributions is 3. Although scores on both exams vary from 1 to 5, there are more scores close to the center on the AP® Statistics exam and more scores at the extremes on the AP® Calculus exam.

1.81 a

1.83 e

1.85 b

## Section 1.3

### Answers to Check Your Understanding

*page 59:* **1.** The mean weight of the pumpkins is

$$\bar{x} = \frac{3.6 + 4.0 + 9.6 + \ldots + 5.4 + 31 + 33}{23} = 9.935 \text{ pounds.}$$

**2.** First we must put the weights in order: 2.0, 2.8, 3.4, 3.6, 4.0, 4.0, 5.4, 5.4, 6.0, 6.0, 6.1, **6.6**, 9.6, 9.6, 11.0, 11.9, 12.4, 12.7, 13.0, 14.0, 15.0, 31.0, and 33.0. Because there are 23 weights, the median is the 12th weight in this ordered list. The median weight of the pumpkins is 6.6 pounds.

3. I would use the median to summarize the typical weight of a pumpkin in this contest because the distribution of pumpkin weights is skewed to the right with two possible upper outliers: 31.0 and 33.0 pounds.

*page 65:* **1.** Range = Max − Min = 23.3 − 21.5 = 1.8 mpg

**2.** The standard deviation of 0.363 mpg tells us that the highway fuel economy of these 25 model year 2018 Toyota 4 Runners typically varies by about 0.363 mpg from the mean of 22.404 mpg.

**3.** $Q_1 = \frac{22.2+22.2}{2} = 22.2$; $Q_3 = \frac{22.6+22.6}{2} = 22.6$; $IQR = Q_3 - Q_1 = 22.6 - 22.2 = 0.4$ mpg

**4.** I would use the interquartile range (*IQR*) to describe variability because there appears to be one upper outlier and one lower outlier. The *IQR* is resistant to outliers, but the range and standard deviation are not.

*page 70:* **1.** Median $= \frac{166+167}{2} = 166.5$ grams; $Q_1 = 163$ grams; $Q_3 = 170$ grams. The $IQR = 170 - 163 = 7$ grams. An outlier is any value below $Q_1 - 1.5(IQR) = 163 - 1.5(7) = 152.5$ grams or above $Q_3 + 1.5(IQR) = 170 + 1.5(7) = 180.5$ grams. This means that the value 152 grams is an outlier. The boxplot is displayed here.

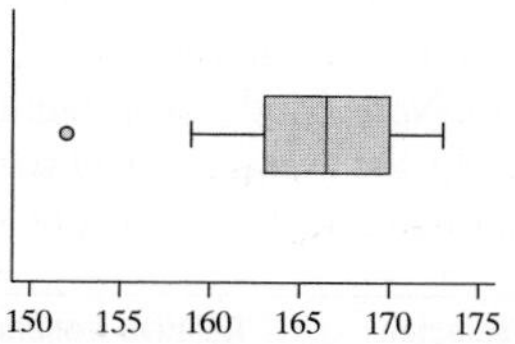


**2.** No, the graph does not support their suspicion. The first quartile of the distribution of weight is 163 grams, which means that at least 75% of the large fries that they purchased weighed more than the advertised weight of 160 grams.

## Answers to Odd-Numbered Section 1.3 Exercises

1.87 **(a)** The mean of Joey's first 14 quiz scores is 85. **(b)** Including a 15th quiz score of 0, Joey's mean would be 79.3. This illustrates the property of nonresistance. The mean is not resistant. It is sensitive to extreme values.

1.89 **(a)** 85 **(b)** Since there are now 15 quiz scores, the median is 84. Notice that the median did not change much. This shows that the median is resistant to outliers.

1.91 **(a)** 8 electoral votes **(b)** Since the distribution of number of electoral votes is skewed to the right, the mean of this distribution is greater than its median. **(c)** A parameter because it is a number that describes the population of all 50 states and the District of Columbia.

1.93 The mean house price is \$276,200 and the median is \$234,200. The distribution of house prices is likely to be quite skewed to the right because of a few very expensive homes, some of which may be outliers. When a distribution is skewed to the right, the mean is typically bigger than the median.

1.95 **(a)** The median is 2 servings of fruit per day. **(b)** $\bar{x} = \frac{194}{74} = 2.62$ servings of fruit per day

1.97 **(a)** Range = Max − Min = 98 − 74 = 24. The range of Joey's quiz grades after his unexcused absence is 98 − 0 = 98. **(b)** The range may not be the best way to describe variability for a distribution of quantitative data because the range can be heavily affected by outliers.

1.99 The standard deviation is 2.52 cm. The foot lengths of these 14-year-olds from the United Kingdom typically vary by about 2.52 cm from the mean of 24 cm.

1.101 **(a)** The size of these 18 files typically varies by about 1.9 megabytes from the mean of 3.2 megabytes. **(b)** If the music file that takes up 7.5 megabytes of storage space is replaced with another version of the file that only takes up 4 megabytes, the mean would decrease slightly. The standard deviation would decrease as well because a file of size 4 megabytes will be closer to the new mean than the file of 7.5 megabytes was to the former mean.

1.103 Variable B has a smaller standard deviation because more of the observations have values closer to the mean than in Variable A's distribution. That is, the typical distance from the mean is smaller for Variable B than for Variable A.

1.105 $Q_1 = 1.9$ megabytes; $Q_3 = 4.7$ megabytes; $IQR = Q_3 - Q_1 = 4.7 - 1.9 = 2.8$ megabytes

1.107 An outlier is any value below $Q_1 - 1.5(IQR) = 1.9 - 1.5(2.8) = -2.3$ megabytes or above $Q_3 + 1.5(IQR) = 4.7 + 1.5(2.8) = 8.9$ megabytes. There are no files in the data set that have fewer than −2.3 megabytes or more than 8.9 megabytes, so there are no outliers in the distribution.

1.109 **(a)** The distribution is skewed to the right because the mean is much larger than the median. Also, $Q_3$ is much farther from the median than $Q_1$. **(b)** The amount of money spent typically varies by about \$21.70 from the mean of \$34.70. **(c)** The first quartile is 19.27 and the third quartile is 45.40, so the *IQR* is $45.40 - 19.27 = 26.13$. Any points below $19.27 - 1.5(26.13) = -19.925$ or above $45.40 + 1.5(26.13) = 84.595$ are outliers. Because the maximum of 93.34 is greater than 84.595, there is at least one outlier.

1.111 **(a)** The median is 9, the first quartile is 3, and the third quartile is 43. The *IQR* is $43 - 3 = 40$. An outlier would be any value below $3 - 1.5(40) = -57$ or above $43 + 1.5(40) = 103$. This means the value of 118 is an outlier. The boxplot is shown here.

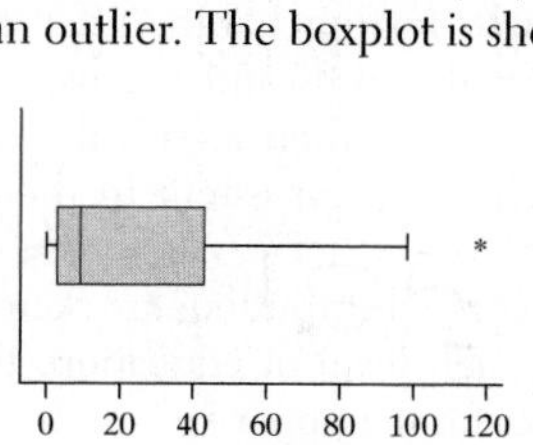


**(b)** The article claims that teens send 1742 texts a month, which works out to be about 58 texts a day (assuming a 30-day month). Nearly all of the members of the class (21 of 25) sent fewer than 58 texts per day, which seems to contradict the claim in the article.

1.113 **(a)** The median and *IQR* would be a better choice for summarizing the center and variability of the distribution of electoral votes than the mean and standard deviation because the boxplot reveals that there are three outliers in the data set. The mean and standard deviation are not resistant measures of center and variability, so their values are sensitive to these extreme values. **(b)** The stemplot reveals that the distribution has a single peak, which cannot be discerned from the boxplot. Also, the stemplot reveals that there are actually four upper outliers rather than three. The value of 29, which is an outlier, gives the number of electoral votes for two states. In the boxplot, this appears as one asterisk. However, there are two states that have that many electoral votes, not one.

1.115 *Shape:* The distribution of energy cost (in dollars) for top freezers looks roughly symmetric. The distribution of energy cost (in dollars) for side freezers looks roughly symmetric. The distribution of energy cost (in dollars) for bottom freezers looks skewed to the right. *Outliers:* There are no outliers for the top or side freezers. There are at least two bottom freezers with unusually high energy costs (over \$140 per year). *Center:* The typical energy cost for the side freezers (median $\approx$ \$75) is greater than the typical cost for the bottom freezers (median $\approx$ \$69), which is greater than the typical cost for the top freezers (median $\approx$ \$56). *Variability:* There is much more variability in the energy costs for bottom freezers ($IQR \approx$ \$20) than for side freezers ($IQR \approx$ \$12) or for top freezers ($IQR \approx$ \$8).

1.117 **(a)**

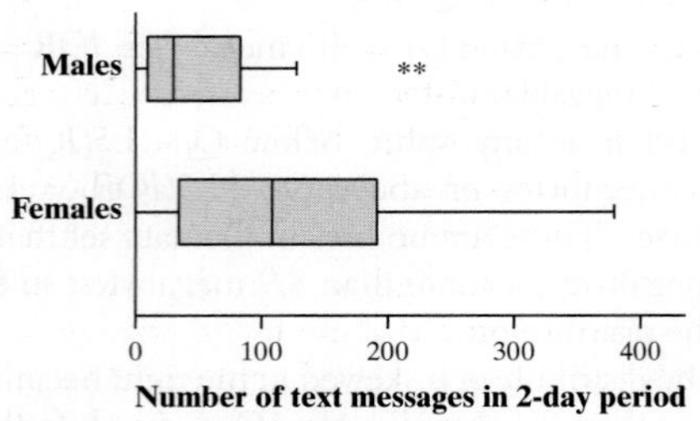


**(b)** Numerical summaries are given here.

| | $n$ | $\bar{x}$ | $s_x$ | Minimum | $Q_1$ | Median | $Q_3$ | Maximum |
|---|---|---|---|---|---|---|---|---|
| Male | 15 | 62.4 | 71.4 | 0 | 6 | 28 | 83 | 214 |
| Female | 16 | 128.3 | 116.0 | 0 | 34 | 107 | 191 | 379 |

These values are statistics because they are numbers that describe the samples of students.

**(c)** The data from this survey project give very strong evidence that male and female texting habits differ considerably at the school. A typical female sends and receives about 79 more text messages in a 2-day period than a typical male. The males as a group are also much more consistent in their texting frequency than the females.

1.119 **(a)** All five income distributions are skewed to the right. This tells us that within each level of education, there is much more variability in incomes for the upper 50% of the distribution than for the lower 50% of the distribution. **(b)** Even though the boxplots are not modified (visually showing outliers), it is clear that the upper whisker for the advanced degree boxplot is much longer than 1.5 times its *IQR*, indicating that there is at least one upper outlier in the group that earned an advanced degree. **(c)** As education level rises, the median, quartiles, and extremes rise as well—that is, every value in the 5-number summary gets larger. This makes sense because one would expect that individuals with more education tend to attain jobs that yield more income. **(d)** The variability in income increases as the education level increases. Both the width of the box (the *IQR*) and the distance from one extreme to the other increase as education levels increase.

1.121 **(a)** One possible answer is 1, 1, 1, and 1. **(b)** 0, 0, 10, 10 **(c)** For part (a), any set of four identical numbers will have $s_x = 0$. For part (b), however, there is only one possible answer. We want the values to be as far from the mean as possible, so the squared deviations from the mean can be as big as possible. Our best choice is two values at each extreme, which makes all four squared deviations equal to 25.

1.123 d

1.125 e

1.127 A histogram is given here.

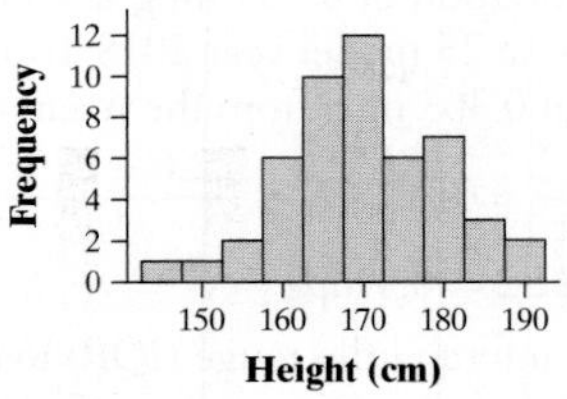


This distribution is roughly symmetric with a single peak at 170 cm. There do not appear to be any outliers. The center of the distribution of heights can be described by the mean of 169.88 cm or the median of 169.5. The heights vary from 145.5 cm to 191 cm, so the range is 45.5 cm. The standard deviation of heights is 9.687 cm and the *IQR* is $177 - 163 = 14$ cm.

## Answers to Chapter 1 Review Exercises

R1.1 **(a)** The individuals are buyers. **(b)** The variables are zip code (categorical), gender (categorical), buyer's distance from the dealer (in miles) (quantitative), car model (categorical), model year (categorical), and price (quantitative). *Note:* Model year might be considered quantitative if we want to know the average model year.

R1.2 First, a relative frequency table must be constructed.

| Candy selected | Relative frequency |
|---|---|
| Snickers® | 8/30 = 26.7% |
| Milky Way® | 3/30 = 10% |
| Butterfinger® | 7/30 = 23.3% |
| Twix® | 10/30 = 33.3% |
| 3 Musketeers® | 2/30 = 6.7% |

The relative frequency bar graph is given here.

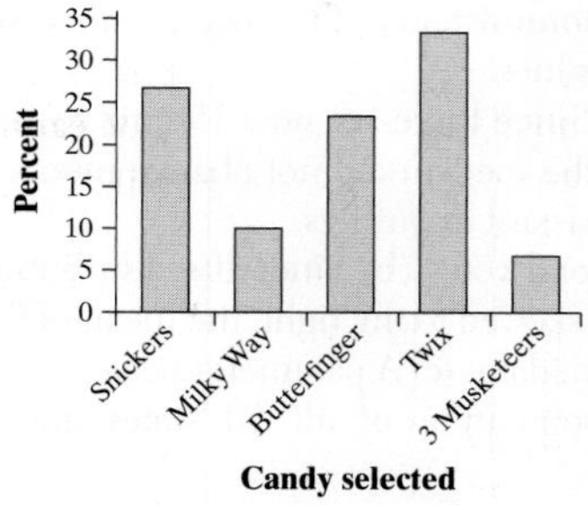


Students preferred Twix the most (one-third of the students chose this candy), followed by Snickers (27% relative frequency), Butterfinger (23% relative frequency), Milky Way (10% relative frequency), and lastly 3 Musketeers (7% relative frequency).

R1.3 **(a)** The graph is misleading because the "bars" are different widths. For example, the bar for "Send/receive text messages" should be roughly twice the size of the bar for "Camera," but it is actually much more than twice as large in area. **(b)** It would not be appropriate to make a pie chart for these data because they do not describe parts of the same whole. Students were free to answer in more than one category.

R1.4 (a) $148/219 = 0.676 = 67.6\%$ were Facebook users. (b) $67/219 = 0.306 = 30.6\%$ were aged 28 and over. (c) $21/219 = 0.096 = 9.6\%$ were older Facebook users. (d) $78/148 = 0.527 = 52.7\%$ of the Facebook users were younger students.

R1.5 (a)

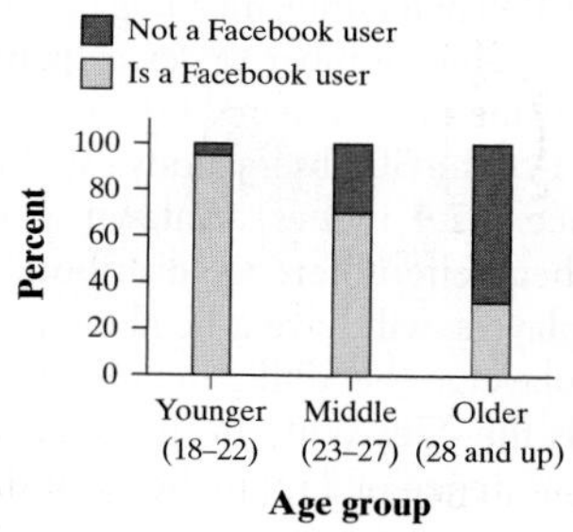


(b) From both the table and the graph, we can see that there is an association between age and Facebook use. As age increases, the percent of Facebook users decreases. For younger students, about 95% use Facebook. That drops to 70% for 23- to 27-year-old students and drops even further to 31.3% for older students.

R1.6 (a) A stemplot is shown here.

```
48 | 8
49 |
50 | 7
51 | 0
52 | 6799
53 | 04469
54 | 2467
55 | 03578
56 | 12358
57 | 59
58 | 5
```

Key: 48 | 8 = 4.88

(b) The distribution is roughly symmetric with one possible outlier at 4.88. The center of the distribution is between 5.4 and 5.5. The densities vary from 4.88 to 5.85. (c) The mean of the distribution of Cavendish's 29 measurements is 5.45. So these estimates suggest that Earth's density is about 5.45 times the density of water. The currently accepted value for the density of the earth (5.51 times the density of water) is slightly larger than the mean of the distribution of density measurements.

R1.7 (a)

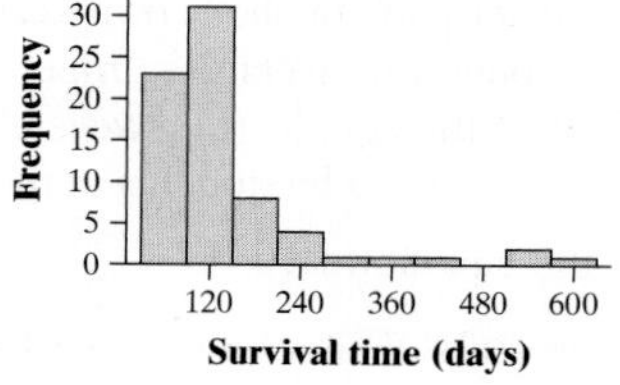


The survival times are right-skewed, as expected.

(b)

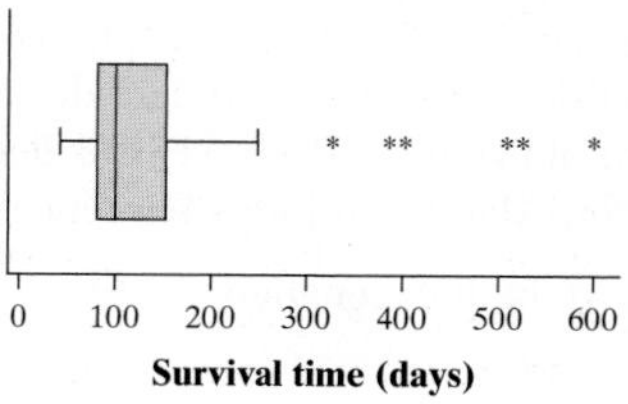


(c) While both graphs clearly show that the distribution is strongly skewed to the right, the boxplot shows that there are several high outliers, and the histogram shows it is single peaked.

R1.8 (a) About 11% of low-income and 40% of high-income households consisted of four or more people. (b) The shapes of both distributions are skewed to the right. However, the skewness is much stronger in the distribution for low-income households. On average, household size is larger for high-income households. In fact, the majority of low-income households consist of only one person. Only about 7% of high-income households consist of one person. One-person households might have less income because they would include many young single people who have no job or retired single people with a fixed income.

R1.9 (a) The amount of mercury per can of tuna will typically vary by about 0.3 ppm from the mean of 0.285 ppm. (b) The $IQR = 0.380 - 0.071 = 0.309$, so any point below $0.071 - 1.5(0.309) = -0.393$ or above $0.38 + 1.5(0.309) = 0.8435$ would be considered an outlier. Because the smallest value is 0.012, there are no low outliers. According to the histogram, there are values above 0.8435, so there are several high outliers. (c) The mean is much larger than the median of the distribution because the distribution is strongly skewed to the right and there are several high outliers.

R1.10 The distribution for light tuna is skewed to the right with several high outliers. The distribution for albacore tuna is more symmetric with just a couple of high outliers. The albacore tuna generally has more mercury. Its minimum, first quartile, median, and third quartile are all greater than the respective values for light tuna. But that doesn't mean that light tuna is always better. It has much larger variation in mercury concentration, with some cans having as much as twice the amount of mercury as the largest amount in the albacore tuna.

## Answers to Chapter 1 AP® Practice Test

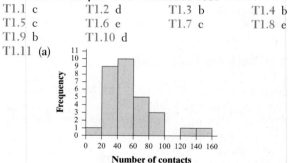

| | | | |
|---|---|---|---|
| T1.1 c | T1.2 d | T1.3 b | T1.4 b |
| T1.5 c | T1.6 e | T1.7 c | T1.8 e |
| T1.9 b | T1.10 d | | |

T1.11 (a)

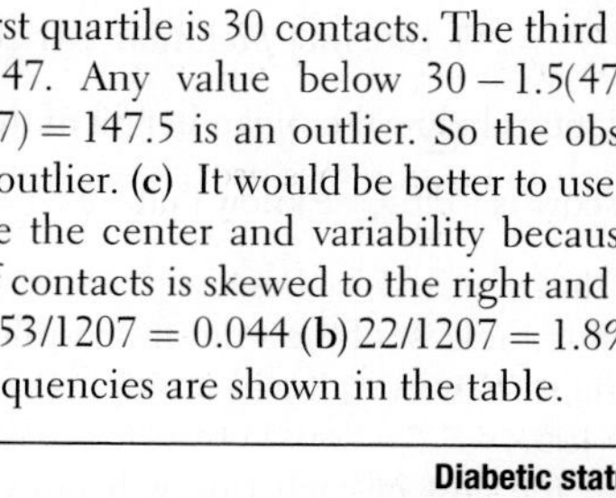


(b) The first quartile is 30 contacts. The third quartile is 77. $IQR = 77 - 30 = 47$. Any value below $30 - 1.5(47) = -40.5$ or above $77 + 1.5(47) = 147.5$ is an outlier. So the observation of 151 contacts is an outlier. (c) It would be better to use the median and $IQR$ to describe the center and variability because the distribution of number of contacts is skewed to the right and has a high outlier.

T1.12 (a) $53/1207 = 0.044$ (b) $22/1207 = 1.8\%$ (c) The conditional relative frequencies are shown in the table.

| | Diabetic status | | |
|---|---|---|---|
| **Number of birth defects** | **Nondiabetic** | **Prediabetic** | **Diabetic** |
| None | 96.1% | 96.5% | 80.9% |
| One or more | 3.9% | 3.5% | 19.1% |

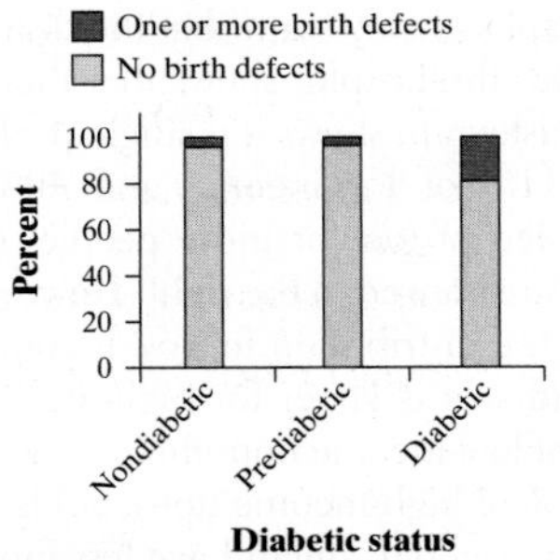


(d) There is an association between diabetic status and number of birth defects for the women in this study. Nondiabetics and prediabetics appear to have babies with birth defects at about the same rate. However, those with diabetes have a much higher rate of babies with birth defects.

T1.13 (a) The longest that any battery lasted was between 550 and 559 hours. (b) Someone might prefer to use Brand X because it has a higher minimum lifetime or because its lifetimes are more consistent (less variable). (c) Someone might prefer Brand Y because it has a higher median lifetime.

T1.14 (a) These numerical summaries are statistics because they are numbers that describe the samples. (b) The distribution of reaction time for the Athlete group is slightly skewed to the right. The distribution of reaction time for the Other group is roughly symmetric with two high outliers. It appears that the Athlete distribution has one high outlier while the Other distribution has two high outliers. The reaction times for the students who have not been varsity athletes tended to be slower (median = 292.0 milliseconds) than for the athletes (median = 261.0 milliseconds). The distribution of reaction time for the Other group also has more variability, as their reaction times have an *IQR* of 70 milliseconds and the athletes' reaction times have an *IQR* of 64 milliseconds.

# Chapter 2

## Section 2.1

### Answers to Check Your Understanding

*page 95:* **1.** c

**2.** For girls of the same age, 87% weigh ≤ Mrs. Munson's daughter and 67% are ≤ her daughter's height.

**3.** About 65% of calls lasted ≤ 30 minutes. This means that about 35% of calls lasted more than 30 minutes.

**4.** $Q_1 = 13$ minutes; $Q_3 = 32$ minutes; $IQR = 32 - 13 = 19$ minutes.

*page 97:* **1.** $z = \dfrac{62-67}{4.29} = -1.166$. *Interpretation:* Lynette's height is 1.166 standard deviations below the mean height of the class.

**2.** Because Brent's *z*-score is −0.85, we know that $-0.85 = \dfrac{74-76}{\sigma}$. Solving for $\sigma$ we find that $\sigma = 2.35$ inches.

*page 103:* **1.** Converting the cost of the rides from dollars to cents will not change the shape. However, it will multiply the mean and standard deviation by 100.

**2.** Adding 25 cents to the cost of each ride will not change the shape of the distribution, nor will it change the variability. It will, however, add 25 cents to the measures of center (mean, median).

**3.** Converting the costs to *z*-scores will not change the shape of the distribution. It will change the mean to 0 and the standard deviation to 1.

### Answers to Odd-Numbered Section 2.1 Exercises

2.1 (a) Because 18 of the 20 students (90%) own ≤ the number of pairs of shoes that Jackson owns (22 pairs of shoes), Jackson is at the 90th percentile in the number of pairs of shoes distribution. (b) 45% of the boys own ≤ the number of pairs of shoes that Raul owns. Raul is at the 45th percentile, meaning that 45% of the 20 boys, or 9 boys, have the same number or fewer pairs of shoes. Raul's response is the 9th value in the ordered list; he owns 7 pairs of shoes.

2.3 (a) Because 11 of the 30 observations (36.7%) are ≤ Antawn's head circumference (22.4 inches), Antawn is at about the 37th percentile in the head circumference distribution. (b) 90% of the 30 players, or 27 players, will have a head circumference that is ≤ this player's. The player at the 90th percentile will have a head circumference that is the 27th value in the ordered list. The player with a head circumference of 23.9 inches is at the 90th percentile of the distribution.

2.5 This means that the speed limit is set at such a speed that 85% of the vehicle speeds are ≤ the posted speed.

2.7 48% of girls her age weigh ≤ her weight and 78% of girls her age are ≤ her height. Because her height is much greater than the median (50th percentile) but her weight is close to the median, she is probably fairly thin.

2.9 (a) No; a sprint time of 8.05 seconds is not unusually slow. A student with an 8.05-second sprint is at the 75th percentile, so 25% of the students took longer than that. (b) The 20th percentile of the distribution is approximately 6.7 seconds. 20% of the students completed the 50-yard sprint in 6.7 seconds or less.

2.11 (a) The first quartile is the 25th percentile. Find 25 on the *y*-axis, read over to the line and then down to the *x*-axis to get about $Q_1 = 4\%$. The 3rd quartile is the 75th percentile. Find 75 on the *y*-axis, read over to the line and then down to the *x*-axis to get about $Q_3 = 14\%$. The *IQR* is approximately $14 - 4 = 10\%$. (b) Arizona, which had 15.1% foreign-born residents that year, is approximately at the 85th percentile. (c) The graph is fairly flat between 20% and 27.5% foreign-born residents because there were very few states that had 20% to 27.5% foreign-born residents that year.

2.13 (a) The *z*-score for Montana is $z = \dfrac{1.9-8.73}{6.12} = -1.12$. Montana's percent of foreign-born residents is 1.12 standard deviations below the mean percent of foreign-born residents for all states. (b) If we let *x* denote the percent of foreign-born residents in New York at that time, then we can solve for *x* in the equation $2.10 = \dfrac{x-8.73}{6.12}$. Thus, $x = 21.582\%$ foreign-born residents.

2.15 (a) The number of pairs of shoes owned by Jackson is 1.10 standard deviations above the average number of pairs of shoes owned by the students in the sample. (b) If we let $\bar{x}$ denote the mean number of pairs of shoes owned by students in the sample, then we can solve for $\bar{x}$ in the equation $1.10 = \dfrac{22-\bar{x}}{9.42}$. Thus, $\bar{x} = 11.64$ is the mean number of pairs of shoes owned by the students in the sample.

2.17 (a) The fact that your standardized score is negative indicates that your bone density is below the average for your peer group. In fact, your bone density is about 1.5 standard deviations below average among 25-year-old women. (b) If we let $\sigma$ denote the standard deviation of the bone density in Judy's reference population, then we can solve for $\sigma$ in the equation $-1.45 = \dfrac{948-956}{\sigma}$. Thus, $\sigma = 5.52\text{g/cm}^2$.

2.19 Eleanor's standardized score of $z = \frac{680 - 500}{100} = 1.8$ is higher than Gerald's standardized score of $z = \frac{29 - 21}{5} = 1.6$.

2.21 **(a)** The shape of the distribution of corrected long-jump distance will be the same as the original distribution: roughly symmetric with a single peak. **(b)** The mean is $577.3 - 20 = 557.3$ centimeters. The median is $577 - 20 = 557$ centimeters. **(c)** The standard deviation is the same, 4.713 centimeters. The *IQR* is the same, 7 centimeters.

2.23 **(a)** The shape of the new salary distribution will be the same as the shape of the original salary distribution. **(b)** The mean and median salaries will each increase by $1000. **(c)** The standard deviation and *IQR* of the new salary distribution will each be the same as they were for the original salary distribution.

2.25 **(a)** The shape of the distribution of corrected long-jump distance in meters will be the same as the distribution of corrected long-jump distance in centimeters: roughly symmetric with a single peak. **(b)** The mean of the distribution of corrected long-jump distance, in meters, is $(577.3 - 20) \div 100 = 557.3 \text{ cm} \div 100 \text{ cm/m} = 5.573$ meters. **(c)** The standard deviation of the distribution of corrected long-jump distance, in meters, is 4.713 centimeters $\div 100 = 0.04713$ meters.

2.27 **(a)** The shape of the resulting salary distribution will be the same as the original distribution of salaries. **(b)** The median will increase by 5% because each value in the distribution is being multiplied by 1.05. **(c)** The *IQR* will increase by 5% because each value in the distribution is being multiplied by 1.05.

2.29 **(a)** The mean temperature reading is $\frac{5}{9}(77) - \frac{160}{9} = 25$ degrees Celsius. **(b)** The standard deviation of the temperature reading is $\frac{5}{9}(3) = 1.667$ degrees Celsius.

2.31 To determine the mean of the lengths of his cab rides in miles, we substitute $15.45 for the mean fare and solve for the mean number of miles: $15.45 = 2.85 + 2.7 \text{ Mean(miles)} \rightarrow \text{Mean (miles)} = 4.667$ miles.

To determine the standard deviation of the lengths of his cab rides in miles, we use the following equation:

$\text{SD(Fare)} = 2.7\text{SD(miles)} \rightarrow 10.20 = 2.7\text{SD(miles)} \rightarrow \text{SD(miles)} = 3.778$ miles

The mean and standard deviation of the lengths of his cab rides, in miles, are 4.667 miles and 3.778 miles, respectively.

2.33 c

2.35 d

2.37 c

2.39 The distribution is skewed to the right. The two largest values appear to be outliers. The data are centered roughly around a median of 15 minutes and the interquartile range is approximately 10 minutes.

## Section 2.2

### Answers to Check Your Understanding

*page 119:* **1.** The graph is shown here.

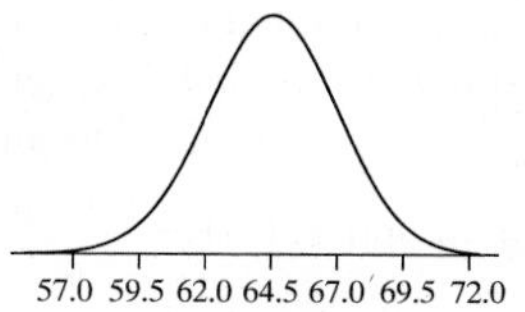


**2.** Because 69.5 inches is 2 standard deviations above the mean, approximately $\frac{100\% - 95\%}{2} = 2.5\%$ of young women have heights greater than 69.5 inches. Therefore, 97.5% of young women have heights less than 69.5 inches.
**3.** Because 62 is 1 standard deviation below the mean, approximately $\frac{100\% - 68\%}{2} = 16\%$ of young women have heights below 62 inches. This is not unusually short.

*page 127:* **1.** (i) $z = \frac{240 - 170}{30} = 2.33$. *Table* A: The proportion of *z*-scores above 2.33 is $1 - 0.9901 = 0.0099$. *Tech:* normalcdf(lower: 2.33, upper: 1000, mean: 0, SD: 1) = 0.0099 (ii) normalcdf(lower: 240, upper: 1000, mean: 170, SD: 30) = 0.0098. About 1% of 14-year-old boys have cholesterol above 240 mg/dl. **2.** (i) $z = \frac{200 - 170}{30} = 1$ and $z = \frac{240 - 170}{30} = 2.33$. *Table* A: The proportion of *z*-scores below $z = 1.00$ is 0.8413 and the proportion of *z*-scores below 2.33 is 0.9901. Thus, the proportion of *z*-scores between 1 and 2.33 is $0.9901 - 0.8413 = 0.1488$. *Tech:* normalcdf(lower: 1, upper: 2.33, mean: 0, SD: 1) = 0.1488 (ii) normalcdf(lower: 200, upper: 240, mean: 170, SD: 30) = 0.1488. About 15% of 14-year-old boys have cholesterol between 200 and 240 mg/dl.

*page 131:* **1.** (i) *Table* A: Look in the body of Table A for the value closest to 0.10. A *z*-score of $-1.28$ gives the closest value (0.1003). *Tech:* invNorm(area: 0.10, mean: 0, SD: 1) = $-1.28$

$-1.28 = \frac{x - 170}{30} \rightarrow -38.4 = x - 170 \rightarrow x = 131.6$

(ii) invNorm(area: 0.10, mean: 170, SD: 30) = 131.6 About 10% of 14-year-old boys have cholesterol levels that are $\leq 131.6$, so a 14-year-old boy who has a cholesterol level of 131.6 would be at the 10th percentile of the distribution.

### Answers to Odd-Numbered Section 2.2 Exercises

2.41 **(a)** The density curve is shown here.

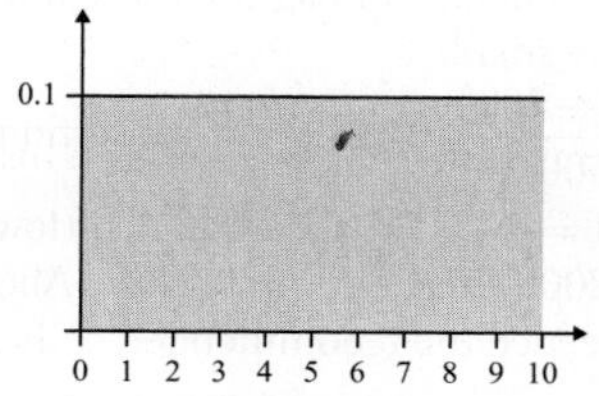


**(b)** Area = (base)(height) = $(5.3 - 2.5)(0.1) = 0.28 = 28\%$. On about 28% of days, Sally waits between 2.5 and 5.3 minutes for the bus. **(c)** The 70th percentile will have 70% of the wait times to the left of it, so the 70th percentile of Sally's wait times is 7 minutes.

2.43 **(a)** The density curve must have the height 0.25 because the area must equal 1. (base)(height) = $(4)(0.25) = 1$. **(b)** Area = (base)(height) = $(5 - 3.75)(0.25) = 0.3125 = 31.25\%$. About 31.25% of the time, the light will flash more than 3.75 seconds after the subject clicks "Start." **(c)** The 38th percentile of this distribution is the time for which 38% of the observations are at or below it.
Area = (base)(height) $\rightarrow 0.38 = (x - 1)(0.25) \rightarrow 1.52 = x - 1 \rightarrow x = 2.52$

Therefore, the 38th percentile of this distribution is 2.52 seconds. Thirty-eight percent of the time the light will flash in 2.52 seconds or less after the subject clicks "Start."

2.45 **(a)** Mean is C, median is B (the right skew pulls the mean to the right of the median). **(b)** Mean is B, median is B (this distribution is symmetric, so mean = median).

2.47 The Normal density curve with mean 9.12 and standard deviation 0.05 is shown here.

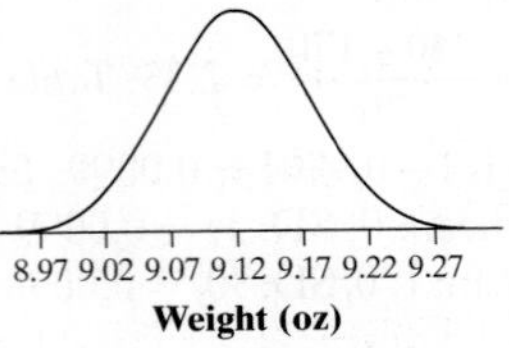


2.49 The mean is at the balance point of the distribution, which appears to be 10. The curve gets very close to the horizontal axis around 3 standard deviations from the mean. Observing the positive side of the curve, the curve shown seems to approach the horizontal axis around the value 16, so 16 should be approximately 3 standard deviations above the mean. Because $16 - 10 = 6$, we estimate 3 standard deviations to be 6 units and therefore 1 standard deviation would be about 2 units.

2.51 **(a)** The value 9.02 is 2 standard deviations below the mean, so approximately $\frac{100\% - 95\%}{2} = 2.5\%$ of bags weigh less than 9.02 ounces. **(b)** The value 9.07 is 1 standard deviation below the mean. About $\frac{100\% - 68\%}{2} = 16\%$ of the bags weigh less than or equal to 9.07 ounces. In other words, 9.07 is approximately the 16th percentile of the weights of these potato chip bags.

2.53 (i) $z = \frac{9 - 9.12}{0.05} = -2.40$; the proportion of $z$-scores below $-2.4$ is 0.0082. (ii) normalcdf(lower: $-1000$, upper: 9, mean: 9.12, SD: 0.05) $= 0.0082$
About 0.82% of 9-ounce bags of this brand of potato chips weigh less than the advertised 9 ounces. This is not likely to pose a problem because the percentage of bags that weigh less than the advertised amount is very small.

2.55 (i) $z = \frac{2400 - 2000}{500} = 0.80$; the proportion of $z$-scores above 0.80 is $1 - 0.7881 = 0.2119$. (ii) normalcdf(lower: 2400, upper: 1000000, mean: 2000, SD: 500) $= 0.2119$. About 21.19% of the meals ordered exceeded the recommended daily allowance of 2400 mg of sodium.

2.57 (i) $z = \frac{1200 - 2000}{500} = -1.60, z = \frac{1800 - 2000}{500} = -0.40$; the proportion of $z$-scores between $-1.60$ and $-0.40$ is $0.3446 - 0.0548 = 0.2898$. (ii) normalcdf(lower: 1200, upper: 1800, mean: 2000, SD: 500) $= 0.2898$. About 28.98% of meals ordered had between 1200 mg and 1800 mg of sodium.

2.59 **(a)** $z = -1.66$
*Table* A: The proportion of $z$-scores above $z = -1.66$ is $1 - 0.0485 = 0.9515$. *Tech:* normalcdf(lower: $-1.66$, upper: 1000, mean: 0, SD: 1) $= 0.9515$. The proportion of observations in a standard Normal distribution that satisfy $z > -1.66$ is 0.9515. **(b)** $z = -1.66$ and $z = 2.85$ *Table* A: The proportion of $z$-scores between $-1.66$ and 2.85 is $0.9978 - 0.0485 = 0.9493$. *Tech:* normalcdf(lower: $-1.66$, upper: 2.85, mean: 0, SD: 1) $= 0.9494$. The proportion of observations in a standard Normal distribution that satisfy $-1.66 < z < 2.85$ is 0.9494.

2.61 **(a)** (i) $z = \frac{3 - 5.3}{0.9} = -2.56$; the proportion of $z$-scores below $-2.56$ is 0.0052. (ii) normalcdf(lower: $-1000$, upper: 3, mean: 5.3, SD: 0.9) $= 0.0053$. About 53 out of every 10,000 times Mrs. Starnes completes an easy Sudoku puzzle, she does so in less than 3 minutes. This is a proportion of 0.0053. **(b)** (i) $z = \frac{6 - 5.3}{0.9} = 0.78$; the proportion of $z$-scores above 0.78 is $1 - 0.7823 = 0.2177$.
(ii) normalcdf(lower: 6, upper: 1000, mean: 5.3, SD: 0.9) $= 0.2183$
About 21.83% of the time, it takes Mrs. Starnes more than 6 minutes to complete an easy puzzle. **(c)** (i) $z = \frac{6 - 5.3}{0.9} = 0.78$, $z = \frac{8 - 5.3}{0.9} = 3$; the proportion of $z$-scores between 0.78 and 3.00 is $0.9987 - 0.7823 = 0.2164$. (ii) normalcdf(lower: 6, upper: 8, mean: 5.3, SD: 0.9) $= 0.2170$. About 21.7% of easy puzzles take Mrs. Starnes between 6 and 8 minutes to complete.

2.63 (i) 0.20 area to the left of $z \rightarrow z = -0.84$. Solving $-0.84 = \frac{x - 5.3}{0.9}$ gives $x = 4.544$ minutes. (ii) invNorm(area: 0.2, mean: 5.3, SD: 0.9) $= 4.543$ minutes. The 20th percentile of Mrs. Starnes's Sudoku times for easy problems is about 4.54 minutes.

2.65 (i) 0.10 area to the left of $z \rightarrow z = -1.28$ and 0.90 area to the left of $z \rightarrow z = 1.28$. (ii) invNorm(area: 0.10, mean: 0, SD: 1) $= -1.282$ and invNorm(area: 0.90, mean: 0, SD: 1) $= 1.282$. The first and last deciles of the standard Normal distribution are $z = -1.282$ and $z = 1.282$, respectively.

2.67 **(a)** (i) $z = \frac{125 - 110}{25} = 0.6$, $z = \frac{150 - 110}{25} = 1.6$; the proportion of $z$-scores between 0.6 and 1.6 is $0.9452 - 0.7257 = 0.2195$. (ii) normalcdf(lower: 125, upper: 150, mean: 110, SD: 25) $= 0.2195$. About 22% of 20- to 34-year-olds have IQ scores between 125 and 150. **(b)** (i) 98% area to the left of $z \rightarrow z = 2.05$. Solving $2.05 = \frac{x - 110}{25}$ gives $x = 161.25$. (ii) invNorm(area: 0.98, mean: 110, SD: 25) $= 161.34$. Scores greater than 161.34 qualify for MENSA.

2.69 **(a)** (i) $z = \frac{3.95 - 3.98}{0.02} = -1.5$; the proportion of $z$-scores below $-1.5$ is 0.0668. (ii) normalcdf(lower: $-1000$, upper: 3.95, mean: 3.98, SD: 0.02) $= 0.0668$. About 7% of the large lids are too small to fit. **(b)** (i) $z = \frac{4.05 - 3.98}{0.02} = 3.5$; the proportion of $z$-scores above 3.50 is approximately 0. (ii) normalcdf(lower: 4.05, upper: 1000, mean: 3.98, SD: 0.02) $= 0.0002$. Approximately 0% of the large lids are too big to fit. **(c)** It makes more sense to have a larger proportion of lids too small rather than too big. If lids are too small, customers will just try another lid. But if lids are too large, the customer may not notice and then spill the drink.

2.71 We are looking for the value of $z$ with an area of 0.15 to the right and 0.85 to the left, and the value of $z$ with an area of 0.03 to the right and 0.97 to the left. We get the values $z = 1.04$ and $z = 1.88$, respectively. Now we need to solve the following system of equations for $\mu$ and $\sigma$: $1.04 = \frac{60 - \mu}{\sigma}$ and $1.88 = \frac{75 - \mu}{\sigma}$. Multiplying both sides of the equations by $\sigma$ and subtracting yields

$0.84\sigma = 15$ or $\sigma = 17.86$ minutes. Substituting this value back into the first equation, we obtain $1.04 = \frac{60-\mu}{17.86}$, or $\mu = 60 - 1.04$ $(17.86) = 41.43$ minutes.

2.73 The distribution of highway gas mileage is not approximately Normal because the distribution is skewed to the right.

2.75 The histogram of these data is roughly symmetric and bell-shaped. The mean and standard deviation of these data are $\bar{x} = 15.825$ cubic feet and $s_x = 1.217$ cubic feet.

- $\bar{x} \pm 1s_x = (14.608, 17.042)$; 24 of 36 observations, or 66.7% of the observations, are within 1 standard deviation of the mean.
- $\bar{x} \pm 2s_x = (13.391, 18.259)$; 34 of 36 observations, or 94.4% of the observations, are within 2 standard deviations of the mean.
- $\bar{x} \pm 3s_x = (12.174, 19.476)$; 36 of 36 observations, or 100% of the observations, are within 3 standard deviations of the mean.

These percentages are quite close to what we would expect based on the empirical rule. Combined with the graph, this provides good evidence that this distribution is approximately Normal.

2.77 The distribution of tuitions in Michigan is not approximately Normal. If it were Normal, then the minimum value would be around 3 standard deviations below the mean. However, the actual minimum has a $z$-score of just $z = \frac{1873 - 10{,}614}{8049} = -1.09$. Also, if the distribution were Normal, the minimum and maximum would be about the same distance from the mean. However, the mean is much farther from the maximum $(30{,}823 - 10{,}614 = 20{,}209)$ than from the minimum $(10{,}614 - 1873 = 8741)$.

2.79 The distribution is approximately Normal because the Normal probability plot is nearly linear.

2.81 The sharp curve in the Normal probability plot suggests that the distribution is right-skewed. This can be seen in the steep, nearly vertical section in the lower left. These numbers were much closer to the mean than would be expected in a Normal distribution, meaning that the values that would be in the left tail are piled up close to the center of the distribution.

2.83 **(a)** A Normal probability plot is shown.

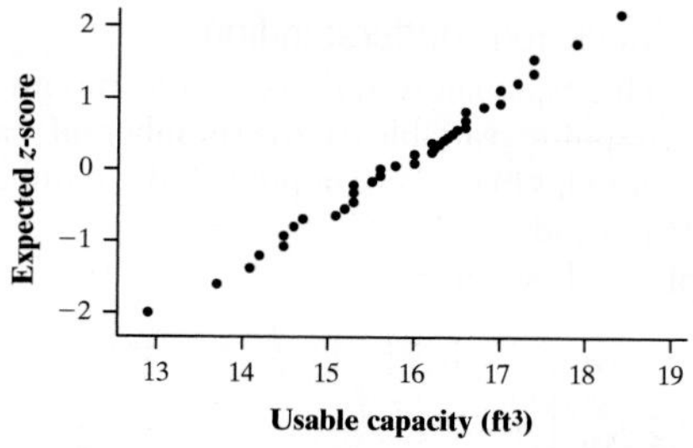


**(b)** The plot is fairly linear, indicating that the distribution of usable capacity is approximately Normal.

2.85 b

2.87 b

2.89 a

2.91 37 of the 38 family incomes in Indiana are at or below \$95,000. Because $37/38 = 0.974$, or 97.4%, this individual's income is at about the 97th percentile. 36 of the 44 family incomes in New Jersey are at or below \$95,000. Because $36/44 = 0.818$, or 81.8%, this individual's income is at about the 82nd percentile.

Indiana: $z = \frac{95{,}000 - 47{,}400}{29{,}400} = 1.62$

New Jersey: $z = \frac{95{,}000 - 58{,}100}{14{,}900} = 0.88$

The individual from Indiana has standardized income of 1.62 while the individual from New Jersey has standardized income of 0.88.

The individual from Indiana has a higher income, relative to others in his or her state because he or she had a higher percentile (97th versus 82nd) and had a higher $z$-score (1.62 versus 0.88) than the individual from New Jersey.

## Answers to Chapter 2 Review Exercises

R2.1 **(a)** $z = \frac{179 - 170}{7.5} = 1.20$; Paul's height is 1.20 standard deviations above the average male height for his age. **(b)** 85% of boys Paul's age are the same height as or shorter than Paul.

R2.2 **(a)** Reading up from 7 hours on the $x$-axis to the graphed line and then across to the $y$-axis, we see that 7 hours corresponds to about the 58th percentile. **(b)** To find $Q_1$, start at 25 on the $y$-axis, move across to the line and down to the $x$-axis. $Q_1$ is approximately 2.5 hours. To find $Q_3$, start at 75 on the $y$-axis, move across to the line and down to the $x$-axis. $Q_3$ is approximately 11 hours. Thus, $IQR = 11 - 2.5 = 8.5$ hours per week.

R2.3 **(a)** If we converted the guesses from feet to meters, the shape of the distribution would not change. The new mean would be $\frac{43.7}{3.28} = 13.32$ meters, the median would be $\frac{42}{3.28} = 12.80$ meters, the standard deviation would be $\frac{12.5}{3.28} = 3.81$ meters, and the $IQR$ would be $\frac{12.5}{3.28} = 3.81$ meters. **(b)** The mean error would be $43.7 - 42.6 = 1.1$ feet. The standard deviation of the errors would be the same as the standard deviation of the guesses, 12.5 feet. **(c)** 62 of the 66 students estimated the width of their classroom to be 63 feet or less, so the student who estimated the classroom width as 63 feet is at the $62/66 = 0.939 \approx 94$th percentile.

R2.4 **(a)** The percent of observations that have values less than 13 is $1 - 0.08 = 0.92 = 92\%$. **(b)** Answers will vary, but the line indicating the median (line A in the graph) should be slightly to the right of the main peak, with half of the area to the left and half to the right. **(c)** Answers will vary, but the line indicating the mean (line B in the graph) should be slightly to the right of the line for the median at the balance point.

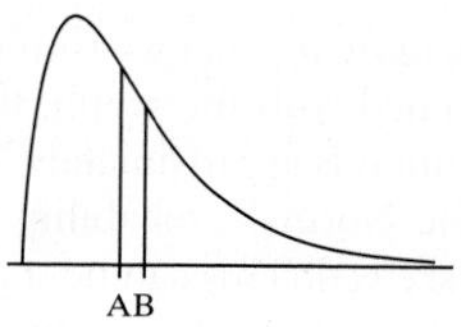


R2.5 **(a)** About 99.7% of the observations will fall within 3 standard deviations of the mean. About 99.7% of the babies had birth weights between <u>2135</u> and <u>5201</u> grams. **(b)** (i) $z = \frac{2500 - 3668}{511} = -2.29$; the proportion of $z$-scores below $-2.29$ is 0.0110. (ii) normalcdf (lower: $-1000$, upper: 2500, mean: 3668, SD: 511) $= 0.0111$. About 1% of babies will be identified as low birth weight. **(c)** (i) The first quartile is the boundary value with 25% of the area to its left $\rightarrow z = -0.67$. The third quartile is the boundary value with 75% of the area to its left $\rightarrow z = 0.67$. Solving $-0.67 = \frac{x - 3668}{511}$

gives $Q_1 = 3325.63$. Solving $0.67 = \frac{x-3668}{511}$ gives $Q_3 = 4010.37$. (ii) invNorm(area: 0.25, mean: 3668, SD: 511) gives $Q_1 = 3323.34$ invNorm(area: 0.75, mean: 3668, SD: 511) gives $Q_3 = 4012.66$. The quartiles are $Q_1 = 3323.34$ grams and $Q_3 = 4012.66$ grams.

R2.6 **(a)** (i) $z = \frac{500-694}{112} = -1.73$, $z = \frac{900-694}{112} = 1.84$; the proportion of $z$-scores below $-1.73$ is 0.0418. The proportion of $z$-scores above 1.84 is $1 - 0.9671 = 0.0329$. (ii) normalcdf(lower: $-1000$, upper: 500, mean: 694, SD: 112) $= 0.0418$ and normalcdf(lower: 900, upper: 100000, mean: 694, SD: 112) $= 0.0329$. The percent of test-takers who earn a score less than 500 or greater than 900 on the GRE Chemistry test is $4.18\% + 3.29\% = 7.47\%$.

**(b)** (i) 99% area to left of $z \to z = 2.33$. Solving $2.33 = \frac{x-694}{112}$ gives $x = 954.96$. (ii) invNorm(area: 0.99, mean: 694, SD: 112) $= 954.55$ The 99th percentile score on the GRE Chemistry test is 954.55.

R2.7 **(a)** (i) $z = \frac{1.2-1.05}{0.08} = 1.88$, $z = \frac{1-1.05}{0.08} = -0.063$; the proportion of $z$-scores between $-0.63$ and 1.88 is $0.9699 - 0.2643 = 0.7056$. (ii) normalcdf(lower: 1, upper: 1.2, mean: 1.05, SD: 0.08) $=$ 0.7036. About 70% of the time the dispenser will put between 1 and 1.2 ounces of ketchup on a burger. **(b)** Because the mean of 1.1 is in the middle of the interval from 1 to 1.2, we are looking for the middle 99% of the distribution. This leaves 0.5% in each tail. (i) 0.005 area to the left of $z \to z = -2.58$. Solving $-2.58 = \frac{1-1.1}{\sigma}$ gives $\sigma = 0.039$. A standard deviation of at most 0.039 ounce will result in at least 99% of burgers getting between 1 and 1.2 ounces of ketchup.

R2.8 The distribution of percent of residents aged 65 and older in the 50 states and the District of Columbia is roughly symmetric and somewhat bell-shaped. The mean and standard deviation of these data are $\bar{x} = 13.255\%$ and $s_x = 1.668\%$.

- $\bar{x} \pm 1s_x = (11.587, 14.923)$; 40 of 51 observations, or 78.4% of the observations, are within 1 standard deviation of the mean.
- $\bar{x} \pm 2s_x = (9.919, 16.591)$; 48 of 51 observations, or 94.1% of the observations, are within 2 standard deviations of the mean.
- $\bar{x} \pm 3s_x = (8.251, 18.259)$; 50 of 51 observations, or 98.0% of the observations, are within 3 standard deviations of the mean.

These percentages are close to what we would expect based on the empirical rule. Combined with the graph, this provides good evidence that this distribution is approximately Normal.

R2.9 The curve in the Normal probability plot suggests that the data are slightly right-skewed. This can be seen in the steep, nearly vertical section in the lower left. These numbers were closer to the mean than would be expected in a Normal distribution, meaning that the values that would be in the left tail are piled up closer to the center of the distribution.

### Answers to Chapter 2 AP® Practice Test

| | | | |
|---|---|---|---|
| T2.1 e | T2.2 d | T2.3 b | T2.4 b |
| T2.5 a | T2.6 e | T2.7 c | T2.8 d |
| T2.9 b | T2.10 c | | |

T2.11 **(a)** A total of 27 of the 40 sale prices were $\leq$ the house indicated in red on the dotplot, so that home is at the $27/40 = 0.675$ $\approx$ 68th percentile. **(b)** $z = \frac{234000-203388}{87609} = 0.35$. *Interpretation:* The sale price for this home is 0.35 standard deviation above the average sale price for the homes in the sample.

T2.12 **(a)** (i) $z = \frac{6-7.11}{0.74} = -1.5$; the proportion of $z$-scores below $-1.5$ is 0.0668. (ii) normalcdf(lower: $-1000$, upper: 6, mean: 7.11, SD: 0.74) $= 0.0668$. About 6.68% of the students ran the mile in less than 6 minutes, which means that $(0.0668)(12000) = 801.6$, or about 802, students ran the mile in less than 6 minutes. **(b)** (i) 0.90 area to the left of $z \to z = 1.28$. Solving $1.28 = \frac{x-7.11}{0.74}$ gives $x = 8.06$ minutes. (ii) invNorm(area: 0.90, mean: 7.11, SD: 0.74) $= 8.06$ minutes. It took about 8.06 minutes for the slowest 10% of students to run the mile. **(c)** If the mile run times were converted from minutes to seconds, the mean would be $(7.11)(60) = 426.6$ seconds and the standard deviation would be $(0.74)(60) = 44.4$ seconds. (i) $z = \frac{400-426.6}{44.4} = -0.60$, $z = \frac{500-426.6}{44.4} = 1.65$; the difference of $z$-scores between $-0.60$ and 1.65 is $0.9505 - 0.2743 =$ 0.6762. (ii) normalcdf(lower: 400, upper: 500, mean: 426.6, SD: 44.4) $= 0.6763$. About 67.6% of students who ran the mile had times between 400 and 500 seconds.

T2.13 No, these data do not seem to follow a Normal distribution. First, there is a large difference between the mean and the median. In a Normal distribution, the mean and median are the same, but in this distribution the mean is 48.25 and the median is 37.80. Second, the distance between the minimum and the median is $37.80 - 2 = 35.80$, but the distance between the median and the maximum is $204.90 - 37.80 = 167.10$. In a Normal distribution, these distances should be approximately the same. Because the mean is larger than the median and the distance from the median to the maximum is larger than the distance from the minimum to the median, the distribution of oil recovered appears to be skewed to the right.

# Chapter 3

## Section 3.1

### Answers to Check Your Understanding

*page 159:* **1.** The explanatory variable is the amount of sugar (in grams). The response variable is the number of calories. The amount of sugar helps to explain, or predict, the number of calories in movie-theater candy.

**2.** A scatterplot is shown here.

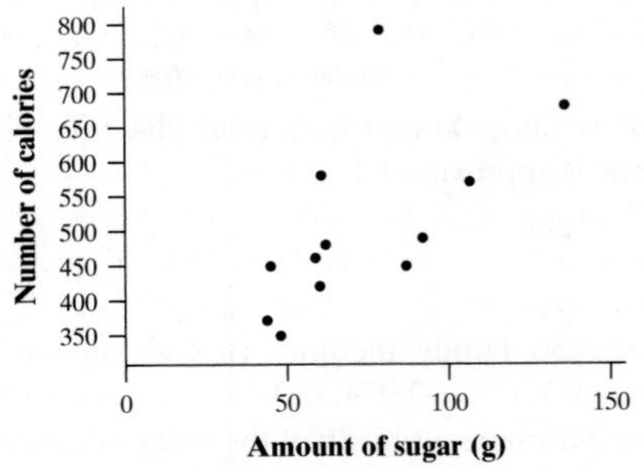


**3.** There is a moderately strong, positive, linear relationship between the amount of sugar contained in movie-theater candy and the number of calories in the candy. The point for peanut

M&M'S®, which has 79 grams of sugar and 790 calories, is an unusual point.

*page 162:* The correlation of $r = -0.838$ confirms that the linear association between dash time and long-jump distance is strong and negative.

## Answers to Odd-Numbered Section 3.1 Exercises

3.1 **(a)** Water temperature is the explanatory variable and weight gain is the response variable. Water temperature may help predict or explain changes in the response variable, weight gain. Weight gain measures the outcome of the study. **(b)** Either variable could be the explanatory variable because each one could be used to predict or explain the other.

3.3 A scatterplot is shown here.

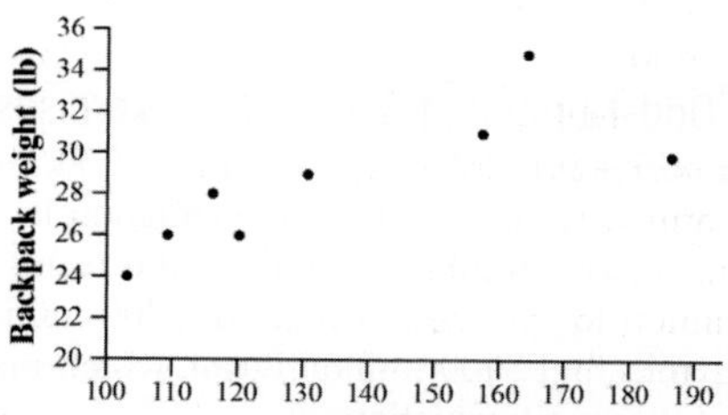


3.5 Other than several athletes who weigh much more than other athletes of the same height, there is a moderately strong, positive, linear relationship between height and weight for these athletes.

3.7 There is a moderately strong, positive, linear association between backpack weight and body weight for these students. There is an unusual point in the graph—the hiker with body weight 187 pounds and pack weight 30 pounds. This hiker makes the form appear to be nonlinear for weights above 140 pounds.

3.9 **(a)** A scatterplot with speed as the explanatory variable is shown.

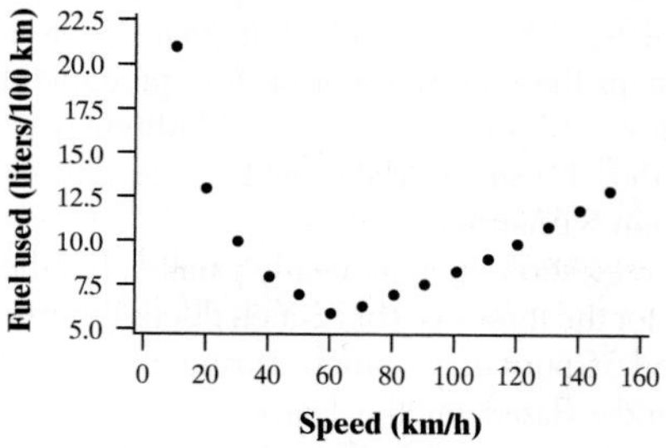


**(b)** There is a strong, nonlinear relationship between speed and amount of fuel used. The relationship is negative for speeds up to 60 km/h and positive for speeds beyond 60 km/h. There are no unusual points.

3.11 For both groups of athletes, there is a moderately strong, positive, linear association between height and weight; however, athletes who participate in the shot put, discus throw, and hammer throw tend to weigh more than other track and field athletes of the same height.

3.13 The relationship is positive, so $r > 0$. Also, $r$ is closer to 1 than to 0 because the relationship is strong.

3.15 The correlation of 0.92 indicates that the linear relationship between number of turnovers and number of points scored for players in the recent NBA season is strong and positive.

3.17 Probably not. Although there is a strong, positive association, an increase in turnovers is not likely to cause an increase in points for NBA players. It is likely that both of these variables are changing due to several other variables, such as time played.

3.19 **(a)** The correlation of 0.87 indicates that the linear relationship between amount of sodium and number of calories is strong and positive. **(b)** The hot dog with the lowest calorie content increases the correlation. It falls in the linear pattern of the rest of the data.

3.21 **(a)** The scatterplot shows a strong, positive, linear relationship between the femur lengths and humerus lengths. It appears that all 5 specimens come from the same species.

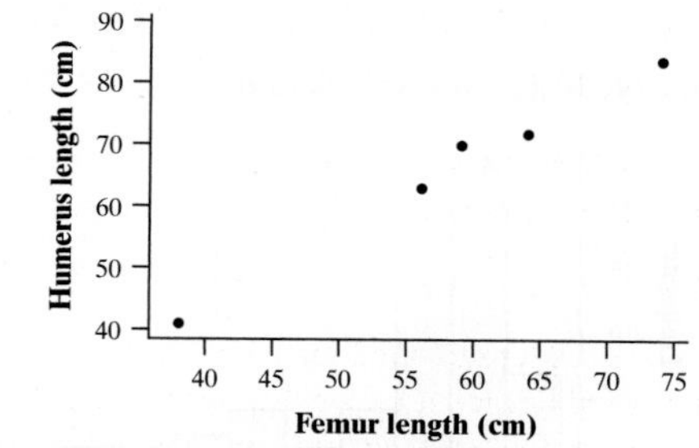


**(b)** The femur measurements have mean of 58.2 cm and a standard deviation of 13.2 cm. The humerus measurements have a mean of 66 cm and a standard deviation of 15.89 cm. The table shows the standardized measurements (labeled *z*femur and *z*humerus) and the product (*z*femur × *z*humerus) of the standardized measurements.

| Femur | Humerus | *z* femur | *z* humerus | Product |
|---|---|---|---|---|
| 38 | 41 | −1.53030 | −1.57332 | 2.40765 |
| 56 | 63 | −1.16667 | −0.18880 | 0.03147 |
| 59 | 70 | 0.06061 | 0.25173 | 0.01526 |
| 64 | 72 | 0.43939 | 0.37760 | 0.16591 |
| 74 | 84 | 1.19697 | 1.13279 | 1.35591 |

The sum of the products is 3.97620, so the correlation coefficient is $r = \frac{1}{4}(3.97620) = 0.9941$. The very high value of the correlation confirms the strong, positive linear association between femur length and humerus length in the scatterplot from part (a).

3.23 **(a)** Correlation is unitless. **(b)** The correlation would stay the same. Correlation makes no distinction between explanatory and response variables. **(c)** If sodium was measured in grams instead of milligrams, the correlation would still be 0.87. Because $r$ uses the standardized values of the observations, $r$ does not change when we change the units of measurement of $x$, or $y$, or both.

3.25 We would expect the height of women at age 4 and their height as women at age 18 to be the highest correlation because it is reasonable to expect taller children to become taller adults and shorter children to become shorter adults. The next highest would be the correlation between the heights of male parents and their adult children because they share genes. Tall fathers tend to have relatively tall sons, and short fathers tend to have relatively short sons. The lowest correlation would be between husbands and their wives. Some tall men may prefer to marry tall women, but this isn't always the case.

3.27 Answers will vary. Here is one possibility.

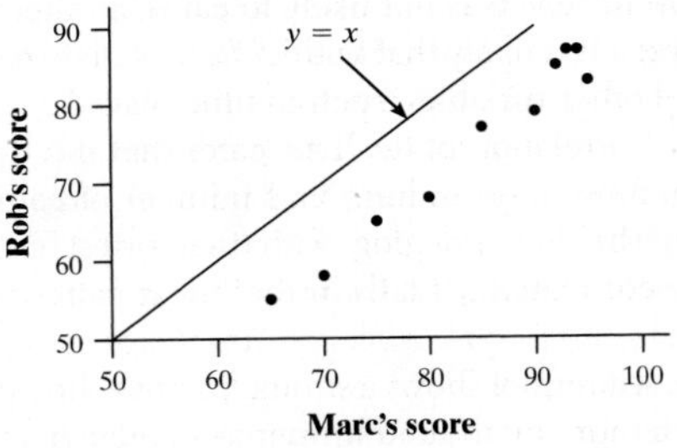


3.29 a

3.31 d

3.33 b

3.35 One possible histogram is shown here.

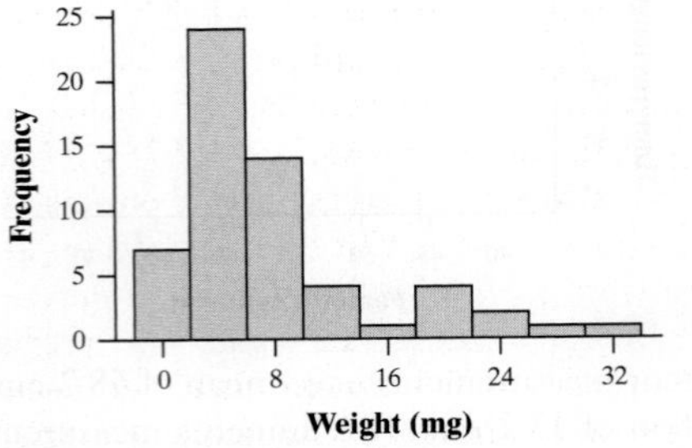


The distribution of weight is skewed to the right with several possible high outliers. The median weight is 5.4 mg and the *IQR* is 5.5 mg.

## Section 3.2

### Answers to Check Your Understanding

*page 181:* **1.** $\hat{y} = 100 + 40(16) = 740$ grams

**2.** The residual = actual $y$ − predicted $y$ = $700 - 740 = -40$ grams. *Interpretation:* This rat weighed 40 grams less than the weight predicted by the regression line with $x = 16$ weeks old.

**3.** The time is measured in weeks for this equation, so 2 years converts to 104 weeks. We then predict the rat's weight to be $\hat{y} = 100 + 40(104) = 4260$ grams, which is equivalent to 9.4 pounds (about the weight of a large newborn human). This is unreasonable and is the result of extrapolation.

*page 182:* **1.** The slope is 40. *Interpretation:* The predicted weight goes up by 40 grams for each increase of 1 week in the rat's age. **2.** Yes, the $y$ intercept has meaning in this context. The $y$ intercept, 100, is the predicted weight (in grams) for a rat at birth ($x = 0$ weeks).

*page 188:* **1.** The equation of the least-squares regression line is $\hat{y} = 16.2649 + 0.0908x$, where $x$ = body weight and $\hat{y}$ = the predicted backpack weight.

**2.** A residual plot is given.

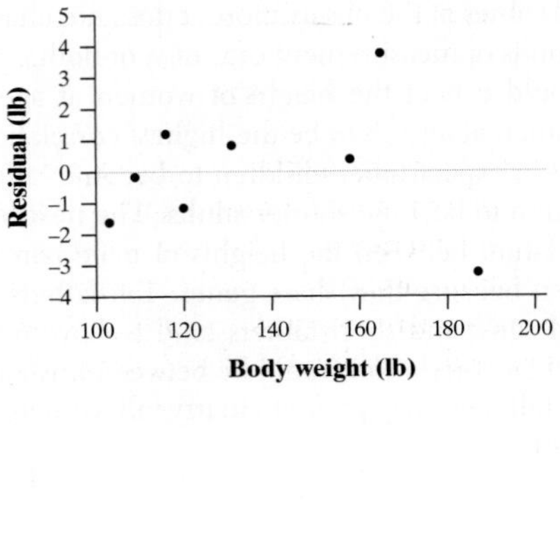


**3.** Because there appears to be a negative–positive–negative pattern in the residual plot, a linear model is not appropriate for these data.

*page 194:* **1.** $\hat{y} = 33.347 + 13.2854x$, where $x$ = duration of the most recent eruption (min) and $\hat{y}$ = predicted interval of time until the next eruption (min).

**2.** The predicted interval of time until the next eruption goes up by 13.2854 minutes for each increase of 1 minute in the duration of the most recent eruption.

**3.** The actual interval of time until the next eruption (min) is typically about 6.49 minutes away from the time predicted by the least-squares regression line with $x$ = duration of the most recent eruption (min).

**4.** $r^2 = 85.4\%$ of the variability in interval is accounted for by the least-squares regression line with $x$ = duration.

### Answers to Odd-Numbered Section 3.2 Exercises

3.37 **(a)** $\hat{y} = 60.7 + 0.139(200) = 88.5$ **(b)** $\hat{y} = 60.7 + 0.139(400) = 116.3$ **(c)** I am more confident in the prediction in part (a) than the prediction in part (b). The payroll for the teams varied from about \$75 million to about \$275 million. \$200 million is in this interval of payrolls, but \$400 million is not, which makes the prediction in part (b) an extrapolation.

3.39 The predicted number of wins for the Chicago Cubs, who spent \$182 million on payroll, is $\hat{y} = 60.7 + 0.139(182) = 85.998$ wins. The residual = actual $y$ − predicted $y$ = $103 - 85.998 = 17.002$. *Interpretation:* The Chicago Cubs won 17.002 more games than the number of games predicted by the regression line with $x$ = \$182 million.

3.41 **(a)** The slope is 0.139. *Interpretation:* The predicted number of wins goes up by 0.139 for each increase of \$1 million in payroll. **(b)** The $y$ intercept does not have meaning in this context. It is not reasonable for a team to have a payroll of \$0.

3.43 **(a)** The predicted number of steps for Kiana, who is 67 inches tall, is $\hat{y} = 113.6 - 0.921(67) = 51.893$ steps. The residual = actual $y$ − predicted $y$ = $49 - 51.893 = -2.893$. *Interpretation:* Kiana took about 2.893 fewer steps than the number of steps predicted by the regression line with $x$ = 67 inches. **(b)** Since Matthew is 10 inches taller than Samantha, I expect Matthew to take $(10)(-0.921) = 9.21$ fewer steps than Samantha.

3.45 **(a)** The regression lines are nearly parallel, but the $y$ intercept is much greater for the throwers. **(b)** 72-inch discus thrower: $\hat{y} = -115 + 5.13(72) = 254.36$ pounds; 72-inch sprinter: $\hat{y} = -297 + 6.41(72) = 164.52$ pounds. Based on the least-squares regression lines computed from the data, we expect a 72-inch discus thrower to weigh about 89.84 pounds more than a 72-inch sprinter.

3.47 No; there is an obvious negative–positive–negative pattern in the residual plot so a linear model is not appropriate for these data. A curved model would be better.

3.49 The predicted mean weight of infants in Nahya who are 1 month old is $\hat{y} = 4.88 + 0.267(1) = 5.147$ kg. From the residual plot, the mean weight of 1-month-old infants is about 0.85 kg less than predicted. So the actual mean weight of the infants when they were 1 month old is about $5.147 - 0.85 = 4.297$ kg.

3.51 **(a)** See part (b). **(b)** $\hat{y} = 300.04 + 2.829x$, where $\hat{y}$ = the predicted number of calories and $x$ = the amount of sugar (in grams).

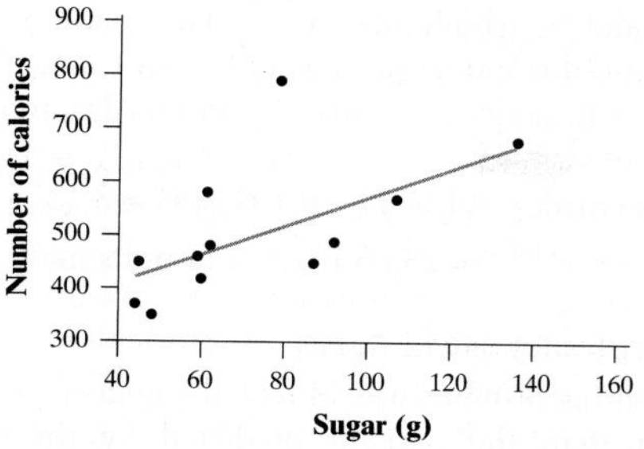


(c) The line calculated in part (b) is called the "least-squares" regression line because this is the line that makes the sum of the squares of the residuals as small as possible.

3.53 A residual plot is shown.

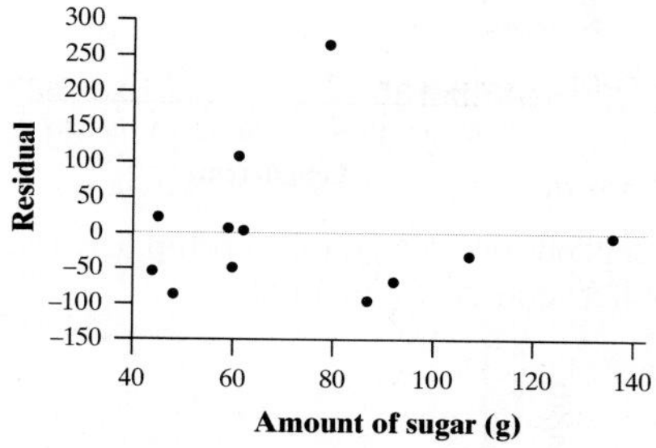


The linear model relating the amount of sugar to the number of calories is appropriate because there is no leftover pattern in the residual plot. The residuals look randomly scattered around the residual = 0 line.

3.55 **(a)** The actual number of strides required to walk the length of a school hallway is typically about 3.50 away from the number predicted by the least-squares regression line with $x$ = height of a student (in inches). **(b)** About 39.9% of the variability in number of steps required to walk the length of a school hallway is accounted for by the least-squares regression line with $x$ = height of a student (in inches).

3.57 **(a)** The predicted free skate score is $\hat{y} = -16.2 + 2.07(78.5) = 146.295$. The residual is $y - \hat{y} = 150.06 - 146.295 = 3.765$. *Interpretation:* Yu-Na Kim's free skate score was 3.765 points higher than predicted based on her short program score. **(b)** The slope is 2.07. *Interpretation:* The predicted free skate score increases by 2.07 points for each additional 1-point increase in the short program score. **(c)** The actual free skate score is typically about 10.2 points away from the score predicted by the least-squares regression line with $x$ = short program score. **(d)** About 73.6% of the variability in free skate score is accounted for by the least-squares regression line with $x$ = short program score.

3.59 **(a)** Yes; there is no leftover pattern in the residual plot, so a linear model is appropriate for these data. **(b)** $r^2 = 60.21\%$ and the slope is positive, so the correlation is $r = \sqrt{0.6021} = 0.776$. **(c)** $\hat{y} = 1.0021 + 0.0708x$, where $x$ is the number of Mentos and $\hat{y}$ is the predicted amount expelled. **(d)** The value of $s = 0.067$ cup. *Interpretation:* The actual amount expelled is typically about 0.067 cup away from the amount predicted by the least-squares regression line with $x$ = number of Mentos. The value of $r^2$ is 60.21%. *Interpretation:* About 60.21% of the variability in amount expelled is accounted for by the least-squares regression line with $x$ = number of Mentos.

3.61 **(a)** $\hat{y} = 11.898 - 0.041(42) = 10.176$ mph. The residual = actual $y$ − predicted $y = 2.2 - 10.176 = -7.976$ mph. *Interpretation:* The actual average wind speed was 7.976 mph less than the average wind speed predicted by the regression line with $x = 42$°F. **(b)** The slope is −0.041. *Interpretation:* The predicted average wind speed decreases by 0.041 mph for each additional 1-degree increase in average temperature (in degrees Fahrenheit). **(c)** The actual average wind speeds typically vary $s = 3.66$ mph from the values predicted by the least-squares regression line using $x$ = average temperature. **(d)** $r^2 = 4.8\%$ of the variability in average wind speed is accounted for by the least-squares regression line using $x$ = average temperature.

3.63 **(a)** The slope is $b = 0.5\left(\frac{2.7}{2.5}\right) = 0.54$. The $y$ intercept is $a = 68.5 - 0.54(64.5) = 33.67$. So the equation for predicting $y$ = husband's height from $x$ = wife's height is $\hat{y} = 33.67 + 0.5x$. **(b)** If the value of $x$ is one standard deviation below $\bar{x}$, the predicted value of $y$ will be $r$ standard deviations of $y$ below $\bar{y}$. So, the predicted value for the husband is $68.5 - 0.5(2.7) = 67.15$ inches.

3.65 **(a)** $\hat{y} = x$, where $\hat{y}$ = predicted grade on final and $x$ = grade on midterm. **(b)** A student with a score of 50 on the midterm is predicted to score $\hat{y} = 46.6 + 0.41(50) = 67.1$ on the final. A student with a score of 100 on the midterm is predicted to score $\hat{y} = 46.6 + 0.41(100) = 87.6$ on the final. **(c)** These predictions illustrate regression to the mean because the student who did poorly on the midterm (50) is predicted to do better on the final (closer to the mean), whereas the student who did very well on the midterm (100) is predicted to do worse on the final (closer to the mean).

3.67 **(a)** Because Jacob has an above-average height and an above-average vertical jump, his point increases the positive slope of the least-squares regression line and decreases the $y$ intercept. **(b)** Jacob's vertical jump is farther from the least-squares regression line than the other students' vertical jumps. Because Jacob's point has such a large residual, it increases the standard deviation of the residuals. Also, because Jacob's vertical jump is farther from the least-squares regression line than the other students' vertical jumps, the value of $r^2$ decreases. The linear association is weaker because of the presence of this point.

3.69 **(a)** There is a moderate, positive, linear association between HbA and FBG. Subject 15 (near the top) is a possible outlier and Subject 18 (to the far right) is a high-leverage point.

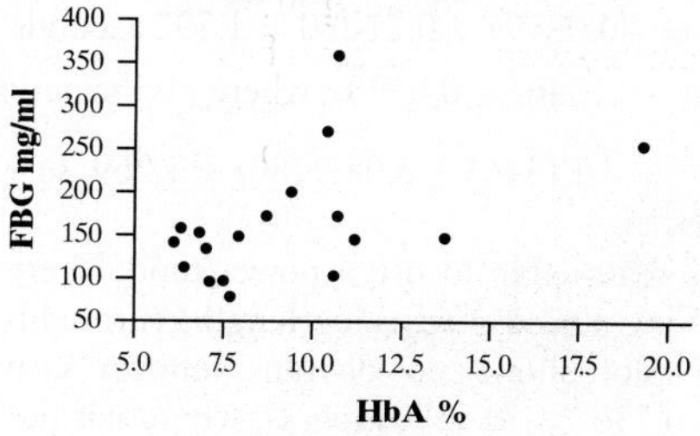


**(b)** Subject 18 is a high-leverage point because it has a much larger $x$-value than the other points. Because the point for subject 18 is in the positive, linear pattern formed by most of the data values, it will make the correlation closer to 1. Also, because the point is likely to be below the least-squares regression line, it will "pull down" the line on the right side, making the slope closer to 0. Without the outlier, the correlation decreases from $r = 0.4819$ to $r = 0.3837$, as expected. Likewise, without the outlier, the equation of the line changes from $\hat{y} = 66.4 + 10.4x$ to $\hat{y} = 52.3 + 12.1x$. **(c)** Subject 15 is an outlier because it has a large residual. The point for subject 15 makes the correlation closer to 0 because it decreases the strength of what would otherwise be a moderately strong positive association. Because this point's $x$ coordinate is very close to $\bar{x}$, it won't influence the slope very much. However, it will make the $y$ intercept increase because its $y$ coordinate is so large compared to the rest of the values. Without the outlier, the correlation increases from $r = 0.4819$ to $r = 0.5684$, the slope changes from 10.4 to 8.92, and the $y$ intercept increases from 66.4 to 69.5.

3.71 a

3.73 c

3.75 d

3.77 b

3.79 **(a)** (i) $z = \dfrac{25 - 18.7}{4.3} = 1.47$; 0.9292. (ii) normalcdf(lower: $-1000$, upper: 25, mean: 18.7, SD: 4.3) $= 0.9286$. About 93% percent of vehicles get worse combined mileage than the Chevrolet Malibu. **(b)** *Table A:* Look in the body of Table A for the value closest to 0.90. A $z$-score of 1.28 gives the closest value (0.8997). Solving $1.28 = \dfrac{x - 18.7}{4.3}$ gives $x = 24.2$. *Tech:* invNorm(area: 0.9, mean: 18.7, SD: 4.3) $= 24.2$. The top 10% of all vehicles get at least 24.2 mpg.

## Section 3.3

### Answers to Check Your Understanding

*page 228:* **1.** Option 1: $\widehat{\text{premium}} = -343 + 8.63(58) = \$157.54$

Option 2: $\widehat{\ln(\text{premium})} = -12.98 + 4.416(\ln 58) = 4.9509 \rightarrow \hat{y} = e^{4.9509} = \$141.30$

Option 3: $\widehat{\ln(\text{premium})} = -0.063 + 0.0859(58) = 4.9192 \rightarrow \hat{y} = e^{4.9192} = \$136.89$

**2.** The exponential model (Option 3) best describes the relationship because this model produced the most randomly scattered residual plot with no leftover curved pattern. This model also has the greatest $r^2$ value of the three options.

### Answers to Odd-Numbered Section 3.3 Exercises

3.81 **(a)** $\hat{y} = -0.08594 + 0.21\sqrt{x}$, where $y$ is the period and $x$ is the length. **(b)** $\hat{y} = -0.08594 + 0.21\sqrt{80} = 1.792$ seconds

3.83 **(a)** $\widehat{y^2} = -0.15465 + 0.0428x$, where $y$ is the period and $x$ is the length. **(b)** $\widehat{y^2} = -0.15465 + 0.0428(80) = 3.269$, so $\hat{y} = \sqrt{3.269} = 1.808$ seconds.

3.85 **(a)** It is reasonable to use a power model here because the scatterplot of log(period) versus log(length) is roughly linear. Also, the residual plot shows no obvious leftover curved patterns. **(b)** $\widehat{\log y} = -0.73675 + 0.51701\log(x)$, where $y$ is the period and $x$ is the length. **(c)** $\widehat{\log y} = -0.73675 + 0.51701\log(80) = 0.24717$, so $\hat{y} = 10^{0.24717} = 1.77$ seconds.

3.87 $\widehat{\log y} = 1.01 + 0.72\log(127) = 2.525$, so $\hat{y} = 10^{2.525} = 334.97$ grams is the predicted brain weight of Bigfoot.

3.89 **(a)** Because the scatterplot of ln(count) versus time is fairly linear, an exponential model would be reasonable. **(b)** $\widehat{\ln y} = 5.97316 - 0.218425x$, where $y$ is the count of surviving bacteria (in hundreds) and $x$ is time in minutes. **(c)** $\widehat{\ln y} = 5.97316 - 0.218425(17) = 2.26$, so $\hat{y} = e^{2.26} = 9.58$ or 958 bacteria.

3.91 **(a)** Model 3, which uses $x$ = distance and $y$ = ln(percent made), because this model produced the most randomly scattered residual plot with no leftover curved pattern. Also, model 3 has the largest value of $r^2$.

**(b)** $\widehat{\ln(\text{percent made})} = 4.6649 - 0.1091(14) = 3.1375$

$\widehat{\text{percent made}} = e^{3.1375} = 23.05$ percent of putts made from 14 feet away

Residual $= 31\% - 23.05\% = 7.95\%$

When the putting distance was 14 feet, the golfers' percent made is 7.95 greater than the percent predicted by the model using $x$ = distance and $y$ = ln(percent made).

3.93 **(a)**

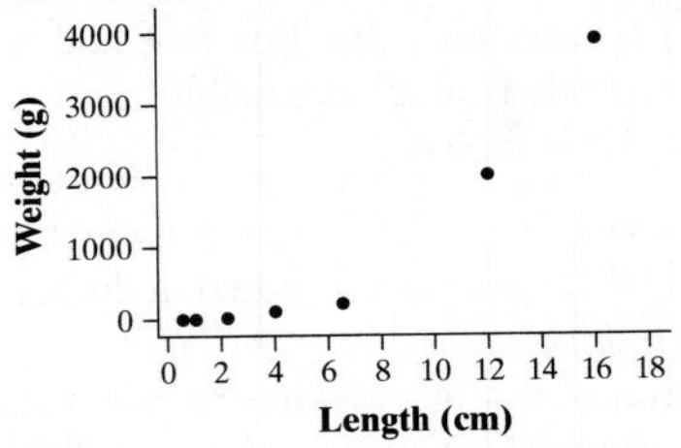


There is a strong, positive curved relationship between heart weight and length of left ventricle for mammals.

**(b)**

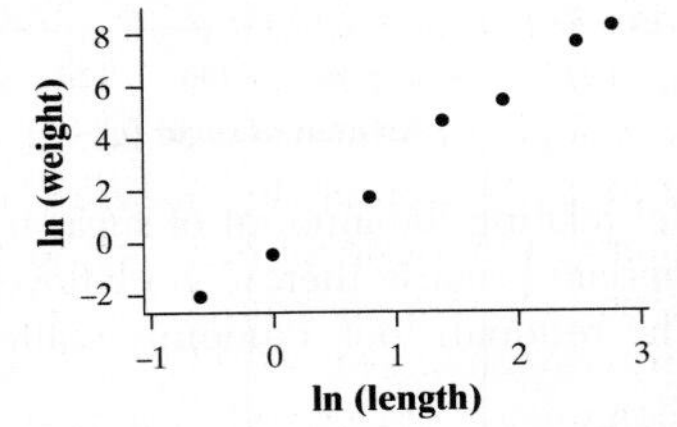


Because the relationship between ln(weight) and ln(length) is roughly linear, heart weight and length seem to follow a power model. An exponential model would not be appropriate because the relationship between ln(weight) and length is clearly curved. **(c)** $\widehat{\ln y} = -0.314 + 3.1387\ln x$, where $y$ is the weight of the heart and $x$ is the length of the cavity of the left ventricle. **(d)** $\widehat{\ln y} = -0.314 + 3.1387\ln(6.8) = 5.703$, so $\hat{y} = e^{5.703} = 299.77$ grams.

3.95 c

3.97 **(a)** $z = \dfrac{3 - 4.5}{0.9} = -1.67$ and $z = \dfrac{6 - 4.5}{0.9} = 1.67$

*Table A:* $0.9525 - 0.0475 = 0.9050$

*Tech:* normalcdf(lower: 3, upper: 6, mean: 4.5, SD: 0.9) $= 0.9044$

There is a 0.9044 probability that Marcella's shower lasts between 3 and 6 minutes.

**(b)** A point is considered to be an outlier if it is more than $1.5IQR$ above $Q_3$, so we need to find the values of $Q_1$ and $Q_3$ for Marcella. $Q_1$ is the boundary value $x$ with 25% of the distribution to its left and $Q_3$ is the boundary value with 75% of the distribution to its left. *Table A:* Look in the body of Table A for the value closest to 0.25 and 0.75. A $z$-score of $-0.67$ gives the closest value to 0.25 and a $z$-score of 0.67 gives the closest value to 0.75. Solving $-0.67 = \dfrac{Q_1 - 4.5}{0.9}$ gives $Q_1 = 3.897$ minutes. Solving $0.67 = \dfrac{Q_3 - 4.5}{0.9}$ gives $Q_3 = 5.103$ minutes.

*Tech:* invNorm(area: 0.25, mean 4.5, SD: 0.9) gives $Q_1 = 3.893$ minutes invNorm(area: 0.75, mean: 4.5, SD: 0.9) gives $Q_3 = 5.107$ minutes. An outlier is any value above $5.107 + 1.5(5.107 - 3.893) = 6.928$. Because $7 > 6.928$, a shower of 7 minutes would be considered an outlier for Marcella.

## Answers to Chapter 3 Review Exercises

R3.1 **(a)** There is a moderate, positive, linear association between gestation and life span. Without the unusual points at the top and in the upper right, the association appears moderately strong, positive, and curved. **(b)** The hippopotamus makes the correlation closer to 0 because it decreases the strength of what would otherwise be a moderately strong positive association. Because this point's $x$ coordinate is very close to $\bar{x}$, it won't influence the slope very much. However, it makes the $y$ intercept higher because its $y$ coordinate is so large compared to the rest of the values. Because it has such a large residual, it increases the standard deviation of the residuals. **(c)** Because the Asian elephant is in the positive, linear pattern formed by most of the data values, it will make the correlation closer to 1. Also, because the point is likely to be above the least-squares regression line, it will "pull up" the line on the right side, making the slope larger and the $y$ intercept smaller. Because this point is likely to have a small residual, it decreases the standard deviation of the residuals.

R3.2 **(a)** A positive relationship means that dives with larger values of depth also tend to have larger values of duration. **(b)** A linear relationship means that when depth increases by 1 meter, dive duration tends to change by a constant amount, on average. **(c)** A strong relationship means that the (dive depth, dive duration) data points fall close to a line. **(d)** If the variables are reversed, the correlation will remain the same. However, the slope and $y$ intercept will be different.

R3.3 **(a)** A linear model is appropriate for these data because there is no leftover pattern in the residual plot. **(b)** Because $r^2 = 0.837$ and the slope is positive, the correlation $r = +\sqrt{0.837} = 0.915$. *Interpretation:* The correlation of $r = 0.915$ confirms that the linear association between the age of cars and their mileage is strong and positive. **(c)** $\hat{y} = 3704 + 12{,}188x$, where $x$ represents the age and $\hat{y}$ represents the predicted mileage of the cars. **(d)** For a 6-year-old car, the predicted mileage is $\hat{y} = 3704 + 12{,}188(6) = 76{,}832$. The residual for this particular car is $y - \hat{y} = 65{,}000 - 76{,}832 = -11{,}832$. *Interpretation:* The actual number of miles this teacher has driven was 11,832 less than the number of miles predicted by the regression line with $x = 6$ years. **(e)** The value of $s = 20{,}870.5$ miles. *Interpretation:* The actual number of miles is typically about 20,870.5 miles away from the number of miles predicted by the least-squares regression line with $x$ = age (in years). The value of $r^2 = 83.7\%$. *Interpretation:* About 83.7% of the variability in mileage is accounted for by the least-squares regression line with $x$ = age (in years).

R3.4 **(a)** Average March temperature is the explanatory variable because changes in March temperature probably have an effect on the date of first blossom. Also, we are predicting the date of first blossom from temperature.

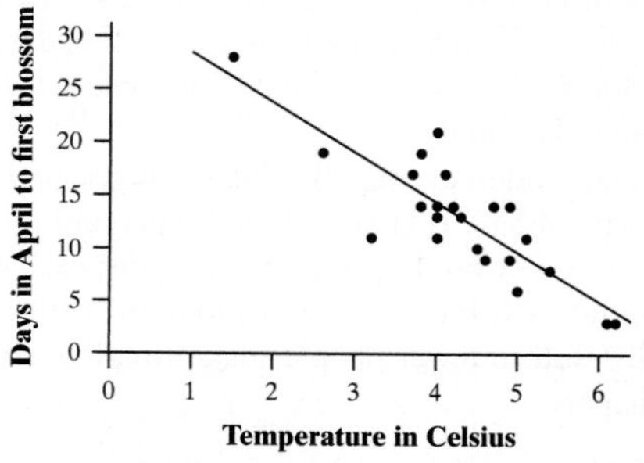


**(b)** The correlation is $r = -0.85$; $\hat{y} = 33.12 - 4.69x$, where $x$ represents the average March temperature and $\hat{y}$ represents the predicted number of days. The slope is $-4.69$. *Interpretation:* The predicted number of days in April to the first blossom decreases by 4.69 days for each additional 1-degree increase in average March temperature (in degrees Celsius). The $y$ intercept tells us that if the average March temperature was 0 degrees Celsius, the predicted number of days in April to first blossom is 33.12 (May 3). However, $x = 0$ is outside of the range of data, so this prediction is an extrapolation and may not be trustworthy. **(c)** No, $x = 8.2$ is well beyond the values of $x$ we have in the data set (1.5 to 6.2). This prediction would be an extrapolation. **(d)** The predicted number of days until first blossom when the average March temperature was 4.5°C is $\hat{y} = 33.12 - 4.69(4.5) = 12.015$. The residual is $y - \hat{y} = 10 - 12.015 = -2.015$. *Interpretation:* The actual number of days until first blossom was 2.015 days less than the number of days predicted by the regression line with $x = 4.5$°C.
**(e)**

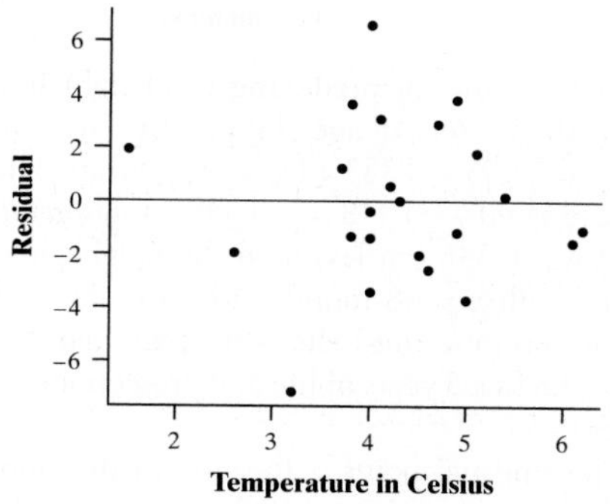


There is no leftover pattern in the residuals, indicating that a linear model is appropriate.

R3.5 **(a)** $b = 0.6\left(\frac{8}{30}\right) = 0.16$; $a = 75 - 0.16(280) = 30.2$; $\hat{y} = 30.2 + 0.16x$, where $\hat{y}$ = the predicted final exam score and $x$ = total score before the final examination. **(b)** $\hat{y} = 30.2 + 0.16(300) = 78.2$ **(c)** The least-squares regression line is the line that minimizes the sum of squared distances between the actual exam scores and predicted exam scores. **(d)** Because $r^2 = 0.36$, only 36% of the variability in the final exam scores is accounted for by the linear model relating final exam scores to total score before the final exam. More than half (64%) of the variation in final exam scores is *not* accounted for by the least squares regression line, so Julie has a good reason to think this is not a good estimate.

R3.6 Even though there is a high correlation between number of calculators and math achievement, we shouldn't conclude that increasing the number of calculators will *cause* an increase in math achievement. It is possible that students who are more serious about school have better math achievement and also have more calculators.

R3.7 **(a)** The predictions of the price of a diamond of this type that weighs 2 carats are:

Model 1: $\widehat{\text{price}} = -98666 + 105932(2) = \$113{,}198$

Model 2: $\widehat{\ln(\text{price})} = 9.7062 + 2.2913\ln(2) = 11.2944$, therefore $\widehat{\text{price}} = e^{11.2944} \approx \$80{,}370$.

Model 3: $\widehat{\ln(\text{price})} = 8.2709 + 1.3791(2) = 11.0291$, therefore $\widehat{\text{price}} = e^{11.0291} \approx \$61{,}642$.

**(b)** The model that does the best job of summarizing the relationship between weight and price is Model 2, which uses $x = \ln(\text{price})$ and $y = \ln(\text{weight})$, because this model produced the most randomly scattered residual plot with no leftover curved pattern. Also, model 2 has the largest value of $r^2$.

### Answers to Chapter 3 AP® Practice Test

| | | | |
|---|---|---|---|
| T3.1 e | T3.2 d | T3.3 e | T3.4 a |
| T3.5 c | T3.6 b | T3.7 e | T3.8 b |
| T3.9 e | T3.10 d | T3.11 c | |

T3.12 (a) There is a strong, positive linear association between Sarah's age and her height.

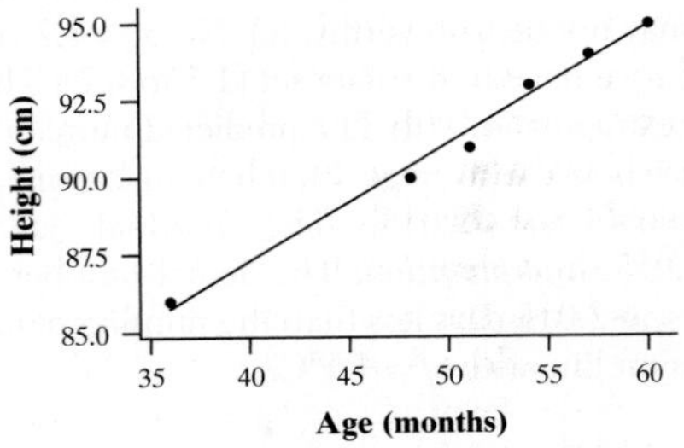


(b) The regression line for predicting $y$ = height from $x$ = age is $\hat{y} = 71.95 + 0.3833x$. (c) At age 48 months, we predict Sarah's height to be $\hat{y} = 71.95 + 0.3833(48) = 90.348$ cm. The residual for Sarah is $y - \hat{y} = 90 - 90.348 = -0.348$. *Interpretation:* Sarah's actual height was 0.348 cm less than the height predicted by the regression line with $x = 48$ months. (d) No; obviously, the linear trend will not continue until she is 40 years old. Our data were based only on the first 5 years of life and predictions should only be made for ages 0–5.

T3.13 (a) The unusual point is the one in the upper righthand corner with isotope value about −19.3 and silicon value about 345. This point is unusual in that it has such a high silicon value for the given isotope value. (b) (i) If the point were removed, the correlation would get closer to −1, because it does not follow the linear pattern of the other points. (ii) Because this point is "pulling up" the line on the right side of the plot, removing it will make the slope steeper (more negative) and make the $y$ intercept smaller. Note that the $y$-axis is to the *right* of the points in the scatterplot. (iii) Because this point has a large residual, removing it will make the size of the typical residual ($s$) a little smaller.

T3.14 (a) Yes; because there is no obvious leftover pattern in the residual plot, a linear model is appropriate for describing the relationship between wildebeest abundance and percent of grass area burned. (b) $\hat{y} = 92.29 - 0.05762x$, where $x$ represents the number of wildebeest and $\hat{y}$ represents the predicted percent of the grass burned. (c) The slope $= -0.05762$. *Interpretation:* The predicted percent of grassy area burned decreases by about 0.058 percent for each additional 1000 wildebeest. The $y$ intercept does not have meaning in this context, as making a prediction for 0 wildebeest is a big extrapolation. It is impossible to know what would happen going from some wildebeest to no wildebeest. (d) The value of $s = 15.988\%$. *Interpretation:* The actual percentage of burned area is typically about 15.988% away from the percent predicted by the least-squares regression line with $x$ = number of wildebeest (1000s). The value of $r^2 = 64.6\%$. *Interpretation:* About 64.6% of the variability in percentage of burned area is accounted for by the least-squares regression line with $x$ = number of wildebeest (1000s).

T3.15 (a) There is clear curvature evident in both the scatterplot and the residual plot.

(b) Option 1: $\hat{y} = 2.078 + 0.0042597(30)^3 = 117.09$ board feet.
Option 2: $\widehat{\ln y} = 1.2319 + 0.113417(30) = 4.63441$ and $\hat{y} = e^{4.63441} = 102.967$ board feet.

(c) The residual plot for Option 1 is much more scattered, while the residual plot for Option 2 is curved, meaning that the model relating the amount of usable lumber to cube of the diameter is more appropriate. Thus, the prediction of 117.09 board feet seems more reliable.

# Chapter 4

## Section 4.1

### Answers to Check Your Understanding

*page 253:* 1. Convenience sample; this could lead him to overestimate the quality if the farmer puts the best oranges on top or if the oranges at the bottom of the crate are damaged from the weight on top of them.
2. Voluntary response sample; those who are happy that the United Nations has its headquarters in the U.S. already have what they want and so are less likely to worry about responding to the question. This means that the proportion who answered "No" in the sample is likely to be higher than the true proportion in the U.S. who would answer "No."

*page 256:* 1. Number the pieces of wood from 1 to 1000. Use the command randInt(1,1000) to select 5 different integers from 1 to 1000. Inspect the corresponding 5 pieces of wood.
2. Number the pieces of wood from 000 to 999. Move along a line of random digits from left to right, reading three-digit numbers, until 5 different numbers between 000 and 999 have been selected. Inspect the corresponding 5 pieces of wood.

*page 261:* 1. Because the quality of the pencils might be the same within each shift, but differ across shifts, use shifts as strata. At the end of each 8-hour shift, label all the pencils produced during that shift from 1 to 5000. Generate 100 different random integers from 1 to 5000 and select those pencils for inspection.
2. The boxes of pencils could be used as clusters because it would be relatively easy to select boxes. At the end of the day, label all the boxes of pencils from 1 to 1500. Generate 30 different random integers from 1 to 1500 and inspect all the pencils in the selected boxes.
3. Select every 15,000/300 = 50th pencil that comes off the production line. As a starting point, randomly select a number from 1 to 50. Select that pencil and every 50th pencil thereafter until 300 pencils have been selected.
4. *Stratified:* We are guaranteed to inspect 100 pencils from each of the three shifts. This will lead to a more precise estimate of overall quality if quality is consistent within each shift but differs across the three shifts. *Cluster:* Simplifies the sampling process. Rather than having to label every pencil produced, only the boxes of 10 pencils would need to be labeled. *Systematic:* It is easier to find the selected pencils if they are selected as they come off the production line. Also, we are guaranteed to obtain a sample of pencils that were manufactured at regular intervals over the course of the day. If we inspect the selected pencils throughout the day, this would allow us to address any manufacturing issues before too many faulty pencils are manufactured.

*page 263:* 1. (a) Undercoverage (b) Nonresponse (c) Convenience
2. The estimate of 84% is greater than the percent of all people in the population who would oppose banning disposable diapers. By making it sound as if they are not a problem in the landfill, this question will result in fewer people suggesting that we should ban disposable diapers.

### Answers to Odd-Numbered Section 4.1 Exercises

4.1 *Population:* The 1000 envelopes stuffed during a given hour. *Sample:* The 40 randomly selected envelopes.

4.3 *Population:* All local businesses. *Sample:* The 73 businesses that return the questionnaire.
4.5 **(a)** A convenience sample. **(b)** The estimate of 7.2 hours is probably less than the true average because students who arrive first to school had to wake up earlier and may have gotten less sleep than those students who are able to sleep in.
4.7 Voluntary response sample; it is likely that those customers who volunteered to leave reviews feel strongly about the hotel, often due to a negative experience. As a result, the 26% from the sample is likely greater than the true percentage of all the hotel's customers who would give the hotel 1 star.
4.9 Voluntary response sample; it is likely that the true proportion of constituents who oppose the bill is less than 871/1128.
4.11 **(a)** Number the 40 students from 01 to 40 alphabetically. Moving left to right along a line from the random digit table, record two-digit numbers, skipping any numbers that are not between 01 and 40 and any repeated numbers, until you have 5 different numbers between 01 and 40. Select the corresponding 5 students. **(b)** Johnson (20), Drasin (11), Washburn (38), Rider (31), and Calloway (07).
4.13 **(a)** Number the plots from 1 to 1410. Use the command randInt(1,1410) to select 141 different integers from 1 to 1410. Select the corresponding 141 plots. **(b)** Answers will vary.
4.15 Stratified random sampling might be preferred in this context because the employees' opinions might be the same within each type of employee (servers, kitchen staff), but differ across employee types. Using employee type as the strata will help provide a more precise estimate of the overall proportion who approve of the no-tipping policy. Select an SRS of 15 employees who are servers and 15 employees who work in the kitchen to form the overall sample.
4.17 No; in an SRS, each possible sample of 250 engineers is equally likely to be selected. However, the method described restricts the sample to having exactly 200 males and 50 females.
4.19 **(a)** Cluster sampling **(b)** In an SRS, the company would have to visit individual homes all over the rural subdivision. With the cluster sampling method, the company has to visit only 5 locations, saving time and money.
4.21 **(a)** Because satisfaction with the property is likely to vary depending on the location of the room, we should stratify by floor and view. From each floor, randomly select 2 rooms with each view. Using a stratified random sample would ensure that the manager got opinions from each type of room and would provide a more precise estimate of customer satisfaction. **(b)** Using floors as clusters, survey the registered guest in every room on each of 3 randomly selected floors. This would be a simpler option because the manager would need to survey guests on only three floors instead of all over the hotel.
4.23 **(a)** *SRS:* It is not practical to number every tree along the highway and then search for the trees that are selected. **(b)** This convenience sample is not a good idea because these trees are unlikely to be representative of the population. **(c)** Because $5000/200 = 25$, randomly select one of the first 25 trees along the highway and then select every 25th tree thereafter.
4.25 Students who do not live in the dorms cannot be part of the sample. Some would live off campus and therefore be less likely to eat on campus than those students who live in the dorm, so the director's estimate for the percent of students who eat regularly on campus will likely be too high.
4.27 People who did not lose much weight (or who gained weight) after participating in the course may be less likely to respond to the survey. This will likely produce an estimated weight loss that is too large, as the people who responded to the survey probably lost more weight than those who did not respond.
4.29 We would not expect many people to claim they have run red lights when they have not, but some people will deny running red lights when they have. Thus, the proportion of drivers obtained in the sample who admitted to running a red light is likely to be less than the proportion who have actually run a red light.
4.31 When asked in person, the boys may claim that they have never cried during a movie (when in reality, they have) because they are embarrassed or ashamed to admit it to the girls asking the survey question. Boys who were given an anonymous survey are more likely to be honest about their experiences.
4.33 **(a)** The wording is clear, but the question is slanted in favor of warning labels because the first sentence states that some cell-phone users have developed brain cancer. **(b)** The question is clear, but it is slanted in favor of national health insurance by asserting it would reduce administrative costs and not providing any counter-arguments. The phrase "do you agree" also pushes respondents toward the desired response. **(c)** Not clear; for those who do understand the question, it is slanted because it suggests reasons why one should support recycling. It could be rewritten as: "Do you support economic incentives to promote recycling?"
4.35 c
4.37 d
4.39 d
4.41 **(a)** The predicted number of points scored decreases by 4.084 points for each additional turnover. **(b)** The actual number of points scored is typically about 57.3 points away from the number of points predicted by the least-squares regression line with $x =$ number of turnovers. **(c)** $\hat{y} = 460.2 - 4.084(17) = 390.772$, so the residual is $y - \hat{y} = 238 - 390.772 = -152.772$. The number of points scored by the San Francisco 49ers was 152.772 points less than predicted based on their number of turnovers. **(d)** Because their point falls below the least-squares regression line and is to the left of the mean number of turnovers, their point decreases the $y$-intercept and increases the slope, making the slope closer to 0 (i.e., less negative). Because the 49ers' point is farther from the line than the rest of the points in the data set, this point increases the standard deviation of the residuals.

## Section 4.2

### Answers to Check Your Understanding

*page 273:* **1.** Experiment because a treatment (brightness of screen) was imposed on the laptops.
**2.** Observational; students were not assigned to eat a particular number of meals with their family per week.
**3.** *Explanatory:* The number of meals per week teens ate with their families. *Response:* GPA (or some other measure of their grades).
**4.** Students who have part-time jobs may not be able to eat many meals with their families and may not have much time to study, leading to lower grades. So we can't conclude that not eating with their family is the cause.
*page 278:* **1.** The patients and the researchers know who is receiving which treatment. This knowledge could motivate some people to take other measures (e.g., exercising more or eating better in general) that would also influence their heart health.
**2.** Use a double-blind experiment; the patients would not know which treatment they received, nor would the researchers know what treatment each patient received.
*page 283:* **1.** A control group would show how much electricity customers tend to use naturally. This would serve as a baseline to

determine how much less electricity is used in each of the treatment groups.
**2.** Number the houses from 1 to 60. Write the numbers 1 to 60 on slips of paper. Shuffle well. Draw out 20 slips of paper (without replacement). Those households will receive a display. Draw out another 20 slips of paper (without replacement). Those households will receive a chart. The remaining 20 households will receive only information about energy consumption.
**3.** To create groups of households that are roughly equivalent at the beginning of the experiment; this will ensure that the effects of other variables (e.g., the thrifty inclination of some households) are spread evenly among the three groups.
*page 289:* **1.** Number the volunteers 1 to 300. Use a random number generator to produce 100 different random integers from 1 to 300 and show the first advertisement to the volunteers with those numbers. Generate 100 additional different random integers from 1 to 300 and show the second advertisement to the volunteers with those numbers. The remaining 100 volunteers will view the third advertisement. Compare the effectiveness of the advertisements for the three groups.
**2.** Because the effectiveness of the ads may depend on a volunteer's familiarity with Jane Austen, block by whether or not the individuals are familiar with the works of Jane Austen. The volunteers in each block would be numbered and randomly assigned to one of three treatment groups: One-third of the volunteers in each block would view the first advertisement, one-third would view the second advertisement, and one-third would view the third advertisement. After viewing the advertisements, the researchers would gauge the effectiveness of the advertisements.
**3.** A randomized block design accounts for the variability in effectiveness that is due to subjects' familiarity with the works of Jane Austen. This makes it easier to determine the effectiveness of the three different advertisements.

## Answers to Odd-Numbered Section 4.2 Exercises

4.43 Although eating seafood may decrease the risk of colon cancer, it is possible that the physicians who ate seafood were also more likely to exercise. Because exercise might decrease colon cancer risk, perhaps the exercise caused the decrease in colon cancer risk, not eating seafood.
4.45 *Explanatory:* Type of program the people watched. *Response:* Number of calories consumed. This was an experiment because the treatments (20 minutes of *The Island*, 20 minutes of *The Island* without sound, and 20 minutes of the interview show) were deliberately imposed on the students.
4.47 **(a)** *Explanatory:* Amount of time in child care from birth to age 4½. *Response:* Adult ratings of their behavior. **(b)** A prospective observational study; no treatments were assigned and the researchers followed the children through their 6th year in school, asking adults to rate their behavior several times along the way. **(c)** No, this is an observational study, so we cannot make a cause-and-effect conclusion. For example, children who spend more time in child care probably have less time with their parents and get less instruction about proper behavior.
4.49 *Experimental units:* Pine seedlings; *Treatments:* Full light, 25% light, and 5% light.
4.51 **(a)** The factors are (1) information provided by interviewer, which had 3 levels, and (2) whether the caller offered survey results, which had 2 levels. **(b)** 6 treatments **(c)** Answers may vary. Here are two treatments: (1) giving name/no survey results; (2) identifying university/no survey results.
4.53 *Explanatory:* Type of snack that was eaten (berries or candy). *Response:* Amount of pasta consumed (measured by the number of calories consumed). *Experimental units:* Women; *Treatments:* 65 calories of berries or 65 calories of candy.
4.55 The control group can be used to show how changes in pain and joint stiffness progress over the three years naturally. If a control group was not used, the treatment could possibly be deemed ineffective at improving the symptoms, when in reality it may be effective at preventing the symptoms from worsening.
4.57 There was no control group. We do not know if this was a placebo effect or if the flavonols actually affected the blood flow. To make a cause-and-effect conclusion possible, we need to randomly assign some subjects to get flavonols and others to get a placebo.
4.59 Yes; if the treatment (ASU or placebo) assigned to a subject was unknown to both the subject and those responsible for assessing the effectiveness of that treatment. If subjects knew they were receiving the placebo, their expectations would differ from those who received the ASU. Then it would be impossible to know if a decrease in pain was due to the difference in expectations or the ASU. It is important for the experimenters to be blind so that they will be unbiased in the way that they interact and assess the subjects.
4.61 Because the experimenter knew which subjects had learned the meditation techniques, he or she is not blind. If the experimenter believed that meditation was beneficial, he or she may subconsciously rate subjects in the meditation group as being less anxious.
4.63 **(a)** Write each name on a slip of paper, put them in a container and mix thoroughly. Pull out 40 slips of paper and assign these subjects to Treatment 1. Then pull out 40 more slips of paper and assign these subject to Treatment 2. The remaining 40 subjects are assigned to Treatment 3. **(b)** Assign the students numbers from 1 to 120. Using the command RandInt(1,120), generate 40 unique integers from 1 to 120, and assign the corresponding students to Treatment 1. Then generate an additional 40 unique integers from 1 to 120 and assign the corresponding students to Treatment 2. The remaining 40 students are assigned to Treatment 3. **(c)** Assign the students numbers from 001 to 120. Pick a spot on Table D and read off the first 40 unique numbers between 001 and 120. The students corresponding to these numbers are assigned to Treatment 1. The students corresponding to the next 40 unique numbers between 001 and 120 are assigned to Treatment 2. The remaining 40 students are assigned to Treatment 3.
4.65 The coach's plan does not include random assignment. Perhaps the more motivated players will choose the new method. If they improve more by the end of the study, the coach cannot be sure if it was the exercise program or player motivation that caused the improvement.
4.67 **(a)** Researchers used a design that compared infants who were assigned to one of three treatments. **(b)** Random assignment helps to create three groups of infants who are roughly equivalent at the beginning of the study. This ensures that the effects of other variables (e.g., genetics) are spread evenly among the three groups of infants. **(c)** Birthweight and whether or not the baby was born prematurely; it is beneficial to control these variables, because otherwise they would provide additional sources of variability that would make it harder to determine the effectiveness of the treatments. **(d)** Having about 17 infants in each group makes it easier to rule out the chance variation in random assignment as a possible explanation for the differences observed in intelligence, language, and memory by age 4.

4.69 **(a)** Expense and condition of the patient; if a patient is in very poor health, a doctor might choose not to recommend surgery because of the added complications. Then, if the non-surgery treatment has a higher death rate, we will not know if it is because of the treatment or because the initial health of the subjects was worse. **(b)** Write the names of all 300 patients on identical slips of paper, put them in a hat, and mix them well. Draw out 150 slips and assign the corresponding subjects to receive surgery. The remaining 150 subjects receive the new method. At the end of the study, count how many patients survived in each group.

4.71 **(a)** Because smaller trees are likely to have fewer oranges than medium or large trees, form blocks based on the size of the trees (small, medium, large). Then randomly assign one-third of the small trees to fertilizer A, one-third to fertilizer B, and one-third to fertilizer C. Do the same for the other two blocks. In the end, measure each tree's orange production. **(b)** In a completely randomized design, the differences in tree size will increase the amount of variability in number of oranges for each treatment. However, a randomized block design would help us account for the variability in number of oranges that is due to the differences in tree size. This will make it easier to determine if one fertilizer is better than the others.

4.73 **(a)** The blocks are the different diagnoses (e.g., asthma) because the treatments (doctor or nurse-practitioner) were assigned to patients within each diagnosis. **(b)** Using a randomized block design allows us to account for the variability due to differences in diagnosis by comparing the results within each block. In a completely randomized design, the variability due to differences in diagnoses will be unaccounted for and will make it harder to determine if there is a difference in health and satisfaction due to the difference between doctors and nurse-practitioners. **(c)** *Advantage:* There would be no variability in the response variables introduced by differences in diagnosis. *Disadvantage:* We would only be able to make conclusions about health and satisfaction with their medical care for diabetic patients like the ones in the experiment.

4.75 **(a)** If all rats from litter 1 were fed Diet A and if these rats gained more weight, we would not know if this was because of the diet or because of initial health. Initial health is a confounding variable because it is related to both diet and weight gain. **(b)** For each litter, randomly assign half of the rats to receive Diet A and the other half to receive Diet B. Number the rats in the first litter from 1 to 10. Use the command randInt(1,10) to select 5 different integers from 1 to 10. Select the corresponding 5 rats and assign them to Diet A. The other 5 rats in this litter will be assigned to Diet B. Repeat the same process for the second litter of rats. **(c)** Answers will vary.

4.77 **(a)** Write the names of all 30 students on identical slips of paper, put them in a hat, and mix them well. Draw out 15 slips and assign the corresponding students to take the online SAT preparation program. The remaining 15 students will take the in-person SAT preparation class. At the end of the study, compare the average improvement in SAT score for each group. **(b)** Create pairs of similar students based on their previous SAT score (the two students with the highest SAT scores are paired together, and so on). Within each pair, one student is randomly assigned to the online preparation class and the other is assigned to the in-person preparation class. For each student, record the SAT improvement after receiving the treatment. **(c)** The matched pairs design is preferred. By matching students by initial score, we can account for the variability in improvement that is introduced by the variability in student ability, making it easier to determine which program is more effective.

4.79 **(a)** This is a matched pairs design because each subject was assigned both treatments. **(b)** Some students are more distractible than others. In a completely randomized design, the differences between the students will add variability to the response variable, making it harder to detect if there is a difference caused by the treatments. In a matched pairs design, each student is compared with himself or herself, so the differences between students are accounted for. **(c)** If all the students used the hands-free phone during the first session and performed worse, we would not know if the better performance during the second session was due to the lack of phone or to learning from their mistakes. By randomizing the order, some students will use the hands-free phone during the first session and others will use it during the second session. **(d)** The simulator, route, driving conditions, and traffic flow were all kept the same for both sessions. That way, the researchers are preventing these variables from adding variability to the response variable.

4.81 **(a)** If the students find a difference between the two groups, they will not know if it is due to gender or due to the deodorant. **(b)** Use a matched pairs design. In this case, each student would have one armpit randomly assigned to receive deodorant A and the other to receive deodorant B. Because each person uses both deodorants, there is no longer any confounding between gender and deodorant. Also, the paired design accounts for the variability between individuals, making it easier to see any difference in the effectiveness of the two deodorants.

4.83 c

4.85 b

4.87 c

4.89 b

4.91 **(a)** (i) $z = -023$; $1 - 0.4090 = 0.5910$ (ii) normalcdf(lower: 500, upper: 10000, mean: 525, SD: 110) $= 0.5899$ **(b)** (i) Solving $-1.28 = \dfrac{x - 525}{110}$ gives $x = 384.2$. (ii) invNorm(area: 0.10, mean: 525, SD: 110) $= 384.0$

## Section 4.3

### Answers to Check Your Understanding

*page 305:* Because the individuals were not randomly selected and this is an observational study and not an experiment, we can only conclude that, for the athletes in the study, those who were removed from play immediately recovered more quickly, on average, than athletes who continued to play.

### Answers to Odd-Numbered Section 4.3 Exercises

4.93 **(a)** Because different random samples will include different students and produce different estimates, it is unlikely that the sample result will be the same as the proportion of all students at the school who use Twitter. **(b)** An SRS of 100 students; estimates tend to be closer to the truth when the sample size is larger.

4.95 **(a)** Yes; it is plausible that the true proportion could be as small as $0.37 - 0.031 = 0.339$ or as large as $0.37 + 0.031 = 0.401$ and 0.50 is not in this interval. **(b)** Increase the number of adults in the sample.

4.97 **(a)** If we repeatedly take random samples of size 124 from a population of couples that have no preference for which way they kiss, the number of couples who kiss the "right way" in a sample varies from about 45 to 76. **(b)** Yes; in the study, 83 couples kissed the "right way"—much higher than what we would expect to happen by chance alone. In the simulations, the largest number of couples kissing the "right way" was 76.

4.99 (a) 4.68 − 4.21 = 0.47 days (b) Assuming the type of clipboard does not matter, there was one simulated random assignment where the difference (A − B) in the mean number of days for the two groups was 0.72. (c) Because a difference of means of 0.47 or higher occurred 16 out of 100 times in the simulation, the difference is not statistically significant. It is quite plausible to get a difference this big simply due to chance variation in the random assignment.
4.101 (a) To make sure that the two groups were as similar as possible before the treatments were administered. (b) The difference in the percent of women who received acupuncture and became pregnant and those who lay still and became pregnant was large enough to conclude that the difference was not likely due to the chance variation created by the random assignment to treatments. (c) Because the women were aware of which treatment they received, we do not know if their expectations or the treatment was the cause of the increase in pregnancy rates.
4.103 Because this study involved random assignment to the treatments (foster care or institutional care), we can infer that the difference in being in foster care or institutional care caused the difference in response. However, the results should only be applied to children like the ones in the study, because these 136 children were not randomly selected from a larger population.
4.105 Because the subjects were not randomly assigned to attend religious services (or not), we cannot infer cause and effect. However, this study involved a random sample of adults so we can make an inference about the population of adults. It appears that adults who attend religious services regularly have a lower risk of dying, but we do not know that attending religious services is the cause of the lower risk.
4.107 Because this study does not involve random assignment to the treatments (amount of anthocyanins consumed), we cannot infer that the difference in blueberry and strawberry intake caused the difference in heart attack risk. In addition, we should only apply the results to women like the ones in this study because these 93,600 women were not randomly selected from a larger population.
4.109 Answers will vary.
4.111 Those Facebook users involved in the study did not know that they were going to be subjected to treatments, and they did not provide informed consent before the study was conducted.
4.113 The responses to the GSS are confidential. The person giving the survey knows who is answering the questions because s/he is at that person's home, but will not share the results with anyone else.
4.115 In this case, the subjects were not able to give informed consent. They did not know what was happening to them, and they were not old enough to understand the ramifications in any event.
4.117 d
4.119 (a)

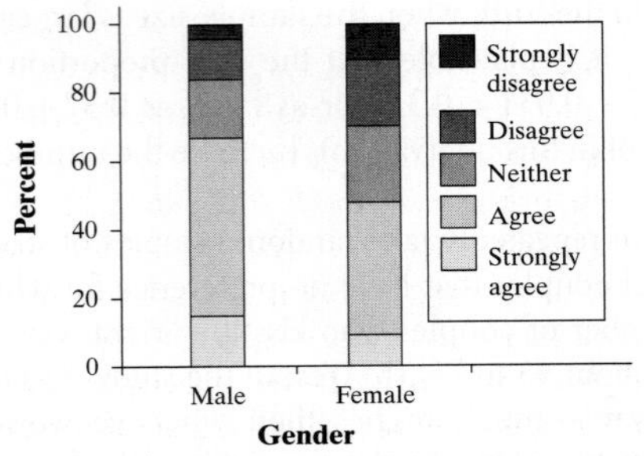


(b) Yes, there is an association between gender and opinion. Men are more likely to view animal testing as justified if it might save human lives: over two-thirds of men agree or strongly agree with this statement, compared to slightly less than half of the women. The percentages who disagree or strongly disagree tell a similar story: 16% of men versus 30% of women.

## Answers to Chapter 4 Review Exercises

R4.1 (a) *Population:* All adult U.S. residents. *Sample:* The 805 adult U.S. residents interviewed. (b) Even though the sample size is very large, it is unlikely that the percentage in the entire population would be exactly the same as the percentage in the sample because of sampling variability. (c) A larger random sample is more likely to get a sample result close to the true population value.
R4.2 (a) One possible answer: Announce in the daily bulletin that there is a survey concerning student parking available in the main office for students who want to respond. Because generally only those who feel strongly about the issue respond to a voluntary response survey, the opinions of respondents may differ from the population as a whole, resulting in an inaccurate estimate. (b) One possible answer: Personally interview a group of students as they come in from the parking lot. Because these students use the parking lot, their opinions may differ from the population as a whole, resulting in an inaccurate estimate. (c) Write the names of all 1800 students on identical pieces of paper. Place the slips of paper into a hat. Shuffle well. Draw out 50 names. Those students will form the SRS of 50 students from the school. An SRS reduces bias by selecting the sample in such a way that every group of 50 individuals in the population has an equal chance to be selected as the sample. (d) Select every 1800/50 = 36th student that enters the school. Select a random number between 1 and 36 to identify the first student, then select every 36th student thereafter. A systematic random sample selects the students on the spot as they enter the building, so we would not have to track down students all over the school as in an SRS.
R4.3 (a) You would have to identify 10% of the seats, go to those seats in the arena, and find the people who are sitting there. (b) It would be better to use the lettered rows as the strata because each lettered row is the same distance from the court and so would contain only seats with the same (or nearly the same) ticket price. Within sections, ticket prices vary quite a bit. (c) Survey all fans in several randomly selected sections (clusters); people in a particular numbered section are in roughly the same location, making it easy to administer the survey. Furthermore, the people in each cluster reflect the variability found in the population, which is ideal.
R4.4 (a) When the interviewer provides the additional information that "box-office revenues are at an all-time high," the listeners may believe that they contributed to this fact and be more likely to overestimate the number of movies they have seen in the past 12 months. Eliminate this sentence. (b) A sample that uses only residential phone numbers is likely to underrepresent younger adults who use only cell phones. If younger adults go to movies more often than older adults, the estimated mean will be too small. (c) People who do not go to the movies very often might be more likely to respond to the poll because they are at home. Because the frequent moviegoers will not be at home to respond, the estimated mean will be too small.
R4.5 (a) The data were collected after the anesthesia was administered. Hospital records were used to "observe" the death rates, rather than imposing different anesthetics on the subjects. (b) *Explanatory:* Type of anesthetic. *Response:* Whether or not a patient died. (c) One variable that might be confounded with choice of anesthetic is type of surgery. If anesthesia C is used more

often with a type of surgery that has a higher death rate, we would not know if the death rate was higher because of the anesthesia or the type of surgery.

R4.6 **(a)** *Experimental units:* Potatoes. The factors are the storage method (3 levels) and time from slicing until cooking (2 levels). There are six treatments: (1) freshly picked and cooked immediately, (2) freshly picked and cooked after an hour, (3) stored at room temperature and cooked immediately, (4) stored at room temperature and cooked after an hour, (5) stored in refrigerator and cooked immediately, (6) stored in refrigerator and cooked after an hour. **(b)** Using 300 identical slips of paper, write "1" on 50 of them, write "2" on 50 of them, and so on. Put the papers in a hat and mix well. Then select a potato and randomly select a slip from the hat to determine which treatment that potato will receive. Repeat this process for the remaining 299 potatoes, making sure not to replace the slips of paper into the hat. **(c)** *Benefit:* The quality of the potatoes should be fairly consistent, reducing a source of variability. *Drawback:* The results of the experiment could then only be applied to potatoes that come from that one supplier rather than to potatoes in general. **(d)** Use a randomized block design with the suppliers as the blocks. For each supplier, randomly assign potatoes to the 6 treatments. Doing so would allow the researchers to account for the variability in color and flavor due to differences in the initial quality of the potatoes from different suppliers, making it easier to estimate how the treatments affect color and flavor of the French fries.

R4.7 **(a)** No; the 1000 students were not randomly selected from any larger population, so we should apply the results only to students like those in the study. **(b)** Yes; the students were randomly assigned to the three treatments, so we can conclude that the reduction in cold symptoms was caused by the masks.

R4.8 **(a)** If all of the patients received the St. John's wort, the researchers would not know if any improvement was due to the St. John's wort or to the expectations of the subjects (the placebo effect). By giving some patients a treatment that should have no effect at all but that looks, tastes, and feels like the St. John's wort, the researchers can account for the placebo effect by comparing the results for the two groups. **(b)** To create two groups of subjects that are roughly equivalent at the beginning of the experiment. **(c)** The subjects should not know which treatment they are getting so that the researchers can account for the placebo effect. Also, the researchers should be unaware of which subjects received which treatment so they cannot consciously (or subconsciously) influence how the results are measured. **(d)** In this context, "not statistically significant" means that the difference in improvement between the St. John's wort and placebo groups was not large enough to rule out the variability caused by the random assignment as the explanation.

R4.9 **(a)** Use 30 identical slips of paper and write the name of each subject on a slip. Mix the slips in a hat and select 15 of them at random. These subjects will be assigned to hard mazes. The remaining 15 will be assigned to easy mazes. After the experiment, compare the time estimates of the two groups. **(b)** Each student does the activity twice, once with the easy mazes and once with the hard mazes. For each student, randomly determine which set of mazes is used first. To do this, flip a coin for each subject. If it's heads, the subject will do the easy mazes followed by the hard mazes. If it's tails, the subject will do the hard mazes followed by the easy mazes. After the experiment, compare each student's "easy" and "hard" time estimate. **(c)** The matched pairs design would be more likely to detect a difference because it accounts for the variability between subjects.

R4.10 **(a)** No, because the subjects did not know the nature of the experiment before they agreed to participate. **(b)** All individual data should be kept confidential and the experiment should go before an institutional review board before being implemented.

## Answers to Chapter 4 AP® Practice Test

| | | | |
|---|---|---|---|
| T4.1 c | T4.2 e | T4.3 d | T4.4 c |
| T4.5 d | T4.6 d | T4.7 e | T4.8 d |
| T4.9 d | T4.10 e | T4.11 d | |

T4.12 **(a)** *Experimental units:* Acacia trees. **(b)** This allows the researchers to measure the effect of active hives and empty hives on tree damage compared to no hives at all. **(c)** Assign the trees numbers from 01 to 72 and use a random number table to pick 24 different two-digit numbers between 01 and 72. Those trees will get the active beehives. The trees associated with the next 24 different two-digit numbers will get the empty beehives and the remaining 24 trees will remain empty. Compare the damage caused by elephants to the trees with active beehives, those with empty beehives, and those with no beehives.

T4.13 **(a)** *Population:* All U.S. students in grades 7 – 12. *Sample:* The 1673 U.S. students in grades 7 – 12 surveyed. **(b)** Random selection reduces the effects of bias due to self-selection and also allows the results to be inferred to a larger population. In this case, students who respond to an online survey might be more interested in computer science, leading to an estimate that is too high. **(c)** No, because of sampling variability. **(d)** The estimate of 54% because it was based on a sample size of 1673, whereas the other two results were based on smaller sample sizes. Larger sample sizes typically yield results that are closer to the truth about the population.

T4.14 **(a)** *Explanatory:* The treatment given (caffeine capsule or placebo). *Response:* The time it takes to complete the tapping task. **(b)** Each of the 11 individuals will be a block in a matched pairs design. Each participant will take the caffeine tablets on one of the two-day sessions and the placebo on the other. The blocking was done to account for individual differences in dexterity. **(c)** After the first trial, subjects might practice the tapping task and do better the second time. If all the subjects got caffeine the second time, the researchers would not know if the increase was due to the practice or the caffeine. **(d)** Yes, if neither the subjects nor the people who come in contact with them during the experiment (including those who record the number of taps) have knowledge of the order in which the caffeine or placebo was administered.

## Answers to Cumulative AP® Practice Test 1

| | | | |
|---|---|---|---|
| AP1.1 d | AP1.2 e | AP1.3 b | AP1.4 c |
| AP1.5 e | AP1.6 c | AP1.7 e | AP1.8 e |
| AP1.9 d | AP1.10 d | AP1.11 d | AP1.12 b |
| AP1.13 a | AP1.14 a | | |

AP1.15 **(a)** The distribution of gains for subjects using Machine A is roughly symmetric, while that for subjects using Machine B is skewed to the left (toward the smaller values). Neither distribution appears to contain any outliers. The center of the distribution of gains for subjects using Machine B (median = 38) is greater than that for subjects using Machine A (median = 28). The distribution of gains for subjects using Machine B (range = 57, $IQR = 22$) is more variable than that for subjects using Machine A (range = 32, $IQR = 15$). **(b)** Machine B because its median gain (38) is greater than it is for Machine A (28), as is the mean ($\bar{x}_B = 35.4$ versus $\bar{x}_A = 28.9$). **(c)** Machine A because it exhibits less variation in gains than does Machine B. The *IQR* for Machine A (15) is less than the *IQR* for Machine B (22). Additionally, the standard deviation for

Machine A (9.38) is less than the standard deviation for Machine B (16.19). **(d)** The experiment was conducted at only one fitness center. Results may vary at other fitness centers in this city and in other cities. If the company wants to broaden their scope of inference, they should randomly select people from the population they would like to draw an inference about.

AP1.16 **(a)** Randomly assign 30 retail sales districts to the monetary incentives treatment and the remaining 30 retail sales districts to the tangible incentives treatment [see part (b) for method]. After a specified period of time, record the change in sales for each district and compare the mean change for each of the treatment groups. **(b)** Number the 60 retail sales districts with a two-digit number from 01 to 60. Using a table of random digits, read two-digit numbers until 30 unique numbers from 01 to 60 have been selected. These 30 districts are assigned to the monetary incentives group and the remaining 30 to the tangibles incentives group. Using the digits provided, the districts labeled 07, 51, and 18 are the first three to be assigned to the monetary incentives group. **(c)** Matching the districts based on their size accounts for the variation among the experimental units due to their size on the response variable—sales volume. Pair the two largest districts in size, the next two largest, down to the two smallest districts. For each pair, pick one of the districts and flip a coin. If it's heads, this district is assigned to the monetary incentives group. If it's tails, this district is assigned to the tangible incentives group. The other district in the pair is assigned to the other group. After a specified period of time, record the change in sales for each district and compare within each pair.

AP1.17 **(a)** There is a very strong, positive linear association between sales and shelf length. **(b)** $\hat{y} = 317.94 + 152.68x$, where $\hat{y}$ = predicted weekly sales (in dollars) and $x$ = shelf length (in feet). **(c)** $\hat{y} = 317.94 + 152.68(5) = 1081.34$. **(d)** The actual weekly sales (in dollars) is typically about \$22.9212 away from the weekly sales predicted by the least-squares regression line with $x$ = shelf length (in feet). **(e)** About 98.2% of the variation in weekly sales revenue can be accounted for by the least-squares regression line with $x$ = shelf length (in feet).

AP1.18 **(a)** (i) $z = \dfrac{-5-0}{22.92} = -0.22$ and $z = \dfrac{5-0}{22.92} = 0.22$; $0.5871 - 0.4219 = 0.1652$ (ii) normalcdf(lower: $-5$, upper: 5, mean: 0, SD: 22.92) = 0.1727 **(b)** (i) Solving $-1.96 = \dfrac{x-0}{22.92}$ gives $x = -\$44.92$. Solving $1.96 = \dfrac{x-0}{22.92}$ gives $x = \$44.92$. (ii) invNorm(area: 0.025, mean: 0, SD: 22.92) = $-\$44.92$ and invNorm(area: 0.975, mean: 0, SD: 22.92) = \$44.92

The middle 95% of residuals should be between $-\$44.92$ and \$44.92.

If 5 linear feet are allocated to the store's brand of men's grooming products, the weekly sales revenue can be expected to be between \$1036.42 and \$1126.26 ($1081.34 \pm 44.92$).

# Chapter 5

## Section 5.1

### Answers to Check Your Understanding

*page 332:* **1.** *Interpretation:* If you take a very large random sample of Pedro's commutes, about 55% of the time the light will be red when Pedro reaches the light.

**2.** **(a)** This probability is 0. If an outcome can never occur, then it will occur in 0% of the trials. **(b)** This probability is 1. If an outcome will occur on every trial, then it will occur in 100% of the trials. **(c)** This probability is 0.001. An outcome that occurs in 0.1% of the trials is very unlikely, but will occur every once in a while. **(d)** This probability is 0.6. An outcome that occurs in 60% of the trials will happen more than half the time. Also, 0.6 is a better choice than 0.99 because the wording suggests that the event occurs often but not nearly every time.

**3.** The doctor is wrong because the sex of the next baby born is a random phenomenon that is unpredictable in the short run, even though it has a regular predictable pattern in the long run. So while approximately 50% of all babies born will be male, even after a couple has seven girls in a row, the probability of the eighth child being a girl is still 50%.

*page 335:* **1.** To carry out a simulation to estimate the probability that a 50% shooter who takes 30 shots in a game would have a streak of 10 or more made shots, begin by assigning digits to the outcomes. Let 1 = make a shot and let 2 = miss a shot. Generate 30 random integers from 1 to 2 to simulate taking 30 shots. Record whether or not there are a series of at least 10 "makes" in a row among the 30 shots. Perform many trials of this simulation. See what percent of the time there is a streak of at least 10 "makes" among the 30 shots.

**2.** There were two trials that resulted in 9 consecutive "makes" among the 30 shots.

**3.** Based on the dotplot, there was only 1 trial out of 50 simulated games in which the player had at least 10 "makes" in a row among the 30 shots. If the player's shot percentage truly is 50% and if the result of a shot truly does not depend on previous shots, then in 50 trials we would expect this player to have a streak of 10 or more makes only about 2% of the time (1 out of 50). Because this phenomenon is unlikely to happen strictly due to chance alone, and the announcer observed the player making 10 shots in a row in a recent game, the announcer is justified in claiming that the player is streaky.

### Answers to Odd-Numbered Section 5.1 Exercises

5.1 **(a)** If you take a very large random sample of times that Aaron tunes into his favorite radio station, about 20% of the time a commercial will be playing. **(b)** No, random behavior is unpredictable in the short run, but has a regular and predictable pattern in the long run. The value 0.20 describes the proportion of times that a commercial will be playing when Aaron tunes in to the station in a very long series of trials.

5.3 **(a)** If you take a very large random sample of women with breast cancer, about 10% of the time the mammogram will indicate that the woman does not have breast cancer when, in fact, she does have breast cancer. **(b)** A false negative is a more serious error because a woman with breast cancer will not get potentially life-saving treatment. A false positive would result in temporary stress until a more thorough examination is performed.

5.5 In the short run, there was quite a bit of variability in the percentage of made 3-point shots. However, the percentage of made 3-point shots became less variable and approached 0.30 as the number of shots increased.

5.7 **(a)** The wheel is not affected by its past outcomes—it has no memory. So on any one spin, black and red remain equally likely. **(b)** The gambler is wrong again. Removing a card changes the composition of the remaining deck. If you hold 5 red cards, the deck now contains 5 fewer red cards, so the probability of being dealt another red card decreases.

5.9 **(a)** To carry out a simulation to estimate the score that Luke will earn on the quiz, begin by assigning digits to the outcomes. For each question, let 1 = guessed correctly and let 2, 3, and 4 = guessed incorrectly. Generate 10 random integers from 1 to 4 to simulate the result of Luke's guess for each of the 10 questions. Record the number of correct guesses (1's). Repeat this three times and record the highest of the three scores. **(b)** There was one trial that resulted in a highest quiz score of 1 correct response out of 10. **(c)** Based on the simulation, the probability that Luke passes the quiz (scores at least a 6 out of 10) is 5/50 = 10%. Five of the 50 trials show a score of at least 6. **(d)** Yes; it is unlikely that Doug would score higher than 6 out of 10 based on randomly guessing the answers. Since Doug scored 8 points out of 10, there is reason to believe that Doug does understand some of the material.

5.11 **(a)** To use a table of random digits to carry out the simulation, number the first class passengers 01–12 and the other passengers 13–76. Ignore all other numbers. Moving left to right across a row, look at pairs of digits until you have 10 unique numbers (no repetitions, because you do not want to select the same person twice). Count the number of two-digit numbers between 01 and 12. Perform many trials of this simulation. Determine what percent of the trials showed no first class passengers selected for screening. **(b)** The numbers, read in pairs, are: **71 48 70** 99 84 **29** <u>**07**</u> 71 48 **63 61 68 34** 70 **52**. The bold numbers indicate people who have been selected. The other numbers are either too large (over 76) or have already been selected. There is one person among the 10 selected who is in first class in this sample (underlined). **(c)** No, there is not convincing evidence that the TSA officers did not carry out a truly random selection. If the selection is truly random, there is about a 15% chance that no one in first class will be selected. So it is not surprising that a single random selection would contain no first class passengers.

5.13 **(a)** To carry out this simulation using slips of paper, label 10 identically sized pieces of paper 0–9. Let 0–5 represent hitting the center of the target and let 6–9 represent not hitting the center of the target. Place the slips of paper into a hat and mix well. Draw out one slip of paper at a time and record the digit. Replace the paper and mix well. Repeat this process until Quinn misses the center of the target. Determine whether she remained in the competition for at least 10 shots. **(b)** To carry out this simulation using a random digits table, let 0–5 represent hitting the center of the target and let 6–9 represent not hitting the center of the target. Moving left to right across a row, read single digits and record whether she hit the center of the target or not. Repeat this process until Quinn misses the center of the target. Determine whether she remained in the competition for at least 10 shots. **(c)** To carry out this simulation using a random number generator, let 0–5 represent hitting the center of the target and let 6–9 represent not hitting the center of the target. Generate a random integer from 0 to 9 and record whether she hit the center of the target or not. Repeat this process until Quinn misses the center of the target. Determine whether she remained in the competition for at least 10 shots.

5.15 This is a legitimate simulation. By letting 1 = feel addicted and 2 = doesn't feel addicted, the simulation would model the claim that 50% of U.S. teens with smartphones feel addicted to their devices.

5.17 This is not a legitimate simulation. When simulating the selection of homework assignments, the selection process would occur without replacement. The slips of paper should not be placed back in the hat.

5.19 To conduct a simulation to estimate the probability that we would have to randomly select 20 or more U.S. adult males to find one who is red–green color-blind, let 0–6 = color-blind and let 7–99 = not color-blind. Use technology to pick a random integer from 0 to 99. Continue picking random integers until we get a number between 0 and 6. Count how many numbers there are in the sample. Repeat this process many times. Estimate the percent of the time we would have to randomly select 20 or more U.S. adult males to find one who is red–green color-blind.

5.21 **(a)** In the simulation, 43 of the 200 samples yielded a sample proportion of at least 0.55. Obtaining a sample proportion of 0.55 or higher is not particularly unusual when 50% of all students recycle. **(b)** Only 1 of the 200 samples yielded a sample proportion of at least 0.63. This means that if 50% of all students recycle, we would see a sample proportion of at least 0.63 in only about 0.5% of samples. Because getting a sample proportion of at least 0.63 is very unlikely, we have convincing evidence that the percentage of all students who recycle is larger than 50%.

5.23 c

5.25 b

5.27 c

5.29 **(a)** This survey displays undercoverage because people who are younger than 50 years old did not have a chance to participate in the survey. Younger people generally don't have as many prescription drugs and might be less willing than older people to support a program like this. The 75% who answered "Yes" is likely an overestimate of the true proportion of Americans who support the program. **(b)** Including the additional information of "can be helped" might have encouraged respondents to say "Yes" to the program because they liked the idea of helping people. This would cause the 75% estimate to be higher than if the question were worded more neutrally.

## Section 5.2

### Answers to Check Your Understanding

*page 346:* **1.** Events A and B are mutually exclusive because a person cannot have a cholesterol level of both 240 or above and between 200 and less than 240 at the same time.

**2.** The event "A or B" is a person who has either a cholesterol level of 240 or above or they have a cholesterol level between 200 and less than 240. $P(\text{A or B}) = P(\text{A}) + P(\text{B}) = 0.16 + 0.29 = 0.45$.

**3.** $P(\text{C}) = 1 - P(\text{A or B}) = 1 - 0.45 = 0.55$

*page 349:* **1.** The probability that the person is an environmental club member is $(212 + 77 + 16)/1526 = 305/1526 = 0.1999$.

**2.** $P(\text{not a snowmobile renter}) = (445 + 212 + 297 + 16)/1526 = 970/1526 = 0.6356$.

**3.** $P(\text{environmental club member and not a snowmobile renter}) = (212 + 16)/1526 = 228/1526 = 0.1494$.

**4.** The probability that the person is not an environmental club member or is a snowmobile renter is 85.06%, which is $(445 + 497 + 279 + 77)/1526 = 1298/1526 = 0.8506$.

### Answers to Odd-Numbered Section 5.2 Exercises

5.31 **(a)** The following table shows the possible outcomes in the sample space. Each of the 16 outcomes will be equally likely and have probability $\frac{1}{16}$.

| | | First roll | | | |
|---|---|---|---|---|---|
| | | 1 | 2 | 3 | 4 |
| Second roll | 1 | (1,1) | (2,1) | (3,1) | (4,1) |
| | 2 | (1,2) | (2,2) | (3,2) | (4,2) |
| | 3 | (1,3) | (2,3) | (3,3) | (4,3) |
| | 4 | (1,4) | (2,4) | (3,4) | (4,4) |

**(b)** $P(A) = 0.25$. There are four ways to get a sum of 5 from these two dice: (1, 4), (2, 3), (3, 2), (4, 1). The probability of getting a sum of 5 is $P(A) = \frac{4}{16} = 0.25$.

5.33 **(a)** The sample space is: Connor/Declan, Connor/Lucas, Connor/Piper, Connor/Sedona, Connor/Zayne, Declan/Lucas, Declan/Piper, Declan/Sedona, Declan/Zayne, Lucas/Piper, Lucas/Sedona, Lucas/Zayne, Piper/Sedona, Piper/Zayne, and Sedona/Zayne. Each of these 15 outcomes will be equally likely and will have probability 1/15. **(b)** There are 9 outcomes in which Piper or Sedona (or both) get to go to the show: Connor/Piper, Connor/Sedona, Declan/Piper, Declan/Sedona, Lucas/Piper, Lucas/Sedona, Piper/Sedona, Piper/Zayne, Sedona/Zayne. Define Event A as Piper or Sedona (or both) get to go to the show. Then $P(A) = 9/15 = 0.60$.

5.35 **(a)** This is a valid probability model because each probability is between 0 and 1 and the probabilities sum to 1. **(b)** $P$(student won't win extra homework) $= 1 - 0.05 = 0.95$. There is a 0.95 probability that a student will not win extra homework. **(c)** $P$(candy or homework pass) $= 0.25 + 0.15 = 0.40$. There is a 0.40 probability that a student wins candy or a homework pass.

5.37 **(a)** The probabilities of all the possible outcomes must add to 1. The given probabilities add up to $0.25 + 0.32 + 0.07 + 0.03 + 0.01 = 0.68$. This leaves a probability of $1 - 0.68 = 0.32$ for $P$(3 or 4). Because the probability of finding 3 people in a household is the same as the probability of finding 4 people (and they are mutually exclusive), each probability must be $0.32/2 = 0.16$. **(b)** $P$(more than 2 people) $= 1 - P$(1 or 2 people) $= 1 - (0.25 + 0.32) = 0.43$. This could also be found using the addition rule for mutually exclusive events. $P$(more than 2 people) $= P$(3 or 4 or 5 or 6 or 7+) $= 0.16 + 0.16 + 0.07 + 0.03 + 0.01 = 0.43$.

5.39 **(a)** The given probabilities have a sum of 0.72 and the sum of all probabilities should be 1. Thus, the probability that a randomly chosen young adult has some education beyond high school but does not have a bachelor's degree is $1 - 0.72 = 0.28$. There is a 0.28 probability that a young adult has some education beyond high school but does not have a bachelor's degree. **(b)** Using the complement rule, $P$(at least a high school education) $= 1 - P$(has not finished high school) $= 1 - 0.13 = 0.87$. There is a 0.87 probability that a young adult has completed high school. **(c)** $P$(young adult has further education beyond high school) $= P$(young adult has some education beyond high school but does not have a bachelor's degree) $+ P$(young adult has at least a bachelor's degree) $= 0.28 + 0.30 = 0.58$. The probability rule used is the addition rule for mutually exclusive events.

5.41 **(a)** $P(B^C) = P$(student does not eat breakfast regularly) $= \frac{295}{595} = 0.496$ **(b)** $P(F \text{ and } B^C) = P$(student is a female and does not eat breakfast regularly) $= \frac{165}{595} = 0.277$. *Interpretation:* If we select a student from the school at random, the probability that the student will be female and does not eat breakfast regularly is 0.277. **(c)** $P(F \text{ or } B^C) = P$(student is a female or does not eat breakfast regularly) $= \frac{110 + 165 + 130}{595} = 0.681$

5.43 **(a)** $P$(not age 18 to 34 and not own at iPhone) $=$
$\frac{189 + 100 + 277 + 643}{2024} = \frac{1209}{2024} = 0.597$

**(b)** $P$(age 18 to 34 or own iPhone) $= \frac{169 + 214 + 134 + 171 + 127}{2024}$
$= \frac{815}{2024} = 0.403$

5.45 **(a)**

| | Black | Not black | Total |
|---|---|---|---|
| Even | 10 | 10 | **20** |
| Not even | 8 | 10 | **18** |
| **Total** | **18** | **20** | **38** |

**(b)** $P(B) = \frac{18}{38} = 0.474; P(E) = \frac{20}{38} = 0.526$ **(c)** The event "B and E" would be that the ball lands in a spot that is black and even. $P(\text{B and E}) = \frac{10}{38} = 0.263$ **(d)** The probability of the event "B or E" means the probability of landing in a spot that is either black, even, or both. If we add the probabilities of landing in a black spot and landing in an even spot, the spots that are black and even will be double counted because events B and E are not mutually exclusive. $P(\text{B or E}) = \frac{18}{38} + \frac{20}{38} - \frac{10}{38} = \frac{28}{38} = 0.737$.

5.47 $P$(own a dog or a cat) $= P$(own a dog) $+ P$(own a cat) $- P$(own a dog and a cat) $= 0.40 + 0.32 - 0.18 = 0.54$

5.49 **(a)** We are given the following information: $P(M) = 0.6$, $P(G^C) = 0.67, P(G \text{ and } M) = 0.23$. Therefore, $P(G) = 1 - 0.67 = 0.33$. $P(G \cup M) = P(G) + P(M) - P(G \cap M) = 0.33 + 0.6 - 0.23 = 0.70$. *Interpretation:* The probability of randomly selecting a student from among those who were part of the census that is a graduate student or uses a Mac is 0.70. **(b)** $P(G^C \cap M^C) = 1 - P(G \cup M) = 1 - 0.7 = 0.3$

5.51 **(a)** A Venn diagram is shown here.

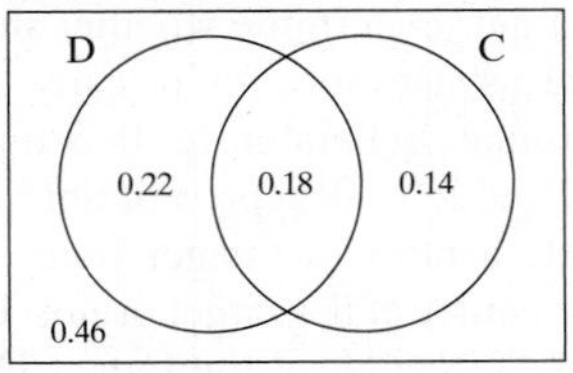


**(b)** $P(D \cap C^C) = 0.22$

5.53 The general addition rule states that $P(A \cup B) = P(A) + P(B) - P(A \cap B)$. We know that $P(A) = 0.3$, $P(B) = 0.4$, and $P(A \cup B) = 0.58$. Therefore, $0.58 = 0.3 + 0.4 - P(A \cap B)$. Solving for $P(A \cap B)$ gives $P(A \cap B) = 0.12$.

5.55 e

5.57 e

5.59 **(a)** The scatterplot for the average crawling age and average temperature is given here.

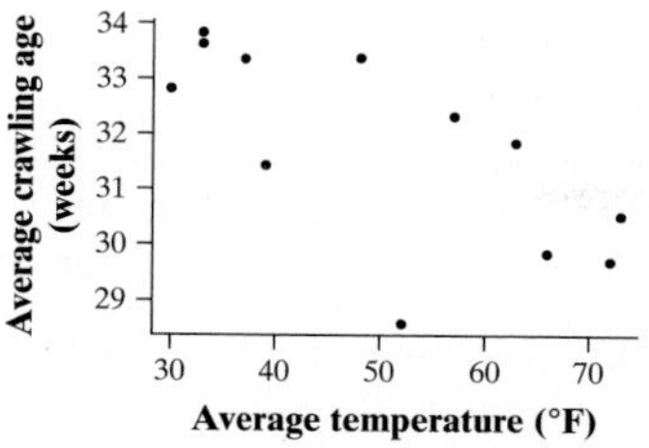


In this scatterplot, there appears to be a moderately strong, negative linear relationship between average temperature and average crawling age. **(b)** The equation for the least-squares regression line is $\hat{y} = 35.68 - 0.0777x$, where $\hat{y}$ is the predicted average crawling age and $x$ is the average temperature in degrees Fahrenheit. **(c)** The slope $= -0.0777$. *Interpretation:* The predicted average crawling age decreases by 0.0777 weeks for each additional 1-degree increase in the average temperature in degrees Fahrenheit. **(d)** We cannot conclude that warmer temperatures 6 months after babies are born causes them to crawl sooner because this was an observational study and not an experiment. We cannot draw conclusions of cause and effect from observational studies.

## Section 5.3

### Answers to Check Your Understanding

*page 362:* **1.** $P(\text{N} \mid \text{E}) = \frac{212}{305} = 0.695$. Given that a survey respondent belongs to an environmental organization, there is about a 69.5% chance that he or she never used a snowmobile.

**2.** $P(\text{E} \mid \text{S}^C) = \frac{212 + 77}{657 + 574} = \frac{289}{1231} = 0.235$

**3.** $P(\text{E}^C \mid \text{N}) = \frac{445}{657} = 0.677$. $P(\text{E}^C \mid \text{renter}) = \frac{497}{574} = 0.866$. $P(\text{E}^C \mid \text{owner}) = \frac{279}{295} = 0.946$. The chosen person is more likely to not be an environmental club member if he or she owns a snowmobile.

*page 365:* **1.** Events A and B are independent. Because we are putting the first card back and shuffling the cards before drawing the second card, knowing what the first card was will not help us predict what the second card will be.

**2.** Events A and B are not independent. Once we know the suit of the first card, then the probability of getting a heart on the second card will change depending on what the first card was.

**3.** The two events, "Female" and "Got an A," are independent. Once we know that the chosen person is female, this does not help us predict if she got an A or not. Overall, 4/28 or 1/7 of the students got an A on the quiz. And, among the females, 3/21 or 1/7 got an A. So $P(\text{got an A}) = P(\text{got an A} \mid \text{female})$.

*page 371:* **1.**

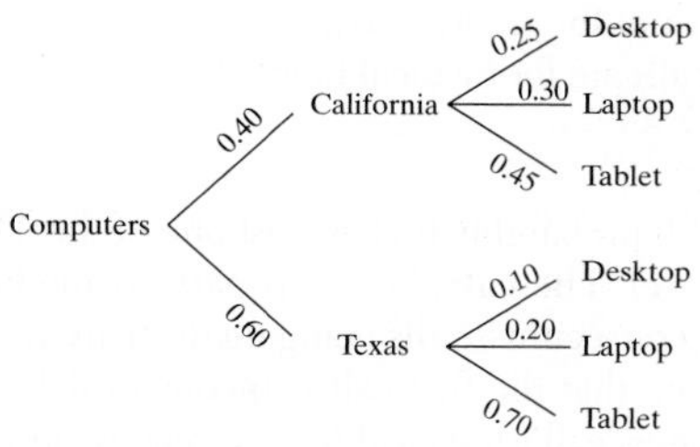


**2.** $P(\text{tablet}) = P(\text{California} \cap \text{tablet}) + P(\text{Texas} \cap \text{tablet}) = (0.40)(0.45) + (0.60)(0.70) = 0.60$. The probability of selecting a tablet is 0.60.

**3.** $P(\text{California} \mid \text{tablet}) = \frac{P(\text{California} \cap \text{tablet})}{P(\text{tablet})} = \frac{(0.4)(0.45)}{0.60} = \frac{0.18}{0.60} = 0.30$. Given that a tablet was selected, there is a 0.30 probability that it was made in California.

*page 375:* **1.** $P(\text{train arrives late on Monday and on time on Tuesday}) = P(\text{Arrives late on Monday}) \cdot P(\text{Arrives on time on Tuesday}) = (0.10)(0.90) = 0.09$

**2.** $P(\text{train arrives late at least once in a 5-day week}) = 1 - P(\text{train never arrives late in a 5-day week}) = 1 - (0.90)^5 = 0.41$

**3.** $P(\text{full-time college student and age 65 or older}) \neq (0.08)(0.15)$ because "Full-time college student" and "Adults age 65 or older" are not independent events. Knowing that a person is a full-time college student decreases the likelihood that he or she is an adult age 65 or older.

### Answers to Odd-Numbered Section 5.3 Exercises

5.61 **(a)** $P(\text{T} \mid \text{E}) = \frac{44}{200} = 0.22$. Given that the child is from England, there is a 0.22 probability that the student selected the superpower of telepathy.

**(b)** $P(\text{E} \mid \text{S}^C) = \frac{54 + 52 + 30 + 44}{99 + 96 + 67 + 110} = \frac{180}{372} = 0.484$

Given that the child did not choose superstrength, there is a 0.484 probability that the child is from England.

5.63 **(a)** $P(\text{female} \mid \text{about right}) = \frac{560}{560 + 295} = 0.655$. Given that the person perceived his or her body image as about right, there is a 0.655 probability that the person is a female. **(b)** $P(\text{not overweight} \mid \text{F}) = \frac{560 + 37}{560 + 163 + 37} = \frac{597}{760} = 0.786$. Given that the person selected is female, there is a 0.786 probability that she did not perceive her body image as overweight.

5.65 **(a)** $P(\text{is studying other than English}) = 1 - P(\text{none}) = 1 - 0.59 = 0.41$. There is a 0.41 probability that the student is studying a language other than English.

**(b)** $P(\text{Spanish} \mid \text{other than English}) = \frac{0.26}{0.41} = 0.6341$. Given that the student is studying some language other than English, there is a 0.6341 probability that he or she is studying Spanish.

5.67 $P(\text{B}) < P(\text{B} \mid \text{T}) < P(\text{T}) < P(\text{T} \mid \text{B})$. There are very few professional basketball players, so P(B) should be the smallest probability. If you are a professional basketball player, it is quite likely that you are tall, so $P(\text{T} \mid \text{B})$ should be the largest probability. Finally, it's much more likely to be over 6 feet tall (B) than it is to be a professional basketball player if you're over 6 feet tall $(\text{B} \mid \text{T})$.

5.69 *Method 1:* $P(\text{D} \mid \text{C}) = \frac{P(\text{D} \cap \text{C})}{P(\text{C})} = \frac{0.18}{0.32} = 0.563$. Given that a household owns a cat, there is a 0.563 probability that the household owns a dog. *Method 2:* Use a Venn diagram. $P(\text{D} \mid \text{C}) = \frac{0.18}{0.18 + 0.14} = \frac{0.18}{0.32} = 0.563$

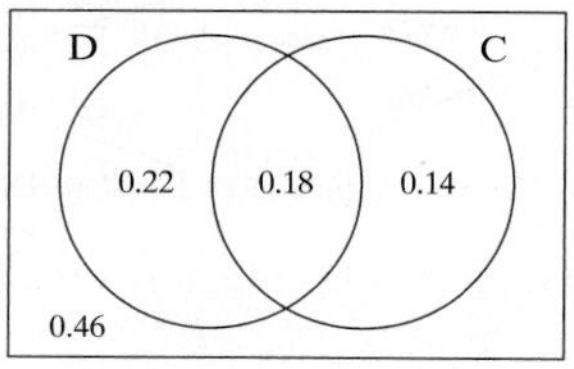

5.71 $P(\text{homeowner}) = 340/500 = 0.68$; $P(\text{homeowner} \mid \text{high school graduate}) = 221/310 = 0.713$. Because these probabilities are not equal, the events "Homeowner" and "High school graduate" are not independent. Knowing that the person is a high school graduate increases the probability that the person is a homeowner.
5.73 **(a)** $P(\text{iPhone} \mid 18-34) = 169/517 = 0.327$. Given that the adult is aged 18–34, there is a 0.327 probability that the person owns an iPhone. **(b)** First we determine $P(\text{iPhone}) = 467/2024 = 0.231$. Because these probabilities are not equal, the events "Own an iPhone" and "Aged 18–34" are not independent. Knowing that the person is aged 18–34 increases the probability that the person owns an iPhone.
5.75 There are 36 different possible outcomes of the two dice: (1, 1), (1, 2), … , (6, 6). Let's assume that the second die is the green die. There are then 6 ways for the green die to show a 4: (1, 4), (2, 4), (3, 4), (4, 4), (5, 4), (6, 4). Of those, there is only one way to get a sum of 7, so $P(\text{sum of } 7 \mid \text{green is } 4) = 1/6 = 0.1667$. Overall, there are 6 ways to get a 7: (1, 6), (2, 5), (3, 4), (4, 3), (5, 2), (6, 1). So $P(\text{sum of } 7) = 6/36 = 0.1667$. Because these two probabilities are the same, the events "Sum of 7" and "Green die shows a 4" are independent. Knowing that the green die shows a 4 does not change the probability that the sum is 7.
5.77 $P(\text{download music}) = 0.29$, $P(\text{don't care} \mid \text{download music}) = 0.67$, $P(\text{download music} \cap \text{don't care}) = P(\text{download music}) \cdot P(\text{don't care} \mid \text{download music}) = (0.29)(0.67) = 0.1943$. About 19.43% of Internet users download music and don't care if it is copyrighted.
5.79 $P(\text{all three candies have soft centers}) = P(\text{1st soft}) \cdot P(\text{2nd soft} \mid \text{1st soft}) \cdot P(\text{3rd soft} \mid \text{1st and 2nd soft}) = \frac{14}{20} \cdot \frac{13}{19} \cdot \frac{12}{18} = \frac{2184}{6840} = 0.319$.
There is a 0.319 probability that 3 randomly selected candies from a Gump box will all have soft centers.
5.81 **(a)** A tree diagram is shown here.

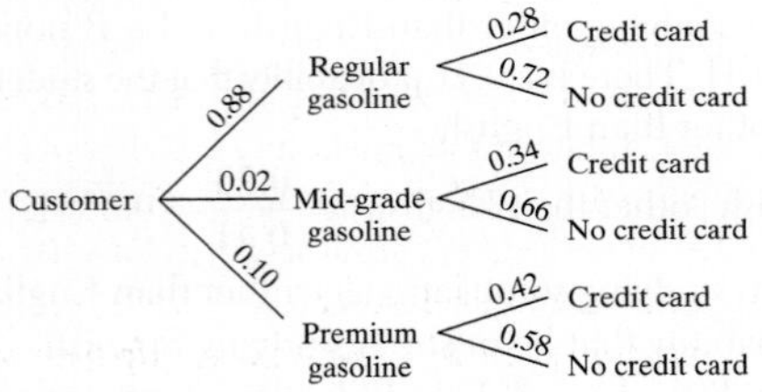


**(b)** $P(\text{credit card}) = (0.88)(0.28) + (0.02)(0.34) + (0.10)(0.42) = 0.295$. There is a 0.295 probability that the customer paid with a credit card. **(c)** $P(\text{premium gasoline} \mid \text{credit card}) = \frac{P(\text{premium gasoline} \cap \text{credit card})}{P(\text{credit card})} = \frac{(0.10)(0.42)}{0.295} = \frac{0.042}{0.295} = 0.142$.
Given that the customer paid with a credit card, there is a 0.142 probability that the customer bought premium gas.
5.83 We need to determine $P(\text{missed his first serve} \mid \text{won the point})$. Here is a tree diagram.

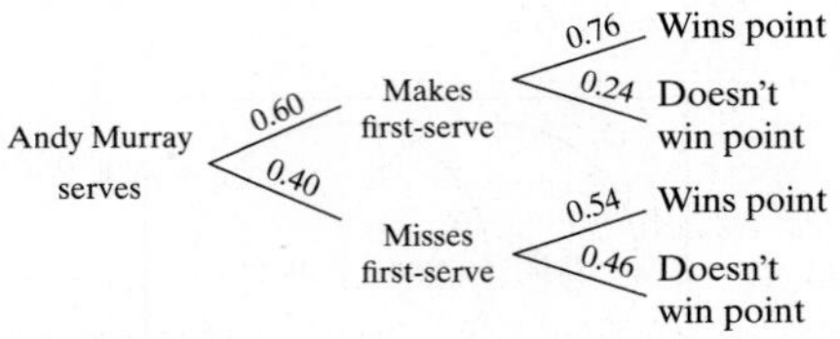


$P(\text{missed his first serve} \mid \text{won the point}) = \frac{P(\text{missed his first serve} \cap \text{won the point})}{P(\text{won the point})} = \frac{(0.40)(0.54)}{(0.60)(0.76) + (0.40)(0.54)} = \frac{0.216}{0.672} = 0.32$. Given that he won the point, there is a 0.32 probability that he missed his first serve.
5.85 We need to determine $P(\text{antibody} \mid \text{positive})$. Here is a tree diagram.

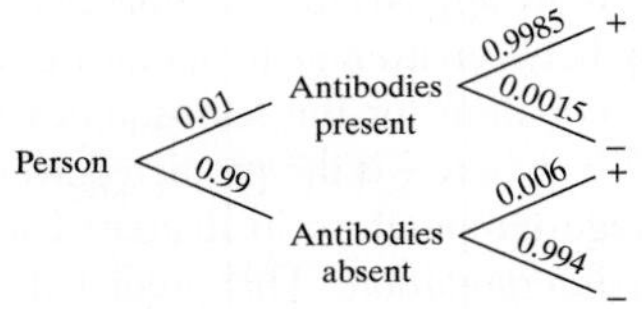


$P(\text{antibody} \mid \text{positive}) = \frac{P(\text{antibody} \cap \text{positive})}{P(\text{positive})} = \frac{(0.01)(0.9985)}{(0.01)(0.9985) + (0.99)(0.006)} = 0.6270$. Given that the EIA test is positive, there is a 0.627 probability that the person has the antibody.
5.87 **(a)** $P(\text{contributed}) = (0.5)(0.4)(0.8) + (0.3)(0.3)(0.6) + (0.2)(0.1)(0.5) = 0.16 + 0.054 + 0.01 = 0.224$. There is a 0.224 probability that a potential donor contributed to the charity. **(b)** $P(\text{recent donor} \mid \text{contribute}) = \frac{P(\text{recent donor} \cap \text{contribute})}{P(\text{contribute})} = \frac{0.16}{0.224} = 0.7143$. Given that the person contributes to the charity, there is a 0.7143 probability that the person was a recent donor.
5.89 The probability that the string of lights will remain bright for 3 years is $(0.98)^{20} = 0.6676$.
5.91 $P(\text{none are late})$
$= P(\text{1st is not late and 2nd is not late and} \ldots \text{and 20th is not late})$
$= P(\text{1st is not late}) \cdot P(\text{2nd is not late}) \cdot \ldots \cdot P(\text{20th is not late})$
$= (0.90)(0.90)\ldots(0.90)$
$= (0.90)^{20}$
$= 0.1216$
$P(\text{at least 1 late}) = 1 - P(\text{none are late}) = 1 - 0.1216 = 0.8784$
5.93 We cannot simply multiply the probabilities together. If a woman is over 55 years old, it is unlikely that she is pregnant. These events are not independent.
5.95 **(a)** $P(\text{all 4 calls are for medical help})$
$= P(\text{1st is medical and 2nd is medical and 3rd is medical and 4th is medical})$
$= P(\text{1st is medical}) \cdot P(\text{2nd is medical}) \cdot P(\text{3rd is medical}) \cdot P(\text{4th is medical})$
$= (0.81)(0.81)(0.81)(0.81) = (0.81)^4 = 0.430$
There is a 0.430 probability that all 4 calls are for medical help. **(b)** $P(\text{at least 1 not for medical help})$
$= 1 - P(\text{all 4 calls are for medical help})$
$= 1 - 0.430$
$= 0.570$
There is a 0.570 probability that at least one of the 4 calls is not for medical help. **(c)** The calculation in part (a) might not be valid because the 4 consecutive calls being medical are not independent events. Knowing that the first call is medical might make it more likely that the next call is medical (e.g., several people might call for the same medical emergency).

5.97 **(a)**

| | | Gender | | |
|---|---|---|---|---|
| | | Male | Female | Total |
| Eye color | Blue eyes | 0 | 10 | 10 |
| | Brown eyes | 20 | 20 | 40 |
| | Total | 20 | 30 | 50 |

The events "Student is male" and "Student has blue eyes" are mutually exclusive because they don't occur at the same time.
**(b)**

| | | Gender | | |
|---|---|---|---|---|
| | | Male | Female | Total |
| Eye color | Blue eyes | 4 | 6 | 10 |
| | Brown eyes | 16 | 24 | 40 |
| | Total | 20 | 30 | 50 |

If the event "Student is male" and the event "Student has blue eyes" are independent, then:

$P(\text{male}) = P(\text{male} \mid \text{blue}) \rightarrow \frac{20}{50} = \frac{x}{10} \rightarrow x = 4$

**(c)** Answers may vary.

| | | Gender | | |
|---|---|---|---|---|
| | | Male | Female | Total |
| Eye color | Blue eyes | 5 | 5 | 10 |
| | Brown eyes | 15 | 25 | 40 |
| | Total | 20 | 30 | 50 |

The events "Student is male" and "Student has blue eyes" are not mutually exclusive because they can occur at the same time (for 5 students). The events "Student is male" and "Student has blue eyes" are not independent because $P(\text{male}) \neq P(\text{male} \mid \text{blue})$. $P(\text{male}) = 20/50 = 0.4$; $P(\text{male} \mid \text{blue}) = 5/10 = 0.5$.
5.99 Two events are independent if $P(A \cap B) = P(A) \cdot P(B)$. It is given that $P(A) = 0.3$, $P(B) = 0.4$, and $P(A \cap B) = 0.12$. Because $0.12 = (0.3)(0.4)$, events A and B are independent.
5.101 **(a)** There are 6 ways to get doubles out of 36 possibilities so $P(\text{doubles}) = 6/36 = 1/6 = 0.167$. There is a 0.167 probability of getting doubles on a single toss of the dice. **(b)** Because the rolls are independent, we can use the multiplication rule for independent events.
$P(\text{no doubles first} \cap \text{doubles second}) =$
$P(\text{no doubles first}) \cdot P(\text{doubles second}) = (5/6)(1/6) = 0.139$.
There is a 0.139 probability of not getting doubles on the first toss and getting doubles on the second toss.
**(c)** $P(\text{first doubles on third roll}) =$

$P(\text{no doubles}) \cdot P(\text{no doubles}) \cdot P(\text{doubles}) = \frac{5}{6}\left(\frac{5}{6}\right)\left(\frac{1}{6}\right) = \frac{25}{216} =$

0.116. There is a 0.116 probability that the first doubles occurs on the third roll. **(d)** For the first doubles on the fourth roll, the probability is $\left(\frac{5}{6}\right)^3\left(\frac{1}{6}\right)$. For the first doubles on the fifth roll, the probability is $\left(\frac{5}{6}\right)^4\left(\frac{1}{6}\right)$. The probability that the first doubles are rolled on the *k*th roll is $\left(\frac{5}{6}\right)^{k-1}\left(\frac{1}{6}\right)$.

5.103 c
5.105 e
5.107 **(a)** $z = \frac{18.5 - 26.8}{7.4} = -1.12$
(i) The proportion of *z*-scores below $-1.12$ is 0.1314.
(ii) normalcdf(lower: $-1000$, upper: 18.5, mean: 26.8, SD: 7.4) = 0.1314. About 13.14% of young women are underweight by this criterion. **(b)** Note that $P(\text{not underweight}) = 1 - P(\text{underweight}) = 1 - 0.131 = 0.869$. $P(\text{at least one is underweight}) = 1 - P(\text{none are underweight}) = 1 - 0.869^2 = 0.2448$. There is a 0.2448 probability that at least one of the two women will be classified as underweight.

## Answers to Chapter 5 Review Exercises

R5.1 **(a)** If you take a very large random sample of pieces of buttered toast and dropped them from a 2.5-foot-high table, about 81% of them will land butter side down. **(b)** No; if four dropped pieces of toast all landed butter side down, it does not make it more likely that the next piece of toast will land with the butter side up. Random behavior is unpredictable in the short run, but it has a regular and predictable pattern in the long run. The value 0.81 describes the proportion of times that toast will land butter side down in a very long series of trials.
R5.2 **(a)** To use a table of random digits to perform the simulation, begin by letting 00 – 80 = butter side down and 81 – 99 = butter side up. Moving left to right across a row, look at pairs of digits to simulate dropping one piece of toast. Read 10 such pairs of two-digit numbers to simulate dropping 10 pieces of toast. Record the number of pieces of toast out of 10 that land butter side down. Determine if 4 or fewer pieces of toast out of 10 landed butter side down. Repeat this process many times. **(b)** Bold numbers indicate that the toast landed butter side down.
*Trial 1:* **29 07 71 48 63 61 68 34 70 52.** Did 4 or fewer pieces of toast land butter side down? **No.**
*Trial 2:* **62 22 45 10 25** 95 **05 29 09 08.** Did 4 or fewer pieces of toast land butter side down? **No.**
*Trial 3:* **73 59 27 51** 86 87 **13 69 57 61.** Did 4 or fewer pieces of toast land butter side down? **No.**
**(c)** Assuming the probability that each piece of toast lands butter side down is 0.81, the estimate of the probability that dropping 10 pieces of toast yields 4 or less butter side down is $1/50 = 0.02$. There is convincing evidence that the 0.81 claim is false. Maria does have reason to be surprised, as these results are unlikely to have occurred purely by chance.
R5.3 **(a)** The sample space is: rock/rock, rock/paper, rock/scissors, paper/rock, paper/paper, paper/scissors, scissors/rock, scissors/paper, and scissors/scissors. Because each player is equally likely to choose any of the three, each of these 9 outcomes will be equally likely and have probability 1/9. **(b)** There are three outcomes—rock/scissors, paper/rock, and scissors/paper—where player 1 wins on the first play. Define event A as player 1 wins on the first play. $P(A) = 3/9 = 0.33$.
R5.4 **(a)** The probability that it is a crossover is $1 - 0.46 - 0.15 - 0.10 - 0.05 = 0.24$. The sum of the probabilities must add to 1. **(b)** $P(\text{vehicle is not an SUV or a minivan}) = 0.46 + 0.15 + 0.24 = 0.85$. The probability that the vehicle is not an SUV or a minivan is 0.85 **(c)** $P(\text{pickup truck} \mid \text{not a passenger car}) =$

$\frac{0.15}{0.15 + 0.10 + 0.24 + 0.05} = \frac{0.15}{0.54} = 0.278$

Given that the vehicle is not a passenger car, there is a 0.278 probability that the vehicle is a pickup truck.
R5.5 **(a)** $P(\text{drives an SUV}) = 39/120 = 0.325$. The probability that the person drives an SUV is 0.325. **(b)** $P(\text{drives a sedan or exercises}) = \frac{25 + 20 + 15 + 12}{120} = 0.60$. The probability that the person drives a

sedan or exercises for at least 30 minutes four or more times per week is 0.60.

(c) $P(\text{does not drive a truck} \mid \text{exercises}) = \frac{25+15}{25+15+12} = \frac{40}{52} = 0.769$

Given that the person exercises for at least 30 minutes four or more times per week, there is a 0.769 probability that the person does not drive a truck.

R5.6 (a) The events "Thick-crust pizza" and "Pizza with mushrooms" are not mutually exclusive. There are two pizzas that have thick-crust and have mushrooms. (b) Let's organize the given information in a table.

| | Thick crust | Thin crust | Total |
|---|---|---|---|
| Mushrooms | 2 | 4 | 6 |
| No mushrooms | 1 | 2 | 3 |
| **Total** | 3 | 6 | 9 |

$P(\text{mushrooms}) = 6/9 = 0.667$ and $P(\text{mushrooms} \mid \text{thick crust}) = 2/3 = 0.667$. Because $P(\text{mushrooms}) = P(\text{mushrooms} \mid \text{thick crust})$, the events "Mushrooms" and "Thick crust" are independent. (c) Use the general multiplication rule. $P(\text{both randomly selected pizzas have mushrooms}) = P(\text{first has mushrooms}) \cdot P(\text{second has mushrooms} \mid \text{first has mushrooms}) = (6/9)(5/8) = 0.417$. If you randomly select 2 of the pizzas in the oven, the probability that both have mushrooms is 0.417.

R5.7 (a) A tree diagram is shown here.

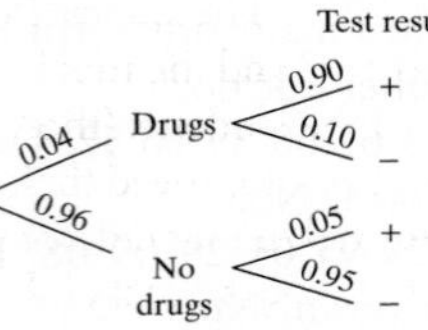


(b) $P(\text{drug test result is positive}) = (0.04)(0.90) + (0.96)(0.05) = 0.084$. The probability that the drug test result is positive is 0.084. (c) $P(\text{uses illegal drugs} \mid \text{positive result}) =$

$$\frac{P(\text{uses illegal drugs} \cap \text{positive result})}{P(\text{positive result})} = \frac{(0.04)(0.90)}{0.084} = 0.429$$

Given that the drug test result is positive, there is a 0.429 probability that person uses illegal drugs.

[illegible] elieves that finding and picking up a penny is [illegible] none of the 10 people believe this) [illegible] 8177

The probability of selecting 10 U.S. adults and finding at least 1 who believes that finding and picking up a penny is good luck is 0.98177.

## Answers to Chapter 5 AP® Practice Test

| | | | |
|---|---|---|---|
| T5.1 d | T5.2 b | T5.3 c | T5.4 b |
| T5.5 c | T5.6 a | T5.7 b | T5.8 a |
| T5.9 e | T5.10 e | | |

T5.11 (a) $P(\text{student eats regularly in the cafeteria and is not a 10th grader}) = \frac{130+122+68}{805} = \frac{320}{805} = 0.398$. There is a probability of 0.398 that a randomly selected student eats regularly in the cafeteria and is not a 10th grader. (b) $P(\text{10th grader} \mid \text{eats regularly in the cafeteria}) = 175/495 = 0.354$. Given that the student eats regularly in the cafeteria, there is a 0.354 probability that he or she is a 10th grader. (c) $P(\text{10th grader}) = 209/805 = 0.260$. The events "10th grader" and "Eats regularly in the cafeteria" are not independent because $P(\text{10th grader} \mid \text{eats regularly in the cafeteria}) \neq P(\text{10th grader})$.

T5.12 Let's organize this information using a tree diagram.

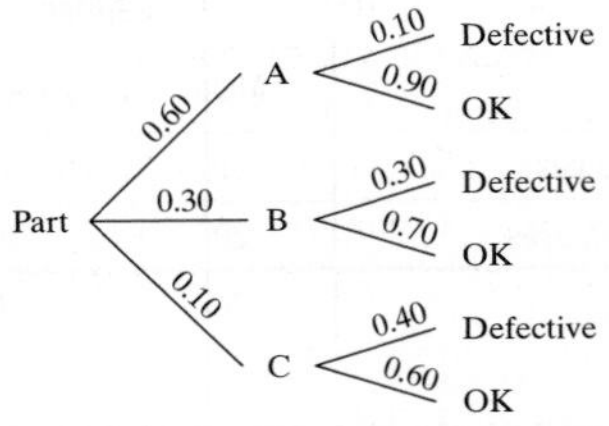


(a) To get the probability that a part randomly chosen from all parts produced in this factory is defective, add the probabilities from all branches in the tree that end in a defective part. $P(\text{defective}) = (0.60)(0.10) + (0.30)(0.30) + (0.10)(0.40) = 0.06 + 0.09 + 0.04 = 0.19$. (b) $P(\text{Machine B} \mid \text{defective}) = \frac{(0.30)(0.30)}{0.19} = \frac{0.09}{0.19} = 0.474$.

Given that a part is inspected and found to be defective, there is a 0.474 probability that it was produced by Machine B.

T5.13 (a) $P(\text{customer will pay less than the usual cost of the buffet}) = 23/36 = 0.639$. There is a 0.639 probability that the customer will pay less than the usual cost of the buffet. (b) $P(\text{all 4 friends end up paying less than the usual cost of the buffet}) = (23/36)^4 = 0.167$. There is a 0.167 probability that all 4 of these friends end up paying less than the usual cost of the buffet. (c) $P(\text{at least 1 of the 4 friends ends up paying more than the usual cost of the buffet}) = 1 - 0.167 = 0.833$. There is a 0.833 probability that at least 1 of the 4 friends ends up paying more than the usual cost of the buffet.

T5.14 (a) To carry out the simulation using a table of random digits, let 00 – 16 = cars with out-of-state plates and let 17 – 99 = other cars. Moving left to right across a row, look at pairs of digits from a random number table until you have found two numbers between 00 and 16 (repeats are allowed). Record how many two-digit numbers you had to read in order to get 2 numbers between 00 and 16. Repeat this process many times. (b) The first sample is 41 **05 09**. The bold numbers represent cars with out-of-state plates. In this sample, it took three cars to find two with out-of-state plates. The second sample is 20 31 **06** 44 90 50 59 59 88 43 18 80 53 **11**. In this sample, it took 14 cars to find two with out-of-state plates. The third sample is 58 44 69 94 86 85 79 67 **05** 81 18 45 **14**. In this sample, it took 13 cars to find two with out-of-state plates.

# Chapter 6

## Section 6.1

### Answers to Check Your Understanding

*page 393:* **1.** Using probability notation, the event "Student got a C" is written as $P(X = 2)$. $P(X = 2) = 1 - 0.011 - 0.032 - 0.362 - 0.457 = 0.138$

**2.** $P(X \geq 3)$ is the probability that the student got either an A or a B. This probability is $P(X \geq 3) = 0.362 + 0.457 = 0.819$.

**3.** The histogram is left skewed. This means that higher grades are more likely, and there are a few lower grades.

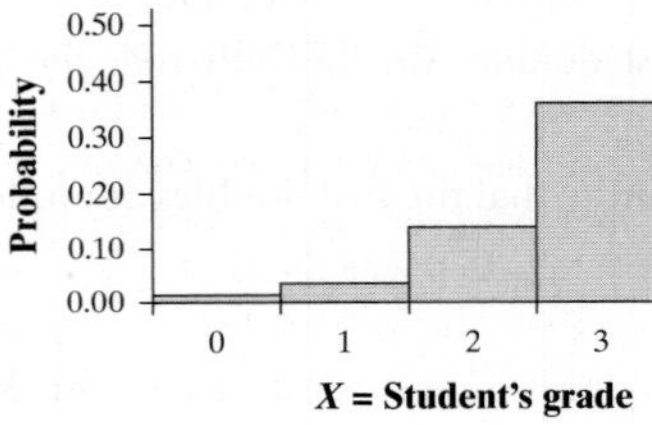

*page 399:* **1.** $\mu_X = (0)(0.011) + (1)(0.032) + (2)(0.138) + (3)(0.362) + (4)(0.457) = 3.222$. If many, many students are randomly selected, their average grade will be about 3.222.
**2.** $\sigma_X^2 = (0-3.222)^2(0.011) + (1-3.222)^2(0.032) + (2-3.222)^2(0.138) + (3-3.222)^2(0.362) + (4-3.222)^2(0.457) = 0.7727$. So $\sigma_X = \sqrt{0.7727} = 0.879$. If many, many students are randomly selected, their grades will typically vary by about 0.879 point from the mean of 3.222.

## Answers to Odd-Numbered Section 6.1 Exercises

6.1 **(a)** $X = 5; P(X = 5) = 0.11$ **(b)** $P(X \leq 3) = 0.72$
6.3 **(a)** $P(Y < 20) = 0.31$; there is a 0.31 probability that the amount of money collected on a randomly selected ferry trip is less than \$20. **(b)** $Y \geq 20; P(Y \geq 20) = 0.69$
6.5

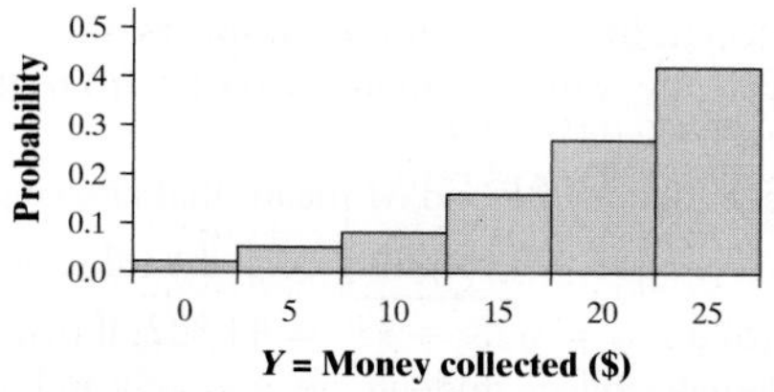


Skewed to the left with a single peak at \$25 collected.
6.7 $\mu_Y = 19.35$; if many, many ferry trips are randomly selected, the average amount of money collected will be about \$19.35.
6.9 **(a)** Right-skewed distribution; the most likely first digit is 1, and each subsequent digit is less likely than the previous digit. **(b)** $\mu_X = 3.441$; if many, many legitimate records are randomly selected, the average of the first digit would be about 3.441.
6.11 **(a)**

| Money collected | 0 | 5 | 10 | 15 | 20 | 25 |
|---|---|---|---|---|---|---|
| Probability | 0.02 | 0.05 | 0.08 | 0.16 | 0.27 | 0.42 |
| Cumulative Probability | 0.02 | 0.07 | 0.15 | 0.31 | 0.58 | 1 |

**(b)** The median of Y is \$20. **(c)** The mean of Y is less than the median of Y because the probability distribution is skewed left.
6.13 $\sigma_Y = 6.429$; if many, many ferry trips are randomly selected, the cost will typically vary by about \$6.43 from the mean of \$19.35.
6.15 $\sigma_X = 2.462$; if many, many records are randomly selected, the value of the first digit of the record will typically vary by about 2.462 from the mean of 3.441.
6.17 **(a)** The company has collected \$1250 and must pay out \$100,000. The company earns \$1250 − \$100,000 = −\$98,750. **(b)** $\mu_Y = \$303.35$; if many, many 21-year-old males are insured by this company, the average amount the company would make, per person, will be about \$303.35. **(c)** $\sigma_Y = \$9707.57$; if many, many policies are randomly selected, the amount that the company earns will typically vary by about \$9707.57 from the mean of \$303.35.
6.19 **(a)** Both distributions are skewed to the right. The center for the "household" distribution is less than the center for the "family" distribution, but the variability of the household distribution is greater than the variability of the family distribution. Also, the event $X = 1$ has a much higher probability in the household distribution. **(b)** $\mu_H = 2.6$ and $\mu_F = 3.14$; the household distribution has a slightly smaller mean than the family distribution. **(c)** $\sigma_H = 1.421$ and $\sigma_F = 1.249$; the standard deviation for the household distribution is slightly larger than for the family distribution.
6.21 **(a)** $P(Y > 6) = 0.333; P(X > 6) = 0.155$; if an expense report contains more than a proportion of 0.155 of first digits that are greater than 6, it may be a fake expense report. **(b)** The mean is 5 because this distribution is symmetric. **(c)** To detect a fake expense report, compute the sample mean of the first digits and see if it is closer to 5 (suggesting a fake report) or near 3.441 (consistent with a truthful report).
6.23 Length = 3600; height = 1/3600; the probability that the request is received by this server within the first 5 minutes (300 seconds) after the hour = 300/3600 = 0.083.
6.25 **(a)** $P(-1 \leq Y \leq 1) = 0.4$; the probability that Mr. Shrager dismisses class within a minute of the end of class is 0.4. **(b)** $\mu_Y = 1.5$; the mean is 1.5 because this distribution is symmetric. **(c)** $k = 2.75$; to find this value, we solve the equation $(4-k)(1/5) = 0.25$ for $k$.
6.27 (i) $z = -1.50; P(Z < -1.50) = 0.0668$.
(ii) $P(Y < 6) =$ normalcdf(lower: −1000, upper: 6, mean: 7.11, SD: 0.74) = 0.0668. There is about a 6.68% chance that this student will run the mile in under 6 minutes.
6.29 **(a)** (i) $z = -1.83$ and $z = 1.5; P(-1.83 < Z < 1.5) = 0.9332 - 0.0336 = 0.8996$
(ii) $P(325 < X < 345) =$ normalcdf(lower: 325, upper: 345, mean: 336, SD: 6) = 0.8998. There is about an 89.98% chance that a randomly selected horse pregnancy will last between 325 and 345 days.
**(b)** (i) 0.80 area to the left of $c \rightarrow z = 0.84$.

$$0.84 = \frac{c - 336}{6} \rightarrow 5.04 = c - 336 \rightarrow c = 341.04$$

(ii) invNorm(area: 0.80, mean: 336, SD: 6) = 341.05. About 20% of horses will have pregnancies that last more than 341.04 days.
6.31 b
6.33 c
6.35 **(a)** Yes; if we look at the differences (Post – Pre) in the scores, the mean difference was 5.38 and the median difference was 3. This means that at least half of the students (though less than three-quarters because $Q_1$ was negative) improved their reading scores. **(b)** No, we do not have a control group that did not participate in the chess program for comparison. It may be that children of this age improve their reading scores for other reasons (e.g., regular school) and that the chess program had nothing to do with their improvement.

## Section 6.2

## Answers to Check Your Understanding

*page 414:* **1.** The probability distribution of X and the probability distribution of Y are shown here.

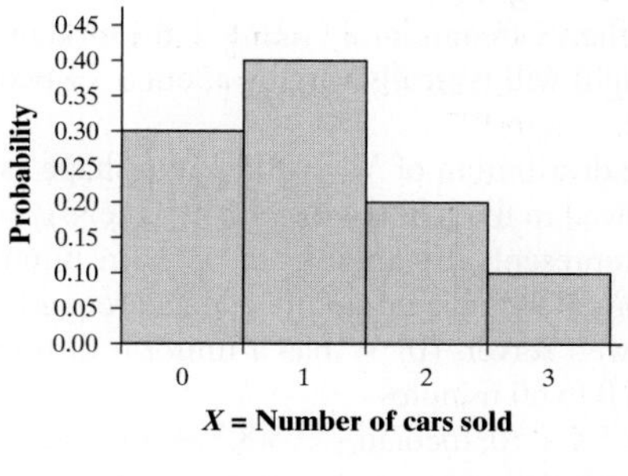


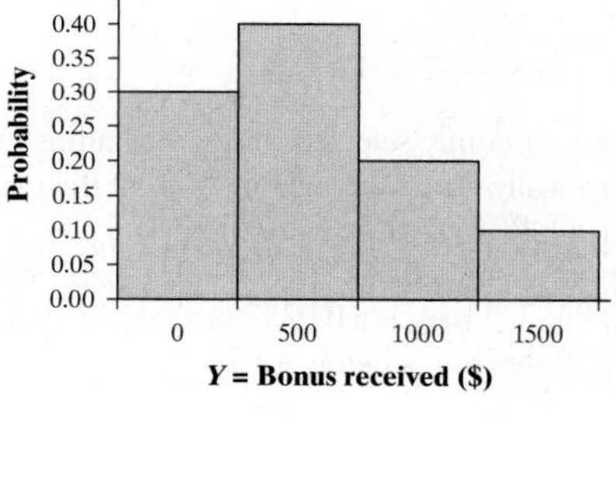

Both distributions are slightly skewed to the right with a single peak. Their shapes are identical.
**2.** $\mu_Y = 500(\mu_X) = 500(1.1) = \$550$
**3.** $\sigma_Y = 500(0.943) = \$471.50$. If many, many Fridays are randomly selected, the bonus earned in the first hour of business will typically vary by about \$471.50 from the mean of \$550.
**4.** Note that $T = Y - 75$. *Shape:* The shape of the probability distribution of $T$ will be the same as the shape of the probability distribution of Y: skewed right with a single peak.
*Center:* $\mu_T = \mu_Y - 75 = 550 - 75 = \$475$.
*Variability:* $\sigma_T = \sigma_Y = \$471.50$.
*page 422:* **1.** $\mu_T = \mu_X + \mu_Y = 1.1 + 0.7 = 1.8$; if many, many Fridays are randomly selected, this dealership expects to sell or lease about 1.8 cars, on average, in the first hour of business.
**2.** Because X and Y are independent, $\sigma_T^2 = \sigma_X^2 + \sigma_Y^2 = (0.943)^2 + (0.64)^2 = 1.2988$, so $\sigma_T = \sqrt{1.2988} = 1.14$. If many, many Fridays are randomly selected, the total number of cars sold or leased in the first hour will typically vary by about 1.14 cars from the mean of 1.8 cars.
**3.** The total bonus is $B = 500X + 300Y$. This means that $\mu_B = 500\mu_X + 300\mu_Y = 500(1.1) + 300(0.7) = \$760$. Because X and Y are independent, $\sigma_B^2 = (500\sigma_X)^2 + (300\sigma_Y)^2 = (500)^2(0.943)^2 + (300)^2(0.64)^2 = 259{,}176.25$. Therefore, $\sigma_B = \sqrt{259{,}176.25} = \$509.09$.

## Answers to Odd-Numbered Section 6.2 Exercises

6.37 Approximately Normal; $\mu_T = 20$ minutes; $\sigma_T = 6.5$ minutes
6.39 **(a)**

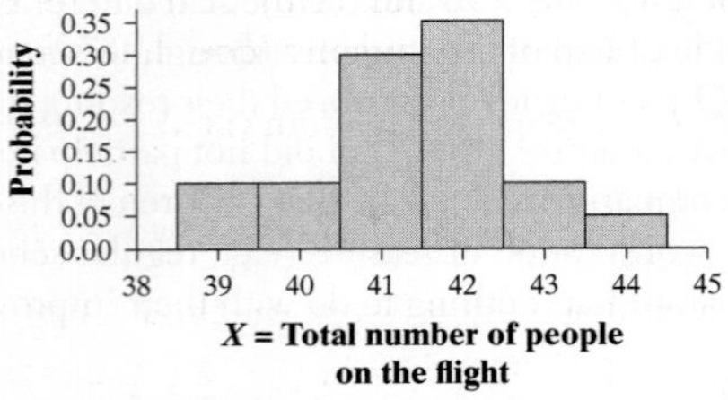


The distribution of total number of people on the flight is roughly symmetric with a single peak at 42 people. **(b)** $\mu_X = 41.4$; if many, many flights are randomly selected, the average of the total number of people on the flight will be about 41.4 people. **(c)** $\sigma_X = 1.24$; if many, many flights are randomly selected, the total number of people on the flight will typically vary by about 1.24 people from the mean of 41.4.
6.41 **(a)** The distribution of $M$ has the same shape as the distribution of X: skewed to the left. **(b)** $\mu_M = \$19.35$ **(c)** $\sigma_M = \$6.45$
6.43 **(a)** $W$ represents the amount of time (in minutes) after the start of an hour at which a randomly selected request is received by a particular web server. **(b)** $W$ has a uniform distribution on the interval from 0 to 60 minutes.
6.45 **(a)** $G = 5X + 50$; $\text{median}_G = 5(8.5) + 50 = 92.5$
**(b)** $IQR_G = 5(9 - 8) = 5$

6.47 **(a)** $\mu_Y = \frac{9}{5}(8.5) + 32 = 47.3$ **(b)** $\sigma_Y = \frac{9}{5}(2.25) = 4.05$; if many, many nights are randomly selected, the temperature in the cabin at midnight will typically vary by about 4.05°F from the mean of 47.3°F. **(c)** (i) $z = -1.80$; $P(Z < -1.80) = 0.0359$
(ii) $P(Y < 40) =$ normalcdf(lower: $-1000$, upper: 40, mean: 47.3, SD: 4.05) $= 0.0357$. There is a 0.0357 probability that the midnight temperature in the cabin is below 40°F.

6.49 $\mu_S = \mu_X + \mu_Y = 732.50 + 825 = \$1557.50$; the average of the sum of the tuitions would be about \$1557.50 for many, many randomly selected pairs of students from each of the campuses.
6.51 $\mu_{F-M} = 120 - 105 = 15$; the average of the difference would be about 15 points, if you were to repeat the process of selecting a single male student, selecting a single female student, and finding the difference (Female – Male) in their scores many times.
6.53 $\mu_{X/50-Y/55} = \mu_{X/50} - \mu_{Y/55} = 14.65 - 15 = -0.35$
6.55 **(a)** Yes; the mean of a sum is always equal to the sum of the means. **(b)** No; the variance of the sum is not equal to the sum of the variances because it is not reasonable to assume that X and Y are independent.
6.57 Because X and Y are independent, $\sigma_S = \sqrt{103^2 + 126.50^2} = \$163.13$. If many, many pairs of students are randomly and independently selected from each of the campuses, the sum of the tuitions will typically vary by about \$163.13 from the mean of \$1557.50.
6.59 **(a)** Independence of $F$ and $M$ means that knowing the value of one student's score does not help us predict the value of the other student's score. **(b)** $\sigma_{F-M} = \sqrt{28^2 + 35^2} = 44.822$; if many, many pairs of male and female college students are randomly and independently selected, the difference (Female – Male) of their SSHA scores will typically vary by about 44.822 points from the mean of 15 points. **(c)** No, we do not know the shapes of the distributions. We cannot assume that the distributions are Normal without additional information.

6.61 $\sigma_D = \sigma_{X/50-Y/55} = \sqrt{\left(\frac{103}{50}\right)^2 + \left(\frac{126.50}{55}\right)^2} = 3.088$; the standard deviation of the difference $D$(Main – Downtown) in the number of units that two randomly selected students take is 3.088.

6.63 $\mu_{X_1+X_2} = \$606.70$, and because the variables are independent, $\sigma_{X_1+X_2} = \sqrt{(9707.57^2 + 9707.57^2)} = \$13{,}728.58$. $W = \frac{1}{2}(X_1 + X_2)$, so $\mu_W = \$303.35$ and $\sigma_W = \$6864.29$.
6.65 **(a)** Normal distribution with mean $= 11 + 20 = 31$ seconds, and because the variables are independent, standard deviation $= \sqrt{2^2 + 4^2} = 4.4721$. **(b)** (i) $z = -0.22$; $P(Z < -0.22) = 0.4129$
(ii) $P(X + Y < 30) =$ normalcdf(lower: $-1000$, upper: 30, mean: 31, SD: 4.4721) $= 0.4115$. There is a 0.4115 probability of completing the process in less than 30 seconds for a randomly selected part.
6.67 Let $D = L - H$; $\mu_D = 105 - 98 = 7$. Because the random variables are independent, $\sigma_D = \sqrt{(10)^2 + (15)^2} = 18.03$. $D$ is approximately Normally distributed with a mean of 7 and a standard deviation of 18.03. We want to find $P(-5 < L - H < 5)$ or $P(-5 < D < 5)$. (i) $z = -0.11$ and $z = -0.67$; $P(-0.67 < Z < -0.11) = 0.4562 - 0.2514 = 0.2048$ (ii) $P(-5 < D < 5) =$ normalcdf(lower: $-5$, upper: 5, mean: 7, SD: 18.03) $= 0.2030$. There is a 0.203 probability that Lamar and Hareesh finish their jobs within 5 minutes of each other on a randomly selected day.
6.69 Let $T = X_1 + X_2 + X_3 + X_4$; $\mu_T = 55.2 + 58.0 + 56.3 + 54.7 = 224.2$. Because the random variables are independent, $\sigma_T = \sqrt{(2.8)^2 + (3.0)^2 + (2.6)^2 + (2.7)^2} = 5.56$. $T$ is approximately Normally distributed with a mean of 224.2 and a standard deviation of 5.56. We want to find $P(T < 220)$. (i) $z = -0.76$; $P(Z < -0.76) = 0.2236$ (ii) $P(T < 220) =$ normalcdf(lower: $-1000$, upper: 220, mean: 224.2, SD: 5.56) $= 0.2250$. There is a 0.2250 probability that the total team time is less than 220 seconds in a randomly selected race.

6.71 Let $D = X_1 - X_2; \mu_D = 1.4 - 1.4 = 0$. Because the random variables are independent, $\sigma_{X_1-X_2} = \sqrt{0.3^2 + 0.3^2} = 0.4243$. $D$ is Normally distributed with a mean of 0 and a standard deviation of 0.4243. We want to find $P(D < -0.8 \text{ or } D > 0.8)$.
(i) $z = 1.89$ and $z = -1.89$; $P(Z < -1.89 \text{ or } Z > 1.89) = 0.0588$ (ii) $P(D < -0.8$ or $D > 0.8) = 1 -$ normalcdf(lower: $-0.8$, upper: 0.8, mean: 0, SD: 0.4243) $= 1 - 0.9406 = 0.0594$. There is a 0.0594 probability that difference is at least as large as the attendant observed.

6.73 c

6.75 **(a)** Yes; the girls who were assigned to receive the fluoride varnish may expect to get fewer cavities and, as a result, might take better care of their teeth. **(b)** This experiment could be double-blind if the girls who are not selected to receive the fluoride varnish received a placebo varnish that would neither help nor hurt their teeth. Also, the dental hygienist who checks the girls' teeth for cavities at the end of the 4 years should not know which girls received which treatment. **(c)** The random assignment should help to make the two groups as similar as possible before treatments are administered.

## Section 6.3

### Answers to Check Your Understanding

*page 434:* **1.** *Binary?* "Success" = get an ace; "Failure" = don't get an ace. *Independent?* Because you are replacing the card in the deck and shuffling each time, the result of one trial does not tell you anything about the outcome of any other trial. *Number?* $n = 10$. *Same probability?* $p = 4/52$. This is a binomial setting, and X has a binomial distribution with $n = 10$ and $p = 4/52$.
**2.** *Binary?* "Success" = over 6 feet; "Failure" = not over 6 feet. *Independent?* Because we are selecting without replacement from a small number of students, the observations are not independent. *Number?* $n = 5$. *Same probability?* The (unconditional) probability of success will not change from trial to trial. Because the trials are not independent, this is not a binomial setting.
**3.** *Binary?* "Success" = roll a 5; "Failure" = don't roll a 5. *Independent?* Because you are rolling a die, the outcome of any one trial does not tell you anything about the outcome of any other trial. *Number?* $n = 100$. *Same probability?* No; the probability of success changes when the corner of the die is chipped off. Because the (unconditional) probability of success changes from trial to trial, this is not a binomial setting.
*page 440:* **1.** *Binary?* "Success" = question answered correctly; "Failure" = question not answered correctly. *Independent?* Mr. Miller randomly determined correct answers to the questions, so knowing the result of one trial (question) should not tell you anything about the result on any other trial. *Number?* $n = 10$. *Same probability?* $p = 0.20$. This is a binomial setting and X has a binomial distribution with $n = 10$ and $p = 0.20$.
**2.** $P(X = 3) = \binom{10}{3}(0.2)^3(0.8)^7 = 0.2013$. There is about a 20% chance that Hannah will answer exactly 3 questions correctly.
**3.** $P(X \geq 6) = 1 - P(X < 6) = 1 - P(X \leq 5) = 1 -$ binomcdf (trials: 10, $p$: 0.2, $x$ value:5) $= 1 - 0.9936 = 0.0064$. There is only a 0.0064 probability that a student would get 6 or more correct, so we would be quite surprised if Hannah was able to pass.

*page 446:* **1.**

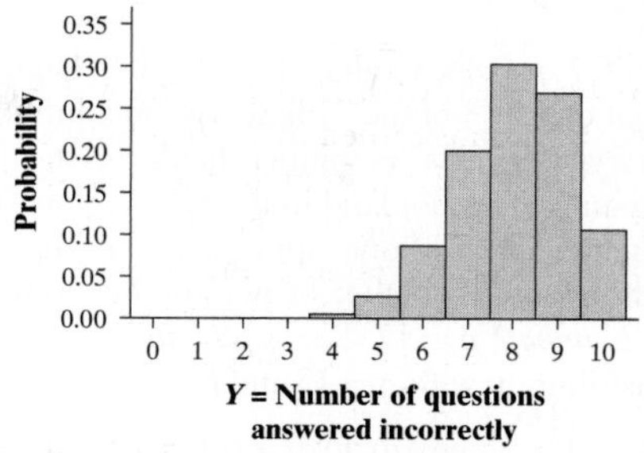


The probability distribution is skewed to the left with a single peak at Y = 8.
**2.** $\mu_Y = np = 10(0.80) = 8$. If many students took the quiz, the average number of questions students would answer incorrectly is about 8 questions.
**3.** $\sigma_Y = \sqrt{np(1-p)} = \sqrt{10(0.80)(0.20)} = 1.265$. If many students took the quiz, we would expect individual students' scores to typically vary by about 1.265 incorrect answers from the mean of 8 incorrect answers.
**4.** The shape of X is skewed right and the shape of Y is skewed left, both with a single peak. The center (mean) of the distribution of X is 2, which is less than the center (mean) of the distribution of Y, which is 8. (Note that the sum of the means is 10 because X + Y = 10.) The variability of both probability distributions is the same (standard deviation = 1.265).
*page 454:* **1.** Die rolls are independent, the probability of getting doubles is the same on each roll (1/6), and we are repeating the random process until we get a success (doubles). This is a geometric setting and $T$ is a geometric random variable with $p = \frac{1}{6}$.
**2.** $P(T = 3) = \left(\frac{5}{6}\right)^2\left(\frac{1}{6}\right) = 0.1157$. The probability is 11.57% that you will get the first set of doubles on the third roll of the dice.
**3.** $P(T \leq 3) = \frac{1}{6} + \left(\frac{5}{6}\right)\left(\frac{1}{6}\right) + \left(\frac{5}{6}\right)^2\left(\frac{1}{6}\right) = 0.4213$. The probability is 42.13% of getting doubles in 3 or fewer rolls.
**4.** $\mu_T = \frac{1}{(1/6)} = 6$. If this random process is repeated many times, the average number of rolls it will take to get doubles is about 6.
$\sigma_T = \frac{\sqrt{1-(1/6)}}{1/6} = 5.477$. If this random process is repeated many times, the number of rolls it will take to get doubles will typically vary by about 5.477 rolls from the mean of 6 rolls.

### Answers to Odd-Numbered Section 6.3 Exercises

6.77 *Binary?* "Success" = survive; "Failure" = does not survive. *Independent?* Yes; knowing the outcome of one elk shouldn't tell us anything about the outcomes of other elk. *Number?* $n = 7$. *Same probability?* $p = 0.44$. X has a binomial distribution with $n = 7$ and $p = 0.44$.

6.79 *Binary?* "Success" = hits; "Failure" = does not hit. *Independent?* Yes, the outcome of one shot does not tell us anything about the outcome of other shots. *Number?* No, there is not a fixed number of trials. Y does not have a binomial distribution.

6.81 X has a binomial distribution with $n = 7$ and $p = 0.44$.

$P(X = 4) = \binom{7}{4}(0.44)^4(0.56)^3 = 0.2304$. *Tech:* $P(Y = 4) =$ binompdf(trials: 7, $p$: 0.44, $x$ value: 4) = 0.2304. There is a 23.04% probability that exactly 4 of the 7 elk survive to adulthood.

6.83 **(a)** *Binary?* "Success" = spinner lands in the blue region; "Failure" = spinner does not land in the blue region. *Independent?* Knowing whether or not one spin lands in the blue region tells you nothing about whether or not another spin lands in the blue region. *Number?* $n = 12$. *Same probability?* $p = 0.80$. X has a binomial distribution with $n = 12$ and $p = 0.80$.

**(b)** $P(X = 8) = \binom{12}{8}(0.80)^8(0.20)^4 = 0.1329$; *Tech:* $P(Y = 8) =$ binompdf(trials: 12, $p$: 0.80, $x$ value: 8) = 0.1329. There is about a 13.29% probability that the spinner will land in the blue region exactly 8 of the 12 times.

6.85 X has a binomial distribution with $n = 7$ and $p = 0.44$. $P(X > 4) = 0.1402$. *Tech:* $P(X > 4) = 1-$ binomcdf(trials: 7, $p$: 0.44, $x$ value: 4) $= 1 - 0.8598 = 0.1402$. This probability isn't very small, so it is not surprising for more than 4 elk to survive to adulthood.

6.87 X has a binomial distribution with $n = 12$ and $p = 0.80$. *Tech:* $P(X \leq 7) =$ binomcdf(trials: 12, $p$: 0.80, $x$ value: 7) = 0.0726. Assuming the website's claim is true, there is a probability of about 7.26% that there would be 7 or fewer spins landing in the blue region.

6.89 **(a)** X has a binomial distribution with $n = 22$ and $p = 0.20$. *Tech:* $P(X \geq 14) = 1-$ binomcdf(trials: 22, $p$: 0.20, $x$ value: 13) = 0.00001. Assuming participants don't have a special preference for the last thing they taste, there is an almost 0% probability that there would be at least 14 people who choose the last kiss. **(b)** Because this outcome is very unlikely, we have convincing evidence that participants have a preference for the last thing they taste.

6.91 **(a)** *Binary?* "Success" = red; "Failure" = not red. *Independent?* Knowing whether or not the light is red for one randomly selected passenger tells you nothing about whether or not the light is red on another randomly selected passenger. *Number?* $n = 20$. *Same probability?* $p = 0.30$. R has a binomial distribution with $n = 20$ and $p = 0.30$. **(b)** The shape is fairly symmetric with a single peak at $R = 6$ passengers.

6.93

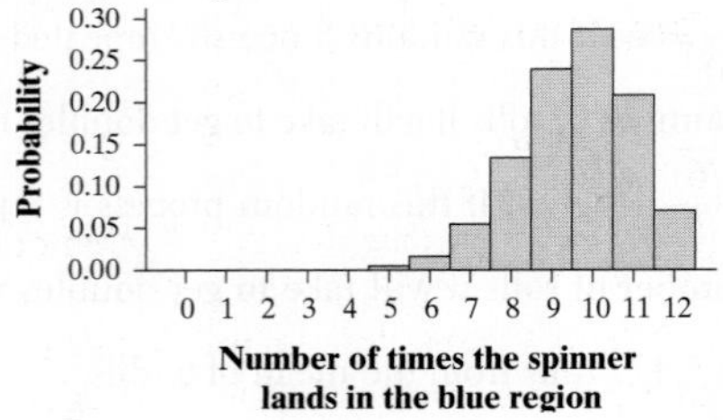


The shape is left-skewed with a single peak at $X = 10$.

6.95 **(a)** $\mu_R = 20(0.3) = 6$; if many groups of 20 passengers were selected, the average number of passengers who would get a red light would be about 6 passengers. **(b)** $\sigma_R = \sqrt{20(0.3)(0.7)} = 2.049$; if many groups of 20 passengers were selected, the number of passengers who would get a red light would typically vary from the mean (6) by about 2.049.

6.97 **(a)** Let Y = number of calls *not* completed. Y has a binomial distribution with $n = 15$ and $p = 0.91$. *Tech:* $P(Y > 12) = 1-$ binomcdf(15, 0.91, 12) = 0.8531. There is an 85.31% probability that more than 12 calls are not completed. **(b)** $\mu_X = 15(0.09) = 1.35$; if we watched the machine make many sets of 15 calls, the average number of calls that would reach a live person would be about 1.35 calls. **(c)** $\sigma_X = \sqrt{15(0.09)(0.91)} = 1.11$; if we watched the machine make many sets of 15 calls, we would expect the number of calls that reach a live person to typically vary by about 1.11 calls from the mean of 1.35 calls.

6.99 **(a)** As long as $n = 100$ is less than 10% of the size of the population (the entire high school), $L$ can be modeled by a binomial distribution even though the sample was selected without replacement. **(b)** *Tech:* $P(L \geq 15) \approx 1-$ binomcdf(100, 0.11, 14) = 0.1330.

6.101 No; we are sampling without replacement and the sample size (10) is more than 10% of the population size (76). We should not view the observations as independent.

6.103 **(a)** $np = (100)(0.11) = 11 \geq 10$ and $n(1-p) = 100(1-0.11) = 89 \geq 10$. Because the expected number of successes and the expected number of failures are both 10 or more, $L$ can be approximated by a Normal distribution. **(b)** We want to find $P(L \geq 15)$. $\mu_L = (100)(0.11) = 11$ and $\sigma_L = \sqrt{(100)(0.11)(0.89)} = 3.129$.
(i) $z = 1.28$; $P(Z \geq 1.28) = 1 - 0.8997 = 0.1003$
(ii) $P(L \geq 15) =$ normalcdf(lower: 15, upper: 1000, mean: 11, SD: 3.129) = 0.1006. There is a 0.1006 probability that 15 or more students in the sample are left-handed.

6.105 When sampling without replacement, the trials are not independent because knowing the outcomes of previous trials makes it easier to predict what will happen in future trials. However, if the sample is a small fraction of the population (less than 10%), the makeup of the population doesn't change enough to make the lack of independence an issue.

6.107 **(a)** X is a geometric random variable with $p = 0.20$; $P(X = 3) = (0.8)^2(0.2) = 0.128$. *Tech:* geometpdf($p$: 0.20, $x$ value: 3) = 0.128. There is a 0.128 probability that it takes Rita exactly 3 pulls to start the mower. **(b)** $P(X > 6) = 1 - P(X \leq 6)$.

$$= 1 - [0.20 + (0.8)(0.2) + (0.8)^2(0.2) + \cdots + (0.8)^5(0.2)]$$
$$= 0.2621$$

*Tech:* 1 − geometcdf($p$: 0.20, $x$ value: 6) = 0.2621. There is a 0.2621 probability that it takes Rita more than 6 pulls to start the mower.

6.109 **(a)** This is not a geometric setting because we can't classify the possible outcomes on each trial (card) as "success" or "failure" and we are not selecting cards until we get 1 success. **(b)** Games of 4-Spot Keno are independent, the probability of winning is the same in each game ($p = 0.259$), and Lola is repeating a random process until she gets a success. X is a geometric random variable with $p = 0.259$.

6.111 **(a)** X is a geometric random variable with $p = 0.097$; $\mu_X = \frac{1}{p} = \frac{1}{0.097} = 10.31$. If this random process is repeated many times, the average number of invoices we would expect to examine in order to find the first one with an amount that begins with an 8 or 9 is about 10.31 invoices.

**(b)** $\sigma_X = \frac{\sqrt{1-0.097}}{0.097} = 9.797$

If this random process is repeated many times, the number of invoices it will take to get the first invoice with an amount that begins with an 8 or a 9 will typically vary by about 9.797 invoices from the mean of 10.31.

6.113 *Tech:* $P(X \geq 40) = 1 - P(X \leq 39) = 1 -$ geometcdf($p$: 0.097, $x$ value: 39) $= 1 - 0.9813 = 0.0187$. Because the probability of not getting an 8 or 9 before the 40th invoice is small, we may begin to worry that the invoice amounts are fraudulent.

6.115 b
6.117 d
6.119 c
6.121 The standard deviations from smallest to largest are: B, C, A.

## Answers to Chapter 6 Review Exercises

R6.1 (a) $P(Y=5)=0.1$; there is a 0.1 probability that a randomly selected patient would rate his or her pain as a 5 on a scale of 1 to 5. (b) $P(Y \le 2)=0.3$ (c) $\mu_X = 3.1$; $\sigma_X = 1.136$
R6.2 (a) Temperature is a continuous random variable because it takes all values in an interval of numbers. (b) $\mu_D = 550 - 550 = 0°C$, and the standard deviation stays the same, $\sigma_D = 5.7°C$, because subtracting a constant does not change the variability. (c) $\mu_Y = \frac{9}{5}(550)+32 = 1022°F$ and $\sigma_Y = \left(\frac{9}{5}\right)(5.7) = 10.26°F$.
R6.3 (a)

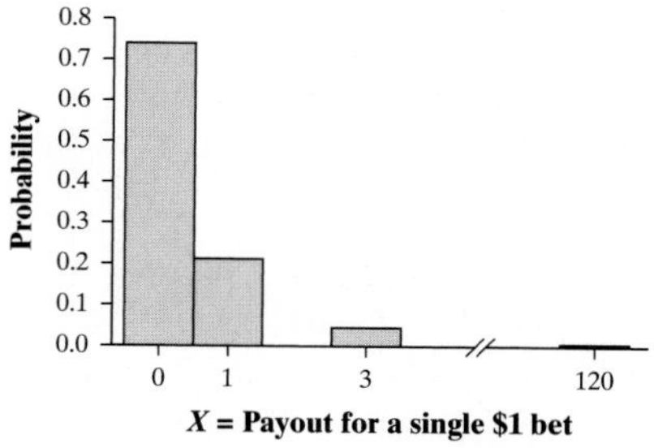


The distribution is skewed right with a single peak at the \$0 payout. The mean of the distribution is \$0.70 and the standard deviation is \$6.58. (b) Payout averages \$0.70 per game. If you were to play many games of 4-Spot Keno, the payout amounts would typically vary by about \$6.58 from the mean of \$0.70. (c) $Y = 5X$; $\mu_Y = 5(0.70) = \$3.50$; $\sigma_Y = 5(6.58) = \$32.90$ (d) Let W be the amount of Marla's payout, $W = X_1 + X_2 + X_3 + X_4 + X_5$; $\mu_W = 5(0.70) = \$3.50$ and $\sigma_W = \sqrt{(6.58^2 + 6.58^2 + 6.58^2 + 6.58^2 + 6.58^2)} = \$14.71$.
R6.4 (a) (i) $z = 0.83$; $P(Z > 0.83) = 1 - P(Z \le 0.83) = 0.2033$ (ii) $P(C > 11) =$ normalcdf(lower: 11, upper: 1000, mean: 10, SD: 1.2) $=$ 0.2023. There is a 0.2023 probability that a randomly selected cap has strength greater than 11 inch-pounds. (b) It is reasonable to assume the cap strength and torque are independent because the machine that makes the caps and the machine that applies the torque are not the same. (c) $C - T$ is Normal with mean $10 - 7 = 3$ inch-pounds and standard deviation $\sqrt{0.9^2 + 1.2^2} = 1.5$ inch-pounds.
(d) (i) $z = -2$; $P(Z < -2) = 0.0228$
(ii) $P(C - T < 0) =$ normalcdf(lower: $-1000$, upper: 0, mean: 3, SD: 1.5) $= 0.0228$. There is a 0.0228 probability that a randomly selected cap will break when being fastened by the machine.
R6.5 (a) *Binary?* "Success" = orange; "Failure" = not orange. *Independent?* The sample of size $n = 8$ is less than 10% of the large bag, so we can view the outcomes of trials as independent. *Number?* We are choosing a fixed sample of $n = 8$ candies. *Success?* The probability of success remains constant at $p = 0.205$. This is a binomial setting, so X has a binomial distribution with $n = 8$ and $p = 0.205$. (b) $P(X = 3) = 0.1532$. *Tech:* $P(X = 3) =$ binompdf (trials: 8, $p$: 0.205, $x$ value: 3) $= 0.1532$. There is a 15.32% probability that exactly 3 of the 8 M&M'S® will be orange. (c) *Tech:* $P(X \ge 4) = 1 -$ binomcdf(trials: 8, $p$: 0.205, $x$ value: 3) $= 0.0610$. There is a 6.1% probability that at least 4 of the 8 M&M'S will be orange. (d) No; it is not very unusual to receive 4 or more orange M&M'S in a sample of 8 M&M'S when 20.5% of M&M'S produced are orange. This happens 6.1% of the time due to chance alone, so this result does not provide convincing evidence that Mars, Inc.'s claim is false.
R6.6 (a) $\mu_X = 8(0.205) = 1.64$; if we were to select many random samples of size 8, we would expect to get about 1.64 orange M&M'S, on average. (b) $\sigma_X = \sqrt{8(0.205)(0.795)} = 1.14$; if we were to select many random samples of size 8, the number of orange M&M'S would typically vary by about 1.14 from the mean of 1.64.
R6.7 (a) Y is a geometric random variable with $p = \frac{3}{12} = 0.25$; $P(Y \le 3) = 0.5781$. *Tech:* geometcdf($p$: 0.25, $x$ value: 3) $= 0.5781$. There is a 0.5781 probability that it takes 3 or fewer spins to get a wasabi bomb. (b) $\mu_Y = \frac{1}{0.25} = 4$
If this random process is repeated many times, the average number of spins it would take to get the first wasabi bomb is 4 spins.
(c) $\sigma_Y = \frac{\sqrt{1-0.25}}{0.25} = 3.464$
If this random process is repeated many times, the number of spins it would take to get the first wasabi bomb would typically vary by about 3.464 spins from the mean of 4 spins.
R6.8 (a) Although the trials are binary ("success" = use and "failure" = does not use) and there are a fixed number of trials ($n = 200$), the residents were selected without replacement, violating the condition of independence. However, since the sample size ($n = 200$) is less than 10% of the population size (all residents in a large city), $T$ has approximately a binomial distribution with $n = 200$ and $p = 0.34$. (b) $n = 200$ and $p = 0.34$, so $np = 68 \ge 10$ and $n(1-p) = 132 \ge 10$. Since the expected number of successes and the expected number of failures are both 10 or more, $T$ can be approximated by a Normal distribution. (c) We want to find $P(T \le 60)$. $\mu_T = (200)(0.34) = 68$
and $\sigma_T = \sqrt{np(1-p)} = \sqrt{(200)(0.34)(0.66)} = 6.699$.
(i) $z = -1.19$; $P(Z \le -1.19) = 0.1170$
(ii) $P(T \le 60) =$ normalcdf(lower: $-1000$, upper: 60, mean: 68, SD: 6.699) $= 0.1162$. There is a 0.1162 probability that at most 60 residents in the sample use public transportation at least once per week.

## Answers to Chapter 6 AP® Practice Test

| | | | |
|---|---|---|---|
| T6.1 b | T6.2 d | T6.3 d | T6.4 e |
| T6.5 d | T6.6 b | T6.7 c | T6.8 b |
| T6.9 b | T6.10 c | | |

T6.11 (a) $P(Y \le 2) = 0.96$; there is a 96% chance that at least 10 eggs are unbroken in a randomly selected carton of "store brand" eggs. (b) $\mu_Y = 0.38$; if we were to randomly select many cartons of eggs, we would expect an average of about 0.38 eggs to be broken. (c) $\sigma_Y = 0.8219$; if we were to randomly select many cartons of eggs, the number of broken eggs would typically vary by about 0.8219 eggs from the mean of 0.38. (d) X is a geometric random variable with $p = 0.11$. We are looking for $P(X \le 3)$. *Tech:* geometcdf($p$: 0.11, $x$ value: 3) $= 0.2950$.
T6.12 (a) *Binary?* "Success" = greets dog first; "Failure" = does not greet dog first. *Independent?* We are sampling without replacement, but 12 is less than 10% of all dog owners. *Number?* $n = 12$. *Same probability?* The probability of success is constant for all trials ($p = 0.66$). X is a binomial random variable with $n = 12$ and $p = 0.66$. (b) $P(X = 6) = 0.1180$ (c) $P(X \le 4) = 0.0213$. *Tech:* binomcdf(trials: 12, $p$: 0.66, $x$ value: 4) $= 0.0213$. Because this probability is very small, it is unlikely that only 4 or fewer owners

will greet their dogs first by chance alone. This gives convincing evidence that the claim by *Ladies' Home Journal* is incorrect.

T6.13 **(a)** Letting $D = A - E$, $\mu_D = 50 - 25 = 25$. Because the amount of time they spend on homework is independent of each other, $\sigma_D = \sqrt{(10^2 + 5^2)} = 11.18$. **(b)** (i) $z = -2.24$; $P(Z < -2.24) = 0.0125$ (ii) $P(D < 0) =$ normalcdf(lower: $-1000$, upper: 0, mean: 25, SD: 11.18) $= 0.0127$. There is a 0.0127 probability that Ed spent longer on his assignment than Adelaide did on hers.

T6.14 **(a)** X has a binomial distribution with $n = 1200$ and $p = 0.13$. $\mu_X = 1200(0.13) = 156$ and $\sigma_X = \sqrt{1200(0.13)(0.87)} = 11.6499$. **(b)** If the sample contains 10% Hispanics, there were $1200(0.10) = 120$ Hispanics in the sample. We want to find $P(X \le 120) = 0.00083$. *Tech:* binomcdf(trials: 1200, $p = 0.13$, $x$ value: 120) $= 0.00083$. If we use the Normal approximation to the binomial distribution, $z = -3.09$ and $P(Z \le -3.09) = 0.001$. Because this probability is small, it is unlikely to select 120 or fewer Hispanics in the sample just by chance. This gives us reason to be suspicious of the sampling process.

# Formulas for the AP® Statistics Exam

*Students are provided with the following formulas on both the multiple choice and free-response sections of the AP® Statistics exam.*

## I. Descriptive Statistics

$$\bar{x} = \frac{1}{n}\sum x_i = \frac{\sum x_i}{n}$$

$$s_x = \sqrt{\frac{1}{n-1}\sum(x_i - \bar{x})^2} = \sqrt{\frac{\sum(x_i - \bar{x})^2}{n-1}}$$

$$\hat{y} = a + bx$$

$$\bar{y} = a + b\bar{x}$$

$$r = \frac{1}{n-1}\sum\left(\frac{x_i - \bar{x}}{s_x}\right)\left(\frac{y_i - \bar{y}}{s_y}\right)$$

$$b = r\frac{s_y}{s_x}$$

## II. Probability and Distributions

$$P(A \cup B) = P(A) + P(B) - P(A \cap B)$$

$$P(A|B) = \frac{P(A \cap B)}{P(B)}$$

| Probability Distribution | Mean | Standard Deviation |
|---|---|---|
| Discrete random variable, X | $\mu_X = E(X) = \sum x_i \cdot P(x_i)$ | $\sigma_X = \sqrt{\sum(x_i - \mu_X)^2 \cdot P(x_i)}$ |
| If X has a **binomial** distribution with parameters $n$ and $p$, then: $P(X = x) = \binom{n}{x} p^x (1-p)^{n-x}$ where $x = 0, 1, 2, 3, \ldots, n$ | $\mu_X = np$ | $\sigma_X = \sqrt{np(1-p)}$ |
| If X has a **geometric** distribution with parameter $p$, then: $P(X = x) = (1-p)^{x-1}p$ where $x = 1, 2, 3, \ldots$ | $\mu_X = \frac{1}{p}$ | $\sigma_X = \frac{\sqrt{1-p}}{p}$ |

## III. Sampling Distributions and Inferential Statistics

Standardized test statistic: $\dfrac{\text{statistic} - \text{parameter}}{\text{standard error of statistic}}$

Confidence interval: statistic $\pm$ (critical value) (standard error of statistic)

Chi-square statistic: $\chi^2 = \sum\dfrac{(\text{observed} - \text{expected})^2}{\text{expected}}$

## III. Sampling Distributions and Inferential Statistics *(continued)*

### Sampling Distributions for Proportions

| Random Variable | Parameters of Sampling Distribution | | Standard Error* of Sample Statistic |
|---|---|---|---|
| For one population $\hat{p}$ | $\mu_{\hat{p}} = p$ | $\sigma_{\hat{p}} = \sqrt{\frac{p(1-p)}{n}}$ | $s_{\hat{p}} = \sqrt{\frac{\hat{p}(1-\hat{p})}{n}}$ |
| For two populations: $\hat{p}_1 - \hat{p}_2$ | $\mu_{\hat{p}_1-\hat{p}_2} = p_1 - p_2$ | $\sigma_{\hat{p}_1-\hat{p}_2} = \sqrt{\frac{p_1(1-p_1)}{n_1} + \frac{p_2(1-p_2)}{n_2}}$ | $s_{\hat{p}_1-\hat{p}_2} = \sqrt{\frac{\hat{p}_1(1-\hat{p}_1)}{n_1} + \frac{\hat{p}_2(1-\hat{p}_2)}{n_2}}$<br>When $p_1 = p_2$ is assumed:<br>$s_{\hat{p}_1-\hat{p}_2} = \sqrt{\hat{p}_c(1-\hat{p}_c)\left(\frac{1}{n_1} + \frac{1}{n_2}\right)}$<br>where $\hat{p}_c = \frac{X_1 + X_2}{n_1 + n_2}$ |

### Sampling Distributions for Means

| Random Variable | Parameters of Sampling Distribution | | Standard Error* of Sample Statistic |
|---|---|---|---|
| For one population $\bar{x}$ | $\mu_{\bar{x}} = \mu$ | $\sigma_{\bar{x}} = \frac{\sigma}{\sqrt{n}}$ | $s_{\bar{x}} = \frac{s}{\sqrt{n}}$ |
| For two populations: $\bar{x}_1 - \bar{x}_2$ | $\mu_{\bar{x}_1-\bar{x}_2} = \mu_1 - \mu_2$ | $\sigma_{\bar{x}_1-\bar{x}_2} = \sqrt{\frac{\sigma_1^2}{n_1} + \frac{\sigma_2^2}{n_2}}$ | $s_{\bar{x}_1-\bar{x}_2} = \sqrt{\frac{s_1^2}{n_1} + \frac{s_2^2}{n_2}}$ |

### Sampling Distributions for Simple Linear Regression

| Random Variable | Parameters of Sampling Distribution | | Standard Error* of Sample Statistic |
|---|---|---|---|
| For slope: $b$ | $\mu_b = \beta$ | $\sigma_b = \frac{\sigma}{\sigma_x\sqrt{n}}$<br>where $\sigma_x = \sqrt{\frac{\sum(x_i - \mu)^2}{n}}$ | $s_b = \frac{s}{s_x\sqrt{n-1}}$<br>where $s = \sqrt{\frac{\sum(y_i - \hat{y}_i)^2}{n-2}}$<br>and $s_x = \sqrt{\frac{\sum(x_i - \bar{x})^2}{n-1}}$ |

*Standard deviation is a measurement of variability from the theoretical population. Standard error is the estimate of the standard deviation. If the standard deviation of the statistic is assumed to be known, then the standard deviation should be used instead of the standard error.

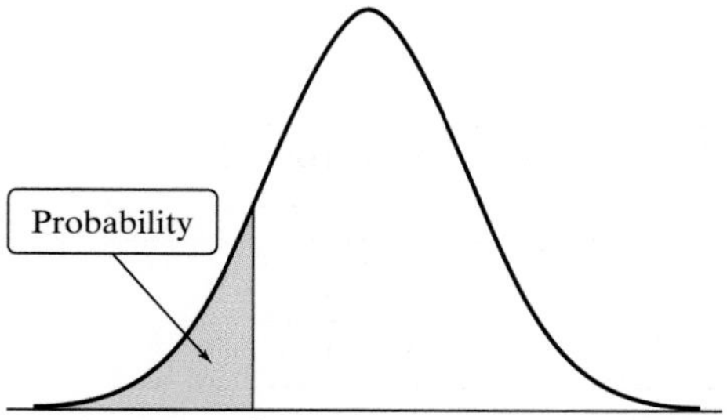


Table entry for $z$ is the area under the standard Normal curve to the left of $z$.

**Table A Standard Normal probabilities**

| $z$ | .00 | .01 | .02 | .03 | .04 | .05 | .06 | .07 | .08 | .09 |
|---|---|---|---|---|---|---|---|---|---|---|
| −3.4 | .0003 | .0003 | .0003 | .0003 | .0003 | .0003 | .0003 | .0003 | .0003 | .0002 |
| −3.3 | .0005 | .0005 | .0005 | .0004 | .0004 | .0004 | .0004 | .0004 | .0004 | .0003 |
| −3.2 | .0007 | .0007 | .0006 | .0006 | .0006 | .0006 | .0006 | .0005 | .0005 | .0005 |
| −3.1 | .0010 | .0009 | .0009 | .0009 | .0008 | .0008 | .0008 | .0008 | .0007 | .0007 |
| −3.0 | .0013 | .0013 | .0013 | .0012 | .0012 | .0011 | .0011 | .0011 | .0010 | .0010 |
| −2.9 | .0019 | .0018 | .0018 | .0017 | .0016 | .0016 | .0015 | .0015 | .0014 | .0014 |
| −2.8 | .0026 | .0025 | .0024 | .0023 | .0023 | .0022 | .0021 | .0021 | .0020 | .0019 |
| −2.7 | .0035 | .0034 | .0033 | .0032 | .0031 | .0030 | .0029 | .0028 | .0027 | .0026 |
| −2.6 | .0047 | .0045 | .0044 | .0043 | .0041 | .0040 | .0039 | .0038 | .0037 | .0036 |
| −2.5 | .0062 | .0060 | .0059 | .0057 | .0055 | .0054 | .0052 | .0051 | .0049 | .0048 |
| −2.4 | .0082 | .0080 | .0078 | .0075 | .0073 | .0071 | .0069 | .0068 | .0066 | .0064 |
| −2.3 | .0107 | .0104 | .0102 | .0099 | .0096 | .0094 | .0091 | .0089 | .0087 | .0084 |
| −2.2 | .0139 | .0136 | .0132 | .0129 | .0125 | .0122 | .0119 | .0116 | .0113 | .0110 |
| −2.1 | .0179 | .0174 | .0170 | .0166 | .0162 | .0158 | .0154 | .0150 | .0146 | .0143 |
| −2.0 | .0228 | .0222 | .0217 | .0212 | .0207 | .0202 | .0197 | .0192 | .0188 | .0183 |
| −1.9 | .0287 | .0281 | .0274 | .0268 | .0262 | .0256 | .0250 | .0244 | .0239 | .0233 |
| −1.8 | .0359 | .0351 | .0344 | .0336 | .0329 | .0322 | .0314 | .0307 | .0301 | .0294 |
| −1.7 | .0446 | .0436 | .0427 | .0418 | .0409 | .0401 | .0392 | .0384 | .0375 | .0367 |
| −1.6 | .0548 | .0537 | .0526 | .0516 | .0505 | .0495 | .0485 | .0475 | .0465 | .0455 |
| −1.5 | .0668 | .0655 | .0643 | .0630 | .0618 | .0606 | .0594 | .0582 | .0571 | .0559 |
| −1.4 | .0808 | .0793 | .0778 | .0764 | .0749 | .0735 | .0721 | .0708 | .0694 | .0681 |
| −1.3 | .0968 | .0951 | .0934 | .0918 | .0901 | .0885 | .0869 | .0853 | .0838 | .0823 |
| −1.2 | .1151 | .1131 | .1112 | .1093 | .1075 | .1056 | .1038 | .1020 | .1003 | .0985 |
| −1.1 | .1357 | .1335 | .1314 | .1292 | .1271 | .1251 | .1230 | .1210 | .1190 | .1170 |
| −1.0 | .1587 | .1562 | .1539 | .1515 | .1492 | .1469 | .1446 | .1423 | .1401 | .1379 |
| −0.9 | .1841 | .1814 | .1788 | .1762 | .1736 | .1711 | .1685 | .1660 | .1635 | .1611 |
| −0.8 | .2119 | .2090 | .2061 | .2033 | .2005 | .1977 | .1949 | .1922 | .1894 | .1867 |
| −0.7 | .2420 | .2389 | .2358 | .2327 | .2296 | .2266 | .2236 | .2206 | .2177 | .2148 |
| −0.6 | .2743 | .2709 | .2676 | .2643 | .2611 | .2578 | .2546 | .2514 | .2483 | .2451 |
| −0.5 | .3085 | .3050 | .3015 | .2981 | .2946 | .2912 | .2877 | .2843 | .2810 | .2776 |
| −0.4 | .3446 | .3409 | .3372 | .3336 | .3300 | .3264 | .3228 | .3192 | .3156 | .3121 |
| −0.3 | .3821 | .3783 | .3745 | .3707 | .3669 | .3632 | .3594 | .3557 | .3520 | .3483 |
| −0.2 | .4207 | .4168 | .4129 | .4090 | .4052 | .4013 | .3974 | .3936 | .3897 | .3859 |
| −0.1 | .4602 | .4562 | .4522 | .4483 | .4443 | .4404 | .4364 | .4325 | .4286 | .4247 |
| −0.0 | .5000 | .4960 | .4920 | .4880 | .4840 | .4801 | .4761 | .4721 | .4681 | .4641 |

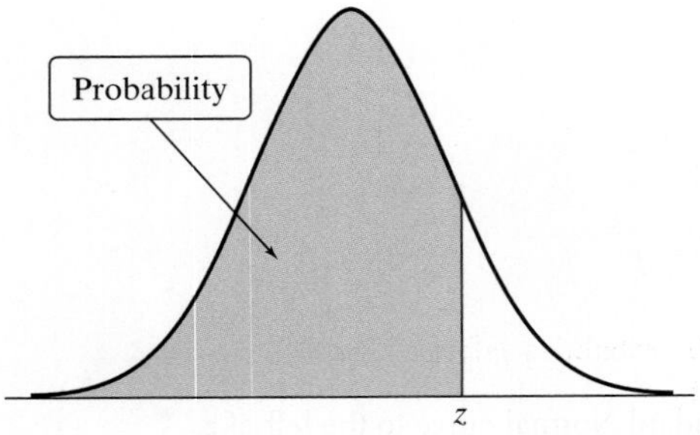


Table entry for $z$ is the area under the standard Normal curve to the left of $z$.

**Table A Standard Normal probabilities (continued)**

| z | .00 | .01 | .02 | .03 | .04 | .05 | .06 | .07 | .08 | .09 |
|---|---|---|---|---|---|---|---|---|---|---|
| 0.0 | .5000 | .5040 | .5080 | .5120 | .5160 | .5199 | .5239 | .5279 | .5319 | .5359 |
| 0.1 | .5398 | .5438 | .5478 | .5517 | .5557 | .5596 | .5636 | .5675 | .5714 | .5753 |
| 0.2 | .5793 | .5832 | .5871 | .5910 | .5948 | .5987 | .6026 | .6064 | .6103 | .6141 |
| 0.3 | .6179 | .6217 | .6255 | .6293 | .6331 | .6368 | .6406 | .6443 | .6480 | .6517 |
| 0.4 | .6554 | .6591 | .6628 | .6664 | .6700 | .6736 | .6772 | .6808 | .6844 | .6879 |
| 0.5 | .6915 | .6950 | .6985 | .7019 | .7054 | .7088 | .7123 | .7157 | .7190 | .7224 |
| 0.6 | .7257 | .7291 | .7324 | .7357 | .7389 | .7422 | .7454 | .7486 | .7517 | .7549 |
| 0.7 | .7580 | .7611 | .7642 | .7673 | .7704 | .7734 | .7764 | .7794 | .7823 | .7852 |
| 0.8 | .7881 | .7910 | .7939 | .7967 | .7995 | .8023 | .8051 | .8078 | .8106 | .8133 |
| 0.9 | .8159 | .8186 | .8212 | .8238 | .8264 | .8289 | .8315 | .8340 | .8365 | .8389 |
| 1.0 | .8413 | .8438 | .8461 | .8485 | .8508 | .8531 | .8554 | .8577 | .8599 | .8621 |
| 1.1 | .8643 | .8665 | .8686 | .8708 | .8729 | .8749 | .8770 | .8790 | .8810 | .8830 |
| 1.2 | .8849 | .8869 | .8888 | .8907 | .8925 | .8944 | .8962 | .8980 | .8997 | .9015 |
| 1.3 | .9032 | .9049 | .9066 | .9082 | .9099 | .9115 | .9131 | .9147 | .9162 | .9177 |
| 1.4 | .9192 | .9207 | .9222 | .9236 | .9251 | .9265 | .9279 | .9292 | .9306 | .9319 |
| 1.5 | .9332 | .9345 | .9357 | .9370 | .9382 | .9394 | .9406 | .9418 | .9429 | .9441 |
| 1.6 | .9452 | .9463 | .9474 | .9484 | .9495 | .9505 | .9515 | .9525 | .9535 | .9545 |
| 1.7 | .9554 | .9564 | .9573 | .9582 | .9591 | .9599 | .9608 | .9616 | .9625 | .9633 |
| 1.8 | .9641 | .9649 | .9656 | .9664 | .9671 | .9678 | .9686 | .9693 | .9699 | .9706 |
| 1.9 | .9713 | .9719 | .9726 | .9732 | .9738 | .9744 | .9750 | .9756 | .9761 | .9767 |
| 2.0 | .9772 | .9778 | .9783 | .9788 | .9793 | .9798 | .9803 | .9808 | .9812 | .9817 |
| 2.1 | .9821 | .9826 | .9830 | .9834 | .9838 | .9842 | .9846 | .9850 | .9854 | .9857 |
| 2.2 | .9861 | .9864 | .9868 | .9871 | .9875 | .9878 | .9881 | .9884 | .9887 | .9890 |
| 2.3 | .9893 | .9896 | .9898 | .9901 | .9904 | .9906 | .9909 | .9911 | .9913 | .9916 |
| 2.4 | .9918 | .9920 | .9922 | .9925 | .9927 | .9929 | .9931 | .9932 | .9934 | .9936 |
| 2.5 | .9938 | .9940 | .9941 | .9943 | .9945 | .9946 | .9948 | .9949 | .9951 | .9952 |
| 2.6 | .9953 | .9955 | .9956 | .9957 | .9959 | .9960 | .9961 | .9962 | .9963 | .9964 |
| 2.7 | .9965 | .9966 | .9967 | .9968 | .9969 | .9970 | .9971 | .9972 | .9973 | .9974 |
| 2.8 | .9974 | .9975 | .9976 | .9977 | .9977 | .9978 | .9979 | .9979 | .9980 | .9981 |
| 2.9 | .9981 | .9982 | .9982 | .9983 | .9984 | .9984 | .9985 | .9985 | .9986 | .9986 |
| 3.0 | .9987 | .9987 | .9987 | .9988 | .9988 | .9989 | .9989 | .9989 | .9990 | .9990 |
| 3.1 | .9990 | .9991 | .9991 | .9991 | .9992 | .9992 | .9992 | .9992 | .9993 | .9993 |
| 3.2 | .9993 | .9993 | .9994 | .9994 | .9994 | .9994 | .9994 | .9995 | .9995 | .9995 |
| 3.3 | .9995 | .9995 | .9995 | .9996 | .9996 | .9996 | .9996 | .9996 | .9996 | .9997 |
| 3.4 | .9997 | .9997 | .9997 | .9997 | .9997 | .9997 | .9997 | .9997 | .9997 | .9998 |

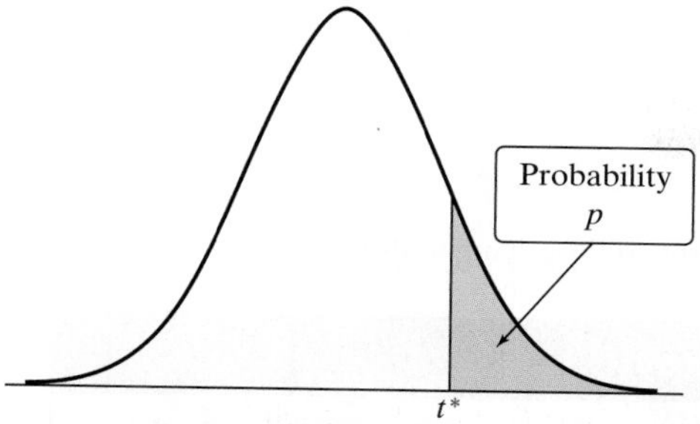


Table entry for $p$ and C is the point $t^*$ with probability $p$ lying to its right and probability C lying between $-t^*$ and $t^*$.

**Table B** ***t* distribution critical values**

| | Tail probability *p* | | | | | | | | | | | |
|---|---|---|---|---|---|---|---|---|---|---|---|---|
| **df** | **.25** | **.20** | **.15** | **.10** | **.05** | **.025** | **.02** | **.01** | **.005** | **.0025** | **.001** | **.0005** |
| 1 | 1.000 | 1.376 | 1.963 | 3.078 | 6.314 | 12.71 | 15.89 | 31.82 | 63.66 | 127.3 | 318.3 | 636.6 |
| 2 | 0.816 | 1.061 | 1.386 | 1.886 | 2.920 | 4.303 | 4.849 | 6.965 | 9.925 | 14.09 | 22.33 | 31.60 |
| 3 | 0.765 | 0.978 | 1.250 | 1.638 | 2.353 | 3.182 | 3.482 | 4.541 | 5.841 | 7.453 | 10.21 | 12.92 |
| 4 | 0.741 | 0.941 | 1.190 | 1.533 | 2.132 | 2.776 | 2.999 | 3.747 | 4.604 | 5.598 | 7.173 | 8.610 |
| 5 | 0.727 | 0.920 | 1.156 | 1.476 | 2.015 | 2.571 | 2.757 | 3.365 | 4.032 | 4.773 | 5.893 | 6.869 |
| 6 | 0.718 | 0.906 | 1.134 | 1.440 | 1.943 | 2.447 | 2.612 | 3.143 | 3.707 | 4.317 | 5.208 | 5.959 |
| 7 | 0.711 | 0.896 | 1.119 | 1.415 | 1.895 | 2.365 | 2.517 | 2.998 | 3.499 | 4.029 | 4.785 | 5.408 |
| 8 | 0.706 | 0.889 | 1.108 | 1.397 | 1.860 | 2.306 | 2.449 | 2.896 | 3.355 | 3.833 | 4.501 | 5.041 |
| 9 | 0.703 | 0.883 | 1.100 | 1.383 | 1.833 | 2.262 | 2.398 | 2.821 | 3.250 | 3.690 | 4.297 | 4.781 |
| 10 | 0.700 | 0.879 | 1.093 | 1.372 | 1.812 | 2.228 | 2.359 | 2.764 | 3.169 | 3.581 | 4.144 | 4.587 |
| 11 | 0.697 | 0.876 | 1.088 | 1.363 | 1.796 | 2.201 | 2.328 | 2.718 | 3.106 | 3.497 | 4.025 | 4.437 |
| 12 | 0.695 | 0.873 | 1.083 | 1.356 | 1.782 | 2.179 | 2.303 | 2.681 | 3.055 | 3.428 | 3.930 | 4.318 |
| 13 | 0.694 | 0.870 | 1.079 | 1.350 | 1.771 | 2.160 | 2.282 | 2.650 | 3.012 | 3.372 | 3.852 | 4.221 |
| 14 | 0.692 | 0.868 | 1.076 | 1.345 | 1.761 | 2.145 | 2.264 | 2.624 | 2.977 | 3.326 | 3.787 | 4.140 |
| 15 | 0.691 | 0.866 | 1.074 | 1.341 | 1.753 | 2.131 | 2.249 | 2.602 | 2.947 | 3.286 | 3.733 | 4.073 |
| 16 | 0.690 | 0.865 | 1.071 | 1.337 | 1.746 | 2.120 | 2.235 | 2.583 | 2.921 | 3.252 | 3.686 | 4.015 |
| 17 | 0.689 | 0.863 | 1.069 | 1.333 | 1.740 | 2.110 | 2.224 | 2.567 | 2.898 | 3.222 | 3.646 | 3.965 |
| 18 | 0.688 | 0.862 | 1.067 | 1.330 | 1.734 | 2.101 | 2.214 | 2.552 | 2.878 | 3.197 | 3.611 | 3.922 |
| 19 | 0.688 | 0.861 | 1.066 | 1.328 | 1.729 | 2.093 | 2.205 | 2.539 | 2.861 | 3.174 | 3.579 | 3.883 |
| 20 | 0.687 | 0.860 | 1.064 | 1.325 | 1.725 | 2.086 | 2.197 | 2.528 | 2.845 | 3.153 | 3.552 | 3.850 |
| 21 | 0.686 | 0.859 | 1.063 | 1.323 | 1.721 | 2.080 | 2.189 | 2.518 | 2.831 | 3.135 | 3.527 | 3.819 |
| 22 | 0.686 | 0.858 | 1.061 | 1.321 | 1.717 | 2.074 | 2.183 | 2.508 | 2.819 | 3.119 | 3.505 | 3.792 |
| 23 | 0.685 | 0.858 | 1.060 | 1.319 | 1.714 | 2.069 | 2.177 | 2.500 | 2.807 | 3.104 | 3.485 | 3.768 |
| 24 | 0.685 | 0.857 | 1.059 | 1.318 | 1.711 | 2.064 | 2.172 | 2.492 | 2.797 | 3.091 | 3.467 | 3.745 |
| 25 | 0.684 | 0.856 | 1.058 | 1.316 | 1.708 | 2.060 | 2.167 | 2.485 | 2.787 | 3.078 | 3.450 | 3.725 |
| 26 | 0.684 | 0.856 | 1.058 | 1.315 | 1.706 | 2.056 | 2.162 | 2.479 | 2.779 | 3.067 | 3.435 | 3.707 |
| 27 | 0.684 | 0.855 | 1.057 | 1.314 | 1.703 | 2.052 | 2.158 | 2.473 | 2.771 | 3.057 | 3.421 | 3.690 |
| 28 | 0.683 | 0.855 | 1.056 | 1.313 | 1.701 | 2.048 | 2.154 | 2.467 | 2.763 | 3.047 | 3.408 | 3.674 |
| 29 | 0.683 | 0.854 | 1.055 | 1.311 | 1.699 | 2.045 | 2.150 | 2.462 | 2.756 | 3.038 | 3.396 | 3.659 |
| 30 | 0.683 | 0.854 | 1.055 | 1.310 | 1.697 | 2.042 | 2.147 | 2.457 | 2.750 | 3.030 | 3.385 | 3.646 |
| 40 | 0.681 | 0.851 | 1.050 | 1.303 | 1.684 | 2.021 | 2.123 | 2.423 | 2.704 | 2.971 | 3.307 | 3.551 |
| 50 | 0.679 | 0.849 | 1.047 | 1.299 | 1.676 | 2.009 | 2.109 | 2.403 | 2.678 | 2.937 | 3.261 | 3.496 |
| 60 | 0.679 | 0.848 | 1.045 | 1.296 | 1.671 | 2.000 | 2.099 | 2.390 | 2.660 | 2.915 | 3.232 | 3.460 |
| 80 | 0.678 | 0.846 | 1.043 | 1.292 | 1.664 | 1.990 | 2.088 | 2.374 | 2.639 | 2.887 | 3.195 | 3.416 |
| 100 | 0.677 | 0.845 | 1.042 | 1.290 | 1.660 | 1.984 | 2.081 | 2.364 | 2.626 | 2.871 | 3.174 | 3.390 |
| 1000 | 0.675 | 0.842 | 1.037 | 1.282 | 1.646 | 1.962 | 2.056 | 2.330 | 2.581 | 2.813 | 3.098 | 3.300 |
| ∞ | 0.674 | 0.841 | 1.036 | 1.282 | 1.645 | 1.960 | 2.054 | 2.326 | 2.576 | 2.807 | 3.091 | 3.291 |
| | **50%** | **60%** | **70%** | **80%** | **90%** | **95%** | **96%** | **98%** | **99%** | **99.5%** | **99.8%** | **99.9%** |
| | Confidence level *C* | | | | | | | | | | | |

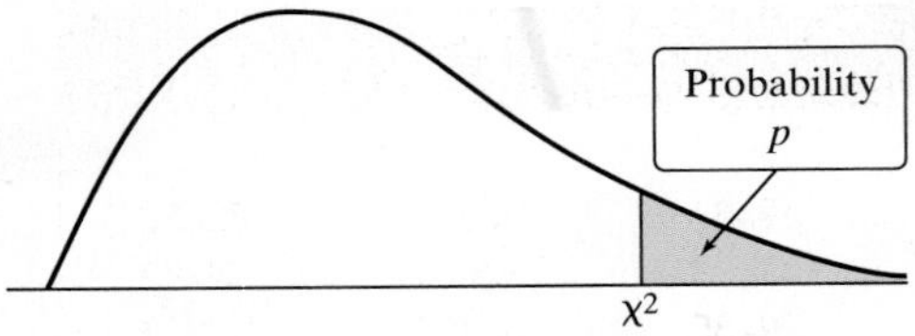


Table entry for $p$ is the point $\chi^2$ with probability $p$ lying to its right.

## Table C Chi-square distribution critical values

| | Tail probability $p$ | | | | | | | | | | | |
|---|---|---|---|---|---|---|---|---|---|---|---|---|
| **df** | **.25** | **.20** | **.15** | **.10** | **.05** | **.025** | **.02** | **.01** | **.005** | **.0025** | **.001** | **.0005** |
| 1 | 1.32 | 1.64 | 2.07 | 2.71 | 3.84 | 5.02 | 5.41 | 6.63 | 7.88 | 9.14 | 10.83 | 12.12 |
| 2 | 2.77 | 3.22 | 3.79 | 4.61 | 5.99 | 7.38 | 7.82 | 9.21 | 10.60 | 11.98 | 13.82 | 15.20 |
| 3 | 4.11 | 4.64 | 5.32 | 6.25 | 7.81 | 9.35 | 9.84 | 11.34 | 12.84 | 14.32 | 16.27 | 17.73 |
| 4 | 5.39 | 5.99 | 6.74 | 7.78 | 9.49 | 11.14 | 11.67 | 13.28 | 14.86 | 16.42 | 18.47 | 20.00 |
| 5 | 6.63 | 7.29 | 8.12 | 9.24 | 11.07 | 12.83 | 13.39 | 15.09 | 16.75 | 18.39 | 20.51 | 22.11 |
| 6 | 7.84 | 8.56 | 9.45 | 10.64 | 12.59 | 14.45 | 15.03 | 16.81 | 18.55 | 20.25 | 22.46 | 24.10 |
| 7 | 9.04 | 9.80 | 10.75 | 12.02 | 14.07 | 16.01 | 16.62 | 18.48 | 20.28 | 22.04 | 24.32 | 26.02 |
| 8 | 10.22 | 11.03 | 12.03 | 13.36 | 15.51 | 17.53 | 18.17 | 20.09 | 21.95 | 23.77 | 26.12 | 27.87 |
| 9 | 11.39 | 12.24 | 13.29 | 14.68 | 16.92 | 19.02 | 19.68 | 21.67 | 23.59 | 25.46 | 27.88 | 29.67 |
| 10 | 12.55 | 13.44 | 14.53 | 15.99 | 18.31 | 20.48 | 21.16 | 23.21 | 25.19 | 27.11 | 29.59 | 31.42 |
| 11 | 13.70 | 14.63 | 15.77 | 17.28 | 19.68 | 21.92 | 22.62 | 24.72 | 26.76 | 28.73 | 31.26 | 33.14 |
| 12 | 14.85 | 15.81 | 16.99 | 18.55 | 21.03 | 23.34 | 24.05 | 26.22 | 28.30 | 30.32 | 32.91 | 34.82 |
| 13 | 15.98 | 16.98 | 18.20 | 19.81 | 22.36 | 24.74 | 25.47 | 27.69 | 29.82 | 31.88 | 34.53 | 36.48 |
| 14 | 17.12 | 18.15 | 19.41 | 21.06 | 23.68 | 26.12 | 26.87 | 29.14 | 31.32 | 33.43 | 36.12 | 38.11 |
| 15 | 18.25 | 19.31 | 20.60 | 22.31 | 25.00 | 27.49 | 28.26 | 30.58 | 32.80 | 34.95 | 37.70 | 39.72 |
| 16 | 19.37 | 20.47 | 21.79 | 23.54 | 26.30 | 28.85 | 29.63 | 32.00 | 34.27 | 36.46 | 39.25 | 41.31 |
| 17 | 20.49 | 21.61 | 22.98 | 24.77 | 27.59 | 30.19 | 31.00 | 33.41 | 35.72 | 37.95 | 40.79 | 42.88 |
| 18 | 21.60 | 22.76 | 24.16 | 25.99 | 28.87 | 31.53 | 32.35 | 34.81 | 37.16 | 39.42 | 42.31 | 44.43 |
| 19 | 22.72 | 23.90 | 25.33 | 27.20 | 30.14 | 32.85 | 33.69 | 36.19 | 38.58 | 40.88 | 43.82 | 45.97 |
| 20 | 23.83 | 25.04 | 26.50 | 28.41 | 31.41 | 34.17 | 35.02 | 37.57 | 40.00 | 42.34 | 45.31 | 47.50 |
| 21 | 24.93 | 26.17 | 27.66 | 29.62 | 32.67 | 35.48 | 36.34 | 38.93 | 41.40 | 43.78 | 46.80 | 49.01 |
| 22 | 26.04 | 27.30 | 28.82 | 30.81 | 33.92 | 36.78 | 37.66 | 40.29 | 42.80 | 45.20 | 48.27 | 50.51 |
| 23 | 27.14 | 28.43 | 29.98 | 32.01 | 35.17 | 38.08 | 38.97 | 41.64 | 44.18 | 46.62 | 49.73 | 52.00 |
| 24 | 28.24 | 29.55 | 31.13 | 33.20 | 36.42 | 39.36 | 40.27 | 42.98 | 45.56 | 48.03 | 51.18 | 53.48 |
| 25 | 29.34 | 30.68 | 32.28 | 34.38 | 37.65 | 40.65 | 41.57 | 44.31 | 46.93 | 49.44 | 52.62 | 54.95 |
| 26 | 30.43 | 31.79 | 33.43 | 35.56 | 38.89 | 41.92 | 42.86 | 45.64 | 48.29 | 50.83 | 54.05 | 56.41 |
| 27 | 31.53 | 32.91 | 34.57 | 36.74 | 40.11 | 43.19 | 44.14 | 46.96 | 49.64 | 52.22 | 55.48 | 57.86 |
| 28 | 32.62 | 34.03 | 35.71 | 37.92 | 41.34 | 44.46 | 45.42 | 48.28 | 50.99 | 53.59 | 56.89 | 59.30 |
| 29 | 33.71 | 35.14 | 36.85 | 39.09 | 42.56 | 45.72 | 46.69 | 49.59 | 52.34 | 54.97 | 58.30 | 60.73 |
| 30 | 34.80 | 36.25 | 37.99 | 40.26 | 43.77 | 46.98 | 47.96 | 50.89 | 53.67 | 56.33 | 59.70 | 62.16 |
| 40 | 45.62 | 47.27 | 49.24 | 51.81 | 55.76 | 59.34 | 60.44 | 63.69 | 66.77 | 69.70 | 73.40 | 76.09 |
| 50 | 56.33 | 58.16 | 60.35 | 63.17 | 67.50 | 71.42 | 72.61 | 76.15 | 79.49 | 82.66 | 86.66 | 89.56 |
| 60 | 66.98 | 68.97 | 71.34 | 74.40 | 79.08 | 83.30 | 84.58 | 88.38 | 91.95 | 95.34 | 99.61 | 102.7 |
| 80 | 88.13 | 90.41 | 93.11 | 96.58 | 101.9 | 106.6 | 108.1 | 112.3 | 116.3 | 120.1 | 124.8 | 128.3 |
| 100 | 109.1 | 111.7 | 114.7 | 118.5 | 124.3 | 129.6 | 131.1 | 135.8 | 140.2 | 144.3 | 149.4 | 153.2 |

**Table D Random digits**

| Line | | | | | | | | |
|---|---|---|---|---|---|---|---|---|
| 101 | 19223 | 95034 | 05756 | 28713 | 96409 | 12531 | 42544 | 82853 |
| 102 | 73676 | 47150 | 99400 | 01927 | 27754 | 42648 | 82425 | 36290 |
| 103 | 45467 | 71709 | 77558 | 00095 | 32863 | 29485 | 82226 | 90056 |
| 104 | 52711 | 38889 | 93074 | 60227 | 40011 | 85848 | 48767 | 52573 |
| 105 | 95592 | 94007 | 69971 | 91481 | 60779 | 53791 | 17297 | 59335 |
| 106 | 68417 | 35013 | 15529 | 72765 | 85089 | 57067 | 50211 | 47487 |
| 107 | 82739 | 57890 | 20807 | 47511 | 81676 | 55300 | 94383 | 14893 |
| 108 | 60940 | 72024 | 17868 | 24943 | 61790 | 90656 | 87964 | 18883 |
| 109 | 36009 | 19365 | 15412 | 39638 | 85453 | 46816 | 83485 | 41979 |
| 110 | 38448 | 48789 | 18338 | 24697 | 39364 | 42006 | 76688 | 08708 |
| 111 | 81486 | 69487 | 60513 | 09297 | 00412 | 71238 | 27649 | 39950 |
| 112 | 59636 | 88804 | 04634 | 71197 | 19352 | 73089 | 84898 | 45785 |
| 113 | 62568 | 70206 | 40325 | 03699 | 71080 | 22553 | 11486 | 11776 |
| 114 | 45149 | 32992 | 75730 | 66280 | 03819 | 56202 | 02938 | 70915 |
| 115 | 61041 | 77684 | 94322 | 24709 | 73698 | 14526 | 31893 | 32592 |
| 116 | 14459 | 26056 | 31424 | 80371 | 65103 | 62253 | 50490 | 61181 |
| 117 | 38167 | 98532 | 62183 | 70632 | 23417 | 26185 | 41448 | 75532 |
| 118 | 73190 | 32533 | 04470 | 29669 | 84407 | 90785 | 65956 | 86382 |
| 119 | 95857 | 07118 | 87664 | 92099 | 58806 | 66979 | 98624 | 84826 |
| 120 | 35476 | 55972 | 39421 | 65850 | 04266 | 35435 | 43742 | 11937 |
| 121 | 71487 | 09984 | 29077 | 14863 | 61683 | 47052 | 62224 | 51025 |
| 122 | 13873 | 81598 | 95052 | 90908 | 73592 | 75186 | 87136 | 95761 |
| 123 | 54580 | 81507 | 27102 | 56027 | 55892 | 33063 | 41842 | 81868 |
| 124 | 71035 | 09001 | 43367 | 49497 | 72719 | 96758 | 27611 | 91596 |
| 125 | 96746 | 12149 | 37823 | 71868 | 18442 | 35119 | 62103 | 39244 |
| 126 | 96927 | 19931 | 36809 | 74192 | 77567 | 88741 | 48409 | 41903 |
| 127 | 43909 | 99477 | 25330 | 64359 | 40085 | 16925 | 85117 | 36071 |
| 128 | 15689 | 14227 | 06565 | 14374 | 13352 | 49367 | 81982 | 87209 |
| 129 | 36759 | 58984 | 68288 | 22913 | 18638 | 54303 | 00795 | 08727 |
| 130 | 69051 | 64817 | 87174 | 09517 | 84534 | 06489 | 87201 | 97245 |
| 131 | 05007 | 16632 | 81194 | 14873 | 04197 | 85576 | 45195 | 96565 |
| 132 | 68732 | 55259 | 84292 | 08796 | 43165 | 93739 | 31685 | 97150 |
| 133 | 45740 | 41807 | 65561 | 33302 | 07051 | 93623 | 18132 | 09547 |
| 134 | 27816 | 78416 | 18329 | 21337 | 35213 | 37741 | 04312 | 68508 |
| 135 | 66925 | 55658 | 39100 | 78458 | 11206 | 19876 | 87151 | 31260 |
| 136 | 08421 | 44753 | 77377 | 28744 | 75592 | 08563 | 79140 | 92454 |
| 137 | 53645 | 66812 | 61421 | 47836 | 12609 | 15373 | 98481 | 14592 |
| 138 | 66831 | 68908 | 40772 | 21558 | 47781 | 33586 | 79177 | 06928 |
| 139 | 55588 | 99404 | 70708 | 41098 | 43563 | 56934 | 48394 | 51719 |
| 140 | 12975 | 13258 | 13048 | 45144 | 72321 | 81940 | 00360 | 02428 |
| 141 | 96767 | 35964 | 23822 | 96012 | 94591 | 65194 | 50842 | 53372 |
| 142 | 72829 | 50232 | 97892 | 63408 | 77919 | 44575 | 24870 | 04178 |
| 143 | 88565 | 42628 | 17797 | 49376 | 61762 | 16953 | 88604 | 12724 |
| 144 | 62964 | 88145 | 83083 | 69453 | 46109 | 59505 | 69680 | 00900 |
| 145 | 19687 | 12633 | 57857 | 95806 | 09931 | 02150 | 43163 | 58636 |
| 146 | 37609 | 59057 | 66967 | 83401 | 60705 | 02384 | 90597 | 93600 |
| 147 | 54973 | 86278 | 88737 | 74351 | 47500 | 84552 | 19909 | 67181 |
| 148 | 00694 | 05977 | 19664 | 65441 | 20903 | 62371 | 22725 | 53340 |
| 149 | 71546 | 05233 | 53946 | 68743 | 72460 | 27601 | 45403 | 88692 |
| 150 | 07511 | 88915 | 41267 | 16853 | 84569 | 79367 | 32337 | 03316 |

# AP® Statistics Course Skills Alignment to UPDATED *The Practice of Statistics,* 6th edition

The table below displays the skills that students should develop during an AP® Statistics course through frequent repetition. **Bold** references indicate the chapter(s) of UPDATED *TPS* 6e where the skill is explicitly addressed. The text provides additional opportunities for students to practice each skill in subsequent chapters.

| Skill 1 | Skill 2 | Skill 3 | Skill 4 |
|---|---|---|---|
| **Selecting Statistical Methods** | **Data Analysis** | **Using Probability and Simulation** | **Statistical Argumentation** |
| *Select methods for collecting and/or analyzing data for statistical inference.* | *Describe patterns, trends, associations, and relationships in data.* | *Explore random phenomena.* | *Develop an explanation or justify a conclusion using evidence from data, definitions, or statistical inference.* |
| **1.A** Identify the question to be answered or problem to be solved. **Chapters 1–12** | **2.A** Describe data presented numerically or graphically. **Chapters 1, 2, 3** | **3.A** Determine relative frequencies, proportions, or probabilities using simulation or calculations. **Chapters 1, 5, 6, 7** | **4.A** Make an appropriate claim or draw an appropriate conclusion. **Chapters 1–12** |
| **1.B** Identify key and relevant information to answer a question or solve a problem. **Chapters 1–12** | **2.B** Construct numerical or graphical representations of distributions. **Chapters 1, 2, 3** | **3.B** Determine parameters for probability distributions. **Chapters 6 and 7** | **4.B** Interpret statistical calculations and findings to assign meaning or assess a claim. **Chapters 1–12** |
| **1.C** Describe an appropriate method for gathering and representing data. Gathering data: **Chapter 4** Representing data: **Chapters 1, 2, 3** | **2.C** Calculate summary statistics, relative positions of points within a distribution, correlation, and predicted response. **Chapters 1, 2, 3** | **3.C** Describe probability distributions. **Chapters 6 and 7** | |
| | **2.D** Compare distributions or relative positions of points within a distribution. **Chapters 1, 2** | | |
| **INFERENCE** | | | |
| **1.D** Identify an appropriate inference method for confidence intervals. **Chapters 8, 10, 12** | | **3.D** Construct a confidence interval, provided conditions for inference are met. **Chapters 8, 10, 12** | **4.C** Verify that inference procedures apply in a given situation. **Chapters 8–12** |
| **1.E** Identify an appropriate inference method for significance tests. **Chapters 9, 11, 12** | | **3.E** Calculate a test statistic and find a *p*-value, provided conditions for inference are met. **Chapters 9, 11, 12** | **4.D** Justify a claim based on a confidence interval. **Chapters 8–12** |
| **1.F** Identify null and alternative hypotheses. **Chapters 9, 11, 12** | | | **4.E** Justify a claim using a decision based on significance tests. **Chapters 9, 11, 12** |

## Technology Corner References

*TI-Nspire and other technology instructions are on the website at highschool.bfwpub.com/updatedtps6e.*